Praise for *Silent Spring Revolution*

"Rich in facts, anecdotes, and ana prehensive and vivid history of ... battles and advances is profound... struggle to conceive of a similarly constructive way for- ward in a time of worsening climate crisis and political gridlock. . . . The scope and importance of this volume is mighty." —Donna Seaman, *Booklist*

"Doug Brinkley has done it again! *Silent Spring Revolution* showcases his mastery of the art of storytelling, deep knowledge of presidential history, and passion for the natural world. Through exhaustive research and beau- tiful writing, Brinkley creates vibrant portraits of the grassroots activists—Rachel Carson, Cesar Chavez, and William O. Douglas—whose work influenced the levers of power inside the White Houses of John Kennedy, Lyndon and Lady Bird Johnson, and Richard Nixon with extraordinary results. This is not only a majestic work of history; it is an urgent call for our time."
—Doris Kearns Goodwin, Pulitzer
Prize–winning historian

"As this story unfolds from Kennedy to Johnson, *Silent Spring Revolution* proves consistently captivating and takes its place alongside the trilogy—*The Wilderness Warrior* and *Rightful Heritage*—as an essential addition to twentieth- century presidential history."
—Steve Nathans-Kelly, *New York Journal of Books*

"A solid addition to the literature of the intersection of environmentalism and politics." —*Kirkus Reviews*

"Absorbing and impassioned narrative that transports readers to a different, more optimistic world of popular support for eco-awareness and collective action. It is a book that sticks with you and implies that the 'highway to climate hell' is far from our only choice."
—Matthew Dallek, *Washington Post*

"Beginning with *The Wilderness Warrior*, his sweeping book on Theodore Roosevelt and the conservation movement, Doug Brinkley has been our foremost chronicler of how presidents shape our environmental policy. In this very important and timely book, filled with fascinating historical revelations, Brinkley describes how three presidents of the Long Sixties embraced the wave of environmental concerns unleashed by Rachel Carson's great work. The sense of common stewardship provides great lessons for today's partisan and poisonous political discourse."
—Walter Isaacson

"*Silent Spring Revolution* is an inspirational tome, one that shows how a few dedicated people can take on the status quo and create meaningful change."
—Curt Schleier, *Minneapolis Star Tribune*

"Douglas Brinkley is among the greatest living American historians. For over a decade, he's been publishing a masterpiece series on the history of the American environmental movement. His *Silent Spring Revolution* is a fascinating deep dive into a pivotal decade for America and its people, told by one of our best historians."
—Sam Matey, Substack

"A work of stunning erudition by one of our most brilliant chroniclers of the American past. By meticulously detailing how courageous activists sparked an environmental revolution that fueled the legislative imaginations of three very different presidents, Douglas Brinkley also renders a vivid portrait of the endangered species of bipartisan cooperation in the effort to save our planet. In this magisterial account, Brinkley proves himself a man for all seasons: he springs into action and refuses to take the summer off in the desperate drive to prevent our fall into an even more destructive winter of environmental discontent."

—Michael Eric Dyson, author of *Entertaining Race: Performing Blackness in America*

"Brinkley is an enthusiastic and engaging guide, pulling together a fascinating narrative explicating the profound effect the environmental movement had on three very different presidents and their administrations."

—Kristofer Collins, *Pittsburgh Magazine*

"Brinkley's wide-ranging account shows how rapidly public policy can evolve when social mobilization reaches a tipping point. In an age of escalating urgency about climate change—and when the politics of the energy transition seem ever more intractable—the history this book relates is of the utmost relevance."

—Jason Bordoff, *Foreign Policy*

"In his *Silent Spring Revolution*, Douglas Brinkley provides an urgently needed history of the American environmental movement, and the prophetic and courageous men and women who led it. Their gripping story is both a warning and a promise of what might still be possible as the struggle for environmental sustainability reaches a crisis point. Every American should read it and ponder why it is that the bipartisan gains of the recent past seem so tragically out of reach today."

—Jane Mayer, chief Washington correspondent, *The New Yorker*, author of *Dark Money: The Hidden History of the Billionaires Behind the Rise of the Radical Right*

"Brinkley's *Silent Spring Revolution* is a deep-dive environmental history that offers a peaceful path forward for a nation consumed by divisiveness."

—Andrew Dansby, *Houston Chronicle*

"Countless superlatives could be used to describe this fascinating book about how the environmental movement came of age during an era known as the Long Sixties, which began in the late 1950s and continued until the early 1970s. Frankly, I can't think of many that would be guilty of hyperbole or embellishment. Brinley's prose is engaging throughout the book, and it keeps building like a sequence of crescendos. What I found utterly brilliant was the depth of his reporting, something that can offer lessons for many journalists."

—Tom Henry, Society of Environmental Journalists

"*Silent Spring Revolution* is a luminous history of the environmental movement that emerged in the tumult of the 1960s. Brinkley's deft mosaic of powerful forces and powerful men puts the main spotlight on a woman,

Rachel Carson. Soft-spoken but fearless, she galvanized an enduring cause by reminding us that the Earth is a natural mansion . . . our living home . . . and we abuse it at our peril."

—William Souder, author of *On a Farther Shore: The Life and Legacy of Rachel Carson*

"If Douglas Brinkley writes it, I will read it. With *Silent Spring Revolution*, he hits another home run: an irresistible tale of courage in the face of stubborn, dangerous ignorance and staggering arrogance, peppered with some of the most unforgettable names in history. I was swept up from page one." —Candice Millard

"Douglas is a master synthesizer. . . . *Silent Spring Revolution* is a huge book, but scale is necessary to tell the story of what happened in the period of intense environmental activism. Many thanks to Douglas Brinkley for this book. . . . There is concern in this digital age that people lack the time and patience to read any books, let alone big ones like *Silent Spring Revolution*. But if one is interested or concerned about the environment today, they will be richly rewarded by making the effort and taking the time to read this book. . . . Brinkley is a master." —John Miles, *National Parks Traveler*

"As the US grapples with climate change, Douglas Brinkley reminds us that a new generation of twenty-first-century environmentalism can save the planet from ruin." —Michael Svoboda, Yale Climate Connections

"Even as it vanishes under concrete and asphalt, wilderness keeps an emotional hold on the collective American heart. Douglas Brinkley's rich and illuminating history of the modern environmental movement shows how tenacious dreamers battled against all odds to save some of our most treasured wild places. It is both a wonder, and a timely inspiration."   —Carl Hiaasen

"With the United States grappling with climate change and resource exhaustion, Douglas Brinkley's meticulously researched and deftly written *Silent Spring Revolution* reminds us that a new generation of the twenty-first century can save the planet from ruin. A terrific book."
—Luke Metzger, Environment America

"Douglas Brinkley's *Silent Spring Revolution* is about my hero Rachel Carson. It's an awesome look at the 1960s and the 1960s environmental movement offering profound insight for the age of climate change."
—Alice Waters author of *We Are What We Eat*

New York Times's "Ezra Klein Show" Book Recommendation for 2023

SILENT SPRING
REVOLUTION

SILENT SPRING

REVOLUTION

John F. Kennedy, Rachel Carson,
Lyndon Johnson, Richard Nixon, and the
Great Environmental Awakening

DOUGLAS BRINKLEY

HARPER ● PERENNIAL

NEW YORK ● LONDON ● TORONTO ● SYDNEY ● NEW DELHI ● AUCKLAND

HARPER ● PERENNIAL

HarperCollins books may be purchased for educational, business, or sales promotional use. For information, please email the Special Markets Department at SPsales@harpercollins.com.

FIRST HARPER PERENNIAL EDITION PUBLISHED 2024.

Library of Congress Cataloging-in-Publication Data has been applied for.

ISBN 978-0-06-321292-3 (pbk.)

24 25 26 27 28 LBC 5 4 3 2 1

Dedicated to my wife, Anne Brinkley . . . Everlasting Gratitude
and the Walden Woods Project

The more clearly we can focus our attention on the wonders and realities of the universe about us, the less taste we shall have for destruction.

—Rachel Carson, accepting the John Burroughs medal (1952)

When will people fully understand and accept the obligation to the future—when will they behave as custodians and not owners of the earth?

—Rachel Carson to Stewart Udall, November 12, 1963

Contents

Part III: The Environmentalism of Lyndon Johnson and Richard Nixon (1964–1973)

CONTENTS { xi }

John F. Kennedy, Coos Bay, Oregon, 1959. On his trip to the Pacific coast, he met with Senator Richard Neuberger, a lobbyist for the establishment of the Oregon Dunes National Seashore.

Preface

As I sit at my office desk at home in Austin, Texas, my bookshelves are packed with conservation histories. But across the room, I see arresting images on TV of California firefighters at Yosemite struggling to prevent some of the world's oldest giant sequoias from burning. Amid the western slopes of the Sierra Nevada, John Muir's hallowed Mariposa Grove, with its 200-foot trees, is right now in the path of the 4,200-acre Washburn Fire and swathed in columns of menacing white smoke. The big trees, as Muir called them, could normally withstand fire due to their thick, moist bark. Now, with warmer annual temperatures and sustained droughts, their bark is thinner, drier, and less able to keep the giant sequoias protected when they're threatened.

Human-caused climate change is now everywhere, evident even in the bark of two-thousand-year-old trees. Massive wildfires have become so routine around Yosemite that visitors arriving at all four entrances are greeted by the charred reminders of the ongoing catastrophe. What California is experiencing isn't a series of freak global-warming events, and Yosemite isn't an anomaly. This is the new normal, courtesy of our nation's—indeed, the world's—addiction to fossil fuels. Here in Austin, it's a brutal 110° Fahrenheit (about 43° Celsius)—while I sit inside, hiding in air-conditioned comfort from the weather, from nature, from my typically active life outdoors. Triple-digit heat has led to poor air quality, which exacerbates my asthma.

Writing *Silent Spring Revolution* during these drastic years of climate danger has been a long, affecting adventure, but one that buoyed my spirits over our current predicament. Exactly sixty years ago, Rachel Carson's *Silent Spring* exposed the dangers of pesticides, turned environmentalism into a public health crusade, and helped galvanize a whole new generation of "green" activists. During the Long Sixties (1960–1973), this *Silent Spring* generation inspired three presidents to heroic environmental action and moved Americans of all stripes to stand up to protect the only planet we have from defilement.

Although it doesn't constitute a major part of this narrative, I document how John Kennedy, Lyndon Johnson, and Richard Nixon all dealt with climate change. The burning of fossil fuels is the main cause of global warming, as greenhouse gases released into the atmosphere contain the sun's heat, raising global average temperatures and fueling vicious weather events, including record-shattering heat. Both JFK and LBJ knew about the climate threat, courtesy of high CO_2 emissions, but at the time it seemed like a distant problem. White House adviser Daniel Patrick Moynihan went so far as to write President Richard Nixon a memo in 1969 about "the carbon dioxide problem," warning that the heated planet could cause ice caps to melt and oceans to rise. "Goodbye, New York," Moynihan wrote. "Goodbye, Washington, for that matter. We have no data on Seattle."

My journey to write *Silent Spring Revolution* began with my book *The Wilderness Warrior: Theodore Roosevelt and the Crusade for America* (2009), which I envisioned as the first of three volumes linking US presidential history (one of my fortes) to three waves of twentieth-century environmental progress and policy: Theodore Roosevelt (1901–1909); Franklin D. Roosevelt (1933–1945); and the Long Sixties triumvirate of Kennedy, Johnson, and Nixon. In *The Wilderness Warrior*, I described how our twenty-sixth president preserved over 234 million acres of wild America between 1901 and 1909. In *Rightful Heritage: Franklin D. Roosevelt and the Land of America* (2016), I documented the progressive second wave, in which Secretary of Interior Harold Ickes, Audubon Society–inspired Eleanor Roosevelt, and the US Forest Service's intrepid mountaineer Bob Marshall acted on FDR's enthusiasm for preserving treasured landscapes in every state.

Sandwiched in between, I wrote *The Quiet World: Saving Alaska's Wilderness Kingdom 1879–1960* (2011), which focused on the storehouse of natural resources situated in the Last Frontier and chronicled the guardians, from John Muir to Dwight Eisenhower, who campaigned to forever protect the wondrous paradise from Glacier Bay to the Arctic Range. These days the Greenland ice sheet is vanishing much faster than projected, which adds to the rising sea level. Alaska's Arctic coastline is exposed to more sunshine, which causes intense warming, leading to unprecedented melting of blue-green glaciers and sprawling ice fields. The wild Alaska of *The Quiet World*, where over 60 percent of our national parks are located, is vanishing. The threatened polar bear has become the symbol for climate change awareness. How will *Ursus maritimus* survive in perpetuity if its Arctic habitat disappears?

Of all the many books I have written since earning my PhD from Georgetown University back in 1989, I consider these histories, taken together, to be a true cornerstone of my work, merging my presidential history focus with a deep-seated passion for national parks, ecology, and wildlife. (I am not including an additional title of mine from 2006—*The Great Deluge: Hurricane Katrina, New Orleans, and the Mississippi Coast*—because it is as much a story of political and civic crises as it is a tale of environmental upheaval.)

Silent Spring Revolution, then, completes what I have envisioned as a presidential trilogy, detailing how, after the radioactive shock of Hiroshima, a network of conscientious postwar anti-nuclear and protoenvironmental activists launched a reform-minded revolution and how three very different presidents drove a cascade of remarkable "green" reform measures into American public law.

During that era, trust in the federal government was sky-high before the Vietnam War brought it crashing down. When Kennedy was in office, three-quarters of the public expressed faith in the government; it is down to 18 percent today. Decades of anti-government zealotry, however, have taken a toll, among other things, on US environmental protection funding. Most of the give-and-take backstories of regulatory laws passed by Congress in the Long Sixties—recounted in these pages—were bipartisan initiatives. Democrats and Republicans boldly united to save the Great American Outdoors from further desecration.

Take, for example, the Clean Air Act of 1970, which passed the

xvi PREFACE

House by 374 to 1 and the Senate by 74 to 0. Likewise, the Endangered Species Act of 1973 was passed by a 355 to 4 vote in Congress and a 92 to 0 vote in the Senate. That's roughly the way visionary environmental laws were enacted throughout the Kennedy, Johnson, and Nixon years. A lot of wrangling, frustration, and tireless negotiating eventually led to the introduction of well-reasoned bills on Capitol Hill, which were passed under the stalwart leadership of such conservationists as Senator Henry M. Jackson (D-Washington) and Thomas Kuchel (R-California).

Just about all of Washington's power players in the Long Sixties seized on any last-gasp opportunities at hand to preserve America's precious natural resources; embrace a high standard for quality of life; and deliver clean air and clean water bills to demanding constituents. Because the public insisted on a greener tomorrow, three presidents embraced environmental regulation as administration hallmarks. "Our resources will not be protected without the concern and help of every private citizen," President Kennedy warned at the National Wildlife Federation (NWF) headquarters opening in Washington, DC, in 1961. "By mobilizing private efforts through the organization, you are helping not only to develop the wildlife resources of our country—but you are helping to create that kind of America of open spaces, of fresh water, a green country—a place where wildlife and natural beauty won't be despoiled—where increasing urbanized population can still go to the country, can still turn back the clock of our civilization and find the material and spiritual strength upon which our greatness as a country depends."

President Kennedy, under the rubric of the New Frontier, negotiated with the Soviet Union and Great Britain for the banning of atmospheric and underwater testing of nuclear weapons after the Cuban Missile Crisis. His administration also thoughtfully embraced the anti-pesticide findings of Carson's *Silent Spring* as the scientific data ended up supporting her assertions. On July 23, 1962, at a famous press conference, Kennedy sought to tame the chemical industry's outrage toward "Miss Carson's book." The words he offered from his bully pulpit were meant to discourage rising corporate attacks on Carson, the brave former US Fish and Wildlife Service employee who had once deemed oceans "the great mother of life" and spoke truth to power about the scourge of pesticides. *Silent Spring* began with a doomsday description of a fictional American town doused in DDT, which eerily could be various climate-affected zones of contemporary times: "Then a strange blight crept over the area, and everything began to change. Some evil spell had settled on the community: mysterious maladies swept to flocks of chickens; the cattle sickened and died. Everywhere was a shadow of death."

That September, when *Silent Spring* triggered a crusade against DDT, there was no *environmental* policy-making in the modern sense. It was still the *conservation* movement of Theodore Roosevelt, John Muir, and Gifford Pinchot that held national sway, in all its preservation and wise-use variations. And while the Cold War conservation community had its bipartisan Wilderness Warriors such as Senator Frank Church (D-Idaho) and Representative John Saylor (R-Pennsylvania) in the 1960s, these lawmakers focused primarily on conserving pristine land for outdoor recreationists. It was Rachel Carson, full stop, who, in an urgent, visceral way, sparked an ecorevolution with *Silent Spring* by connecting Rooseveltian preservation with public health concerns about the pesticide DDT. Carson, the galloping Paul Revere of Earth stewardship, warned Americans that, depending on the communities in which they lived, their children weren't safe playing on grassy lawns or netting crawfish in creeks or even wandering in a field of yellow wildflowers.

In the wake of *Silent Spring*, Lyndon Johnson upgraded New Conservation as a White House priority. With his own Texas Hill Country as his idyllic foundation—sixty miles from where I sit in Austin—and his wife, Lady Bird, urging him onward, he established thirty-five national parks, most within an easy commute of big cities. His environmental protection and beautification efforts were epic: the Land and Water Conservation Fund Act, the Water Quality Act, the Highway Beautification Act, the National Historic Preservation Act, the Clean Water Restoration Act, Air Quality Act, the National Trails Act, and the establishment of the Canyonlands (Utah), North Cascades (Washington), Redwood (California), Indiana Dunes (Indiana), Pictured Rocks (Michigan), and Guadalupe (Texas) as new National Park Service units—all in just five remarkable years.

President Johnson's National Wilderness Preservation System, signed into law on September 3, 1964, saved pristine roadless back-country from the destruction of bulldozers and steam shovels. To-day such wilderness areas constitute about 4.5 percent of the United States' landmass. "We must maintain the chance for contact with beauty," Johnson wrote in Presidential Policy Paper No. 3 on November 1, 1964. "When that chance dies a light dies in all of us. We are the creation of our environment. If it becomes filthy and soiled, then the dignity of the spirit and the deepest of our values are in danger."

Then came President Nixon. To the surprise of the political intelligentsia, the California-bred Republican became a reluctant environmentalist, promoting sewage treatment plants and clean air laws with the consummate skill of a New Dealer. Long before the Watergate scandal destroyed his presidency, the conservative politician shocked liberals by making environmental policy the centerpiece of his remarkable January 22, 1970, State of the Union address. Relying on his White House domestic policy adviser, John Ehrlichman, a smart land and water lawyer from Seattle, Nixon supported the Endangered Species Acts of 1969 and 1973, the Occupational Safety and Health Administration (OSHA), the Marine Mammal Protection Act (MMPA), and many more.

While it's not as visually exciting as Senator Robert F. Kennedy (D-New York) shooting white-water rapids on the Colorado River or canoeist Sigurd Olson paddling the Boundary Waters of Minnesota and praising the "singing wilderness" in his journals—both recounted here in vivid detail—the culminating event of the Long Sixties ecology zeitgeist was President Nixon's signing of the National Environmental Policy Act (NEPA) on New Year's Day 1970 (at the Western White House in San Clemente, California). The act enshrined a legal foundation for United States environmental policy and required that "any major federal action significantly affecting the quality of human environment" would require an evaluation of the public disclosure of potential environmental impact through a required environmental impact statement (EIS). These goals have become the foundational stones of US environmental law in every administration from Nixon to Biden.

The *Silent Spring* vanguard really came into its own on July 9, 1970, when Nixon sent reorganized plans to Congress, creating the Environmental Protection Agency (EPA) and the National Oceanic

Atmospheric Administration (NOAA). Carson, who had passed away from cancer six years before, would have been jubilant: her worries about pesticides, industrial pollution, and ocean protection had finally been taken seriously by the US government. The EPA officially opened its doors that December, with mandates to dispense with the hodgepodge of oversight agencies (the Atomic Energy Commission along with the Departments of Interior; Agriculture; and Health, Education, and Welfare) and streamline federal programs (for radiation standards, air and water pollution, pesticide control, and solid-waste management) under a single umbrella. The EPA was a godsend for a nation struggling with how to juggle the competing interests of economic growth and proper environmental stewardship.

Moreover, it was the EPA that finally banned DDT nationally, ten years after Carson's book warned of its deleterious effects. William Ruckelshaus, the first EPA administrator, deserves the credit for that one. Investigative panels and committees throughout the decade substantiated the danger of pesticides. Ruckelshaus, in banning DDT, knew he was validating Carson for the ages. "The continued massive use of DDT," Ruckelshaus said, "posed unacceptable risks to the environment and potential harm to human health."

Nixon, during his notorious five-and-a-half-year presidency, was a clever Machiavellian chess player when it came to navigating the environment-versus-energy divide. Playing bureaucrats off each other with coldhearted cunning, he was disgruntled about his own EPA ban on DDT because it inflamed cotton growers in the South and ranchers in the West. As Nixon told White House adviser H. R. Haldeman—who recounted the conversation in his unpublished diary entry on June 9, 1972—he *had* to retain Ruckelshaus "to keep the environmentalists happy." But in a countermove, Nixon soon ordered that all the "left-wing liberals" and environmental activists on the EPA's staff be dismissed. "They believe in it [the environmental movement]," he complained to Haldeman. "You can't have an advocate dismantle something they believe in."

Rachel Carson—like her spiritual hero, the Nobel Peace Prize–winning Dr. Albert Schweitzer working as a medical missionary in French Equatorial Africa—was a *believer* that proper Earth stewardship was a holy imperative. She reminded readers in her sea trilogy of *Under the Sea-Wind* (1941), *The Sea Around Us* (1951), and *The Edge of the Sea* (1955) that America's future depended on nature being in balance. Sadly, today that balance is egregiously out of whack. During the decade I spent writing this narrative history, most of the Atlantic seaboard National Park Service units established in the Long Sixties that I explore herein—such as Assateague National Seashore (Maryland and Virginia), Cape Cod (Massachusetts), Cape Lookout (North Carolina), Cumberland Island (Georgia), and Fire Island (New York)—have been mangled by the frightening snarl of climate change. So have human residences and wildlife habitats all over the nation. The poisonous climate cocktail of ravaging wildfires, megafloods, and monstrous blizzards costs American taxpayers billions of dollars every year (soon to climb to the trillions), disrupts lives, causes vast displacement, ruins communities, and kills countless citizens.

Since I began my first draft of chapter 1, extreme weather has been brutalizing our National Park System. Vicious flooding wiped out roads in Yellowstone National Park, forcing the emergency evacuation of thousands of visitors. In rural California near Yosemite, farming communities are contending with arsenic in their water, ten times above allowable limits. Record-setting heat waves are driving temperatures into triple digits along the Lewis and Clark Trail. Water-bombing planes are proving ill equipped to extinguish infernos in Oregon's Cascades. Around Big Bend, the Big Thicket, and Guadalupe—three important National Park System units in Texas—the effects of chronic drought have turned grazing pastures into a parched dust bowl. Soaring temperatures in Colorado's Great Sand Dunes and Black Canyon National Parks have led to bison dying from heat and stress. Drought and wildfires threaten the survival of the prickly-pronged icons *Carnegiea gigantea* of Arizona's Saguaro National Park. Freaky atmospheric rivers are soaking the Olympic Peninsula in Washington State, causing record-setting rainfall and epic floods. Changing rainfall patterns and warming weather are poised to decrease the cloud cover in the Great Smoky Mountains, which could be catastrophic to the lush green forests. No national park is immune to the warming of the planet.

Many of the storied North Cascades National Park peaks that Jack Kerouac swooned over in his novel *The Dharma Bums* (1958) are no longer heavily snow-capped in summer. The blue-green ice sheets in Glacier Bay National Park, Alaska, that Muir wrote about in his classic *Travels in Alaska* (1915) have begun to melt even faster than previously forecast. Montana's Glacier National Park will soon have an obsolete name. Hurricanes, tornadoes, heat waves, nor'easters, and bomb cyclones are increasing with high-octane ferocity as climate change churns wickedly and ransacks our National Park System.

My buddy Terry Tempest Williams recently sent me a lively letter about her summer with her husband, Brooke, in Castle Valley, Utah—a report card of sorts, about how the red rock wilderness near Arches National Park is faring in this time of environmental crisis. She wrote:

> *Drought is here in the American West, especially this beautiful broken landscape where we live. I find such solace here even as it burns. A few weeks ago, it was 114 degrees. Last summer, there were places on the Colorado River you could walk across. I saw a doe and fawn cross the river and I thought it was a heat wave hallucination. And just last week, we had a flash flood. I smelled it before I heard it—I heard it before I saw it—and when the rushing water came like a wall of churning red water, fear was transforming into awe. Water! I had forgotten the sound of roaring water in this decade of drought. Sandstone boulders the size of small cars were storming through the arroyo twenty feet from our house. I stood on the berm and watched the torrents with the groundwater now saturated reaching my ankles, then shins in seconds. One lone monarch fluttered above the chaos, dipping down in graceful intervals to pollinate the penstemons—orange-black wings floating in a blue sky—this days after her species had been declared endangered. Violence and grace have always existed side by side.*

With the twenty-first-century climate challenge weighing heavily on Terry, Brooke, and me, the three of us and many of those in our literary circle have sought solace in the long shadow of Henry David Thoreau as well as the grassroots activists he inspired during the Long Sixties. That story begins on April 23, 1851, when Thoreau spoke at the Concord Lyceum about the interrelationship of God, man, and nature and delved into the spiritual power of wilderness. Thoreau ended his oration with eight words that live on as a mantra: "In Wildness is the preservation of the World." The sentiment became popularized when *The Atlantic*, a month after his death, published Thoreau's essay "Walking" in May 1862, with that line as the centerpiece.

Most Americans know Thoreau from reading *Walden*, with its simple assertion, "I went to the woods because I wished to live deliberately." But it's the "In Wildness" epigram from "Walking" that was embraced as near scripture for many of the environmentalist leaders profiled in this narrative. Toward the end of his life, Thoreau famously called for townships to have "a park, or rather a primitive forest, of 500 or a thousand acres, where a stick should never be cut for fuel, a common possession forever, for instruction and recreation." His posthumously published *The Maine Woods* (1864) went further, calling for "national preserves." Less than ten years later, Congress created Yellowstone as America's first national park, a template for all that followed. Thoreau was crucial to the *Silent Spring* generation not only as an environmentalist forerunner but as a social rebel and dissident, a nineteenth-century visionary bridging the American narrative from the War of 1812 to the Long Sixties—and up through the present day.

Just as Thoreau's 1849 essay "Resistance to Civil Government" nourished nonviolent protests by Cesar Chavez, Norman Cousins, Martin Luther King Jr., and Coretta Scott King (all key actors in this narrative), Thoreau's "In Wildness" precept electrified the soul-searching imaginations of writers such as Marjory Stoneman Douglas, Robert Frost, Wallace Stegner, Joseph Wood Krutch, Nancy Newell, Edward Abbey, N. Scott Momaday, and Gary Snyder during the Kennedy, Johnson, and Nixon years. Likewise, William O. Douglas (Supreme Court justice), David Brower (executive director of the Sierra Club), Howard Zahniser (executive director of the Wilderness Society), and Sigurd Olson (president of the National Parks Association) kept Thoreau in mind when warring against pesticides, water and air pollution, hyperindustrialization, and the despoiling of public lands.

Thoreau's words motivated Rachel Carson in a very personal way. After Carson learned she had breast cancer, she adopted a sentence from Thoreau's *Winter* journal to spur her writing of *Silent Spring*: "If thou art a writer, write as if thy time were short, for it is indeed short at the longest." In a bit of environmental synchronicity, *Silent Spring* was published in 1962, the centenary of Thoreau's death. When Carson died, *Walden* was found on her nightstand at her Silver Spring, Maryland, home.

In 1962, David Brower, executive director of the Sierra Club, published a book titled *In Wildness Is the Preservation of the World*, the introduction of which includes the observation, "To me, it seems that much of what Henry David Thoreau wrote more than a century ago was less timely in his day than it is in ours." In the 1950s and early 1960s, as the postwar automobile and trucking boom surged, America experienced a ghastly environmental crisis: the vexing threat to air quality posed by leaded gas as well as underregulated factory emissions, which left cities such as Los Angeles and New York choking under domes of smog. Grassroots anti-smog groups proliferated across America, with concerned citizens like Hazel Henderson and Mary Amdur warring against poisoned air, armed with persuasive scientific data.

Veering off the mindless grind of consumerism run amok in the Long Sixties was the indomitable William O. Douglas, a Thoreauvian who protested with his leather boot–laced feet, staging high-publicity hikes to forever protect the North Cascades and Olympic Peninsula (Washington), the C&O Canal (Maryland–Virginia), the Red River Gorge (Kentucky), the Allagash Waterways (Maine), Allerton Park (Illinois), and the Buffalo River (Arkansas). When this Supreme Court justice was not busy drafting Court opinions—and voluminous dissents—his hobby (beyond writing books on travel and the environment) was successfully stopping dams from being built. Such "green" victories against dams led other environmentalists, famous or not, to act with equal intensity to save wilderness values during the Kennedy, Johnson, and Nixon years.

Activists defending the ecological integrity of the Hudson River and Lake Erie constitute an essential part of this historical recounting. The intensified argument was that clean water was considered an American birthright. When journalist Hunter S. Thompson went to Wisconsin in 1972, he peered into Lake Michigan and was horrified. "The lake is out there somewhere," he wrote. "A giant body of water full of poison. You can still find a few places that serve 'fresh seafood' in Milwaukee, but they have to fly it in from Maine or Bermuda, packed in dry ice. People still fish in Lake Michigan, but you don't want to eat what you catch. Fish that feed on garbage, human shit, and raw industrial poisons tend to taste a little strange."

The Clean Water Act of 1972 became one of the great legislative achievements in American civic and environmental history. This act inspired other countries to emulate us. Around the same time the Nixon administration established two of America's most ambitious national recreation areas (NRAs)—Golden Gate (California) and Gateway (New York and New Jersey) to help beautify waterways around urban areas. (Today a similar Lone Star Coastal National Recreation Area is urgently needed for the Houston-Galveston area along the Gulf of Mexico.) And the evidence is clear that Nixon was very proud of California's magnificent giant tree groves. When he made his historic trip to China in 1972, the president presented Mao Zedong with California redwood and sequoia saplings as the most culturally appropriate gifts representing the wild glories of America.

Despite Nixon himself lacking a Thoreauvian connection, his administration maintained one in the person of Russell Train, undersecretary of the interior (1969–1970) and head of the Council of Environmental Quality (1970–1973), who told me he had shepherded the Endangered Species Act into law with Thoreau as his "literary guide." In another instance of Thoreau being more timely in an age later than in his own, the act protected nonhuman creatures that Thoreau considered "honest spirits as good as myself, any day," living their lives in "a civilization other than our own." Today, though, even with the foresight of the Endangered Species Act, over one-third of North America's fish and wildlife species are at risk of mass extinction in the coming decades due to threshold effects in the carbon cycle and chemical contamination. The good news, however, is that the rewilding of Yellowstone with gray wolves (*Canis lupus*) has been a success story. Other endangered species—such as the leatherback sea turtle (*Dermochelys coriacea*), Atlantic cod (*Gadus morhua*), and monarch butterfly (*Danaus plexippus*)—can be reversed from the brink of extinction with visionary policy adjustments from Washington lawmakers and biological laboratories working on species salvation issues with emergency sirens going off.

Strange as it may seem, one of my models for this book was Daniel Yergin's *The Prize: The Epic Quest for Oil, Money, and Power* (1991), a masterpiece of energy policy history. Just as Yergin didn't tackle environmental politics head-on in *The Prize*, I haven't sought to analyze the actions of the oil lobby and coal consortiums following World War II. I'm not here to demonize the petroleum industry, the Bureau of Reclamation, the Forest Service, or the Army Corps of Engineers as greedheads—though many of the leaders profiled in this book did. But the stark truth is that many hydroelectric dams built between 1957 and 1975 were unnecessary, especially the enormous one at Glen Canyon, which is killing the Colorado River, as well as the four dams on the lower Snake River that are destroying the migratory salmon runs. The federal government has been sued many times by the Nez Perce, Yakama, and Umatilla tribes for decimating the Columbia River basin of the Pacific Northwest. Hopefully the Biden administration, in the name of "tribal justice," will remove these pork-barrel dams from the Pacific Northwest in the near future. Salmon need cold, swift-flowing streams and rivers to spawn (not humongous dam ladders the fish can't jump). Recent heat waves have also caused parasites to invade Alaska's Yukon River, thereby decimating Chinook and Coho salmon runs.

I largely skirt the counterrevolution that emerged against Rachel Carson, Ralph Nader, Barry Commoner, and the other environmental voices of the Long Sixties: it arose with vengeance in early 1974, around the noisy time of Watergate, just as this narrative ends. The anti-federal-regulation manifesto, which had a galvanizing effect, kicked in when Lewis Powell—a Harvard-trained lawyer and Virginia tobacco lobbyist—wrote a confidential memo to the chair of the US Chamber of Commerce a year before becoming a Supreme Court justice in 1972. Believing that the *Silent Spring* environmentalist threat came from "the campus pulpit, the media, the intellectuals and literary journals, the arts and sciences, and from politicians," he warned that if federal restrictions on industry were to continue or expand, "the *survival* of the free enterprise system" was at stake. Powell elaborated a multitiered strategic plan to counter Rachel Carson and her tribe of trailblazers by organizing an anti-regulatory political knock-down with its own media and corporate leaders seizing control of universities with pro-business boards. Counterintuitively, the Powell memo turned out to be a backhanded tribute to how brazenly effective the *Silent Spring* eco-activists were in rattling Wall Street's cages.

Johnson's White House aide Bill Moyers, who went on to renown at PBS, would later write that big business's counterattack against the Carson revolution was "swift and sweeping—a domestic version of Shock and Awe." And when the Arab oil embargo hit American gas pumps with a wallop in late October 1973, the public pushback gave the bigwigs in the Petroleum Club of Houston new ammo against the green lobby. Nothing disturbed the collective consumer's pocket-book more than high gasoline prices and runaway inflation. The Silent Spring Revolution ended when gasoline hit $0.55 a gallon in 1974 (which in 2022 dollars would be $3.05).

Feeling cornered by lawmakers such as Ed Muskie (D-Maine) and Senator Gaylord Nelson (D-Wisconsin), the number of corporations with public affairs offices in Washington grew from one hundred in 1968 to more than five hundred by 1978. Anti-regulatory rhetoric became institutionalized in chamber of commerce circles and legitimized by new conservative think tanks such as the Cato Institute, the Heritage Foundation, and the American Enterprise Institute. By 1977, five years after its creation, the anti-regulatory, pro-business lobbying group Business Roundtable could boast that 113 of the Fortune 200 companies were members, accounting for nearly half the American economy.

Furthermore, because the United States government owned 47 percent of all land in the American West, a series of Sagebrush Rebellions simultaneously exploded onto the national scene in the 1980s and beyond in states such as Nevada, Oregon, and Idaho. Anti-government protests rose in shrillness and violent intent. Conservative Republicans after Nixon, perhaps the last New Dealer to serve as president, grew increasingly convinced that the Silent Spring Revolution was leading to nothing less than the end of American free-market capitalism. The Business Roundtable didn't mind incremental Theodore Roosevelt–style conservation reform aimed at national park creation (i.e., eco-tourism dollars to invigorate local economies near the federal parks). But Lyndon Johnson– and Richard Nixon–style regulatory overreach against the chemical and extraction industries, they fumed, were a socialist dragon they sought to slay. These post-1974 anti-regulatory forces—both Powellian and Orwellian—have never left the national stage, expanding under President Ronald Reagan, a staunch foe of federal regulation, and reaching their apotheosis during the Trump administration. Today's Supreme Court is the triumph of Powell's pro-corporate stance reversing Douglas's anti-industry movement, which had established environmental protection as fundamental to life, liberty, and the pursuit of happiness as outlined in the Declaration of Independence.

Just now, as I type on my laptop, a news story popped up that the EPA has found cancer-causing PFAS "forever chemicals" in the water supply of my city of Austin, which are linked to cancer, liver damage, fertility issues, and thyroid disease. Who will be the Rachel Carson of the PFAS crisis? Who will hold accountable the derelict companies that manufacture a group of about twelve thousand different chemicals used in stain-resistant foam, waterproof cosmetics, and nonstick cookware? Apparently, environmental experts say it is unwise now to allow rainwater on your tongue in Austin or swim in local watering holes such as Hamilton Pool and Waller Creek. How's that for a brave new world?

The time has come for a Grand Reversal. Legal powerhouses such as the National Resources Defense Fund and state, community, and citizen groups around the nation must sue reckless polluters, chemical dumpers, resource gougers, and land ravagers to save our beautiful nation from ecological ruin. In a country as highly legalistic as the United States, going to court is the only way to hold the defilers accountable.

What hopefully will be instructive about reading *Silent Spring Revolution* is learning how grassroots citizens *demanded* an American life replete with birdsong, sweet waters, fresh air, and green pastures aplenty. Under the Environmental Defense Fund slogan "Sue the Bastards," a string of citizen class-action legal suits were won following the passage of NEPA. The Grand Canyon, Potomac River, North Woods of Maine, and Redwood forests were saved via court battles. Even wealthy moguls who made dark money in the extraction world were engaged in the "not in my backyard" (NIMBY) movement to protect sanctified landscapes like Biscayne Bay, Puget Sound, Cape Cod, the San Juan Islands, and Point Reyes. As philanthropist Laurance Rockefeller used to say, the "biggest privilege" an individual can have is to serve a cause larger than oneself, such as protecting glorious landscapes from bulldozers and chemical flush.

It's also my contention in *Silent Spring Revolution* that the environmental justice movement of the late twentieth century emerged as a set of serious policy considerations with the parsing of the Civil Rights Act of 1964 and of Title VI (which forbade the appropriation of federal money to discriminate on the basis of race, color, or national origin). When Cesar Chavez led strikes against grape growers to draw media attention to the mistreatment of farm workers, he was operating under the double whammy of *Silent Spring* and the Civil Rights Act. When Ralph Abascal of the California Rural Legal Assistance filed lawsuits at the bequest of half a dozen migrant farm workers in 1969, which led to the banning of DDT in California three years before the EPA prohibited the pesticide nationally, he was likewise following in the Carson tradition. And certainly, when Martin Luther King, Jr., struggled on behalf of Black sanitation workers in Memphis, Tennessee, shortly before his murder in April 1968, the environmental justice movement had taken full root.

When writing *Silent Spring Revolution*, I was reminded that every environmental disaster is an opening for much-needed environmental reform. In 1948, after more than twenty people died in Donora, Pennsylvania, from chemical emission sickness, the clean air movement was born. In 1969, Cleveland's Cuyahoga River was so polluted with industrial waste that it actually burst into flames. The public outcry was fierce, forcing Congress to pass the Clean Water Act in 1972. Today, Donora is a lovely town to visit, and the Cuyahoga is in much better ecological shape and is home to healthy populations of steelhead trout, northern pike, and some sixty other fish species.

Echoing something Woody Guthrie once coined about his folk songs, I consider *Silent Spring Revolution* to be a "hope machine," reminding readers that our own times aren't uniquely oppressive and that everyday people and nonprofit groups retain the capacity to fight for a green tomorrow in the United States and around the world. Optimism must remain in our oxygen. And the good news on this sweltering hot July afternoon in Austin is that the Mariposa Grove has narrowly escaped being engulfed in flames. On my television, exhausted firefighters are spraying down trees in Mariposa Grove and wrap sequoias—including the three-thousand-year-old Grizzly Giant—in protective foil. These crusading first responders are our surrogates. We all need to become hands-on Earth stewards (or at least vote for politicians who prioritize protecting our natural resources). And the EPA must proactively regulate methane (a vicious greenhouse gas that leaks from oil and gas wells) and end our coal dependency while we pivot to renewable energy. Most important, the entire United States must follow California's lead in prohibiting the sale of new gasoline-powered cars by 2035.

In some small way, I hope this book illuminates how an engaged citizenry can bring America's natural beauty back from the brink. On the second Earth Day, in 1971, a public service campaign was launched under the rubric "People Start Pollution; People Can Stop It." It can just as easily be said that "People Use Fossil Fuels; People Can Stop It." My fear is that too many Americans don't give a hoot in hell about being good air, water, and land stewards. The ostrich syndrome has become a coping mechanism for scores of dissatisfied Americans. What's clear is that we need a new, bold generation of twenty-first-century wildlife biologists like Archie Carr, forest protectors like Stewart Udall, entomologists like E. O. Wilson, ocean stewards like Sylvia Earle, and environmental justice warriors like Marion Moses, who nursed sick San Joaquin Valley agricultural workers suffering from the ravages of harmful chemicals.

People can reverse the damage of the planet with Carson-like conviction, a thoughtful ecological conscience, an understanding of Dr. Albert Schweitzer's "reverence for life" philosophy, and a reasoned trust in fact-based science. If nothing else, I hope *Silent Spring Revolution* helps readers reconnect with America's public lands and freshwater resources: our lakes, mountains, rivers, seashores, islands, deserts, marshes, and woods—and their myriad inhabitants.

In 1970, John Lindsay, then mayor of New York City, posed a prescient existential question: "Beyond words like environment and pollution, there is a single question: Do we want to live or die?" It is my contention that only direct climate action can save this land, this world, the human race itself. America must have a shared environmental ethic to help arrest the high cost of our fossil fuel addiction. Reducing greenhouse gas emissions is the surest way to mitigate climate impact on our treasured national parks, wildlife refuges, and other sacred public lands.

"We are the most dangerous species on the planet, and every other species has cause to fear our power to exterminate," the novelist Wallace Stegner observed. "But we are also the only species which, when it chooses to do so, will go to great efforts to save what it might destroy."

Austin, Texas
July/August 2022

Part I

Protoenvironmentalists (1945–1959)

CHAPTER 1

The Ebb and Flow of John F. Kennedy

John F. Kennedy was in love with the Atlantic Ocean marine environments. In coming years, he would lead the fight on behalf of the establishment of the Cape Cod National Seashore.

I

Beguiled by the way sea and sky played together, almost always unpredictably, John F. Kennedy was enthralled by the complexity of the Atlantic Ocean: the moody sky, the invisible might of the tides, shifting clouds, and the yaw and pitch of movement. To be on the water in a sailboat, even in a cruel wind, provided him with a profound connection with nature. At the America's Cup dinner in Newport, Rhode Island, in 1962, President Kennedy spoke on behalf of "those of us who regard the ocean as a friend," as he put it.

> I really don't know why it is that all of us are so committed to the
> sea, except I think it is because in addition to the fact that the sea

changes and the light changes, and ships change, it is because we all came from the sea. And it is an interesting biological fact that all of us have, in our veins, the exact same percentage of salt in our blood that exists in the ocean, and, therefore, we have salt in our blood, in our sweat, in our tears. We are tied to the ocean. And when we go back to the sea, whether it is to sail or to watch it we are going back from whence we came.[1]

Oceans were mystical to Jack Kennedy. Though he was not known to surrender emotionally to land and bodies of fresh water, he drew spiritual nourishment from the hypnotic rush of surf on a shoreline at high tide. At such times, a calm peacefulness washed over him, propelled by the fresh scent of the breeze, the glint of the sun setting on the sea, and the primordial sound of water crashing and rippling. Like any good mariner, Kennedy knew that the ocean drives climate and weather, regulates temperature, and holds 97 percent of Earth's water. Given his family lineage and predisposition, his muse and spiritual home was the Atlantic seaboard, from Cape Cod to Newport to Palm Beach. That bond with the sea made him sympathetic with the protoenvironmentalism that swept over the United States in what historians call the "Long Sixties" (1960–1973); indeed, seashore preservation would be a catalyst of the Silent Spring Revolution, which the writer Rachel Carson ignited.

Kennedy was born on May 29, 1917, in Brookline, Massachusetts, a middle-class streetcar suburb of Boston. The solid residence at 83 Beals Street, now a National Park Service site, was a safe harbor in a troubled world with the United States' involvement in the Great War in full swing, only a month old. His mother was a member of one of Boston's most successful political families, the Fitzgeralds. His father, Joseph Kennedy, Sr., a 1912 graduate of Harvard University, parlayed what connections an Irish Catholic might make within Boston's business circles into financial success in Wall Street investment banking and real estate development, culminating in the acquisition of RKO Pictures and Pathé Studios. The senior Kennedy, a fierce competitor, was determined that his four sons wouldn't be spoiled rich kids; he *willed* them to make a mark of distinction on American life, preferably in politics.

Jack Kennedy spent much of his youth in New York, first in the Riverdale section of the Bronx in a house not far from the Hudson River.[2] There he joined a Boy Scout troop to learn Indian lore, wood-craft, canoeing, and camping skills. "When I am a scout," he wrote in a letter, "I have to buy canteens, haversacks, blanket, searchlights, poncho things that will last for years."[3]

Moving north to bucolic Westchester County, the Kennedys later bought a large Georgian house in Bronxville, near Sarah Lawrence College. The property had beautiful, well-manicured grounds replete with rosebushes, hydrangeas, and daffodils. Rose Kennedy was attracted to suburban living. Uncomfortable with the grind and grease of big cities, she encouraged the future president to embrace the natural world as holy. Although her father, John Francis "Honey Fitz" Fitzgerald, was a two-term mayor of Boston, he had moved his family to the historic village of Concord, Massachusetts. Rose had grown up exploring woodlands and regularly swimming in Henry David Thoreau's fabled Walden Pond, considered the birthplace of American environmentalism. She credited Thoreau's 1854 memoir *Walden; or, Life in the Woods* with encouraging her to appreciate the outdoors in a divine way, to simplify her life, and to understand that even a small act, such as planting a flower garden or an elm tree, was a commitment to Earth's future.[4] She considered Thoreau's essay "Walking" was akin to a Catholic catechism.[5] When a 1992 biographer of Rose Kennedy called her "aloof" from her own children, the family issued a protest statement to the *New York Times*. "She took us for walks in our strollers and piled us into the family station wagon to go swimming at Walden Pond."[6]

Summering in Maine as a girl, listening to Atlantic waves breaking, Mrs. Kennedy understood the transcendental intent of New England philosophy. Henry Wadsworth Longfellow's poem "The Tide Rises, The Tide Falls" was dear to her. Old Orchard Beach's salt marshes—rich with cordgrass, seaside goldenrod, and fiddler crabs—was for her a sanctified landscape.[7] Wherever she lived, her shelves were lined with books about the natural world by the likes of John Burroughs and Florence Merriam Bailey. She made it a family tradition to visit Walden Pond, where her children studied her great literary hero, who had also championed self-government over authoritarianism. Rose

Kennedy regularly quoted Thoreau's line "Heaven is under our feet as well as over our heads."[8]

Once on a trip to Russia with her daughter Kathleen, Mrs. Kennedy visited the National Library in Moscow with a singular mission: to find out whether Thoreau's works had survived the censors of the Soviet dictator Joseph Stalin. To her delight, she discovered his entire oeuvre lined up on the shelves. "Her environmentalism aligned with her natural frugality," her grandson Robert F. Kennedy, Jr., recalled in his book *American Values: Lessons I Learned from My Family*. "Never waste anything" was among her favorite mottos, and she even squirreled away uneaten fragments of her beloved York Peppermint Patties.[9] In other ways, Rose indulged herself with the family's wealth, taking long trips and buying expensive clothes, but she did embrace Theodore Roosevelt–style conservation as her ethic, and she inoculated that belief into her children.

In 1926, Joseph Sr., at Rose's urging, rented Malcolm Cottage, a waterfront retreat at 50 Marchant Avenue on the Cape in Hyannis Port, Massachusetts. It didn't take the couple long to realize that their children were happier playing on the unspoiled shore and shell middens of Nantucket Sound than in leafy Westchester County. They called it the "Big House," a many-gabled, white frame clapboard structure with green shutters built in 1904, with a boathouse and a broad two-acre grass lawn right on the shore. After two years, they bought it. "It was a tradition for our family that the moment we arrived at Hyannis Port each summer, before we even entered the house," Jean Kennedy Smith, the eighth child, recalled, "we would run down to the breakwall to say hello to the sea."[10]

Much has been written about the Kennedy family life on Cape Cod. The Big House exuded salt-tinged comfort, and field guides to help identify migratory birds were scattered throughout. Jack took to spending time on the family's sailboat. He learned how to tack in a gathering tide, navigate in difficult breakers, use a signaling kite, and, most important, avoid sandbars. Sunsets were a source of melancholy for him unless the sky promised to be star-filled soon. When he turned fifteen in 1932, his parents gave him a twenty-five-foot Wianno Senior sloop he named *Victura* (Latin for "about to win").

Before long, Jack and his older brother, Joseph Jr., were winning races around New England.[11]

Predictably, Thoreau's lesser-known 1865 book *Cape Cod* was treasured in the Kennedy household. Thoreau had first hiked the Outer Cape's unsullied shoreline in 1849, and he is credited with inventing the name "Great Beach" for the Outer Cape dunes. Jack adopted as his own Thoreau's romantic notion about Cape Cod: "a man may stand there and put all of America behind him."[12]

Another of Rose's literary attachments was Henry Beston's memoir *The Outermost House: A Year of Life on the Great Beach of Cape Cod* (1928). Beston spent a solitary year studying the Outer Cape through the changing seasons.[13] In a section of *The Outermost House* called "The Headlong Wave," he captured the sounds of the surf in lyrical prose and then moved into a sweeping perspective: "Night and day, age after age, so works the sea, with infinite variation obeying an unalterable rhythm moving through an intricacy of chance and law."[14]

Inspired by Beston's book, Mrs. Kennedy occasionally secluded herself in a wind-shorn beach shanty for a reprieve. There she read with the ocean breeze as companion and the scolding of gulls circling overhead as music.[15] Beston's words had taught her the healing art of mindfulness—to be aware always of nuances in natural surroundings. One chapter in *The Outermost House*, the one in which Beston spies on a solitary man swimming naked, reminded Rose of her gregarious father, who once stripped off all of his clothes and wrapped himself in a garland of seaweed.[16] "The sea has many voices," Beston told readers. "Listen to the surf, really lend it your ears, and you will hear in it a world of sounds: hollow boomings and heavy roarings, great watery tumblings and tramplings, long hissing seethes, sharp, rifle-shot reports, splashes, whispers, the grinding undertone of stones, and sometimes vocal sounds that might be the half-heard talk of people in the sea."[17]

Born in the Boston suburb of Quincy on June 1, 1888, to an upper-middle-class Irish American clan, Beston learned from his physician father to love the wild coasts of New England. In 1925, he bought thirty-two acres along the dunes near the Eastham Life Saving Station on the eastern edge of Cape Cod. He constructed a two-room cottage, twenty by sixteen feet with ten windows all facing the Atlantic.[18] That "fo'castle" residence provided the thirty-nine-year-old author a perch from which to observe marine life in all its manifestations. Beston, the Kennedys, and generations of advocates later pressed for federal protection of the Outer Cape. Surely that meeting of the wild shore, sea, and sky, they felt, was equal in its own way to Yellowstone or Yosemite in natural grandeur, brimming with historic, scenic, recreational, and scientific value. Gaining protection would prove to be a decades-long effort, for at the outset there was not even a category for national seashore.

In 1928, Rose Kennedy bought a new book of poems by Robert Frost: *West-Running Brook*. Its title poem became one of her credos. It is set in Derry, New Hampshire, where all rivers flow toward the ocean except West-Running Brook, which, true to its name, runs in the opposite direction. Frost asked:

> What does it think it's doing running west
> When all the other country brooks flow east
> To reach the ocean?[19]

The poem reminded Rose Kennedy of Thoreau's advice to follow the beat of "a different drummer" rather than the conventional mores of society.[20] Frost was smitten with the Darwinian precept that the human race began in a primitive, aquatic environment, then evolved landward—the recurrent theme of "West-Running Brook."

Books inspired by the natural world, including those of Henry Beston and Robert Frost, seemed nostalgic after the stock market crashed on October 29, 1929. It wasn't a great time to be Cape Cod seashore guardians and New England poetic scribes. President Herbert Hoover proved to be unable to curtail the colossal economic downturn and was generally unwilling to find innovative ways to put the unemployed to work. By contrast, the governor of New York, Franklin D. Roosevelt, responded in myriad ways, one of which was the first proof that conservation could be a boon to the economy. In October 1931, he created the Temporary Emergency Relief Administration (TERA), which provided work relief to young men who planted trees,

An intrepid traveler and chronicler of the natural world, Henry Beston is seen here on the front steps of the Fo'Castle, the cottage on the dunes of the Outer Cape where he lived alone for a year. Beston's *The Outermost House* became a favorite book of both John F. Kennedy and Rachel Carson.

protected wildlife, and stocked ponds with fish throughout the Empire State.[21]

When Roosevelt won the White House in 1932, he quickly set up a federal program on the TERA model. The Civilian Conservation Corps (CCC), which from 1933 to 1942 hired millions of unemployed teens and young men to plant more than 3 billion trees and stock lakes with fish across the United States and much more.[22]

The Kennedy boys were far too rich to labor for the CCC. But the very idea of a "tree army" and beautification squad influenced Jack's sense of civic engagement. The teenager wasn't unique in absorbing the New Deal enthusiasm for America's parklands. He came of age in Roosevelt's first two terms, when FDR's passion for conservation affected practically every community, and he absorbed the New Deal attitude that under creative liberal leadership, natural resource management and infrastructure renewal could work seamlessly together.

Throughout his childhood, Jack was prone to serious bouts of illness and was often unable to explore the outdoors.[23] Rose Kennedy remembered that her son was "a very, very sick little boy" during his prep school years, often "bed-ridden and elfin-like."[24] In 1933, the Kennedys bought the La Querida estate in Palm Beach. The Mediterranean Revival house, designed by the architect Addison Mizner and set on two acres, boasted multiple second-floor balconies with sweeping beachfront views of the Atlantic. Throughout his life, Jack often retreated to La Querida to soak up the Caribbean trade winds and the therapeutic sunshine. For decades to come, Christmas and Easter holidays in Palm Beach were a Kennedy family tradition. The Kennedy boys, especially Jack and his younger brother Robert, aimed to be like the bottle-nosed dolphins—"at one" with the Atlantic. "I remember how as teenagers in Florida [Jack] and Bobby, on even the roughest days, would swim miles out into the ocean," Edward "Ted" Kennedy, the youngest of the boys, recalled. "The storm-warning flags would be flapping furiously in the wind and rain, and they'd be frolicking in the surf like a couple of polar bears."[25]

What set the Kennedy family apart from most others in exclusive Palm Beach was that their saltwater swimming pool had been turned into a giant aquarium populated with sea creatures captured in the Atlantic. Neither Joe Sr. nor Rose objected to their children releasing their live catch—pompano, small bonito, and even a nurse shark—into the pool, where, as grandson Robert Kennedy, Jr., recalled, the young people "swam among" the marine species "wearing goggles, and flippers," anxious to study the specimen's biological variations.[26]

II

Joe and Rose Kennedy's inclination toward seashore preservation was elevated by their relationship with a young, talented lawyer named William Orville Douglas. Ten years younger than Joe Sr. and twenty years older than Jack, Douglas was something of a fixture within the family: a legal adviser, weekend naturalist, and traveling companion. "It was as if he [Douglas] were an uncle or second cousin," Jean Kennedy Smith recalled. "He was just always around."[27] His relationship with Joe Sr. was extremely close. Douglas, who was quite liberal, and Kennedy, who was conservative to the same degree, met on the common ground of the New Deal, which inspired both men in two respects. One was the New Deal's emphasis on *action*; the other was that the New Deal offered the most fertile ground in national government since the American Revolution for men of serious ambition. During FDR's first term, Joe Sr. and Douglas were ready for brazen action and anxious for quicksilver advancement in public affairs, and they met at that juncture, remaining loyal, trusting friends ever afterward.

Whenever an opportunity presented itself, Douglas spoke fondly about his hometown of Yakima, Washington, which was enclosed in an arid foothill-rimmed agricultural valley with the rugged Cascade Mountains some forty miles to the west. Often wearing a rumpled sheepherder's hat with a cigarette hole burned on the flap, a thick flannel shirt, and in cold weather a red-and-black lumberman's coat, Douglas was at home on whatever forest trail or sagebrush patch he hiked. Walking through some of the most extensive large conifer forests on Earth turned him into an instinctive nature preservationist. Over time, he hiked trails on Mount Adams, fished the lower Snake River, and snowshoed the Wenatchee Mountains. The berry-laden Willamette Valley, the arid high-desert Lincoln Plateau, and the gently rolling hills of Palouse Valley constituted what Douglas described as the Pacific Northwest's "symphony" of wilderness. "Those who never learned to walk will never know its beauty," he wrote. "Only those who choose to get lost in it, cutting all ties with civilization, can know what I mean. Only those who return to the elemental world can know its beauty and grandeur—and man's essential unity with it."[28]

Douglas was born on October 16, 1898, in a farmhouse on the Otter Tail River in Minnesota. His father was a forthright, rigid Presbyterian minister. In 1901, Orville, as he was then known, was stricken with infantile paralysis and intestinal colic, leaving him weak and with a pronounced limp. The family moved briefly to Estrella, California, before ending up in Cleveland, Washington. After Reverend Douglas died in 1904, his widow, Julia Douglas, moved her family to Yakima, hub of an important fruit-producing valley. All three Douglas children did odd daily labor jobs to help put food onto the kitchen table.

With sheer willpower and defiance, young Orville undertook a grueling rehabilitation regimen to strengthen his leg muscles, taking long walks along the Yakima River even in rain and snow. First he hiked to the low desert hills around Yakima and later into the snow-capped mountains. His Yakima walks were the root source of his "unfenced libertarian spirit," as his biographer Bruce Allen Murphy explained in *Wild Bill*.[29] As Douglas himself wrote, "A people who climb the ridges and sleep under the stars in high mountain meadows, who enter the forest and scale the peaks, who explore glaciers and walk ridges buried deep in snow, these people will give their country some of the indomitable spirit of the mountains."[30]

Standing over six feet tall, with crystal-clear blue eyes, tousled brown hair, a prominent nose, and a taut boxer's frame, Douglas was a self-styled intellectual pugilist. From a young age, he was boldly ambitious, never hiding the fact that amassing power was what motivated him. He wasn't after power for its own sake, though. Out of his occasional experiences as a fruit picker grew an unshakable compassion for society's downtrodden. In 1916, he graduated from high school as class valedictorian and was awarded a scholarship to Whitman College in Walla Walla, Washington. While there, he discovered the nearby Blue Mountain trails with their resinous scent of pines, firs, and spruces in the high-altitude terrain. Before long, he explored such high-altitude peaks as Kloochman, Fifes, and Cleman, all of which were within a five-hour drive of Walla Walla (with its motto, "a town so nice they named it twice," though in the Nez Percé language, it means "waters running together"). The hikes helped him think like the mountain range itself.[31] Over time, those outings led him to explore the entire Cascades range, which extends from Canada through Washington and Oregon and into California to the gap south of Lassen Peak. He also developed a passion for fly-fishing for cutthroat and rainbow trout in the clamors of mountain rivers, including the swift-running Snake and Columbia.[32]

After Douglas earned his degree from Whitman in 1920, he enrolled in Columbia Law School in New York City, despite lacking the requisite funds. To save money, he hopped a Great Northern boxcar on a freight train headed east, befriending along the way an array of freewheeling hobos. Whether he embellished his romantic yarns about his youth to inhabit the Jack London vagabond tradition or stuck to the autobiographical truth has been debated by scholars, but the fact of his steely and colorful personality has not.

Douglas was an outstanding Columbia Law student, growing into one of the most promising constitutional lawyers of his generation and graduating second in a class that included Paul Robeson and Thomas E. Dewey. Intellectually, he gravitated to the "living Constitution" theories, especially as related to the First and Fourth Amendments. In 1925, having married Mildred Riddle of Yakima, he joined a powerhouse Manhattan law firm. In 1931, he moved to New Haven to become a full-time law professor at Yale University.

In 1933, Joe Kennedy, Sr., was rewarded for his contributions to Roosevelt's presidential race with the chairmanship of the new Securities and Exchange Commission (SEC). He, in turn, hired Douglas, who had written an article lambasting the SEC as a weak weapon to solve a treacherous problem. Imbued with a great luminosity of spirit and brazen self-assurance, Douglas met his match in Joe Sr. on both counts.

In September 1934, Douglas entered Kennedy's private office to talk about his new job. "What do I do?" he asked.

"Goddammit, I thought you knew," Kennedy replied. "Maybe I got the wrong man down here."[33]

The intimidated Douglas reassured Kennedy that he'd hired the right lawyer to conduct a thorough study so that the SEC could take action against Wall Street corruption based on hard facts.

Within months, the conservative mogul adopted the feisty Douglas as a protégé. Ever challenged by one another's policy differences, the two headstrong mavericks shared a resolute intensity, each growing stronger with the rebuffs that life hurled at them.

From 1937 to 1939, the Kennedys rented a spacious house in Marwood, Maryland, with bedrooms galore to accommodate their large family and guests.[34] "I knew all the boys—except for Joe Jr. who was the eldest," Douglas recalled, "and their sisters."[35] In Maryland, Douglas began coaching the Kennedy children, Jack (seventeen) and Bobby (nine), on the joys of hiking and rock climbing in Patapsco State Park's forest of big deciduous trees. Rock Creek Park in Georgetown—historically the third unit of the National Park System, created by Congress in 1890—became "Camp Douglas" to the Kennedys; there they learned to identify sycamores, sugar maples, red oaks, and basswoods. To them, he was the western frontiersman "Wild Bill," who crawled into hollow trees like a bear and up live ones like a marten.

Douglas's Pacific Northwest upbringing affected his constitutional viewpoint. Whenever a bulldozer rolled onto undeveloped land, Douglas lamented the ecological desecration. As watchdog for the high-desert Ochoco Mountains near his fishing cabin in Oregon, he worried that the experts—chemists, engineers, hydrologists, mathematicians, and physicists—who were carrying the day in the twentieth century were stacked on the side of mindless progress. Just because somebody was a great scientist, he warned, didn't mean that his or her judgments were aimed at the common good—much less Earth stewardship. Leery of the unforeseen consequences of growth for growth's sake—air and water pollution, loss of agricultural land, and natural resource exhaustion—Douglas felt a sense of what geographer Yi-Fu Tuan called "topophilic" sentimentality (an all-encompassing bond between people with a geographical place) for practically any wild place in Washington and Oregon.[36]

"We must provide enough wilderness areas," he insisted, "so that no matter how dense our population, man—though apartment born—may attend the great school of the outdoors and come to know the joy of walking the woods, alone and unafraid. Once he experiences that joy he will be restless to return over and over again to discover the never-ending glories of God's wilderness."[37]

In addition to Joe Sr., Douglas made a friend and nurtured a senior ally in President Roosevelt, a fellow graduate of Columbia Law School and forestry maven. The wily Douglas regularly played poker and drank scotch with FDR, the two swapping good yarns. The president came to see that he had a "good" heart and a sense of justice based on his fundamental belief in what he called "guaranteed liberty" and the Bill of Rights. In the spring of 1937, when FDR tried to pack the Supreme Court by adding six justices to the sitting nine, so as to enable New Deal measures to pass judicial review, Douglas backed the president's effort, though it failed. Two years later, when Louis Brandeis's seat on the Supreme Court opened up, Douglas was nominated by FDR and easily confirmed.[38] Soon after, Douglas had a private meeting with the eighty-two-year-old outgoing Brandeis, who offered one piece of sage advice: "Whenever you feel stressed or overburdened with workload, leave the court, head to the Chesapeake & Ohio Canal along the Potomac River, and just tramp or canoe along until your mind is free of clutter."[39]

Douglas followed Brandeis's advice; it was in line with his appreciation of the ecological interrelationship of plants, humans and other animals, and the world in which they all coexisted. In coming years, Douglas would influence and help direct scores of lawyers to think about conservation rights and remedies not only in terms practical law, but as one of the natural rights reserved to the people of the United States and protected by the Ninth Amendment.[40]

III

Douglas wasn't the only person to influence the Kennedy boys to explore the outdoors. In May 1936, Joseph Sr. arranged for Jack (then nineteen) and Joe Jr. (twenty) to work on the forty-thousand-acre Jay Six cattle ranch in Benson, Arizona. Lying along the San Pedro River, forty miles southeast of Tucson, it was owned by the entrepreneur John Speiden, a Wall Street financier who had moved west in 1932, seeking a healthier climate. With the help of breeders, he developed one of the finest stock farms in the Southwest. Soon he was inviting guests to show off the Jay Six.

Earning a dollar a day sweating in the broiling heat, Joe Jr. and Jack built corrals and an adobe edifice and learned to punch cows.[41] According to Speiden, they worked "very hard" herding cattle to graze on jojoba bushes but mainly thought about meeting girls across the Mexican border. Jack routinely ended his letters written from the Jay Six with the salutation "Home on the Range."[42]

Jack entered Harvard University in the fall of 1936, taking his latest boat, *Flash II*, with him. He had won the Nantucket Sound Star Class Championship and would go on to win the coveted McMillan Cup, the oldest collegiate sailing event.[43] In addition to the reputation Jack earned at Harvard as an excellent swimmer and sailor, he became known for his constant hijinks and zest for life. Some classmates recalled that around Jack it was always spring; he was all libido. Carefully reading the *New York Times*, *National Geographic*, and history books became one of his quieter daily rituals. He was fashionably at ease with himself, while making grades that were only slightly above average. His senior thesis on British diplomacy in the lead-up to World

War II was titled *Why England Slept*; when published in 1940, it sold well enough (with assistance from Joe Sr.) to give twenty-three-year-old Jack Kennedy a budding reputation as a serious commentator on world affairs.

In the fall of 1941, Jack was sworn in as a naval ensign. After the United States entered the war, he was assigned to the Motor Torpedo Boat Squadrons Training Center in Melville, Rhode Island. The unpretentious, self-deprecating Kennedy emerged as a natural leader. "I never heard him complain about anything," Fred Rosen, his closest friend at Melville, recalled. "The boats were pretty shaky for someone who had a bad back, but I never heard him complain about it."[44]

Kennedy was ordered to the Pacific to take command of a small, speedy patrol torpedo boat, *PT-109*. On August 1, 1943, a Japanese destroyer collided with the boat near the Solomon Islands, slicing it apart and killing two of the thirteen men on board. Combating the agitated sea for hours, Kennedy managed to lead the survivors to a deserted coral island. After the men of *PT-109* were rescued, Kennedy's cool heroism became a seminal part of his biography.

A year later, when President Roosevelt failed to support the renomination of Vice President Henry Wallace in 1944, a competition developed for the open spot on the ticket. Two of the most widely discussed possibilities were Senator Harry Truman of Missouri and Associate Justice William O. Douglas of the Supreme Court. Roosevelt personally preferred Douglas, but in the end, he listened to party leaders and chose Truman. It was Douglas's best chance to gain national office. Joe and Rose Kennedy bemoaned the fact that Douglas hadn't been picked. In terms of presidential ambitions, however, they were getting ready to place their hopes on one of their own: Joe Jr.

But the biggest blow of all was about to hit the Kennedy family. In August, Joe Jr. was killed at the age of twenty-nine while flying a mission over the English Channel. Emotions aside, the horrific loss affected Jack more than anyone else in his family. He was the eldest remaining son, with all that implied to his single-minded father. In early 1945, he headed to Castle Hot Springs, Arizona, thirty miles north of Phoenix in the Sonoran Desert, to recuperate and to rethink his future. He wanted to be a writer of some kind, but knew well that it was politics that awaited him.

CHAPTER 2

Harry Truman

POLLUTED AND RADIATED AMERICA

Supreme Court Justice William O. Douglas (1898–1980) poses by a tree in front of his child-hood home at 111 N. Fifth Avenue in Yakima, Washington. "My love is for what many would put down as the dreariest aspects of the dry foothills of the West," Douglas boasted. "Sagebrush and lava rock."

I

On July 16, 1945, Jack Kennedy was in Europe working as a special correspondent for Hearst Newspapers, covering the Potsdam Conference, when the world's first nuclear bomb was detonated in the Alamogordo Bombing Range in the Jornada del Muerto (Dead Man's Journey) desert near White Sands, New Mexico. Other sites had been considered for the blast—San Nicolas Island, off the southern coast of California, and the sandbars off Texas' Gulf Coast—but for both safety and secrecy, the planners had chosen the flat, barren area owned by the US government, where the nearest human habitation was twenty-seven miles away in the town of Carrizozo. Code-named "Trinity," the test would be the most violent man-made shock to the Earth known up to that time, using about thirteen and a half pounds of plutonium. The plan, if President Harry Truman authorized it, was to attack Japan with an exploding uranium-based atomic bomb, followed, if necessary, by an imploding plutonium-based one. With a minimum amount of uranium available, the first bomb—untested—was already en route to its staging area. The plutonium version would be used in the July test. Called "the gadget" by the Manhattan Project scientists at the New Mexico lab, the implosion-style device was a sphere just under sixty inches across and weighed ten thousand pounds. Even some of those scientists predicted it would fail, so new was atomic technology.

To the physicist Edward Teller, the predawn blast at ground zero was "like opening the heavy curtains of a darkened room to a flood of sunlight."[1] The white flash temporarily blinded the closest observers ten thousand yards away. A huge orange-and-yellow fireball about two thousand feet in diameter rose into a narrow column that flattened into a mushroom shape.[2] Nearby livestock suffered skin burns, bleeding, and loss of hair. Oppenheimer, one of the gadget's chief designers, looked to the Bhagavad Gita, the holy book of Hindu scripture, for fitting words: "Now I am become Death, the destroyer of worlds."[3]

The devastation caused by "the gadget" was beyond what the scientists had imagined. The light from the initial fireball was more brutal than anything ever manufactured on Earth and was so intense that it would have been visible from Mars. The temperature at its core was 50 percent higher than that of the surface of the sun as the bomb exploded with the force of more than twenty thousand tons of TNT or, in another comparison, the destructive power of thousands of fully loaded bombers. Just seconds after the detonation, observers saw "a swirling column of orange and red, darkening as it rose until it looked like flames of burning oil."[4] The steel tower that had held the gadget was completely vaporized.[5]

The rationale for the test was to see if the bomb functioned; it did and Truman authorized the use of atomic weapons on Japan. Virtually no planning had been done to estimate Trinity's lasting effects, either on living things or the environment. At the test site, US military personnel had been given nothing more for protection than suntan lotion and dark welding glasses. Moreover, high winds on the day of the test were disregarded as only a minor inconvenience. The result was that wildlife, people, and livestock were exposed to radioactive fallout carried far beyond the bombing range.[6]

Three weeks after the Trinity test, on August 6, 1945, the uranium bomb was dropped from 31,000 feet onto the Japanese city of Hiroshima. It exploded 1,900 feet above the ground, annihilating 90 percent of the city and its residents in a circle one mile in diameter. The damage from heat, blinding light, and radiation stretched for at least twelve miles from the blast site. In the aftermath, survivors were treated for burns and other injuries, but something then occurred that baffled doctors. The death toll actually began to rise. It didn't spike downward as in every other type of human-induced disaster. During the span from thirty to sixty days after the bombing, tens of thousands of Japanese people died from radiation sickness—something that had never been seen before as a lethal environmental hazard. A plutonium bomb was dropped on Nagasaki three days later. The annihilation of the two cities changed the world forever. A new frightening global order, one in which nuclear bombs were supreme, had been born.[7]

Joe Kennedy, Sr., was horrified that President Truman had ordered the use of the initial atomic bomb on the civilian population in Hiroshima. He rushed to see his friends *Time* publisher Henry Luce and Cardinal Francis Spellman to persuade them to lobby Truman against dropping a second bomb on Japan. Both ignored his appeal.

One person who sided with Joe Sr. was Bill Douglas. The day Hiroshima was bombed, he was climbing toward the summit of Mount Adams (elevation 12,281 feet) in Washington State. Reading reports from the Pacific upon his return filled him with anger. He believed that if FDR had still been alive, he would have demonstrated the atomic bomb on a military target or Pacific islet—not on Japanese civilians. Deeply concerned about the moral and ethical implications of Hiroshima, Douglas determined that Truman wasn't fit for command. Years later, when Douglas was interviewed at his rustic cabin, he was asked by Eric Sevareid of CBS News if history would have been different if he had been president instead of Truman in 1945. The justice snapped, "There would have been no bomb dropped on Hiroshima."[8]

Echoing his father, Jack Kennedy also worried about the humanitarian complications of the atomic bomb. His quiet dissent contrasted with the jubilation in the streets of New York and Boston. Even though he was thrilled that World War II was over and that the United States had defeated Japan, his personal views echoed those of the CBS News reporter Edward R. Murrow, who observed, "Seldom, if ever, has a war ended leaving the victors with such a sense of uncertainty and fear, with such a realization that the future is obscure and that survival is not assured."[9]

An essay that Kennedy read with particular attention was the one Norman Cousins famously wrote on the day he heard the news reports of the atomic attack on Hiroshima. The thirty-year-old Cousins had been editor in chief of the *Saturday Review of Literature* since 1942, with the stated mission to publish authors who tried to "restore to writing its powerful tradition of leadership in crisis." He succeeded in that with his solemn essay, "Modern Man Is Obsolete," which ran in the magazine in August 1945; it was so popular that it was adapted into book form just two months later and would go on to have fourteen printings and be read by millions. By raising moral and ethical concerns about Hiroshima, Cousins's book helped galvanize an antinuclear movement around the world.

Modern Man Is Obsolete was not a work that offered a lot of answers. With the initial essay having been written less than a day after Hiroshima, Cousins couldn't possibly foresee all the global ramifications of the new weapon. He did succeed in posing the kind of questions that would guide—and haunt—the era going forward. Though he didn't discuss environmental damage specifically, he recognized damage to the natural world as an intrinsic aspect of the Atomic Age. "Mankind," he wrote, "has leaped centuries ahead in inventing a new world to live in, but he knows little or nothing about his own part in that world. He has surrounded and confounded himself with gaps—gaps between revolutionary science and evolutionary anthropology, between cosmic gadgets and human wisdom."[10]

The health hazards of radiation fallout from atomic bombs were not a strong concern in 1945. And few connected the weapons with ecological destruction of the Earth. First, as Cousins pointed out, little was really known about the long-term biological effects, even by scientists, let alone the general public. Second, the only issue on most people's minds in August 1945 was winning World War II and then keeping the peace. And third, from the viewpoint of those familiar with battlegrounds in either world war, contamination of the land was a by-product of every modern weapon; the fact that atomic bombs were exponentially worse was a matter of degree difficult to grasp in the wake of Hiroshima.

Cousins also argued in his extended essay that "the quintessence of destruction as potentially represented by modern science must be dramatized and kept in the forefront of public opinion." That proved hard to do during the early Cold War, when anti-Soviet sentiment peaked. But Cousins took up his own challenge. Educating the American public on the lasting effects of atomic activity by humans and keeping those realities "in the forefront" was a job he assumed even as the two Japanese cities were still smoldering from what were, inarguably, the greatest environmental disasters of all time.

Born in West Hoboken, New Jersey, on June 24, 1915, Cousins, like Kennedy, had been sickly as a child, leading him to a deep sympathy for people who suffered illness or hunger. After Columbia University, he joined the staff of the *New York Evening Post* (now the *New York Post*) and was also hired as the book critic for *Current History*. A riveting conversationalist, Cousins was likable beyond description. He was an active member of New York's literary crowd and was an ethicist who aligned himself with Dr. Albert Schweitzer, who would be the primary subject of two of his books.[11]

In 1940, Cousins joined the staff of the *Saturday Review of Literature*, a struggling magazine that, as editor in chief, he built into a veritable handbook of postwar liberalism. In his columns, he stormed against nuclear weapons, often on the theme that even in tests they destroyed life on Earth. He continually analyzed the complexity of the Atomic Age and typically concluded that in exercising the right to change everything but themselves, unethical humans took the Earth for granted altogether too often. Whereas other Americans cheered the demolition of Hiroshima and Nagasaki, Cousins was filled with "a primitive fear, the fear of the unknown, the fear of forces man can neither channel nor comprehend."[12]

Imbued in the Judeo-Christian ethic, Cousins harbored "the deepest guilt" that the atomic bomb had been used on Japanese civilians during World War II. Recalling the New Testament's Book of Revelations, with its prophecies of a looming apocalypse of fire, he contended that the advent of nuclear weapons "overnight" intensified and magnified "the possibility of global destruction in scientific terms." Equally compelling were the harrowing articles by John Hersey in the *New Yorker*, which exposed the horror of radiation sickness among Japanese civilians and which in 1946 became the classic book *Hiroshima*. E. B. White of the *New Yorker* seconded Cousins's consternation, vehemently disapproving of man's "stealing God's stuff."[13]

Jack Kennedy pondered the world-changing events in Japan. That autumn, he told friends that world history would be forever divided into two epochal categories: before and after Hiroshima. On October 8, he spoke about the uncertainty surrounding atomic weapons in an address to the United War Fund in Boston. "In the past years, we have heard much about the horrors of war, but we have always felt that war was preferable to certain alternatives. . . . War has never been the ultimate evil," he explained. "Now, however, that may have changed. We may be forced to make the sacrifices that will insure peace. We can only pray that man's political skill can keep abreast of his scientific skill; if not, we may yet live to see Armageddon."[14]

Kennedy, echoing Cousins, lamented in his Boston speech that science had outstripped other considerations. "The few men who had the scientific knowledge to create this bomb," he said, "are not qualified by experience to make the political decisions on what should be done with it." Significantly, he stepped into that breach, suggesting specific possibilities on those political considerations. His last point was the shortest and simplest, and also the one that would guide his postwar political career: "The atomic bomb," he said, "may be so destructive a weapon that it may force people to keep the peace."[15]

The Cold War with the Soviet Union commenced immediately after World War II. For both Democrats and Republicans, the United States' primary foreign policy objective was to contain Soviet expansion. The same would soon be true of the United States' other wartime ally, China. First, though, Americans who, like Kennedy, were trying to imagine the new order of the Atomic Age suspected that it would be only a short time before the Soviets, too, had "the bomb."

Douglas was one of the few who recognized that even without war—and certainly with one—an atomic world would do grave harm to the environment. His opinion was borne out by Operation Crossroads, the joint Army-Navy tests at Bikini Atoll in the summer of 1946. The radiation hazards of Trinity and the two bombs dropped on Japan strongly influenced the decision to locate the tests in the far-flung Pacific. The Bikini site was a semicircle made up of some dollops of land in the Marshall Islands region of Micronesia, twenty-four hundred miles southwest of Hawaii. The Army-Navy joint force, consisting of forty-two thousand soldiers and sailors, traveled to the vicinity. Their mission was to participate in the July 1 Operation Crossroads nuclear test. The month before, Douglas had been featured on a CBS Radio broadcast to discuss inherent dangers of atomic proliferation. Among those appearing with him were Albert Einstein and Vice Admiral William Blandy, the commander of Operation Crossroads. Douglas and Einstein spoke in favor of immediately ceding control of atomic capability to the United Nations, on the understanding that all powers attaining that nuclear capability in the future would do the same. For Douglas, the global abolition of atomic weapons would be the best solution.

The impetus of Operation Crossroads was quite the opposite; in part, it was meant to intimidate the USSR with a very public, well-documented demonstration of American nuclear prowess.[16] The detonations were a vivid display that nuclear weapons were going to become a cornerstone of Cold War military strategy. The operation began with an atmospheric detonation. Twenty-four days later, on July 24, a second atomic bomb was set off underwater. As part of the operation, servicemen measured the amount of radiation in the air, in lagoons, on nearby land, and on test ships populated by animals without protective gear. Exposure was a serious threat to US servicemen participating in the series of Bikini tests. That specter wasn't taken very seriously by the Truman administration. But Crossroads did presage a significant shift with the president's August 1 signature on the act that removed control of atomic activities from the US military and awarded it to the newly created Atomic Energy Commission (AEC). Though the postwar era held out the promise of perhaps limitless atomic energy for peaceful purposes, the prospect of mass destruction was horrifying. Clearly, some sort of regulatory control of nuclear energy, if not its abolition, was essential. In the conflict that erupted over civilian versus military control of nuclear power, Kennedy favored the former. When authority shifted to the AEC, composed of a five-member civilian board, he was relieved.

Meanwhile, in the Pacific, the chief medical expert, Dr. Stafford Warren, a radiologist, put a stop to a planned third nuclear test scheduled for mid-August. The joint Army-Navy force hurriedly pulled just about every member of the command out of Bikini. In a report filed during the Operation Crossroads tests, Warren charged that the fleet commanders had pursued "a blind, hairy-chested approach to the matter of radiological safety."[17] In an article for *Life* magazine a year later, he described what the first blast had done to the sea:

> As the radioactivity of the fission products lessened, a more insidious hazard was discovered. The area of *slight* contamination was spreading outside the target area. The algae in the water, moreover, had absorbed radioactive particles and passed them on to little fish. These died at the end of the second week and were eaten by larger fish. These died in the third week, their decayed bodies passing the radioactivity back to the algae. When the algae collected on the bottom, the radioactivity was sometimes strong enough to be detected through the steel hulls.[18]

The fact that from 1946 to 1958, the United States detonated sixty-seven nuclear bombs at its Pacific Proving Grounds, based in the trust territory of the Marshall Islands, appalled Cousins and Douglas. That coral atoll, surrounded by a lagoon of well over two hundred square miles, had become poisoned by the fallout. Radiation made the Marshall Islands unsafe for human habitation, an anti-paradise of contamination, illness, and anguish.[19] People in the territory were infected for years afterward. Not until 1978, during congressional hearings, did hundreds of test workers, soldiers, and local residents, most of them Indigenous, have the chance to formally object to being exposed to radiostrontium during the nuclear weapons tests in the late 1940s and 1950s. Cousins and Douglas, from a public health perspective, had been correct from the start.[20]

II

Another anti-nuclear voice at the time of Operation Crossroads was Truman's secretary of the interior, Harold Ickes. Born on March 15, 1874, in Hollidaysburg, Pennsylvania, on the eastern slopes of the Allegheny Mountains, in 1890 Ickes moved to Chicago, where he worked his way through both the undergraduate program and law school at the University of Chicago. Disgusted that industrialization and railroads had turned parts of his adopted city into blighted slums, Ickes sought solace in traveling to Yellowstone, exploring the national park on horseback. Upon his return to Chicago, he became a progressive reporter anxious to slay corporate dragons on behalf of unions and conservation groups. Not only was Theodore Roosevelt, the Bull Mooser, his lifetime role model, but like him, Ickes championed the reform cause of natural resource conservation.

When Franklin Roosevelt announced in 1932 that he was running for president, Ickes formed an independent Republican committee that supported him extravagantly. Once FDR defeated Hoover, he thanked Ickes for his fundraising prowess by appointing him secretary of the interior and public works administrator. As head of Interior for more than twelve years, Ickes expanded the National Park System to include the Mammoth Cave, Great Smoky Mountains, Big Bend, Isle Royale, Joshua Tree, and many other units. Aware that Interior's reputation had been tarnished by the sleazy Teapot Dome affair under President Warren Harding, Ickes sought to transform some national parks from tourist-based cash registers into roadless wilderness areas for aesthetic reasons. "I am not in favor of building any more roads in the national parks," he said. "I do not have patience with people whose idea of enjoying nature is dashing along a hard road at fifty or sixty miles an hour."[21] And he instructed his park superintendents to limit concessionaires in the parks. "I do not want any Coney Island," he said. "I want as much wilderness, as much nature preserved and maintained as possible. . . . I think parks ought to be for people who love to camp and love to hike and who . . . [want] a renewed communion with Nature."[22]

Remaining at Interior at the start of the Truman administration, the indomitable Ickes feared that the president would be a patsy for stockman associations, the oil lobby, and lumber outfits. That hunch seemed to be borne out in 1946, when Truman asked Edwin Pauley, a multimillionaire oilman in California, to be secretary of the navy. Ickes sensed that Pauley, having Truman's support, intended to drill for oil off the coast of Santa Barbara, with energy security as the rationale. As petroleum administrator for the entire war effort, a job that had ended only months before, Ickes knew even more about the oil and gas world than Truman and Pauley combined.

With a crusader's zeal, Ickes charged that Pauley had advised against the federal government's seeking title to offshore oil fields from the states because such action would lead California oilmen (presumably including himself) to withhold contributions to the Democratic Party. Ickes believed that neither private energy companies nor Sacramento lawmakers should be trusted with decisions about drilling off of California's beautiful 840-mile coast because they were apt to be greedy and unconcerned about seashore conservation. He wanted Congress to pass an Outer Continental Shelf Lands Act to oversee drilling in federal waters. Unable to stand the press scrutiny and charges of cronyism, Pauley withdrew his name from consideration for the cabinet post. Hopping mad, Truman blanched at Ickes's insinuation that he, as well as Pauley, was beholden to Big Oil.

In the showdown over Pauley that followed, Ickes resigned his Interior post on principle, leaving in February 1946. The "old curmudgeon," as he happily styled himself, felt vindicated in his view that many oil companies not only were reckless polluters but also sought to snare federal regulators in profit schemes. As he complained in 1948, Truman was the kind of pro–extraction industry president who allowed "the oil companies to get away with murder."[23] Called upon to assess Ickes, who died in February 1952, Truman said that the Interior Department legend "was often irascible and could be intolerant of the opinions of others. . . . His rare gifts of irony and invective made him a formidable opponent in public debate."[24] The law Ickes fought for was passed in 1953, and a system was set up that required the federal government to auction offshore oil leases at staggered intervals.[25]

Replacing Ickes as interior secretary was the New Deal energy expert Julius Krug, who had been the chairman of the War Production Board at the end of World War II. With the exception of banning logging in the Olympic Mountains, Krug did virtually nothing positive for the environment before he resigned in 1949 to become co-founder of the Volunteer Asphalt Company in Knoxville, Tennessee. The reason Krug accomplished little was not solely his fault. His boss, President Truman, thought Interior needed to accommodate tourists, not bottle up wilderness areas for elite hikers who wanted to enjoy solitude. Krug, in fact, was ahead of his time in advocating reduced dependence on fossil fuels and the development of alternative sources of energy, including solar, which was barely known at the time but which he campaigned to develop. President Truman blocked all expensive solar initiatives.[26]

For the most part, Truman spoke only begrudgingly about the preservation of the natural world. When Michael Grunwald, the author of *The Swamp: The Everglades, Florida, and the Politics of Paradise*, spoke at a 2007 symposium on "Truman and Conservation," he began by saying a friend had told him to simply say, "Ladies and gentlemen, Truman didn't care about the environment. Thank you very much."[27]

Grunwald's friend was right. During his White House years, Truman promoted modern engineering and scientific marvels aimed at controlling nature. Determined not to allow the nation to slump into another depression, he directed his administration to apply scale-of-war mobilization to large-scale agriculture in the Midwest, South, and West. Farms were encouraged by the USDA to conglomerate and automate. The beauty of the American countryside—tall-grass pastures, wildflower meadows, crawfish-laden creeks—was supplanted by industrial agriculture relying on chemical fertilizers. Overgrazing began to despoil basins and flatlands. Mountainsides were seen as future coalfields. Lakes were marred by excess fertilizer from agricultural runoff. In the West, Truman allowed stockmen easy access to public lands. For New Dealers, including Ickes, it was painful to see progressive conservation on the wane, as if the Dust Bowl drought of the 1930s had never occurred.

During World War II, oil and gas companies, to the consternation of Ickes, had employed a method called "well stimulation" to produce more fuel. Exploding ordnance at the bottom of a well became viable in Texas and Louisiana. After the war, the oil field service firm Halliburton pioneered a process known as hydraulic fracture, or "fracking" (the injection of fluid chemicals far underground to cause cracks in a rock formation, thereby releasing more oil and gas). The historian Daniel Raimi described Halliburton's first attempt in 1949: "While the operation did not employ conventional explosives, it nonetheless involved some seriously flammable fluids, a mixture of napalm and gasoline pumped roughly 2,500 feet underground into the massive Hugoton gas field in western Kansas."[28]

In 1929, just before the stock market crash, the Forest Service had created "primitive areas" across America where timbering and road building were to be minimized, though not disallowed. At President Roosevelt's behest, tougher regulations forbade logging and motorized vehicles in "primitive areas" while also mandating Forest Service inspections that continued into the early 1960s. The reviews included public hearings and the ecological study of primitive areas in order to evaluate their boundaries, along with lease requests for logging, grazing, mining, and recreation. The Forest Service then adjusted certain boundaries before reclassifying primitive acreage into "wilderness areas" and "wild areas," which was intended to preserve it from the intrusions of loggers and road builders altogether.

Unfortunately, reclassification didn't guarantee permanent protection in a world full of land plunderers and resource gougers. The gestalt of the country after the war was about full employment and industrial development, and Truman's policies followed suit. Once Truman became president, the increased need for lumber, stoked by a housing and construction boom, caused the Forest Service to reduce the size of several "primitive areas" and "roadless areas" in order to open more acreage to commercial enterprises, including tree harvesting. Returning GIs from Europe and the Pacific wanted houses, and timber logged from public forests was a postwar necessity. Logging companies hired sawyers and drivers of mechanical skids to clear-cut Forest Service tracts as quickly as possible. "As the tree-length logs are dragged out of the woods, sometimes straight up or down a steep slope, the trees of the 'next crop' are damaged by the skidder or by the dragged logs," the agrarian poet Wendell Berry of Kentucky complained of this destructive practice. "The woods is left a shambles, for nobody thought of the forest rather than the trees."[29]

Public lands were stressed in another way, as well. The hunger for outdoor recreation activities grew faster than ever before, prompted by suburbanization, road construction, increased automobile ownership, and the end of wartime gasoline rationing.

Furious that Truman opened the public lands to reclamationists and extraction profiteers, the Wilderness Society concluded that only a new federal law would protect wilderness permanently. In June 1947 at a gathering in Rainy Lake, Minnesota, the society launched a campaign for a national wilderness preservation system. "Here is the place of places to emulate, in reverse, the pioneering spirit of Joliet and Marquette," the society's president, Benton MacKaye, told the conservationists. "They came to quell the wilderness for the sake of civilization. We come to restore the wilderness for the sake of civilization. . . . Here is the central strategic point from which to relaunch our gentle campaign to put back the wilderness on the map of North America."[30]

At the time, the federal government owned about one-third of the country's land. Some of those holdings were east of the Mississippi, where deforestation, wetlands drainage, overgrazing, and abysmal farming methods caused soil erosion. But the lion's share was in the West and the territory of Alaska. To administer the federal acreage, Truman established the Bureau of Land Management (BLM) in 1946 by combining the General Land Office with the Grazing Service. As a "multiple-use" public land management agency, the BLM was granted authority to oversee around 420 million acres. Some of the prettiest western landscapes—such as the northern edge of what is now Grand Canyon–Parashant National Monument (Arizona) and the Missouri Breaks (Montana)—were on BLM lands.[31] A unit of the Department of the Interior, the BLM was mandated to "sustain the health, diversity, and productivity of the public lands for the use and enjoyment of present and future generations."[32] Mining, fracking, and exploratory drilling companies could lease BLM lands and not pay royalties until their products reached the market.

From the start, preservationists in the Pacific Northwest distrusted the new agency, which overnight came to oversee more land than any other entity in the United States. Chambers of commerce in the West, in contrast, looked to the BLM as a potential friend. Back east in West Virginia, coal deposits were surface-mined by destroying a mountainside and polluting the nearest waterways with acid drainage. Fracking was encouraged as a way for energy companies to profitably develop oil and gas from shale even on BLM lands. In Truman-era conflicts between mining and clean water, grazing and recreation, and timber and wildlife, the BLM usually sided with quick-dollar moneymaking proposals. Edward Abbey, the author of the nonfiction classic *Desert Solitaire: A Season in the Wilderness*, nicknamed the BLM "Bureau of Livestock and Mining."

As Ickes had predicted, special interests plundered BLM lands, yet the agency's mission was not clearly defined. In columns written for the *New York Post* and *New Republic*, private citizen Ickes defended national parks as wildlife habitat and warned against building unnecessary roads and hydroelectric dams in the West. It disgusted him that national forests established by both Theodore and Franklin Roosevelt for future generations to enjoy were being violated for short-term economic gains. "The omnipotent bulldozer continued its undisputed reign," the married activists Peggy and Edgar Wayburn noted in the *Sierra Club Bulletin*, echoing Ickes, "more good earth was sliced, flattened, and smothered under concrete than ever before."[33]

III

Bill Douglas, as disgusted as anyone with the Truman record on wilderness and most other issues, was with the Kennedy family in Palm Beach when Jack decided he'd run for Congress in Massachusetts's 11th District in 1946. Throughout the year, Douglas sat in on family discussions concerning campaign strategy (including commissioning a hundred thousand copies of Hersey's PT boat article "Survival" to be distributed around the district).[34] When Jack won, carrying 72 percent of the vote, Douglas celebrated by presenting him with a handsome edition of Thoreau's *Cape Cod*.[35] Kennedy saw his election as an endorsement by Massachusetts voters in the new generation entering politics.

But once Kennedy entered the House, he didn't do much of importance, certainly not on issues related to the management of natural resources. He refrained from publicly criticizing Truman's lackluster Fair Deal conservation agenda, even while his mother, Rose, complained openly about the president's indifference to grassroots activists lobbying for Atlantic seashore protection.[36] Green as a new leaf when it came to understanding the intricacies of FDR's scientific forestry versus Truman's extraction mania, Kennedy used Douglas as his off-hours conservation adviser.

Although Jack was not inclined to work exceptionally hard in Congress, he was determined not to be a minor figure in the Cold War being waged between the United States and the Soviet Union. Foreign policy became his strength. He grew increasingly concerned with the Soviet Union's potential development of nuclear weapons, satellites, and intercontinental ballistic missiles. The breathing room between the Second World War and the Cold War was short lived. In March 1946, Winston Churchill, Jack's hero, had warned of an "iron curtain" that was descending on Eastern Europe as the Soviet Union sought to expand its influence there. Closer to home, the AEC was given exclusive control of US deposits of uranium, which had been discovered in Utah, Arizona, and Colorado.[37] As a crusading anti-Communist representing an urban district, Jack gave minimal (if any) thought to the environmental and public health hazards of uranium mining. What did consume him was the fear he shared with millions of a nuclear showdown between the United States and the Soviet Union if atomic weapons proliferated.

As Truman prepared to run for a full White House term in 1948, Douglas was among those talked about as his running mate. Joe Kennedy, along with Harold Ickes, took part in a movement to draft Douglas. The ticket made sense: Truman had low ratings, and the rugged mountain climber Douglas was a media darling with a popular following in western states. After giving the possibility some thought, Douglas—dismayed that Truman had dropped nuclear weapons on Japanese civilian populations, used the BLM to the benefit of large-scale agriculture, and mistreated Ickes—withdrew his name from consideration. Another factor, most assuredly, was that he wanted to retain the lifetime power of his Supreme Court post.

With presidential ambitions fading, Douglas devoted more time to the causes of wilderness and wildlife protection. He wasn't alone. In 1947, a small group of women led by Mary Breck established Defenders of Furbearers, a nonprofit organization based in Washington, DC (renamed Defenders of Wildlife in 1959). Its initial focus was a ban on leg-hold traps. Trappers used those spring-loaded implements because they damage only the leg, but the trapped animal suffers for days before dying, typically of dehydration. Ranchers used the traps, too, to rid their land of coyotes, wolves, and bears. Hoping to promote coexistence among humans, predators, and livestock, the Defenders began working to save predators in North America from extinction.

About the same time, twin brothers Frank and John Craighead made a base at Yellowstone National Park, where for twelve years they studied with biological precision the black and grizzly bear populations, which had been decimated by habitat loss as well as overhunting. Originally from Washington, DC, the Craigheads, experts on Rocky Mountain mammals, worried that the charismatic large mammals of North America were "vanishing" (the term for endangered species before the 1960s) at an astonishing rate. It was exceedingly rare to see a wolverine, a marten, a Florida panther, or a red wolf roam the American landscape anymore. Gifted at generating *National Geographic*–style publicity, the Craigheads feared that if the Texas grizzly could become extinct in 1890, the Rocky Mountain grizzlies, which they sought to protect from overhunting, could easily meet the same fate.[38]

Another voice for wildlife protection that emerged during the Truman years was that of Marjory Stoneman Douglas. She had arrived in Miami in 1915, escaping New England and a bad marriage in order to reunite with her father, whom she hadn't seen since her parents had separated twenty years before. One of the reasons why Marjory's parents had split was that Frank Stoneman was a determined entrepreneur but a sadly unsuccessful one. In Miami, however, he had become a successful publisher, starting first the *News Report* and later the *Miami Herald*. Marjory became not only his friend for life but also a natural-born reporter for the *Herald*.

A frequent visitor to the *Herald* offices was another New England transplant, Ernest Coe. Born the year after the Civil War ended, Coe went to Miami in the 1920s to make money and retire quietly. He didn't do either. Working as a landscape architect, he discovered the Everglades at a time when most Floridians maligned the swamp as a mosquito-thick hellhole. Coe decided that, on the contrary, it was magnificent and needed to be protected as a biologically intact national park. Marjory remembered the days in the mid-1920s when Coe had been the only one who thought so and had been "always talking about his park."[39] She joined his effort early on, not as a naturalist but as a friend. As she explained later, it was extremely hard to enter the Everglades. The ecosystem was too hot and buggy for her. But from afar she appreciated the natural wetlands and exotic birdlife for its biological diversity.

Powered by Coe's unceasing efforts, the movement to create the national park succeeded with the passage of a federal law in 1934. Roosevelt and Ickes led the charge. It succeeded up to a point, that is. The monetary and local requirements attached to the law were not satisfied for thirteen more years—with Coe pushing and pulling the parties all the while. During that time span, Mrs. Douglas made a name for herself as a writer and was invited to publish a book in the *Rivers of America* series that Rinehart & Co. was undertaking; she shocked the editors by choosing the Everglades. Few topographers thought of the swamp as a river, but Mrs. Douglas did. "There are no other Everglades in the world," her celebrated first lines read. "They are, they have always been, one of the unique regions of the earth, remote, never wholly known. Nothing anywhere else is like them; their vast glittering openness, wider than the enormous visible round of the horizon, the racing free saltness and sweetness of their massive winds, under the dazzling blue heights of space."[40]

Mrs. Douglas titled her 1947 effort *The Everglades: River of Grass*. Her gorgeous prose impacted the environmental history of Florida by reimagining the Everglades as a source of free-flowing fresh water essential to wildlife and people of the region. With descriptions of the history, hydrology, geology, and "crowd of changing forms, of thrusting, teeming life," the book, an instant hit, turned the public perception of that ecosystem around.[41] It replaced the fallacy that Florida was glamorous despite the Everglades with the truth that the state was paradise specifically because of the ibises, spoonbills, and herons found in the wetlands.[42]

Many people helped in the final effort to open Everglades National Park, which occurred officially on December 6, 1947. President Truman wasn't one of them, but he did bring attention to the region by attending the dedication ceremony and delivering a fine speech that was broadcasted nationwide:

> Here in Everglades City we have the atmosphere of this beautiful tropical area. Southeast of us lies the coast of the Everglades Park, cut by islands and estuaries of the Gulf of Mexico. Here are deep rivers, giant groves of colorful trees, prairie marshes, and a great many lakes and streams.
>
> In this park we shall preserve tarpon and trout, pompano, bear, deer, crocodiles and alligators—and rare birds of great beauty. We shall protect hundreds of all kinds of wildlife which might otherwise soon be extinct.[43]

Toward the speech's end, Truman boasted that his administration was all about efficient irrigation, building dams, mining coal, and other such "wise use" of natural resources. The best lines of the day was falsely attributed to Marjory Stoneman Douglas: "The Everglades is a test. If we pass it, we get to keep the planet." It was environmentalist Joe Podger who uttered those immortal words.[44]

Harold Ickes was not at the Everglades ceremony. It infuriated him that the new national park had reduced Interior's proposal by almost 15 percent. The acreage kept shrinking as concessions were made to commercial developers and farmers. Though never presenting proof, Ickes intimated that Truman was in the pocket of "fat cat" Florida real estate developers. He also believed that Truman was a tool of the Army Corps of Engineers (ACE), in addition to lumber barons, oil companies, sugar plantations, and public utilities. Likewise, Bill Douglas charged that the Corps was engaged in "rampant vandalism" of the glorious Everglades ecosystem. Sure enough, as they predicted, after the national park dedication, ACE launched a congressionally approved flood control effort known as the Central and Southern Florida Project. It took twenty years, but ACE constructed seventy-five dams and spillways aimed at irrigating 15,200 square miles of south Florida farmland. Furthermore, a mind-boggling 1,300-mile network of canals and levees was constructed, forever stripping much of wild Florida of its innate natural beauty.[45]

IV

During the Truman years, air pollution became a flash point, capturing the attention of public health officials and conservationists alike. Air quality, however, was minimized as being a big-city problem, and lawmakers in Washington left municipalities to make their own regulations, if any, on smokestack emissions. Citizens had a sense that something was off kilter, but they weren't organized and didn't have the legal tools to combat the dirty air that was a threat to their health. Three years into the Truman presidency, the problem came to the fore with a vicious "killer smog," as it was called, that took the lives of citizens in one unfortunate Pennsylvania community.[46]

Starting on October 26, 1948, a temperature inversion occurred in the air above the mill town of Donora, southeast of Pittsburgh. In any other town, it would have generated nothing more than five stuffy, windless days. In Donora, the inversion killed twenty residents and made seven thousand others sick. Donora was backed by hills that rose more than four hundred feet above the Monongahela River, so the horseshoe bend of the river and the hills formed a natural saucer, which served to bottle up the air on the day of the atmospheric inversion. On the same day, as usual, poisonous chemicals, including hydrofluoric acid, carbon monoxide, and sulfur dioxide, were emitted from the town's three factories, which made steel, zinc, and wire products. Donorans had no choice but to breathe those noxious chemicals, and most of the population in the town of 13,000 citizens developed serious respiratory illness. With medical professionals overwhelmed, firemen raced about Donora taking oxygen tanks to the gasping sick. "The Donora incident for the first time established a clear link between air pollution and disease, or even death in America," the historian Scott Hamilton Dewey wrote in *Don't Breathe the Air: Air Pollution and U.S. Environmental Politics, 1945–1970*, "for while earlier news reports and scientific studies had made hesitant suggestions of statistical links between dirty air and disease, after Donora, there was a smoking gun with air pollution fingerprints all over it."[47] It would, however, take time and a dedicated scientist to identify those "fingerprints."

Americans found the situation in Donora alarming, but at first it was chalked up to a natural phenomenon of some sort. In the aftermath, factory owners insisted that kooky weather had been the cause. The locals knew better, and an early academic study out of the University of Cincinnati pointed to air pollution from the United States Steel factory in Donora and the zinc mill owned by the American Smelting and Refining Company (ASARCO). Lawsuits followed on behalf of Donora residents. ASARCO decided to respond with a highly respected, purportedly independent study of its own. It underwrote research by a biochemist at Harvard University, Mary O. Amdur.

Amdur was herself born in Donora on February 18, 1921, and had earned a PhD in biochemistry at Cornell University. Dedicated to science, incorruptible, and curious, she was working on pollution-related issues at the time of the 1948 disaster. After five years of testing, she concluded that sulfuric acid emissions produced by the zinc plant were a public health menace.[48] She presented her findings at a

Dr. Mary Amdur was a public health researcher who spent a lifetime fighting against air pollution. She became interested in the subject after the Donora smog incident of 1948. Amdur studied how the interactions of particles and gases in smog affected the lungs of humans and animals. Her research influenced the development of air pollution standards in the United States.

conference, despite attempted interference from ASARCO operatives in attendance. On her return to Harvard, she learned that ASARCO had made its opinions clear. ASARCO fired Amdur and suppressed her paper on Donora. Undaunted, she found a different position at Harvard and continued the pioneering advances in environmental understanding that earned her the title "mother of air pollution toxicology."[49]

Educating the public about the health risks of dirty air was an uphill battle. Although later Amdur would work at the Massachusetts Institute of Technology and then New York University, she never received tenure, much less a full professorship. That was partly because of sexism in academia, but she was also ostracized for having conducted research resented by industrialists at schools financially beholden to those same titans of—in many cases—chemical pollution. The corporate power of intimidation was especially strong in the late 1940s, when the environmental movement was in an embryonic stage and had yet to find youthful activists in numbers to join the New Deal warhorses such as Douglas and Ickes.[50]

When Congressman Kennedy read about the Donora smog incident in the *New Yorker* in 1950, he understood that the right to clean air was, in fact, a mass public policy concern.[51] Quietly, without a press release or speech to announce it, he joined a cadre of Democratic lawmakers demanding that the federal government at least study air pollution.[52] Whether or not the nascent anti–air pollution movement was aware, it had found one of its future leaders in Kennedy. A healthy environment, he knew, wasn't just something one expected on vacation to Yellowstone or hiking the Appalachian Trail; it was about the air and water resources in the backyards of voters.

As noted, Truman, by contrast, was uninterested in practically any regulatory stance that would stunt postwar US business growth. Nevertheless, Secretary of the Interior Krug persuaded him to hold a United States Technical Conference on Air Pollution, the first-ever American meeting on the subject. The climb was going to be steep. No air pollution legislation would be enacted in Pennsylvania, the site of the Donora horror, for twelve years.[53]

Donora, Pennsylvania, during the infamous smog episode of October 26–31, 1948.

V

Another, new type of atmospheric pollution—radiation—was increasingly poisoning Cold War America. On August 29, 1949, the Cold War grew vastly more serious as the Soviet Union successfully detonated its first atomic bomb. A year later, in June 1950, US troops were deployed to fight Communist forces in Korea. Furthermore, the arrest in England on February 2, 1950, of Klaus Fuchs as an atomic spy who had leaked secrets to the Russians during the waning days of the Manhattan Project increased public anxiety about the weapons capacity of the Soviet Union. Pressured by the military and key members of the AEC, President Truman made two crucial decisions in response to those events: the full-speed-ahead development of the hydrogen bomb and the authorization of a series of nuclear tests in the continental United States. The prospect of the spread of radiation far and wide in America, without assessing public health or environmental considerations, was ominous.

After considering sites including the North Carolina shores near Cape Fear and coastal areas in Texas and Alaska, on December 18, 1950, Truman's AEC chose a five-hundred-square-mile tract sixty-five miles from Las Vegas, which became the Nevada Proving

Grounds (NPG) for nuclear testing. (In 1955, it was renamed the Nevada Test Site, or NTS.) The site would soon grow to nearly fourteen hundred square miles. It consisted of two Air Force bombing ranges in a secluded region that reminded Dr. Norris Bradbury, the director of the Los Alamos National Laboratory, of the stark New Mexico desert where the world's first atomic bomb had been assembled.[54] In truth, the Nevada Test Site was a transitional zone between the Great Basin and Mojave deserts. A vast array of species lived in the ecosystem: kit fox, rattlesnakes, tortoises, golden eagles, striped whipsnakes, mule deer, coyotes, mountain lions, and wild horses. Creosote bush dominated the lower elevations; while the higher mountain altitudes were a vast woodland of piñon, juniper, and sagebrush. Throughout the ecosystem freshwater springs sustained life.[55]

On January 27, 1951, the US government detonated its first atomic device at the Nevada site in a basin known as Frenchman Flat. The blinding flash could be seen a hundred miles away. The AEC knew that the prevailing easterly winds would blow the radioactive debris away from highly populated California cities. In the contemptuous words of one federal bureaucrat, the Nevada tests would affect only "a low-use segment of the population"—meaning sustenance farmers in Nevada towns such as Indian Springs and Cactus Springs, as well as Indigenous tribal communities throughout the region.[56]

With the Korean War locked in a stalemate, the AEC continued tests at the Nevada range. Even though the effects of nuclear fallout in Japan had been well documented by medical experts and AEC technicians knew that people living within sixty miles of ground zero would likely fall ill from radioactive fallout, they persisted in doing the tests anyway. Critics concluded that the agency's guiding principle was nothing must stop the tests. "Jesus, it was bright!" a shepherd, Kern Bulloch, recalled after seeing the first Nevada test. "I put my hands up like that and you could doggone near see your bones. And then that cloud . . . mushroomed right over our camp and our herd. Pretty soon here comes some jeeps with Army personnel, and they said to us, 'My golly, you fellas are in a hot spot.' We didn't even know what they were talking about."[57] The Bulloch family sued the AEC for costing them millions of dollars due to radioactivity in their sheep. They lost their case in 1955. But *Bulloch v. United States* became the opening salvo of a flurry of lawsuits against the AEC for gross negligence.[58]

At the test site, officials allowed military personnel to be as close as twenty-five hundred yards from ground zero—before moving them even closer immediately following detonation. As it was, soldiers stationed as close as 3.9 miles from the explosions suffered from radiation sickness. "Atmospheric testing in Nevada continued along these lines throughout most of the fifties," the historian A. Constandina Titus wrote in *Bombs in the Backyard: Atomic Testing and American Politics.* "Many of the tests were defense oriented and involved troop maneuvers as well as experiments to determine the effects of a nuclear blast on various kinds of military equipment."[59] Besides military personnel, hundreds of Army jeeps and tanks and other forms of military hardware were strewn across the Nevada desert floor at varied distances from ground zero to monitor how they'd fare in a nuclear attack.

During the 1950s, nuclear mushroom clouds from test site detonations could be seen from towns as far away as northwestern Arizona and southeastern California. Fallout from the tests blew through southern Utah, where citizens experienced a sharp spike in cancer rates. The combination of aboveground and underground tests left the landscape a wasteland of radioactive rubble. Between 1945 and 1992, the United States conducted 1,054 nuclear tests, 216 of them in the atmosphere. The key testing sites were in Nevada, the Marshall Islands, New Mexico, and Arizona.

Nonetheless, despite some dissent from ecologists and public health advocates, most US lawmakers believed that nuclear testing was necessary to outmaneuver the Soviets in the Cold War. An "iron triangle" of government, industry, and science—one that played down the public health realities of atomic fallout—had taken firm root across the United States. With such quiescence and narrow perspective, Truman sided with the Pentagon brass, foreign policy elites, irrigation ranchers, and uranium mining companies who claimed that atomic fallout was no big deal. When it came to the American Southwest, Truman had scant interest in desert residents, including Indigenous peoples, or in ecology. His steep priorities on Cold War defense and economic growth reflected those of war-weary America. In response to efforts to protect Native American cliff dwellings in the Southwest from dams and extraction activities, he said, "It has always been my opinion that food for coming generations is much more important than bones from the Mesozoic period."[60]

VI

During the Truman years, no one knew for certain where Jack Kennedy stood on environmental conservation and the Nevada tests, for he exuded a faint boredom as a congressman, "sort of drifting" around for an issue to sink his teeth into, as Bill Douglas recalled in an oral history interview.[61] Douglas used his influence to urge Kennedy to lead the congressional fight for clean air quality and clean water. He saw an opening for Kennedy, in that twenty-one states had adopted water control laws by 1946 while the federal government itself had no unified standard.[62] The fact that American families couldn't swim in Lake Erie, the Mississippi River, and the Potomac River because of contamination did concern Kennedy enough that on June 30, 1948, he voted for the first federal water control program—but he wasn't a leader in that fight on the Hill. The Federal Water Pollution Control Act, passed by Congress, began a new effort by the federal government to regulate water quality. It was one law, not as strong as it should have been, but nevertheless significant. The federal government was assuming some level of responsibility for reducing water pollution.

At Kennedy family dinners, Douglas continued to tutor both Jack and Bobby about the organizations that were pushing to preserve forests, lakes, rivers, and wildlife. He explained how the Sierra Club, founded in 1892, safeguarded national parks; the National Audubon Society, founded in 1940, protected bird life; and the National Wildlife Federation, founded in 1936, taught the public about wildlife protection and pushed for establishing federal agencies to save species. All the groups wanted Rooseveltian preservation to stay alive in the more conservative Truman years. Genuine fellowship and camaraderie brought Douglas close to the two Kennedy brothers, especially Bobby. On weekends they would take day trips to places such as the Antietam battlefield and the forest trails in the Catoctin Mountains.[63] His determination to make Jack Kennedy a modern-day preservationist also grew. At one of those family meals, Douglas learned that seashore preservation—the protection of marine ecosystems, such as Cape Cod and the US Virgin Islands—was Jack's soft spot in the preservation realm. Thoreau's *Cape Cod* swelled him with local Massachusetts pride. Nobody important on the national stage was focused exclusively on seashore protection. That, Douglas believed, would be Jack's opportunity. "Douglas was Jack's compass in pointing out how he could be most useful on the preservation front," Jean Kennedy Smith recalled. "It was all about the seashores."[64]

CHAPTER 3

Rachel Carson and the Shore of the Sea

Rachel Carson, 1962

I

Just twenty-five miles west of Hyannis Port, where the Kennedys and thousands of other families enjoyed the sail-dotted water along the coast, the Woods Hole Oceanographic Institution was home to those fascinated with the world beneath the ocean surface. Established in 1930, Woods Hole operated beside the previously existing Marine Biological Laboratory and the Woods Hole Laboratory of the US Commission of Fish and Fisheries. Rachel Carson, who worked as an ocean science researcher at all three over the years, considered the coastal village of Woods Hole the most "delightful place" to "biologize."[1]

During the summer of 1929, the trim and proper twenty-two-year-old Carson was on a research fellowship at the MBL and living in a cottage run by the lab. Down a path from her quarters was the marvelous Nobska Point Light Station. On a clear day, she would sit there and gaze toward the vacation island of Martha's Vineyard, watching sea ducks bob and collecting spine-studded urchins at the low-tide line. Migrating eels and bottom-dwelling mollusks became her primary interests that summer and later. Though Carson had never seen the ocean until that summer, she had read about the diverse animal societies to be found on shorelines. Chatting with conchologists or ichthyologists at the MBL was a thrill for the budding zoologist. Carson enthused that it "would be very easy to acquire the habit of coming back every summer" at Woods Hole, conducting marine experiments in the huge redbrick Crane Laboratory building and diving into books at the extensive oceanographic library.[2] But those weren't the only attractions. "To understand the shore, it is not enough to catalogue its life," she wrote. "Understanding comes only when standing on a beach, we can sense the long rhythms of earth and sea that sculptured its landforms and produced the rock and sand of which it is composed; when we can sense with the eye and ear of the mind the surge of life beating always at its shores—blindly, inexorably pressing for a foothold."[3]

That summer Carson began a study of the cranial nerve in reptiles. If serious research at the MBL set her course professionally, long walks on the sand and swims in the waves gave her a life-affirming love of the sea. The two strains of Carson's Woods Hole stint combined with her own inborn view of the world. To her, all of God's creatures had a will to live and were worthy of compassion. Though not outwardly philosophical, she did believe that every person had to respectfully repay an eternal debt to nature. At a young age, she understood ecology as a basic notion: *We live in the house of life, and all the rooms connect.* A line by the English poet Francis Thompson nicely summed up her view: "Thou canst not stir a flower / Without troubling of a star."[4]

Rachel Carson was born on May 27, 1907, in Springdale, Pennsylvania, a town northeast of Pittsburgh in the lower Allegheny Valley.

The third and youngest child of Robert Warden Carson and Maria McLean Carson, she grew up on her parents' sixty-five-acre woodland farm. The small white farmhouse sat atop a steep hill that sloped toward the 325-mile-long Allegheny River, not far away. The deep river elbowed sharply at Springdale before flowing its final sixteen miles to converge with the Monongahela at Pittsburgh. Encouraged by her parents, she explored the white pine groves and lush meadows surrounding the farm, treating the ecosystem as a backyard nature laboratory.[5]

Growing up a landlubber in western Pennsylvania, Carson regarded seashells as the most wondrous and elegant collectibles imaginable. Bright eyed and studious, she enjoyed observing wildlife, reading fiction, and writing her own essays. Fortunately, she had a booster in her mother. Maria Carson, who embraced natural history, botany, and bird-watching, passed those hobbies on to Rachel. "I can remember no time when I wasn't interested in the out-of-doors and the whole world of nature," Rachel reminisced years later. "Those interests, I know, I inherited from my mother."[6]

In every butterfly or frog, Carson saw something larger than herself. It saddened her to see the Allegheny River used as a giant sewer, with glue factories in the area polluting the water and disrupting wildlife. There, humans blatantly dominated and destroyed the natural world. Two power plants in Springdale—the West Power Company and Duquesne Light Company—were stark examples of the industrial order blighting the countryside and fouling the air. All around Pittsburgh, there were factories spewing dirty, ash-black smoke into the atmosphere. The waterways where the city's famed "three rivers" converged was foul.[7]

At eleven, Rachel made her debut as a writer with a prizewinning short story, "A Battle in the Clouds," in *St. Nicholas Magazine*. Other stories followed, all in this popular American children's periodical that Scribner's had founded after the Civil War. Under her mother's loving guidance, in 1922 she wrote an essay, "My Favorite Recreation," which described a day in the forests of Pennsylvania. At the finish, she wrote, "The cool of approaching night settled. The wood-thrushes trilled their golden melody. The setting sun transformed the sky into a sea of blue and gold. A vesper-sparrow sang his evening lullaby. We turned slowly homeward, gloriously tired, gloriously happy!"[8]

Carson's devotion to nature writing never waned. At Parnassus High School, classmates raved that she was a marvelous writer with a sharp intellectual bent, like Henry David Thoreau or Herman Melville. The inscription about her in the senior yearbook put their sentiments into rhyme:

> Rachel's like the mid-day sun,
> Always very bright,
> Never stops her studying
> 'Till she gets it right.[9]

Excited by literature and the natural world, Rachel enrolled in the Pennsylvania College for Women (PCW; later Chatham University) in 1925. The elegant, green-leafed college campus was tucked away in a wealthy neighborhood in Pittsburgh, three miles from the Allegheny. Rachel found it idyllic, admiring the well-maintained lawns and hardwood trees. Clad in dark bloomers, stockings, and white tennis shoes, she played goalie on the field hockey team and got along well with the eighty-eight other women in her class. Although not an extrovert, she was witty and always an original. When asked in freshman English to write about herself, she reflected instead on the outdoors. "I love all the beautiful things of nature, and the wild creatures are my friends," she revealed. "What could be more wonderful than the thrill of having some little furry animal creep closer and closer to you, with wondering but unafraid eyes?"[10]

Rachel started out as an English major but soon fell under the spell of Professor Mary Scott Skinker, the head of the biology department. Beautiful, stylishly dressed, and full of self-confidence, Skinker was the star faculty member at PCW. In the 1920s, an undergraduate biology course typically included such topics as food safety, hygiene, agronomy, public health, nutrition, and sanitation—something akin to an "intersection between science and home economics." Skinker was more of a purist. In 1925, in the so-called Scopes Monkey Trial in Dayton, Tennessee, the schoolteacher John Scopes was forbidden to teach Charles Darwin's theory of evolution in high school. Skinker, an admirer of *On the Origin of Species*, taught students about the interconnectedness of life in the lower Allegheny region, depicting the living world from a holistic perspective. "Eventually it dawned on me," Carson commented later, "that by becoming a biologist I would be giving myself something to write about."[11] Before long, she began thinking of combining a literary career with zoology.[12]

After graduating with honors in the spring of 1929, Rachel headed to Woods Hole. As a "beginning investigator" of the Atlantic Ocean, she then entered graduate school in zoology at Johns Hopkins University in Baltimore on a one-year scholarship. Whenever possible, she worked to improve the flow and rhythm of her prose. After earning her master's degree in 1932, she wrote articles about the Chesapeake Bay for the *Baltimore Sun*, focusing on mid-Atlantic fish and wildlife under headlines such as "It'll Be Shad Time Soon" and "Chesapeake Eels Seek the Sargasso Sea."[13] Her *Sun* articles showcased her broad interest in progressive conservation writ large.[14] However, freelancing for the newspaper, though satisfying, didn't earn her much money to live on.

After Carson's father, Robert, died in 1935, she taught part-time at the University of Maryland and Johns Hopkins summer school. With him gone, there was little income with which to support her mother and herself. When Professor Skinker learned that Rachel was in a financial bind, she recommended applying to the Bureau of Fisheries in Washington, DC, which had recently undertaken a series of radio broadcasts entitled *Romance Under the Sun*. It needed a part-time biologist-type who could write. Carson recalled later, "I happened in one morning, when the chief of the biology division [Elmer Higgins] was feeling rather desperate—I think at that point he was having to write the scripts himself. He talked to me a few minutes and then said: 'I've never seen a written word of yours, but I'm going to take a sporting chance.' That little job, which eventually led to a permanent appointment as a biologist, was in its way, a turning point."[15] She was hired as a junior aquatic biologist in the Division of Scientific Inquiry, writing radio scripts about fish.[16]

At about the same time, Carson had a turning point in her freelance writing career, as well, when *Atlantic Monthly* published her "Undersea" essay in its September 1937 issue. In it, she brilliantly explored tidal pools, surface waters, and the seafloor, treating humans as aliens and predators, at a remove from the ocean ecosystem. Only by *thinking* like a sea creature, she felt, could a scientist fully grasp the marine environment. "Who has known the ocean?" she wrote. "Neither you nor I, with our earth-bound senses, know the foam and surge of the tide that beats over the crab hiding under the seaweed of his tide-pool home; or the lilt of the long, slow swells of mid-ocean, where shoals of wandering fish prey and are preyed upon, and the dolphin breaks the waves to breathe the upper atmosphere."[17]

One of Carson's biographers, Linda Lear, believed that it was Maria who introduced her to the teachings of Albert Schweitzer, a world-famous Lutheran medical missionary who Norman Cousins thought was a modern-day saint. Schweitzer taught that "compassion for everything" was the beginning and foundation of morality. In his thinking, there were echoes of others, but Schweitzer spoke to the twentieth century with a sure, simple message that was nonetheless dynamic, guiding the course of at least three generations worldwide. He had been raised in eastern France among the foothills of the Vosges Mountains (equivalent to Carson's Allegheny Mountains), cherishing the farm animals, migratory birds, and above all, his companion dogs and cats. Intuitively channeling the spirit of Schweitzer, Carson also believed that all creatures were her kindred spirits.[18]

In September 1913, Schweitzer, with his unruly gray hair and bushy mustache, was a well-educated physician cruising along the Ogowe River in French Equatorial Africa (now Gabon). He and his wife, Helene, a nurse, were headed upriver to run Lambaréné, a jungle hospital that treated West African patients from nearby villages for such maladies as dysentery, malaria, and leprosy. Attempting to relieve human misery was the couple's lifework. As they floated past the thick jungle forests, Schweitzer pondered existence. At sunset on their third day on the Ogowe, after watching hippopotamuses wallow on a mudbank, he had an epiphany. Feeling a sudden sense of human puniness and disregard for the natural world, a holy notion struck him. It was an attitude of mind, not a rational set of propositions. He described the experience and the understanding it had left in his book *Civilization and Ethics*, calling it *Ehrfurcht vor dem Leben* (reverence for life).

Christian theologians have placed Schweitzer's "reverence for life" philosophy outside of the Orthodox Christian tradition; for Schweitzer, a blade of grass or a sparrow flying was an intricate part of God's grand plan. Two lines of verse by the British poet William Blake—"A Robin Red Breast in a Cage / Puts all Heaven in a Rage"— cut to the core of Schweitzer's philosophy. "In everything you recognize yourself," Schweitzer advised. "The tiny beetle that lies dead in your path—it was a living creature, struggling for existence like yourself, rejoicing in the sun like you, knowing fear and pain like you. And now it is no more than decaying matter—which is what you will be sooner or later, too."[19]

During World War I, while Germany was perfecting death devices such as barbed wire, mustard gas, machine guns, and aerial bombs, Schweitzer was heeding Jesus's call to help the sick and impoverished. What made Schweitzer's Christianity unique was his strong pantheist streak, his extended, soulful compassion for all animals. Borrowing from decades of deep reading in sacred Buddhist, Christian, and Jainist texts, he lamented that the modern industrial-technological existence was too rooted in social Darwinism's "survival of the fittest" thinking. To Schweitzer, the theory of evolution, although indisputable, was wrongheaded in its focus on dominance. Human consciousness, he believed, had an ethical dimension not well considered by Darwin and his followers. At their best, humans were spiritual creatures who needed to develop and maintain an openhearted respect and compassion for life, rejecting death and destruction and embracing universal tenderness. After the brutality of World War I, Schweitzer's viewpoint inspired millions of people around the world. One of them was Rachel Carson. She called Schweitzer "the one truly great individual our modern times have produced."[20]

In practice, Schweitzer had a pecking order for medical treatment at Lambaréné: humans first, even with minor complaints, followed distantly by other creatures.[21] As ethically dubious as that might seem, it was in fact consistent with his views. "Reverence for life" didn't describe a peaceable kingdom in which all was benign. Schweitzer recognized the "will to live" and understood that an individual being of whatever species needed to answer that drive, in some cases by

Albert Schweitzer with his pet pelican in French Equatorial Africa (Gabon) 1951.

killing other species. Borrowing from the Hindu text Bhagavad Gita, Schweitzer's philosophy, in essence, stood for the Creator's "commandment not to kill and not to damage earth unnecessarily."

Schweitzer expressed his ethical standard: "Whenever I injure life of any kind, I must be quite clear as to whether this is necessary or not. I ought never to pass the limits of the unavoidable, even in apparently insignificant cases."[22] Carson lived and worked by that rule. She felt that doing unnecessary harm to wildlife was appalling and she was somewhat unique among field researchers in that she habitually returned specimens under scientific scrutiny to the place from whence they had come, so that they could resume their lives.

In 1938, bolstered by positive feedback from the *Atlantic* essay, Carson spent another summer at Woods Hole gathering material for her first book. She also went on a working vacation at a Bureau of Fisheries station in Beaufort, North Carolina. The sandpits and tidal marshes along the mid-Atlantic Coast, protected by the Outer Banks of barrier islands, provided her with rich material for her project, which would become *Under the Sea-Wind*. Her adoption of sea creatures such as Scomber, the primary character in *Under the Sea-Wind*, resembled Henry Williamson's treatment of otters in his book *Tarka the Otter* (1927), while the musicality of her prose was reminiscent of Beston's *The Outermost House*.[23] For the rest of Carson's life, her main fascination was discovering the exotic creatures who lived in saltwater ecosystems, especially crustaceans, seaweeds, invertebrates, and small fish—what one of her biographers, William Souder, called "an embassy of living things."[24]

Under the Sea-Wind, her first and favorite book, was published in November 1941 by Simon & Schuster. Its artwork was by Howard Frech, an illustrator Carson had worked with in Baltimore. *Under the Sea-Wind* was a masterly dissertation on the life cycles of three ocean creatures, told from the perspective of the animals themselves. The first section, "Edge of the Sea," was about the migratory journey of the sanderling pair Silverbar and Blackfoot from the Outer Banks of North Carolina to the Arctic Circle and back to Patagonia. "The Gull's Way," the second part, centered on Scomber the mackerel, swimming through a gauntlet of near-death experiences from the shores of New England to the Continental Shelf. Finally, "River and Sea" depicted migration as experienced by Anguilla the eel, who journeyed to the Sargasso Sea, south of Bermuda, a spawning ground for American and European eels.[25] All three migration stories were based on the latest research on the life cycles of sea creatures.[26] Carson had anthropomorphized the sanderling, mackerel, and eel to help the reading public see the world through their underwater eyes and so absorb their dramatic survival stories in a more sincere way.

In the preface, Carson, praising Beston's *The Outermost House*, admitted to being awestruck by that Cape Cod memoir's great simplicity, beauty, and reflection of oceanic rhythms.[27] With Beston pointing the way, she vowed "to make the sea and its life as vivid a reality for those who may read the book as it has become for me during the past decade."[28]

About a dozen years later, in the mid-1950s, John F. Kennedy received a copy of *Under the Sea-Wind* as a gift from his mother.[29] Although *Under the Sea-Wind* sold in low numbers, perhaps because Pearl Harbor occurred right after it debuted, it became celebrated by marine biologists and other lovers of the animal kingdom.

Seldom had a scientifically accurate delineation of ocean species been written with such gorgeous literary flare. At the time it was published, glowing reviews poured in; the *New York Times* called it a "beautiful and unusual" book for a naturalist.[30] William Beebe, who had ventured three thousand feet deep in a record-breaking descent in the Bathysphere off the coast of Bermuda, authoritatively declared in the *Saturday Review of Literature* that he couldn't detect a single error in *Under the Sea-Wind*.[31] In a 1945 anthology that Beebe edited, *The Book of Naturalists: An Anthology of the Best Natural History*, an essay by Carson appeared alongside work by such immortals as John James Audubon and Henry David Thoreau. Pleased to be recognized as a defender of the world's oceans and sea life, Carson professed that her ambition was to leave the natural world "a better place to live in."[32]

II

Learning from government biologists at the US Fish and Wildlife Service's Patuxent Wildlife Research Center in Maryland that synthetic pesticides were having a deleterious effect on wildlife, Carson pitched a *Reader's Digest* editor with an idea for an article about DDT (dichlorodiphenyltrichloroethane). "Practically at my backdoor here in Maryland, an experiment of more than ordinary interest and importance is going on," she wrote on July 15, 1945. "We have all heard a lot about what DDT will soon do for us by wiping out insect pests. The experiments at Patuxent have been planned to show what other effects DDT may have if applied to wide areas: what it will do to insects that are beneficial or even essential; how it may affect waterfowl, or birds that depend on insect food; whether it may upset the whole delicate balance of nature if unwisely used."[33] Scientists at Patuxent had begun testing the effects of synthetic chemicals on the environment and wildlife as early as 1941. Carson's DDT proposal was rejected, but her concern about the ubiquitous chemical persisted.

Pesticides had been employed for centuries, but their use had exploded across America during World War II, supported in part by military-funded research. DDT had been synthesized in the late nineteenth century, but it wasn't until 1939 that a Swiss chemist, Paul Hermann Müller, discovered that it could be used to eliminate a vast range of insects via its toxic effect on nerve tissue. The US military used DDT to control mosquitoes and other insects that transmitted malaria, dengue fever, and typhus to troops in the Pacific and Europe. Soldiers were dusted in it. DDT was inexpensive to produce and lethal to bugs, and it seemed innocuous around humans.

One reason DDT was embraced so readily was that chemicals had yielded a plethora of man-made miracles in the first half of the twentieth century. For example, after a long epoch of intermittent tragedies, typhoid was newly controlled by water chlorination, thanks to chemists, reformers, and hydro engineers. It wasn't much of a stretch to believe that DDT was another wonder, one that would stop insects from spreading disease and destroying crops.[34]

After V-J Day, the DDT left over from the war effort was allocated by the Truman administration for use on public lands. Meanwhile, farmers throughout the Midwest considered DDT a miracle panacea against grasshoppers, fruit flies, mosquitoes, beetles, and boll weevils. Workers without protective gear or masks sprayed it on crops from prop planes and trucks. They weren't concerned about whether it was safe—after all, the US government was promoting it. Only the concerned biologists at the US Fish and Wildlife Service were studying its toxicity in other species. And it wasn't just DDT. Union Carbide, DuPont, and Shell developed their own synthetic agrochemicals, such as aldrin, dieldrin, endrin, and 2,4-D. Mixed with fertilizers made from surplus nitrogen no longer needed for wartime explosives, such chemicals intensified the spread of industrial farming throughout the United States.

By the time Carson pitched the DDT story to *Reader's Digest*, she knew there was a growing body of scientific research proving that many bird species, including brown pelicans, cormorants, and peregrine falcons, were at risk when their habitats were contaminated with DDT.[35] In 1945, the federal government released a study that found traces of DDT in the milk of cows that had grazed on grass sprayed with the chemical. It recommended that farmers apply "safe alternative substitute insecticides" to control flies and lice on cattle. In May 1946, Carson wrote and released a Fish and Wildlife report claiming that DDT was harmful to fish, shellfish, eagles, and hawks.

In 1949, Carson became editor in chief at the Fish and Wildlife

Service. At the time, Patuxent housed the USFW wildlife library and so she had access to disturbing reports that chemical corporations chose to ignore. Because the new DDT research was scattered and inaccessible for the average reader, Carson began collecting the Patuxent material, perhaps for a book someday about the perils of insecticides. Around this time, she wrote three press releases alerting fish-processing plants, specifically, to the hazards of using DDT in their facilities. In all three, she advised fisheries to "consult experts about using DDT." Early on, she had understood that pesticides used on the land would eventually drain into the ocean, disrupting or destroying water life.

For much the same reason, Carson opposed atomic testing by the US government in the Pacific Ocean. She learned about the ecological damage it caused from the biologist Roger Revelle, who was attached to the Scripps Institution of Oceanography in California and who was later known for his pioneering work foretelling global warming. Carson had befriended Revelle during the war when he was assigned to the Bureau of Naval Research. In 1946, he was appointed to chair the task force analyzing the radiation fallout from the Bikini Atoll tests. By coincidence, Carson had edited the Biological Survey made of the far-flung islands before the Operation Crossroads detonations. Revelle asked Carson to proofread his Bikini Atoll reports. Looking at the destruction of ocean habitat the joint Army-Navy atomic testing brought, she hoped that future tests would be banned on ecological grounds.[36]

III

Aligning with Carson in her anti-DDT and anti–nuclear testing concerns after World War II was a young biology professor at Washington University in St. Louis, Missouri: Barry Commoner. Born on May 28, 1917, in Brooklyn, New York, Commoner was the precocious child of Russian Jewish immigrants. Though city-raised, he displayed a connection to the natural world from a very young age. At ten years old, he roamed Prospect Park in Brooklyn, collecting leaf and insect specimens to scrutinize under his Sears, Roebuck microscope. A marvelous student, with his intense glare, heavy eyeglasses, thick black eyebrows, and flare for mathematics, he graduated from Columbia University's zoology program with honors in 1937.

Commoner next went to Harvard, where he earned a doctorate in cellular physiology in 1941. He enlisted in the Navy, which assigned him to develop a system for spraying DDT from planes. The plan was to clear islands of insects such as lice and mosquitoes that caused disease among soldiers. Commoner and his crew discovered, to their disgust, that although DDT sprayed from bombers effectively eliminated flies on the beach, new flies were quickly drawn to "feast on the tons of fish that DDT had poisoned, thereby creating a cycle of ecological disaster."[37] The startled Commoner concluded that DDT made wildlife sick—and people, too, for that matter.

After the war, Commoner's adverse experience with DDT in the Pacific led him to espouse the "precautionary principle" that new chemicals and technologies should not be introduced into society if there was any chance that they were detrimental to human health. He argued that chemical products such as DDT shouldn't be approved until after the Food and Drug Administration had proof that they were environmentally safe. Even though he was unfazed by criticism of his opinions, Commoner was surprised at how many well-trained scientists, particularly at land-grant universities, were then making untenable claims about those types of chemicals merely to make money in agribusiness or advance their academic careers.

Like most US servicemen in the Pacific theater, Jack Kennedy had been doused with DDT in order to kill lice and repel mosquitoes. There is no record that he questioned the procedure. In part because of the use of DDT and other pesticides, disease accounted for fewer casualties than enemy fire for the first time in US military history. However, Carson, Commoner, and that group of Patuxent biologists based in Laurel, Maryland, maintained that DDT was harmful to birds, fish, and possibly humans. For the time being, however, their voices were drowned out by postwar enthusiasm for anything that smacked of newfangled promise. "Modern chemistry rubs its Aladdin's lamp, shakes up its test tubes," the *Washington Post* enthused, "and presto!"[38]

Starting in 1946, Carson originated *Conservation in Action*, a series of US Fish and Wildlife Service booklets describing the activities, human and natural, in national wildlife refuges.[39] In all, there would be twelve, averaging about a dozen pages. Carson wrote the first four, delighting in the project because for the first time in almost a decade, her job took her into the field. For the first one, Carson, naturalist Shirley Briggs, and illustrator Katherine Howe visited part of the Chincoteague National Wildlife Refuge at the southern tip of Virginia's Assateague Island. Carson consulted with refuge employees but also made her own observations within the shoreline, marshland, and maritime woodland. In the text she wrote for the resulting booklet, she offered graceful descriptions of nature at work:

> Like islands standing out of the low marsh areas are the patches of firmer, higher ground, forested with pine and oak and carpeted with thickets of myrtle, bayberry, sumac, rose, and catbriar. Scattered through the marshes are ponds and potholes filled with widgeongrass and bordered with bulrushes and other good food for ducks and geese. This is waterfowl country. This is the kind of country the ducks knew in the old days, before the white man's civilization disturbed the face of the land. This is the kind of country that is rapidly disappearing except where it is preserved in wildlife sanctuaries.[40]

After writing two more installments in the *Conservation in Action* series—on Lake Mattamuskeet in North Carolina and Parker River in Massachusetts—Carson coauthored a pamphlet on the Bear River Wildlife Refuge in Utah. As intended, her structure, and to some extent her tone, was used by other writers in the series. Carson also wrote *Guarding Our Wildlife Resources*, which was different from her other Fish and Wildlife publications. Souder, Carson's biographer, explained, "Carson structured it as a serial tragedy—a story of natural wealth repeatedly squandered, differing only in the details from one class of wildlife to another."[41]

In the *Conservation in Action* booklets on specific wildlife refuges, Carson treaded a thin line between encouraging her readers to appreciate the natural world and enticing the kind of tourism that would damage it. She made it clear that Natural Wildlife Refuges weren't merely weekend playgrounds for fishermen and hunters but sacred places designated to honor the totality of ecological existence. "The preservation of wildlife and of wildlife habitat means also the preservation of the basic resources of the earth, which men, as well as animals, must have in order to live," she wrote in that booklet. "Wildlife, water, forests, grasslands—all are parts of man's essential environment; the conservation and effective use of one is impossible except as the others also are conserved."[42]

Between 1945 and 1948, Carson faced two medical problems that were somewhat unusual for a relatively young woman. She underwent an appendectomy and had a cyst surgically removed from her left breast. Though neither of these was alarming by itself, she was also very often fatigued. In the summer of 1950, when she was forty-three, she had a second growth removed from her left breast. For a scientist, Carson was rather dismissive about her illnesses. Her mission, it seemed, was to teach people about oceans and protect the environment, not to feel sorry for herself.[43] Holding to Schweitzer's "reverence for life" philosophy, she found every second she was alive wondrous.

In 1949, Carson went diving in southern Florida waters for the first time to observe coral reefs. With Shirley Briggs as a companion, she wore a helmet underwater and went down only ten feet for fifteen minutes. In a letter to Beebe, she described her first "deep-sea dive": "I finally got down, under conditions that were far from ideal—water murky, the current so strong I could not walk around but hung onto the ladder. But the difference between having dived—even under those conditions—and never having dived is so tremendous that it formed one of those milestones of life, after which everything seems a little different."[44] In September, *Yale Review* published Carson's essay "The Birth of an Island," which won the $1,000 George Westinghouse Award (Norman Cousins was one of the judges). Carson depicted the animal societies she had encountered with documentary exactness.

Everything about the year 1950 was exciting and hopeful for Carson until she was diagnosed with breast cancer while writing her second book, *The Sea Around Us*. A tumor was removed, and she considered the ordeal to be over, given that her doctors prescribed nothing further. Coinciding with her own medical news, the American

Medical Association (AMA) disingenuously warned that the "chronic toxicity" of most new pesticides, including DDT, was "completely unexpected."[45] That, Carson knew, was bunk. If the AMA cared, it could have read her press releases warning of DDT hazards. For the time being, such scientific, fact-based knowledge rarely left the domain of Rachel Carson, her government colleagues, and a few other open-minded researchers. Nevertheless, in the early Cold War a drumbeat of public health and environmental concerns swept across the nation. In 1952, a report by the International Materials Policy Commission, also called the Paley Commission after its head, CBS chairman William Paley, urged the United States to look to renewable energy resources (solar and wind power) and urged ecological prudence to protect natural resources into the twenty-first century. There was no Environmental Protection Agency in the 1950s (that wouldn't come until 1970), just crisis reactive federal and state stopgap regulatory measures; that wasn't good enough.[46]

Not so far from the paneled rooms and good intentions of the Paley Commission offices, a twenty-two-hundred-acre wetland on the west side of Staten Island in New York City was attracting unwanted attention. Known as Fresh Kills, it was the site of a gargantuan landfill that soon started reeking due to the towering piles of raw garbage and industrial debris dumped there. Nearby residents were often dizzy. Rats roamed the dump unchecked. Rotting food attracted swarms of flies. In response, citizens' groups such as the Staten Island Anti-Garbage Organization and Staten Island Citizens for Clean Air formed to combat the stench and health hazards. They denounced the destruction of the beautiful salt marshes, tidal wetlands, and forests that had been on the site.[47]

At its peak, Fresh Kills received thirty thousand tons of trash per day, growing into a pyramid of waste that became one of the biggest human-made structures in the world. "It had a certain nightmare quality," a New York City sanitation engineer recalled. "I can still recall looking down on the operation from a control tower and thinking that Fresh Kills, like Jamaica Bay, had for thousands of years been a magnificent, teeming, literally life-enhancing tidal marsh. And in just twenty-five years, it was gone, buried under millions of tons of New

York City's refuse."[48] Similar urban dumps encroached on the public health of other cities, as well.

Hard-hitting books gave such subjects urgency. *Road to Survival* (1948) by the ecologist William Vogt was a frightening description of how overpopulation and industrial debris were wiping out essential natural resources; it helped trigger a revival of Malthusianism in the 1950s and 1960s. A book by the naturalist Fairfield Osborn, *Our Plundered Planet* (1948), lambasted the human race as a horrific steward of Earth. But it was scientific forester Aldo Leopold's posthumously published book *A Sand County Almanac* (1949) that truly awakened a new "land ethic."[49]

Based on Leopold's observations of the flora and fauna of his Baraboo, Wisconsin, farmland, *A Sand County Almanac* soon joined Thoreau's *Walden* and Carson's *Under the Sea-Wind* as historic benchmarks among books that advanced ecological understanding. Every fall, thousands of sandhill cranes congregated on the Wisconsin River just behind Leopold's writing shack as they prepared for fall migration. Their rituals stirred his heart and inspired in him a desire to heal damaged landscapes and protect wildlife with his land ethic. As his book was being prepared for publication, Leopold died of a heart attack while trying to extinguish a wildfire on a neighbor's farm. Posthumously, his publisher, Oxford University Press, released *A Sand County Almanac*. Conservation-minded citizens flocked to the book. Such lines as "A thing is right when it tends to preserve the integrity, standing, and beauty of a biotic community" and "There are some who can live without wild things, and some who cannot" became pearls of wisdom for people unnerved by multi-lane highways, Fresh Kills, Donora, and the Nevada Test Range.[50]

In 1950, Bill Douglas gave the postwar conservation movement a bestseller with his book *Of Men and Mountains*, a long memoir of his nature outings in the Pacific Northwest: what he had seen, heard, and felt in the Cascades. Climbing high-altitude peaks allowed him time and perspective to ponder the workings of civilization and reemphasize independence over conformity. "Once man leaned that heavily on people he was not wholly free to live," he philosophized. "Then he became moody rather than self-reliant. He was filled with tensions and doubts. He walked in an unreal world, for he did not know the earth from which he came and to which he would return. He became a captive of civilization rather than an adventurer who topped each hill ahead for the thrill of discovering a new world. He lost the feel of his own strength, the power of his own soul to master any adversity."[51]

The next year Douglas traveled to Tibet and stayed at a Buddhist monastery in the Himalayas; that religious experience shifted his thinking even more, from Theodore Roosevelt–styled conservation to the type of highly complex ecological thinking that Leopold and Carson espoused. A hike in the Tibetan mountains brought the justice to an epiphany. "I realized that Eastern thought had somewhat more compassion for all living things," he wrote. "Man was a form of life that in another reincarnation might possibly be a horsefly or a bird of paradise or a deer. So a man of such a faith, looking at animals, might be looking at old friends or ancestors. In the East the wilderness has no evil connotation; it is thought of as an expression of the unity and harmony of the universe."[52]

IV

Three installments of Carson's book *The Sea Around Us* were serialized in the *New Yorker* in June 1951. Though still not a bench scientist, she translated once again ecological wisdom about the ocean into lucid and poetic prose that Leopold and Carson espoused. That July the book was published and was an instant hit. It stayed on the *New York Times* best-seller list for eighteen months. The public was even hungrier than expected for a comprehensive understanding of the oceans. "That was the book that became my mother's new *Walden* and *The Outermost House*," Ted Kennedy recalled. "She just adored it."[53] Praise poured in from all directions. At Norman Cousins's insistence, the *Saturday Review of Literature* published a photo of Carson on the cover of its July issue and commended her brief Florida coral reef diving effort in the name of ocean exploration, even though she went down only ten feet.[54]

Among the fans of *The Sea Around Us* was Theodore Roosevelt's daughter Alice Roosevelt Longworth. She telephoned Carson to rave about her talent as a public oceanographer. Oxford University Press threw a book party for Carson at the National Press Club in Washington, DC. Clad in a white silk summer dress patterned with wildflowers, Carson looked elegant and fashionable, a woman comfortable in her own skin at the club event. *The Sea Around Us* was favorably reviewed in the *New York Times Book Review*, which delighted her beyond words. This sudden burst of fame led Oxford to reissue *Under the Sea-Wind* in a handsome new paperback edition. When a reporter asked Carson what she had hoped to accomplish in writing *The Sea Around Us*, her reply was that she wanted to educate the public. She said, "An ocean voyage, or a trip to the shore means so much more if you know a few things about the sea."[55]

Wherever Carson traveled to promote *The Sea Around Us*, she championed American shoreline preservation. She told the *Washington Post*, "What has taken centuries to develop is being destroyed in a few years."[56] She drove her point home that Earth was "a water world, a planet dominated by its covering mantle of ocean, in which continents are but transient intrusions of land above the surface of the all-encircling sea."[57] It boggled the mind, Carson wrote, that "life evolved in the sea two and a half billion years ago."[58]

In the most notable pages of *The Sea Around Us*, Carson reflected on island life. Like Charles Darwin in the Galápagos, she believed that islands were the ideal outdoor laboratories in which to study evolution and species habits. "In all the world of living things, it is doubtful whether there is a more delicately balanced relationship than that of island life to its environment," she wrote. Yet, she observed, humans were destroying the environment on islands around the world. For instance, the Hawaiian Islands were "a classic example of the results of interfering with natural balances. Certain relations of animal to plant, and of plant to soil, had grown up through the centuries. When man came in and rudely disturbed this balance, he set off a whole series of chain reactions."[59]

Foreboding runs through much of *The Sea Around Us*. The result of Earth's natural changes combined with reckless human stewardship was, Carson knew, the diminishment of seashore wonders and coral reefs. Even though the Outer Cape, which Thoreau, Beston, and the Kennedys so loved, was doomed to vanish in four or five thousand years, she nevertheless believed that an all-out attempt at preservation should be adopted. "Cape Cod is not old, in geologic terms, being the product of the glaciers of the most recent Ice Age," she wrote, "but apparently the waves have cut away, since its formation, a strip of land some two miles wide."[60]

Carson's literary reputation was sealed once *The Sea Around Us* won the National Book Award. It was soon translated into thirty-two foreign languages and praised far and wide as oceanic science graced with rare literary elegance. Upon accepting the award, Carson said, "If there is poetry in my book about the sea, it is not because I deliberately put it there, but because no one could write truthfully about the sea and leave out the poetry."[61] Furthermore, she received the New York Zoological Society's Gold Medal and a Guggenheim Foundation Fellowship for future research on tidal coasts, in preparation for what would be her third book, *The Edge of the Sea*.

The honor that meant the world to Carson, however, came from the Garden Club of America, established in 1913 by women interested in the greening of the nation through botany and horticulture. It presented her with the prestigious Frances K. Hutchinson Medal for *The Sea Around Us*. What made the award so special was that the Garden Club hadn't previously branched out from land-based flora endeavors to contemplate ocean "gardens," such as the coral reefs off the Florida coast or the kelp beds off Monterey, California. Impressed by *The Sea Around Us*, the club was adopting a more holistic way of thinking about Earth stewardship. Carson wrote in accepting the medal:

Perhaps you are remembering that the first plants lived and evolved in the sea, and that after many millions of years some of these same sea plants came out on the bare land of the continents and in time made them habitable for the land creatures that were to follow. If the sea plants had not done this, it is hard to imagine that the continents would ever have been more than bare deserts of rock, or that life could even have evolved beyond the fishes. And so, there are these underlying unities—one thing following upon another—inevitably linking the destinies of earth and sea, of diatoms and palm trees, of trilobites and men. I am convinced that we can learn to live in harmony with each other and with our environment only by recognizing these ancient unities and the links that bind us to our origins in earth and sea.[62]

With the success of *The Sea Around Us*, Carson resigned from the Fish and Wildlife Service to devote time to writing and taking care of her mother, Maria. In 1957, she would become a mother herself, to her orphaned five-year-old grandnephew, Roger Christie, whom she adopted. Flush with success, Carson bought land on Southport Island in southern Maine with the dream of constructing her own Outermost House and becoming a full-time writer. "By next June I am to have a sweet little place of my own built and ready to occupy!" she wrote her agent, Marie Rodell, in September 1952. "The place overlooks the estuary of the Sheepscot River, which is very deep, so that sometimes . . . *whales* come up past the place, blowing and rolling in all their majesty! And lots of seals, and there is a long pool left in the rocks at low tide, where sculpins and other fish sometimes get stranded."[63]

Seashore protection became Carson's full-time job. The wholesale excavation, dredging, and subdividing of America's shorelines throughout the late 1940s and 1950s had forced her into action. Commercial developers were typically unfamiliar or unconcerned with the need for natural barriers against maritime storms and currents that cut into dunes and destroyed vegetation. They foolishly filled in wetlands and constructed buildings close to the ocean edge. "One of my keen interests," Carson wrote, "is the preservation of some natural seashore areas. There are few beaches left that show no scarring traces of man's presence."[64]

Convincing people to protect their seashores—as had been done at Maine's Acadia National Park—was a priority for Carson while she was promoting *The Sea Around Us* and thereafter. "What has taken centuries to develop is being destroyed in a few years," she lamented in July 1951. She announced that her next book would be a "popular guide" to American seashores that were in need of rescue. When asked by a *Washington Post* reporter what her favorite beaches were, she mentioned coastal Maine and North Carolina's Outer Banks, around Cape Lookout. But her favorite seaside site wasn't a sandy beach; it was Solomons Island, Maryland, where the Patuxent River met Chesapeake Bay. "Lots of sponges, and sea squirts, mollusks, and crabs," she said, her eyes brightening.[65]

The rather isolated boat-building village of Solomons Island was home to the University of Maryland's Chesapeake Biological Laboratory. Watermen in the area were engaged in oyster harvesting. Carson thought of Maryland's Solomons in October 1951 when she wrote a favorable review of Gilbert C. Klingel's book *The Bay* for the *New York Times Book Review*. Noting Klingel's warning that Chesapeake Bay was losing its waterman culture and its salt marshes to industrialization, Carson sounded her own alarm. "The truth of this is all too evident to those who have followed the current progress of the bulldozer and the beach cottage and the hot-dog stand," she wrote. "Perhaps Klingel's book will awaken interest in preserving some of the natural shore areas that are left—places where one can still sense the beauty of the earth and find release from the tensions of our difficult times."[66]

CHAPTER 4

William O. Douglas and the Protoenvironmentalists

JFK's favorite photo of himself was taken by Mark Shaw in 1959, on the dunes near Hyannis Port. The photo has been used frequently since JFK's assassination, to symbolize his "New Frontier" seashore preservation leadership.

I

In 1952, nine years after the publication of Carson's *Under the Sea Wind*, thirty-five-year-old Congressman Jack Kennedy decided to run for the Senate from Massachusetts against the Republican incumbent, Henry Cabot Lodge, Jr. With his pearl white teeth, thick auburn hair, and movie star glow, he emerged as a television media darling.[1] Determined to court such groups as the Massachusetts Audubon Society and the Garden Club of America, he spoke eloquently about protecting the commonwealth's shoreline splendor, keeping the nets of fishermen full, and attacking companies that were polluting rivers to the point of ecological ruin. His campaign was run by his twenty-seven-year-old brother, Bobby Kennedy, an avid outdoorsman and skiing enthusiast, who read *National Geographic* religiously, admired Henry David Thoreau, and thought national parks were the crown jewels of America. "It was all wrapped up in a larger family sense of saving America's heritage and beauty," his brother Ted recalled. "Bobby was more into the fight for wilderness, ecosystems, and the like. Jack was about living in a clean, decent country, the United States as Norway or Sweden. He didn't like landscapes marred."[2]

In November, Jack defeated Lodge by three percentage points in the traditionally Republican state. The telegenic JFK's ascendency to the Senate was timed perfectly with the exploding popularity of television. Although he was celebrated as the author of *Why England Slept* and as a World War II hero, it was his super-cool manner that drew photographers into his orbit as if he were a Hollywood star. Never visibly nervous, he mastered the trick of being both friendly and somewhat elusive. A genius at compartmentalizing, simultaneously a brass-knuckles politician and playboy, a world peace idealist who out-hawked many Republicans in Cold War rhetoric, Kennedy was hard to pigeonhole. He was apparently guided by his father's advice: "Can't you get it into your head that it's not important what you really are? The only important thing is what people *think* you are!"[3]

In January 1953, twenty-four-year-old Theodore Sorensen of Lincoln, Nebraska, joined Kennedy's staff as a legislative aide. Ted had a deadpan sense of humor, bookish disposition, progressive Bull Moose political philosophy, and well-honed skepticism of US military authority. To accentuate a point, he would hold both hands above his head to get Kennedy's undivided attention and then proceed to speak eloquently, with a twinkle in his dark brown eyes that indicated he knew his spiel was damn good. Before long, at a glance, Jack and Ted could practically read each other's minds.

As a boy, Sorensen had spent summers camping along Nebraska's Blue River, six miles southwest of Lincoln. For two of those summers, he had been a YMCA camp counselor, teaching youngsters how to swim, fish, and rock climb. His adeptness as a lifeguard and as a guide on the Blue River was noted by locals. Once, however, he almost drowned aboard an "Indian canoe" while traveling down the

Blue in flash-flood season. Accompanying his father on business road trips across Nebraska, soaking up Willa Cather country from the front seat of a Ford, young Ted embraced the "shimmering landscape" of farm country with agrarian delight. Colorfully named towns such as Big Springs and Red Cloud piqued his penchant for wordplay. In his memoir, *Counselor: A Life at the Edge of History*, he wrote, "I could see majestic barns, hear meadowlarks perched on roadside fences; wonder aloud about the optical illusion of rain pools ahead on the dry concrete highways, and simply take in the vastness of the open skies, the neatly cultivated fields, and the sheer grandeur of the Nebraska horizons and sunsets."[4]

Because Sorensen became so closely identified with Kennedy and lived most of his life in Washington and New York City, his intense affection for Great Plains landscapes has not been properly explored by scholars. Fighting to protect natural resources mattered to him. The fall migration of the sandhill cranes to the Flint Hills was an event he cherished for life. His wife, Gillian Sorensen, recalled in 2022, "He would return often to the Nebraska and Kansas prairie to witness the awe-inspiring congregation of the cranes."[5]

As Kennedy's term in the Senate started in January 1953, Dwight D. Eisenhower was sworn in as America's thirty-fourth president. Under his aegis, the geopolitical rivalry between the United States and the Soviet Union showed no sign of abating. Bound up in the confrontation was a superpower tit-for-tat approach to nuclear testing. Throughout 1953, as part of Operation Upshot-Knothole, nuclear weapons were once again detonated in southern Nevada.[6] The secrecy surrounding the United States' atomic developments at the Yucca Flat test site made it difficult for the public to assess them, yet most Americans were relieved that the United States was building a huge nuclear arsenal. The only way to avert a nuclear war with the Soviets, Secretary of State John Foster Dulles believed, was keeping the United States on a permanent nuclear wartime footing. With polls showing general public approval, the result was that the size of the United States' nuclear arsenal increased during Eisenhower's two-term presidency from 1,005 to 20,000 weapons.[7]

There was hope, however, that Cold War tensions with the USSR might be resolved through diplomacy, especially following the death of the Soviet dictator, Joseph Stalin, shortly after Eisenhower was inaugurated. On December 8, 1953, Eisenhower delivered a groundbreaking speech, "Atoms for Peace," before the United Nations, raising the morale of anti–nuclear test activists.[8] Dismayed by the "awful arithmetic of the atomic bomb" and the "probability of civilization destroyed" in a doomsday showdown, Ike proposed that the nuclear nations pool fissionable material from their stockpiles to be overseen by a new International Atomic Energy Agency, explaining that nuclear energy produced by reactors could be the great side benefit of the Manhattan Project's development of the atomic bomb. That plan didn't materialize, but under the banner of "Atoms for Peace," the United States soon exported more than twenty-five tons of highly enriched uranium, intended to fuel research reactors, to thirty nations.

Senator Kennedy stayed out of the "Atoms for Peace" debate. On the conservation front, he had a predictable regional agenda, expressing dismay that the Merrimack River and Boston Harbor were heavily polluted. Occasionally he ventured into national park–related issues such as joining Justice Douglas's effort to ban a tramway up Washington's Mount Rainier.[9] Because the US government owned 47 percent of all land in the western states, Kennedy, like most New England Democrats, supported Army Corps of Engineers and Bureau of Reclamation dams to aid arid western states such as Colorado and Arizona. Such big dams generated both electricity for cities and good-paying jobs. In some cases, they also controlled flooding and made water available for ranch irrigation. Kennedy was agnostic when it came to the big-dam arguments erupting over Idaho's Hells Canyon or Oregon's Bonneville Dam. By contrast, his vote in 1954 for the construction of the St. Lawrence Seaway was considered brave in the midst of a loud, long debate, as the engineering feat offering a smooth shipping route from the Great Lakes to the Atlantic Ocean would hurt the Port of Boston economically. It was mainly in the realm of seashore preservation and the protection of marine mammals that he exhibited the commitment to national parks that Douglas had encouraged.

Jack Kennedy and Rachel Carson were among the many enthralled by *The Silent World*, a book by Jacques-Yves Cousteau published in 1953 that introduced Americans to the living things that the French explorer had discovered through underwater diving.[10] "My brother was all about *The Silent World*," Jean Kennedy Smith recalled. "It had his name written all over it. It's one of the few books I remember him encouraging me to read."[11] Carson claimed that the book had "riveted her from beginning to end." Cousteau, she wrote, had "revolutionized the means of human access to the undersea world."[12] Three years later, *The Silent World* became an Academy Award–winning documentary film, helping the oceans to grow in stature as a conservationist cause in the eyes of world governments.

In the summer of 1953, Kennedy married Jacqueline Bouvier, who'd grown up in New York City and on a country estate in East Hampton, Long Island. She instantly established a deep and warm friendship with her parents-in-law, Joe and Rose. The Kennedy wedding in Newport, Rhode Island, that September was attended by six hundred guests, including Bill Douglas.

After a honeymoon to California and Mexico, the newlyweds moved into a rented house in the leafy Georgetown section of Washington. But they always considered Hyannis Port home. Before their marriage, *Life* magazine had featured the couple sailing off Cape Cod in Jack's old sloop *Victura*, windblown and carefree.[13] Sometimes when Jack sailed, Jackie stayed on land, painting watercolors of the windswept Cape Cod seashore. Smitten with Jack's young bride, Joe Sr. hung her seascape watercolors prominently around the house. Jackie also composed a poem embracing Jack's indomitable spirit:

> But now he was there with the wind and the sea
> And all the things he was going to be.
> He would build empires
> And he would have sons
> Others would fall
> Where the current runs
>
> He would find love
> He would never find peace
> For he must go seeking
> The Golden Fleece

All of the things he was going to be
All of the things in the wind and the sea.[14]

II

In the fall of 1953, courses on the science of ecology were starting to be offered in US colleges and universities. Advanced degrees in the emergent field were growing in number. The catalyst was a ground-breaking textbook, *Fundamentals of Ecology*, published that year by Eugene P. Odum (with his brother, Howard T. Odum, contributing substantially). With its clear, erudite prose, fine photos, and easy-to-comprehend graphs, *Fundamentals of Ecology* had a galvanizing effect on university science programs.[15] Odum's book emphasized ecosystems (synecology), as well as energetics. Both were unifying concepts that provided ample room for observation and hypothesis testing. In the book, the two were combined by the Odums into "systems ecology."[16]

Eugene Odum explained that the word *ecology* came from the Greek word *oikos*, meaning "house." Even though the great German zoologist Ernst Haeckel had used the term *oekologie* to signify "econ-omy of animals and plants" in 1866, it wasn't until the mid-twentieth century that the study of organisms as part of a community—biology in context—garnered widespread scientific acceptance and entered the popular lexicon as *ecology*. In 1935, the British botanist Arthur George Tansley, writing in the journal *Ecology*, coined the term *eco-system*. He maintained that an environment and all its living organ-isms should properly be understood as a single interactive entity.[17] Odum was consumed by the new science. The message of *Funda-mentals of Ecology*, as well as Carson's *The Sea Around Us*, for that matter, was, in essence, "Beware: humans can impact one aspect of an ecosystem and destroy the entire thing, a process that can thereby destroy Earth."[18]

Fundamentals of Ecology was adopted for academic courses far and wide.[19] It was subsequently translated into twelve languages. "Truly general or basic courses in ecology have often not been available in our largest universities because the subject is fragmented by departmental lines, which traditionally are drawn along morphological or taxo-nomic rather than physiological lines," Odum wrote in his second edi-tion. "Such lines need not be a barrier because, as I have tried to show in this book, any ecologist can present the general principles to stu-dents, regardless of departmental affiliations or whether his training and specialization may be in botanical, zoological, or microbiological sciences."[20]

Eugene Odum, anointed "the father of modern ecology," was born in Newport, New Hampshire, on September 17, 1913, but grew up in North Carolina. His father, Howard Washington Odum, was a pioneering sociologist of the American South; his younger brother, Howard T., would also become a famed ecologist. What all three Odum men shared was a strong belief in the holistic approach to studying both the natural world and human civilization. The Odum men thought that scientists had a societal obligation to inform the world about poisonous chemicals, reckless waste disposal, stationary sources of pollution, and ecosystem destruction, along with resto-ration strategies.[21]

As a young boy, Eugene Odum wrote a column on birds for the *Chapel Hill Weekly*. His delight in backyard naturalism led him to zoology and graduate study at the University of Illinois, which offered pioneering degree programs in ecology. For his dissertation, he stud-ied the cardiovascular systems of birds, winning praise from his peers for inventing a means of monitoring the avian heartbeat. Four years later, Odum earned his doctorate at Illinois in ornithology and ecol-ogy and he was selected as the first resident naturalist at the Edmund Niles Huyck Preserve in Rensselaerville, New York, twenty-eight miles southwest of Albany. His task was to discover how its ecosystem worked as a unified physical, chemical, and biological process.

Using the five-hundred-acre Huyck Preserve as his own Walden Pond, Odum devoted himself to the scientific study of bird variations, vegetation systems, and beech-hemlock climax forests. His belief was that ecology drove the planet with evolution being secondary, a view-point that many other scholars disputed. But he held his own in aca-demic debates. From the study of soil microorganisms to the role of pollinators and from research into the water cycle to comprehending Earth's climate system, Odum peered into every corner of his field.

In 1940, the University of Georgia hired Odum to teach biology. After about eight years in Athens, he pitched a course on ecology, even though it still wasn't then a widely accepted academic endeavor. In a 1998 *Natural History* interview, he said:

> They looked at me and laughed. They thought that ecology was just going out and finding animals and describing and collecting them. They said, "There are no principles. It is just organized natural history. It's not an important subject."
>
> They were right about what it was then, but I got mad and walked out. Later they said, "We didn't mean to hurt your feelings, but what is ecology?" Then I realized that nobody had written a general book about ecology. So with help from my brother Howard, I started to write it.[22]

While Kennedy was serving in the Senate, Odum established an Institute of Ecology at the University of Georgia, serving as its first director. Students in Odum's classes conducted primary research in the freshwater Okefenokee Swamp on the Georgia-Florida border; ripe for study, it is an alligator and water moccasin habitat and palmetto country, with groves of gnarled live oaks bearded with gray Spanish moss. On Sapelo Island, Georgia, he founded a marine institute and led explorations there. Most impressive, he set up the Savannah River Ecology Laboratory (SREL), a 310-square-mile marsh zone considered one of the world's first outdoor science classrooms.[23]

Eventually, Odum was the author of a dozen important ecology books. "Gene was a proselytizer of holism," Karen Porter, a colleague of Odum in the Institute of Ecology, recalled, "and his message of interconnectivity inspired a generation of ecologists."[24]

While Eugene Odum promoted *Fundamentals of Ecology* on college campuses, Carson worked on the third book of her trilogy, *The Edge of the Sea*. Her early plan had been to write a string of entries (like a guidebook) on what could be discovered along the ocean's edge; the book, in fact, was originally titled *A Guide to Seashore Life on the Atlantic Coast*. "I decided that I have been trying for a very long time to write the wrong kind of book," she confided to her editor, Paul Brooks, in 1953, going on to explain that she was changing the structure from "one little thumbnail biography after another" to something more the-matic. "In the beginning I thought of this as a book very different in kind from *The Sea Around Us*, now it seems to me a sort of sequel or companion volume, the former dealing with the physical aspects of the sea, this with the biological aspects of at least part of it."[25]

Throughout 1953, Carson continued her research on tidal life along the Atlantic seaboard states of Maine, North Carolina, and Florida. Often kneeling on the wet sand or wading through a gentle surge, she studied horseshoe crabs, razor clams, and lugworms. That year, RKO released a documentary based on *The Sea Around Us*, which increased her fame. In July, as planned, she moved into the new, pine-paneled cottage that she built in coastal Maine. She named the retreat "Silver-ledges" and was immediately befriended by her neighbors, Stanley and Dorothy Freeman. On summer nights, with the windows open, Carson listened to waves crashing against boulders and voracious her-ring gulls making a commotion overhead. When collecting sea spec-imens such as the darkness-loving ghost crabs and soft-shell clams in buckets, Carson favored a place called Ocean Point, just a mile away along the coast. There, the shifting tide marooned the shells and sand dollars. Many of the shells she found had been left by crustaceans, especially crabs, that molted in the growth process. According to the environmental historian Robert K. Musil, "She saw them all as woven into a larger pattern of life in which each part, each piece is connected and worthy of our reverence and our awe."[26]

In the spring of 1954, Carson, accompanied by Dorothy Freeman, made a pilgrimage to Nobleboro, Maine, to visit Henry Beston and his wife, the poet Elizabeth Coatsworth. The four spent hours sharing stories about exploring the seashores of New England. *The Outermost House*, Rachel told the sixty-five-year-old Beston, was a treasured, al-most sacred text to her.[27] Furthermore, when Beston had glowingly reviewed *Under the Sea-Wind*, back in 1952, she had been overjoyed.[28] Working together as friends, Carson and Beston hoped to preserve parts of Maine and Cape Cod: literary ecologists with a shared value system.

III

Despite a hectic schedule as an associate justice, Bill Douglas didn't neglect his other interests, notably his passion for wilderness protection. Quite the opposite. Between cases, he plotted campaigns to preserve Washington's North Cascades and Alaska's Arctic, Wyoming's Wind River region, and Maine's Allagash waterways. From his Supreme Court chambers, he investigated the howling range of coyotes from aria to chorus and how the ruffed grouse survived winters around Mount Adams, Washington. Well versed in botany, he could identify wildflowers without a guidebook. He knew why witch hazel bears bright yellow flowers in September and where the flower buds of silver maples bloomed in March. On a few occasions, he remarked that he would gladly trade the honor of being in the Supreme Court for that of having collected ferns on the Lewis and Clark Corps of Discovery expedition. Propped up around his chamber desk were pressed wildflowers that he had collected on hikes and then matted.[29] There was also a laminated card with four lines from the poem "Auguries of Innocence" by British poet William Blake.[30]

The so-called green justice sometimes wore his lug-soled hiking boots to the Supreme Court as a nod to Pacific Northwest forest culture. After hours, with a backpack strapped on, binoculars draped around his neck, his shirtsleeves rolled up, and a floppy hat on his head, Douglas would heartily explore the unfenced backcountry of Virginia or Maryland, sometimes with Bobby Kennedy as trail mate. A legal genius of the first rank, yet touched with a Luddite strain, he regularly denounced "machines" that were "waiting to chew up" forest lands.[31] When chauffeured to the Supreme Court, Douglas would sometimes ask his driver to detour to the Potomac River so that he could check the water for traces of manganese, a heavy metal used in industrial processes that can cause neurological damage to humans and wildlife.

Douglas was particularly concerned that pent-up demand for cars after World War II would scar the landscape in myriad ways. It was as if the wartime slogan "We're all in this together" had been replaced by "Fill'er up!" Sales of US-built motor vehicles skyrocketed. Jack Kerouac, the Beat Generation writer, summed up the postwar era this way in his novel *On the Road* (1957): "Whither goest thou, America, in thy shiny car in the night?"[32] When Truman became president, 26 million cars were in service in the United States. By 1954, under Eisenhower, the number had soared to 40 million.

The sharp uptick in automobiles, all of which used leaded gasoline in that era, exacerbated the vexing postwar air quality crisis. Smog was noticeable in most big cities. "Virtually no one in the oil industry was prepared for the explosion of demand for all oil products," Daniel Yergin wrote in his book *The Prize: The Epic Quest for Oil, Money, and Power.* "Gasoline sales in the United States were 42 percent higher in 1950 than they had been in 1945, and by 1950, oil was meeting more of America's total energy needs than coal."[33] The growth trend kept increasing throughout the decade and beyond.

Starting in 1954, Douglas had another complaint, blasting the Army for selling surplus jeeps in the arid West; there, when driven off road, they damaged soil, destroyed native plants, and wreaked havoc on wildlife by destroying habitats, crushing burrows, and damaging streams. Having grown up with pockets of roadless wilderness around him in Washington State, Douglas considered public wild lands an American birthright as unimpeachable as the First Amendment or any other. On the issue of roads, off-roading, and other encroachments, he encouraged fellow protoenvironmentalists to use lawsuits as a weapon against the Forest Service, Bureau of Reclamation, and extraction companies in the American West. Playing the role of matchmaker, Douglas would help both the Sierra Club and the Wilderness Society acquire *pro bono* attorneys for their preservationist crusades.[34] Seemingly unafraid of conflict-of-interest charges, Douglas had essentially turned his Supreme Court office into a wilderness lobby way station, encouraging elite conservation groups to send him their newsletters, field notes, and various preservation campaign literature.

Richard Schwartz, a Yale law professor, recalled that his friend Douglas was deeply distraught over the way the Department of Agriculture and the Corps of Engineers operated without ecological accountability. He distrusted the Forest Service, a huge part of Agriculture, carping that the days of visionaries there such as Gifford Pinchot and Henry Graves had vanished, replaced by a bureau that thought it was a law unto itself. From 1954 until 1970, Douglas would point out that the Forest Service secured budget increases for timber sale administration of around 66 percent of its requests but only 20 percent of its requests for wildlife protection and recreation.[35] Douglas scoffed at its wild lands management policies under Eisenhower as craven philistinism.[36] He especially distrusted the Forest Service's land reclassification process, which in practice opened wild areas to road building, mineral extraction, and timber cutting. Douglas told Schwartz, "I'm ready to bend the law in favor of the environment and against the corporations."[37]

At the time, Douglas was making big changes in his personal life. He divorced his first wife, Mildred, in 1953, after twenty-nine years of marriage while having an affair with Mercedes Davidson, whom he married in early 1954. In midcentury America, divorce was still frowned upon, so with the breakup of his marriage, Douglas by and large surrendered his longtime dream of a run for the White House. And in 1963, he divorced again and immediately married twenty-three-year-old Joan Martin. Divorcing her in 1966, he married during the same year twenty-two-year-old Cathleen "Cathy" Heffernan, an Oregonian who shared his passion for America's backcountry. With Cathy as soulmate, he found domestic happiness right up until his death in 1980.

Collegiality with the "fraternity of fishermen"—what he considered the "most congenial brotherhood the world around"—mattered more to Douglas, he boasted, than close relationships with fellow Supreme Court justices or other Washington lawmakers.[38] Spending as many months in Oregon as possible at his tamarack cabin on the Lostine River, camping on sandy Cannon Beach looking offshore at the basalt monolith Haystack Rock, or climbing the peaks of the California Coast Ranges brought spiritual balance to Douglas's hyperactive mind. Whenever JFK had a question about nesting seabirds or pollution abatement during his Senate years, he looked to Douglas, who often teased him, "The trouble is, Jack, that you've never slept on the ground."[39]

In early January 1954, Douglas's concern about the explosion in road building struck close to home. He grew incensed over a proposed motorway along the bed of the abandoned Chesapeake & Ohio Canal. The canal bed stretched from the Georgetown section of Washington to Cumberland, Maryland. Refuting the derogatory notion that the C&O had become an "ugly ditch," Douglas charged that a paved road would create abominable traffic and ruin the historical, natural, and recreational features of the 184.5-mile towpath, which generally paralleled the Potomac River. The noise of highway traffic, he fumed, would destroy the tranquility of a hike along the old canal, a project first championed by George Washington. Modeled after the Erie Canal, the C&O had been constructed with a total of seventy-four locks between 1828 and 1850; it was instrumental in moving coal out of the Allegheny Mountains, as well as in enhancing the overall development of the mid-Atlantic states. It ceased to operate in 1924. The National Park Service came out in favor of the parkway, ameliorating its position by proposing a ban on commercial activity along the route. In contrast, Douglas envisioned a scenic, *roadless* Chesapeake & Ohio Canal National Historical Park, where visitors could discover the woods at the back door of the nation's capital, as his birdwatching friends Louis Halle, Roger Tory Peterson, and Rachel Carson already had.

After Maryland lawmakers originated the idea of building a highway over or next to the old C&O, a January 1954 *Washington Post* editorial written by associate editor Merlo Pusey endorsed the plan.[40] Douglas fired off a letter to the *Post*, arguing vehemently against a multiple-lane road there. A few months later, the Audubon Society of the District of Columbia published Shirley Briggs's book *Washington: City in the Woods*. It also denounced the proposed C&O parkway as being "based on false premises" aimed at destroying natural splendor.[41] Carson had worked with Briggs at the Fish and Wildlife Service and supported her preservationist manifesto about wildlife in the backyard of the capital. As the battle heated up, a sign was posted by local preservationists at Shepherdstown, West Virginia, which is on the canal: "Justice Douglas, keep to the right. Booby traps to the left are for the *Post* editors."[42]

With characteristic verve, Douglas challenged the *Post* editorial board to hike the entire C&O Canal towpath with him. He wanted them to experience firsthand the natural beauty of the smallest scenes en route, as well as the biggest: Great Falls and Mather Gorge, which the proposed parkway would destroy.[43] "One who walked the canal its full length could plead that cause with the elegance of a John Muir," Douglas wrote. "He would get to know muskrats, badgers, and fox; he would hear the roar of wind in thickets; he would see strange islands and promontories . . . the whistling wings of ducks would make silence have new values for him. Certain it is that he could never acquire that understanding going 60, or even 25, miles an hour."[44]

To their credit, Robert Estabrook, the *Post*'s editorial page editor, and his associate editor Pusey accepted the challenge to make an eight-day hike. On March 20, a party of thirty-seven, including Dr. Olaus Murie and Harvey Broome of the Wilderness Society, joined the Supreme Court justice on the towpath.

Within Fish and Wildlife circles, Olaus Murie epitomized professionalism in field observation. Douglas flat-out claimed that nobody in North America knew more about mammals than his friend Olaus. Born on March 1, 1889, in Moorhead, Minnesota, the son of Norwegian immigrants, Murie did his undergraduate work at the University of the Pacific (where John Muir's papers are housed). In 1927, he earned a master's degree from the University of Michigan and began his life as a scientific conservationist. Before long, he was roaming all over Alaskan and British Columbian wilderness, befriending Eskimo and Cree communities. After studying elk herds in Jackson Hole, Wyoming, he concluded that the traditional approaches to wildlife management by the Fish and Wildlife Service were destabilizing ecosystems. Refusing to see wilderness from a predominately utilitarian viewpoint, he wrote about the "moral fabric" in the relationship between humans and other living creatures.[45] With his brother, Adolph, he spent six years in Alaska studying caribou migrations.

After Murie resigned from Fish and Wildlife in 1945, he devoted his time to keeping hydroelectric dams from destroying western rivers and wilderness areas. His biggest win in that regard was stopping the construction of the Glacier View Dam in Montana. Regularly, Murie and his wife, Margaret, corresponded with Douglas about dam projects. Their home in Moose, Wyoming, became a conservation workshop, a meeting place of great importance for outdoor enthusiasts from all over the world. It would be impossible to exaggerate how beneficial Douglas's alliance with the Muries would prove to be in the late 1950s and early 1960s.

Also along for the C&O Canal expedition was Sigurd Olson, recently named director of the National Parks Association. The gentle, warmhearted Olson was already a legendary advocate of roadless wilderness protection in the upper Midwest. Through tireless, high-level lobbying, he had convinced the Truman administration to create an airspace reservation over the roadless areas of Minnesota's Superior National Forest. Born in Chicago, on April 4, 1899, Olson moved to Door County, Wisconsin, in the early 1900s as a youngster. His father was a Baptist minister who taught him the secrets of Lake Michigan's shoreline. Inspired by Henry David Thoreau, he attended the University of Wisconsin at Madison to study zoology and botany. After receiving a BS degree in 1920, he attended the University of Illinois, eventually earning an MS degree in the new field of ecology. Shortly thereafter, he was hired to teach biology and geology at a junior college in Ely, Minnesota. That put him near the Quetico Superior wilderness along the US-Canada border.[46]

Just like Eugene Odum, Olson preferred teaching classes outdoors and asked many old-time north country wilderness guides to give guest lectures. The more experience he had canoeing the vast boundary waters between the United States and Canada, the greater kinship he felt with the voyageurs, the eighteenth-century French trappers and fur traders who had journeyed into the far northern interior of North America. Although not a cofounder, Olson joined the Wilderness Society in 1935 and would rise to become president in the late 1960s. But it was the role of National Parks Association leader that brought him into Bill Douglas's C&O Canal protest orbit.

Reporters from *Time, Life,* and CBS Radio followed the hike along the Potomac. Nightly news television networks ran favorable segments on the protest. The *Washington Post,* of course, covered the adventure daily; the Potomac Appalachian Trail Club of Washington provided the delegation with food. Enjoying the blush of buds blooming in late winter and avoiding briar tangles, the group followed Douglas's pace of 112 steps per minute, marching almost twenty-three miles the first day.[47] The *New York Times* dubbed them the "blister brigade."[48] For fun, the group, with Olson, their "poet laureate," taking the lead, composed "The C&O Song," a thirty-one-stanza ditty set to the tune of "Low Bridge, Everybody Down" (also known as "The Erie Canal

The C&O Canal troupe, pushing through a snowstorm, passes beneath a Western Maryland Railroad trestle on March 21, 1954. Justice William O. Douglas is on the right.

Song"). Driving snow and frigid wind didn't hamper the nature lovers, who encouraged gaggles of Boy Scouts, fishermen, and even equestrians to join their anti-road cause.

Walking long miles, pausing only for something like an examination of fungi networks along the forest floor, Douglas's merry band of hikers encountered raccoons, woodchucks, muskrats, and fox. The melodies of wood thrushes lent a lovely musical cadence to the outing. Hawks were spotted circling for prey. Round-pad otter prints were discovered along the riverbank. One evening, a joke was played on Murie, considered North America's expert on animal tracks. A Mexican burro was brought clandestinely to the trail to leave hoofprints in the dirt. Early the following morning, Murie rubbed his face, seemingly perplexed by the imprints. Then, to the stunned group, he nailed it, saying, "If it weren't for the fact that there aren't any for hundreds of miles, I'd say it was a Mexican burro."[49]

Near the end of the long trek, the *Post* editors, Estabrook and Pusey, quit the march. Behaving like an Oregon Trail pioneer, Douglas, despite a rending cough, forged onward with eight other finishers, though they climbed onto a boat for the last few miles. Once back in

Georgetown, his "blister brigade" was congratulated by Douglas Mc-Kay, secretary of the interior, on their intrepid anti-parkway activism.

The *Post* retracted its original position, suggesting that parts of the canal indeed should be saved as a federal park.[50] Soon afterward the National Park Service withdrew its support for the highway. According to National Park Service historian Barry Mackintosh, "Vocal public sentiment ran strongly against the canal parkway in the months after the Douglas hike."[51] Douglas's widely publicized style of direct action was adopted around the country in local conservation battles to save treasured landscapes.

Douglas then took another step, campaigning to establish the Chesapeake & Ohio Canal Association as a permanent watchdog group to protect the historic waterway. "For one who hikes the towpath, every quarter-mile or so brings a new view—a sharp cliff, a flat meadow, a rough hillside, a narrow gorge," he wrote.[52] From there, his efforts continued and gathered momentum. President Eisenhower, who quietly admired Douglas's feistiness, surprised conservationists by designating the C&O Canal a national monument in January 1961, just days before he left the White House. Because of Douglas's persistence, in 1971 the C&O Canal was designated a National Historical Park, affording a further layer of preservation.

A precedent had been set. By "hiking and hollering," as Douglas put it, he proved that a "Gandhian protest" against a highway could establish greenbelts, nature preserves, and open spaces in and around large metropolitan areas.[53] And he had formed a deep friendship with Sigurd Olson. Soon they would hike in Olympic National Park (to protest salvage logging) and canoe in the Quetico Superior wilderness, hoping to save Minnesota's Boundary Waters Canoe Area as a wilderness.[54] Douglas admired Olson for writing conservation-themed prose in the vein of Aldo Leopold about Minnesota's north country. Olson's first book, *The Singing Wilderness*, published in 1956, followed by *Listening Point* in 1958 and *The Lonely Land* in 1961, convinced Douglas to write his own *My Wilderness* books to celebrate the Olympics, the Wind River, Allagash, and other ecosystems that he so deeply loved.

Friendship with Douglas likewise inspired Olson to press for a national park in Minnesota. In the summer of 1956, he organized his first canoe expedition into Saskatchewan, following a five-hundred-mile route of the Churchill River that French voyagers had long before used. The trip not only garnered publicity, it was a spiritual revelation for Olson that reflected his belief that the hand of God was manifest in nature. Like Carson, Olson produced nature writing that was both lyrical and scientifically exact. In *The Singing Wilderness*, he asked readers to understand the essential role that caribou moss plays in North Country ecosystems. "Those silvery little tufts before me are the shock troops of the north, the commandos with which the plant kingdom made a beachhead on a barren, rocky ridge. Surviving where other types would die, needing nothing but crystalline rocks and air, they prepare the way for occupation and for the communities to come."[55]

For Olson, hunting for wild berries, building campfires, trout fishing, and other outdoor activities in Minnesota had a "ritualistic significance" that was an essential part of America's heritage.[56] Nobody before had written as beautifully as Olson did on the ancient art of paddling in 1961.[57] "The movement of a canoe is like a reed in the wind. Silence is part of it, and the sounds of lapping water, bird songs, and wind in the trees. It is part of the medium through which it floats, the sky, the water, the shores."[58]

Even though Kennedy didn't participate in Douglas's C&O Canal protest hike, he embraced the preservation of the tidal Potomac (sometimes referred to as "the nation's river") as a worthy cause. "My family was deeply invested in keeping the Potomac scenic and unpolluted from the 1950s onward," Robert F. Kennedy, Jr., Bobby's son, recalled. "We were 110 percent behind what Bill Douglas was doing to raise awareness about this incredible American river."[59] On weekend drives from his Georgetown home, Jack Kennedy traveled to the white water rapids at Great Falls to picnic and throw a football. Kennedy treasured the Potomac as a scenic wonder and was infuriated that by the 1950s, it had become a polluted sewer with algae choking the riverbed. Under the grim conditions, oyster and crab populations in the Chesapeake Bay, into which the Potomac flowed, were dying off while rubbish flourished.[60]

Taking his protest a leap further, Douglas, backed by the Kennedy family, demanded that the Potomac, already a veritable crime scene, not be obstructed by a proposed Seneca Dam near Great Falls. "It isn't too late to save the Potomac and the Chesapeake Bay from ruin," he would assert.[61] Great Falls, visible from the C&O Canal, was located about fifteen miles from Washington.

Political reporters came to understand that Jack Kennedy was an ally of progressive conservationists on issues such as saving the C&O Canal and restoring the Potomac, establishing national seashores, and advancing federal anti-pollution laws. In 1954, John B. Oakes of the *New York Times*, the newspaper's first real environmental reporter, called the two top conservation-attuned Democrats on Capitol Hill, Kennedy and Representative Lee Warren Metcalf of Montana.[62] (Metcalf's pet cause was the protection of the United States' rivers and streams.)[63] Although he was not prone to grandstanding on conservation and anti-pollution measures, Kennedy was a reliable supporter of the National Parks Association, the Izaak Walton League, and the Audubon Society. As congressman and senator he backed legislation conserving ever-shrinking wild places, fighting smog, and cleaning up polluted waterways. Most visible was his campaign for national seashores. Among Republicans, Oakes wrote, Representatives John P. Saylor (wilderness) and Leon Gavin (migratory bird protection), both Pennsylvanians, were the most forceful champions.

Oakes thought Jack was a serious conservationist; perhaps that was chiefly because Bobby, his brother, even more than Douglas, had his ear. The most gentle and sensitive of the Kennedy children, Bobby had an affinity for animals much like that of two of his heroes, St. Francis of Assisi and Albert Schweitzer. After serving in the Naval Reserve from 1944 to 1946, he returned to Harvard University, where he exhibited an interest in presidential history. After graduating, he traveled to Europe and the Middle East. In 1948, with Bill Douglas as his mentor, he enrolled at the University of Virginia Law School, convincing the justice to guest lecture in one of his classes. Two years later, Bobby married Ethel Skakel in Greenwich, Connecticut. It was a happy marriage; both enjoyed hiking, skiing, horseback riding, and living with a menagerie of animals.

In 1955, when Douglas received a visa to travel around the Soviet Union, a rare privilege for an American, he took the twenty-nine-year-old Bobby along with him. For seven weeks, they traveled Russia, including the Siberian backcountry, studying the eco-system and meeting with villagers. The justice wrote later that Bobby "was never overawed by greatness and was always at ease whether his companions were dukes or peasants."[64]

In 1957, Bobby and Ethel bought Hickory Hill, an estate in the northern Virginia countryside of McLean, where they would raise their children. Ethel's brother Jimmy traveled the world on wilder-ness treks and along the way collected animals to give to his nieces and nephews: rhesus and spider monkeys, marsupials, and a Central American honey bear. It was the kind of menagerie that Theodore Roosevelt would have enjoyed, though it certainly wasn't fair to the wild animals that were transplanted. One Christmas, Uncle Jimmy sent a mysterious crate. When Robert and Ethel removed the ply-wood, out waddled a California sea lion, which they named Sandy. The children were enthralled by the barking animal, which resided in the backyard swimming pool and became a beloved part of the family. "Sandy ate mackerel by the barrel, devouring everything but the eyeballs, which we found scattered like marbles across the pool, patio, and lawn," Bobby Kennedy, Jr., recalled. "Sandy rode in the car, accompanying my mother when she picked us up after school, and my sister Kathleen taught him to jump off the diving board through a hula hoop."[65]

IV

Any hope that President Eisenhower, in the spirit of "Atoms for Peace," would ban nuclear testing in the atmosphere or underwater vanished when he authorized more atomic tests at Bikini Atoll in the Mar-shall Islands. The first of seven detonations over the course of seven weeks was on March 1, 1954. Though Bikini's population had been permanently moved after the 1946 tests there, all eighty-six residents of Rongelap Island, ninety-eight miles away, became sick from radio-active fallout in 1954. In addition, the first test rained ash on a Japa-nese fishing boat, *Lucky Dragon*, that was trolling ninety miles away.

The crew became violently ill, and one fisherman died. Aboveground testing of hydrogen bombs had been denounced by scientists worldwide, but the Eisenhower administration persisted, causing the AEC scientist Willard F. Libby to claim a few years later that the risk from fallout was minimal and irrelevant to "our pleasures, our comforts, and our material progress."[66]

When Dr. Linus Pauling, a biochemist at the California Institute of Technology in Pasadena, suggested that ten thousand people had died of leukemia because of aboveground nuclear testing, a few Washington lawmakers vowed to investigate the health hazards of radiation. Others accused Pauling of being a Communist rabble-rouser. The State Department wouldn't grant him a new passport. But Pauling, who won the Nobel Prize in Chemistry in 1954, had global influence nonetheless.[67] Rachel Carson, a staunch admirer, corresponded with Pauling, as well as Barry Commoner and the Greater St. Louis Citizens' Committee for Nuclear Information, incorporating their data into her writings in coming years.[68] By then, polls showed that more than 50 percent of Americans believed, as did Carson, Pauling, and Commoner, that nuclear fallout was a "real danger" to life on the planet.[69] Not surprisingly, an even higher proportion worried that nuclear war could occur at any moment.

Soon, a marching song was written by the Campaign for Nuclear Disarmament:

> Men and women stand together!
> Do not lead the men to war
> Make your minds up now or never
> Ban the bomb forever more.[70]

That abiding fear of the bomb, not merely of dying but of sacrificing the planet to temporal hatreds, pressed a sector of vehement anti-nuclear activists forward. To differentiate themselves from conservationists (a group looking to properly manage natural resources), those ecology-minded citizens spoke in terms of the interdependent components of nature and the irreversible damage that atomic bombs could do to planet Earth. As the historian Michael Egan noted, "The total human exposure to fallout was clearly much greater than the AEC had predicted." Adding fuel to the anti–nuclear testing movement were the lies told in AEC publications in 1953–54 about fallout being evenly distributed around the world. On the contrary, meteorologists had discovered that fallout was concentrated in the North Temperate Zone band.[71]

V

Fallout sickness from nuclear testing in the Marshall Islands and the Nevada Proving Grounds were just two of the catastrophes that protoenvironmentalists rallied against during the Eisenhower years. Smog was another. In Los Angeles and Pasadena, citizens were increasingly feeling the effects of it. The problem had started during World War II, when dust fall from industrial smokestacks and incinerators had increased in metropolitan Los Angeles from about 100 tons to nearly 400 tons per day. Figures showed that 60 percent of the state's smog came from automobile exhaust, 35 percent from industry, and the rest from incinerators. To combat the dangerous air, Los Angeles implemented some preliminary measures, installing dust precipitators and prohibiting open burning. The dust fall was reduced to 200 tons a day, but Angelenos couldn't celebrate the reduction because, as they were aware, population growth in southern California—and with it, more and more cars—outstripped initial efforts against smog. They also focused on the fact that sulfur dioxide, produced by the burning of coal and fuel oil, was still blanketing the metropolitan region. "Despite a vigorous and successful effort to control industrial emissions of hydrocarbons," Commoner wrote, "Los Angeles was still in the grip of smog."[72]

In December 1954, Governor Goodwin Knight declared a state of emergency in California because people were developing respiratory illnesses caused by the contaminated air throughout the Los Angeles basin. Stephen Royce, the owner of the Huntington Hotel in Pasadena, was an outspoken leader in the anti-smog effort. Operating on blind intuition, not data, he complained that the tourist industry was in danger economically because of poisoned air. Instead of seeing blue skies above Mount Baldy and pure sunlight on Grauman's Chinese Theatre, visitors were coughing from the brown smog, eyeballs shot red. Royce founded a Los Angeles Citizen's Committee on Air Pollution and led the fight for air-quality laws in the state capital, Sacramento. The most dangerous culprit of all, Royce believed, was sulfur dioxide. "Air pollution was affecting our sunshine," he told the *Los Angeles Times*. "We have a greater natural asset than any tourist community and it is being ruined by smog."[73]

Joining forces with Royce was a group of Pasadena mothers who demanded better air quality for their children. Wearing high-fashion dresses, jewelry, and high heels, they put on gas masks and took to the streets. Dubbing themselves the Smog-a-Tears (a play on Walt Disney's Mouseketeers), and looking like Hollywood starlets, they marched through downtown Pasadena carrying placards that read HELP US FIGHT THIS BLACK DEATH, WE WANT A CLEAN SWEEP OF SMOG, and GIVE US PURE AIR.[74] The campaign was visual, hard-hitting, and a brilliant way to publicize the environmental menace.

The effort to combat air pollution in southern California was shifting into high gear, and reporters started covering it as an important public policy issue. Throughout the mid-fifties, the *Los Angeles Times* published interviews with physicians explaining the health dangers connected to smog. According to serious evidence, smog exacerbated or even caused chronic obstructive pulmonary disease (COPD). "Industry and population have increased more rapidly than we have been able to increase our technological control of the pollutants," Dr. Francis Pottenger, Jr., the secretary of the Los Angeles County Medical Commission on Smog Research, told the *Times*. "If it weren't for the efforts of the Air Pollution Control District probably none of us would be able to live in this area at this time."[75]

Under extreme pressure from citizens, Los Angeles County organized the groundbreaking Air Pollution Control District (APCD) to curtail emissions from industry. It made perfect sense, but the measure was to little avail, since the smog persisted. With that, the APCD turned to Arie Haagen-Smit, a biochemistry professor at Caltech.

When the Dutch-born scientist visited Pasadena in the late 1930s, he could see the nearby San Bernardino Mountains; a decade later, he couldn't. In 1950, Haagen-Smit undertook innovative research that proved that the pollutants weren't necessarily sulfur-based, as in the Donora, Pennsylvania, poison-air disaster. Instead, the components were mainly oxides of nitrogen and hydrocarbons. "Each one of these emissions by itself would be hardly noticed," he wrote. He discovered, however, that in "the presence of sunlight, a reaction occurs, resulting in products which give rise to the typical smog symptoms."[76]

The oil industry, on the defensive, slammed Haagen-Smit's conclusions, reporting that hydrocarbons and nitrous oxides produced by gasoline combustion in cars, buses, and trucks were to blame for California's smog. Relying on its long-standing ties to Stanford University, the industry arranged for counter research, leading to a 1954 square-off between the brain trusts of Caltech and Stanford. Haagen-Smit won his point through meticulous chemical notations and public experiments that anyone could understand. In one, for example, he put a box of rubber bands outdoors in the hot California sunshine and let people see that, within four minutes, they began disintegrating. Professor Haagen-Smit could trace exactly what was occurring in the atmosphere, and it led back to LA's millions of cars.[77]

Dirty air wasn't only a California phenomenon. The American Cancer Society blamed poisonous factory air for a 1950s spike in cancer deaths in five major cities. Smog was also triggering respiratory illnesses, such as asthma and COPD, in Baltimore and Detroit. In New York City, residents were getting ill from the estimated 92,500 pounds of soot that blanketed the five boroughs annually. The menace of poison air caused E. B. White of the *New Yorker* to assert caustically that soot was "the topsoil of New York."[78] As the columnist Elsie Robinson wrote in the White Plains, New York, *Journal News*, referring to smog, "You may not have breathed its gray-white blanketing or choked over its acrid fumes but, sure as God made apples, your life is affected by the endless contamination of our modern problem life."[79]

Bad air in New York City forced some senior citizens to move to Sunbelt states, notably Arizona—where air pollution barely existed at the time. Throughout the 1950s, there was a clamor for news organizations in metropolitan New York City to provide air quality reports on a daily basis. An increasing number of citizens demanded regulatory efforts to combat air pollution by state and federal governments.

The more immediate horror of environmental mismanagement gained worldwide attention in December 1952. The place was London, England, where the "Great Killer Fog of 1952" was caused by the deadly combination of industrial smoke and an inversion leading to becalmed weather.[80] It was the Donora smog incident of 1948 writ large. After five days of darkness caused by the combination of emissions and fog, more than four thousand people perished. That figure rose to twelve thousand in the aftermath. The BBC warned citizens during the air crisis to stop emissions of smoke from cars, factories, and homes in order to help rid the atmosphere of the smog.[81]

Almost a year later, in November 1953, New York City had its own bout with smoke/fog: smog. An inversion trapped the exhaust from industrial facilities, chimneys, and automobiles. Because New Yorkers had grown used to dirty air, they generally shrugged it off at first. After two days, though, newspapers started reporting on the air pollution crisis. One breathless front-page *New York Times* headline read "Smog Is Really Smaze; Rain May Rout It Tonight; Four-Day Concentration of Smoke and Haze Causes Optical Illusions and Discomfort—Two Airports Close as Fog Is Added; Heavy, Heavy Hangs Over: Smaze, a Blend of Smoke and Haze, Smothers City Area with a Grayish Pall; RAIN MAY WASH OUT 4-DAY SMOG TONIGHT."[82] Thousands of New Yorkers were reporting asthma symptoms, burning eyes, and heavy coughing. In the end, at least 240 people died.[83]

In 1955, in the aftermath of the poisonous smoke in Donora, Pennsylvania, and federal studies by the US surgeon general of New York, Los Angeles, and other cities, Congress passed the Air Pollution Control Act, the first federal legislation to address the issue. However, it only provided funds for government research and technical assistance relating to air pollution at its source. Even though the word *control* was in the act's title, it didn't empower the federal government to punish polluters. Kennedy thought that the legislation didn't do enough to abate the air pollution crisis but that it at least acknowledged that the smog problem would eventually require federal regulatory intervention. The act, he believed, would provide a baseline of data for that future action and also inform the public about the health hazards of smog. It was a logical first step. Yet, polluting industries instantly balked at being "watched" by the federal government, insisting that smoke and dust were simply the price of being an advanced industrial nation. Kennedy pushed back, warning his constituents that manufacturing cities such as Lowell and Lawrence, both in Massachusetts, could easily become the next Donora or London.

In September 1957, Kennedy delivered a speech at the US Conference of Mayors gathering in New York City titled "Our American Cities and Their Second-Class Citizens," in which he insisted that tougher environmental standards had to become law. He spoke of "a growing need to control the pollution of air and water, for the good not only of a single city but of a much broader area. . . . But municipal governments have too often been required to attack these problems alone, without the assistance of other governments with a stake in their solution." In his appeal for what would, in the 1980s, be labeled "environmental justice," he lamented the "shame of our cities," the tragedy of poor neighborhoods becoming dumping grounds for industrial waste, the "blight and decay" of neighborhoods, "slum housing," "congested traffic," "declining and deficient mass transit," and "inadequate parking, recreational, and trash collection facilities."[84] Kennedy complained specifically against the Salem Harbor Power Plant, which relied on coal to produce electricity. It had recently opened to great fanfare in the coastal city of Salem,[85] a charming town that was otherwise the essence of New England's historical, cultural, and maritime heritage.[86]

The severe environmental and public health impacts of sulfur dioxide, mercury, and nitrogen oxide emissions from the Salem Harbor Power Plant were disturbing to Kennedy and other liberal commonwealth politicians. Particulate matter, or soot, which was composed of very fine particles—particles so small they were inhaled deep into the lungs and directly entered people's bloodstream—turned Salem into a public health and environmental disaster zone.[87] There were coal sludge pools along the harbor that leaked into the city of Wenham's water supply. "Jack had seen the Salem Harbor plant," his brother Ted recalled. "It infuriated him."[88]

The owner of Salem Harbor Power Plant was Dominion Resources, one of the biggest energy companies in the United States, known for environmental carelessness.[89] At Dominion's Yorktown Power Station in Yorktown, Virginia, another historic town, the utility company started in 1957 disposing of more than 500,000 tons of coal ash near a tidal tributary of Chesapeake Bay. Known in history as the Chisman Creek poisoning, the dumping went on until 1974, leaking nickel, vanadium, arsenic, and selenium into the groundwater, nearby ponds, and the estuary.[90] Few people knew what was going on with Dominion's coal-combustion residual pits until 1980, but the specter of such dumping was beginning to be a grave concern more than twenty years before. "We must pay increased attention to the problem of water pollution," Kennedy said with alarm in 1959. "The availability of clean water is of importance not only to our health and living standards but to our industrial development as well. The early growth of New England was possible because there was plenty of good water available. Today pollution is the largest single destroyer of New England water resources, hindering the economic health and prosperity of the region."[91]

CHAPTER 5

Wilderness Politics, Dinosaur National Monument, and the Nature Conservancy

David R. Brower, the first executive director of the Sierra Club (1952–1969), taking photographs of Utah's threatened Escalante River. An ardent wilderness defender, he was deemed by the *New York Times* as "the most effective conservationist in the world."

I

From 1953 to 1957, Senator Kennedy joined fellow Democrats in rebuking Eisenhower's first secretary of the interior, Douglas McKay, a former Oregon governor who had built a fortune in Portland selling insurance and automobiles. McKay, a fierce anti–New Dealer, was considered a front man for business interests and the timber, oil, gas, and mining industries. "Giveaway" McKay, as critics called him, was known as a "gospel of growth" philistine who wasn't much interested in the preservation of the natural world. He was the type of market-oriented profiteer who looked at a pristine old-growth forest and saw only lumber. Upon his appointment to the cabinet, he had received a note from Thomas E. Dewey, the Republican nominee for president in 1948, in appreciation of "the wonderful job I know you will do in slaying the socialist dragon of the Interior Department."[1]

Intent on downsizing Interior and gutting the legacy of Harold Ickes, McKay abolished five divisions, fired four thousand workers, and set up obstacles to budget appropriations. Out of the gate, he dismissed Albert M. Day, the director of the Fish and Wildlife Service and the former boss of Rachel Carson. "The action against Mr. Day," she protested, "is an ominous threat to the cause of conservation and strongly suggests that our national resources are to become political plums."[2]

Carson was right. McKay made the entire system of refuges that Fish and Wildlife administered—which FDR and Ickes had established in 1940—vulnerable to sweetheart oil and gas leases.[3] Bill Douglas complained that under McKay's stewardship of the department, primeval forests were being illegally logged in Olympic National Park for the "greatest dollar return."[4] McKay seemed uninterested in the aesthetic value of the national forests. True, he saw some benefit in maintaining wildlife refuges for sportsmen and Bureau of Land Management lands for livestock grazing. But his paramount goal was the privatization of public lands in the American West. As long as McKay was perceived as a political asset, helping to keep western states in the Republican fold, Eisenhower gave him a wide berth to undo Ickes's New Deal preservation measures. Publicly, McKay sympathized with miners, loggers, and ranchers who railed against "federal colonialism in the West."[5]

McKay was excited about the uranium boom in the western states, as were defense strategists and a great many newly minted prospectors. In 1950, a Navajo shepherd unearthed uranium in the Grants mineral belt in northwest New Mexico. Four years later, California miners found uranium deposits northeast of Bakersfield. Lodes were found in Oregon in 1955 in the White King and Lucky Lass mines.

Nevada and Utah also had strikes. The largest deposit of uranium in North America was discovered at Coles Hill in Virginia, but citizens there worried that mining it would contaminate the water; eventually, the commonwealth enacted a ban on extracting the mineral.[6]

Arid western states worried less about water contamination. Encouraged by a world market that kept prices high, prospectors combed remote terrain with Geiger counters. When they found the ore, as did a Texan named Charlie Steen, who reaped a $130 million windfall in Utah, big companies and complex extraction operations soon followed. A battle royal ensued between western legislators, both Republicans and Democrats, who favored uranium mining for its economic benefits and protoenvironmentalists who opposed it, warning about a rise in local cancer rates and tainted water systems. The anti-nuclear advocates spoke mostly about exposure to radiation making mine workers sick, but many of the victims weren't miners. In coming years, the AEC would be forced to compensate people in rural areas, many of them people of color, whose health had been adversely affected by contaminated debris from uranium mining operations nearby.

The earliest organized extraction of uranium in the United States had taken place during World War II on the Navajo reservation in the Four Corners region, where Colorado, Utah, Arizona, and New Mexico met. A team from a front company, secretly representing the Manhattan Project, arrived at Utah's Diné Bikéyah, the Navajo name of the huge reservation, meaning "the people's sacred lands," requesting permission to mine uranium. Seeing it as a patriotic obligation, the Navajo cooperated. Operations continued in the postwar era, when mining proliferated throughout the Four Corners and Colorado Plateau regions. Laborers from the Navajo Nation, as well as the Hualapai, Havasupai, and Hopi, received high wages but no protection from exposure to radon gas or radioactive dust. "Like the pesticide and lead exposure issues for other people of color, the uranium issue for thousands of Southwestern Indians became a legacy of victimization," the historian Robert Gottlieb wrote. "All through the uranium fever period Native American miners were recruited without the AEC or the nuclear industry providing information about potential hazards."[7] Notably, when the US Public Health Service studied the impact of uranium mining on the health of the workers, it "initially focused its attention on white miners."[8] The results of medical studies were shrouded or purposefully incomplete during the 1950s, but later research showed that the incidence of lung cancer was four and a half times higher among uranium workers—of any race—than in the general population.

From a short-term economic perspective, uranium mining and nuclear testing brought prosperous new industries to Nevada, in particular. In Clark County, in fact, where Las Vegas is located, the official seal was an atomic mushroom cloud. The cover of the Las Vegas High School yearbook, *The Wildcat Echo*, during the Eisenhower years was also a mushroom cloud.[9] In 1953, *National Geographic* published an article by Samuel Matthews, "Nevada Learns to Live with the Atom," reporting on the pride that residents felt that the state was home to nuclear testing and the uranium boom. Even though mining companies left behind thousands of tapped-out mines, uranium tailings, and spills, few residents cared. "While blasts teach civilians and soldiers survival in an atomic war, the Sagebrush State takes the spectacular tests in stride," he wrote. According to Matthews, Nevadans deemed the atomic tests "only one more superlative in a state endowed with already spectacular history and scenery."[10]

The influential historian Bernard DeVoto came to the defense of western public lands with a series of scathing essays written in the H. L. Mencken tradition. DeVoto, originally from Utah, won a massive audience and a Pulitzer Prize for his trilogy of books about the West: *The Year of Decision: 1866* (1943), *Across the Wide Missouri* (1947), and *The Courage of Empire* (1952). In addition, he was well known in the 1950s for his "Easy Chair" column in *Harper's Magazine*. There, he condemned the Interior Department under McKay for collusion with business interests. An arch critic of the Forest Service and the BLM who wanted to raise the temperature of dissent, he savaged the Western Cattlemen's Association and the oil and gas lobby for turning the American West into a "plundered province."[11]

Almost twenty years after DeVoto died of a heart attack in November 1955, the novelist Wallace Stegner published a biography of the man he called "the nation's environmental conscience and liberty's watchdog, the West's most comprehensive historian and most affectionate spokesman and most acid critic."[12] The conservation movement in the mid-1950s minus DeVoto's outspoken leadership was defanged. His "Easy Chair" column had been a one-man tribunal in which land skinners and corporate polluters had been exposed for crimes against nature.[13] "He was the first conservationist in nearly half a century," the historian Arthur M. Schlesinger, Jr., observed, "to command a national audience."[14]

New protectors of public lands were about to step up, however, and in even greater numbers than before. David Brower of the Sierra Club, for one, was ready to take center stage in the brutal national debate over public lands and rivers. Born on July 1, 1912, in Berkeley, California, he'd spent his childhood exploring Grizzly Peak and its gorgeous ridgeline, which ran straight down to San Francisco Bay. His boyhood home was only a few blocks from the University of California, where his father, Ross, taught mechanical drawing and where the idea of a National Park Service was hatched in 1915. David's parents joked that there was an almost religious glint in David's fierce blue eyes when he was outdoors. When he was fifteen, he found a butterfly that was black with white wings and green undersides. Insisting that the subspecies weren't in any butterfly guides, he declared it a discovery. And indeed it was; in 1932, it was named *Anthocharis sara reakirti broweri* after him.

Mountain climbing became young Brower's passion. In the spirit of exploration and bonhomie, he eventually shimmied his six-foot, one-inch frame up such landmarks as the Ship Rock formation in the Utah Diné Bikéyah in New Mexico and Mount Waddington in the Coast Range of British Columbia. Enamored of John Muir's dangerous feat of climbing towering sequoias in the High Sierra during a wind squall or rainstorm to feel the raw power of nature, Brower wanted to become part of the giant tree, not separate from it. "The most fun I had was talking about it afterwards," he recalled of one such climb. "In the tree I found plenty of pitch and on the descent very prettily scratched my arm. However, if I find another easy tree I'll do it again."[15]

In 1935, Brower was hired to promote tourism to the Yosemite Valley. There he encountered the celebrated photographer Ansel Adams, whose black-and-white images regularly appeared in the *Sierra Club Bulletin*. Adams, a sparkling raconteur, enthused about America's national parks and told Brower how to look properly at "stone, space, and sky." After serving in the Army in World War II, Brower took over the editorship of the *Bulletin* and then, in 1952, became the first executive director of the club. At the time, the San Francisco–based Sierra Club was a rather exclusive society of seven thousand people, mainly well-to-do white men hungering for weekend outings in Yosemite. Brower, with encouragement from Adams, nudged the Sierra Club toward a broader-based activism and a part in the fight to protect public lands. He had made the same transition he expected of the club, going into the Army ten years before as an outdoorsman and nature lover and coming out a protoenvironmentalist.

Recalling his World War II years as an intelligence officer in the Army's 10th Mountain Division, he explained why. Rising in the ranks from private to major, he had taught ten thousand soldiers survival skills essential for combat in the European Alps. He also served there, in combat. After having witnessed how war and industrialization had affected Europe's natural world, he returned to Berkeley, driven to save the vanishing North American wilderness from the wrecking ball of slash-and-burn modernity. "In such parts of the mountains of Italy, Austria, Switzerland, and Jugoslavia as I have been able to observe are the shattered remains of what must have been beautiful wildernesses," he wrote in the *Sierra Club Bulletin*. "These wild places had their one-time inaccessibility to defend them—their precipices, mountain torrents, their glaciers and forests. But they lost their immunity; they felt the ravages of a conqueror. And now they're dead."[16]

Assuming the mantle of John Muir, who had cofounded the Sierra Club in 1892 and died in 1914, wasn't easy for Brower. Looking like an Old Testament prophet, Muir had been an ecstatic evangelist for national parks; Brower was more buttoned down and reserved in his presentation. Unfortunately, throughout the second half of the twentieth century, reporters regularly intimated that Brower was Muir resurrected. According to his son Kenneth, "My father himself never expressed any enthusiasm for this notion. But the lives of the two conservationists did run parallel in remarkable ways. Both were autodidacts. Both were hopelessly in love with the Sierra Nevada. Both loved and nurtured the Sierra Club. Both hated dams."[17]

With his flowing gray beard and expert knowledge of glaciers, Muir had been an inextinguishable legend whose proudest accomplishment had been campaigning to establish Yosemite National Park in 1890. However, his biggest defeat followed when the O'Shaughnessy Dam was built in the Hetch Hetchy Valley in the park, flooding it in 1923.

Likewise, Brower's first triumph—his leading role in the successful fight against a pair of dams that would have flooded Dinosaur National Monument in Utah and Colorado—was followed immediately by his greatest defeat: his failure to stop the construction of the Glen Canyon Dam on the Colorado River. Kenneth Brower wrote, "It did not escape my father, as he fought those Colorado dams in the 1950s, that O'Shaughnessy Dam in Hetch Hetchy and Muir's long struggle against it was a foreshadowing of his own tribulations; that the dead water of Hetch Hetchy Reservoir made a potent symbol for what he and his movement were up against, not just on the Colorado and its tributaries, and not just with rivers, but with threatened species and ecosystems and landscapes everywhere on earth."[18]

Brower's campaign to stop the dams in Dinosaur National Monument became the center of the fight to protest the legitimacy of the National Park System. Related to the struggle was Brower's perception that the proposed Echo Park Dam project threatened not merely a single valley floor but the sanctity of the Dinosaur ecosystem. The campaign also reassessed in a public way the role of perhaps the most powerful entity in the American West: the Bureau of Reclamation, the part of the Interior Department that didn't protect wild land and rivers; it built dams on them.

The Bureau of Reclamation was created in 1902 to manage water in the West; to that end, it planned and built dams. Its work undoubtedly allowed the region to grow, and during the Great Depression, it provided jobs and new opportunities that supported the western economy. By the 1950s, protoenvironmentalists believed that enough dams had been built and that the bureau needed to be reined in from simply damming the western states into a shadow of their wild former glory. That resistance hardened when the Bureau of Reclamation sought to construct a huge dam near Echo Park inside Dinosaur National Monument, a rather obscure unit of the National Park System.

The main construction, the Echo Park Dam, would be on the southeast flank of the Uinta Mountains, on the border between Colorado and Utah, and at the confluence of the Green and Yampa Rivers.

In addition, there would be a Split Mountain Dam further downriver, both to be components of the multipurpose Colorado River Storage Project, a well-supported reclamation effort to harness the waters of the upper Colorado River and its tributaries for urban and agricultural use and, of course, for electrical power.

Dinosaur National Monument was only eighty acres when President Woodrow Wilson established it in 1915 to protect Allosaurus, Deinonychus, and Abydosaurus fossils. Franklin Roosevelt expanded it by more than 200,000 acres in 1938, specifically to save the scenic canyons of the Green and Yampa Rivers.[19] But in the late 1940s, the Bureau of Reclamation announced plans to construct a network of dams along the upper Colorado River and its tributaries in four western states. In 1950, Harry Truman's secretary of the interior, Oscar Chapman, approved the construction of the Echo Park Dam on the Green River and the Split Mountain Dam on the Colorado.

After Eisenhower entered the White House, he increased the pressure to start construction, mindful of the need for power and water storage in an arid part of the West. In 1954, Brower became central to the debate, which pitted the inherent worth of the national monument, a symbol of the entire National Park System, against the value of the electricity that the dams would produce. Brower and his congressional champions drew a line at the integrity of the national park. They not only wanted to defend Dinosaur from destruction but save other national parks and monuments from the same fate. Brower's protoenvironmentalists demanded: (1) removal of the Echo Park and Swift Mountain Dams from the large package of Colorado River system dams being advocated by the Bureau of Reclamation and (2) insertion of a policy statement that read "It is the intention of Congress that no dam or reservoir constructed under the authorization of this Act shall be within any national park or monument."[20]

Each side argued that its vision for the canyons would result in a higher benefit for the public. Working in favor of Brower's side of the argument was the fact that the park held more than eight hundred paleontological zones. A strong ally was Rachel Carson, who had joined the Wilderness Society in 1950 and was decidedly against the Echo Park and Swift Mountain Dams.[21]

Meanwhile, the White House had other federal dam projects

within National Park Service sites queued up behind the Echo Park project. Sensing that precedent was at stake in the Echo Park/Split Mountain Dam projects, Brower turned pugilist. In his mind, something was illogical, if not downright anti-American, about infringing on the integrity of the National Park System because of the West's ever-increasing need for more cheap federal electricity.

Brower and the Sierra Club weren't alone in fighting to preserve the sanctity of Dinosaur National Monument. Howard Zahniser of the Wilderness Society; Ira Gabrielson, president of the Wildlife Management Institute; and Sigurd Olson of the National Parks Association all threw their weight into the cause.[22] A power struggle ensued between western waterpower enthusiasts and National Park System purists. In Washington, DC, committees in both the House and Senate held public hearings on the strategy of the Colorado River Storage Project. "I wonder in our mad rush to dam every river, chop down every tree, utilize all resources to the ultimate limit," Olson testified at the Senate hearing, "if we are not destroying the very things that have made life in America worth cherishing and defending?"[23]

On May 13, 1955, Brower drove to Hetch Hetchy with his 16-millimeter movie camera and the talented photographer Philip Hyde at his side. Together they spent hours filming and photographing Hetch Hetchy Valley (dammed) and then Yosemite Valley (wild). The Sierra Club used Hyde's marvelous photos to illustrate the book *The Battle for Yosemite*; Brower's film footage became the documentary *Two Yosemites*. The desired effect was to show what would happen to Dinosaur National Monument if the Echo Park and Split Mountain Dams were built.[24]

On Halloween in 1955, an unusual full-page advertisement was published in the *Denver Post*, signed by the Council of Conservationists. The anti-dam coalition's goal was to intimidate the promoters of the Colorado River Storage Project. The ad threatened that if the dams were approved, a unified national parks protection lobby would stop them by any means necessary. Newspapers hadn't been used in quite that way before. Something new was brewing in the American West. An all-out showdown over Dinosaur National Monument was at hand.[25]

Following that thunderbolt of an ad, an energized Brower leapt into the Dinosaur fight with a kit bag full of public relations tricks. Using the controversy to raise awareness about the sanctity of national monuments, pamphlets, lectures, movies, and postcards were all woven into the Sierra Club blitz. He organized trips down the Green River to show politicians and the press firsthand the devastation the dams would render to the Dinosaur National Monument. Brower's Sierra Club mission statement was unambiguous: "We're not blindly opposed to progress; we're opposed to blind progress."[26]

As part of his campaign, Brower recruited Wallace Stegner to edit and contribute to *This Is Dinosaur: Echo Park Country and Its Magic Rivers*, an illustrated book designed to display the natural splendors of the area to the general public. As a child, Stegner had moved with his restless family all around the West.[27] His troubled father couldn't seem to settle down. This transient childhood instilled in Stegner the desire to be a novelist of the true American West. By the time of the Dinosaur battle, Stegner, a professor of English and the director of the creative writing program at Stanford University, had won critical praise for such novels as *The Big Rock Candy Mountain* (1943). He had recently published a well-received biography of John Wesley Powell, the first documented European American surveyor to explore the Colorado River through the Grand Canyon. Depicting the powerful link between the western landscape and its people was his literary forte.

The Colorado River is 1,400 miles long, the seventh longest river in the United States, and Stegner thought damming it to provide hydroelectric power at the expense of the National Park Service to be a moral affront.[28] The Colorado River Compact of 1922 had allocated the river's annual flow among the seven western states in its basin. The upriver states had a hard time meeting their obligations to guarantee a set amount of water for use by downriver states. They needed a way to make the flow predictable. Stegner was concerned that the Bureau of Reclamation didn't take into consideration changing conditions in terms of weather and population. Tapping the river and its tributaries too much could turn a dry year into a Dust Bowl–like drought. Growing cities in Utah, Colorado, and Arizona would then face serious water shortages. The point of his *This Is Dinosaur* essay was that the Colorado River was in danger of being overallocated.

Other contributors to *This Is Dinosaur* were the publisher Alfred A. Knopf and the naturalist Olaus Murie, who wrote essays explaining that the Echo Park Dam threatened historic, archaeological, and ecological resources.[29]

Brower and company won the battle. On April 11, 1956, Congress removed the Echo Park and Split Mountain Dams from the Colorado River Storage Project. Victory for the sanctity of national parks and national monuments was achieved. Eisenhower signed legislation that designated alternative sites for the Colorado River Storage Project but forbade dams from marring any national park. The Dinosaur battle turned out to be the test case for the preservation of the noneconomic value of wilderness in public land-use policy.

Even though the Dinosaur battle was a victory for the Sierra Club, Brower felt hoodwinked. On the upside, the permanent protection of national parks and monument ecosystems had become law. The Sierra Club alliance had turned a blind eye to the Bureau of Reclamation's alternative: the Glen Canyon Dam, to be built farther down the Colorado River, just outside of the national monument. Brower and company realized too late that Glen Canyon possessed stupendous natural attributes worthy of permanent preservation. Sadly, they would be destroyed. Edward Abbey described the loss in *Desert Solitaire: A Season in the Wilderness*: "To grasp the nature of the crime that was committed imagine the Taj Mahal or Chartres Cathedral buried in mud until only the spires remain visible. With this difference: those man-made celebrations of human aspiration could conceivably be reconstructed while Glen Canyon was a living thing, irreplaceable, which can never be recovered through any human agency."[30]

Upon reflection, Brower was embarrassed that he'd missed the big picture. Trading the serenity of Dinosaur for allowing the Glen Canyon Dam to be built, he realized, wasn't morally sound. When he embarked on a boat trip through Glen Canyon before construction began, he was taken with the tapestried walls, slide rock arches, and elegant willows of the canyon, all of which would soon be underwater. The pang of the impending loss radicalized him.[31] He would become fond of saying "What we save in the next few years is all that will ever be saved."[32] Another shift also compelled him. No longer would the Sierra Club be oriented solely toward southwest wild lands: Brower was ready to fight for natural ecosystems all across America, from Washington's North Cascades to Maine's Allagash watershed to Georgia's Cumberland Island.

In the mid-1950s, Brower's scope expanded in a different sense as he became a leading national voice against air and water pollution. From his home in the hills of Berkeley, he witnessed smog obscuring the view down to San Francisco Bay. The black butterfly with white wings and green undersides that he had identified as a boy could no longer be found around Grizzly Peak. "We built here for the view of San Francisco Bay and its amazing setting," he wrote in December 1956. "But today there is no beautiful view; there is hideous smog, a sea of it around us. 'It can't happen here,' we were saying just three years ago. Well, here it is."[33] In response, as the 1960s approached, the irascible Brower drove "the transformation of conservation into environmentalism in the United States," as the prominent historian Hal K. Rothman put it.[34]

II

Even though Brower received most of the media attention for leading the Dinosaur campaign, Howard Zahniser of the Wilderness Society played the key behind-the-scenes role, working from his modest office in Washington, DC. Soft-spoken and alert to political nuance, "Zahnie," as he was affectionately known, knew the corridors of Capitol Hill intimately. Since the 1930s, he had been writing wilderness essays and articles and he had attracted a fan base. Like Rachel Carson, he had worked as a writer for the US Biological Survey (and later the Fish and Wildlife Service), promoting national wildlife refuges. What differentiated Zahniser from Brower and Olson was that he didn't regularly canoe, rock climb, or ski, but he did love to walk. A faith-based Christian moralist, Zahniser, always calm and reasonable in demeanor, insisted that national parks uplifted the spirit of citizens in America's mechanized and urban society even if, like him, they were places seen mainly in pictures and the imagination. Thus, nothing about Zahniser was deceitful when he studded his sentences about the North Cascades and Boundary Waters with superlatives. He

Howard Zahniser was president of the Wilderness Society from 1945 to 1964. He fought to save the sanctity of Dinosaur National Monument from intrusive hydroelectric dams.

felt what he couldn't see. As a strategist, he was polished at recruiting grassroots foot soldiers, while winning over Washington lawmakers and assembling coalitions of conservation-oriented groups.[35]

Zahniser was born on February 25, 1906, in Franklin, Pennsylvania, yet another committed environmentalist raised near the Allegheny River. Growing up, he would explore the hemlock bottoms and hardwood groves of Minister Creek, taking in the diverse plant life of the Allegheny Plateau. Protecting the heavily forested countryside in his home state would be a lifetime passion for Zahniser. Pennsylvania was literally "Penn's Woods," named after the state's heroic Quaker founder, William Penn, whose 1681 edict proclaimed that one in five acres of forest must be spared the ax. Unfortunately, he wasn't heeded. Over the next two centuries, successors to his utopia for religious freedom would raze not just 80 percent, but 99.9 percent of the primeval forest there.[36]

After graduating from Greenville College in Illinois in 1928, Zahniser moved to Washington, DC, to do graduate work at American

University and George Washington University. Led by an intense love
of Concord transcendentalism, he made a name for himself writing for
conservation groups and was among the eight cofounders of the Wilderness Society in 1935. Starting in 1945 as executive secretary of the
society, Zahniser edited the *Living Wilderness*, a hard-hitting magazine filled with warnings of landscapes pillaged in the interest of speculation, "progress," and commercialism. Always present, with a quick
mind and polite demeanor, he never self-aggrandized or fell prey to
diversions. Having served as head of the Thoreau Society in 1956–57,
his mantra was: "In Wildness is the preservation of the world."[37] The
seasonal pageants of nature—in which he believed the Kingdom of
God was visible—gave his *Living Wilderness* writings a religious cast.
His prose descriptions of the flowering dogwood, the chokecherry,
and the sugar maple were as lovely as those of Thomas Wolfe on the
Great Smoky Mountains or John Burroughs on the Catskills.[38] Less
artistically, Zahniser took it upon himself to lobby Congress to establish roadless primitive areas within national forests.

The wilderness movement with its modern imperative had co-
alesced in 1949 when a hundred conservationists and researchers
gathered in San Francisco; they were there for the Sierra Club's first
biennial Wilderness Conference. Some activists were seeking ways to
protect more of the High Sierra, while others wanted Washington's
North Cascades and California's Coast Ranges saved for the ages.
There were concerns about limiting the number of cattle on public
grazing lands in Montana and Wyoming.[39] Zahniser, perennially focused on saving habitats in their most natural form, insisted that the
attendees think more broadly and deeply and define the shrinking
American wilderness everywhere in both legal and moral terms. "We
saw that safeguarding wilderness involves the wilderness of ourselves
and of other visitors to the wilderness, for we all have an inborn ten-
dency to make over wilderness rather than adapt ourselves to it," he
said.[40] At the second conference, in 1951, the attendees strengthened
their resolve to lobby for roadless public lands.

Like Douglas, Zahniser considered the "wilderness experience" to
be an American birthright. However, there was a fine line between
them. Douglas saw wilderness as a sphere that humans could visit
as long as they didn't leave a trace, hiking, camping, and fishing as

he loved to do. Zahniser, deep down, would have preferred that humans stay out of designated wilderness and simply leave it alone. He was realistic about the need to compromise on that point. His conviction was twofold: only landscapes that were purposefully set aside would be saved in perpetuity in pristine shape, and wilderness areas in private hands would likely be developed commercially in the future. Given that, he thought that the federal government, as an opening salvo, needed to save around 10 million acres of wilderness quickly before it vanished. And he knew that it was incumbent on conservation groups such as the Wilderness Society and the National Parks Association to launch an aggressive battle to bring about a new public lands designation.[41]

Zahniser rose to the forefront of the wilderness lobby during the fight to keep dams out of Dinosaur National Monument and other national parkland. Unlike others in the movement, including Brower and Olson, he thought primarily in terms of the vote count on Capitol Hill. Using his ability to persuade people with his courtly kindness, he was credited with personally persuading 120 congressmen to change their votes during the Dinosaur National Monument debate, writing many of the speeches delivered in Congress against the Colorado River Storage Project and, in 1956, coordinating the final compromise that protected the national monument's integrity.[42]

After the Dinosaur showdown, President Eisenhower remained supportive of private utilities but began denouncing some public dam proposals. When the Army Corps of Engineers tried to build two federal dams on the Buffalo River in Arkansas, he vetoed the bills, calling them a waste of money.[43] He was stalwart in proceeding with the Glen Canyon Dam, however.

By the time construction started on Glen Canyon, most of the finest river sites in America had already been manipulated either by the Bureau of Reclamation, which operated primarily in the West, or by the Army Corps of Engineers (whose motto was "Controlling Nature"). By the Eisenhower years, few major rivers in the continental United States flowed freely. The Colorado River, as just one example, had by 1964 an array of nineteen large dams and reservoirs that contained four times the river's annual flow and gave the US government systematic management over its watershed. "No longer did the river remotely resemble the wildly surging, unpredictably flooding river explored by John Wesley Powell almost a century earlier," Steven Solomon wrote in his book *Water: The Epic Struggle for Wealth, Power, and Civilization*. "Each drop was measured, every release calculated, and every event on the river planned by its central managers. It was the lifeblood of the entire southwestern United States."[44]

The building of the Glen Canyon Dam spurred Zahniser and other protoenvironmentalists to fight harder for a wilderness protection law, by which certain public lands would be protected in their natural condition. Unlike Yellowstone National Park, with its massive lodge-hotels, restaurants, curio shops, and observation decks, true wilderness zones wouldn't allow man-made structures—not even gravel roads, horse stables, or restrooms. Given that land within national forests had been opened to logging, grazing, and mining since the Great Depression, the idea faced a steep road of its own.[45] In 1956, at the first Biennial Northwest Wilderness Conference in Portland, Oregon, Zahniser got started, drafting a bill to create just such a preservation mandate. Before long, Representative John Saylor, an outdoors-loving Republican from Pennsylvania, introduced it in the House, and Hubert Humphrey, a Democrat from Minnesota, did the same in the Senate.

With his horn-rimmed glasses, studious demeanor, and calm speaking voice, John Saylor generally dedicated his public career to protecting the natural world. He was born on June 23, 1908, in Somerset County, Pennsylvania, in the heart of the Allegheny Mountains. In fact, he grew up with the Gallitzin State Forest almost in his backyard. Saylor entered Congress in 1949 with civil rights and conservation as his top concerns. His parents had placed a high value on nature, one that Saylor adopted while seeking undeveloped areas for himself canoeing the Allegheny River near home, as well as Minnesota's gorgeous Saint Croix River, and rafting down Utah's Green River. Dubbed "St. John" by the Sierra Club and a "Theodore Roosevelt conservationist" by legislative colleagues, Saylor strongly opposed impounding or damming rivers in rich habitats. At the end of his twenty-four years in Congress, he was cited by the Izaak Walton League as "one of the first to stand up and really fight for clean air and water, the establishment of parks, and our other environmental needs."[46]

In the mid-1950s, Saylor dedicated himself to a goal that was as difficult as it was epochal: he wanted Congress to pass the Wilderness bill and protect pristine land from any disturbance by humans. The system, as Saylor and Zahniser envisioned it, would be managed by four federal agencies: the National Park Service, the Forest Service, the Fish and Wildlife Service, and the Bureau of Land Management. Sigurd Olson of Minnesota, riding the literary success that year of *The Singing Wilderness*, was delegated to ask Senator Humphrey to sponsor the bill.[47] "I have worked closely with Howard Zahniser and others for some time on this measure and feel that in view of the mounting pressures of population, commercialization, and industrial expansion, the only way to assure future generations that there will be any wilderness left for them to enjoy is to give such areas Congressional sanction now," he wrote Humphrey. "I feel strongly that this is the last chance to preserve the wilderness on this continent, for we are on the verge of an era where the pressures to destroy or change it will become greater than anything we have ever experienced."[48]

The notion that wilderness was facing its "last chance" united leaders in conservation with the new wave of protoenvironmentalists. They saw their time as a defining moment in US history. The frontier that Americans had long taken for granted closed in the late nineteenth century. In the mid-1950s, the wild lands that had been equally taken for granted were in serious danger of disappearing forever. "The founders of the wilderness preservation movement shared a passionate conviction that wilderness is not some luxury but a vital bulwark in our individual lives and the bedrock of our distinctive American culture," the historian Doug Scott wrote. "Their breakthrough concept was to link the existence of ecological wholeness most perfectly exemplified in wilderness areas to human well-being and our very civilization."[49]

On June 7, 1956, Humphrey introduced the bill in the Senate; a month later, Representative Saylor introduced the companion bill in the House.[50] Critics pounced on both lawmakers as socialist Luddites. Fred Childers, the publisher of the *Ely Miner* in Minnesota, was outraged that Humphrey was spearheading an effort that would make the National Forest Service's huge Superior Roadless Primitive Area, containing hundreds of pristine lakes, miles of undeveloped rivers, and the largest virgin forests east of the Rocky Mountains off-limits. The North Country was a treasure chest of resources to extraction companies, which were important to Ely. The new bill threatened to make the tract, renamed Boundary Waters Canoe Area in 1958, a true wilderness. Keeping it open to timber and mining interests, Childers argued, was essential for rural families trying to eke out a living. Smearing Humphrey as a "Brutus or Judas," Childers urged readers of the *Ely Miner* to write the senator a sharply worded protest letter, stating that Ely would become a ghost town if mining, lumbering, and motorboat tourism were driven out of town. He challenged Humphrey to hold a town hall meeting in Ely and get an earful from his constituents about how damaging the proposed Wilderness bill would be to the local economy. "This fellow Childers is a reactionary editor in Ely," Humphrey wrote to a staff member. "He hates my guts, and has been after me for years. He feels he has a good issue now, so I want to take him on—head on—so let's give it to him."[51]

Humphrey, a former mayor of Minneapolis, who entered the Senate in 1949, only dabbled in conservation politics. Civil rights, humanitarian foreign aid, and the nuclear test ban were his liberal calling cards. In 1955, he heard Zahniser give a speech in Washington, titled "The Need for Wilderness Areas," and he was hooked. Believing that wilderness was a good national issue for Democrats, one that had a solid base of backers in the Twin Cities and rustic resorts in the Boundary Waters region, Humphrey took the lead on the Wilderness bill the next year. His sponsorship of the legislation was his full baptismal dunk in the waters of environmental policy. He was the single most persistent backer of the bill in the Senate during the early years of what would be a long legislative fight.[52] "He is a climber, and wants *national publicity*, and fast," Olson enthused to a friend about Humphrey's preservationist leadership.[53]

Joining Humphrey's quest for the bill in the Senate was fellow Democrat Richard L. Neuberger of Oregon. A son of German Jewish restaurant owners with a passion for the natural beauty of the Pacific Northwest, Neuberger had graduated from the University of Oregon in 1935. Two years later, he wrote *Integrity: The Life of George W. Norris*, a biography of the progressive Republican senator from Nebraska (a favorite of Ted Sorensen). The *New York Times* took notice and hired Neuberger to become its roving correspondent in the Pacific Northwest. Elected to the Oregon House of Representatives in 1940, he continued to write occasionally for the *Times*. In World War II, he served in the army for three years in the Pacific, before returning to Oregon and writing the important *The Lewis and Clark Expedition* (1951). A well-rounded, versatile intellectual steeped in western history, Neuberger became in 1954 the first Democrat to win a Senate seat in Oregon in forty years.

Easily recognized by his trademark polka-dotted bow tie, Neuberger had an encyclopedic knowledge of America's rivers. In Washington, he struck up a fast friendship with Bill Douglas, occasionally hiking with him on the C&O Canal. Long before the word *biodiversity* came into widespread use in the 1980s, Neuberger and Douglas were concerned about the loss of wildlife species in the Pacific Northwest. A member of the Wilderness Society, Neuberger was a strong opponent of Echo Park Dam and regularly attacked Eisenhower's Bureau of Reclamation for being in cahoots with the private power interests in the Pacific Northwest.[54] "He loved the outdoors and used it over and over again as a theme for many of his articles," Douglas recalled of Neuberger. "He was a frequent visitor at my tamarack cabin on the Lostine River in the Wallowas of eastern Oregon."[55]

Neuberger and Douglas once joined a couple of Forest Service employees on a three-day, hundred-mile trip by horseback across the Wallowas. Neuberger later wrote about it in the *New York Times*.[56] Inspired by Humphrey's Wilderness bill, he would also introduce a bill in 1959 to save the hundred-thousand-year-old Oregon Dunes, along with the Sea Lion Caves, in his home state. With great drive, Neuberger also led the campaign to establish the Fort Clatson National Memorial near Astoria; it had been the temporary winter quarters built by the Lewis and Clark expedition in 1805. The locale, a compound reconstruction, helped increase tourism along the Oregon route of the expedition.

As Humphrey, Saylor, and Neuberger soon learned, opposition to the Wilderness bill by such lobbying groups as the American Pulpwood Association, the American Cattlemen's Association, and the American Mining Association was wolverine fierce. The National Park Service and the US Forest Service opposed it, too, because they saw it as an attempt by Congress to diminish their authority over federal land.[57] Furthermore, the idea of banning human encroachment on millions of acres was a hard sell. An ethos of extracting riches from natural resources continued to prevail in western states. Substantive or procedural objections were constantly being issued by reclamationists and developers. Year after year, Zahniser was forced to revise the bill's language. Senator James E. Murray of Montana said in support of the legislation, "In my opinion, it is very unfortunate that some of the commercial interests have exaggerated the effect of this splendid wilderness proposal. The timber people are not going to be 'closed down' or left without resources to grow. The bill is not going to terminate any grazing permits held by livestock men."[58]

As chairman of the Interior and Insular Affairs Committee in the 1950s, Murray favored multipurpose dams and was a virtual rubber stamp for the federal development of hydroelectric power in the West. His fingerprints were on such large Montana dams as Canyon Ferry on the Missouri River, Yellowtail on the Bighorn River, and Libby on the Kootenai River. Because Murray was from Butte, a mining vortex, he was a hero to timber workers, the railroad brotherhoods, and scores of lesser unions. So it's somewhat surprising to realize that he worked hard on passage of the Wilderness bill. Indeed, before his death in 1961, four months after leaving the Senate, he suggested that America needed a far-reaching National Environmental Policy Act. He believed that such a comprehensive law would require federal agencies, which often polluted and plundered, to adhere to certain national environmental values and goals.

Murray argued that without specific, far-reaching environmental policy, federal agencies were neither able nor inclined to consider the environmental impact of their actions. He also sought the creation of an executive-level board or council that would gather information regarding the state of the environment and provide policy advice based on it to the White House. In 1959, he introduced the Resources and Conservation Act in Congress. Almost all of what Murray sought to accomplish with that bill would finally be enacted ten years later as President Richard Nixon's National Environmental Policy Act (NEPA).[59]

III

In the 1950s, another group emerged in the cause of land preservation. This one didn't look to the federal government for help. Instead, in 1951, a bipartisan coalition of activists formed the Nature Conservancy (TNC) to bypass the government and buy beautiful spaces before cement trucks and bulldozers arrived. The TNC sought to protect ecologically important lands and waters from Florida's mangrove thickets to Alaska's fragile tundra, from North Dakota's potholed prairies to Louisiana's cypress bayous. The nonprofit actively bought, traded, leased, and otherwise gained control of the land most at risk. Today its real estate portfolio includes around 120 million acres of land and 5,000 miles of rivers worldwide. With more than one million members, TNC also had notable successes in conservation easements on land that it didn't own. Democrats and Republicans both thought TNC was a clever way to sidestep Washington politics and save land.

Richard Pough, an engineer who had written a seminal article for the *New Yorker* in 1945 about the peril of DDT for wildlife, was among the founders of the group; he persuaded Lila Acheson Wallace, a cofounder of *Reader's Digest*, to seed TNC with $100,000. With that, the fledging group was able to buy its first property: sixty acres of old-growth forest in the Mianus River Gorge in Westchester County, New York, was saved from commercial development. TNC recognized that the moist, chilly microclimate along the Mianus River was the ideal environment for the growth of eastern hemlock, American beech, oak, and various species of ferns.

The Mianus River Gorge, with its 350-year-old trees and gorgeous waterfalls, was the perfect acquisition with which to launch TNC. The growth and power of TNC was remarkable. By 1970, the conservancy had purchased the entire island of St. Vincent off Florida's Gulf Coast for $1 million and the Sunken Forest on Fire Island off the south shore of Long Island for $2 million. By the twenty-first century, it was buying 1,000 acres a day in more than seventy countries.[60]

For years, Katharine Ordway of Minnesota, born on April 3, 1899, was TNC's most generous financial contributor. Her father was the principal owner of the Minnesota Mining and Manufacturing

Company (now known as 3M). She graduated from college at the usual age but returned to academia in her fifties to study land use, as well as biology, at Columbia. At the time, she'd recently inherited a sizable fortune from her father. Soon she was using her money to preserve the kind of land use she liked best: utterly natural and free from encroachment by humans. In the mid-1960s, she gave TNC $750,000 to buy 1,500 acres of forest at a site called Devil's Den. It became the largest tract of protected land in Fairfield County, Connecticut.

At home in high society or slogging across a prairie holding an umbrella for shade, Ordway eventually donated more than $64 million to TNC. Though she lived in Weston, Connecticut, she was personally involved in saving tall-grass prairie land in the Midwest. Sounding like Bill Douglas, she once asked TNC's president, "Can't you get me a prairie that we can go out on and not see *anything* else—no houses, no power lines, anything?"[61] With Ordway's donations, TNC was able to purchase, among other plains parcels, the Konza Prairie reserve in Geary County, Kansas, where big bluestem, purple prairie clover, and pasqueflower bloomed among the sea of grass.[62]

TNC had a keen understanding of ecosystem wholeness. Working from its Washington, DC, headquarters, and through its local chapters, it was largely responsible for halting development that would exterminate such diverse living things as the fringe-toed lizard in California's Coachella Valley and a patch of Knowlton's cactus (a species so rare that TNC biologists have refused to divulge its location). Before the conservancy acted to protect vast tracts of nutrient-rich shallows along the Platte River in Nebraska, there were fewer than fifty breeding pairs of sandhill cranes left in the upper Midwest. A decade after TNC made its purchases, nearly ten thousand of the crimson crowned cranes graced the Central Flyway skies. (The Crane Foundation also joined TNC in the rehabilitation project.) In addition to land purchases, the conservancy highlighted the importance of ecological diversity, pointing out that a quarter of all drugs prescribed in the United States incorporated plant extracts; that aspirin was derived from willows; and that oil extracts from nuts had been found to ward off bacterial infections. Even bee venom, it pointed out, had applications for arthritis. Such examples were numerous.[63]

IV

For many Native American tribes, the Nature Conservancy's ethic was an ancient custom; land was dominated only by itself. As Chief Luther Standing Bear of the Oglala Lakota once explained of the prairie lands that Katharine Ordway strove to protect through TNC, by tradition his people didn't consider the Great Plains and Dakota Badlands as being *wild*. "Only to the white man was nature a 'wilderness' and . . . the land 'infested' with 'wild animals' and 'savage' people," he said.[64] Nevertheless, many tribal leaders in the 1950s agreed with Howard Zahniser's notion that "We are part of the wilderness of the universe."[65] To many Indigenous people, mountains, lakes, soil, earth, skies, wind, and waterways were sacred.

Although tribal customs varied in the Eisenhower years, most Native Americans maintained a traditional reverence for the interconnectedness of all life-forms. Though it was sheer romanticism to call Indigenous people the first environmentalists, their sense of land stewardship was, in most regards, ahead of that of white Euro-American culture. Long before Aldo Leopold's *A Sand County Almanac*, Chief Seattle's farewell speech of 1854 to his Suquamish and Duwamish people was an ecological homily against the evil of "nature conquest."[66] Native American tribes treated special land formations such as Oklahoma's Wichita Mountain, New Mexico's Blue Lake, and Arizona's Monument Valley as sacred sources of life. To many Native Americans, a grove of trees, the wind, a meadow, or stone pillars were to be respected as gifts from the Creator. Their ancient cultures were rooted spiritually and emotionally in the earth. As the novelist Jack Kerouac wrote in *On the Road*, "The earth is an Indian thing."[67]

Kerouac encouraged postwar baby boomers to don backpacks and explore America's public lands. In his 1958 novel, *The Dharma Bums*, Kerouac positioned Japhy Ryder (modeled on the poet Gary Snyder) as the leader of a "rucksack revolution" about to take place, in which nonconformist youths would be drawn to wild rivers, green mountains, desert mesas, remote islands, and rain-drenched forests. Since 1956, when Kerouac worked for the Forest Service as a forest fire lookout on Desolation Peak in the North Cascade Range of northwest Washington, he had championed wilderness as a spiritual salve. His fire lookout quarters, which had no electricity or plumbing, served as his Walden Pond retreat. In that cloudy blue-green wilderness, he looked for wildfires and studied Buddhism. Being in the deep back-woods, out of range, away from TV screens and washing machines, liberated him, at least temporarily, from the crushing conformity of American culture.[68] With his romantic backcountry prose, the Beat writer caused legions of die-hard followers, most of them barely out of their teens, to think about wilderness in spiritual terms.

In a very different way, Walt Disney, too, aroused widespread pub-lic interest in protecting wildlife habitats. His 1946 animated short film *Willie the Operatic Whale* brought a new awareness of the in-telligence and grandeur of real whales.[69] Disney then launched a se-ries called *True-Life Adventures*, documentaries starring charismatic wolves, prairie dogs, seals, and other creatures. At first, the films were thirty-minute short subjects, but when such titles as "The Beaver Valley" and "Water Birds" won Academy Awards, Disney decided to expand the format into full-length features. Remarkably, those com-plex nature films were hits, bringing Disney even more Oscars. The footage of a buffalo giving birth to a tremulous calf, a rattlesnake and tarantula in a death battle, and mallard ducks sliding on ice was rivet-ing. According to Steven Watts, "Smoothly blending education with entertainment, the True-Life Adventures were full of information about the natural world, and they carried a prescient message with the warnings about the environmental dangers of human invasion of the natural world."[70]

With Disney films, *The Dharma Bums*, and the legacy of Aldo Leopold stoking popular interest in wilderness, the time was right for visionary legislation.[71] Working in tandem with Brower, in April 1959, Zahniser distributed a fresh draft of his Wilderness bill to Forest Ser-vice and National Park Service directors, seeking their support. To his dismay, Conrad Wirth, the director of the NPS, dismissed the bill as a ludicrous attack on his agency's ability to build roads and restrooms in parks under his jurisdiction. It conflicted directly with "Mission 66," a ten-year NPS plan to rearrange national parks for car-borne tourists and to expand visitor services within them, all by 1966, the fiftieth anniversary of the National Park Service.

Everybody in the conservation orbit, it seemed, had an opinion about how to improve Zahniser's drafts. One, in particular, gave Zahniser something especially valuable: the perfect word. When Zahniser visited Polly Dyer, a Seattle-based activist, to discuss saving the Washington State seashore from intrusive roads, she described the area as "untrammeled." The adjective jumped out at him. To his mind, that meant areas not "subjected to human controls or manipulations."[72] He used it in all subsequent drafts to explain exactly what he meant.

"Untrammeled" nature wasn't the aim of every conservation group, and corralling them all to work together on battles such as saving Dinosaur National Monument and the Wilderness bill was no easy matter. Moreover, conservation activism was a crisis discipline, and all of the conservation groups were competing for donations, so it was a tricky dance.[73] The main goal, in reality, was getting money with which to operate. Despite that, an unwritten rule took hold among national groups addressing environmental activism: mock other conservation-oriented groups behind their backs, but *never* criticize them in public.[74]

As Zahniser and Brower saw it, by the late 1950s, the term *conservation*, so closely identified with Theodore and Franklin Roosevelt, had lost its power to move people to action. They were protoenvironmentalists, postwar Americans who had a new ecological awareness that the Earth was being ransacked, polluted in myriad ways and at an alarming rate. Not all activists, no doubt, agreed on tactics. There was the wilderness faction of preservationists, who wanted to ban roads in pristine parts of federal lands. There was the Mission 66 national parks crowd, who wanted new units added to the Interior Department system, with plenty of amenities provided for tourists. The so-called hook-and-bullet outdoors types, exemplified by the Izaak Walton League, the Boone and Crockett Club, and the National Wildlife Federation, were determined to protect wildlife so humans could continue to hunt and fish. Some forester types continued to latch on to Gifford Pinchot's maxim that sustainable natural resource development meant "the greatest good for the greatest number of people." It was between 1965 and 1975, when mass concern in the United States, Canada, and western Europe arose, that the inclusive terms *environmentalism, environmentalist,* and *the environment* took firm hold and became commonplace in the public square.[75] All types of people on the range of "green" activism used them, with the core understanding that an *environmentalist* cared about the Earth—a lot.

The fight against air pollution, in particular, created a slew of regional single-issue nonprofits, such as Stamp Out Smog in Los Angeles. In fact, concerns about the speed with which air quality was deteriorating made air pollution the predominant environmental issue for most Americans. In 1955, Congress authorized a three-year study of air pollution—just a study, not vigorous reform—that focused on exhaust fumes in New York City, the metropolitan area with the greatest concentration of automobiles in the world. During Eisenhower's two-term presidency, automobile exhaust was responsible for 70 percent of the lead contamination of the environment. But the administration shrugged off pollution as the province of state governments, even as it connected the states with a huge federal highway system that added insult to hypocrisy by encouraging long-distance motor vehicle use.[76]

Just as infuriating to conservation groups, in 1960 Eisenhower vetoed a clean-water bill for much the same reason; he insisted that water pollution was "a uniquely local blight."[77] It was a response that might have satisfied Americans ten years before, but the threats represented by nonbiodegradable products, automobile exhaust from leaded fuel, dioxins released by incineration, cement superhighways, and the sprawl of suburbia were catapulting protoenvironmentalist activism into the public eye with vehemence. Giving the protest backbone was a new breed of scientists, alarmed by the Nevada nuclear tests and unregulated chemical-manufacturing facilities, by the Donora and New York City smog incidents, by unnecessary blocking of western rivers by dams, and the alteration of the Everglades ecosystem by a network of canals, levees, and pump stations. According to the Office of the Surgeon General, "The quantity and quality of America's water supplies are being endangered by 'synthetic' living and 'push button' industrial processing . . . and the problems threaten to increase in the future."[78] That had hit crisis proportions as the 1950s ended.

Brower and Zahniser conceived and led the environmental preservation movement, which had room for outdoor recreationists, ecologists, protoenvironmentalists, anti-pollution advocates, sportsmen, and anti–nuclear testing protesters. Coalitions of nonprofit activists worked in unison to win high-stakes battles over the dark forces of what Eisenhower himself would call the military-industrial complex. Bursting with ideas and energy, they charted an activist political course with a priority on lobbying on Capitol Hill and in statehouses nationwide on a wide array of conservation concerns.

Nobody summarized the shared outlook of the protoenvironmentalists and anti–nuclear testing activists as eloquently as Rachel Carson did in the spring of 1954 before a thousand young women in Columbus, Ohio. In some ways, environmentalism was a reaction to the rampant consumerism of the postwar years, a point that she recognized before just about anyone else. The author of *Under the Sea-Wind* denounced the madcap rush of postwar technology, contrasting it with the inherent divinity of the natural world. As she put it, "I'm not afraid of being thought a sentimentalist when I stand here tonight and tell you that I believe natural beauty has a necessary place in the spiritual development of any individual or any society. I believe that whenever we substitute something man-made and artificial for a natural feature of the earth, we have retarded some part of man's spiritual growth."

Carson went on to decry the trend toward "a perilously artificial world," which to her was epitomized by Levittown tract houses, shopping centers with massive asphalt parking lots, and a proposed six-lane highway through Rock Creek Park in Washington, DC. She asked, "Is it the right of this our generation, in its selfish materialism, to destroy these things because we are blinded by the dollar sign?" Unlike the single-minded, money-obsessed "Giveaway" McKay types looking for the next deal, she hoped that American stewards would focus on the "wonders and realities of the universe about us," which would lead to "less taste" for reckless ecological destruction.[79]

CHAPTER 6

Saving Shorelines

Senator John F. Kennedy and fiancée Jacqueline Bouvier go sailing while on vacation at the Kennedy compound in Hyannis Port, Massachusetts, in 1953.

I

The National Park Service's *A Report on a Seashore Recreation Area Survey of the Atlantic and Gulf Coasts,* issued in 1955, echoed Rachel Carson's environmental concerns, as expressed so directly in her Columbus address.[1] Looking at the 3,700 miles of coastline from Maine to Texas, the study revealed that only 240 miles of that long stretch was preserved for public use—and half of that was on North Carolina's beloved Cape Hatteras, which in 1937 became the United States' first national seashore.[2] The study identified fifty-four other seashores worthy of local, state, or federal preservation, among them Cape Cod (Massachusetts), Fire Island (New York), Bogue Banks

(North Carolina), Marco Island (Florida), Padre Island (Texas), and Cumberland Island (Georgia). The report became the basis of a widely distributed pamphlet issued by the NPS, "Our Vanishing Shoreline."

The pamphlet started by bemoaning the ubiquity of such signs in coastal areas as PRIVATE PROPERTY, NO TRESPASSING, and SUBDIVISION: LOTS FOR SALE. It warned bluntly that seashores were being mauled by excessive commercial and residential developments. "Our Vanishing Shoreline" also made clear that oceans were being jeopardized on and off shore by oil and gas exploration, engineering projects, mining, housing, motel strips, and high-volume tourism. Tidal pools and dune system habitats of shorebirds and nesting turtles were being destroyed by reckless commercial developers.

Kennedy obviously understood why people loved living at the water's edge: his own family's houses were on the Atlantic. The reality, however, was that the country was in grave danger of losing the entirety of its natural shoreline habitat. The pace of building was unsustainable. What had been clusters of cottages and bungalow colonies in New England, New Jersey, and Florida were growing crowded with high-rises and hotels. Wetlands behind beachfront were filled in for tract homes. In New Jersey, unbroken, undisturbed beaches supporting once-pristine ecosystems were disappearing. New York City was dumping its garbage from barges into the Atlantic just a few miles out at sea. The magnificent Jersey shore was blighted by flotsam of the nation's biggest city: medical waste and tons of smelly garbage, along with the occasional dead body, washed up onto the beaches and made its way into the state's other marshes.

Washington lawmakers were jarred by the NPS report and began to talk about how to stop the New Jerseyization of coastlines along the Atlantic and Gulf coasts. According to Jean Kennedy Smith, her brothers Jack and Bobby "recoiled" from the depiction in the report of the engineering debacle along the Jersey shore.[3] The report noted that once lovely New Jersey beaches had deteriorated or effectively died due to the effect of overengineering, with hundreds of groins and miles of seawalls, all designed to keep the beaches safe for humans and their property, no matter the cost to shore life. A seawall running from Sea Bright to Monmouth Beach, two communities on a New Jersey barrier beach, constituted a particularly egregious example.

"Our Vanishing Shoreline" identified numerous vulnerable sites as priceless oceanfronts; of them, Cape Cod's Great Beach captured Kennedy's immediate attention. Since the early European settlement on the Cape, residents had built houses on high ground, purposefully away from the stormy seas and the thin soil at the coastline. As a result, Great Beach, along the easternmost coast, contained the longest undeveloped sweep of sand dunes in New England.[4] The dunes hadn't been bulldozed to enlarge the view; sand hadn't been pumped or trucked to the beach to "improve" it; and cement walls hadn't yet been engineered to block waves. "The beach is excellent in places, quite variable, and backed by cliffs 150 feet high in some sections," the NPS report noted. "The dunes are spectacular, some more than 50 feet high, the vegetative cover is seminatural but varied. The geology of the area is interesting and the history outstanding."[5]

Spurred on by "Our Vanishing Shoreline," Kennedy fought to establish Cape Cod National Seashore as a 28,645-acre park, stretching from Provincetown at the end of the cape to the tip of Nauset Beach at the elbow, a distance of about forty miles. He envisioned it as a permanently protected strip along the Atlantic side of the Cape, averaging a mile in width. There were, however, some serious hurdles to overcome: 18,000 of the acres were privately owned; 7,000 (including twenty ponds) were state-owned; 1,000 were owned by towns; and 2,500 were owned by the Department of Defense. Not all of those entities were interested in the federal park idea.

Like most of the potential national seashores, the site on Cape Cod was complicated by the presence of homes and businesses. In the case of land closest to the shore, the federal government would need to buy parcels from private citizens and companies; in addition to finding the funds, backers needed to win over suspicious New Englanders who thought that a Cape Cod National Seashore was a federal land grab.[6] The bigger problem was convincing defiant residents of affected towns to vote in favor of ceding tracts owned or controlled by those towns. As far as many of them were concerned, the establishment of a national seashore would kill the local economy.

About the same time as the NPS report, Kennedy was convalesc-ing from back surgery in Palm Beach, and drafting chapters with Ted Sorensen for *Profiles in Courage*, a book about eight US senators that would win the Pulitzer Prize in Biography in 1957. Jackie Kennedy believed the history project had "saved his life" by directing his at-tention away from his excruciating back pain.[7] After weeks of being bedridden, Kennedy was finally able to return home from Florida on crutches and then, with the help of Jackie and his aide David Powers, make his way into the Atlantic waters of Cape Cod. "He stood there feeling the warm saltwater on his bare feet," Powers recalled, "and broke into a big smile."[8]

II

In 1955, Rachel Carson finished *The Edge of the Sea*, the third volume of her ocean trilogy; in it, she explored marine mysteries along the Atlantic seaboard from Maine to Key West. That August, the *New Yorker* published two excerpts. "The edge of the sea is a strange and beautiful place," Carson's book began. "All through the long history of Earth it has been an area of unrest where waves have broken heavily against the land, where the tides have pressed forward over the con-tinents, receded, and then returned. For no two successive days is the shore line precisely the same."[9]

Three of the most memorable places Carson described in *The Edge of the Sea* are islands along the shore: Beaufort on Port Royal Island, South Carolina; Woods Hole, Massachusetts; and St. Simons Island, Georgia. At St. Simons, the curved sea bottom at low tide was a wide tidal flat. Eugene Odum, author of the popular ecology textbook, believed that that feature made St. Simons Island the ideal place to examine the biology of the seafloor. Carson's notebooks from St. Si-mons exhibited a kind of euphoria. "Walking back across the flats of that Georgia beach, I was always aware that I was treading on the thin rooftops of an underground city," she wrote. "Of the inhabitants themselves little or nothing was visible. There were the chimneys and stacks and ventilating pipes of underground dwellings, and various passages and runways leading down into darkness."[10]

To the surprise of Carson, *The Edge of the Sea* was a bestseller. Once again critics raved about her literary artistry and scientific prowess. Carson's most personal book was a new kind of ecology writing, infectious and informative. The esteemed oceanographer Henry Bigelow of Harvard University, whom Carson admired, wrote her a fan letter, saying that "plenty of people write about the sea . . . but no one else writes with the feeling of the sea that shows in your every sentence. And no one else equals the delightful way in which you express yourself."[11]

Time magazine praised Carson's prose as an "underwater ballet" brimming with "the life breath of science on the still glass of poetry."[12] Because Anne Morrow Lindbergh's book *Gift from the Sea* was number one on the *New York Times* best-seller list and Carson's effort logged in at number eight, the women were dubbed "the First Ladies of the Sea."[13] None of the adulation went to Carson's head. When Dorothy Freeman wrote her about weeping over the gushing *Time* review, Carson wrote back, "But dear—truly—I am too conscious of the many flaws in my work to accept some of these high estimates without strong reservations. They are something to try to be worthy of in the future and perhaps that is the only way I can regard them."[14] Carson told her friends William Curtis and Nellie Lee Bok that the royalties she earned from *The Edge of the Sea* would be donated to the future of preservation "for untouched oases of natural beauty that remain in the world."[15] Working with the Nature Conservancy reflected her commitment to such places.

On a rainy spring day in 1956, Carson joined a handful of concerned conservationists for a summit at the elegant Wiscasset Inn, not far from her Southport Island home. Concerned that the extensive forests along the Maine coast were being destroyed by unchecked commercial developers and that the nearby Mason Station power plant on the Sheepscot River, which was coal fired, was a menace, the group plotted an environmental counteroffensive. With an air of authority, Carson recommended that they establish a chapter of the Nature Conservancy in Maine. "It is the only group I know which is doing something practical about actually preserving areas," she told her fellow activists.[16]

With Carson leading the charge, a new TNC chapter was founded in Maine that fall. Large topographical maps of the Pine Tree State were unrolled at her home. Information was gathered about saving what became the 5,000-acre Big Reed Forest Reserve, the largest contiguous old-growth forest in New England, and the salt ponds on the shores of Muscongus Bay, where Carson had studied blue mussels and green crabs.

As a conservation ecologist, Carson took a particular interest in acquiring land for TNC within a transition zone between the eastern deciduous forests of New England and the northern boreal forest. But she was also involved in land purchases ranging from the Allagash River in the north to Casco Bay in the south. In the group's first five years, Carson raised important questions that have influenced TNC thinking ever since: What is the correct balance between habitat protection and economic development? How should the chapter set priorities? How would the chapter's land stewardship differ from that of state and federal agencies?

For years, Carson had toyed with the idea of writing a book tentatively titled *The Air Around Us*. Just months after founding the TNC chapter, she did the next best thing, writing a show on clouds and the earth's atmosphere for *Omnibus*, a public television series. "Hidden in the beauty of the moving clouds is a story that is as old as the earth itself," her text began. "The clouds are the writing of the wind on the sky. They carry the signature of masses of air drifting across sea and land. They are the aviator's promise of good flying weather, or an omen of furious turbulence hidden within their calm exterior. But most of all, they are cosmic symbols, representing an age-old process that is linked with life itself."[17]

Making the documentary increased Carson's concern about the danger of nuclear weapons testing in Nevada. The wind about which she wrote for the show was blowing fallout all over the American Southwest and beyond. She would say later, "It is surprising that so little thought seems to have been given to the biological cycling of materials in one of the most crucial problems of our time: the understanding of the true hazards of radiation and fallout."[18] People who felt that way, typified by Carson, Commoner, Douglas, and even Dr. Schweitzer, couldn't look to President Eisenhower for positive action on nuclear testing.

Rachel Carson in her favorite setting, the edge of the sea, near her home in Southport, Maine. Her 1955 book, *The Edge of the Sea*, was researched there.

It became clear during Eisenhower's 1956 presidential campaign that a reduction of testing was not even to be discussed; nor did he fancy anti-pollution laws. Quite the opposite: that June he launched his signature National System of Interstate and Defense Highways. A major rationale for pouring so much concrete from coast to coast was to provide city residents with escape routes from a possible atomic attack. Furthermore, Eisenhower was a military logistics expert; he saw the superhighway as a way to deploy military equipment proficiently within the United States. Congress allocated $26 billion to pay for the limited-access Eisenhower roadways. The Interstate Highway System was also designed to eliminate traffic congestion, replacing what one booster called "undesirable slum areas" with ribbons of concrete, and to make coast-to-coast trucking more efficient. Road engineers paid no attention to how the superhighways might adversely damage the environment or open land in low-income neighborhoods.

"All that concrete encouraged flooding, and salts and oils carried in runoff poisoned nearby ponds and streams and fostered the growth of invasive weeds," historian Earl Swift wrote in *The Big Roads*. "Rural interstates presented insurmountable barriers to small mammals, turtles, and amphibians, one study concluding that a four-lane divided highway was as much a barrier to small creatures as a body of fresh water twice as wide. The slaughter of game by auto approached, and would soon exceed, that by hunting."[19]

Around that time, Keep America Beautiful was founded in New York City by such businesses as the Owens-Illinois Glass Company and the American Can Company. Concerned that the construction of the Interstate Highway System was going to increase the amount of garbage being thrown out of car windows, Keep America Beautiful popularized the term *litterbug* to discourage the practice. In coming years, it would launch an effective TV campaign, "Every Litter Bit Hurts." The corporations worked in tandem with conservation groups such as the Izaak Walton League and the Garden Club of America to distribute recycling containers to the public. With postwar consumer products flooding the country, the well-financed Keep America Beautiful sought nothing less than garbage-free streets, parking lots, roadways, and parklands. The organization's America Recycles Day led to the adoption by myriad states of bottle-return regulations. Oregon and the New England states adopted recycling with gusto.

III

For the 1956 Democratic National Convention in Chicago, Kennedy had been asked to narrate the party's campaign film, "The Pursuit of Happiness"; he also delivered a nominating speech for former Illinois governor Adlai Stevenson.[20] That introduced him on a national stage. Stevenson, influenced by Norman Cousins, ran his campaign on a nuclear test ban platform. With his detail-oriented mind and moral sincerity, he tried to educate the public about why nuclear tests in Nevada were health hazards and he called for an outright moratorium on testing in the atmosphere.[21] After Stevenson secured the Democratic nomination, he left the choice of a vice presidential nominee to the convention. Kennedy came close to winning the frenzied competition, but bowed out smoothly when the momentum shifted to Senator Estes Kefauver of Tennessee.

The unexpected conservation policy star in Chicago was Harry Truman. After being out of the White House for almost four years, he criticized Eisenhower sharply for abusing public lands. The extraction philosophy of Interior Secretary Douglas "Giveaway" McKay, in particular, irked the ex-president. "For three and a half years the Eisenhower administration has been using every trick and device," Truman charged in his speech, "to pry our water power and our forests, our parks and oil reserves out of the hands of the people and into the pockets of a few selfish corporations."[22]

The Stevenson-Kefauver ticket was defeated in November, but the campaign had elevated Kennedy's brand. In the years leading up to the 1960 race, he generated excitement on a national level on par with Lyndon Johnson and Hubert Humphrey. Thinking ahead to that election, Kennedy pushed a "third-wave" conservation strategy, one that built on the White House records of the two Roosevelts. He proposed expanding the National Park Service (especially by adding seashore units) and designating millions of acres as protected wilderness areas, mainly in the West. Though it made sense for a Massachusetts senator to promote a national seashore on Cape Cod, his plan was at once a foray into national leadership and the politics of western water rights.

Eisenhower replaced Douglas McKay at the beginning of his second term. The new interior secretary was a Nebraska newspaper publisher, Fred Seaton, who had formerly been a senator and assistant secretary of defense. Of medium build, with blondish hair and piercing blue eyes, Seaton was an affable man, interested in outdoor recreation. There was no mistaking the fact that he favored the Bureau of Reclamation, but he also sought to protect the National Park Service.[23] Even as the Alaska territory vied for statehood, Seaton hoped to establish large national wildlife refuges there.

Conrad Wirth, an Interior Department leader who had been promoting national seashores since the New Deal era, had become the NPS director in 1951. Increasingly thereafter, he made shoreline protection a priority. He turned to Paul Mellon, heir to the Mellon bank fortune, to financially seed preservation programs. Mellon, more of a "beautification" philanthropist than a wilderness warrior, had largely funded the establishment of the White House Rose Garden, the landscaping of nearby Lafayette Park, and the preservation of Cumberland Island, Georgia. Prodded by Wirth, Mellon used his Old Dominion Foundation and Avalon Foundation to fund the NPS's crucial report on the seashores in 1955. His priority was establishing new national parks along the Atlantic seaboard, but he subsequently agreed to pay as well for an aerial survey of the entire 5,500-mile American shoreline of the Great Lakes.

The Great Lakes shoreline survey was finished in late 1958, resulting in a report called *Remaining Shoreline Opportunities in Minnesota, Wisconsin, Illinois, Indiana, Ohio, Michigan, Pennsylvania, New York*. It was a game changer. When the survey was undertaken, there were only three national parks in the region: Isle Royale National Park (Michigan), Grand Portage National Monument (Minnesota), and Perry's Victory National Memorial (Ohio). The visionary report made thirteen specific recommendations for new parks while arguing that 15 percent of the Great Lakes shoreline be permanently protected, by whatever means.

Two of the recommendations were for areas in the Great Lakes region of Michigan: Pictured Rocks, on the shore of Lake Superior in the Upper Peninsula, and Sleeping Bear Dunes, not far from Traverse City in the Lower Peninsula. The colorful sandstone cliffs at Pictured Rocks, along with its sublime waterfalls, rock arches, sea caves, and forty-three miles of shoreline, were among the most gorgeous natural features in the lower forty-eight states, the report said. When frothing water tumbled into Lake Superior over the cliffs of the Munising Formation in the spring or across snow-blanketed Beaver Creek in the winter, Pictured Rocks was magical.

Sleeping Bear Dunes, situated along Lake Michigan, was just as gorgeous, according to the report. With its thirty-five miles of pristine shoreline, the proposed park also included two islands, North and South Manitou, replete with forests, sandy beaches, and ancient glacial wonders. Among Sleeping Bear's most phenomenal features were its dunes, which towered four hundred feet above the lake. Formation of the dunes varies with the weather, as *Shoreline Opportunities* explained, giving the proposed park a dynamic appeal. The wind off Lake Michigan picks up grains of sand and carries them along until irregularities in the surface—perhaps grass—slow the wind and they come to rest. At that juncture the accumulated pile of sand acts as an obstacle to the wind and captures more sand, incrementally building a dune.

The leading promoter of the establishment of Sleeping Bear Dunes and Pictured Rocks was Michigan's junior senator, Phil Hart. Born on December 10, 1912, to a wealthy family in Bryn Mawr, Pennsylvania, he earned a bachelor's degree from Georgetown University in the 1930s and then a law degree from the University of Michigan. For a few years, he practiced law in Detroit. In 1941 he joined the Army as a lieutenant colonel in the infantry. He was wounded on D-Day at Utah Beach in Normandy. After the war, he returned to Michigan and held numerous state government positions. Hart's wife, Jane, came from an old Detroit family, strengthening his ties to his adopted state. A skilled aviator, Jane Hart was the first female helicopter pilot in Michigan and later qualified as a Mercury 13 astronaut (though due to gender bias women were denied further advancement in the space program).

Phil Hart became a senator in 1959, a social justice intellectual whose heroes were Martin Luther King, Jr., and Franklin Roosevelt. Four of Hart's Senate colleagues—Clinton Anderson of New Mexico, Paul Douglas of Illinois, James Murray of Montana, and Richard Neuberger of Oregon—sponsored S. 2460, a shoreline preservation bill, promoted under the slogan "Save Our Shorelines" (or simply "SOS"). Those conservation-minded lawmakers proposed a $50 million appropriation to establish ten shoreline recreation areas, on both sea and lake. Topping the list was Kennedy's beloved Cape Cod. Others were Padre Island, Texas; Point Reyes, California; the Oregon Dunes; the Indiana Dunes; and both Pictured Rocks and Sleeping Bear Dunes in Michigan. "Nearly all the great National Parks of the United States are in mountain ranges," Senator Neuberger told the *New York Times*. "In the process of setting aside these magnificent upland reserves, the nation has neglected another realm which is equally alluring to the tourist and the seeker of outdoor recreation. This realm consists of the seacoasts and shorelines of the United States, which are among the most beautiful on earth."[24]

IV

On August 19, 1958, Bill Douglas set out on yet another high-profile hike, this time to a secluded Pacific Ocean beach in Olympic National Park. His mission was to stop a future roadway planned for the quiet

northern coastal corner of his home state. Douglas led thirty conservation luminaries on a three-day, twenty-two-mile environmental
protest hike from Cape Alava, the westernmost point in the forty-
eight contiguous United States, south to the Quillayute River. Looking like a weather-beaten tramp in his thrift-store walking clothes,
including a shapeless cap that was so floppy that it slopped over on
one side of his face, Douglas made for good newspaper copy. If the
highway extension was built, Douglas and company argued, it would
wreck the longest unbroken coastline in the continental United States,
while also endangering the habitat of the northern spotted owl and
the marbled murrelet. The *Seattle Post-Intelligencer* reported, "The
hiking party hopes to demonstrate the irreplaceable value of the land
and advocate its preservation."[25]

As a Pacific northwesterner, Douglas had every right to protect the
rushing rivers, temperate rain forests, tidelands, and saltwater beaches
of the Ocean Strip portion of Olympic National Park. Writing about
this windblown coast, he said, "I like to lose myself in the solitude of
this beach, the solitude that no automobile can puncture."[26] At a press
conference, he spoke reverentially about his cabin on Quillayute River
near Rialto Beach. Keeping the beach roadless was the "last chance
to save a piece of our seashore free of the noise, smell, and sight of
automobiles."[27] Douglas's hike and salmon barbecues seized public
support. The roadway plan was abandoned.

Along the way, Douglas had used the protest to let the press know
that a group of Democratic senators, Kennedy and Neuberger among
them, supported his roadless preservation cause in Washington State.
In November, Kennedy won reelection, carrying every city in Massachusetts and winning 73 percent of the overall vote. Conservationists
were thrilled that Kennedy had matured into a leader of the Cape
Cod effort and the Wilderness bill. Another scion of a rich eastern
family, Laurance Rockefeller, also emerged with the will to confront
environmental problems. Born May 26, 1910, he was the fourth child
of John D. Rockefeller, Jr. On the coffee table in the Kennedy family's
Cape Cod living room for decades to come was the 1957 book by
Nancy Newhall, *A Contribution to the Heritage of Every American: The
Conservation Activities of John D. Rockefeller, Jr.* It documented how the
famous oil mogul's son had donated many millions for the Sugar Pines

addition to Yosemite; interpretive museums at Yellowstone, Grand Canyon, and Mesa Verde, and expansion of national parks at Acadia, Grand Teton, Jackson Hole, the Great Smokies, and the Shenandoah Valley. In 1954, California, after three decades of effort, acquired the South Grove Trail of Calaveras Big Trees, and it was the Rockefeller gift of $1 million that made that possible. Newton B. Drury wrote in the *Sierra Club Bulletin,* "Those who had the privilege of dealing with John Rockefeller Jr., in these enterprises were impressed not only by his generosity but also by his deep understanding of the importance of this cause, by his sensitivity to landscape beauty, and by the meticulous care that he took personally to master all the details involved."[28]

All six of Rockefeller Jr.'s children cared about the preservation of American landscapes and cultural resources. But none more than Laurance "Larry" Rockefeller, who was born in New York City on May 26, 1910. After graduating from Princeton in 1932, he married Mary Billings French, in Woodstock, Vermont. (In 1992, they donated their summerhouse and farm there to the National Park Service, dedicating it to US conservation history.) The couple made saving American landscapes the pillar of their philanthropy. "Iconic American landscapes spoke to him," his business associate Douglas Home recalled. "His idea of happiness was to get friends on horseback, fish for trout, and get back to the wonders of nature."[29] It was Rockefeller who spurred the Interior Department to conduct three seminal Eisenhower-era surveys aimed at establishing new parks, *A Report on the Seashore Recreation Area Survey of the Atlantic and Gulf Coasts, Our Fourth Shore: Great Lakes Shoreline Recreation Area Survey,* and *Pacific Coast Recreation Area Survey.*[30]

In many ways, the eclectic LR—a well-liked, moderate Republican, venture capitalist, and rangy outdoorsman with a lively sense of humor—invented the idea of ecotourism. Soft-spoken with a narrow birdlike nose, he was suspicious of technological hubris and sprawl. The gist of his philosophy was that ecosystems needed to be preserved for wildlife to flourish and recreationists to enjoy. "What you try to do is to share with other people your own personal experiences without destroying the things you want to share," he said.[31] At his JY Ranch in Wyoming, visitors were given cabins with stunning views of the Tetons across grounds teeming with moose, deer, and raptors in the sky.

Laurance Rockefeller's most ambitious project focused on the US Virgin Islands. Enthralled by their lush tropical beauty yet concerned about the industrialization taking place there, he bought five thousand acres on St. John and built a luxury resort with seven beaches on Caneel Bay. On the property was Mary's Trail, named after his wife, which is considered among the finest hikes in the Caribbean. Home to abundant wildlife as well as healthy coral reefs that teemed with marine life, Caneel Bay offered a new kind of luxury: not ritzy but simple and becalmed, so that guests did not disturb the paradisiacal island atmosphere. On December 1, 1956, he donated all of his holdings on St. John to the National Park Service. In coming years, he would make more donations, in the form of NPS inholdings (private lands within park boundaries).

When Congress established the Outdoor Recreation Resources Review Commission (ORRRC) in early 1958, Rockefeller was asked to serve as chairman.[32] Though the commission was saddled with a ponderous, bureaucratic name, conservationists placed exultant hope on the report it was preparing, to be presented to the president and Congress on January 31, 1962. Key members of the commission were four senators: Clinton P. Anderson, Democrat of New Mexico; Henry C. Dworshak, Republican of Idaho; Henry M. "Scoop" Jackson, Democrat of Washington; and Jack R. Miller, Republican of Iowa. The conservation-driven congressman John Saylor was also a member. The task force sought to find innovative ways to protect natural resources and maintain open spaces near urban areas. Rockefeller was the ideal leader for the assignment. Open to fresh ideas, an amateur nature photographer, and an enthusiastic equestrian, he got along with everybody. He understood from his family's Standard Oil wealth that petroleum companies actually *loved* national parks: tourists, driving long distances to remote places, generated filling-station business. With ardor and a sense of grand purpose, Rockefeller and the commission undertook an inventory of the United States' national outdoor recreational resources to determine what the needs would be by 1976 and 2000; then they recommended conservation policies to meet those anticipated needs.[33]

With homes in Woodstock, Vermont, and Jackson Hole, Wyoming, Rockefeller knew both eastern and western wilderness-related issues. He was never keen on the damming of rivers, and his trademark characteristic was his refusal to be impetuous; he liked to weigh things before deciding on them.

Jack and Jackie Kennedy visited the Caneel Bay Resort, and Jack plotted with Rockefeller as to how the Interior Department could acquire Buck Island, a star of the Caribbean adjacent to St. John.[34] Meanwhile, Save Our Shorelines (SOS) was becoming a full-fledged movement. The Sierra Club joined the effort to save shorelines, devoting the September 1958 issue of the organization's *Bulletin* to protecting a gorgeous stretch of the Marin County coast just north of San Francisco.[35] The burgeoning interest in ocean life continued to be promoted by Rachel Carson and Jacques Cousteau. Archie Carr's 1959 book, *The Windward Road*, which documented the plight of green sea turtles in the Caribbean and Central American waters, likewise garnered media attention. Carr lobbied for laws to protect sea turtles on US shores by banning dune buggies from Florida's coastline in order to protect turtles' nests.[36]

Carr was born on June 16, 1909, in Mobile, Alabama. Like Carson, his academic focuses were zoology and marine biology. At the University of Florida, he earned his undergraduate and master's degrees and, in 1937, a doctorate degree. He stayed in Gainesville to teach ecology, with a specialty in sea turtles. Carr's Community Ecology course was extremely popular because students went on field trips to exotic places such as the Sand Pine scrub near Ocala, Florida; the Okefenokee Swamp in Georgia; and even Costa Rica. Regularly, Carr took students to watch sea turtles return to land for the spawning season.

Carson admired such books by Carr as *Handbook of Turtles: The Turtles of the United States, Canada, and Baja California* (1952) and *High Jungles and Low* (1953). She learned of the primary spawning sites of sea turtles, including Padre Island, Texas, which "Our Vanishing Shoreline" had recommended for national-seashore status. In *The Edge of the Sea*, Carson wrote about loggerhead, green, and hawksbill turtles with awe, describing how they emerged from the ocean and lumbered "over the sand like prehistoric beasts to dig their nests and bury their eggs."[37]

Joining the SOS cause, Carson published an article titled "Our

Ever-Changing Shore" for *Holiday* magazine, expressing her deep concern that Americans were abusing their beaches and estuaries. Lamenting that only 6 percent of the Atlantic and Gulf coastline was in federal or state hands, she took up the cause of saving the "wilderness of beach and high dunes where Cape Cod, after its thirty-nine miles thrust into the Atlantic, bends back toward the mainland" and the "rocky cliffs and headlands" of Oregon. "In every outthrust headland, in every curving beach, in every grain of sand, there is a story of the earth," she wrote.[38]

Jack Kennedy spoke about preservation before the California legislature in Sacramento in the spring of 1959. "It is an impressive experience to come from the heart of Massachusetts to the heart of California," he said. "Much is different—much is the same. We border the smoky North Atlantic—you border the blue Pacific. You send us tuna—we send you lobster. Our rivers flow unchallenged into the sea—your rivers are being harnessed to the benefit of all your citizens." Although Cape Cod National Seashore was first on his wish list, he championed California's Point Reyes as second in line.[39]

In early June, he spoke out on behalf of the Potomac River, infuriated that a river so celebrated in American history had been turned into a stinkhole dump for raw sewage and industrial sludge. Kennedy argued that the unregulated discharges were damaging the ecological balance of the waterway and potentially led to toxic algae blooms. He favored a bill authorizing a $75 million appropriation to build sewer lines and filtration facilities. "The Potomac is the most polluted river west of the Nile," he fumed.[40] Starting in 1959, he demanded swift federal action to clean up more of America's polluted waterways. Calling rivers a "most precious treasure," much of which "we have failed to use properly," he publicly asserted that "we must pay increased attention to the problem of water pollution."[41]

V

During the 86th Congress, the first bill to establish a Cape Cod National Seashore was filed by representatives Edward P. Boland,

Thomas P. "Tip" O'Neill, and Philip J. Philbin, all of Massachusetts. Kennedy planned to shepherd the bill through the Senate. On September 3, 1959, he officially cosponsored the bill with his Republican colleague from Massachusetts Leverett A. Saltonstall. If established, Cape Cod National Seashore would be the first national seashore of any size designated within a populated area. Kennedy helicoptered around the Outer Cape to highlight the land issues, stating his hope that Thoreau's "white, sandy, and very bold shore" would be saved from coastal development for eternity.[42] Saltonstall's legislative assistant believed that Kennedy, with eyes on the White House, was "playing to a national gallery."[43] That was probable, but Kennedy was entangling himself in an issue that was highly controversial among his constituents on the Outer Cape. It would take all of his political magic and muscle to win their favor for the park.

While Kennedy lobbied in Washington for the NPS designation, Henry Beston, working as a lecturer at Dartmouth College, toured the nation, advocating for the protection of beachfront in Massachusetts. At the same time, he donated his Outermost House to the Massachusetts Audubon Society.[44] (It was destroyed in a storm in 1978.)

Throughout 1959 and 1960, public meetings of officials, boards of selectmen, and residents were held across the Cape. Critics insisted that a Cape Hatteras–style federal park would stunt commercial development in coastal Massachusetts. Local builders Charles Frazier of Eastham and Tony Duarte of Truro claimed that the federal designation would adversely affect the economies of towns within the proposed park, notably Provincetown, Truro, and Wellfleet. It was one thing to save the remote sandstone cliffs, deep canyons, and rugged mountains in the West, they said. Cape Cod, by contrast, boasted a string of fast-growing towns of more than ten thousand people each. Real estate values were up and the locals liked it that way. They perceived the national seashore as economic colonialism, elitism, and insensitive to real estate developers.

Kennedy took part in Senate negotiations that addressed the number of acres to be turned over to the Interior Department, the park boundaries in each town that would be affected, and land parcels that towns hoped to exclude from the proposed seashore. When he himself bought a white frame house on Irving Avenue in Hyannis Port, not far from his parents' compound, the press happily noted that Cape Cod was his home, in every sense. Posed photographs of the presidential aspirant sailing in Nantucket Sound and golfing in Hyannis Port appeared in national periodicals. At Kennedy's behest, Brower testified before a Senate subcommittee in October 1959. "There is not really too much choice before us," he stated in a press release that Kennedy undoubtedly appreciated. "We adults are but stewards of America's wilderness and natural scenery. If we want to pass this heritage on to our children, then we must act now to save the beauty of Cape Cod."[45]

When the Senate subcommittee hearings on the proposed Cape Cod National Seashore ended that December, the expectation was that the Kennedy-Saltonstall bill would pass. Even so, drafting various land easements and completing myriad other legal details remained tricky business. Determining who owned the constantly shifting sand dunes was legally complex. If Kennedy and the Audubon Society, which ran a bird sanctuary on Massachusetts Bay close to Wellfleet, hadn't kept pushing for its passage, the bill might have stalled.

Somewhat mystically, another force coming to Kennedy's aid for passage of the national seashore act was the ghost of Henry David Thoreau. The *New York Times* columnist Caroline Bates renewed the book's currency in "Walking in Thoreau's Footsteps on Cape Cod," in which she embraced the national seashore idea. "Development is now reaching out to this section of the Cape, but if it becomes a National Seashore Recreational Area, as proposed, one of the country's most magnificent stretches of unspoiled beach will be saved," she wrote. "Few coastal areas have escaped the encroachments of civilization, and even fewer retain the primitive grandeur of the Great Outer Beach. The adventurer in search of wild America visits this coast for the same reason that Thoreau did."[46]

Bill Douglas was especially proud of JFK in 1959, as the young senator championed Cape Cod National Seashore, the National Wilderness Preservation System, and a Clean Air Act. Douglas had assessed his young friend as "a playboy in public office" up to 1958."[47] Now Kennedy was a conservation leader. In the coming months, the Massachusetts senator would consult with his father and Douglas about whether the timing was right for a run for the White House in 1960. The day Jack decided to run, Joe Sr. walked across Capitol Hill from the Senate Office Building to Douglas's Supreme Court chambers and said, "Bill, the great decision has finally been made. Jack is good to go for the presidency. You can bet your ass on that."[48]

CHAPTER 7

Protesting Plastics, Nuclear Testing, and DDT

Barry Commoner working in his Washington University office in St. Louis in the early 1960s. He became one of America's top anti–nuclear testing and anti-DDT activists.

I

To the consternation of Barry Commoner, World War II had caused a dangerous uptick in the chemical-manufacturing business. DDT was the most notorious, but innovations in synthetic rubber, plastics, other pesticides, and artificial fibers such as rayon and nylon were being introduced on an accelerating basis. Coal tar, from which the first such synthetics had been derived, was demoted in favor of the by-products of cracking procedures used to process petroleum, part of the era's enormous expansion of the petrochemical industry. Back then, the media, by and large, didn't question whether the innovations had any possible negative side effects. Instead, postwar culture celebrated "Better things for better living through chemistry," a catchphrase coined by a DuPont chemist in 1938 after he discovered the compound called polytetrafluoroethylene, or PTEF, which became Teflon.[1] If there hadn't been vigorous Food and Drug Administration studies on the long-term health consequences of these new synthetic chemicals, they would have gone entirely unregulated. As it was, the cascade was twofold: new chemicals and new processes to produce them. Either one or both could be detrimental to the natural world.

One particularly harmful postwar industrial ingredient arrived under the group name "phosphates." An innovation in making detergent, they are derived from bone. Since the time of Julius Caesar, soap had been made by mixing animal fat (organic) with wood ashes (inorganic) to produce a water-soluble substance (soap) able to clean grease-coated surfaces. That kind of soap was biodegradable and safe to use. By contrast, phosphates, compounds containing phosphorous, oxygen, and sometimes hydrogen, are effective detergents but wreak havoc on lakes and rivers by turning waterways green with thick algae blooms that kill aquatic plants. When phosphates run off from detergents, fertilizers, and other man-made substances, the result is excessive plant/algae growth, which sets off a chain of events that eventually suffocates fish, other animals, and even the native underwater plants. When the nutrient level is too high in that way, the system is called *eutrophic.*

Plastic production, another postwar growth industry that troubled Commoner, had negative environmental effects, too, but received little scientific attention at first.[2] During the Great Depression, plastic had been used in only a few consumer products, such as living room furniture and radios. After Pearl Harbor, new ways to employ plastics were developed rapidly as part of the war effort. Soon plastics were used to make nylon parachutes, lightweight airplane parts, toothbrushes, and combs. Following V-J Day, US companies discovered clever new ways to mass-market plastic products. Some, such as syringes and test tubes, had social benefits. The new wave of postwar children, dubbed baby boomers, took to plastic-molded toys: popular items such as hula hoops, Frisbees, Barbie dolls, and Lego blocks. Cloth diapers, ceramic plates, and water glasses were replaced by single-use items. Formica-topped tables and counters were the fad in diners and cafes. Baby bottles were made with polyethylene. Plastic straws became ubiquitous. Vinyl kitchen floors were the homeowners' new preference. Garbage pails and laundry baskets made of plastic were mass-marketed as consumer musts to help alleviate the drudgery of housekeeping.[3] *Life* magazine, in a sexist celebration of "throwaway living," gleefully raved about the "liberation of the American housewife from drudgery" thanks to the miracle of plastics.[4]

Unfortunately, some of the modern compounds grouped under the umbrella term *plastic* aren't biodegradable. Though it's true that many plastics are biodegradable over a long half-life, others aren't; it depends on the availability of microbes that can break them down. Some plastics won't disappear for about 450 years. Millions of tons of plastic ended up in lakes and the ocean annually, killing multitudes of dolphins, whales, seabirds, seals, fish, and sea life of all kinds. No matter how arduously Rachel Carson, Henry Beston, Archie Carr, and Jacques Cousteau sought to educate citizens in preventive science regarding the oceanic world, they were shouting into a maelstrom of commercialism. Americans yearned—or were conditioned to yearn—for quicker, easier disposable products, while luxuriating in what the novelist John Updike called "that sweet tangy plastic new-car smell."[5]

A challenge facing postwar protoenvironmentalists was how to persuade consumers to think like a part of nature and thus question the true desirability of chemicals such as polychlorinated biphenyls (PCBs), for example, and petroleum distillates. How to convince the Department of Defense that the dispersal of ammonium perchlorate, an additive mixed into rocket fuel and munitions, threatened groundwater systems? How to convince fire departments that flame retardants, known as polybrominated diphenyl ethers, were poisoning mothers and newborns throughout the United States? Commoner and his colleagues weren't opposed to these new products per se, but they thought that increased scientific research was needed to determine how the chemicals and plastics could be made and used safely.

Parallel with consumer pollutants, the atmospheric testing of nuclear weapons grew as the arms race between the United States and the Soviet Union accelerated. Citizens around the globe, as well as members of Eisenhower's inner circle, harbored increasing doubts about the sanity of atmospheric and underwater testing. A burgeoning "Science of Survival" movement led by Commoner claimed that humans were absorbing quantities of strontium-90 and iodine-131 from fallout that were significant and potentially lethal.[6]

The great awakening occurred on April 26, 1952, when physicists monitoring radioactivity at the Rensselaer Polytechnic Institute in Troy, New York, detected an unexpected surge in radiation. "The surge, associated with a deluge of rain, was determined to be radioactive debris—fallout—from nuclear tests in Nevada thirty-six hours earlier that had blown across the country and been brought to earth by heavy rain," the historian Michael Egan explained in *Barry Commoner and the Science of Survival*. That meant that fallout was descending on the Earth much faster than the AEC was admitting, Commoner warned.

By 1957, scientists like Commoner knew that strontium-90, a radioactive isotope produced during hydrogen bomb detonations, had entered human bodies through milk consumption. Dairy farmers in Wisconsin and Iowa suddenly feared that their agribusiness was going to be destroyed. The AEC's nonchalant response only worsened the public health scare. Nobody could agree on what constituted a safe level of strontium-90. The fear was that if you drank milk regularly—as most children do—you were being poisoned with a cancer-producing substance that would accumulate in your bones and teeth and eventually lead to premature death. Given that hazard, many people began to believe that a global ban on nuclear atmospheric and underwater testing was desperately needed.[7]

An anti–nuclear testing movement spontaneously sprang to life. With Norman Cousins leading the charge, peace activism merged easily with other concerns motivating ecologists, preservationists, and public health officials. Borrowing against his stock in the *Saturday Review* magazine, Cousins contributed $50,000 to establish the National Committee for a Sane Nuclear Policy (SANE), whose mission was to ban atomic testing globally.[8] Its genius lay in seeking direct citizen involvement all over the world. Cousins lobbied Eisenhower, Stevenson, Kennedy, and other Washington power players to impose a ban. In the late 1950s, the only thing holding up a test ban treaty involving the United States, Great Britain, and the Soviet Union was

Close-up of *Saturday Review* editor and anti-nuclear activist Norman Cousins. His organization SANE led to the effort to ban the nuclear weapons testing in the atmosphere and underwater.

the number of on-site inspections that the Kremlin would allow. A suspicious Nikita Khrushchev was convinced that inspections would allow the United States to spy.[9]

In April 1957, when the first US nuclear power plant opened in Fort Belvoir, Virginia, there was little national cheering. Quite reasonably, many citizens didn't want nuclear power plants in their backyards, no matter how much energy the facilities produced. (A decade after SANE was founded, fourteen nuclear power plants were operating in the United States, most constructed with a heated "not in my back yard"—NIMBY—environmental protest as a backdrop.) While SANE's primary aim was to ban nuclear testing, Cousins dreamed of stuffing the whole nuclear genie back into the bottle.[10]

Dr. Martin Luther King, Jr., took the side of Cousins and even signed a SANE advertisement. Asked in early December 1957 for his views about nuclear weapons, King didn't mince words, saying:

> I definitely feel that the development and use of nuclear weapons should be banned. It cannot be disputed that a full-scale nuclear

war would be utterly catastrophic. Hundreds and millions of people would be killed outright by the blast and heat, and by the ionizing radiation produced at the instant of the explosion. . . . Even countries not directly hit by bombs would suffer through global fall-outs. All of this leads me to say that the principal objective of all nations must be total abolition of war. War must be finally eliminated or the whole of mankind will be plunged into the abyss of annihilation.[11]

Not too coincidentally, Dr. King and Carson had the same literary agent at the time: Marie Rodell, an ardent liberal and protoenvironmentalist who believed fervently in a ban on nuclear weapons. Like any agent, she sought out people with something to say, but in her case, she wanted the cause to be something with which she wholeheartedly agreed. She would have influenced both clients, if they needed it, to think in terms of banishing the bombs.

If there were a founding document for the emergence of what came to be known as the Science Information Movement, it was Barry Commoner's sobering speech in 1957 at a meeting of the American Association for the Advancement of Science in Indianapolis. It presaged the preference of protoenvironmentalists for making their own decisions and for basing them on raw facts, rather than on childlike trust in government. Commoner discussed the duty and obligation that scientists had to inform the public about the danger of strontium-90 pollution from nuclear fallout. In many ways, it was the much-needed refutation of a famous and erroneous statement by the Manhattan Project physicist Edward Teller: "Fallout from testing is dangerous only in the imagination."[12]

Refuting Teller, who is considered to be "the father of the hydrogen bomb," Commoner turned his Indianapolis speech into an article in the magazine *Science* the next year.[13] In "The Fallout Problem," he not only outlined the long-term hazards of global radiation from nuclear weapons testing but also skillfully connected public policy advocacy with scientific expertise. Yet he shrewdly claimed that it wasn't the job of scientists alone to argue for the banning of nuclear testing; instead, a coalition between socially engaged scientists and an informed public was humanity's best hope to "conserve the environment and preserve life on Earth."[14] Commoner took steps to provide the kind of information the public needed. Convinced that strontium-90 was responsible for a spike in leukemia, bone cancer, and other diseases, he launched a program called the Baby Tooth Survey at Washington University in St. Louis. Dr. Louise Reiss, assisted by her husband, also a physician, partnered with the school of dentistry at the university to start collecting children's baby teeth. Scientists had come to suspect that radioactive fallout from nuclear testing was embedded in America's food supply and eventually wound up in human teeth and bones. The St. Louis group was collecting tangible data to settle the question. "Dr. Louise Reiss and her assistants have had extraordinary success in getting local schools—public, private, parochial—to help in the teeth collection," the *Nation* reported in June 1959. "Some 250,000 forms had been distributed to reach all lower-grade students."[15]

Baby teeth arrived at Louise Reiss's office on West Pine Boulevard at the rate of fifty a day. Each arrival was treated like a rare jewel. It was public science—not politics—at its very best. To encourage children to donate baby teeth (before the tooth fairy flew off with them), Dr. Reiss had a button made with a smiling Dennis the Menace–like kid missing his front teeth, the words I GAVE MY TOOTH TO SCIENCE encircling him. Dentists would give their young patients the button. Reiss reflected in 1996, "I continue to be moved by the knowledge that a group of organized people can effectively pressure government if they come up with data instead of rhetoric."[16]

While Dr. Reiss collected baby teeth with an eye on strontium-90, Linus Pauling, a Nobel laureate in physics and opponent of nuclear testing, believed that human tissue also had to be studied for exposure.[17] Nevertheless, in the Red Scare climate of the McCarthyite 1950s, Pauling was attacked as a Communist for warning about the health perils of fallout. The Senate Subcommittee on Internal Security demanded that Pauling turn over the names of fellow scientists who had helped him circulate an anti-nuclear petition.[18]

The big question was how to force the United States, the USSR, and Great Britain to end nuclear testing. In January 1957, Norman Cousins journeyed to French West Africa to visit Albert Schweitzer. Cousins's companion on the trip was Clara Urquhart, a photographer who later collaborated with him on the 1960 book *Dr. Schweitzer of*

Lambaréné. They arrived at Schweitzer's jungle hospital in early 1957 carrying with them letters of "good wishes" from President Dwight Eisenhower and Indian prime minister Jawaharlal Nehru. It would prove to be a fruitful pilgrimage.

For starters, Schweitzer had manuscripts locked in a trunk and he allowed Cousins to read them. "Dr. Schweitzer had written his book [*The Kingdom of God and Primitive Christianity*] in longhand on the reverse side of miscellaneous papers," Cousins recalled. "Some of them were outdated tax forms that had been donated to Lambaréné by the French colonial administration. Some were lumber requisition forms used by a lumber mill not far away on the Ogowe River. Some came from old calendars."[19] Schweitzer agreed to let Cousins publish a section of the manuscript in the *Saturday Review.*

Their negotiations didn't end there. Cousins knew that the Nobel Peace Prize laureate was, according to a Gallup Poll, the fifth most admired man in the world. He told Schweitzer that if he should denounce nuclear testing by the great powers, every country in the United Nations would have to take notice. Schweitzer was reluctant. "All my life I have carefully stayed away from making pronouncements on public matters," he said. ". . . I have tried to relate myself to the problems of all humankind rather than to become involved in disputes between this or that group."[20]

But Cousins, charming and persistent as ever, his heart full of goodwill, stayed on it. Feeding baby goats bleating outside his thatch-roofed headquarters and pondering how he could be most effective as a peace activist, Schweitzer suddenly burst into a wide smile. He had made a decision. Instead of writing an article or issuing a press release, he would speak to the world against continued nuclear testing on Radio Oslo. Beside himself with gratitude, Cousins had achieved the two African objectives: Schweitzer had agreed both to publish an article in the *Saturday Review* and to be the high-profile leader in the global anti–nuclear testing movement.

When Schweitzer delivered the Radio Oslo appeal on April 24, 1957, he took his audience by the hand and explained, in what he called a "Declaration of Conscience," the horrors of open-air nuclear testing. Referring to Marie Curie's pioneering late nineteenth-century experiments on radioactivity, which eventually led to her death from radiation poisoning, he stated that strontium-90, descending in rain and snow in fallout from nuclear tests, was lethal to all God's creatures. "The radioactive elements in grass, when eaten by animals whose meat is used for food, will be absorbed and stored in our bodies," he explained. "In the case of cows grazing on contaminated soil, the absorption is affected when we drink their milk. In that way, small children run an especially dangerous risk of absorbing radioactive elements." As Cousins had hoped, Schweitzer stated flat out that open-air testing by the United States, Great Britain, and the Soviet Union had to end. "We must muster the insight, the seriousness, and the courage to leave folly and to face reality," he implored.[21]

Clearly, Schweitzer didn't believe that the qualitative leap of post-war technology was a victory for human civilization over nature.[22] Reverence for any creature, no matter how seemingly insignificant to humans, was the starting point and foundation of his moral code. People had to respect every living thing, from ant to zebra, because every member of every species, in the end, is one of God's creatures. The US tests of nuclear weapons in the Nevada desert and the Soviet tests in Novaya Zemlya were unethical, demonic, and foolish. "We stroll around in thoughtlessness," Schweitzer said on Radio Oslo. "It must not be that we will not pull ourselves together in good time and summon up the reason, the seriousness, and the courage to renounce it and concern ourselves with reality."[23]

Schweitzer dubbed nuclear testing the "elixir of death"; he felt the same way about the adverse effect of DDT on wildlife. But he could lead only one global crusade at a time. At his hospital, however, he had banned using chemical pesticides even though insects were a problem. Anti-DDT signs were posted throughout the grounds. One read, "Do not use insecticides for killing the poor creatures. Invite them to take a walk in nature. Insecticides are dangerous for your health."[24]

Concern about DDT was growing in the United States in 1957. Some months after Schweitzer's broadcast, landowners in Long Island, New York, brought a far-reaching lawsuit seeking a ban on the aerial spraying of a blend of DDT and fuel oil that affected over 12,000 square miles in the Northeast. The practice was an attempt to eradicate the gypsy moth, whose caterpillars feed on trees, killing them in alarming numbers. To fight the scourge of the insects, crop planes would spray up to fourteen times a day. Carson telephoned sources and friends to find out what was happening on Long Island, in particular. The leader of the grassroots anti-spraying action—with whom she connected through her agent as an intermediary—was Marjorie Spock, a respected teacher and follower of Rudolf Steiner's theories of biodynamic gardening. Her older brother, Dr. Benjamin Spock, was the most famous pediatrician in the United States; his book *The Common Sense Book of Baby and Child Care* would sell more than 50 million copies.

Marjorie Spock was born on September 8, 1904, in New Haven, Connecticut. Blessed with an open mind and a penchant for alternative education, she moved to Switzerland to study and work with Steiner. In her thirties, she returned to the United States and earned degrees in education from Columbia University. Spock was living on Long Island with her friend Mary "Polly" Richards, who suffered from multiple chemical sensitivity disorder, which forced her to eat only simply grown fruits and vegetables—those that we today call "organic." Marjorie grew chemical-free produce in her garden. The aerial spraying of DDT contaminated the organic produce and threatened Richards's health in a serious way. In full protest mode, Spock and Richards signed a petition to halt the US government from poisoning their backyard from the air. Spock charged that she had a constitutional right to garden as she liked without government chemical intervention. Throughout the three-year legal ordeal, she wrote impressive reports of the court proceedings and became a spokesperson for banning DDT. She was joined in the suit by Robert Cushman Murphy, an ornithologist retired from the American Museum of Natural History in Manhattan, and Archibald Roosevelt, a son of President Theodore Roosevelt.

The first indication of Carson's discontent with DDT had come in 1945, when she had suggested an article (to no avail) on the subject for *Reader's Digest* after scrutinizing wildlife research done by Clarence Cottam and Elmer Higgins. Now the problem of synthetic pesticides rekindled her interest. Initially, she opposed the use of the chemical because it was terrible for birds and fish. Spock influenced her to understand the magnitude of DDT's effects on food and the environment overall.

In a letter to the *Washington Post* on April 10, 1959, Carson charged that the use of highly poisonous hydrocarbons and organic phosphates had led to the mass killing of robins. The spraying of pesticides, she knew from research conducted by George J. Wallace, an ornithologist at Michigan State University, and Joseph J. Hickey, a wildlife biologist at the University of Wisconsin at Madison, on the effects of DDT on peregrine falcons, was decimating birdlife. It thinned the eggshells of eagles, peregrines, pelicans, and ospreys.[25] "To many of us, this sudden silencing of the song of birds, this obliteration of the color and beauty and interest of bird life, is sufficient cause for sharp regret," she wrote. "To those who have never known such rewarding enjoyment of nature, there should yet remain a nagging and insistent question: If this 'rain of death' has produced so disastrous an effect on birds, what of other lives, including our own?"[26] According to Carson's biographer Linda Lear, that letter provided the public with the first clue that Carson was studying the impact of synthetic chemicals. Furthermore, the letter brought praise from Agnes Meyer, whose family owned the *Washington Post*, and the activist Christine Stevens, the founder and president of the Animal Welfare Institute. Both women became huge public boosters of Carson's book *Silent Spring* when it was published.

In 1960, the Spock-inspired lawsuit went all the way to the Supreme Court; however, the Court denied a relatively minor writ, effectively refusing to hear the case. Bill Douglas was the only justice who opposed that action; he was more than willing to hear the DDT case. In fact, he wrote a brilliant dissent explaining his position. His words reached far beyond a legal audience when both *Saturday Review* and *U.S. News & World Report* published it. His dissent ended:

The need for adequate findings on the effect of DDT is of vital concern not only to wildlife conservationists and owners of domestic animals but to all who drink milk or eat food from sprayed gardens.

We are told by the scientists that DDT is an insoluble that cows get from barns and fields that have been sprayed with it. The DDT enters the milk and becomes stored by people in the fatty tissues of the body. Because it is a potential menace to health the Food and Drug Administration maintains that any DDT in milk in interstate commerce is illegal.

The effect of DDT on birds and on their reproductive powers and other wildlife, the effect of DDT as a factor in certain types of disease in man such as poliomyelitis, hepatitis, leukemia, and other blood disorders, the mounting sterility among our bald eagles have led to increasing concern in many quarters about the wisdom of the use of this and other insecticides. The alarms that many experts and responsible officials have raised about the perils of DDT underline the public importance of this case.[27]

Marjorie Spock and Polly Richards, with help from Carson and Douglas, awakened the media to DDT's dangers, both to the environment and to human health. Spock always said of the Long Island case, "We lost the battle but won the war."[28] She believed that the federal government ran roughshod over anybody who was opposed to pesticides, as did state and county agriculture commissioners. "They brushed us off like fleas," she complained. From the summer of 1957 to early 1960, Spock circulated detailed reports about her days in and out of the courts. Carson was a regular reader of the dispatches. Once the last trial ended, Carson also procured a transcript of the hearings.

Following the anti-DDT case, Spock collaborated with Ehrenfried Pfeiffer, a German soil scientist and advocate of biodynamic agriculture. She was also a large-scale compost- and farm-adviser for the growing biodynamic movement. In 1957, Spock moved to Maine, where she worked for decades as a "look to the land" biodynamic farmer. Luckily for history, she recorded her first in-person encounter with Carson at a restaurant. "Upon landing we went to meet her at the inn's parking lot," she recalled. "As we approached it, a slightly built woman came around the bend walking unhurriedly. Seeing us she smiled, but did not change her pace. When we knew Rachel better, we realized how typical it was of her to keep to her own way in everything. Neither at this nor further meetings did she strike us as an exuberant, outgoing nature. But there was no heaviness in her somewhat grave demeanor, no lack of warmth in her reserve, or unease in her incapacity for chit-chat. Rather did she seem so disciplined to concentration, so given to listening and looking and weighing impressions as to be unable to externalize."[29]

The Long Island legal case was a historic watershed, one of the first modern environmental actions led by a community of environmental activists. Consumers of agricultural products took notice. In 1959, newspaper stories were published regarding cranberries grown in the Pacific Northwest containing high levels of the weed killer aminothiazole, which had induced cancer in laboratory rats even in low doses. The Department of Health, Education, and Welfare, founded in 1953, advised consumers to not buy cranberries for Thanksgiving. Grocers all over the country withdrew cranberry products from their shelves. The American Cyanamid Company, the primary maker of aminothiazole, tried to defend its herbicide, but the damage was done. The combination of cancer and Thanksgiving cranberries awoke the nation to the danger that might exist on the grocery shelf.[30]

Health and environmental issues related to pesticides began to come to light more often. "The more I learned about the use of pesticides," Rachel Carson wrote, "the more appalled I became."[31] In 1957, the Department of Agriculture launched a program to eradicate the fire ant, a stinging non-native species that had entered the country during World War II through the port of Mobile, Alabama. Dieldrin and heptachlor were sprayed across 1.7 million acres in the Southeast. Such compounds were more lethal than DDT, yet the USDA didn't seem worried about the effect on humans or the environment in general. In 1960, the US Fish and Wildlife Service affirmed that the spraying was responsible for mass deaths among fifty-nine species of animals, including waterfowl.

A range of pesticides were developed and distributed with little or no testing to avoid such disasters. But it was DDT that was, by far, the most common one in use. In Michigan, DDT was sprayed to combat a bark beetle that spread Dutch elm disease; as a result, some bird populations were wiped out.[32] Fly fishermen along the Yellowstone River in Montana found the once pristine waterway's bank littered with the carcasses of dead brown trout and suckers; their carcasses were found to contain DDT.

In 1958, as Carson was learning more about the USDA's fire ant eradication project, she encountered a life-changing statement of Schweitzer's in a bulletin of the International Union for Conservation of Nature and Natural Resources. In only twenty-eight words extracted from a letter to a French apiarist, the good doctor encapsulated the crux of Carson's concern about both DDT and atomic fallout. It was Marjorie Spock who brought his words to her attention. Schweitzer had written, "Modern man no longer knows how to foresee or to forestall. He will end by destroying the earth from which he and other living creatures draw their food."[33]

Carson tacked the quote on a note card at her desk in Silver Spring, Maryland. It motivated her to work late into the night. Borrowing from Schweitzer, she planned to call her anti-DDT book *Man Against the Earth* until her agent, Marie Rodell, said it was inappropriate and confusing. Carson wanted a title like that to indict nuclear fallout and pesticide poisoning as the primary postwar scourges destroying the web of life. "In my floundering I keep asking myself what I would call it if my theme concerned radiation, having some illogical feeling that would be easier," she wrote Paul Brooks, her Houghton Mifflin editor. "As you have seen in the cancer chapter, I keep hammering away at the parallel. Whether radiation or chemicals are involved the basic issue is the contamination of the environment."[34]

From 1958 to 1960, in her letters to friends and scientists, Carson constantly deplored the fact that innocent farmers and gardeners simply didn't know that dieldrin, chlordane, malathion, and other backyard chemicals were potentially cancer causing. Commoner was right: the citizenry had to be properly informed of the perils. Determined to be scientifically exact, Carson had cell biologists and agronomists review chapters of *Silent Spring*. She pondered whether regenerative agriculture could be implemented in the near future, if a proper public education campaign were implemented.

Another danger for the environment arose from species brought into a country accidentally or irresponsibly. A 1958 book by the British biologist Charles S. Elton, *The Ecology of Invasions by Animals and Plants*, warned of the destruction wreaked on ecosystems in that manner. The gypsy moth, imported into the United States from Europe by an etymologist in the 1860s, was a prime example. A more recent one is the python population in the Everglades. Elton's book is considered the first study of the dangers that invasive species pose to an ecosystem. "The time had come when it must be written," Carson said of the subject of his book. "We have already gone very far in our abuse of this planet."[35]

That same fall of 1958, Carson received a letter from Edward O. Wilson, a young Harvard entomologist, about pesticides altering insect life cycles. Like Archie Carr, he had been raised in Mobile, adopted *Walden* as his "bible" (particularly the section on black ants versus red ants as battle gladiators), and knew the fire ant situation in Alabama intimately; in fact, he was the ranking expert on ants of all types. Wilson, who would win two Pulitzer Prizes for his books *The Ants* (coauthored with Bert Hölldobler) and *On Human Nature*, encouraged Carson to forge ahead with a cautionary book on DDT. "The subject is a vital one," he wrote, "and needs to be aired by a writer of your gifts and prestige."[36] In response, Carson wrote, "I am eager to build up, in every way I can, the positive alternatives to chemical sprays, for I feel that a book that is wholly against something cannot possibly be as effective as one that points the way to acceptable alternatives."[37]

Gathering a wide array of data and staying current with scientific research had always been Carson's strong suit. She drew on the expertise of a legion of scientist friends and nature-loving admirers. Like E. O. Wilson, forward-thinking North American biologists, both inside and outside of government, wanted to help her with *Silent Spring*. Her telephone stayed busy as environmental activists and top-tier public health scientists called with field data reports. Carson's timing for questioning DDT could hardly have been more precipitous. As Robert Musil related in his book *Rachel Carson and her Sisters: Extraordinary Women Who Have Shaped America's Environment*, "heavy-handed attempts to eradicate fire ants and gypsy moths with pesticides had gone awry."[38]

Martin Luther King, Jr., linked the civil rights movement to ecological concerns and the threat of nuclear war at many junctures in his activist career. In 1959, five months after being stabbed in Harlem at a book signing for *Peace and Freedom*, he spoke at the War Resisters League. There King fully supported the league's mission of universal disarmament. "Not only in the South, but throughout the nation and the world, we live in an age of conflicts, an age of biological weapons, chemical warfare, atomic fallout and nuclear bombs," he said. "Every man, woman and child lives, not knowing if they shall see tomorrow's sunrise." Then he asked a question that was an opening salvo for the merging of the peace, civil rights, and environmental movements of the 1960s: "What will be the ultimate value of having established social justice in a context where all people, Negro and White, are merely free to face destruction by strontium 90 or atomic war?"[39]

Part II

John F. Kennedy's New Frontier (1961–1963)

CHAPTER 8

Forging the New Frontier

STEWART UDALL AND LYNDON JOHNSON

Stewart Udall gave the US conservation movement a stalwart friend in Washington during the Kennedy and Johnson administrations. He was secretary of the interior from 1961 to 1969.

I

There were those close to Jack Kennedy who felt that he'd been running for president during most of his political life: attending meetings and giving speeches all over the country, year in and year out, in election and nonelection years alike. All the while, he was building a framework of support for the future. That future arrived on January 2, 1960, when Senator Kennedy read a statement at the Capitol in Washington, announcing his candidacy. "I believe," he said, "that the Democratic Party has a historic function to perform in the winning of the 1960 election, comparable to its role in 1932."[1]

For most of the rest of 1960, Kennedy followed the traditional politician's handbook, going to cookouts and coffee klatches and plant gates. Having a private plane, courtesy of the family fortune, helped him crisscross the nation and avoid the vagaries of commercial air travel. As both a stump speaker and crowd mingler, JFK—the shorthand that the press had taken up—was exemplary. Touting public service, he championed legislation in the Senate for a Youth Conservation Corps, modeled after Franklin Roosevelt's Civilian Conservation Corps; its mission would be to beautify neglected public lands. Garnering endorsements from a wide variety of protoenvironmentalists and conservationists, he embraced the Wilderness bill as the flagship of his progressive conservation agenda.[2]

Other serious contenders for the Democratic presidential nomination in 1960, such as Lyndon Johnson, Stuart Symington, and Adlai Stevenson, had yet to emphasize Rooseveltian conservation to the extent Kennedy had with his support of the Cape Cod National Seashore. Among top Democrats, only Hubert Humphrey, the original Wilderness bill sponsor in the Senate, saw conservation as a significant voter issue. At that point, Johnson thought of conservation in terms of reclamation projects, mineral rights, and scientific forestry. Believing that public lands had to generate job growth, he opposed efforts to establish Padre Island National Seashore in his home state of Texas and was for the controversial Hells Canyon Dam on the Snake River in Idaho and Oregon.[3] There wasn't a Bureau of Reclamation dam proposal in the West that Johnson didn't like.

The 1960 campaign shaped up as the first of many presidential contests in which the environment was *an* issue, but not *the* issue. Oddly enough, most Americans had to be reminded that there was a chronic problem, even though, in the case of water pollution, it was staring them right in the face. At Onondaga Lake, in New York, executives of Solvay Process, the chief corporate polluter in the country, lived in swank houses overlooking the dead body of water, even as their company's waste, including PCBs, gushed into it. There was no shame, there was no shock. Effectively, every city had a horribly polluted body of water attached to it, yet it was almost as though residents had grown numb to the discharges. They stopped seeing the filth or at least stopped looking. When the Standells had a hit song about Boston in 1966, it was called "Dirty Water," in reference to the Charles River and Boston Harbor. The condition of the Hudson River in New York was also "celebrated" in song, when a character in Stephen Sondheim's *Company* promises in jest that "By Monday I'll be floating in the Hudson with the other garbage." As to the Potomac in Washington, Bill Douglas lambasted the Eisenhower administration for turning it into a "Chantilly Cocktail," an unhealthy mixture of water and processed sewage that made Washingtonians afraid to drink their tap water or even brush their teeth with it.

Such examples were legion all across the country and Douglas could recite the facts—as far as they were known by the science community. The Navy, he charged, was poisoning San Francisco Bay by allowing its fleets to dump raw sewage there. He said that cities in Alaska, including Anchorage and Juneau, didn't have any proper sewage treatment facilities, leading to daily dumping of human waste into pristine waters. Kodiak Harbor was so polluted by a seafood cannery that the harbor waters couldn't even serve as holding tanks for fish. Douglas lacerated NASA for poisoning Ohio's Rocky River with mercury. In Texas, he lamented, toxic metals such as mercury, arsenic, barium, and boron could be discharged into waters freely, provided that a state permit was obtained. He was infuriated that seemingly nobody in the federal government knew the precise quantity of chemicals being flushed into US waters by industrial plants. Nor had anyone measured the capacity of particular rivers and lakes to absorb a particular waste without causing ecological harm.[4]

Kennedy believed that waterways could no longer be repositories of untreated industrial sludge and human waste. Proximity to otherwise attractive residential areas and farmland didn't help. The runoff from fertilizers, herbicides, and pesticides collected in waterways, as did household chemicals, including phosphates from laundry detergents. In addition to the diminished potability of clean water itself, many freshwater fish species were rendered inedible. Raritan Bay in New Jersey, adjacent to New York Bay, was an ecological catastrophe owing to the sins of the many refineries and factories in the vicinity. The once lush wetlands area was full of begrimed salt marshes and trash-heaped mudflats.

Kennedy had intended for pollution to be an important issue in the 1960 campaign, but it was dwarfed by foreign policy concerns, particularly the so-called missile gap with the Soviet Union. As a Cold War hard-liner, JFK supported closing the gap (which was later shown to be a figment of the military's imagination), but he was unique in that he also supported a nuclear test ban treaty with the USSR and Great Britain. For that he was lacerated by Republicans, including the moderate Nelson Rockefeller of New York, who was seeking the GOP nomination for president.

During the 1960 campaign, conservation commanded JFK's attention only in spurts. "Yet my dad would say Kennedy always sided with us, no matter the issue," David Brower's son Kenneth recalled. "It was just a given that JFK would be wherever the preservationists were at. That was enough."[5] Bill Douglas had the same opinion, maintaining that Kennedy's "instincts" were good on environment and wilderness issues, even though he didn't have the green touch of Theodore and Franklin Roosevelt.[6]

After Adlai Stevenson's defeat in 1956, Democratic Party intellectuals had started making environmental concerns and nuclear test ban part of the liberal policy agenda. At a SANE (National Committee for a Sane Nuclear Policy) rally in Madison Square Garden in May 1960, more than twenty thousand attendees cheered Eleanor Roosevelt, Norman Cousins, and Walter Reuther as they called for ending nuclear testing in the atmosphere. The economist John Kenneth Galbraith, in his 1958 book *The Affluent Society*, laid out the pitfalls of postwar America's achievements, and environmental degradation was one. "The family which takes its mauve and cerise, air-conditioned, power-steered, and power-braked automobile out for a tour passes through cities that are badly paved, made hideous by litter, blighted buildings, billboards, and posts for wires that should long since have been put underground," he wrote. "They pass on into a countryside that has been rendered largely invisible by commercial art. . . . They picnic on exquisitely packaged food from a portable icebox by a polluted stream." Galbraith's prototypical midcentury family continued their rural excursion by camping overnight. "Just before dozing off on an air mattress, beneath a nylon tent, amid the stench of decaying refuse, they may reflect vaguely on the curious unevenness of their blessings. Is this, indeed, the American genius?"[7]

The Affluent Society jarred Congressman Stewart Udall of Arizona into evaluating environmental conservation as a public policy priority.[8] With his close-cropped hair, chiseled features, hawk nose, gentlemanly demeanor, and rugged build, he looked like a test pilot straight out of a NASA astronaut recruitment brochure. He told Kennedy that with the help of lawyer friends, he could deliver Arizona's delegates to his presidential campaign. (Arizona didn't hold a Democratic primary for the 1960 election.) The senior US senator from Arizona, Carl Hayden, a conservative Democrat first elected in 1926, by contrast, was all out for Johnson of Texas. When Hayden learned that Udall was backing Kennedy, he derided the congressman as a naive upstart.[9] Kennedy welcomed Udall's endorsement, though, especially when the "upstart" outmaneuvered Hayden and delivered the Arizona delegation to him.

Stewart Udall was born on January 31, 1920, in St. Johns, Arizona, a sagebrush town of 3,700 founded in the 1880s by his grandfather David King Udall. As a child, Stewart learned about land and water conservation from his father, Levi Stewart Udall, a farmer and attorney who eventually served as chief justice of the Arizona Supreme Court. Levi Udall was one of many Mormons in western politics then. The Church of Latter-Day Saints placed a high religious value on ethical land stewardship: if one Mormon town in Arizona polluted the Little Colorado River, it would adversely affect the neighboring Mormon community downstream. St. Johns, a tight-knit community sandwiched between the Fort Apache and San Carlos Indian Reservations, was perched high atop the Colorado Plateau. Surrounded by US public lands punctuated by arroyos and tablelands in their bright rust-orange colors, the arid region was an outdoors adventurescape for any bright, rock-climbing child to explore.

Like the Udalls, farmers in St. Johns cultivated maize and alfalfa and raised livestock. "You had to be a conservationist to survive," D. Burr Udall, a sibling, recalled. "We were raised to never throw anything away. We were the Make-Do Udalls. We recycled things over and over again."[10] During the Great Depression, the family had minimal electricity, and most homes had outhouses instead of indoor plumbing. "Modern civilization rode a slow horse to St. Johns," Stewart Udall recalled.[11]

As a teenager, Stewart, often accompanied by his family dog, would scour the desert for arrowheads, pottery shards, and snake-skins. After sunset, the Udall family would tune in to the radio signal from Albuquerque and listen to Beethoven, and they would some-times read passages from the Book of Mormon aloud. "I grew up, not in the wilderness, but, from the back door of the house that I lived in, you could shoot ducks in season and you could, some mornings if you listened real hard, hear coyotes," Udall recalled of his childhood. "You could hike off into places where at least you could imagine nobody else had been."[12]

Along with his brother Morris, two years his junior, Stewart milked cows, slaughtered hogs, and tended poultry. The Udall chil-dren erected chicken wire as a backstop and outfield fence for baseball games. During rodeo season, their homemade diamond made a conve-nient holding pen for the calves. Stewart's other siblings, Inez, Elma, Eloise, and Burr, likewise performed ranch duties. The outdoors was an all-seasons school in which the Udalls learned to ford the Little Colorado River, identify migrant birds, attract hummingbirds, and ride horses.

Whereas young Jack Kennedy had daydreamed about the Pilgrims' sea voyage to the New World, the Udall children fantasized that they were part of the expedition of Francisco Vásquez de Coronado, the Spanish conquistador who had trekked through the White Mountains in 1580 looking for the Cities of Cibola (the mythical Seven Cities of Gold). The Udalls listened to old-timers' stories of the explorers and settlers who had traveled there by pack mule, train, stagecoach, and covered wagons. "I was told that Coronado's trail came by St. Johns as it paralleled the wagon road that later ran from Fort Apache to the Zuni Indian Reservation," Stewart Udall wrote in his book *To the In-land Empire: Coronado and Our Spanish Legacy.* "To a child growing up in a lonely land, these revelations were like a fairy tale dropping out of a book into life. Sometimes when the spring winds were roar-ing, I would stand and listen. In my imagination I could hear the clang of Spanish armor rising across the hills from the West."[13]

Physically fit, a lean and agile six feet, three inches in height, Udall was a model St. Johns high school student, entering livestock in county fairs, earning literary prizes, and playing quarterback in football. Considered a natural leader, he attended the University of Arizona in Tucson, preparing to become a lawyer even as he excelled as a star point guard on the varsity basketball team. But religion and World War II interrupted his plans. In 1940, he left college to serve as a Mormon missionary in upstate New York and western Pennsylvania, and in 1942, he enlisted in the Army Air Corps, which assigned him to the Fifteenth Air Force in Italy. As a waist gunner on B-24 bombers, he went on thirty-five bombing missions and rose to the rank of captain.[14]

In 1946, after the war, Udall returned to the University of Arizona, where his best friend was his brother Morris, known as "Mo." Known for his basketball prowess (he played a year with the Denver Nuggets), Mo was a cutup, even writing a memoir later in life titled *Too Funny to Be President*.[15] Eventually, the brothers opened a law firm in Tucson. Stewart, in his spare time, traveled about the Southwest studying geographical features and ancient petroglyphs. The Tucson Mountains, west of the city, became a favorite spot to freshen his mind. An amateur nature poet, he rhapsodized about the immortal sound of time in the tranquil sea of desert. Udall became a "friend" of Saguaro National Monument, located in the Sonoran Desert. "Right after the Second World War, throughout the 1940s and 1950s, I began hunting for ways to enlarge the national monument," he recalled. "To my mind the Saguaro was amongst the prettiest scenery in America."[16]

While at law school Stewart fell in love with Ermalee "Lee" Webb; they married in 1947. Smart, vibrant, and self-confident, Lee was from the Phoenix suburb of Mesa and knew the Mazatzal Mountains backcountry well. Lee vigorously promoted Native American arts, becoming a supporter of the Institute of American Indian Arts in Santa Fe and the Heard Museum in Phoenix. The husband-and-wife team loved Thoreau's *Journals*, collected Navajo and Hopi artifacts, and befriended the Pulitzer Prize–winning poets Robert Frost and Carl Sandburg. Together they had six children—Tom, Scott, Denis, Jay, Lynn, and Lori.[17]

After Udall was elected to Congress in 1954, he was appointed to the House Interior and Insular Affairs Committee, which dealt with the western issues that he knew so well. The committee's chairman, Wayne Aspinall, a Democrat from Colorado, believed that "junior members are to be seen and not heard."[18] Because Udall wouldn't be muzzled, his relationship with Aspinall was tense from the start; the rivalry would continue for fifteen years as they sparred over what constituted wilderness, reclamation, and wild rivers.

When Congress was in session, the Udalls lived in McLean, Virginia, along the Potomac River. They were stunned to find that the muddy river in their backyard was ravaged by rank pollution. Though Cumberland, Maryland, had built the Potomac's first wastewater treatment plant in 1957, it wasn't enough.[19] Both Stewart and Lee Udall joined Douglas's crusade to clean up the river from persistent organic pollutants (POPs) and raw sewage.

During the Eisenhower years, both Kennedy and Udall were young Turks, but it wasn't until 1959 that they became political allies; that year they worked on the Labor-Management Reporting and Disclosure Act, also known as the Landrum-Griffin Act, which safeguarded employees' rights to organize, negotiate collectively, and choose a union. Both men were pleasantly surprised that Walter Reuther's progressive United Auto Workers union was working in sync with conservation groups such as the Izaak Walton League and the National Wildlife Federation. "I liked the way his mind worked, the way [Kennedy] attacked a problem and stuck with it," Udall told reporter William V. Shannon.[20]

Because by 1960 Udall had served on the Interior and Insular Affairs Committee for six years, Democrats surmised that he aspired to be secretary of the interior. They were probably right. "My own state is in one way a microcosm of the Interior Department," Udall would observe. "I have had, as a Congressman, more Indians, more Indian reservations, more national park areas than any other Congressman in the United States."[21]

With his love of a rugged outdoors life, Udall was a seasoned conservationist in the style of Theodore Roosevelt. And like Roosevelt, who in 1903 had dammed the Salt River to bring critical power to metropolitan Phoenix, Udall was a champion of reclamation. He consistently campaigned to increase the allocation of more water from the Colorado River to his growing state's needs. He helped convince Kennedy to support a proposed 336-mile network of canals and pumping stations aimed at delivering water to Phoenix and Tucson.[22] When Kennedy gave speeches west of the Mississippi, he would cite his support for public hydropower, saying that it would reclaim western lands, supply energy for cities and industry, provide opportunities for recreational pleasure at newly formed reservoirs, and meet the needs of the rapidly expanding western population. He learned that from Udall.

On June 18, 1960, Kennedy arrived in Durango, Colorado, to speak at the state's Democratic Convention, which was choosing delegates. The other candidates skipped the convention or sent surrogates. JFK left Boston at midnight on his plane and arrived in Durango at 8:00 a.m. He looked tired, according to reporters, but he was there. He could've delivered perfunctory remarks, but instead, he spoke spontaneously to "Westerners highly conscious of the resource issue," as a local reporter put it.[23] No doubt many, if not most, of the attendees had a stake in the extraction industries or water-dependent agriculture. In a long, significant speech, Kennedy didn't pander to the convention voters he needed for the Democratic nomination who may have been more interested in using natural resources than in saving them. On the attack, of course, against Eisenhower and his vice president, Richard Nixon, who was vying for the head of the Republican ticket, Kennedy derided the failure of the Republicans to "protect our great abundance of natural wealth" and "to control the wasteful and destructive pollution of our waterways." Sounding like Franklin Roosevelt, he warned that America's "priceless stock of natural beauty is being eroded—our wildlife going unprotected—our fish stocks destroyed." He did speak of increased reclamation efforts, but his theme was imbued with respect for nature. As always, he spoke of the future: "Wherever we look—from forest, to mountain, to river, to the very air we breathe we can see America's priceless heritage of natural wealth being dissipated by Republican despoilment, under-development and neglect."[24]

That emphasis on preservation was courageous, and it drew respect from the attendees, who had been divided on their choice for a presidential nominee. After that speech, they quickly decided to give him all twenty-one of Colorado's delegates. One other piece of news emerged from the campaign stop in Durango: Kennedy promised that, if elected, he'd appoint a westerner as secretary of the interior.[25]

Kennedy's growing identity as a liberal on the environment

mirrored that of his generation. Throughout the 1950s, he had focused on other concerns, as had they, by and large. By 1960, however, prophetic voices, including those of Bill Douglas, Jacques Cousteau, Rachel Carson, David Brower, and Howard Zahniser, were being heard, and arguments about wilderness preservation and clean air were resonating. Young adults could see ecological devastation with their own eyes, as could Kennedy, one of the best-traveled people in the country.

Lake Erie, with the stupendous Bass Islands just offshore from Port Clinton, Ohio, had long been a renowned vacation playground. A major transportation route with boat traffic from four major cities—Buffalo, Erie, Cleveland, and, by way of the eponymous river, Detroit—the lake, however, had become largely unswimmable. By the time Kennedy announced his run for the presidency, it was literally a dump, filling with sewage effluent, phosphates, and industrial discharge. The lake's ecosystem had been ruined; what remained was decaying fish and clumps of algae. More than 2,600 square miles of the lake had been depleted of oxygen and were thus unable to support fish life or aquatic plant growth. The federal government believed that 5 million to 9 million gallons of oil and other petroleum products were dispersed yearly into the Detroit River, which as a result had suffered huge winter duck kills.

While campaigning, Kennedy listened to the concerns of biologists, hydrologists, and ichthyologists regarding Great Lakes pollution. They worried that Lake Erie was becoming so polluted that it had tainted local tap water. Dirty water carried the additional risk of waterborne diseases such as dysentery and cholera. If elected president, Kennedy promised, he would spearhead federal abatement action against Michigan and Ohio factories and other polluters on the Detroit River.[26] Walter Reuther agreed that oil tankers and terminals needed to be fiercely regulated by the federal government if states weren't up to the task. With Reuther at his side, Kennedy spoke passionately about establishing national lakeshores and national migratory waterfowl refuges in the ecologically struggling Great Lakes region.[27] Saving the lakes was a gargantuan job that would need presidential leadership and bipartisan political will from enlightened conservation-minded lawmakers.

II

On a spring day, Rachel Carson interrupted her work on *Silent Spring* to go to a Washington hospital to have two small lumps on her left breast removed. Over the previous twelve years, she had had other such growths removed without serious complications. But as she returned to writing that spring, she felt depleted and fatigued, struggling to focus on the manuscript. In fact, her doctor was suspicious of the growth, believing that Carson had "a condition bordering on malignancy."[28] Tests proved him right.

Determined not to be derailed from writing, Carson plodded onward, when she wasn't too weak, meticulously editing and perfecting the book. With forebodings about the severity of her condition, she was on a race against the clock. All over her home in Silver Spring were note cards, reports, and clippings about DDT, each with the information she wanted to include in her narrative. During practically every waking hour, she tackled the stacks of research documents and wondered how she could complete her book.[29]

Carson was also involved in politics for the first time as a member of the Democratic Advisory Council. The party chose her as an environmental thought leader, the ideal person to introduce it to a fresh vocabulary: *habitat, reverence for life, ecosystem, endangered species, DDT poisoning,* and *strontium-90*.[30] She delivered her recommendations that June. The key conservation topics the Democratic Party needed to tackle head-on, she believed, were pollution control, curtailment of the use of pesticides harmful to the environment, saving national seashores, and passage of the Wilderness bill. She urged Democrats to criticize the Eisenhower administration for its gypsy moth and fire ant extermination programs, using, as they did, DDT.

One of Carson's suggested policies showed progress when the Cape Cod National Seashore bill advanced to the House and Senate subcommittees in June. "From both the public and expert response which we have had to this legislation," Kennedy explained at the Senate hearings, "we have reason to believe that our legislation may help to set a pattern for future seashore park development in other parts of the country."[31] Kennedy's appearance at the hearing demonstrated the value he placed on the Cape Cod bill. Most of that summer of 1960, he was of course on the campaign trail—as was most of his family. So many members of the Kennedy family campaigned for him that his rival, Hubert Humphrey, complained that running against JFK was like "an independent merchant running against a chain store."[32] Two politicians who were drawn to Kennedy were intellectual liberal Democratic senators from the Pacific Northwest: Henry M. Jackson of Washington and Frank Church of Idaho. Because both preferred Kennedy to Stevenson or Johnson in 1960, they were soon dubbed "New Frontiersman," after Kennedy's campaign promise of a "New Frontier."

Born in Everett, Washington, on May 31, 1912, the son of Norwegian immigrants, Jackson was contemplative and polite from an early age. His sister called him "Scoop" because he resembled a cartoon character of the day, one that managed to make other people do his chores. That did not describe the hardworking Jackson, but the name stuck. His appreciation of nature was instilled into him by camping trips in the Cascades of northern Washington. "I think it was . . . here that he got his love for the mountains and streams and wildlife that he fought to protect during his long life; he was an environmentalist long before it became fashionable," revealed Jackson's boyhood friend, Robert M. Humphrey, who became a newspaper columnist in Everett.[33]

In 1940, after earning a law degree at the University of Washington, Jackson ran for Congress and became its youngest member, at twenty-eight. He joined the Army the next year but returned to Congress on Roosevelt's order.[34] A Cold War hawk, he was reelected to the House five times before challenging Republican senator Harry Cain in 1952. Jackson won and would serve in the Senate for decades. His two prime areas of interest were military affairs and conservation. Passing the Wilderness bill and establishing North Cascades National Park in Washington were two of his chief conservation objectives.[35]

Frank Church, a thirty-two-year-old Stanford University–trained lawyer, was elected to the Senate in 1956. Though most Idaho politicians thought forestlands were something to exploit for lumber, not to save in wilderness condition, Church was an exception. Like Lyndon Johnson, he was in favor of dams in the West (though he'd change that position by the end of the decade). Nonetheless, he devoted enormous political capital to saving landscapes such as Idaho's Sawtooth National Forest from development. It was the kind of land that he had hoped the Wilderness bill could protect from commercial activities. Debating the bill in the Senate, he said, "The Wilderness bill is of primary importance to westerners. The vanishing wilderness is yet a part of our western heritage. We Westerners have known the wilds during our lifetimes, and we must see to it that our grandchildren are not denied the same rich experience during theirs."[36]

Church, a third-generation Idahoan, was born on July 25, 1924, in Boise, where his parents operated an outdoor gear store. Hunting and fishing were a big part of his youth, as was horseback riding in the Sawtooth Mountains. During the summers months, he would fish for native cutthroat and rainbow trout in the Middle Fork of the Salmon River. Known as the River of No Return, it cut through dense forests to form the Salmon River Canyon, the second deepest canyon in North America after the Grand Canyon. Seeing a mule deer there, head erect and tense, large antlers looking polished in the declining September sun, was for Church "magical."[37] Determined that Idaho needed to shed its reputation as one big potato field, he beat the drum for increased tourism to Selway Falls, Bruneau Dunes, Crater of the Moon, and Clarendon Hot Springs. "I never knew a person who felt self-important in the morning," Church said, "after spending the night in the open on an Idaho mountainside under a star-studded summer sky."[38]

Kennedy and Church became fast friends in the Senate in 1957. Because Kennedy was seven years older, Church adopted him as something of a political big brother. The influence, though, went both ways. Behaving like fraternity brothers, they jested, teased, and pranked each other. Once when JFK visited Church in Idaho, his plane almost crashed near Twin Falls. "He teased Frank about it unmercifully," Church's wife, Bethine, recalled. "Jack would say, 'My God, you tried to kill me when I visited your state,' or jokingly tell people, 'Don't get in a plane that Frank booked.'"[39]

That spring of 1960, Bill Douglas found himself on a discomfiting airplane ride of another sort. He joined a delegation of senators, led by Majority Leader Lyndon Johnson, flying to Portland for the funeral of Senator Richard Neuberger, who'd died of cancer at forty-seven. President Eisenhower had lent his official jet to the Democratic delegation. "So in that manner I crossed the continent and paid my respects to Dick's memory," Douglas wrote. "We were in the air nine hours that day—five going West and four returning."

During the flight to Oregon, Johnson talked with Douglas about his presidential bid while ridiculing Kennedy and Stevenson. Hungering for Douglas's support, the cagey Johnson dangled the vice presidency before the justice; Douglas didn't take the bait. Somehow Johnson didn't realize how committed Douglas was to JFK (or perhaps he did and was trying to flip him). "During his eight-hour talk, his theme was 'My First Hundred Days in the White House,'" Douglas recalled. "He was as sure of the Democratic nomination as he was that the plane would make a successful round trip."[40]

III

Kennedy was nominated for president on the first ballot on July 15 at the Democratic National Convention in Los Angeles. In his acceptance speech, he declared that the "New Frontier" had arrived. That served as a reminder that Richard Nixon, the GOP nominee for president on the ticket with Henry Cabot, was part of the old era, having served as Ike's vice president for the past eight years.[41] The Democratic Party platform, which Kennedy embraced, contained language directly addressing the "poisoned earth and sky."[42] It held that the upsurge in population, the expansion of industry, the testing of nuclear weapons, and unregulated pollution would devastate the country if strong environmental laws weren't enacted. Putting the oil, gas, synthetic chemicals, and coal lobbies on notice, the Democrats promised controls on pollution caused by stationary emissions and automobile exhaust. The platform also called for the proper disposal of chemical and radioactive waste. By contrast, the Republican National Convention in Chicago—following Nixon's lead—downplayed conservation.

Richard Nixon was born on January 9, 1913, the second of five boys, on his family's lemon farm in Yorba Linda, California. Because his parents were Quakers, he was taught to refrain from drinking alcohol and swearing, behaviors that he would become famous for in his long and storied political career. The family ran a grocery store during most of Nixon's youth, making a scant living, but enough to allow for participation in class trips to the Grand Canyon and Yosemite. Both national parks enthralled him. The great adventure of his life was hiking the rim of the Grand Canyon with high school friends from Yorba Linda.

The Pacific Ocean, however, was Nixon's most personal tie to nature. His parents would take the children for outings in gorgeous beach towns such as Dana Point and Laguna Beach. "While my father had little time off from school and family chores, growing up in California's Orange County meant growing up swimming in the nearby Pacific Ocean," his daughter Tricia Nixon Cox recalled. "One of my special childhood memories was riding the surf with my father. He taught me how to dive under, and not be rolled by waves. Having learned that lesson, he then taught me how to wait for the right wave to ride back to shore."[43] Like Kennedy, Nixon chose in 1969 to make his primary home next to the ocean; his beach house was in San Clemente, south of Laguna Beach, while Kennedy's was in a similar oceanside situation on the other side of the country in Cape Cod.

A brilliant student, Nixon turned down a scholarship to Harvard for family reasons, graduating from Whittier College and then Duke University Law School. After distinguished service in the Navy, deployed, as Kennedy was, in the South Pacific, Nixon entered politics. He was elected to the House of Representatives in 1946—the same year Kennedy was. The two were friendly in their new jobs. A rising star of the GOP, Nixon was elected to the Senate in 1950 and then from 1953 to 1961 was President Eisenhower's vice president. Nowhere along the way did conservation or ecology become a talking point for him. He didn't seize upon air quality as a political issue, even though he represented California, the state that had made more progress establishing air pollution control programs than any other.

In contrast, Kennedy believed that California could serve as a national model for curbing air pollution. So did other Democrats. Their platform listed "Water and Air Pollution" as a separate section, one clause of which stated that "Federal action is needed." Specifically, research would be done on "the rapidly growing problem of air pollution from industrial plants, automobile exhausts, and other sources" and "disposal of chemical and radioactive wastes, some of which are now being dumped off our coasts." Another section, "Outdoor Recreation," addressed the imperative for the inclusion of shorelines along with wilderness areas in federal park management. Kennedy benefited from the environmental planks in the platform. The approach of the Nixon Republicans, in contrast, looked old-fashioned. Without ever mentioning pollution, their platform suggested that a citizens' board make recommendations regarding the use of public lands. The Democrats were clearly thinking in terms of environmental conservation, while the Republicans were content to stick with Eisenhowerian utilitarian conservation and with a "Giveaway" McKay tone at that.[44]

That August, *Life* magazine asked both presidential candidates to delineate what the national purpose should be. Nixon said nothing about the environment. Kennedy did, and framed the challenge in terms of social justice: "Even in material terms, prosperity is not enough when there is no equal opportunity to share in it, when economic progress means overcrowded cities, abandoned farms, technological unemployment, polluted air and water, and littered parks and countrysides."[45]

Kennedy succeeded in highlighting the distinctions between himself and his opponent when it came to the environment. Oddly enough, he could have drawn the same distinctions between himself and his running mate, Lyndon Johnson. Given his druthers, Kennedy would have preferred Scoop Jackson, Frank Church, or Minnesota governor Orville Freeman on the ticket.[46] All three men were his friends. For his part, Bobby Kennedy pushed for Walter Reuther, the president of the United Automobile Workers, environmentalist extraordinaire and ally of Martin Luther King, Jr. But Texas was too big a prize to lose, so JFK picked Lyndon Johnson. Disappointed at losing the top spot, Johnson reluctantly agreed to be on the ticket.

The dark-eyed and rugged Johnson, a teacher by training and a politician to the core, was born on August 27, 1908, in a three-room farmhouse along the emerald Pedernales River in Texas's Hill Country. His hometown of Johnson City had only three to four hundred residents at the time, lacked electrification, and had no sewage system. As a boy, he climbed limestone bluffs, hunted in mesquite thickets, collected salamanders, pickled peaches, and tended to cattle grazing in the tall grass on the Edwards Plateau. "Those hills and that river

Lyndon Baines Johnson at his ranch in the Texas Hill Country. His political heroes were Theodore and Franklin Roosevelt.

were the only real world that I really had in those years," he recalled. "So I did not know much about how much more beautiful it was than that of many other boys, for I could imagine nothing else from sky to sky. Yet the sight and the feel of that country somehow or other burned itself into my mind."[47]

Johnson studied at Southwest Texas State Teachers College (now Texas State University) in San Marcos and then took a job teaching Mexican American children in Cotulla, Texas. Entering the world of politics as a congressional aide, he was tapped by President Roosevelt to direct the Texas division of the National Youth Administration, a job at which he excelled. In 1937, he won his own seat in Congress, using his office to bring electricity to the Hill Country. During World War II, Johnson served in the Navy, then returned to the House and won a seat in the Senate in 1948. He became minority leader in 1953, while still in his first term—the youngest leader in Senate history. Although he was hungry for power he didn't conform with the status quo, and cared deeply about the opportunity for Americans, regardless of color or ethnicity. He sponsored the Civil Rights Act of 1957. As his party's leader in the Senate, Johnson was effective, but demanding on a minute level. New York's Emanuel Celler, the chairman of the House Judiciary Committee, remarked, "I heard Johnson say one time that he wanted men around him who were loyal enough to kiss his ass in Macy's window and say it smelled like a rose."[48]

Johnson's backroom political style didn't appeal to the suave Kennedy. The Texas leader believed he was the hardest working lawmaker in Washington, which justified his uncouth acts, vindictiveness, and violations of personal space as a persuasive tool.[49] Bobby Kennedy thought the Senate majority leader was an uncivil "animal," unfit for service in the New Frontier.[50] But with Johnson on the ticket, JFK's chance of winning the White House improved. The Kennedy brothers would have to make the best of aligning themselves with Johnson. As a consolation prize to Reuther, Kennedy embraced the union president's vision of sending young Americans to help the poor in developing nations—a Peace Corps idea that Reuther had been promoting since 1950.

On the conservation front, Johnson probably should have been aware of the importance of intelligent land stewardship. His favorite novel—John Steinbeck's *The Grapes of Wrath* (1939)—was based on a man-made ecological catastrophe during the Great Depression that destroyed ranches and farms in Texas, New Mexico, and Oklahoma. It was soil depletion that had led to the Dust Bowl. Johnson, however, was left wanting to help the rural poor, without setting an equal priority on nurturing the natural world around them.

Johnson liked Bureau of Reclamation dams, because he knew firsthand how electricity lit up forgotten regions, literally and figuratively. On the campaign stump, he highlighted FDR's Tennessee Valley Authority and how it had transformed an entire rural region by providing electricity. Protoenvironmentalists liked electricity, too, but thought it was long past the time to stop desecrating scenic wild rivers such as the Colorado and Guadalupe in Texas provide it. Nor was Johnson excited about Kennedy's national seashore and lakeshore advocacy; it smacked, he felt, of the federal government bottling up potentially lucrative real estate. Even with economic incentives for locals, the prospect wasn't popular in Texas business circles.

Johnson *did* love the CCC notion that spending time camping under the stars in places like Lake Brownwood or Bastrop developed character in a young person. His promotion of hunting made him popular with the Boone and Crockett Club and the National Wildlife Federation. The hint that he could someday become an environmentalist, however, was betrayed by his own greatest loyalty, which was to the Hill Country expanse where he was born and where, when all was said and done, his spirit soared.

Johnson's wife, Claudia, shared his love of the outdoors and sympathy for people who were struggling. A cook once said Claudia was "as purty as a ladybird," referring to the ladybird beetles (ladybugs) of the Caddo Lake region in east Texas. The nickname "Lady Bird" stuck to her for life. She was from Karnack, a hamlet near the Louisiana border. As a little girl, she was spellbound by the mysterious Caddo Lake, a twenty-three-mile-long waterway near her home. Adorned with a labyrinth of sloughs and low semi-submerged islands, thick with bald cypress trees carrying Spanish moss, the lake was replete with alligators, paddlefish, egrets, owls, and beaver.[51]

Growing up, Lady Bird "loved to paddle in those dark bayous, where time itself seemed ringed around by silence and ancient cypress trees, rich in festoons of Spanish moss. . . . It was a place for dreams."[52] Wandering around the perimeter of Caddo Lake, enthralled by the natural world, Lady Bird would study the daffodils, daisies, and crimson clover fields as if studying rare heirlooms. She would declare the first flower that bloomed every spring "queen" in a Cinderella-esque ceremony.[53]

At an early age, Lady Bird's mother, Minnie Lee, who had struggled with depression, fell down a staircase and died. Netting Black Swallowtail butterflies and collecting loblolly pinecones helped Lady Bird's grieving heart heal. "I grew up listening to the wind in the pine trees of the East Texas woods," she recalled.[54] In 1930, when she was eighteen years old, she was accepted into the University of Texas. "I fell in love with Austin the first moment that I laid eyes on it, and that love has never slackened," she said later.[55] At UT, she majored in journalism, writing about the poetry of John Keats in the *Daily Texan* and working as publicity manager for the university's Texas Sports Association, which oversaw the women's athletics program. "From the time I was seventeen until I left the university, I had all the beaus I could handle," she recalled late in life. "I had a lot of fun. Crazy, wild, city fun. I think I fell in love every April."[56]

After Lady Bird graduated, she met Lyndon, who asked her out for a breakfast date in Austin. She remembered him as being "excessively thin but very, very good looking, with lots of black hair, and the most outspoken, straightforward, determined manner I had ever encountered. I knew I had met something remarkable, but I didn't quite know what."[57] The twenty-six-year-old Johnson proposed to her after only one date. She hesitated, but not for long. When Lady Bird took Johnson home to Karnack, her father said matter-of-factly, "You've been bringing home a lot of boys. This one looks like a man."[58]

On November 17, 1934, Lyndon and Lady Bird were married. After honeymooning in Mexico, they moved into a modest Washington apartment. The Roosevelt administration had created the National Youth Administration to help 45 million Americans between the ages of sixteen and twenty-five find jobs, receive vocational education, and achieve higher standards of education; Johnson was offered the position of Texas administrator. According to Monroe Billington in the *Journal of Negro History*, "Johnson's concern was to help the youth of Texas—regardless of the Great Depression." He was active in affording African American youth the same opportunities offered to Whites. At first, Black leaders were almost confused; such things didn't happen in Texas then. As Billington reported, "One black contemporary concluded that Johnson helped blacks not only because he wanted to do a good job, but also because he 'cared for people.'"[59]

Proud of the Alamo heritage and natural beauty of Texas, Lyndon and Lady Bird Johnson were true conservationists, enamored with such landscapes as the Davis Mountains, Palo Duro Canyon, and Enchanted Rock. In 1951, they bought a ranch along the Pedernales River, surrounded by raised limestone strata with incised caves and canyons, near where Lyndon had grown up. Never comfortable around the rich, Johnson felt most at ease with humble people in rural America. Picnic parties—barbecues and basket lunches in state parks—became mainstays in the way the Johnsons cultivated friendships. Porches, swings, and half-open leaning gates were his milieu. The sweet familiarity of his Pedernales spread, the ranch hands in shirtsleeves working to feed the cattle, and the hens clucking before nightfall made the place his oasis from the grind of Washington politics. Yet the dichotomy of the man meant that when Washington came to call at the ranch, he played his part to the hilt.

LBJ relished telling visitors how his ancestors drove cattle down the Chisholm Trail, which ran past the Hill Country. Identifying himself with this independent frontier place was an important part of his political persona. Playing catch-up with the Kennedys, he bought a JetStar aircraft and built visitor quarters where politicians, journalists, and world figures could stay overnight. A *New York Times* reporter noted that Johnson cultivated the image of "the breezy, two-fisted, overpowering range king who rules from horizon to horizon and from can-see to can't-see with iron will and fast gun."[60] That was exactly the plan. "Sell the Johnson image," Lyndon once instructed a press secretary. "You know, like Marshal Matt Dillon . . . big, six-feet three, good-looking—a tall, tough Texan coming down the street."[61] He even played the part for Kennedy.

Late in 1960, after the Democrats captured the White House, JFK visited the Ranch.

Early one morning, he was sound asleep when Johnson excitedly woke him to show him the sunrise. Dressed in camouflage hunting gear, LBJ insisted that the president-elect join him in a drive around his spread. They made the tour in a white Cadillac convertible supplied with high-powered rifles in case they found game to shoot. Kennedy's biographer, William Manchester, was the first to tell the story. JFK, who wasn't a hunter, didn't care about the prospect of killing deer and he was too sleepy to appreciate the limestone outcroppings. Suddenly Johnson spotted a deer and handed Kennedy a rifle. "Shoot! Shoot!" he urged. Kennedy shot the deer from the car, to his regret, and then asked to end the outing. Later, he told Ted Sorensen that he felt ashamed. He thought hunting warm-blooded mammals a grotesque endeavor. In a letter to Senator George Smathers of Florida, he wrote, "That will never be a sport until they give the deer a gun."[62]

Johnson, however, years later, told a different version of the hunting excursion in private to his aides: "Forcing that poor man to go hunting? Hell, he not only killed one deer; he insisted on killing a second! It took three hours and I finally gave up. I said, 'Mr. President, we just can't do it.'" JFK might have shot the deer and later, as it all sunk in, regretted it. Johnson defended his version of the story with an incontrovertible point, saying, "I think it is the greatest desecration of his memory that an 'impotent' vice president could force this strong man to do a goddamned thing."[63]

Just prior to the Democratic National Convention, Kennedy received a copy of *This Is the American Earth*, a Sierra Club book by Ansel Adams and Nancy Newhall intended to convince readers that natural settings were invaluable to the nation. The cover image was Adams's masterpiece, "Aspens, Northern New Mexico, 1958," a picture of a stand of quaking aspens. The images contributed by Adams and other photographers were supplemented by Newhall's Thoreau-style questions, such as "What is the price of exaltation?" and "What is the value of solitude?" Biblical quotes such as "the tongues of angels," "evils as old as man," and "Famine and Pestilence" were scattered throughout the oversized volume. Adams dubbed the interplay between the photos and prose "synesthetic."[64]

Adams's luminous landscape photography was a potent tool in the effort to persuade Americans to save wilderness areas. Born on February 20, 1902, in San Francisco, Adams fell in love with nature as a boy. A gifted pianist but an indifferent student, he was given his first camera in 1916 and immediately trained his eye on California's landscapes, mastering the ever-changing sunlight along the wild Pacific Coast. Throughout the 1920s, he worked as the custodian of the Sierra Club's lodge in Yosemite National Park, taking astonishing photographs of John Muir's treasured landscapes.

He joined the Sierra Club's board of directors during the Depression and championed the emotional aspects of the wilderness. By 1960, Adams was lobbying for new parks and seashores, for the Wilderness bill, and the preservation of the Big Sur coast of central California. At the same time, he reviled Eisenhower's NPS Mission 66 infrastructure effort.[65]

This Is the American Earth delivered the jolt that Adams, Newhall, and Brower had hoped for. It also proved to be a powerful catalyst that brought the wilderness movement into the Kennedy campaign. Douglas had already praised the book as "one of the great statements in the history of conservation."[66] The *New York Times* praised its "magnificent photographs and eloquent words."[67] The *Deseret News*, a Salt Lake City daily owned by the Church of Latter-Day Saints, said: "The Sierra Club has been in the forefront of virtually every effort to lock up the West's scenic resources, and this book is a most skillful, persuasive part of the campaign."[68] In the coming years, other large-format books published by David Brower and the Sierra Club would become tangible tools for lobbying on Capitol Hill, helping to save treasured landscapes. With the success of the books, some people even misconstrued the role of the Sierra Club, thinking it was mainly a book publisher instead of an environmental nonprofit.

IV

As Rachel Carson finished her volume on DDT, she was calling it *Dissent in Favor of Man*, a phrase inspired by a magazine account of Douglas's dissent in the Supreme Court's action during the Long Island DDT case. No matter how brilliant and timely the book, it might not have sold eight copies with that title, but such is the way of authors and working titles. The final push on the work was extremely trying. Not only was Carson fighting cancer, she was suffering from viral pneumonia and ulcers. Only her closest friends and associates knew how debilitating her health problems were. That was her choice. She sought to conceal her illness, wearing a wig when her hair started falling out during chemotherapy, for fear that chemical companies would attack her *Silent Spring* research by saying "She's sick with cancer and wants to blame the pesticides."[69]

The thought of Kennedy in the White House lifted Carson's hopes. Spending most of her days at home in either Maine or Maryland, propped up on pillows, trying to heal, and working on *Silent Spring*, Carson somehow managed to find the energy to volunteer for his campaign. In the last weeks before his nomination, she joined the Women's Committee for the New Frontier, which met at the Kennedys' N Street home in Georgetown.[70] The group consisted of forty-seven prominent women who backed Kennedy. The committee's leaders included former labor secretary Frances Perkins and Eleanor Roosevelt. Only Carson represented the cause of environmental protection within the influential group. She was grateful to receive a thank-you note afterward from JFK, whom she had never met.[71] Carson then hunkered down, finishing *Silent Spring*. In a letter to her friend Dorothy Freeman, she wrote, "I suppose as I grow older and become more aware that life is not only uncertain but short at best, the sense of urgency grows to press on with the things I need to say."[72] Others felt the same urgency—about her book.

Within a closed circle of ecologists and nature enthusiasts, word spread that Carson was going to publish her anti-DDT findings. To augment her Fish and Wildlife data, Douglas and other conservationists sent her documents showing the negative effects that pesticides were having on living things. In *Silent Spring* she cited the scientific findings of Dr. Robert Rudd, Dr. John George, Professor Joseph Hickey, Professor George Wallace, and many others who helped her bring national attention to their fact-based papers about the dangers of spraying DDT.[73]

Carson also found the energy to write a new preface for *The Sea Around Us*, the book Kennedy so loved. Drawing from her DAC work, she warned that the sea—"The last frontier of Earth"—was being abused as a dumping ground for the contaminated garbage of the Atomic Age. The new preface was a stark call for government action against nuclear testing and ecological degradation. From a literary perspective, the preface was a warm-up exercise, a glimmer of the concerns that would be explored more authoritatively in *Silent Spring*. "It is a curious situation that the sea, from which life first arose, should now be threatened by the activities of one form of that life," she wrote. "But the sea, though changed in a sinister way, will continue to exist; the threat is rather to life itself."[74]

Much like Carson, Douglas was anticipating better days for the environment under a Kennedy administration. Somehow, even as a member of the Supreme Court and as a very busy author, he found the time to go on a multitude of nature trips every year, in addition to spending summers in the backwoods of the Northwest. Just traveling to the places he sought to explore took days, as was the case of his September 1960 canoe trip down the Allagash River in north-central Maine. It was a long way from the nation's capital. In fact, it was a long way from almost everywhere. Sections of the wild Allagash were still wide whitewater—which was, incidentally, perfectly safe to drink. The gorgeous waterway had been a birch-bark canoe route long before Christopher Columbus arrived in the New World. It sickened Douglas to think that without prohibitive laws, it was only a matter of time before roads, dams, and lumber operations destroyed the ecosystem. As was his wont, Douglas began campaigning on his return for a "green" corridor ranging from one to four miles wide on each side of the river and also including the lakes and tributaries that lay in the Allagash watershed. Deeming the area the largest primitive area east of the Mississippi, with the exception of the Everglades, the justice believed that Kennedy would possess the vision to create an Allagash National Riverway, a new type of designation Douglas foresaw.[75]

After the Democratic National Convention, Stewart and Lee Udall took two of their sons on a four-day adventure on the upper Colorado River at Glen Canyon in Utah. The first concrete was poured only two months before for a massive dam in the canyon, putting the Udalls among the last beings of any species to see the canyon walls with its petrographs, put there by both nature and Indigenous people. Rumors were circulating that Udall was a front-runner to be secretary of the interior if Kennedy won.[76] Udall carped that fall that if groups Nixon favored—the Western States Petroleum Association, the American Sheep Industry Association, and the National Taxpayers Union—had their way, there would be no national parks or wilderness areas. Both Udall and Douglas warned that "what was classified as national forests today can be logged tomorrow—if a few men in Washington so decided."[77]

On November 8, 1960, the Kennedy family convened in Hyannis Port to watch the election returns on television, dining on Maryland crab washed down by cocktails.[78] In the end, Kennedy beat Nixon in the Electoral College by a count of 303 votes to 219, but in the popular vote, he won by only 113,000 votes out of the 68 million cast.

After the excitement of election night, Jack and Jackie Kennedy took separate routes. Jackie, who was about eight months pregnant, went to Washington to rest, in anticipation of delivery by cesarean section in December. Jack, meanwhile, also needed rest, but he chose the quieter surroundings of his parents' estate in Palm Beach. Flush with victory, the president-elect began vetting cabinet appointments at what the media were calling the Winter White House. Kennedy golfed at the Palm Beach Country Club, ate lunch at Hamburger Haven, attended St. Edward Catholic Church, beachcombed, and worked on his inaugural address, looking up to stare out at the intoxicating blue-green Atlantic for inspiration. Writing from Florida to the Society of American Foresters, he promised that a "surge forward in the conservation of forest soils and wildlife" was at hand.[79]

Just a week after the birth of his son John Jr., on November 25, the president-elect invited Stewart Udall to his N Street home in Georgetown to offer him the post of secretary of the interior. "The Kennedy crowd thought a lot about the West in 1960," recalled Udall's son, Tom, who served in the Senate from 2008 to 2020. "They sent Ted Kennedy to work with Stewart in trying to pull in southwest voters. And their grassroots effort worked. Arizona Democrats, against all odds, went for Kennedy in 1960. My dad got a lot of credit for that."[80] Jackie remembered it the same way. She later explained to the historian Arthur Schlesinger, Jr., that her husband had admired Udall tremendously and felt he "owed him" Interior for delivering Arizona to his campaign.[81]

The elated Udall immediately accepted Kennedy's offer. After a polite chat, they agreed that their first New Frontier conservation priorities would be passing the Wilderness bill and establishing new national seashores. As Udall was preparing to leave the residence, he asked Kennedy to consider having the poet Robert Frost participate in the inauguration by reading an original poem. "Oh no," the president-elect joked. "You know that Robert Frost always steals any show he is part of. If I did that the same thing would happen to me that happened to Lincoln at Gettysburg, with Edward Everett."[82] Then he burst into a full smile and agreed that having the esteemed New England poet read a poem was a fine idea.

Rose Kennedy had read Frost's "West-Running Brook" to her children on trips to Walden Pond. Two of the American writers the president-elect knew best were Thoreau and Frost. At campaign speeches throughout 1960, JFK quoted Frost's lines "But I have promises to keep / And miles to go before I sleep / And miles to go before I sleep." There was poetry in the politician, and Frost was his guiding light. For his part, Frost, on his eighty-fifth birthday, had endorsed Kennedy for president.[83] Kennedy felt that Frost's art could be appreciated on various levels with its imagery related to New England landscapes, both in their natural state and as farmland.

CHAPTER 9

Wallace Stegner's "Wilderness Letter"

Pulitzer Prize–winning American author and environmentalist Wallace Stegner (1909–1993) photographed at his home in Menlo Park, California, August 1976.

I

The Rockefeller Commission, officially the Outdoor Recreation Resources Review Commission (ORRRC), had been busy studying the importance of parks and preserves since its inception in 1958. "Man has a basic need for the outdoors," its chair, Laurance Rockefeller, philosophized. "He needs it for physical recreation, for contemplation, for stimulation of his senses and sensibilities. It is more than a matter of looking at scenery—though this, too, has its elemental value. It is a matter of assuring, for man, that measure of contact with nature that is vital in developing the whole man."[1]

A key member of ORRRC's team was James Gilligan, a prominent forester who had earned his PhD from the University of Michigan in 1953. Rockefeller had tapped Gilligan to lead the commission's Forest Service policy recommendations for the Wilderness bill. His able research assistant was twenty-five-year-old David Pesonen, who was attending the renowned School of Forestry at the University of California, Berkeley.[2] An outdoors enthusiast since childhood, Pesonen had read Wallace Stegner's eye-opening *Beyond the Hundredth Meridian: John Wesley Powell and the Second Opening of the West* (1953) about the famous Colorado River expedition of 1869. Courtesy of Stegner, Pesonen understood that the arid West didn't lend itself to large-scale farming without obtrusive dams to irrigate property.[3]

In June 1960, Pesonen wrote to Wallace Stegner, who was working on a novel and teaching at Stanford University, with an important request to help the ORRRC determine whether the Wilderness bill, as currently circulating around Congress, was rational conservation policy. Should the next president, be it Kennedy or Nixon, embrace roadless wilderness within national forests as essential for the future of American heritage. Was that nondevelopment designation fair to local economies? These were the types of questions Pesonen hoped Stegner, a very busy man, would help him with. It was a Hail Mary pass to the literary hero of the Dinosaur National Monument campaign. "I waited for about five months for a reply," Pesonen recalled "and then one day, shortly after Kennedy was elected, the six-page 'Wilderness Letter' arrived."[4]

Stegner's "Wilderness Letter" was a 2,492-word plea to the Department of Agriculture to defend pristine roadless areas in America's national forests. It became one of the foundational documents of the environmental movement, stating in part:

If I may, I should like to urge some arguments for wilderness preservation that involve recreation, as it is ordinarily conceived, hardly at all. Hunting, fishing, hiking, mountain-climbing, camping, photography, and the enjoyment of natural scenery will all, surely, figure in your report. So will the wilderness as a genetic reserve, a scientific yardstick by which we may measure the world in its natural balance against the world in its man-made imbalance. What I want to speak for is not so much the wilderness uses, valuable as those are, but the wilderness idea, which is a resource in itself. Being an intangible and spiritual resource, it will seem mystical to the practical-minded—but then anything that cannot be moved by a bulldozer is likely to seem mystical to them.

I want to speak for the wilderness idea as something that has helped form our character and that has certainly shaped our history as a people. It has no more to do with recreation than churches have to do with recreation, or than the strenuousness and optimism and expansiveness of what the historians call the "American Dream" have to do with recreation. Nevertheless, since it is only in this recreation survey that the values of wilderness are being compiled, I hope you will permit me to insert this idea between the leaves, as it were, of the recreation report.

Something will have gone out of us as a people if we ever let the remaining wilderness be destroyed; if we permit the last virgin forests to be turned into comic books and plastic cigarette cases; if we drive the few remaining members of the wild species into zoos or to extinction; if we pollute the last clear air and dirty the last clean streams and push our paved roads through the last of the silence, so that never again will Americans be free in their own country from the noise, the exhausts, the stinks of human and automotive waste. And so that never again can we have the chance to see ourselves single, separate, vertical and individual in the world, part of the environment of trees and rocks and soil, brother to the other animals, part of the natural world and competent to belong in it.[5]

The ORRRC thought the letter was magnificent. Career employees of Interior and some in the Forest Service loved Stegner's missive. His letter provided a moral underpinning for the Wilderness bill, just as Albert Schweitzer's "Declaration of Conscience" on Radio Oslo had done for the anti–nuclear testing crusade. In the December 1980 issue of *Living Wilderness* magazine, Stegner looked back and observed that, "Altogether, this letter, the labor of an afternoon, has gone farther around the world than other writings on which I have spent years."[6] When a friend joked to Stegner that he had become a conservation hero, the author scoffed at the idea. He was no Howard Zahniser or David Brower. "I am a paper tiger, typewritten on both sides," he said.[7]

What Stegner would call the "geography of hope"—the protection of public lands—summoned the fighting spirit of New Frontier conservationists. "We simply need," Stegner argued, "that wild country available to us, even if we never do more than drive to its edge and look in."[8] That lifted the concept of "use" from the domain of the measurable to an American ideal and gave the ORRRC the argument for roadless wilderness on public lands that it needed. Conservation was not a business, to be judged by "success" in drawing the most people. Describing one of his favorite places, a summit in Wayne County, Utah, Stegner explained how the unsullied wilderness would exist: "Save a piece of country like that intact, and it does not matter in the slightest that only a few people every year will go into it. That is precisely its value. Roads would be a desecration, crowds would ruin it." Those without the physical strength or the proximity to take a walk within a roadless wilderness "can simply contemplate the idea, take pleasure in the fact that such a timeless and uncontrolled part of earth is still there."[9]

Emboldened by Stegner, the mild-mannered Howard Zahniser kept pushing his idea that wilderness helped Americans preserve their capacity for wonder: the possibility of experiencing a herd of caribou, a black bear on the prowl, a stream of trout, or a condor flying over a lonely canyon. He emphasized the scientific value to humans to a greater degree than Stegner did, but both thought that wilderness areas could be conceived of as outdoor laboratories for botanists, ichthyologists, ornithologists, and mammalogists. The favor went the other way, too. In 1960, the American ecologist Nelson G. Hairston and his colleagues Frederick E. Smith and Lawrence B. Slobodkin published a seminal paper entitled "Community Structure, Population Control, and Competition," which Stegner believed proved the important function in ecosystems of top-down forces (predation) and bottom-up forces (food supply).[10] In practical terms, that meant, for example, that the eradication of wolves in Yellowstone would lead to the overgrazing of mountainsides by deer, and attendant problems thereafter. It was a scientific justification for the preservation of roadless, "uncontrolled" wilderness.

With Stegner's stirring arguments circulating in conservation nonprofit circles and Zahniser promoting the Wilderness bill on Capitol

Hill, Udall prepared to join the fight from his new cabinet post. Introduced as the next secretary of the interior at the Carlyle Hotel in New
York City, with the president-elect at his side,[11] he told reporters that
the next few years would be "our last chance" to protect seashores,
lakes, rivers, and old-growth forests.[12] Kennedy then affirmed that
the New Frontier was, in essence, the "third wave" in conservation,
following the advances made by Presidents Theodore and Franklin D.
Roosevelt.[13] "We haven't got much more time to get land for parks,"
he warned. "We have one at Cape Cod that we are anxious to do, and
time is running out. These areas are all being built up. I would say this
is probably one of the last chances we would have."[14]

For preservationists, the arrival of Udall to head Interior was like
the second coming of Harold Ickes. The *St. Louis Post-Dispatch* greeted
the new secretary with the headline "Udall Expected to Halt Trend,
Restore Interior Department as Champion of Natural Resources."[15]
Most media outlets understood that Udall wanted the Wilderness
bill passed and many new national park units created. What reporters didn't know was Udall's views relating to Native American sovereignty. The Bureau of Indian Affairs, after all, was part of his Interior
portfolio. When one reporter asked Udall whether "Indians would get
an extension of human rights," the Arizonan gave a positive but vague
response: "We haven't done enough for our Indian citizens. . . . We can
and we intend to do a better job."[16] That, inexplicably, was as specific
as he got. He didn't express strong views thereafter, either, which was
extremely disappointing, in view of the passion he and Mrs. Udall
displayed for amassing Indian artwork. But then, collectors tend to
respond to items, quite separate from the creators. An interest in artwork, held forth as a credential for the new secretary of the interior,
didn't guarantee bold new policy initiatives toward the rank federal
mistreatment of Indigenous citizens.

Often wearing cowboy boots and a bolo tie, Udall, a clothes hound
like Kennedy, enjoyed being depicted as the new western man. As
the *New York Post* noted, he was "a happy cross-breed of Western
outdoorsman and Eastern-style intellectual."[17] Jacqueline Kennedy
adored Udall; later in life, she rafted with him down the Colorado in
the Grand Canyon. Lady Bird Johnson enjoyed the company of the
rugged, debonair Udall when they rafted together in the following

years down the Rio Grande and Snake River. Right away, Udall publicly embraced the Sierra Club, National Wildlife Federation, and the Wilderness Society as groups that could help him set the policy agenda.

Such Washington power players as FBI director J. Edgar Hoover, Montana senator Mike Mansfield, and Secretary of Labor–designate Arthur J. Goldberg all sent Udall heartfelt congratulatory notes.[18] Senator Paul Douglas of Illinois, who was eager to establish an Indiana Dunes National Lakeshore along southern Lake Michigan, was also quick to encourage Udall. "Greatly pleased that you are saving the scenic shorelines of America for recreational purposes," he wrote. "We don't have any time to lose."[19] Even Republican senator Barry Goldwater, the Arizonan known as "Mr. Conservative," felt that Kennedy had scored a success in choosing Udall. Not only did Goldwater send a congratulatory telegram, but he volunteered to help Udall in any way possible. "Regardless of our political differences," he wrote, "I feel you are most qualified to make an excellent Secretary of the Interior."[20]

Another sign that President-elect Kennedy was prioritizing preservation was his selection as secretary of agriculture of Orville Freeman, a rarity inasmuch as he was well educated in both forestry and farm problems. The governor of Minnesota since 1955, Freeman had championed soil and water conservation, earning him high marks from the Minnesota Conservation Federation and the Izaak Walton League.[21] He was a passionate supporter of the Wilderness bill, because he wanted permanent protection for the Boundary Waters along the Canadian border as well as Pictured Rocks in Michigan and the Apostle Islands in Wisconsin.

Freeman, forty-two, was the youngest secretary of agriculture in history.[22] Diversified farming of plants and animals—growing a variety of both on one farm, as opposed to the monocultures favored by agribusiness—was his policy métier. At Agriculture, "Orv," as Kennedy affectionately called him, would ably oversee a budget of $6.4 billion, only $180 million of which went to the US Forest Service. Generally distrustful of the federal bureaucracy, Kennedy felt that Freeman might be able to shrink the sprawling Agriculture Department while making sure that the Forest Service maintained or improved its conservation course. From Udall's perspective, Freeman was an ideal counterpart because he wasn't dominated by agribusiness conglomerates—and it didn't hurt that he was a fellow outdoors enthusiast. "If there was such a thing as a conservation cabal, Freeman was a member of it," Udall recalled. "That bode well for Interior and Agriculture to collaborate. Together we were determined to overcome ancient and deep-seated bureaucratic rivalries."[23]

The big conservation surprise of December 1960 came from President Eisenhower. In a highly significant parting gesture, he announced three new, enormous national wildlife refuges in Alaska: the Arctic, in the northeast corner of the state, the home of the last great caribou herds; Izembek in the southwest, an essential migration stopover for migratory birds; and Kuskokwim along the western coast, the greatest waterfowl breeding ground in North America. Saving pristine wilderness had become, to a degree, a bipartisan cause, and that was crucial for New Frontier conservation efforts. The impetus for Eisenhower's directives in Alaska was Bill Douglas, who had visited the Sheenjek River camp of Olaus and Mardy Murie in 1956; on his return he regaled Ike with reasons to protect Alaskan public lands.[24]

II

On January 20, 1961, at the Eastern Portico of the US Capitol, forty-three-year-old John F. Kennedy was sworn in as the country's thirty-fifth president. Battling frigid temperatures, he delivered an inaugural address of only 1,355 words, a masterpiece of oratory that heralded a new era. Although the address did not champion natural resource management, it brimmed with the promise of new ideas and new ways. Implicit in the renewal of America about which Kennedy spoke was the third-wave conservation reform movement that he had been discussing. He had written the speech together with Ted Sorensen, and it was perfect. The evergreen line, a model of elevated language and memorable rhetoric, was a clarion call for public service: "Ask not what your country can do for you—ask what you can do for your country."[25]

Robert Frost did almost steal Kennedy's shining moment. Ushered to the podium by Udall, the white-haired sage began reading his new poem, "Dedication." After reciting only a few lines, however, he stopped, struggling with the glaring sun. "No, I'm not having good light here at all," he confessed. Despite a strenuous, desperate effort by Vice President Lyndon Johnson to gallantly block Frost's typescript from the sun with his top hat, Frost couldn't see his page properly. Refusing to be stymied, he improvised, reciting a poem he knew by heart—"The Gift Outright"—which began "The land was ours before we were the land's."[26] At the end, the crowd of more than twenty thousand people roared its approval.

Kennedy enjoyed that afternoon's inaugural parade, but he was embarrassed by the litter and blight along the route. On a whim, he established the President's Advisory Council on Pennsylvania Avenue to beautify that major boulevard. He also asked Udall to work with Nathaniel Owings, a nationally known architect with Skidmore, Owings & Merrill, to redesign the National Mall and nearby areas. The beautification of Washington was at hand.[27]

By contrast to Eisenhower, who had treated his interior secretaries as minor cabinet members, Kennedy immediately elevated Udall,

Stewart Udall with poet Robert Frost walking together at Dumbarton Oaks, in Washington, DC, while attending a Henry David Thoreau celebration.

looking on him as the lightning rod of the New Frontier's third-wave conservation. More than that, he trusted the new secretary. The White House didn't supervise Udall closely. "I was a wheeler and dealer, and it was a wonderful time," he said later.[28] Above all else, he inherited the stewardship of 760 million acres of federally owned land, ranging from Arctic Alaska and the North Woods of Maine to the Marshall Islands and the Everglades. He oversaw 55,000 employees in managing national parks, historic sites, fish and wildlife habitats, land and water management facilities, reclamation and hydroelectric power development, mine safety, Indian affairs, and overseas territorial administration. The *New York Times* noted, "It is, in other words, a big job."[29]

The seven-story Interior Building spans two large city blocks. Aiming for a southwestern feel, Udall decorated his high-ceilinged office there as if it were a natural history library at the El Tovar Hotel on the Grand Canyon rim. He hung an enlarged color photo of Rainbow Bridge, the majestic 290-foot-high sandstone arch in southern Utah. First People's antiquities were curated by Udall's staff and displayed in museum cases. A series of Ansel Adams's landscape photos of national parks from the 1940s and 1950s were likewise displayed in the department's corridors. Employees were immediately pleased with their upgrade from bureaucratic cave dwellers to curators and custodians of national historical relics.

To reach people outside the department, Udall began authoring articles on camping, fly-fishing, canoeing, and mountain climbing. Consciously, he was emulating Bill Douglas, who regularly contributed articles to the *Sierra Club Bulletin* and the *Living Wilderness* as a way to jump-start environmental preservation campaigns. In Udall's first essay, "National Parks for the Future," which appeared in the *Atlantic Monthly*, he praised Eisenhower's Mission 66 infrastructure program, which aimed to restore the national parks by fixing roads and building lodges, but he also issued a caution: the NPS, a victim of its own success, was short of rangers, field biologists, and even concessionaires.[30]

In the article, Udall worried that some sites, such as the Great Smoky Mountains National Park, a prime example, were being overwhelmed by tourism. Starting in the 1950s, Gatlinburg, Tennessee, gateway town for the park, was commercialized like Coney Island or Las Vegas. In an effort to attract midwestern tourists on their way to and from Florida, Gatlinburg became a sprawl of curio shops, minia-ture golf courses, fudge makers, and amusement rides. Traffic down-town was bumper to bumper. Inside the park, heavy pollution was visible from scenic lookouts. Overflowing garbage cans in the park became smorgasbords for black bears and raccoons. From Udall's per-spective, Great Smoky was being over-loved and under-managed. His article proved that he understood the parks' complex problems.

Udall liked being called "Stewart in chief" of public lands.[31] Hoping to be the face of the NPS in the public imagination, he soon pushed a plan to preserve Canyonlands (Utah), Flint Hills (Kansas), North Cascades (Washington), Great Basin Park (Nevada), and Santa Cruz Island (California), along with the three federal seashores: Cape Cod, Padre Island, and Point Reyes.[32] The only one Udall promoted that was not eventually designated was Kansas's Flint Hills (which be-came a national preserve in 1996). In the spirit of Franklin Roosevelt, he hoped to rehabilitate drained wetlands and transform certain areas into rest stops and watering holes for migratory birds on their long seasonal journeys. Udall wanted to increase the number of national wildlife refuges in Florida, Texas, North Carolina, and California. Fortunately, leading ecology-minded Republicans on Capitol Hill— especially Senator Thomas Kuchel of California and Representative John Saylor of Pennsylvania—were willing to collaborate. "President Kennedy has already struck the keynote," Udall wrote. "If we seize the opportunity and act to save the spaciousness and grandeur of our land, later generations may record this period as one of the most significant in the American conservation movement."[33]

With such a busy National Park Service agenda, Udall needed a crackerjack assistant. In early 1961, he hired Sharon Francis (née Fair-ley) for the job. Born on March 17, 1937, in Seattle, she grew up ex-ploring the forests of the Cascade and Olympic Mountains, beguiled by the trees—their awe-inspiring size, age, and canopy. At thirteen, she joined the Mountaineers, founded in 1906 to build a commu-nity of nature lovers who enjoyed climbing, hiking, and photography among the peaks of the region. As a Mountaineer, she studied the tenets of modern ecology with zeal. She also saw firsthand what hap-pened when those tenets were flouted. "At one point I was coming down off Boston Peak into a Douglas fir forest," she recalled. "They were logging there. It was one of those moments that changed my life. Got off the mountain and called Polly Dyer of the Conservation Committee of the Mountaineers. I said, 'Give me something meaningful to do, I want to save forests!'"[34]

Francis had telephoned the right person. An intrepid hiker and conservation proponent, Polly Dyer was a driving force for decades in preserving Pacific northwest wilderness. It was she who suggested the word *untrammeled* for use in the Wilderness bill. Born in 1920, she spent most of her childhood traveling the East Coast as the family followed her father, a government engineer, to various jobs. At a Girl Scout camp on the shore of the Chesapeake Bay, she discovered in herself a reverence for nature.[35] As an adult, she moved to Washington State and began campaigning to stop clear-cutting in a number of locales, including the Stehekin Valley; a copper mine in the shadow of Glacier Peak; and a ski resort, tramway, and golf course on Mount Rainier.[36]

While still a teenager, Francis took Dyer's suggestion to investigate illegal lumbering in Olympic National Park. Pacific Northwest timber barons, working in sync with the US Forest Service, dominated the forests like a colony of termites, illegally cutting down trees in the park. In 1952, working undercover, Francis began documenting illegal harvesting in the Olympic forestlands. "I did a couple of things," she recalled. "I hiked logging roads taking notes and measured how wide the access road had been built. A wider road was evidence of use by wider, log-hauling trucks. I quietly interviewed park employees who were afraid to be quoted for fear of losing their jobs. A lot of the park workers showed me protest letters they had covertly written to the National Park Service in Washington, to no avail. It was a wonderful experience for a girl like me to put my teeth into something real like that."[37]

Francis's report, disseminated widely by Dyer, became a crucial document in the Mountaineers' successful effort to save Olympic National Park from further cutting. That same year, she won a scholarship to Mount Holyoke College in South Hadley, Massachusetts, to study under Paul Sears, a celebrated conservationist. A political science major, she ended up writing her senior thesis in 1959 on the American wilderness preservation movement. She was by then a frontline fighter in the effort to establish North Cascades National

Park. When the club asked Scoop Jackson, the chairman of the Senate Interior Committee, to support the North Cascades effort, he replied, "If you get up a big enough parade, and I'll step out front and lead it."[38]

In mid-January 1961, Francis interviewed for the position of assistant to the secretary of the interior. By then she was married. Seven months pregnant, she didn't know if a nine-to-five government job would even be possible for her. Udall started the interview by praising a recent article by Francis, "Once Upon a Mountain," in the *Living Wilderness*.[39]

"Those North Cascades," he asked her. "Should they be a national park?"

"Yes," Francis said firmly, describing the rugged mountain peaks, dense evergreen forests, alpine lakes, and wildflower-filled valleys. With that answer, Francis began her stint as Udall's special assistant, with an office on the fifth floor. She was tasked with generating regional enthusiasm for the North Cascades National Park effort and helping Udall draft magazine articles. There was a familial feeling about the Udall office from the get-go. Francis's son, Christopher, was born that March. Within weeks she was back at work, sometimes bringing her infant along. "Stewart was very much a dad," she recalled. "He liked that I was a mom. He came by my office with my baby in the file drawer crib. He really enjoyed it."[40]

Family and outdoors recreation was in vogue in the New Frontier. While Kennedy was running for president in 1960, he and Jackie leased Glen Ora, a well-manicured farm in Middleburg, Virginia, horse country. Jackie loved equestrian activities such as steeplechases, and it proved the perfect place for the first couple to go for a gallop or an easy afternoon trot. Glen Ora was second only to Hyannis Port as the Kennedys' favorite place to get away from official Washington. Between 1960 to 1963, they spent thirty-six weekends at Glen Ora.[41] He couldn't enjoy riding, due to his back problems, but he thrived at the farm, surrounded by the horses Jackie so loved. Being in the pastoral setting around thoroughbreds taught JFK how to be far less sensitive to press criticism. "He became so tolerant—like a horse you see in the field in summer—the flies have annoyed him at first," Jackie wrote Ted Sorensen years later. "But there are long months before they will go away—so he does his work—which is eating grass—and just flicks his tail. Whenever I was upset by something in the papers—he always soothed me and told me to be more tolerant."[42]

Two weeks after taking office, Kennedy dropped the flag and launched the New Conservation agenda in earnest, sending his comprehensive *Special Message to the Congress on Natural Resources* to Congress. In impressive detail, it covered the concerns surrounding forests, air and water quality, oceans, energy, recreation, and public land stewardship.[43] Although his televised presidential speeches received vastly more media attention, his "special messages" were more pertinent, being specific.[44] Praising the accomplishments of Theodore Roosevelt and Gifford Pinchot, Kennedy firmly prodded Congress to establish Cape Cod (Massachusetts), Padre Island (Texas), and Point Reyes (California) as national seashores. He also spoke out against the overharvesting of high-grade timber, abuse of topsoil in the Great Plains, and the overmining of minerals in Appalachian states such as Kentucky and West Virginia; all needed to cease.[45] "The Nation's remaining underdeveloped areas of great national beauty are being rapidly pre-empted for other uses," he warned.

CHAPTER 10

The Green Face of America

Frank Church of Idaho seen here as a young man in the White Clouds Wilderness, circa 1940. As a US senator from 1957 to 1981, he became a driving force behind the Wilderness Act of 1964 and Wild and Scenic Rivers of 1968.

I

The amount of traveling that Stewart Udall did around the country in the spring of 1961 was staggering; he was always on the move. Some of the many backdrops may have become a blur after a while, but not on the trip he took in late April to a conference at Grand Canyon National Park.[1] Flying to the park, he happened to pass over the heart of the Colorado Plateau in southeastern Utah and had a bird's-eye view of the maze sculpted hoodoos, austere arches, sinuous canyons, towering spires, and standing rocks, all set amid stair-step benches and wilderness highlands. It was the Canyonlands at the confluence of the Green and Colorado Rivers. Udall learned that the historical back-story of Canyonlands dated back twelve thousand years, most of it in mystery, from the presence of ancestral pueblos to the whisper of dis-appeared civilizations. He knew with certainty that the area screamed for permanent protection. According to his son Tom, "My dad . . . saw the convergence of rivers and said, 'That's not a dam site but a national park in the waiting.'"[2]

The interior secretary rewrote the beginning of the speech he would give at the conference, paying homage to the Utah landscape he'd just seen: "No valley or basin in the nation has such grandeur and variety of beauty. I suspect there could be two or three more national parks in the area we flew over. There are still some areas and I hope we save some of them before it's too late."[3] Once back in Washing-ton, he teased Sharon Francis that the North Cascades she loved were green but the Canyonlands, which he preferred, were pink.[4]

Udall moved forward with a campaign to make the Canyonlands a national park. There were already grassroots efforts to save more than three hundred thousand acres of red rock by making a national park with four districts—the Island of the Sky, the Needles, the Maze, and the Confluence, referring to the two rivers that had carved two in-tricate canyons into the rock. The superintendent of nearby Arches National Monument, Bates Wilson, had been hoping for just such a designation. Wilson was an outdoorsman of the old-frontier variety, who slept under the stars, handled rattlesnakes, and went on week-long hikes in the slickrock wilderness. On board with the plan, he rou-tinely hosted activists and lawmakers whom Udall sent his way, taking them on memorable jeep rambles to the confluence of the banks of the Green and Colorado Rivers, the Maze District, and Horseshoe Canyon.

In Washington, Udall told Kennedy that he would make a formal inspection tour of Canyonlands that summer. If, as he thought, it was worthy of national park status, he would fight to persuade Congress to establish a unit there before the 1964 presidential election. He knew he'd have a powerful adversary, however, one with both deep pockets and importance to national security: the uranium industry in Utah and Arizona.

When Udall resigned from the House of Representatives to join

Kennedy's cabinet, his thirty-eight-year-old brother, Morris, narrowly won a special election to fill the seat. The Udalls—"Big Brother Stew" and "Little M," as the national press corps dubbed them—acquired a certain glamour in the public eye, similar on a far lesser scale to that of the Kennedy brothers.[5] Being a "liberal egghead" carried a negative stigma in southwest ranch country, but the Udalls were largely given a pass—perhaps because they were rock-solid Arizonans who rode horses, wore cowboy boots, hiked flattop mesas, and, most of all, were pro-dam. Mo Udall would go on to serve fourteen full terms in the House.[6] Like his brother, he prioritized conservation, reclamation, and public lands issues.

Anti-Udall blowback did ensue inside Arizona and elsewhere. Udall was more of a Harold Ickes liberal than Republicans had anticipated. Instead of being a reliable Bureau of Reclamation pro-dam booster—as his votes for the Glen Canyon project had indicated—he was showing signs of resisting some efforts by the Bureau of Reclamation to harness waterpower for energy. Bill Douglas, initially a skeptic, was pleased to find Udall receptive to preventing the Rankin Rapids Dam on the Allagash. In days to come, Douglas spoke out against multipurpose dams on the Buffalo River (Arkansas), Sangamon River (Illinois), Sunfish Pond (New Jersey), Little Tennessee River, and many more. All the while, Udall's sentiments seemed to be on the wild and scenic river side—except when it came to his home state of Arizona.

Part of the reason for Udall's partial transformation was his education in riverine ecology, courtesy of George B. Hartzog, Jr., the superintendent of the Jefferson National Expansion Memorial in St. Louis. In 1961, Udall and Hartzog took a float trip on the Current River in the Ozarks. Hartzog, a South Carolinian who had been ordained a minister as a teenager, had scant formal education. Nevertheless, he passed the bar exam on the third try—without ever going to law school—and served as a US Army captain during World War II. In the late 1940s, he became an NPS ranger and rose in rank to become superintendent of the memorial in St. Louis. "Beyond his skill as a park planner, it was the winning, masterful touch George had in dealing with all kinds of people that made him an unforgettable person," Udall recalled. "From that first encounter there was a brotherly relationship between us; we were the same age, and we had both learned about life growing up in the Great Depression in small towns."[7]

For the next few years, Udall and Hartzog worked in tandem to push for passage of the Ozark National Scenic Riverways bill, which sought to preserve the beautiful, hundred-mile-long Current River and its Jacks Fork tributary. The bill also sought to protect the river's surroundings, which encompassed thirteen caves, eleven geological sites, and more than forty archeological sites. Those rivers, in the heart of the Ozarks, were popular for canoeing and tubing. The Ozark National Scenic Riverways would be created in 1964.

Shortly after Udall's Canyonlands inspection, Douglas was out fighting against another proposed dam, the Riverbend project, which would flood forty-two miles of the Potomac Valley above Great Falls to a point just below Harpers Ferry, West Virginia.[8] Douglas recruited protesters to hike with him along the C&O Canal—a one-day, sixteen-mile reenactment of the 1954 protest. Stewart and Lee Udall, along with Sigurd Olson, were among the walkers. "We deal with values that no dollars can measure," Douglas declared.[9]

Once again, Douglas's hike drew glowing press coverage; power brokers in baggy clothes and boots reciting Robert Frost poems such as "Spring Pools" and "The Bear" made for colorful copy. An unexpected downpour forced Justice Douglas and his fellow hikers to seek shelter in the Old Angler's Inn in Potomac, Maryland. After draping their wet ponchos over the tables and ordering beverages, they unwrapped their homemade sandwiches and sat down to eat. "I run this place to make money, not to serve tramps," the proprietor's wife, Olympia Reges, scolded. "Get off that rug. Get over there with the rest of the wet ones," she ordered Senator Paul Douglas. And, turning to Stewart Udall, she shouted, "You look like a bum. Get out!" He complied and ate his lunch outdoors. When one of the hikers asked Reges if she knew whom she was bossing around and then told her he was a cabinet secretary, she replied indifferently, "Well, is he going to clean up the mess you make?"[10] The wet hikers still inside burst out in laughter and the story was carried in dozens of newspapers, garnering more publicity for the protest. The *Guardian* in Manchester, England, headlined the story "Minister Taken for a Tramp."[11]

Douglas's environmental activism truly knew no bounds. He even wrote a biography of John Muir for children in 1961, hoping to inspire a new generation.[12] The same year, incredibly, he issued yet another book on the conservation theme, *My Wilderness: East to Katahdin*—a follow-up to his 1960 book, *My Wilderness: The Pacific West*. Each described a number of extraordinary areas; the ones in *East to Katahdin* were all east of the Cascades and Sierra Nevadas. The *New York Times* titled its glowing review, "The Predator Is Man," paraphrasing a line from the book. Douglas, with his books, marches, articles, and buttonholing of practically anyone he met, was not only unceasing in his efforts but one of the most energetic people ever documented. "If I were to name one true hero of the sixties environmental movement, it was Bill Douglas," Udall said decades later.[13] After *East to Katahdin* came out, with a chapter on the Allagash waterway, the Maine ecosystem became famous for the first time, and Douglas stayed busy trying to protect it permanently. Conrad Wirth, the NPS director, stalled serious consideration for it, because of resistance from locals who wanted paved roads to enhance commercial timber development and those who favored damming the St. John River (of which Allagash was a tributary) for hydroelectric power. None of that stopped Douglas. By dint of his job title and connections, especially with the Kennedy family, he was perhaps the most powerful New Frontiersman *not* in the administration.

"I would like to see a big public hearing of the Park Service put somewhere in northern Maine," Douglas wrote Wirth in late 1961. "Wintertime would not be the best, but come early Summer or late Spring, it could be arranged. Fort Kent would be one possible place. They have a college there and hearing facilities might be available."[14]

Douglas next made an ally of Maine senator Edmund Muskie. They set their sights on Udall.

Muskie joined Douglas in urging Udall to tour the Allagash region. He did. "Senator Muskie got a little floatplane and we went up to canoe the Allagash," Udall recalled. "He was anxious to leave the river alone, and it was threatened by the Rankin Rapids Dam. So the Allagash was identified as one of the finest wild rivers in the eastern part of the country, and the fact that people in Maine wanted to save it left an impression on me, particularly when I canoed it. I had never been canoeing, really; we didn't have that many rivers in Arizona. Senator Muskie knew that Maine couldn't afford to buy land, yet the state wanted to do the job itself, so we eventually matched their funds with federal funds. We preserved the river for about $3 million."[15]

Udall was looking forward to the administration's first landmark accomplishment: national seashore status for Cape Cod. Even as he helped shepherd it through Congress, he was also maneuvering it through the media, cleverly striking a deal for good press coverage with the young *New York Times* reporter David Halberstam, who summered in Truro, a town on the Cape.[16] Both knew that setting aside forty miles of beach to create a national seashore from Provincetown Harbor to Chatham was a fantastic idea. But difficult. "We made a pact," Udall said later. "I'd keep him in the loop on all the administration's big Interior Department battles on things like oil import reductions and bills. He'd sneak a Cape Cod article into the paper. A [New York] *Daily News* reporter named [Peter] Edson figured this out and tried to damage both Halberstam and myself . . . to no avail."[17]

After a few months in office, Udall had bonded with several other *New York Times* writers, especially John B. Oakes and Russell Baker. Oates, a quiet leader in Hudson River Valley conservation efforts, had just joined the editorial board of the *Times* when Kennedy became president. He elevated the Hudson River and Cape Cod on the paper's opinion page. Russell Baker was a leading reporter for the paper and would write a well-known, even beloved column starting in 1962.

For all the support that the Cape Cod bill had in newspaper offices, the only opinions that mattered were in Congress. Many elected officials thought that the rights of homeowners and small businesses were about to be trampled upon. However, those objections were successfully countered by Udall. In March, when pressed by the House Subcommittee on National Parks, Forests, and Public Lands about how the national seashore designation would affect commerce in Cape Cod towns such as Chatham and Wellfleet, he pointed to a 1960 economic development report that the NPS had commissioned; it predicted that the park would create a boom in year-round visitation to the area.[18] In addition, Senator Leverett Saltonstall, a Massachusetts Republican, devised a way to allow many existing houses within the proposed park to stay in private hands. That and the support of two

Massachusetts Republican congressmen, Hastings Keith and Silvio Conte, bolstered Udall's case for the park. To solidify it, he had to address tedious details about land easements, exemptions, and buyouts. He said, "Let there be no doubt about it, we are in a race between those who would develop the last, best segments of the seashore for industrial and other commercial purposes and those who would preserve these limited areas."[19]

To complement the national seashores initiative, the Kennedy administration moved to establish a string of national lakeshores around the Great Lakes. That initiative coincided with the longtime dream of nature writer Sigurd Olson to gain some sort of federal protection for Boundary Waters adjacent to Lake Superior National Forest in Minnesota as a national park. Olson was receiving good press in 1961 for his book *The Lonely Land*, about a canoe trip he took in Canada that mimicked, as accurately as possible, the kind of travel undertaken by the famed *voyageurs* two hundred years before. Across much of Canada and the northern United States, *voyageurs* were expert in navigating any body of water, from a convenient pond to a rushing river to the Great Lakes. For that and other reasons, Olson, the modern *voyageur*, was the right man at the right time to advise Udall on the candidate lakeshores.

Often smoking a pipe, his ears attuned to the cries of Minnesota loons and screech owls, Olson was "the personification of the wilderness defender," according to former Sierra Club president Dr. Edgar Wayburn.[20] Having earned a science degree from the University of Wisconsin in 1920, he understood upper Midwest ecosystems as both an outdoorsman and a limnologist. Starting with *The Singing Wilderness* (1956), Olson wrote elegant prose about the ecology of Minnesota's freshwater lakes, which numbered 11,842. Traveling deep into the backcountry, an incorruptible and upbeat spirit, he became a master of the canoe—conquering currents, whirlpools, and shallows. "The way of a canoe is the way of the wilderness, and of a freedom almost forgotten," he wrote. "It is an antidote to insecurity, the open door to waterways of ages past and a way of life with profound and abiding satisfactions. When a man is part of his canoe, he is part of all that canoes have ever known."[21]

In the spring of 1961, Olson was working with the Wilderness

Sigurd Olson pours lake water into a can holding smallmouth bass fingerlings in a bay on Crooked Lake, Minnesota.

Society to protect two Minnesota sites, the Boundary Waters and another site, which didn't yet have a name. The two sites were both contiguous with the Superior National Forest, which bordered the Great Lake of the same name. In June 1962, Olson toured the region with the director of the National Park Service, Conrad Wirth, and the Minnesota governor, Elmer Andersen. All agreed that the water-rich region should be a national park. While canoeing on Crane Lake they brainstormed to think of a name. Olson spoke up on one of his favorite topics: the eighteenth-century *voyageurs* who had conducted commerce on the lakes of the area. He suggested "Voyageurs" as a name for the park. Wirth slapped his knee. "That's it!" he shouted. Establishing the park wasn't as easy. It took time to iron out the details, but at least the park had a vivid name.

Olson was intrigued by Udall, especially after receiving a letter about him from a mutual friend. "I don't think I have ever known anyone to move so far in so many directions in such a short period of time as the secretary [Udall]," Frank Masland, the chairman of the National Parks Advisory Board, wrote Olson in March. "If anything, he moves faster than Kennedy."[22]

The intrepid Olson emerged as one of Udall's top consultants on

wilderness-related issues and he was even tapped to scout for potential areas to add to the NPS. "In Alaska," Olson's biographer David
Backes wrote, "Olson played an important role in selecting millions of
acres of land that ultimately received congressional protection under
the Alaska National Lands Conservation Act of 1980."[23]

II

After being marginalized during the Truman and Eisenhower administrations, the Sierra Club and the Wilderness Society were raring to
take center stage in the New Frontier. In a sign of pent-up energy,
one thousand protoenvironmentalists from both inside and outside
government attended the Sierra Club's 7th Wilderness Conference in
San Francisco on April 7 and 8, 1961. The star attraction was Ansel
Adams, who was still on a roll with *This Is the American Earth*. At
fifty-nine, Adams was preparing to move from San Francisco to the
Carmel Highlands, near Monterey, along the rugged Central Coast of
California. The house he built, overlooking the Pacific, would soon
become a way station for California environmental activists. On any
given day in the coming years, Brower, Stegner, or the Wayburns could
be found sitting by Adams's fireplace or on the patio, plotting ways to
protect pristine California seascapes such as Point Reyes and remnant
coastal redwood groves. At the 1961 conference, Adams lectured on
"The Artist and the Ideals of Wilderness," a talk that summoned the
ghosts of Henry David Thoreau and John Muir.

Wearing a khaki shirt and bolo tie, Adams put on his tortoiseshell-
framed eyeglasses to read his text. Lobbying on behalf of the Wilderness bill, he invoked traces of transcendentalism as he observed that
the "Glory of God" had to be embraced if a third wave of ecological-
minded conservation were to sweep the land following in the footsteps of TR and FDR. "We have the vast and luminous evidence of
God in the realities of the cosmos in which we live," he told the audience. "We are at the threshold of a new revelation, a new awakening.
But what we have accomplished up to this time must be multiplied a
thousand-fold if the great battles are to be joined and won."[24]

Adams good-naturedly chastised his cohorts for being "too

clannish and self-centered." He said that conservationists needed to dismount from their high horses and talk to union workers, farmers, and small-business owners. He calculated that for every enthusiastic environmentalist at the San Francisco conference, there were 450,000 indifferent US citizens. Recruitment efforts to cultivate relationships with elementary school teachers, blue-collar communities, and people of color had to increase.[25]

Another intellectual star at the San Francisco conference was Joseph Wood Krutch. Less well known now, in 1961 he was revered by the conference attendees for his environmental musings and natural history books about Arizona ecosystems. Born on November 25, 1893, in Knoxville, Tennessee, he fell in love with the ethereal Great Smokies at an early age. As a student at the University of Tennessee in Knoxville, he adopted the Irish playwright George Bernard Shaw and the North Carolina novelist Thomas Wolfe as his literary heroes.[26]

The intellectually ambitious Krutch earned master's and doctorate degrees in drama from Columbia University. For almost thirty years starting in 1924, he was the drama critic of the *Nation*. After the war, his fascination with Manhattan super-sophistication receded. Imbued with a pastoral sensibility, the urbane Dr. Krutch longed for the days when the transcendentalism of Ralph Waldo Emerson and Margaret Fuller had been the defining pulse of the American intelligentsia.

Understandably, then, in 1948 Krutch chose to write a biography of Henry David Thoreau; the book launched a revival of interest in Thoreau's ideas.[27] Thoreau's 1862 essay "Walking," with the immortal line "In Wildness is the preservation of the World," touched Krutch deeply. Thoreau posed a radical challenge to his readers: "to regard man as an inhabitant, or a part and parcel of Nature, rather than a member of society."[28] Inspired, Krutch visited the Sonoran Desert and among the creosote bushes and saguaros, he had an epiphany. Tired of New York's brutal pace and worried that his own agitated heart would soon give out, he relocated with his wife to Tucson.

Nobody else has written about the Sonoran ecosystem with the tenderness of Krutch. His masterful 1952 nonfiction book *The Desert Year* successfully intertwined nature writing with serious philosophy.[29] Compassion for all God's creatures, even rattlesnakes and Gila monsters, was the Schweitzeresque philosophy that Krutch professed.

In *The Voice of the Desert: A Naturalist's Interpretation*, published three years later, he wrote unpretentious essays about arthropods—tarantulas, black widow spiders, scorpions, and centipedes—and small desert mammals, such as kangaroo rats, with unobtrusive wit and self-effacing humor.[30]

In 1958, Krutch published *Grand Canyon: Today and All Its Yesterdays*, a nonfiction homage that argued that "the wilderness and the idea of wilderness is one of the permanent homes of the human spirit."[31] In doing his geological fieldwork, Krutch had learned that whereas the majestic buttes of the Grand Canyon might seem unchanging, in truth, small changes slowly altered the landscape every hour. Combing the whole of the Grand Canyon and studying the myriad geological ages of the red-rust cliffs left him spellbound. To Krutch, the park was a sublime composition of endless themes and variations. "To say something is solid as the ground under one's feet is not to say much," he wrote. "That ground may be lifted miles into the air."[32]

Krutch felt that Udall was a politician in bed with the reclamation boys. During the floor debate in the House over the Glen Canyon Dam in 1954, an enlightened Democratic congressman from Florida, James Haley, dropped a hunk of canyon shale into a glass of water, and his colleagues had watched the rock disintegrate. His point was that the geology at the site wouldn't maintain a reservoir. Udall, also a congressman then, dramatically countered Haley by dropping his own piece of shale into a glass and saying he'd drink the water. That wasn't quite the point (and as it turns out, Haley was right; the reservoir at Glen Canyon loses 15 percent of its water to seepage, a high and very counterproductive amount). It was a bizarre demonstration by Udall, but it showed just how close an ally he was with the Bureau of Reclamation when it came to constructing the Glen Canyon Dam. A person with a sense of decency, Krutch believed, wouldn't murder a natural wonder like Glen Canyon.

In his eloquent speech in San Francisco, Krutch rejected the wise-use conservation tradition observed by the US Forest Service under Gifford Pinchot's direction as dangerously antiquated: "The premise of most conservationists today is the old premise that only man and his utilitarian needs count," he noted.

During most of American history that premise was workable. There was so much on our continent that it seemed inexhaustible, and the problem was how to tame and to use the wilderness. A generation ago we woke up to the fact that the frontier was disappearing and that there might conceivably be an end to resources. So, we began to talk about conservation—but usually only in the narrow sense—talked about it without changing our fundamental premise. Therefore, conservation meant no more than careful exploitation with some concern for the needs of the future as well as of the present. It is my conviction that this sort of conservation is not enough to ensure our welfare, either materially or spiritually.[33]

At the time of the conference, Krutch had just finished his autobiography, *More Lives than One*. Published in 1962, it concluded with a plea to control growth in Tucson.[34] The five-acre property that the Krutches shared was on the far outskirts of Tucson—where a bedridden Joseph Krutch was propped up on the last day of his life on May 22, 1970, so that he could see the beloved Sonora Desert with wildflowers blooming.[35]

Krutch drew inspiration from Thoreau's notion of "township parks," the idea of promoting urban wilderness as a New Frontier policy. Why not save the New Jersey Meadowlands as a bird refuge, one that New York City residents could also enjoy? Why not protect the San Gabriel Mountains winding through greater Los Angeles? Couldn't Houston have a Galveston Bay federal park? It wasn't too late.[36] As he told his audience in San Francisco, Thoreau's Harvard graduation speech of 1837 was still relevant in 1961, and he quoted it: "This curious world which we inhabit is more wonderful than it is convenient, more beautiful than it is useful. It is more to be admired than to be used."[37]

With one voice, the conference urged the Kennedy administration to keep its early conservation promises and build on them, while there was still time. It became clear in San Francisco that the Sierra Club had become the most powerful nonprofit conservation organization. Some skeptics regarded the California-based nonprofit as a gaggle of Berkeley alumni, a white-privilege yodeling cult that acted as though Yosemite was where Jesus had given the Sermon on the Mount.

Others mocked the club as predatory, always hungry for donor checks to finance the board of directors' junkets to the North Cascades and Lake Tahoe. But with the publication of *This Is the American Earth*, and with distinguished elders such as Douglas, Adams, and Stegner among the club's rank and file, it was hard to dismiss those in the long shadow of Muir as a mere alpine social club. With Brower in charge, the Sierra Club, in truth, was the vortex of the wilderness preservation movement coming to power in the United States. Under his guidance, it grew from 2,000 to 77,000 members from 1952 to 1968. And its reach across the continent made Washington lawmakers take notice.

Recognizing the political power of the Sierra Club, Secretary Udall, who had been invited to speak at the conference, responded to the call to action inherent in the room. He aimed his comments directly at Brower, mentioning his name a half-dozen times. He praised *This Is the American Earth* and promised to implore the Departments of the Interior and Agriculture to collaborate wholeheartedly to protect the United States' dwindling natural resources. It became apparent from Udall's performance that he wanted to be perceived as a man having an ecological conscience who, like the busy man in Robert Frost's poem "To the Thawing Wind," had escaped from his "narrow stall" in Washington, even with reports on the floor, to be turned "out of door" to golden California. To win the audience over, especially anti-dam skeptics such as Krutch, he read lines from Frost's poem "Dust of Snow":

> The way a crow
> Shook down on me
> The dust of snow
> From a hemlock tree
> Has given my heart
> A change of mood
> And saved some part
> Of a day I had rued.[38]

Udall then endeared himself to his listeners even further by reading aloud Stegner's "Wilderness Letter." Thunderous applause greeted

Stegner's admirable words: "Without any remaining wilderness we are committed wholly, without chance for even momentary reflection and rest, to a headlong drive into our technological termite-life, the Brave New World of a completely man-controlled environment."[39] When Udall finished, the crowd offered a standing ovation. That was the moment when protoenvironmentalists understood that the Kennedy administration, with the exception of backing too many Army Corps of Engineers and Bureau of Reclamation hydroelectric dams, tilted to Muir's side of the long-standing conservation versus preservation debate.

Shortly after his address in San Francisco, Udall gave an in-depth interview to the *Deseret News*. Instead of talking about work on the dam at Glen Canyon, which naturally interested readers in Utah, he focused on open space and the Wilderness bill:

> The glory of America has always been its green face . . . its spaciousness. The whole character of the American people has been shaped by living on a virgin continent where men could test themselves against the wilderness. But our land is changing before our eyes. The bulldozers are eating away the last of the wild areas in the East, and even in the West, rapid population growth is exerting pressure on open spaces.[40]

The conference, with its emphasis on equalizing the relationship of humankind and the natural world, reflected an ecological trend arising across the country. Kennedy, as president, may not have mounted the bully pulpit as a preservationist, but he didn't have to. His secretary of the interior, clearly, was taking up the fight, as well as the mantle of Harold Ickes. And when the administration promised to work toward passing the Wilderness bill, young people, in particular, were drawn to it. So it was that after his inauguration, the conservation movement broadened. Folk singers, movie actors, novelists, and historians spoke out about ecology. The grand old men of poetry, such as Robert Frost and Carl Sandburg, pointed the way to a new generation of writers exploring their own connection to the physical world. In the Bay Area, the Beat poets Lawrence Ferlinghetti and Michael McClure launched the cutting-edge *Journal for the Protection of All*

Beings, which had an ecological theme. Jackie Kennedy had read Jack Kerouac's account in *The Dharma Bums* about the Beat writer's Forest Service stint on Desolation Peak in the North Cascades, during which, influenced by Buddhist texts, he realized that seeking solitude in nature leads to spiritual enlightenment.[41] A literary hero of the emerging youth counterculture, Kerouac drew his readers to respect wilderness and look for places to experience it for themselves, thus expanding the potential constituency for backcountry conservation.

III

On one plane ride in early June, Udall fretted in his diary that he wasn't part of Kennedy's inner circle; he felt like an appendage of the New Frontier instead of a trusted adviser. On the big issues of the first half of 1961—the Freedom Riders, NASA's Project Apollo, and Cold War diplomacy with the USSR—he wasn't consulted by the president. "Are parks really important, I ask myself, as compared with Laos and Castro" he pondered. "Yes, if one takes the long-haul as I must and [Senator] Clint Anderson restores my confidence in the importance of our sector of the front." He bolstered his morale by realizing that saving priceless landscapes would gain support for Kennedy "in the way missiles in place cannot."[42]

Anderson, a Democrat from New Mexico, harbored an extravagant belief: that the 1960s would come to be known as the environmental decade. Born in Centerville, South Dakota, he nearly died of tuberculosis while growing up at the turn of the nineteenth century. He was admitted to an Albuquerque sanitarium, and due to a combination of expert doctors, fresh air, and sheer willpower, his health rebounded. For the rest of his life, he insisted that the natural world held curative powers. Plagued with health problems throughout his life, he never faltered, becoming a Washington conservation infighter, first as a congressman and then as Truman's secretary of agriculture. Influenced by Aldo Leopold's land ethic, he campaigned to protect such treasured New Mexican landscapes as Organ Mountains–Desert Peaks and Kiowa Grasslands.

When Anderson won a Senate seat in 1948, he became a national conservation thought leader. Although he was an ardent supporter of the Bureau of Reclamation's damming the Colorado River to produce hydroelectric power, he reiterated his long-standing belief that saving wilderness heritage was a top priority. Throughout the 1950s and into the 1960s, he continued to champion a National Wilderness system, writing that, "Wilderness is a demonstration by our people that we can put aside a portion of this which we have as a tribute to the Maker and say—this we will leave as we found it."[43] Anderson was a crucial and very close ally of JFK's conservation policies from his Senate seat.

In July 1961, when Udall embarked on the inspection of Utah's Canyonlands that he had promised Kennedy to undertake, he took his wife and children along. Accompanying them were Frank Masland, Orville Freeman, and a few others. The group rafted the rapids of Cataract Canyon, swam in the calmer parts of the Colorado River, and rode in Jeeps to see the surrounding prickly-pear countryside. "This rugged country has much of the grandeur of the Grand Canyon, but has a wider variety of outdoor resources including sandstone formations which have a beauty that rivals anything in the National Park System today," Freeman wrote in his diary. "In time the splendor and great variety of the scenic beauty of this area will make it one of the gems of the National Park Service."[44]

That prospect, however, wasn't a given. Uranium excavations were encroaching, drawing near with the accompanying drilling equipment, pipelines, pumpjacks, towering storage tanks, and a sprawling network of roads built to accommodate heavy truck traffic. Udall, Freeman, and Masland swore that they would use all their combined power and influence with JFK and Congress to establish Canyonlands National Park.

Through the summer, the Kennedy administration worked hard for the Wilderness bill. With Anderson as the lead sponsor, it passed the Senate by a 73–12 vote in September. Frank Church, the Idaho Democrat, and Thomas Kuchel, the California Republican who was the minority whip, dined together to toast the accomplishment. Yet it wasn't time to declare victory. In the coming months, the House Interior Committee, chaired by Aspinall, would approve a completely rewritten substitute that was unacceptable both to the Sierra Club and to the Kennedy administration. Representative John Saylor derided

Aspinall's version as a "perversion" of the Senate's bill. "It was indeed a substitute bill," he fumed. "For the preservation of wilderness, it substitutes protection for exploiters of our wilderness areas."[45] For Howard Zahniser, it was back to the drawing board to find a sweet-spot formula, a quid pro quo of some kind, to get Aspinall to soften his stance.

While leaving the Wilderness and Cape Cods bills to Udall, Kennedy thought that he had scored a diplomatic coup in June 1961 by persuading Khrushchev to include on the agenda of their upcoming summit a discussion of ending nuclear weapons testing in the atmosphere, in outer space, and underwater. Ever since 1956, Kennedy had supported a global ban on nuclear weapons and he'd favored the temporary moratorium on atmospheric nuclear testing, which had gone into effect in November 1958. He wanted to make it permanent. That view was in sync with the United Nations Disarmament Commission, which wanted the nuclear powers of the time—the United States, the United Kingdom, France, and the Soviet Union—to end atmospheric testing. After traces of radioactive deposits were found in wheat and milk in the northern United States, millions of Americans agreed.

At the vaunted summit with Khrushchev in Vienna, Austria, the Soviet leader was recalcitrant, and there was no substantive talk about a test ban. Shortly after the summit, he declared his willingness to halt western access to Berlin and he threatened war if the United States tried to stop him. With tensions rising precipitously, Kennedy's foreign policy advisers pleaded with the president to resume testing of nuclear weapons to prove that the United States wasn't afraid of the Kremlin's brinksmanship tactics. He was reluctant.

Then in August 1961, as the Berlin Wall was being constructed, Khrushchev announced his decision to restart atmospheric nuclear testing. Over the next three months, the USSR conducted thirty-one nuclear tests, including the detonation of the biggest nuclear bomb in history: 58 megatons, four thousand times as powerful as the bomb dropped on Hiroshima.[46] Outraged by the Soviet tests, Kennedy pursued a last-ditch diplomatic effort to put the moratorium back into place. But in the end, he felt compelled to resume US testing, and he did so on April 25, 1962. Nevertheless, he was determined to find a diplomatic way to forever ban atmospheric and underwater nuclear testing during his first term. It seemed insane to him that radioactive fallout should be blowing around the world and that the superpowers' leaders, sitting on opposite sides of the world, could destroy civilization with the push of a button.

IV

On August 7, 1961, in the midst of the Berlin Crisis, Kennedy had one of his proudest moments, signing the legislation that established the 26,666-acre Cape Cod National Seashore. At the White House ceremony, surrounded by Democratic and Republican lawmakers who had supported the project, Kennedy pronounced his hope that Cape Cod's "marshes, the seascapes, and the sea itself should remain inviolate for all time for all men."[47] At last, Massachusetts had its own national park on a grand scale, like those in the western states. The park increased the NPS's acreage in the eastern United States by 38 percent. "This act makes it possible for the people of the United States through their Government to acquire and preserve the natural and historic values of a portion of Cape Cod for the inspiration and enjoyment of people all over the United States," Kennedy boasted. "This is a wise use of our natural resources, and I am sure that future generations will benefit greatly from the wise action taken by the Members of the Congress who are here today."[48] According to Paul Schneider in *The Enduring Shore: A History of Cape Cod, Martha's Vineyard, and Nantucket*, "the most important factor in the creation of the first national seashore was that the summer White House at the time was located in Hyannis Port, and its resident was convinced of the importance of saving Thoreau's Great Beach."[49]

Encompassed in Kennedy's Cape Cod National Seashore were forty miles of priceless sandy beach, salt marshes, and upland forests facing the Atlantic. Lighthouses and wild cranberry bogs gave the landscape an old-style New England charm. Parts of six towns, including Truro, Wellfleet, and Provincetown, were included in the federal park.

In the summer of 1961, Cape Cod was overrun with tourists determined to spot a real live Kennedy, the same way visitors to

Yellowstone clamor to see grizzly bears from their cars. They hoped to get a glimpse of Jack and Jackie cruising on their yacht, enjoying the sea breezes at Katama Bay, or perhaps snap a photograph of Robert and Ethel Kennedy playing touch football with their children, buying ice cream cones in town, or arriving at St. Francis Xavier church. More and more, Sunset Hill Drive, which took visitors around Hyannis Port by way of Kennedy's neighborhood, suffered from traffic congestion—and a Secret Service contingent to look everyone over.

The success of the Cape Cod National Park effort enthralled a wide range of preservationists. Horace Albright, a cofounder of the National Park Service and its director from 1929 to 1933, used the event to plead for additional federal shorelines. In a letter to the *New York Times*, he promoted Point Reyes, Oregon Dunes, and Padre Island. It was also incumbent on Kennedy, he wrote, to establish three national lakeshores in the Great Lakes region: Indiana Dunes, Pictured Rocks, and Sleeping Bear Dunes. Those treasured places were all being studied. Albright made a good point about the American spirit, an entity much bandied about among conservationists, but not usually in a practical way. "Our pioneer American instincts turn our thoughts to building," Albright wrote, "we have always taken land for granted. But now we must give thought to preserving some of the land, which can only decrease."[50] Arriving Americans, bringing the concept of landownership with them (it was unknown to Indian nations), started marking the land by building on it from the earliest days. As Albright perceived, that habit had to change when ecosystem preservation was needed in places like shores, where so much as building a house on a half acre was detrimental to the fragile environment.

Kennedy designated a slew of important national wildlife refuges in his first year in office. Among them were spots of extreme importance to specific species: Wapanocca (Arkansas), a migration stopover for warblers and other neotropical birds along the Mississippi River Flyway; Washita (Oklahoma), a shelter area for white-tailed deer and the endangered Texas horned lizard; Ottawa (Ohio), the home of some of the last remnants of the Great Black Swamp era of the Lake Erie marshes; and Wyandotte (Michigan), the prime stopover for migratory diving ducks of the Anatidae family along the Detroit River. The national press wasn't very interested in new wildlife refuges; the deskbound Rachel Carson, however, cheered for each new refuge as she was finishing *Silent Spring*. Perhaps, she thought, some ambitious young Fish and Wildlife Service employee would oversee a new *Conservation in Action* series to document the Kennedy refuges.[51]

Shortly after the Cape Cod National Seashore victory, Stewart and Lee Udall visited Robert Frost at his bucolic homestead, the Homer Noble Farm in Ripton, Vermont. Together they rambled the thickly forested countryside, even wandering around the Bread Loaf writers' compound together. As in Utah, it was hard for the interior secretary to hike without reciting Frost verses. Because Frost's poems were rooted in the green glens, small farms, and country stores of rural New England, Udall returned to Washington convinced that the three-room cabin where Frost slept and wrote, along with the woodlands around the town of Ripton, needed to be saved as a park. That plan, however, went nowhere. (Middlebury College would later acquire Frost's home in Ripton.)

Udall's obsession with Frost had its detractors who charged that he was trying to ride the poet's coattails to his own literary fame. That insinuation—fair or unfair—was repeated when Udall hired Wallace Stegner in the fall of 1961 to work at Interior. Other top writers joining Udall's staff included Native American historian Alvin M. Josephy, Jr.; Harold Gilliam of the *San Francisco Chronicle*; and Peter Matthiessen, the author of *Wildlife in America*.

One afternoon, Udall complained to Stegner about the bad press he was getting for being a media hog. "How do you shake this sort of thing?" he asked. Stegner suggested he do what "outlaws used to do": ride far upstream to shake the posse. "Just lower your profile, disappear for a while, and write a book."

"About what?" Udall asked.

"About what you are doing here and what you think needs to be done."[52]

Clearly, Udall, who did enjoy the sunshine of fame, wasn't about to disappear from the public eye. But the idea of writing a book struck his fancy. Out of his conversation with Stegner was born *The Quiet Crisis*. Stegner was the foreman of the project, corralling chapter drafts from the department's stable of writers. Udall inevitably received most of the credit, which Sharon Francis said he "deserved, as he participated in draft after draft." Stegner concurred and bent over backward to support the secretary. "The [book's] big and unique strength ought to be that it comes from you," he wrote in a letter to Udall, "it flows out of the source of political power, and therefore has the authority and the hopefulness of practical accomplishment that a lot of conservation writing lacks! We can sing all we want to, but when a man with his hand on the lever sings the same song, something new has been added."[53]

Before the end of the year, Kennedy used the Antiquities Act to establish Buck Island Reef National Monument, much to the elation of Laurance Rockefeller. Buck Island and its rare elkhorn coral reef are situated just off the northeast coast of St. Croix Island in the US Virgin Islands. *National Geographic* magazine called Buck Island's beach one of the world's most beautiful. The NPS designation meant that spotted eagle rays, nurse sharks, and other rare creatures were safe and that snorkelers and divers from around the world could see them. In signing the proclamation, JFK stressed that Buck Island Reef was of "great scientific interest and educational value to students of the sea and to the public."[54] Seventy-three-year-old Joseph Kennedy, Sr., knowing how much his son Jack loved the Virgin Islands, made a present for him of the mural *Christiansted Harbor at Saint Croix, the Virgin Islands* by Bernard Lamotte. It was soon displayed alongside the White House swimming pool.

Kennedy also resorted to the Antiquities Act to establish Russell Cave National Monument in northeastern Alabama. The Russell Cave was used by hunter-gatherers as many as twelve thousand years ago. Named for the local family that had owned the site for decades, the cave was not a claustrophobic warren of dripping rooms and cramped passages; instead, it's wide—and often sunny near the mouth. Native Americans used the larger parts of Russell Cave as a communal shelter in cold weather. Unfortunately, in the 1950s, thieves stole ancient fishing hooks fashioned from bone, flint spearheads, and decorative pottery pieces from the cave. The National Geographic Society, which had managed to acquire the site, believed that federal protection of Russell Cave was imperative, both for its own sake and because the southern states had too few national monuments. While campaigning for support of the monument, Udall invited Alabama's senators on

Potomac River cruises, excursions which Udall called "cultivating for conservation."

To the delight of Joseph Wood Krutch, Udall convinced Kennedy to use the Antiquities Act to add 15,360 acres of federally owned land in Arizona's Tucson Mountain Park to the Saguaro National Monument.[55] Kennedy and Udall pursued the enlargement to stop mining in the park and preserve the site's magical mix of mesquite and palo verde trees and prickly pear, cholla, saguaro, and hedgehog cacti.[56] The park was close enough to Tucson to be considered urban wilderness. Though a native Arizonan, Udall was learning to observe more carefully the saguaro cactus, the pale green symbol of the Southwest with its slim fingers reaching skyward. "An apartment building of the desert, the saguaro provides living space for several species of birds," he enthused, mimicking Krutch. "The Gila woodpecker, elf owl, gilded flicker, sparrow hawk, purple martin, and flycatcher nest inside the stems, while larger birds such as the red-tailed hawk and great horned hawk live among the branches."[57]

Buck Island Reef, Russell Cave, and the Saguaro expansion proved that President Kennedy wasn't afraid to use his executive power for the sake of nature. Although those national monuments weren't huge, they functioned as momentum builders in subsequent New Frontier battles. At the end of the administration's first calendar year, Udall was also successful in adding two Old West forts, Fort Smith (Arkansas) and Fort Davis (Texas), to the National Park System.

As 1961 closed, Udall received his dream present from Stegner: an annotated outline of his book—and a title, *The Quiet Crisis*.[58] The facile author had also drafted four chapters of the book to get it started. Stegner, whose government employment was ending, praised Udall in a letter. "I think I have not yet said, for the record, what an exhilarating experience Washington was," Stegner told his boss. "As my grandmother used to say, I enjoyed it. I also got a good inside glimpse of your vision of the next few years in conservation, and I'm sold. I'll help any way I can."[59]

CHAPTER 11

Rachel Carson, the Laurance Rockefeller Report, and Kennedy's Science Curve

A helicopter sprays DDT in the Wallowa-Whitman National Forest in Oregon and Idaho. The pesticide was used by the Forest Service to combat the tussock moth, which damaged millions of Douglas firs during one of its periodic plagues.

I

By early January 1962, Rachel Carson had completed all but the last chapter of *Silent Spring*. She had already started distributing chapters to a wide range of influential and ecologically concerned New Frontiersmen, including Stewart Udall and Jerome Wiesner, Kennedy's science adviser. While preparing the book for publication, Carson contracted iritis, a painful eye inflammation. The ailment limited her work, already hampered by her poor health, to a few hours a day. Stress consumed her. Everything seemed urgent, from the need for libel insurance to worries about publicity for the book.

What encouraged Carson the most as she finished *Silent Spring* was President Kennedy's State of the Union address on the eleventh of January. Calling for "a new long-range conservation," Kennedy forcefully sought the expansion of national parks and forests and the permanent "preservation of our authentic wilderness areas." Written by Sorensen, the words vibrated with ecological overtones. The New Frontier got it: nothing in nature was independent of all else (and that included the American people).[1]

That February and March, working long hours as best she could, Carson made her final changes to the manuscript and delivered it to Houghton Mifflin. Another copy went to William Shawn at the *New Yorker*, who enthused that Carson had turned pesticide use into "literature."[2] Then the drive for publicity began. Marie Rodell reminded the Houghton Mifflin team, Carson's *The Sea Around Us* had sold around 2 million copies.[3] That didn't make the road for *Silent Spring*, a different kind of book, much easier, though. "This is not an easy book to tell people about," Brooks warned. "We are going to have to work up something of a crusade—on a local level—if we are to reach a really wide audience."[4]

With the *New Yorker* slated to run an excerpt in the June 16 issue, the nervous Carson sought meet-and-greet opportunities. One of them occurred on May 21, 1962, when she attended a dinner speech by William O. Douglas before the National Parks Association in Washington. His topic was the plan of the Army Corps of Engineers to dam the Potomac: "A matter of life and death for one of America's most beautiful and most historic valleys," he said.[5] The threat extended beyond any one river, though: the rapid acceleration of engineering technology made multipurpose dams easier to construct throughout the United States.[6]

While gunning for the Corps of Engineers, Douglas unexpectedly evoked *Silent Spring* before the packed auditorium of nature lovers. He had read portions of the manuscript and had been mightily impressed. To him, the lesson of *Silent Spring* wasn't just the perils of DDT; it was far bigger. "The chemical specialist knows what effect an insecticide will have on a weevil, but it takes a person who is broader gauged and who is not on the payroll of a chemical company to advise whether the men who use the insecticide will be harmed or when the insecticide, washed by irrigation water or by rain into the rivers, will injure plant life and fish," he said.[7]

Speaking in general terms, Douglas pointed out that the Army Corps of Engineers and the Manufacturing Chemists' Association, whose membership included DuPont, Monsanto, and Dow, were reading from the same playbook. First, they offered surface appeal to the public: DDT keeps those pesky mosquitoes away from your backyard barbecue; the new dam will cause your electricity bill to drop. What the industrialists didn't mention was that the sprays and hydroelectric power could be disastrous in the long run. To Douglas, those producers of synthetic chemicals, like the public works engineers, were hoodwinking Americans. "Those who doubt this thesis," he said, "should read Rachel Carson's forthcoming book *Silent Spring* and learn how dangerously far the expert chemists have taken us toward ruination of the earth and its waters through poison."[8]

Carson was impressed that the prosecutorial Douglas minced no words in his lecture about riverine destruction. The Corps of Engineers hadn't offered the public an analysis of the environmental damage that the proposed Seneca Dam on the Potomac would do. Their justification in May 1962 had been "an insult to the layman's intelligence," Douglas charged. That blueprint contained, for example, no data on the project's impact on wildlife. Douglas aimed to get the National Parks Association behind the fight to save the Potomac Basin. "It is as irresponsible as a court would be if a case were set for trial without the issues being drawn and with no witnesses available except those selected by one side," he insisted. "What the Corps of Engineers has done in the case of Seneca Dam is a discredit to responsible agency action."[9]

Douglas was trying to shame the Corps into holding public hearings. If he could pull that off, he could likely kill the Seneca Dam project, driving a stake straight through the blueprints. Town halls, he believed, were the best friends of the unharnessed river movement. If fair hearings were held, the citizens of Virginia, Maryland, and the

District of Columbia would at least learn what lands—and wildlife—the dam would inundate.

Douglas admitted that the Army engineers were first-class builders and insisted that they weren't inventors of destruction, as some protoenvironmentalists painted it. But Douglas asserted that their engineering expertise didn't mean that their ethics were correct. Somebody had to speak out on behalf of the aesthetic and spiritual values of nature. To his everlasting consternation, the Mississippi River had been overbuilt with a string of dams and nonstop levees. Similarly, the Ohio River was "back-to-back impoundments" for nearly a thousand miles. On the Colorado, dams had destroyed scenic canyons and the river didn't run freely into the Gulf of California anymore. By the time the Colorado River got to Mexico it was a mere trickle. "As knowledge becomes more and more compartmentalized," he said, "the expert in one sense becomes more and more a threat to society. For his expertise, though lacking the value judgment necessary for the common good, tends to carry the day."[10]

After his National Parks Association lecture, Douglas buttonholed Carson to tell her face-to-face how much he appreciated *Silent Spring*. Writing to Dorothy Freeman afterward, Carson said: "The justice and Mrs. D. were at dinner. He came to me and said 'Your book is tremendous. I'm selling it every place I go.' In his talk at the Forum, which was about Potomac River dams and the Army Engineers, he digressed to speak of the havoc wrought by 'experts' in other fields, and added, 'Everyone should read Rachel Carson's forthcoming book, *Silent Spring*, to learn what the chemical engineers are doing to our world.' "[11]

In Douglas's mind, *Silent Spring* was a warning flare not only about DDT but also about nuclear fallout, water pollution, smog—the entire industrial-military attack on the natural world. The unfortunate resumption the previous year of atmospheric nuclear testing by the Soviets and in turn by the United States was, if anything, even more worrisome to him in the context of Carson's description of the ecological ripple effects of the chemical contaminants. Fallout from Soviet testing had blown over the United States, and radiation readings in US cities such as Minneapolis, Des Moines, and St. Louis were on the verge of surpassing the federal guidelines for "safe" limits. Milk suppliers on dairy farms throughout the Great Plains and Midwest were detecting radiation in the milk.[12]

What Carson did brilliantly, Douglas thought, was to connect pesticides and radioactive fallout in terms of the ecological doom they both threatened. Only in the twentieth century, Douglas and Carson agreed, had human beings gained the capacity to destroy the earth. Douglas felt that the opening fable Carson told in her book about a spring minus birdsong was the perfect metaphor for environmental degradation writ large. She had written, "No witchcraft, no enemy action had silenced the rebirth of new life in this stricken world. The people had done it themselves."[13]

If one were to pick a moment when the modern environmental movement was born, it was when the American public read this paragraph in *Silent Spring*:

In this now universal contamination of the environment, chemicals are the sinister and little-recognized partners of radiation in changing the very nature of the world—the very nature of its life. Strontium 90, released through nuclear explosions into the air, comes to earth in rain or drifts down as fallout, lodges in soil, enters into the grass or corn or wheat grown there, and in time takes up its abode in the bones of a human being, there to remain until his death. Similarly, chemicals sprayed on croplands or forests or gardens lie long in soil, entering into living organisms, passing from one to another in a chain of poisoning and death.[14]

Carson, for her part, had acknowledged in *Silent Spring* the literary inspiration she had drawn from Douglas's 1961 *My Wilderness: East to Katahdin*.[15] In the book, Douglas had rhapsodized about the sun and wind in his face in places such as the White Mountains of New Hampshire and Mount Katahdin in Maine, recounted the unmatchable pleasure canoeing the wilderness of Quetico-Superior in Minnesota, and expressed his despair over the aerial spraying of DDT to kill insects. There is no better literary example of the synchronicity between the movements to ban DDT and to pass the Wilderness bill than Douglas's book. At times, his plea for preservation surpassed that of Thoreau or Muir in terms of advocacy. "We need a restatement of national purpose," he wrote. "We need to bring to our educational programs a new ethic. Man is capable of care as much as he is of destruction. Preservation of beauty, tenderness in relation to other life, communication with nature—these too can be awakened and given a powerful thrust. If we make conservation a national cause, we can raise generations who will learn that earth itself is sacred."[16]

Because Douglas was certain that chemical companies were culpable of poisoning the environment, he became Carson's greatest encourager.[17] Likewise, she quoted him on four separate occasions in *Silent Spring*. Drawing from a chapter of *East to Katahdin*, she told how he had visited Bridger National Forest in Wyoming in 1959 and been appalled by the irrecoverable ecological destruction he saw here. In that once verdant Wyoming landscape, the US Forest Service, "yielding to pressure of cattlemen for more grasslands," had sprayed ten thousand acres to clear it of sagebrush, killing off the area's willow trees as well. Douglas arrived the next year, Carson reported, and became almost ill with rage. "The moose were gone and so were the beaver," Carson lamented in *Silent Spring*. "Their principal dam had gone out for want of attention by its skilled architects, and the lake had drained away. None of the large trout were left. None could live in the tiny creek that remained, threading its way through a bare, hot land where no shade remained. The living world was shattered."[18]

Stunned by the depth of Carson's research on the perils of DDT to wildlife, Douglas would declare in the fall that *Silent Spring* was "the most revolutionary book" since Harriet Beecher Stowe's *Uncle Tom's Cabin*. He wrote a favorable recommendation to the Book-of-the-Month Club, saying "This book is the most important chronicle of this century for the human race."[19]

It had taken Carson over four hard years to write *Silent Spring*. It had required a different kind of digging from her sea trilogy. No longer was she glorying in the beauty of starfish or the wonders of the laboratories of Woods Hole. Physically savaged by "a whole catalog of illnesses," in her words, she had studied scientific papers and reports to aim damning prose at chemical companies that poisoned the environment. Upon finishing the last chapters in early 1962, Carson wrote to a wildlife conservationist friend, Lois Crisler, about the literary-scientific high-wire act she had performed: "The beauty of the living world I was trying to save has always been uppermost in my mind—that, and anger at the senseless, brutish things that were being done. . . . Now I can believe I have at least helped a little."[20]

II

Coincident with the completion of *Silent Spring* was the release, in early 1962, of the Outdoor Recreation Resources Review Commission (ORRRC) final report. It wasn't a slapdash affair. With its well-written text, detailed investigatory findings, coherent charts, and useful tables, the report offered rock-solid recommendations for the Kennedy administration and Congress. To help the New Frontier's third-wave conservation effort succeed, the president drew on the ORRRC report for his State of the Union address a few weeks later: "We also need for the sixties—if we are to bequeath our full national estate to our heirs—a new long-range conservation and recreation program—expansion of our superb national parks and forests—preservation of our authentic wilderness areas."[21]

Citing a dire shortfall in public parks, the ORRRC report insisted that the opportunity to visit a wondrous natural setting was not merely a pleasant possibility but an American birthright. Concerned about the rapidly growing population, the report suggested that the federal government acquire new recreation sites or help states and municipalities to do so, and it recommended the establishment of a Bureau of Outdoor Recreation to coordinate the other agencies working on the issue. Chairman Rockefeller had used his bipartisan influence to make sure the ORRRC report was read in government offices at all levels. The timing of the widely disseminated report was right and the press couldn't ignore it.

Many protoenvironmentalists, however, resented the report's overwhelming theme: that nature was a playscape. David Pesonen, the forester—an investigator for ORRRC—was among those who considered the report a compromise between conservation and commercialism, saying that it "lean[ed] wearily on the obvious, the indisputable, the conventionally wise, the irrelevant."[22] Importantly, the report exposed a dividing line between preservation as an end in itself and environmentalism as attention to a complex web of ecological interactions. Kennedy recognized the difference and was intent on being the first president to cross into the new way of thinking about Earth stewardship.

Weeks later, on March 1, Kennedy reinforced the science-driven New Frontier "green" agenda. He submitted a *Special Message to the Congress on Conservation* in which he asserted that as the decade took shape, the environment was rightly assuming its towering importance with other major national concerns. He termed the Cape Cod National Seashore a "path-breaker," mentioned similar candidates for that status, and stressed that "fast-vanishing public shorelines of this country constitute a joint problem for the Federal Government and the States."[23] Kennedy urged Congress to enact many of the legislative measures the ORRRC report recommended. The presidential message also dealt with the threats of pollution and the outright depletion of natural resources.

A landmark communication to Congress, Kennedy's *Special Message* introduced to the nation the notion of ecology—the understanding that nothing in nature, including humanity, is independent of all else. "Our national conservation effort must include the complete spectrum of resources, air, water, and land; fuels, energy and materials; soils, forests, and forage; fish and wildlife," he said. "Together they make up the world of nature which surrounds us."[24]

Most members of the Wilderness Society and the Sierra Club thought Kennedy's special message (and the Rockefeller report) erred on the side of caution.[25] "Recreation," which addressed human needs, was not the same as "preservation," which addressed natural ones. Nevertheless, members of the groups cheered the fact that Kennedy wanted to expand the National Park System with the addition of Point Reyes, Padre Island, Canyonlands, Ozark Rivers, and Sleeping Bear Dunes, among other sites. For the first time, parts of the Great Basin, from the tip of the Sierra Nevada to the peaks of the Wasatch Range (across California, Nevada, Utah, Oregon, Idaho, and Wyoming), 200,000 square miles, were included on the Interior Department's wish list. Nevada's two senators, Alan Bible and Howard Cannon, both Democrats, reflecting the attitude of the ORRRC report, believed that the basin would become a major tourist attraction if the new site incorporated the 13,000-foot Wheeler Peak, the famous Lehman Caves, and stands of 4,800-year-old bristlecone pines said to be the oldest living things on Earth.[26] Kennedy also proposed Prairie National Park, comprising some 57,000 acres of the bluestem country in Pottawatomie County, Kansas.

National parks tended toward sites of dramatic scenery: mountains, deserts, and, more recently, shorelines. Prairie land was nothing more than reassuringly gentle, by contrast. A hundred years before, it had reached as far as a thousand miles in every direction from Kansas. Since then, wheat had largely replaced the wildflowers. When Udall had first brought up the heartland park idea, the president rolled his eyes. "Kansas?" he said. "Why do you want a national park in Kansas? What the hell have they ever done for us [Democrats]?"[27] He was cognizant of the fact that Kansas had two Republican senators and had gone for Nixon in 1960. But Udall had shown Kennedy tallgrass prairie photos of the Flint Hills. That had impressed the president enough to keep the Great Plains park idea alive.

The opposition to Prairie National Park was as fierce as any that had ever greeted a national park proposal. "I suggest that Mr. Udall take a trip from here to Oklahoma and look at the bluestem grass which paves the way," one Kansas resident carped. "How do you expect tourists to drive through 12 million acres of the same grass to look at 60,000 acres of the same grass which you say would be unique?"[28] Some who rejected the idea were Kansas farmers who resented the loss of any open land. Some of the objectors didn't like national parks, and some just plain didn't like the federal government. "I wonder if it ever occurred to our energetic Secretary of the Interior that we can't just make one big federal park out of the United States," wrote an Arizonan about the proposed prairie park. If Udall were to get his way, she said, "then he and the tourists can put on their Bermuda shorts, tramp down the grass, and the government can spend millions of our dollars putting in toilets, camping facilities, highways, electricity, hiring rangers, and in general make the property useless to everyone but the federal government."[29]

In December 1961, when Udall visited the site at the confluence of the Blue and Kansas Rivers, he actually encountered armed resistance.

Carl Bellinger, the rancher then leasing the land, approached him holding a gun, accused the secretary of trespassing, and demanded that he leave immediately.[30] Udall was shocked; he complied but redoubled his support for the park on his return to Washington.[31] So did Kennedy and so it was that the establishment of Prairie National Park became part of the president's *Special Message* the following spring.

In Kennedy's first year as president he had made New Frontier conservation activism a morally compelling endeavor. "Kennedy opened the door a bit," Udall explained. "He said, 'Let's have a Wilderness bill.' He said, 'Let's have a couple or three national seashores, Cape Cod and so on.' He opened the door and people rushed in."[32]

Kennedy did the same on a more profound scale in civil rights. In several ways that were considered bold at the time, Udall expressed his commitment to racial equality in deeds as well as words. He knew he'd be backed by the White House. Having zero tolerance for racial bigotry, Udall was disgusted by Washington's professional football team, then known as the Redskins, which as late as 1961 refused to hire Black football players. Udall told the team's owner, George Preston Marshall, that if he didn't immediately integrate the Redskins, they'd be banished from their home stadium, which was administered by the Interior Department. Marshall capitulated. With the addition of five Black athletes, including Ron Hatcher and Bobby Mitchell, an eventual Hall of Famer, the Redskins became a far stronger franchise.[33] Some sportswriters comically suggested that Udall should be named NFL coach of the year.

Udall was just as incensed that the National Park Service employed Black rangers only in the Virgin Islands. In 1962, he implemented an Interior Department program to increase the number of Blacks employed as rangers and supervisors. He started by asking the presidents of Historically Black Colleges (HBCs) for help in identifying candidates. In the summer of 1962, Robert Stanton, a twenty-two-year-old student at Huston-Tillotson College in Austin, Texas, was hired to work at Grand Teton National Park. "I was very excited about the opportunity," Stanton recalled. "I borrowed money to buy a train ticket to Jackson Hole. Taking a train from Texas to Wyoming to work in the Tetons was like going to the moon."[34] The Grand Teton employees embraced him with open arms; to his surprise, he encountered no racism. (By contrast, in Stanton's hometown of Fort Worth, Black residents were allowed in only one public park.) Thanks to Udall's racial integration employment edict, Stanton soaked up the jagged-peak beauty of the Tetons and came to know Laurance Rockefeller on a first-name basis.

Stanton had never even visited a national park when he took the summer job in Wyoming. His father was a hay baler, and his mother worked as a short-order cook in a restaurant in Fort Worth's Mosier Valley, one of the oldest African American communities in Texas, founded shortly after the Civil War. Stanton struggled to pay his college bills, so the summer job in the Tetons was especially welcome. The following year, he graduated from Huston-Tillotson with a bachelor of science degree, and then he worked another summer in Wyoming for the NPS.[35]

In 1966, after several years as an administrator at his alma mater, Stanton started a career with the NPS. Working in the capital for Udall, he recruited African Americans, Hispanics, Asian Americans, and Native Americans into the park service. Through the years, he also helped to create three new national parks focused on African American history: the houses of George Washington Carver in Diamond, Missouri, and Booker T. Washington in Franklin County, Virginia, and a memorial to Mary Mcleod Bethune in Washington, DC. In addition, he was largely responsible for opening Frederick Douglass's house in the Anacostia neighborhood of Washington, DC, as an NPS site.

President Bill Clinton appointed Stanton director of the NPS in 1997, the first African American in that post.[36] Under Stanton's leadership, the NPS launched an impressive recruiting initiative to bring women and minorities into superintendent positions while increasing staff diversity.

Udall was also determined that the Interior Department take the plight of Native Americans more seriously. In a clever move, he arranged for JFK to write the introduction to the 1961 book *The American Heritage Book of Indians*. The president admitted in the foreword that "American Indians remain probably the least understood and most misunderstood Americans of us all."[37] Josephy, in his post at Interior, often testified before Congress about Native American issues

Robert G. Stanton was the first African American National Park Service ranger at Wyoming's Grand Teton National Park, in 1962. A beloved figure with Interior Department employees, Stanton went on to become director of the NPS under President Bill Clinton (1997–2001).

such as returning Alcatraz Island (California) and Blue Lake (New Mexico) to the Taos Pueblo. Through it all, Udall said Josephy always fought for "Indian rights and justice."[38]

Udall generally bypassed the fact that citizens living in high-poverty pockets, especially racial minority groups, were disproportionately exposed to toxic air and water. Though he was thrilled to work with Sigurd Olson to protect the upper Mississippi River watershed in Minnesota, he turned a blind eye to the eighty-five-mile stretch of the river between Baton Rouge and New Orleans known as "Cancer Alley."[39] Residents of impoverished African American communities in southern Louisiana who were getting sick from industrial pollution had no Sierra Club or the Wilderness Society to represent them. In Miami's West Grove, thousands of poor people fell ill from the carcinogenic emissions of a municipal trash incinerator called Old Smokey. Such health problems caused by incinerators were common.

Native Americans during the Kennedy years and later were

victims of systemic environmental injustice: Navajo and other nations in Udall's beloved Southwest were suffering abuses attributed to uranium mining. Church Rock, New Mexico, was the site of the longest continuously running uranium mine on Navajo land. Throughout Udall's tenure as secretary, the tribe leased land to mining companies that did not obtain consent from Navajo families or report the health consequences of their activities. The companies dramatically depleted the limited water supply. Worse, they contaminated what was left with uranium tailings. And they got away with it. Kerr-McGee and United Nuclear Corporation, the two largest mining companies, argued that the Federal Water Pollution Control Act didn't apply to their industry and maintained that Native American land wasn't subject to environmental protection. The courts didn't force them to comply with US clean water regulations until 1980.[40]

As the historian Marc Reisner wrote in *Cadillac Desert: The American West and Its Disappearing Water*, Udall presented himself as a vigorous preservationist, climbing Alaskan peaks, touring a national park with David Brower, and rafting swift-running rivers with US senators. But away from the cameras, he lobbied to construct a nuclear-powered desalination plant off Long Beach "to slake Los Angeles' giant thirst"; conspired to have water-carrying aqueducts built from the Columbia River to the Southwest; stayed mum about the Everglades ecosystem being channelized by the Army Corps of Engineers; and was part of the US government contingent who wanted to dam the Colorado River near the Grand Canyon during the Kennedy years.[41] He undoubtedly thought of himself as racially enlightened, with his Hopi doll collection and his initiative to hire African American rangers. In truth, he took only modest steps in behalf of minorities, seeing his job as serving the mainstream, which was, to judge by his actions, the middle and upper classes of white America.

III

Perhaps what most distinguished President Kennedy from his predecessors, Harry Truman and Dwight Eisenhower, in the environmental realm was the abundant faith he placed in scientists. In 1961, *Time* magazine had named US scientists as a group "Men of the Year," reasoning that they were the "true 20th century adventurers, the explorers of the unknown, the real intellectuals of the day."[42] Because Kennedy kept his Addison's disease secret, the public never realized that their young president had a special reason to be intrigued by the advent of new antibiotics and surgical procedures. To gather the highest type of technical knowledge on the nuclear issue, the president relied on Jerome Wiesner, his special assistant for science and technology. Having worked at Los Alamos National Laboratory in the early Cold War years, Wiesner could lend a key voice on limiting nuclear proliferation and banning nuclear testing. He would advise Kennedy on establishing the Arms Control and Disarmament Agency. "Not since Merlin has any head of state made greater use of, or relied more on, his Chief Science adviser than John F. Kennedy relied on Jerry Wiesner," the White House speechwriter, Ted Sorensen, quipped, referring to King Arthur and his wizard.[43] On issues pertaining to environmental policy as well as nuclear nonproliferation, Wiesner was Kennedy's expert.[44]

If Wiesner understood, to at least some extent, the many considerations Kennedy had to keep in mind, Linus Pauling, an even more accomplished scientist, did not. His advice to Kennedy was pointed and impatient. On March 1, infuriated by the anticipated renewal of atmospheric atomic tests, he sent a searing telegram to Kennedy that started by asking if the president was, by his actions, going to "go down in history as one of the most immoral men of all time and one of the greatest enemies of the human race?"[45]

Pauling didn't realize, or perhaps care, that Kennedy was personally opposed to nuclear testing and working hard to protect the world's oceans from industrial abuse. In early 1961, Kennedy asked Wiesner for a memorandum on the health of the oceans with input from Jacques Cousteau. The White House examined everything from seafood harvesting to mapping the ocean floor to cost-effective desalination. In keeping with the Kennedy mandate to learn more about that "entirely new dimension," the ocean, Udall recruited Roger Revelle—the director of the Scripps Institution of Oceanography in La Jolla, California, whose 1936 PhD dissertation had focused on understanding how climate affects marine environments—for guidance. During

World War II Revelle was a navy commander and predicted wave and surf action for amphibious landings. After the war, he studied the grim effects that atomic tests had had on the Bikini Atoll. Starting in 1950, he directed Scripps, which became part of the University of California at San Diego. There he pioneered the study of the influence of carbon dioxide on oceans and the atmosphere. After conducting numerous experiments, he concluded that societal use of fossil fuels was resulting in excessive carbon dioxide, which would heat the planet in a disturbing way. The very first academic paper directing attention to the impact of emissions of carbon dioxide on global climate was published by Revelle and his Scripps colleague Hans Suess in 1957 in the journal *Tellus*.[46]

That *Tellus* article led to the creation of a celebrated station at Mauna Loa, Hawaii, for the measurement of a variety of atmospheric data. According to Revelle's observations, CO_2 was heating the planet. He embraced the chance to take his global warming message to Washington: a prescient climatologist struggling to get his research taken seriously by Cold War–era lawmakers.

Kennedy became the first president to be warned about climate change owing to the human use of fossil fuels. Both Wiesner and Revelle gave him the facts. Time has proved, of course, that they were right. Senator Clinton Anderson also raised the alarm, telling JFK that climate change could "affect the level of the seas and hence the habitability of the continental coastal shelves." One of Anderson's most pressing concerns was weather modification—whether due to atomic testing, factories, automobile exhaust, or any other man-made cause the planet was heating up.[47]

It wasn't all doom and gloom in the ocean science world. Upbeat stories about marine mammals also made news during the Kennedy years. At Woods Hole Oceanographic Institution, William Schevill and William Watkins documented for the first time that sperm whales make a medley of loud clicks, a phenomenon that sailors had reported since the days long before Herman Melville. As founders of contemporary cetology, both experts in bioacoustics, Schevill and Watkins released in 1962 six thousand field recordings of seventy marine mammal species. It was clear from their tapes that whales and dolphins are able to socialize and communicate with sounds. Ten years later, the songs of whales were captured on tape and analyzed for the first time. On the subject of communication, the biologist E. O. Wilson declared that "the most elaborate *single* display known in any animal species may be the song of the humpback whale."[48]

In 1961, the physician and neuroscientist John C. Lilly published *Man and Dolphin*, in which he postulated that the bottlenose dolphin possesses a degree of cognition and awareness not unlike that of humans.[49] From his Coconut Grove laboratory in Florida, he led the way for marine mammals to be treated ethically because of their super intelligence. Lilly went so far as to authorize an experiment in which researcher Margaret Howe Lovatt taught a dolphin called Peter to speak some English, including greeting her by name every morning.[50] In 1963, the movie *Flipper* depicted a dolphin as an intricate part of a Florida family, interacting and certainly communicating with it. It led to a television show that helped to convince Americans of the charismatic value of marine mammals.[51]

As Albert Schweitzer had warned, however, there was something inherently cruel and selfish about the way humans related to animals. After the funding for the experiment with Peter ended, the dolphin was transported to an unfamiliar habitat, with no further contact with Lovatt. Bewildered and abandoned, he committed suicide. Dolphins are capable of that, too, so another researcher found.[52]

CHAPTER 12

The White House Conservation Conference

(MAY 24–25, 1962)

Senator John Kennedy and labor leader Walter Reuther at a United Auto Workers Convention, in 1959. Until his death in 1970, in a plane crash, Reuther was a fierce environmental justice advocate.

I

Just how much of a revival the work of Henry David Thoreau had enjoyed in America's environmental community became clear when Stewart Udall presided over a grand party—outdoors, of course—to celebrate the hundredth anniversary of the writer's death. Udall had asked Sharon Francis to plan the event, which was to be cohosted by the Department of the Interior and the Wilderness Society. It was held at Dumbarton Oaks Park, the mansion in Georgetown at which the United Nations was conceived. Udall, true to his usual leanings, was determined that Robert Frost and Bill Douglas speak to the gathering.

Douglas was an especially dedicated fan. Wherever he hiked, he carried in his backpack a copy of Thoreau's *Walden*, which he had practically memorized. If colleagues weren't well versed in Thoreau, Douglas didn't respect or trust them as genuine environmentalists. Working with Francis to organize the party was Howard Zahniser, the executive director of the Wilderness Society. Not only did Frost and Douglas agree to speak on Thoreau, but the invitation list included other famous people, including Chief Justice Earl Warren, Appalachian Trail founder Benton MacKaye, poet Robert Lowell, and India's ambassador to the United States, B. K. Nehru.

Francis had never seen her boss so excited to host an event. "I commented to Stewart that he had seemed a 'serene coil of energy' as he escorted Frost around the grounds," she recalled. "Under a tree canopy, Udall stood and spoke of Thoreau as 'having lasting power to mold character and govern nations.'"[1] When it was Frost's turn to speak, he praised Thoreau's *Walden* as one of the three greatest storybooks of all time (the other two were *The Voyage of the Beagle* and *Robinson Crusoe*).

In front of a few hundred guests, Udall introduced Bill Douglas as one who shared Thoreau's concern for the estrangement of humans from their natural surroundings. Not missing a beat, drinking scotch, Douglas began by quoting a man who'd recently taken a walk around Walden Pond, where he had encountered "One hundred and sixteen beer cans, 21 milk bottles, 7 Coca-Cola bottles, the remains of 14 campfires, a shoe box, eggshells, soap, half-eaten sandwiches, Dixie cups . . ." and the list of garbage went on. Douglas then warned about the unnecessary turnstile commercialization of sacred places like Walden: "The problem of mass invasion of wilderness areas presents serious problems of this character all over the country."

Lamenting the fact that Walden Pond was used as a dump site in Concord, he hoped that Thoreau's classic would become a foundational text for future environmentalists. "I have traveled with Thoreau everywhere he went in New England," he added. "Thoreau's curiosity was about the wonders of creation, including man, but mostly about those wonders which are at our feet and yet which we seldom see."[2]

At Dumbarton Oaks, Udall was all abuzz about the upcoming White House Conference on Conservation, the first since Theodore

Roosevelt had hosted one for America's governors in 1908. Five hundred conservation decision-makers were invited to the White House event, from nonprofit preservationists to extraction industry representatives.[3] Rachel Carson asked Ruth Jury Scott, a Pennsylvania conservationist, and Adele "Nicki" Wilson, an Interior Department publicist, to accompany her to the event, to be held in the State Department auditorium. Owing to Carson's fame, any environmental leader would gladly have accompanied her to the conference, but she asked two people who were on the front lines, not in the evening news. Like Carson, both believed that the indiscriminate use of pesticides was poisoning Earth to the point of no return.

Ruth Scott was the newest member of Carson's informal circle of environmental allies. During the summer of 1961, Carson had received a fan letter from Scott, saying she would be attending an Audubon camp in Maine soon and hoped to meet her. Scott, a landscape designer, was the state bird chairman of the Garden Club Federation of Pennsylvania. An organic farming pioneer, she was launching a Roadside Vegetation Management Project centered on teaching people about the hazards of pesticides.[4] She and her husband had built a house overlooking Powers Run above the Allegheny River, only five miles from Carson's girlhood home in Springdale. Scott hoped to protect that stretch of the Allegheny from industrial ruin, and to that end was working closely with the National Audubon Society and the Nature Conservancy. When Carson and Scott met, they instantly became friends. Together, they plotted how to place warnings about pesticides in Audubon publications. Scott and her friend Nikki Wilson frequently worked together on environmental causes. They were a highly effective team and thus excellent candidates to accompany Carson, who was weakened by cancer and always somewhat shy.

David Brower used the White House occasion to press for the Sierra Club's campaign to establish the North Cascades National Park. Sigurd Olson came full of news about the Voyageurs National Park preservation efforts near his hometown of Ely, Minnesota.[5] Zahniser chatted up potential allies for the Wilderness bill. Senator Phil Hart of Michigan distributed photo-laden pamphlets promoting Pictured Rocks and Sleeping Bear Dunes as national lakeshores. Laurance Rockefeller, who chaired the first day's proceedings, emphasized establishing new federal parks near big cities, such as New York, Chicago, and Houston. To that end, the administration was willing to prioritize Indiana Dunes, near Chicago; Fire Island, near New York City; and Assateague Island, a three-hour drive from Washington, DC.

Hubert Humphrey, the original sponsor of the Wilderness bill, chaired the second day of the White House conference. Most Democrats in attendance accepted whatever Udall thought needed to be prioritized, and that was a long list. As he told the press covering the conference, a third-wave conservation movement of "Rooseveltian proportions" needed to coalesce in the New Frontier.[6] Republican lawmakers, by contrast, were divided into two camps: one led by Representative John Saylor, the supreme wilderness-proponent, the other led by Utah governor George Clyde, who favored extraction industries such as mining and logging. Clyde said, "We live under a system of free enterprise, and private initiative must be given maximum latitude and responsibility for resource development."[7]

A hotly contested topic between conservationists was the impending fate of the Wilderness bill. The ORRRC recommended that pristine public lands in the United States measuring over 100,000 acres "containing no roads usable by the public," and showing "no significant ecological disturbance from on-site human activity," be eligible for a new "wilderness" designation. As of the time of the conference, that designation would forbid any commercial activity, so it was easy to see where Clyde and his side stood on the classification issue.

When President Kennedy strode to the podium on the second day, cameras snapped and applause erupted. In what was described later as an "informal" speech, JFK failed to make any dramatic announcements. The strongest policy statement heard by the attendees picked up on the question of easy accessibility, as the president expressed interest in a fund to establish parks near eastern cities—and not just in the public land-rich West. His most quoted line from that day was "We do not want . . . this eastern coast to be one gigantic metropolitan area stretching from north of Boston to Jacksonville, Florida, without adequate resources for our people to participate and see some green around them."[8]

Because the oceans contain 97 percent of the Earth's water, Kennedy also turned to a pet idea, challenging American scientists to develop an inexpensive means of water desalinization. "I have felt that whichever country can do this in a competitive way will get a good deal more lasting benefit than those countries that may be even first in Space," he said.[9] That was quite a statement from the US president who, a year before to the day, had prioritized putting a NASA astronaut on the moon.

Encouraging the attendees to keep championing conservation, Kennedy told a story about General Hubert Lyautey, who had been a French colonial administrator in Morocco. As Kennedy recounted it, Lyautey asked his gardener to plant a tree. "But this won't flower for a hundred years," the gardener warned. Lyautey replied, "In that case, plant it this afternoon."[10]

Kennedy's speech, or at least his presence, raised the morale of those who cared deeply about preserving America's forests, parks, rivers, and shorelines, but there were complaints from all sides of the conservation debate. Those who opposed Kennedy's third-wave conservation movement accused the White House of using the conference as "an all-Administration show," with no dissenting voices or real debate. Protoenvironmentalists complained that it "failed to deal with major conservation problems," and that "no major new programs emerged."[11] Senators like Frank Church and Clinton Anderson were frustrated that the Wilderness bill was stuck in Congress, even though the Senate had vigorously backed the legislation. The administration, they noted, seemed unable to get it across the finish line. Holding the line, liberal Democratic lawmakers in both the House and the Senate refused to gut the bill of vestigial language about wilderness as being a sacred place for "solitude or a primitive and unconfined type of recreation."[12] *Unconfined* meant no hotels or cottages and not even a lean-to.

The chief obstacle to passage was Wayne Aspinall, the powerful chairman of the House Interior and Insular Affairs Committee. Cranky, obstructionist, and buttressed by western ranchers and the oil and gas lobby, Aspinall had been largely blocking the wilderness movement agenda for years. As a Truman Democrat, he was an unbending advocate of a multiple-use policy for public land. For that reason, he insisted that mining and timbering be allowed even in designated primitive areas of national forest and BLM land. Aspinall, a master congressional strategist, didn't like the way the Wilderness bill allowed the executive branch to pick the roadless areas without consulting Congress. When asked if he would ever compromise on his position, Aspinall retorted, "Not on your life."[13] His obstruction was no small matter; as chairman of the committee, he could simply neglect to schedule a hearing on the Wilderness bill. By refusing a hearing, he kept his committee from voting on it, and by doing that, he kept it well away from the full House.

Aspinall's stance made him popular with pro-development voters of both parties throughout the West. In his home state, the Grand Junction *Daily Sentinel* praised the fact that he "single-handedly took on the powerful conservation lobby this year in stopping the steamroller on the Wilderness bill. He was the only speaker at the White House Conference on Conservation who underscored the importance of multiple use of public land. His speech emphasizing the importance of managing resources 'for the benefit of the many and not the few' was stonily received by the audience."[14]

The Wilderness lobby had begged sympathetic politicians to steamroll Aspinall, but none could. While chairman, Aspinall answered only to himself and his voters back home. Because of that, frustrated environmentalists descended on his district in western Colorado to tell the people there just exactly what was going on. "I was thrown off one ranch," Michael McCloskey, who later became the executive director of the Sierra Club, remembered. "Another rancher fingered his pistol and said, 'There's about to be a shooting.'" Aspinall complained about outside agitators like McCloskey. But the next spring, congressional hearings were held on the wilderness designation. Thanks to the tireless crusade of environmentalists and political opportunities opening in Washington because of Kennedy, Aspinall started leaning toward compromise.[15]

Those who complained that the White House conference was one-sided had a point. Though opposing opinions were allowed, Udall had stacked the deck in favor of those taking the Laurance Rockefeller–Clinton Anderson approach over the Wayne Aspinall pro-extraction industry line. The *New York Times* under John Oakes, the *Washington Post*, the *St. Louis Post-Dispatch*, and the Louisville *Courier Journal* under Barry Bingham, Jr., all praised the New Frontier's preservation leadership. "A narrower viewpoint was expressed by those who insisted on conservation exclusively for 'use' and decried the establishment of natural areas set aside solely as quiet places," the *Times* noted. ". . . Such a position misses the point. Everyone favors more efficient management . . . but the 'preservationists' believe—and fortunately there are now a good many of them in high places in this Administration—that steps must be taken now to protect the remnants of our wilderness."[16]

Around the time of the White House conference, Wiesner complained in the White House that the Fish and Wildlife Service, the Bureau of Land Management, and the Forest Service didn't understand that relationships among living things constitute an ecosystem. For example, one afternoon in 1963, Roger Revelle "burst into" Sharon Francis's office at Interior. Distressed, he asked what in hell had gone wrong to cause a massive die-off of fish in the reservoir behind the recently constructed Flaming Gorge Dam on the Green River, continuing downstream into Dinosaur National Monument in Utah and Colorado. Rachel Carson called Udall, wanting to know the same thing.

Francis immediately investigated. What she soon called the "fisheries fiasco" had been triggered after the Wyoming and Utah fish and game departments decided that the filling of the new reservoir provided a once-in-a-lifetime opportunity to rid hundreds of miles of tributary streams of "trash fish" such as humpback, chub, bonytail, razorback sucker, and pikeminnow.[17] Ignorant of ecology, the fishing managers intended to replace them with millions of alluring rainbow trout, which would become a magnet for fishermen and the money they would spend on their sport. The poisons used for the biological transformation were rotenone and potassium permanganate. Three days of application in 1962 resulted in 450 tons of dead fish, scores of disgusted eyewitnesses, photographers along riverbanks, and, most seriously, dying fish downstream in Dinosaur National Monument, a unit of the National Park System. Francis's investigation of what had transpired revealed that the Bureau of Sport Fisheries and Wildlife had allocated $150,000 of federal funds to the project, but neither those in the bureau nor their state counterparts understood the consequences of what they were doing. "They mis-estimated the mixing of rotenone and potassium permanganate, and did not understand the effects of water temperature or siltation on the chemicals they used," Francis wrote in an unpublished memoir. "While they acknowledged that the new reservoir was going to create significant changes in riverine ecology, they expanded the damage several-fold by exterminating biological life for several hundred miles upstream."[18]

II

The White House conference reenergized the forces behind *Our Fourth Shore: Great Lakes Shoreline Recreation Area Survey*, which the NPS had issued three years before. And if the effort to save Great Lakes shorelines as public spaces gained even a little momentum, Senators Paul Douglas of Illinois and Philip Hart of Michigan were front and center, willing to wage an all-out fight for them.

Indiana Dunes, located on a strip of undeveloped shoreline of Lake Michigan in Porter County, Indiana, less than an hour's drive east of Chicago, was a prime example of the potential. The Save-the-Dunes movement had been ongoing since 1917, when NPS director Stephen Mather published *Report on the Proposed Sand Dunes National Park, Indiana*; it stated "the advisability of securing, by purchase or otherwise," the Lake Michigan beachfront in northwest Indiana, "with a view to creating a national park."[19] The 1959 survey's tone was more urgent, declaring that the 15,000 acres were in need of immediate preservation, for without such protection Bethlehem Steel and National Steel, in need of ports, were poised to desecrate the pristine shoreline. The dunes were a biological wonder, the home of an unusual number of species, some of which were left over from earlier climatic eras. More than a hundred types of birds nested in the area, and another 250 flocked to trees during migration seasons.

An influential supporter of the Indiana Dunes was the poet Carl Sandburg, who became the éminence grise for the cause. Much of Sandburg's poetry was inspired by Chicago; it was he who had christened the city "Hog Butcher for the World, / Tool Maker, Stacker of Wheat, / Player with Railroads and the Nation's Freight Handler; /

Carl Sandburg near his home east of Chicago on Lake Michigan. An ally of Senator Paul Douglas (Democrat-Illinois), the populist poet was the Grand Old Man of the Indiana Dunes National Lakeshore movement.

Stormy, husky, brawling, / City of the Big Shoulders."[20] Having won the Pulitzer Prize three times, once for his four-volume Abraham Lincoln biography, he was especially revered throughout the Midwest and Appalachia as a huckleberry sage. If Yosemite had John Muir as its publicist and the Everglades had Marjory Stoneman Douglas, the Great Lakes states had Carl Sandburg, who called the world's largest freshwater system "the heart of North America."[21]

Udall knew that Sandburg supported the national lakeshore designations and the two bonded over the cause. Udall, of course, never met a literary lion he didn't like. Before long he was writing to Kennedy, almost in the tone of a bobbysoxer hoping to invite a movie star to the prom. "When you are reelected and we have the Inauguration in January 1965," he told Kennedy, "it would be a wonderful gesture (if he is still with us), to invite Carl to fill the artist's role in the official ceremonies. If you were to ask him to participate—and perhaps to sleep the night before in Lincoln's bed at the White House—this tribute to our only poet-historian would, I am sure, receive a warm and wide acclaim."[22]

For his part, Sandburg was also in touch with Senator Douglas, having contacted him for the first time in 1958. "I have known those dunes for more than forty years and I give my blessing and speak earnest prayers for all who are striving for this project," he wrote Douglas. "Those dunes are to the Midwest what the Grand Canyon is to Arizona and Yosemite to California. They constitute a signature of time and eternity: once lost the loss would be irrevocable."[23]

The heartland poet had written to the right Democratic politician. Born on March 26, 1892, Douglas developed a lifelong commitment to conservation as a young man. After earning degrees from Bowdoin and Columbia, he became a professor of economics at the University of Chicago. He advised Gifford Pinchot, TR's ex–chief of forestry, then Pennsylvania's governor, and FDR's scrappy interior secretary, Harold Ickes. Serving in World War II when he was in his fifties, he rose to the rank of lieutenant colonel in the Marines, earning a Bronze Star and two Purple Hearts in the Pacific theater. With his slicked-back silver hair, a loose smile, and wire-rimmed glasses, Douglas resembled a handsome Main Street businessman out of a Sinclair Lewis novel. But looks can be deceiving, for he was more interested in nature preservation than in capitalizing on the military-industrial complex for Chicago.

In 1948, Douglas was elected to the Senate from Illinois, where he gained a reputation as an advocate of honest government, civil rights, and conservationism. He would stand for the Save-the-Dunes movement in the late 1950s and 1960s, and even in the early 1970s, when he was no longer in office. Along the way, he embraced the grassroots activism of Dorothy Buell, the head of the Save the Dunes Council, and Hazel Hannell. "The worst features of our civilization threatened to invade what had been an earthly paradise," Douglas wrote in defense of the Dunes. "The gross national product from the area, as measured in dollars, would probably increase. But so would smoke, polluted water, crime and license ugliness would replace beauty."[24]

Lobbying on Capitol Hill with force and urgency, Douglas argued that Chicago was at the core of a "great industrial complex which stretches from Gary, Indiana to Milwaukee, Wisconsin." It didn't need to grow northward from Gary into the celebrated Dunes. The senator was specific, citing the plans of US Steel and Bethlehem Steel to turn the area over to mills, with a new deep-water port to be dredged along the shoreline. He said, "If we do not act quickly, all the places of beauty will be taken over and destroyed. We will have steel mills, cement factories, slag piles and more asphalt jungles and these, I suppose, will be hailed as signs of progress. But we will also have more psychiatric hospitals."[25] The industrial towns of Michigan City and Gary had already marred the sand dunes' perimeter with steel mills, smokestacks, and rusting storage tanks. "There are plenty of sites for steel mills," he implored. "But there is only one Dunes. Do not destroy them."[26]

To win the Indiana Dunes fight and other preservation causes, Paul Douglas employed Lee Botts as his one-woman environmentalist brain trust. Born Leila Carman in Mooreland, Oklahoma, she earned degrees at Oklahoma A&M and the University of Chicago. Based in Chicago during the 1950s and 1960s, she was an admirer of Barry Commoner. Working out of Douglas's Senate office, she spearheaded the campaign to build widespread support for Indiana Dunes National Lakeshore. Her preservation soon expanded to the entire Great Lakes sand dunes systems. Another of her passions was agitating to eliminate the discharge of phosphates and other pollutants in Lake Michigan. And she regularly wrote opinion columns and gave speeches about the irreplaceable value of the freshwater Great Lakes. In Chicago, she became known as a tireless activist for open spaces within the city. If Douglas was one of the most effective senators on landscape preservation, it was because he had Botts on his staff. When Botts died in 2019 at ninety-one, her *New York Times* obituary said, "She became a writer, a grass-roots organizer, an educator, and a municipal and federal government official whose work would touch practically every drop of water and every mile of shoreline in the Great Lakes basin while educating tens of thousands of people in its ecology."[27]

III

Along with lakeshore preservation, Kennedy and Udall remained interested in riverine ecology, a pet cause of Justice Douglas and Sigurd Olson. There are more than 250,000 rivers in the United States stretching over 3.5 million miles, all with features that conservationists were saying deserved protection. Kennedy asked Orville Freeman, his agriculture secretary, to establish a wild and scenic river study team at the department. Since the late fifties, the NPS had embraced the Current (Missouri), Allagash (Maine), and Suwannee (Florida and Georgia) to stop hydroelectric dams from destroying their ecosystems. Udall believed that if he could get Freeman on board, their departments could conspire against the Army Corps of Engineers to establish a visionary wild-and-scenic-river system. That didn't mean that Udall had changed his mind about damming the Colorado River for the Central Arizona Project, but he was willing to designate some eastern rivers as out of bounds to the reclamationists.

After the White House conference, Sigurd Olson took Udall on a canoe trip on the Saint Croix River, along the Wisconsin-Minnesota border, in the hope of having it designated a national river of some kind. Olson had convinced the Northern States Power Company to transfer to the US government seventy miles of frontage along both sides of the scenic river where a huge power dam had once been slated for construction. It would take even more than that to secure protection of a river hundreds of miles long, but getting Udall's agreement with the idea was a step in the right direction.

River preservation grew into a hot-button issue in wilderness protection circles. While planners at the Interior and Agriculture Departments workshopped the idea of a wild-and-scenic-river system, the Kennedy administration specifically pressured the Senate Select Committee on National Water Resources to permanently protect portions of the Current River and the Jacks Fork River in Missouri as free-flowing streams. Eventually, the Interior-Agriculture team, led by Stanford Young, developed a list of 650 rivers for possible federal preservation. It was winnowed down to 70 and then to 22. Money was allocated for detailed studies of the finalists. Eight would be chosen for the first round of protection. In 1965, a Wild and Scenic Rivers bill was introduced in Congress. The bill specified several of the rivers to be preserved in the initial round, including Idaho's Middle Fork of the Clearwater and its stupendous tributaries the Lochsa and the Selway, as well as the rollicking Rogue in Oregon.[28]

Within the Interior Department, the National Park Service was also working to protect complex river systems. One place the battle was joined was the free-flowing, 150-mile-long Buffalo River in northern Arkansas, a tributary of the White River considered a sacred waterway by the Osage, Caddo, and Quapaw peoples. The Army Corps of Engineers was planning to harness it with at least one dam. If it was built, the river's smallmouth bass, bubbling springs, frothy water, and rocky limestone cliffs would be lost forever. Harold Alexander, an Arkansas Game and Fish Commission conservationist, contacted such environmental groups as the Sierra Club and the National Audubon Society for help. "A stream is a living thing," he wrote. "It moves, dances and shimmers in the sun. It furnishes opportunities for enjoyment and its beauty moves men's souls."[29]

The message, though well worded, didn't persuade the groups to help. The Sierra Club, in particular, was just not focused on southern concerns, so Alexander decided to establish his own nonprofit: the Ozark Society. Its mission was the preservation of environmental quality in the Ozark-Ouachita Highlands. Bill Douglas, to his credit, was one force in the conservation movement with no regional predilections or prejudices. Cleverly, Alexander lured him to canoe the Buffalo River in 1962. After the trip, the justice declared the river "a national treasure worth fighting to the death to preserve."[30] With Douglas on board, Alexander had found the way to reach the national audience that was needed. With the aid of Arkansas senator J. William Fulbright, Douglas convinced Udall to draft a plan to save 132 miles of the Buffalo.[31] How it would be done wasn't clear, but the Outdoor Recreation Resources Review Commission had promoted the idea of national rivers, and it seemed like a good place to start.

Neil Compton was a northeastern Arkansas native, a graduate of the University of Arkansas Medical School. Outside of his obstetrics practice, he was interested in preserving the scenic, scientific, and recreational wonders of the Ozark ecosystem, and he gladly helped Alexander found the Ozark Society. They had the great luck to have the painter Thomas Hart Benton on their side. Since 1926, Benton had been visiting Arkansas from his home in Missouri to fish the Buffalo for smallmouth bass; he also used his boat to navigate the upper river, looking for vistas to paint. His paintings of the Buffalo River—*Welch Bluff* and *Bat House (Skull Bluff) near Woolum*—are masterpieces of

American landscape art. Benton also wrote a broadside that was circulated all over the Ozarks, offering a heartfelt plea for the preservation of his favorite river:

> As a lover of the great scenic beauty of the Buffalo River I would like to add my name to those others which are lined up in protest, against plans to put a dam across its waters. . . .
>
> Man, hog tied as he largely is, with the steel tentacles of an increasingly mechanistic world, and with the prospect of being tied ever tighter, needs some areas of escape, of escape to the natural world from which he came. He'll need these all the more in the future.
>
> The Buffalo River provides one of these areas.
>
> I say, and I intend it emphatically, let the river be.[32]

Many rural Arkansans were furious that the Interior Department was trying to take ownership of the Buffalo River. The situation escalated into a violent showdown. Canoeists were fired at with rifles, were entrapped by barbed wire blocking their way downriver, and were harassed by local police for being Communists. Even the peaceful part of the designation debate was starting to split down party lines. The Buffalo River preservation effort aggravated pro-dam, small-government Republicans; they feared that the precedent in the Ozarks would encourage other states to follow suit. Conservation, which had been a bipartisan affair since Theodore Roosevelt's days, was slowly drifting into becoming more of a liberal Democratic endeavor. There was an expanding chasm between environmental regulators and industrial laissez-faire capitalists. In Michigan, labor leader Walter Reuther promoted environmental quality as a union right and led the UAW to become the most vocal organization in America demanding pollution control and nature conservation for workers.[33]

Reuther attributed his love of fishing and planting trees to growing up in rural West Virginia. All workers, he said, deserved clean rivers and lakes. During the Eisenhower years, he had chosen to live in an exurb outside Detroit because the Paint River that flowed through town was full of trout and bream. When the septic tank system broke and raw sewage entered the creek, Reuther sprang into action, organizing neighbors into a grassroots conservation association that brought about the construction of a new sewage treatment plant.[34] Reuther was also an active proponent of establishing Sleeping Bear Dunes and Pictured Rocks National Lakeshores in Michigan. Reuther would tell the annual conference of the Water Pollution Control Federation: "If we continue to destroy our living environment by polluting our streams and poisoning our air . . . we put the survival of the human family in jeopardy. . . . We may be the first civilization in the history of man that will have suffocated and been strangled in the waste of its material affluence—compounded by social indifference and social neglect."[35]

The UAW maverick was a strong advocate of New Frontier preservation efforts. The outdoor recreation industry never had a better friend in the labor world. He favored every national wildlife refuge proposed in the 1960s. Since the founding of the Fish and Wildlife Service as a bureau of the Interior Department in 1940, refuges had been established primarily to protect migratory waterfowl, but in July 1962, the Kennedy administration, under Reuther's sway, reached farther, calling for the designation of federal lands to protect endangered species such as bald and golden eagles. That year's Refuge Recreation Act made wildlife protection a priority—a move that, of course, both Reuther and Carson supported.[36] Some conservation scholars believe that is the root of what became the Endangered Species Acts of 1966, 1969, and 1973.[37] A committed environmentalist, Reuther supported the efforts of Paul Douglas, Philip Hart, and Hubert Humphrey after the three senators lobbied for a string of stopover preserves for migratory birds in the upper Midwest.[38]

When Kennedy authorized the establishment of the new refuges after the White House Conference on Conservation, he also approved plans to increase the size of five existing ones. Both the Interior and Agriculture departments worked in sync on the effort. "This is a heartening beginning of a conservation program that was given unprecedented stimulus a year ago when Congress passed the new wetlands bill," Udall proclaimed to the press. "It is a banner day for hunters, conservationists, and others who are materially assisting our States and Federal agencies in the scientific management of our wildlife resources."[39]

One of Kennedy's new 1962 refuges was Alamosa National Wildlife Refuge in the San Luis Valley in southern Colorado. Alamosa has a permanent water supply, and it compares with the best of the North Dakota pothole prairies in duck proliferation. It is also an essential wintering habitat for other migratory birds. Another new unit was Davis Island National Wildlife Refuge on the site of the old Jefferson Davis Brierfield Plantation near Vicksburg, Mississippi. With many wetlands along the Mississippi having been drained, Davis Island proved to be a marvelous sanctuary for a plethora of wildlife.

Another of Kennedy's picks was Eastern Neck National Wildlife Refuge, an island at the conjunction of the Chesapeake Bay and the Chester River on Maryland's Eastern Shore, a miraculous place to be in mid-November, when more than ten thousand great white tundra swans winter in nearby waters. Over the years, the 2,286-acre island had struggled to produce adequate food for wintering waterfowl. The Kennedy administration returned the overused landscape to its natural state through a combination of federal regulations, impoundments, and woodland management.

Cross Creeks National Wildlife Refuge in northwestern Tennessee attracts migrating eagles every winter. Consisting of 8,862 acres of rich Cumberland River bottomlands surrounded by forested hills and rocky cliffs, it "sets the birding hotline buzzing from time to time with such rarities as roseate spoonbills, wood storks, snowy owls, and cinnamon teals," according to *Guide to the National Wildlife Refuges*.[40]

President Kennedy also took a personal initiative in creating Prime Hook National Wildlife Refuge, a preeminent Delaware stopover site for snow geese that nest in the northern Arctic tundra. The goose was the lovable emblem for US Fish and Wildlife Service refuges. Arriving in October, the geese sometimes number sixty thousand; by year's end, they cover the waterways. In the spring, other shorebirds blanket the beach like a moving carpet.[41]

The preservation of Prime Hook, situated just over eighty miles south of Wilmington, drew the opposition of Elbert Carvel, Delaware's Democratic governor. Where JFK saw a migratory bird sanctuary, Carvel envisioned commercial and industrial development. The incensed Kennedy, who had sailed the area, wrote Carvel a tough letter, requesting that the governor "retract his objections to save the Atlantic Flyway sanctuary, whose ocean frontage and diversity of habitat made it a magnet for migratory birds."[42] Carvel capitulated, allowing the Prime Hook refuge to go forward.

By the late spring of 1962, Kennedy was consciously wrapping himself in Theodore Roosevelt's conservation mantle. Prodded by Udall, he established the Theodore Roosevelt Birthplace (New York City) and Sagamore Hill (Roosevelt's home) in Oyster Bay, New York, as national park sites. Both edifices soon became Interior Department venues for policy discussions on subjects such as public parks and wildlife protection.[43] Udall had just finished work on a chapter in his book *The Quiet Crisis* about Theodore Roosevelt and how he had saved more than 230 million acres of American scenery for future generations to enjoy. By equating the legacy of the Rough Rider with the Wilderness bill, Udall was engaging in smart conservation politics. Democrats, Republicans, and independents alike admired the twenty-sixth president.[44]

IV

In 1962, in the spirit of New Frontier conservation, Attorney General Bobby Kennedy, his wife, and four of their children ventured to Olympic National Park in Washington for a four-day camping trip with Bill Douglas. Keeping a brisk pace, they hiked ten or eleven miles a day, fished for trout in the Elwha River, and slept in sleeping bags under the stars at campsites they called "The Olympic Hilton."[45]

On the hikes, Douglas and RFK talked about world problems. It was as if they were once again tramping around Siberia together. In a remote area called "the Last Frontier," Douglas told stories about his off-road travels in the Brooks Range and the North Cascades, both of which he wanted saved as wilderness areas. Douglas also taught RFK's sons, David and Bobby Jr., the art of fly-fishing. "The Elwha is a cold, clear stream, ideal for trout, and I did my best with David, who was an instinctively excellent fisherman," Douglas recalled. "I taught Dave how to float a dry fly and how to stimulate a nymph with a wet fly. I taught him to study the water for bug hatches and try to match the hatch with a fly of his own. I even tried to teach him the roll-cast, which the fisherman casts forward without bringing the line over his shoulder."[46] Douglas billed the seminars as "master classes."

CHAPTER 13

Rachel Carson's Alarm

Rachel Carson examines sea urchins and shells near her house in Southport, Maine, a year before her ground-breaking book *Silent Spring* was published. The boy in the photograph is Carson's grandnephew and adopted son, Roger Christie.

I

Two days after the White House Conference on Conservation ended, Carson, at Udall's suggestion, met with his assistant Paul Knight to prepare for attacks that would inevitably greet the publication of *Silent Spring*. Some of them would come from agencies of the federal government. In particular, the Department of Agriculture was home to many bureaucrats tasked with eradicating certain destructive insects. They weren't used to outside criticism or the need to consult other government agencies.[1] Even though Secretary Freeman was a reasonable person, he'd be under pressure to side with the large-scale pesticide applications his department was doing. In the face of

Carson's book, he instructed underlings not to be "trigger happy" but to build a fact-based defense of continued USDA insecticide use.[2]

An example of the siloed approach, habitual at USDA, is the high-profile campaign the department had initiated against the fire ant in the mid-1950s before the Fish and Wildlife Service had an opportunity to investigate its potential harm to the environment. The ecological damage in the Deep South was considerable. It took three years for quail populations to rebound from DDT. After that controversy, a less toxic pesticide started being deployed against the fire ant, and by 1962, the Fish and Wildlife Service's insecticide research budget leapt from nothing to $883,000. "There is need for intensive study of alternative methods of pest control," the *Washington Post* opined. "In some cases, as with alfalfa, resistant hybrid forms may provide better protection than the spray gun."[3]

One outgrowth of the fire ant disaster was the Federal Pest Control Review Board, established in 1961 in response to Kennedy's special message directing federal agencies engaged in pesticide programs to coordinate their application efforts. The new guard brought the Departments of Agriculture, Defense, the Interior, and Health, Education, and Welfare together to evaluate their programs in terms of their effectiveness, economy, and safety. Recognizing the threat the board posed to their lucrative business, pesticide companies lined up eminent scientists to fend off Carson's fact-based attacks.[4] Scientists and administrators in the federal government—in Agriculture and Defense, in particular—liked DDT and were ready to fight for it, too. Concern about the impact of DDT on the environment and public health was being featured in periodicals and on TV. Synthetic pesticides, air pollution, strontium-90, and sewage contamination were all national worries. But the federal government downplayed the environmental concerns about what Linda Lear termed the "gospel of technological prowess" on which *Silent Spring* was an attack.[5]

Carson, the reluctant crusader, knew it. Knight did, too, and was anxious to help her prepare for the onslaught of criticism after *Silent Spring* was published. "At that time," he wrote later, "Interior stood almost alone among federal agencies in considering the persistent pesticides as serious and possibly permanent contaminants."[6] It was a yeoman's job to prepare Carson for assaults by the lucrative pesticide world. Already the chemical lobby was ridiculing Carson for not having a PhD or academic affiliation. In their minds, she was a popular writer of beach books, not a scientist.

Shortly after Carson met with Knight, she spoke at the annual convention of the Biological Sciences Group of the Special Libraries Association. Introduced as the noted author of *The Sea Around Us*, she warned the librarians that the indiscriminate use of insecticides was killing birds and wiping out "whole runs of salmon." Convinced that pesticides caused chromosome damage to wildlife, she speculated that humans likewise weren't safe. Since World War II, she said— obviously referring to the use of nuclear weapons—man had possessed a "truly frightening power to change the world around him. . . . The price . . . might well be the destruction of man himself."[7]

Carson apparently had four primary aims in writing *Silent Spring*: creating an enduring work of literature on par with *The Sea Around Us*, alerting the public to the health dangers of indiscriminate pesticide usage, forcing the government to regulate the synthetic chemical industry more stringently, and strongest of all, helping to save the planet from destruction. That was no small order. When one considers her ambitions, one should remember that she didn't write *Silent Spring* in a bubble. Just as she was holding chemical manufacturers responsible for DDT poisoning, Frances Oldham Kelsey, a pharmacologist and physician at the Food and Drug Administration, was preparing to square off against the pharmaceutical lobby over the drug thalidomide, which she blamed for causing thousands of babies in Europe and Canada to be born with stunted limbs.[8]

Following the advice of Dr. George Crile, a Cleveland Clinic oncologist who read a few chapters, Carson refrained from writing that pesticides are always carcinogenic (even though a top Mayo Clinic cancer expert was convinced that they were). Crile worried that Carson was being selective, that a different writer could easily mine the existing scientific literature and research and come to an opposite conclusion.[9] Following his guidance, Carson asserted in *Silent Spring* only that DDT *possibly* caused cancer, but also admitted that a significant number of people who had been exposed to it hadn't contracted the dreaded disease.[10]

Carson made a bold point in comparing DDT to radioactive nuclear fallout; both were government-sanctioned poisons that people couldn't see in the air they breathed. Using *Silent Spring* as a vehicle to teach the public about ecology, she lamented that only since World Wars I and II had a single species, *Homo sapiens*, grotesquely mauled the evolution of life on Earth. After eons in balance, the natural order was being attacked in a crude and thoughtless way. The novelist Vladimir Nabokov, a self-taught butterfly expert, had urged his students to exude "the passion of a scientist and the precision of a poet"; Carson hoped she had approached that lofty objective in *Silent Spring*.[11]

On June 12, Carson delivered a commencement address at Scripps College in Claremont, California, that echoed her fears of the "ominous problem" of the by-products of atomic fission in her preface to the revised edition of *The Sea Around Us*, which was published the previous year.[12] The trip entailed her first cross-country flight. Just gazing down at Earth's topography mesmerized her. Her speech at Scripps was titled "Of Man and the Stream of Time"; in it, she postulated, that before Hiroshima, she had never thought that people could destroy the natural world order. Postwar nuclear testing—in the Marshall Islands, Siberia, Nevada, and elsewhere—had awakened her to the death knell of planetary destruction. "The once beneficent rains are now an instrument to bring down from the atmosphere the deadly products of nuclear explosions," she said. "Water, perhaps our most precious natural resource, is used and misused at a reckless rate. Our streams are fouled with an incredible assortment of wastes—domestic, chemical, radioactive—so that our planet, though dominated by seas that envelop three-fourths of its surface, is rapidly becoming a thirsty world."[13]

The next morning, Carson learned from her editor Paul Brooks that *Silent Spring* was the main selection of the Book-of-the-Month Club for October: towering news in the world of publishing back then. Moreover, she learned that Bill Douglas had agreed to write the catalog copy and that was even more stupendous news.[14] Meanwhile, on June 16, the *New Yorker* began its three-part serialization of the book. The magazine's editor, William Shawn, said that Carson had turned the DDT problem into enduring "literature."[15] An enormous number of letters, mostly enthusiastic, poured into the *New Yorker* office.[16]

The opening chapter of *Silent Spring* was called "A Fable for Tomorrow." Readers were introduced to a village that bore, by no coincidence, a distinct resemblance to Carson's hometown of Springdale, Pennsylvania, during her youth. "There was once a town in the heart of America where all life seemed to live in harmony with its surroundings," she wrote "The town lay in the midst of a checkerboard of prosperous farms, with fields of grain and hillsides of orchards, where, in spring, white clouds of bloom drifted above the green fields."[17]

The allegorical fable that followed, with a doomsday message at which some scientists balked, would become, in the words of Carson's biographer William Souder, "one of the great set pieces in American literature."[18] Carson next imagined a "strange blight" attacking the terrain like some evil spell in a horror movie. A granular white powder could be seen on rooftops, and sheep died en masse. The streams lost their fish, and colorful wildflowers turned a sere brown. With the disappearance of the robins and wrens, a weird stillness pervaded the community. "The few birds seen anywhere were moribund; they trembled violently and could not fly," she wrote. "It was a spring without voices."[19]

In the next chapter, "The Obligation to Endure," Carson described the state of unbalance between nature (which had taken eons to develop) and the poisons distributed by human beings in Cold War America, notably the twin demons of radiation and DDT. "It is not my contention that chemical insecticides must never be used," she wrote. "I do contend that we have put poisonous and biologically potent chemicals indiscriminately into the hands of persons largely or wholly ignorant of their potentials for harm."[20] Paul Brooks told Carson that if a reader of *Silent Spring* could make it through chapter 3, "Elixirs of Death," the rest of the book would be "smooth sailing."[21] That's because Carson broke down, without dumbing down the scientific descriptions, the two primary classes of pesticides: chlorinated hydrocarbons (which disintegrated slowly and include DDT, chlordane, dieldrin, aldrin, endrin, and heptachlor) and organic phosphates (which are extremely toxic but don't linger in the environment for too long, including parathion and malathion). Anyone hoping for *Silent Spring* to be Florida or California beach reading would be dissuaded upon encountering her prose in that scientifically dense chapter.[22]

Unlike Carson's sea trilogy, *Silent Spring* was written without gladness. The fact that one species, *Homo sapiens*, was destroying the planet made her heartsick. Other naturalists, John Muir and Aldo Leopold among them, had written about the ravaging of landscapes: the ruination of rivers, the clear-cutting of woods and consequent erosion of watersheds, the overgrazing of the western range, and the ill-thought-out agriculture that birthed dust bowls. Those activities were often local. That wasn't the case with nuclear testing and pesticides (or what Carson called "biocides"), such as DDT. They indiscriminately contaminate all ecosystems. Because the threat of DDT was so dire, Carson had abandoned her rhapsodic observations of nature, especially the sea, to tackle the high-stakes issue of the global poison in the earth, air, and water. Fourteen further chapters in *Silent Spring* described the use of DDT and the ramifications in blunt, clear language.

As Carson was finishing *Silent Spring*, she wrote a dear friend, "No, I myself never thought the ugly facts would dominate, and I hope they don't. The beauty of the living world I was trying to save has always been uppermost in my mind—that, and anger at the senseless brutish things that were being done. I have felt bound by a solemn obligation to do what I could—if I didn't at least try I could never again be happy in nature. But now I can believe I have at least helped a little. It would be unrealistic to believe one book could bring a complete change."[23]

Carson headed north to rock-ribbed Maine, to enjoy a quiet summer at Silverledges before the book was officially published. With her friends Dorothy and Stan Freeman, she relaxed, watching the advancing and retreating tides from their oceanfront deck. With her grand-nephew Roger Christie in tow, she enjoyed the diving terns, nesting parula warblers, and scavenging gulls of the seashore more than ever before, even though radiation treatments had ravaged her body and shrunk her frame. When summer ended, the spiritually reenergized Carson headed back to Silver Spring, Maryland.

That summer, Carson's network of supporters grew. But so did the line of detractors, who prepared to pounce on her research. Companies such as American Cyanamid and Monsanto could see that *Silent Spring* set up Big Chemical as the planet's worst environmental desecrator. The Velsicol Chemical Corporation, headquartered in Illinois—the sole producer of the chlorinated hydrocarbons chlordane and heptachlor—threatened to sue Houghton Mifflin for libel if it published *Silent Spring*.[24] Velsicol and other chemical manufacturers worried that the book would trigger lawsuits by citizens claiming that they had been poisoned in their own backyards. They were outraged with Carson's contention that, "It is not possible to add pesticides to water anywhere without threatening the purity of water everywhere."[25] It infuriated chemical companies that Carson had written in grandstanding fashion, "If the Bill of Rights contains no guarantee that a citizen shall be secure against lethal poisons distributed either by private individuals or by public officials, it is surely only because our forefathers, despite their considerable wisdom and foresight, could conceive of no such problem."[26]

The *New York Times* published its first editorial on *Silent Spring*— "Rachel Carson's Warning"—on July 2, 1962. If Carson's work succeeded in raising public concern, according to the paper's editors, "the author will be as deserving of the Nobel Prize as was the inventor of DDT."[27] That reference was to Paul Hermann Müller, the Swiss chemist credited with developing DDT; he had won the Nobel in Physiology or Medicine in 1948. A few weeks later, the *Times* ran a story about Carson cleverly titled "'Silent Spring' Is Now Noisy Summer." The uproar over the book was indeed ferocious. Unsurprisingly, the chemical industry's reaction focused even more media attention on *Silent Spring*, thus helping to promote Carson and her conclusions. For Velsicol, which also produced polybrominated biphenyls, cattle feed additives, and other chemicals, Carson posed a threat to the company's very existence.

The stage was set for a public relations war, once the book was officially released in late September, between protoenvironmentalists on one side and Big Chemical on the other. Most Americans were caught in the middle. Carson had never asked readers to take her word for the things she described, which was why she included pages of references.[28] Working against her was the fact that DDT was credited with having controlled the spread of malaria to a degree that saved countless lives. Carson argued that that victory had in fact been short-lived because mosquitoes were already developing resistance to biocides— including DDT—and were poised to return in even greater swarms.

Many conservation groups tried to straddle the issue by refraining from taking sides: not an admirable but a politick course. The National Audubon Society's president, Carl Buchheister, who had read the *New Yorker* excerpts, was a case in point. Lawyers from Velsicol lobbed veiled threats at John Vosburgh, the *Audubon* editor, and Charles H. Callison, Buchheister's assistant, warning them to beware of associating with Carson. Vosburgh and Callison ignored them, although they did fear lawsuits. *Audubon* then published an excerpt of *Silent Spring* along with an editorial criticizing Velsicol's pesticide program. Incredibly, however, Buchheister refused to permit the National Audubon Society to endorse the book. His was not the only conservation group to have reservations about Carson's research. The National Wildlife Federation was also against it—and in favor of pesticides. The Sierra Club gave it an official endorsement, but when members implied that Carson had gone too far, the organization's board was forced into a period of serious self-examination. The themes of *Silent Spring* were game-changing to many of the blue-chip conservation nonprofits.

As Udall and Knight had feared, Carson ran into serious trouble with Orville Freeman. His Department of Agriculture was scouring the *New Yorker* for libelous material. The USDA's premier constituents—farmers and agribusinesses—were already up in arms about Carson's shocking language and one-sided picture. A twenty-six-year-old investigative reporter, Robert A. Caro, later a respected author, wrote a five-part series in the Long Island paper *Newsday* about the threat *Silent Spring* posed to the tight relationship between the chemical industry and the USDA. "The lid is about to blow off this behind-the-scenes controversy," he wrote in *Newsday*, "overselling scientific evidence that chemical pesticides, enthusiastically promoted by the United States Agriculture Department despite 16 years of warnings, have decimated species of wildlife and now threaten man with cancer, leukemia, and abnormal gene development."[29]

Caro had been following DDT since Marjorie Spock and Bill Douglas had made headlines about the chemical having killed wildlife on Long Island. He had also followed the cranberry scare of 1959, the London smog tragedy, and the FDA's emergency retraction of its approval of the use of thalidomide by pregnant women to prevent nausea. Caro, who would go on to write Pulitzer Prize–winning biographies of LBJ and the New York urban planner Robert Moses, had styled himself as an environmental reporter before the term came into vogue. In *Newsday*, he reported how the USDA jumped onto the DDT bandwagon in the 1940s and never reconsidered the decision.[30] For his *Newsday* stories, he interviewed Freeman, who said that on balance, DDT was more beneficial than dangerous. Quite disingenuously, Freeman told Caro that Carson had been an ally in getting the truth out to the American people but hadn't in *Silent Spring* offered alternative ways to keep insects in check. That, of course, wasn't her job. With solid evidence, Caro showed how deeply intertwined the USDA was with the synthetic pesticides industry.

II

On August 29, 1962, at a late-afternoon press conference at the State Department, President Kennedy opened the session by announcing that Felix Frankfurter, Douglas's nemesis on the Supreme Court, was retiring. Then he was peppered with questions about Nikita Khrushchev, the United Nations, and Cuban shipping traffic. Toward the end, he fielded a question about DDT. The Caro *Newsday* series had ended five days earlier and pesticides were being discussed nationally. "Mr. President," a reporter asked, "there appears to be a growing concern among scientists as to the possibility of dangerous long-range side effects from the widespread use of DDT and other pesticides. Have you considered asking the Department of Agriculture or the Public Health Service to take a closer look at this?"

After reading the *New Yorker* excerpts along with the First Lady, Kennedy knew that Carson would receive an onslaught of abuse from the Big Chemical companies. At the White House press conference he was ready. And that alone put the issue on a high policy level. What Kennedy said next put *Silent Spring* there, too. The president replied, "Yes, and I know that they already are. I think particularly, of course, since Miss Carson's book, but they are examining the matter."[31]

It was a crisp exchange, only a few lines. But it proved to be a turning point. New Frontier conservation began its transformation into what would become environmentalism, linking together public health, the balance of nature, wildlife protection, and objections to nuclear testing. By referring to the existence of a government investigation into the poisoning of the environment by products that unassuming consumers routinely used—and that Fortune 500 companies were manufacturing with the approval of the USDA and the Public Health Service—Kennedy's utterance was brave. One would be hard pressed to find a precedent for it.

For Kennedy, both DDT and nuclear radiation were issues involving human existence.[32] Perhaps his daring evocation of "Miss Carson's book" was the public service equivalent of Theodore Roosevelt's embrace of Upton Sinclair's novel *The Jungle*, a searing indictment of unsanitary conditions in Chicago meatpacking plants; it had led to the passage of the Pure Food and Drug Act in 1906. Overall, Kennedy believed that the information contained in *Silent Spring* warranted a study of pesticide use and perhaps federal regulation of chemical companies. By accepting and even promoting the importance of Carson's work, he joined forces with the protoenvironmentalists. Sticking with the conservation status quo or staying mum on a public health crisis wasn't an option, at least not for him.

Without necessarily embracing Carson's controversial conclusions, Kennedy made it clear that his administration took *Silent Spring* very seriously. However, the government investigation he had cited to the press was not, in fact, under way that day. But on the day after, it was. On August 30, Kennedy announced that the federal government's approach to pesticide use would be reviewed by the Federal Council for Science and Technology and a panel of the President's Science Advisory Committee to be headed by Jerome Wiesner. (Its report would be made public in May 1963.) "What Kennedy did say was sufficient to galvanize the chemical industry to take action," Carson's biographer Linda Lear wrote. "It was not an accident that the day after this news conference, the National Agricultural Chemicals Association issued its own pro-pesticide propaganda booklet, 'Fact and Fancy.'"[33]

The gist of the chemical counterattack was that Mr. Fancy (aka Kennedy) was an East Coast elitist who yachted frivolously around Cape Cod, his treasured national seashore, while allowing DDT manufacturers to be unjustly vilified in the public square. It asked why Kennedy didn't remind the public that without DDT, many US troops during World War II, especially in the Pacific theater, would have died from malaria. Was Kennedy likewise going to turn against other World War II–era technological marvels such as synthetic rubber, microwave electronics, jet propulsion, and orange juice concentrate?[34] Furthermore, the association warned, chemical factory shutdowns would mean thousands of lost jobs in Illinois, Michigan, and Ohio.

Udall's Interior Department had helped publicize *Silent Spring* while simultaneously protecting the president from embarrassment if Carson's research collapsed under critical peer review. Where detractors and fans alike agreed was that *Silent Spring* was going to be a terrifying book to read.

The media hullabaloo over *Silent Spring* tickled Kennedy's sense of irony.[35] That one bold author could upend the entire chemical industry was proof that the grand American reformist movement was alive and well. *Silent Spring* jibed with Kennedy's New Frontier reliance on a scientific approach anchored by experts to solve public policy problems. In a commencement address at Yale University that June, he warned the students not to let truth-telling become elusive in their lives: "Too often we hold fast to the clichés of our forebears. We subject all facts to a prefabricated set of interpretations. We enjoy the comfort of opinion without the discomfort of thought."[36] Kennedy, as president, didn't have to have an instantaneous opinion on DDT. He only had to keep his mind open to the truth, while the Rachel Carsons of the world led the reform charge on what constituted environmental and public health hazards.

It was quite a turnaround. In 1958, Senator Kennedy had coauthored legislation to authorize $100 million for a World Health Organization malaria project based on using DDT dissolved in oil or kerosene in developing countries. Four years later, owing to "Miss Carson's book," he feared that the scientific data revealing that DDT was potentially deadly for the full range of living creatures warranted deep investigation.[37] He took to heart Carson's richly detailed writings on the effects of DDT on sentient beings besides humans. In one example from the book, two zoologists at the University of Michigan reported that spraying DDT on campus had wiped out Dutch elm disease, a fungus spread by bark beetles. Tree surgeons had done so to save the

Fogger truck spraying DDT on beaches at Long Island, New York. In 1967, the Environmental Defense Fund (EDF) grew out of the fight to ban DDT.

elms, but the entire population of ravens on the 110-acre campus had been destroyed.[38] That case proved, as Carson had written, that a synthetic pesticide poisoned "all life with which it comes in contact: the cat beloved of some family, the farmer's cattle, the rabbit in the field, and the horned lark out of the sky. These creatures are innocent of any harm to man."[39]

The spate of controversies that followed the publication of *Silent Spring* that September whipsawed it with alternating praise and condemnation. The president of Montrose Chemical, the primary manufacturer of DDT, described Carson as "not a scientist, but a fanatic defender of the cult of the balance of nature." His proof was that her opening chapter was a hybrid of nature writing and fiction. The American Cyanamid Company charged that *Silent Spring* was a compendium of factual distortions. Louis McLean, the general counsel for Velsicol Chemical Corporation, intimated that Carson was a Communist sympathizer.[40]

Carson had fully expected Velsicol and American Cyanamid to challenge her work. But she wasn't prepared for the onslaught of vitriolic put-downs in popular periodicals she read. The *Economist* denounced *Silent Spring* as a "shrill tract" of "propaganda written in white-hot anger with words tumbling and stumbling all over the page."[41]

In addition to the harsh treatment that Carson's work was receiving, the fifty-five-year-old author faced a barrage of *ad hominem* attacks simply for being a woman. Trying to silence *Silent Spring*, Big Chemical painted her as a hysterical pseudoscientist and nature-loving freak. Some made up lies that were supposed to smear Carson, calling her a lesbian and a faddist. William J. Darby of the Vanderbilt University School of Medicine attacked her in *Chemical and Engineering News* for "reversion to a passive social state devoid of technology, scientific medicine, agriculture, sanitation, or education."[42]

Time magazine browbeat Carson for "putting literary skill second to the task of frightening and arousing her readers" and shopping "over simplifications and downright errors." In full gender-bias mode, it also mocked her findings as "hysterically overemphatic" and an "emotional and inaccurate outburst."[43] The *Catholic* likewise resorted to the knee-jerk means of undercutting female intellectuals, slamming Carson for "emotionalism." Tacitly defending the status quo, *Life* magazine thought the book was as exaggerated as the chemicals industry claimed in response to it (which described a world unlivable without pesticides).[44] The personal attacks flew at Carson from all quarters in such derogatory phrases as "priestess of nature," a "hysterical woman," and a "member of a mystical cult." A former secretary of agriculture, Ezra Taft Benson, even got into the act. "She's a spinster, isn't she?" he wrote to his old boss Dwight Eisenhower. "Why is she so worried about genetics?" Besides insults, chemical companies threatened the *New Yorker*, Houghton Mifflin, and Carson herself with lawsuits, though none was ever filed.[45]

In August, Kennedy awarded Frances Oldham Kelsey the President's Award for Distinguished Federal Service in recognition of her discovery that thalidomide causes deformities in babies.[46] Carson told the *New York Post*, "It is all of a piece, thalidomide and pesticides— they represent our willingness to rush ahead and use something new without knowing what the results are going to be."[47] For every critical article, though, there were more appreciative ones. For example, the celebrated anthropologist Loren Eiseley, writing in the *Saturday*

Review, praised the book as "a devastating, heavily documented, relentless attack upon human carelessness, greed, and irresponsibility."[48]

Silent Spring stayed on the *New York Times* bestseller list for six months; it would eventually sell 6 million copies. Unaccustomed to immense fame, Carson again retreated to coastal Maine. Even though many journalists wanted to interview her, she kept her public appearances to a minimum. Only if she thought her presence would help a conservation group would she agree to speak or sign books. At the time, she was too busy fighting cancer to worry much about blowback. More often, *Silent Spring* was her voice and soul: it spoke for her. With the Atlantic as her breezy muse, she finished a long essay for the *World Book Encyclopedia Yearbook* about ocean resources.[49] Staying focused on the sea allowed her to remain detached from the roar that *Silent Spring* had triggered.

David Brower read *Silent Spring* on a flight from Salt Lake City to San Francisco on its publication day. It was a eureka moment: instinctively he saw how ecology and public health would be forever interlinked. He wanted the Sierra Club to be part of the zeitgeist, but it wasn't that easy. He'd have to go up against conservative members of the Sierra Club board, who thought Carson was an environmental extremist or an irresponsible scientist. One member, Thomas Jukes, a founder of the club's Atlantic States chapter, actually allowed a TV crew to film him swallowing a teaspoonful of DDT at a press event to prove how safe the chemical was for humans.[50]

Brower looked for new board members for the Sierra Club, people who were sympathetic to Carson's environmental thinking, preferably with access to the media. In short order, he recruited Bill Douglas; the *New York Times* editorial page editor and columnist John Oakes; the Houghton Mifflin (and *Silent Spring*) editor Paul Brooks; the *Sunset* magazine travel editor Martin Litton; the photographer Eliot Porter; and the novelist and wilderness advocate Wallace Stegner. Old-time board members, many of whom were outdoors recreationists, were livid at what they took to be a guerrilla tactic by Brower: packing the board with his powerful friends. Brower stuck to his guns. A new way of environmental thinking was on the rise, thanks to the Carson juggernaut, and he meant to put the Sierra Club into its vanguard.

CHAPTER 14

Point Reyes (California) and Padre Island (Texas) National Seashores

President John F. Kennedy (seated) signs S. 476, a bill to establish the Point Reyes National Seashore in California, in 1962. *Standing, left to right*: Representative Wayne N. Aspinall (Colorado); Representative J. T. Rutherford (Texas); Senator Hubert H. Humphrey of Minnesota; Secretary of the Interior, Stewart Udall; Representative John P. Saylor (Pennsylvania); Senator Alan Bible (Nevada); Senator Clair Engle (California); Representative Clem Miller (California); unidentified (*in back, mostly hidden*); Representative Jeffery Cohelan (California); unidentified (*in back*); executive director of the Sierra Club, David Brower.

I

On August 17, 1962, President Kennedy left Washington to embark on a whirlwind tour of three western states: South Dakota, Colorado, and California. Udall wanted the president's presence to bring attention to the work that the New Frontier was doing in the realm of natural resource management, especially with multipurpose dams, which, as Udall put it in an interview eight years later, "still had some magic then."[1] To also prove that Kennedy was a nature enthusiast, Udall had arranged for a Yosemite National Park trip. JFK meandering around Tuolumne Meadows, an open landscape dotted with creeks and pools, ringed by what Ansel Adams called "sculptures in stone," with colorful names such as Cathedral Peak and Unicorn Peak was guaranteed to generate great press copy.[2]

Only four days before leaving for the West, Kennedy was sailing and sunbathing on Johns Island off the Maine coast. Yet that was different from hiking to the top of Vernal Falls in Yosemite for a man ravaged with Addison's disease. "The President with his health problem," Udall explained, "with his back and everything, never was able to do the things that Bobby would have done as president, you know, of getting out and running a river—the vigorous life things—which I always regretted. But that was not the case and this, in a way, and the fact that his big love was the sea—you know, he was tied to the seacoast and the sea (that's where he spent his life)—he didn't have the kind of earth feeling that Bobby Kennedy developed, for example. So this always was part of it."[3]

As 1962 was an election year for the House and many seats in the Senate, various Democrats seeking office glommed onto the president's trip for a day or so at a time. One of them was Wayne Aspinall, who flew west with JFK and stayed on through the stop in Colorado. Aspinall liked Kennedy personally. But he was a tough bird for whom friendship made little difference when it came to his desire for reclamation dollars. In 1956, Aspinall had hurt JFK by refusing to help him win the vice presidency. Far from alienating him, Kennedy had cultivated him and finally, in about 1958, had asked him straight out why he'd thrown his support to Lyndon Johnson. Aspinall told Kennedy point-blank that he mistrusted Massachusetts Democrats on the issue of western dams. Kennedy had countered with his voting record. "Well," Aspinall had replied, "it never has appeared to me that you're as interested as much in the West as in even Europe."[4]

From that frank exchange, Kennedy embraced Aspinall as a mentor on western politics. One example of their working relationship was the Fryingpan-Arkansas Project, a complex plan to pull water from Colorado's western slope to the farmland and residential areas of the eastern slope.[5] In a later conversation, when Kennedy flaunted his loyalty to the West by saying that he'd voted for "Fry-Ark" in the Senate, Aspinall failed to be impressed. Still trying to teach Kennedy how a master of western water politics looked at the world, he replied that the simple yes-no votes in the Senate had left the House (namely, Aspinall's committee) to work out the nearly unending details of the complicated project. It took ten years for Aspinall to be satisfied that the bill would ensure a successful future for Fry-Ark. He wanted Kennedy to understand the specific irrigation and electricity considerations that dams entailed, which, he feared, were foreign to a sailor from Cape Cod and which made New Englanders think of the arid American West as a simple yes-no type of region.

The trip began in Pierre, South Dakota, with George McGovern, a preservationist at heart then campaigning for a Senate seat, at Kennedy's side. The stop lasted only two hours. There was no opportunity to explore woods that sustained a cross-section of northeastern wildlife: beaver, muskrat, wood duck, great blue heron, or deer, to name the most obvious. At the Oahe Dam, only eleven miles away, Udall introduced the president to a sprawling steel and concrete complex built by the Army Corps of Engineers. In the process he called South Dakota "Teddy Roosevelt country." Kennedy was there to dedicate the massive project, which he proudly noted was one of six major dams built or planned to harness the Missouri River. He praised the hydroelectric power produced on the Missouri River for northern Great Plains communities. No word was uttered about wilderness preservation, national parks, or pollution-free waterways. "This dam will produce enough electric energy, this one dam, to light the city of Edinburgh, Scotland," he enthused at Oahe. "This dam and the rest of the dams on this river, which 30 years ago would have provided only floods and darkness, now provide irrigation and light."[6]

When Kennedy arrived in Pueblo, Colorado, more than a hundred thousand people lined the sides of Highway 50 to see the motorcade pass. The timing was hardly coincidental, inasmuch as the front-page news of the *Denver Post* was his signing the previous day of the Fryingpan-Arkansas Project into law. The future of Colorado's southeastern section was suddenly bright, with water on its way in a steady flow to irrigate farms, as well as growing western cities. As

the president's jet approached Pueblo, it circled over the future location of a dam that was part of the huge and contentious Fry-Ark
project. Kennedy came back from his quarters on the plane to look
out the windows at the dam site with Aspinall, an enormous moment for the congressman. On the ground, where temperatures hit
the high nineties, Kennedy met an enthusiastic audience at the local
football stadium. In his speech there, he praised the Fry-Ark project,
which would take water from around nine thousand feet, sometimes
through mountains, to irrigate an entire valley in eastern Colorado.[7]
Mountain farmers who had struggled with erosion and those ranchers in the lower counties who suffered dust bowl drought conditions
couldn't wait for Fry-Ark to start delivering water. Many others had
been resisting the project for over a decade.[8]

Deviating from the prepared Fry-Ark script, Kennedy spontaneously pivoted to a preservationist theme, advocating national seashores and passage of the Wilderness bill. With Aspinall nodding
in agreement, he also pushed for his proposed Youth Conservation
Corps, aimed at beautifying American landscapes. "I would rather
have those unemployed boys and girls who hang around on street corners today working in our parks and forests and making something
of this country and their lives than staying at home and wondering
what's going to become of them," he said.[9]

As noted, Udall had strategically included a visit to Yosemite on
Kennedy's crowded itinerary to pay homage to the National Park System. The sheer majesty of El Capitan and Half Dome, the secretary
reckoned, would kindle the president's enthusiasm for large-acreage
national park sites in Washington's North Cascades and Utah's Canyonlands.[10]

Experiencing the Yosemite wilderness had been a rite of passage
for presidents for generations. Abraham Lincoln had signed legislation
preserving the Sierra treasure in 1864. Just as Theodore Roosevelt had
John Muir at his side in 1903 when he camped in the park, Kennedy
had Udall with him on August 17, 1962—when he stayed at a lodge.
Udall had originally wanted Virginia and Ansel Adams to escort the
president around Yosemite, but prior commitments had made that
impossible for them. (To express his regret, Adams sent the president
a copy of the out-of-print *My Camera in Yosemite Valley*.) To express

a different kind of regret, Adams had written Udall on August 14: "The local paper (San Joaquin Valley *Fresno Bee*, I think) noted that the president would visit Yosemite, enjoy the Ahwahnee Hotel and the world-famous Firefall. This is typical of the emphasis given to the artificial impositions! The Ahwahnee is an excellent hotel, but the Firefall remains, for me, a very sad concession to commercialism and misdirected publicity. I hope the Press will note that the President *did* see the cliffs and forests, and what is left of the waterfalls at this time of year."[11]

The letter suggests that Adams's scheduling conflicts may have been produced, in part, by his reluctance to endorse the commercial side of Yosemite the president was going to see. By the time of the Yosemite visit, Adams's arresting photographs of the park had become iconic. Having sat on the Sierra Club board since 1934, with an association with Yosemite that went back even further, Adams thought landscape preservation mattered more than economic considerations when it came to federal parks. The Yosemite that Kennedy was to experience was, to his mind, partially ruined by slob tourism; there were simply too many gas-guzzling cars causing chronic congestion, too many air-conditioned getaway lodges, and too many ticky-tacky concession shops. And then there was the Firefall, in which humans improved on Yosemite's natural beauty, as though training a brown bear to dance improved on the majesty of such a grand animal.

With Kennedy as he reached Yosemite were California's Democratic governor, Edmund Brown, and Thomas Kuchel, its super-liberal Republican senator, both thoughtful conservationists. Upon arrival, Kennedy hopped into a candy apple red convertible so he could cruise to the historic Ahwahnee Hotel in the fresh mountain air. That evening, he ventured onto a second-floor balcony, where he had a spectacular view of Glacier Point. As Adams had resented, the president witnessed the Yosemite Firefall show that evening: hot embers pitched from the summit cascaded three thousand feet into Yosemite Valley, creating the effect that the flow of water was on fire. Ever since Kennedy had seen the Firefall featured in the 1954 movie *The Caine Mutiny*, he had wanted to witness the spectacle for himself. It was a natural inclination; the effect was unforgettable. Thousands of other

Governor Edmund "Pat" Brown of California explains to President Kennedy and Stewart Udall certain geological features of Yosemite National Park.

spectators were thrilled that the debonair JFK was in attendance with them that August evening to make the experience historic.

The next day, Kennedy explored the Yosemite Valley, accompanied by Udall and Douglas Hubbard, the chief naturalist at Yosemite National Park. JFK's ever-observant face looked ruddy and healthy, but the puffy eyes behind his Ray-Ban sunglasses showed the toll of his time in office. Although the weather was hot, the president wore a dark Brooks Brothers suit. At a small ceremony, he was made an honorary ranger and presented with a souvenir redwood plaque. "I'm sure all Americans who come here will be proud of our park service, and will be happier about America after seeing this valley," he said in gratitude.[12]

At one point during the morning tour, JFK announced that he wanted to drive up to Mirror Lake and hike back down. Ironically, it was his very presence that made that impossible. The road was jammed with people hoping to see him. His car couldn't get to Mirror Lake, so the hike was scrubbed.[13] The change of plans meant he could see much more of the park, though, in the short time available. There were some pleasant, spontaneous moments en route. When he journeyed to Yosemite Falls from Inspiration Point, he encountered a young Japanese couple, Mr. and Mrs. Sachiko Won of Tokyo, who waved with unrestrained excitement. "We're visiting from Japan," Mrs. Won called out. Like any good politician, Kennedy shook hands. "How long do you plan to stay in this country?" he asked. "Forever, now," gushed Mr. Won, who, slightly overwhelmed, hugged his wife.[14]

Eli Setencich, a columnist for the *Fresno Bee*, followed Kennedy's progress at Yosemite, impressed most of all by the president's friendly humor as he mixed with a wide array of awestruck tourists; in his typical fashion, he didn't merely wave or say hello, he asked most of those who greeted him a question. On only one occasion, Setencich wrote, did Kennedy want the last word. Before a gathering of reporters, Governor Brown pointed to the president and said in a jocular way, "He's coming back next year to climb Half Dome." Kennedy didn't lose a beat: "No, Udall's going to climb it."[15]

Udall was pleased that the Yosemite visit had gone well. He wrote Adams, "The President loved *your* valley, the natural beauty, I mean, not the Firefall!"[16] In fact, the secretary was ecstatic that Kennedy had enjoyed himself so much, not merely at Yosemite but on all his western stops; that set Udall thinking out loud about another western tour in 1963.

On the western trip Kennedy also spoke in Los Banos, California, to break ground for the San Luis Dam on the creek of the same name. A cooperative effort between the Golden Bear State and national government agencies, the dam was important chiefly as a reservoir that would gather water from a number of sources. As Kennedy toured the site, he seemed to be thinking of the criticism that his former adversary Aspinall had made years before. He told the crowd at Los Banos, "What this country needs is a broad, new conservation effort, worthy of the two Roosevelts, Theodore and Franklin, who lived in New York, and who helped build the West; an effort to build up our resource heritage so that it will be available to those who come after us."[17]

All-important California, with its forty-five electoral votes, was the most environmentally conscious state in America. Of course, the degree varied across the big state. At the time of Kennedy's visit, the residents of the Central Coast of California, with its small river valleys and beach dunes, led the way as home to the most eco-aware citizens in all of America. Across the state, a public education campaign had been set into motion for schoolchildren to learn about local botany, sustainable living, ecology, endangered species, and the philosophy of bioregulation. In 1962, California had a progressive environmental ethic and Kennedy was embraced as a leader who understood it. His administration was then partnering with Sacramento lawmakers to keep Lake Tahoe, one of America's largest mountain lakes, "forever blue."

Because the Sierra Club was based in San Francisco, the nonprofit adopted the Point Reyes Peninsula, only forty miles northwest of the Golden Gate Bridge, as a prime preservationist concern. The club's president, Edgar Wayburn, made it a high priority. With San Francisco Bay Area commercial and residential real estate expanding in all directions, he knew that a window of opportunity was closing. Wayburn, a physician by day, was a dogged behind-the-scenes environmental agitator. He differed from many conservation leaders in that he didn't seek personal publicity. On the Reyes effort, Wayburn teamed with *San Francisco Chronicle* reporter Harold Gilliam, a Udall ally, to awaken the Bay Area to the importance of saving the prominent headland. The Sierra Club's executive director, David Brower, along with conservation legends such as Horace Albright and Sigurd Olson, also lent support to the preservation effort, as did Ansel Adams.

For years, Adams—wearing a work shirt, pens usually protruding from his pocket, Keds tennis shoes on his feet—had hiked Point Reyes with preservation in mind. With his trusty Leica in hand, he took stunning photos of the peninsula overlooking Drakes Bay. Two of his finest art photos, *Barn, Point Reyes, California, 1940* and *Road and Fence, Point Reyes* (ca. 1939), were taken on such hikes. Captivated by the open hills, windswept beaches, bishop pines, perennial grasslands, and thick willow scrub, Adams wrote hundreds of letters lobbying to protect the unspoiled Marin County shoreline from development.[18]

On the political front, Thomas Kuchel helped the Kennedy administration gather bipartisan support for the Point Reyes National Seashore designation. As the Senate minority whip, Kuchel asserted that scenic oceanfront landscape such as Cape Cod was essential to the National Park Service's strategy of establishing national seashores and lakeshores. An avid bird-watcher, Kuchel had already successfully fought to designate Tule Lake (California) and Upper Klamath (Oregon) as federal migratory bird refuges. As the leading Republican conservationist in the Senate, Kuchel lent political clout and a bipartisan glow to the effort at Point Reyes.[19]

The Point Reyes campaign had much going for it. In 1962, Harold Gilliam's book *Island in Time: The Point Reyes Peninsula* appeared under the Sierra Club large-format imprint.[20] In emulation of *This Is the American Earth*, the text was accompanied by photos taken by the masterful Eliot Porter of shipwreck debris, northern elephant seals, coastal redwoods, and picturesque lighthouses. Udall wrote the foreword to Gilliam's book, and in it he observed that Point Reyes "owes its wide uncluttered dimensions to ranchers—western husbandmen— who have maintained its land for a century and owes its public recognition to quiet citizens in private life who know what it is to love the earth and who care about working for it."[21]

The work of Wayburn, Gilliam, and Porter paid off, as did the fact that Point Reyes would be a recreation oasis for city dwellers, in keeping with the ORRRC report. On September 13, 1962, in a ceremony at the White House, President Kennedy signed a bill authorizing the purchase of 53,000 wild acres and allocating $13 million for Point Reyes National Seashore. Standing at the president's side during the Oval Office ceremony was David Brower, who had given everybody in attendance a complimentary signed copy of *Island in Time*. (There is a photo of Kennedy surrounded by lawmakers and activists clutching copies of the oversized book.) As at Cape Cod National Seashore, NPS management had pushed to allow some private landowners to remain; in the case of Point Reyes, a limited number of dairy operations were grandfathered in. A coastal radiotelegraph marine station was refurbished as a historical attraction and the Tule elk (*Cervus canadensis*) was reintroduced into the grasslands to replace cattle, for, as the naturalist Lois Crisler wrote, "wilderness without animals is mere scenery."[22]

No sooner did Kennedy create the Point Reyes National Seashore than Wayburn orchestrated a new grassroots Bay Area movement to protect fifty-nine miles of noncontiguous bay and ocean shoreline-adjacent acreage as part of what in 1972 would become the Golden

Gate National Recreation Area, a remarkable collection of points of interest from southern San Mateo County to northern Marin County.

The red-hot controversy in California conservation in the 1960s was the fate of coastal redwood stands. On one hand, Americans took bedrock pride in the fact that the enormous trees were among the very oldest living things on Earth. On the other, they were buying deck furniture made of those same trees. Logging old-growth redwoods was a highly profitable business, operated by big, stubborn companies perfectly willing to argue over every tree. The issue, realistically, was not an effort to save the redwood forests as a whole but the designation of just exactly what groves should be spared the chainsaw. For more than fifty years, the Save the Redwoods League had been buying tracts of ancient trees, making trade-offs with lumber companies when necessary. More recently, starting in the 1950s, the Sierra Club had advanced wholeheartedly into the preservation battle, working for total protection of groves through the federal government. Each nonprofit had its own opinion about which big trees to save. The two groups were so adversarial that objective observers felt that their mutual animosity was working against the goal of saving the giants—as soon as possible. The National Geographic Society entered the fray in what it hoped would be an equitable way: by sponsoring a feasibility study on a national redwoods park. The two motivating factors, with which the society assumed everyone would agree, were that the redwoods were worth preserving and that at the present rate of redwood harvesting, about 900 million board feet annually, all old-growth redwoods not protected in parks would be gone by the year 2000 and probably twenty years sooner.[23]

II

While President Kennedy was making Point Reyes a national seashore, another candidate for that designation was nearing success: Padre Island, Texas. Saving the barrier strip of clean white sand and shifting dunes was tougher than saving Point Reyes. For starters, establishing national parks in Texas was never easy. Lone Star State residents prided themselves on the state government's administration of public lands, and they largely discouraged the Interior and Agriculture Departments from doing so. However, in 1941, Franklin Roosevelt had successfully pushed Congress to establish—and Texas to accept—Big Bend National Park, along the border with Mexico, as well as two crucial federal migratory bird reservations elsewhere. Still, compared with other western states such as California and Arizona, Texas contained limited federal acreage for anything but military bases.

As a senator, Vice President Lyndon Johnson had opposed national park status for Padre Island, just south of Corpus Christi in the very south part of Texas's Gulf Coast. Friendly with powerful commercial developers, Johnson thought that locking up valuable beachfront acreage was unreasonable. Democratic Texas senator Ralph Yarborough, LBJ's frequent adversary, supported the park proposal as a means of maintaining the balance with nature in Texas. Yarborough feared that the Gulf Coastal Zone, where fresh and salt waters intermingle, had already been abused by navigation, oil and gas exploration, military operations, and rank commercialization. In June 1958, when Yarborough introduced the Padre Island bill in the Senate, he quoted the Interior Department's persuasive pamphlet "Our Vanishing Shoreline." It termed the barrier island "a place of undying historic charm, . . . one of the most desirable semitropical rest spots in the world."[24] At the time, the Army was holding bombing practices on the island. Yarborough also requested that an article in the *Texas Observer* by Ronnie Duggan be read into the *Congressional Record*; it claimed that the entire island could be purchased for $35 million and then made into a vast seashore park.[25]

As the discussion was continuing, the Army Corps of Engineers was cutting a channel across Padre Island, dividing it in two.[26] The south part, stretching thirty miles, was to be open for development. Typically, it was called "South Padre Island," while the unmolested northern part was just "Padre Island." Padre offered a seventy-mile stretch of beachfront, separated from the mainland by Laguna Madre, one of the saltiest bodies of water in the world and one of the most biologically diverse marine habitats in North America. Sparkling sands blowing from the beachfront created an ever-changing dune system that was a critical species habitat because the dunes formed a natural dike, which prevented storm tides from inundating and destroying essential grasslands. On that big, quiet barrier island, more than six hundred species of plants and wildflowers thrived. Depending on the season, astronomical numbers of heron, ibis, egrets, spoonbills, pelicans, cormorants, ducks, and geese claimed Padre Island as a temporary home. Sand dollars, sea stars, whelks, clams, and snails washed in on the daily tides. One bit of good news was that the channel carved by humans through the island to facilitate boat traffic actually enhanced the habitat for wildlife, with ocean water circulating into Laguna Madre to refresh it. Rumors have long circulated that Padre has another type of treasure, as well as the birds and plants: the buried contraband booty of Gulf pirate Jean Lafitte.

Despite Yarborough's efforts, Padre Island National Seashore stalled in the Senate during the Eisenhower years. The advent of the Kennedy administration presented new opportunities, especially because Yarborough could present a new, fact-filled book by a Corpus Christi resident, Vernon Smylie: *The Secrets of Padre Island: An Informal History of America's Most Fascinating Island*. In January 1961, he reintroduced his bill in the Senate, saying on the floor, "In urging the establishment of Padre Island Seashore, I am not speaking for the New Frontier. Instead, I plead for what many conservationists, many Americans, consider and what really is, geographically, the last frontier. Unless we act now, it will be the lost frontier."[27]

Another important proponent of the park designation was the owner of the *Corpus Christi Tribune*, Edward H. Harte. Nonetheless, protests erupted throughout Nueces County over the bottling up of so much of the beachfront on behalf of sea turtles, shell collectors, and stargazers. Real estate developers from the Corpus Christi area claimed that multimillion-dollar fortunes of future entrepreneurs were being squandered.[28] Lyndon Johnson would have agreed with them. In reply, Yarborough defended the preservation action as visionary.

It wasn't a secret that Kennedy had a chilly relationship with Lyndon Johnson. Behind his vice president's back, he'd unleash acid jests and cynical appraisals. By contrast, Yarborough was Kennedy's type, a modern-day liberal in sync with fellow wilderness warriors Frank Church of Idaho and Clinton Anderson of New Mexico. As a lawyer and Texas land expert, Yarborough knew how to properly acquire acreage for preservation and, when necessary, stop unnecessary road construction he deemed pork.

One of the primary reasons Yarborough sought federal protection for Padre Island was to save nesting areas required by the endangered Kemp's ridley sea turtles, whose eggs were being overharvested. Between late May and early August, the females of the relatively small turtles crawled onto Padre Island to deposit clutches of eggs into the white sand, digging cavities into the beach with their rear flippers to bury them. Afterward, they covered the nest with sand again and returned to the sea. Once the hatchlings emerged, they were on their own. Yarborough identified with the innocent young creatures, living on the edge of a powerful ocean that could smash an animal but that also supported life and helped create new, sentient beings. Padre Island wasn't the only hatching grounds for the gulf species, but it was probably the biggest.

In a strange move, LBJ tried to derail the plan for a Padre Island National Seashore by encouraging two Democratic senators, Robert S. Kerr of Oklahoma and George Smathers of Florida, to reject the bill. Breaking ranks with Democratic conservationists such as Scoop Jackson and Clinton Anderson, Johnson seemed to be awaiting a better deal. At Yarborough's request, Udall made an inspection tour of Padre Island in June 1961 with Anderson and a bevy of Texas officials. Most important, Johnson went along, seeing Padre Island for himself both by air and on the ground. After that junket, the Kennedy administration pushed even harder for the national seashore designation. Johnson continued to dissent, but without his former rigidity.

Yarborough later pointed to the trip as the turning point for the park. To see it in person was, indeed, to understand that it was a breathtaking scenic and recreational treasure. LBJ still tried to horse-trade, offering to name a small beachfront park on the island after Yarborough if some of the gulf coastal lands could be sold to commercial interests, including oil and gas companies. That was dubious behavior for a vice president. Yarborough refused.[29] Beholden to Houston construction firms, Johnson was publicly tight-lipped about Padre Island. The loudest enemy of the park idea was Texas's powerful land commissioner, Jerry Sadler. He contended that the "coastal area was home to untapped producing oil wells and represented great mineral wealth" and characterized the transfer of the tract as "a giveaway." Commenting on the plan to bar commercial development from the island, he said, "That may suit poets and bird watchers, it suits neither Jerry Sadler, Commissioner of the General Land Office, nor Jerry Sadler, citizen and taxpayer."[30]

National momentum was clearly on Kennedy's side. Within months, Sadler reversed himself, becoming a supporter of Padre Island. It took a little longer, but LBJ followed. The sand beaches of Padre Island were something of a maritime miracle no one could resist forever.

On April 11, 1962, the Senate passed the Padre Island National Seashore bill—Kerr and Smathers were among the thirty-nine "nay" votes—and the bill moved to the House, where it eventually passed. On September 28, 1962, just two weeks after the Point Reyes victory, President Kennedy signed the bill creating Padre Island National Seashore at a White House ceremony with Ralph Yarborough leaning over his shoulder.[31] Compared to the excitement that had accompanied the Cape Cod and Point Reyes bills, national enthusiasm for the Padre Island National Seashore was low key. Nonetheless, approximately seventy miles of the most extensive natural barrier island along the entire Gulf of Mexico was protected for the ages. As a footnote, Corpus Christi's newspaperman, Ed Harte, would establish a $46 million endowment in 2000 at Texas A&M's Corpus Christi campus to create a marine institute like Woods Hole specifically for the study of the Gulf of Mexico. One of many initiatives of the Harte Research Institute is the conservation of the Kemp's ridley sea turtle on Padre Island.

Building on his conservation victories, President Kennedy signed a bill amending FDR's Soil Conservation and Domestic Allotment Act of 1936, which helped landowners, especially farmers, take better care of their holdings and reduce erosion. Udall was concerned that the Department of Agriculture was paying to have wetlands drained for agricultural purposes and Interior was buying such ecosystems to make them Fish and Wildlife Service refuges. Kennedy had mentioned that situation in his Special Message on Conservation earlier in 1962. In the compromise he'd encouraged, Interior would have first dibs on wetlands designated for draining. If it didn't acquire them as wildlife refuges, Agriculture could proceed with its usual reclamation program. The beneficiaries would be migratory birds, which need wetlands as way stations. The amendment applied to the states crucial to such birds: Minnesota, North Dakota, and South Dakota. Under Kennedy's leadership, new national wildlife refuges such as Moody (Texas), Harris Neck (Georgia), Davis Lake (Mississippi), and Delevan (California) were also established.

Yarborough, flush with victory in the fight for Padre Island, began a campaign to establish Guadalupe National Park, an area of ancient limestone rocks and high desert peaks in west Texas's Trans-Pecos region.[32] His ally was the peripatetic conservationist Bill Douglas, who promised to help the cause by writing a book titled *Farewell to Texas*, about how the state was on the verge of losing its wilderness assets. Douglas recommended to Yarborough that he market Guadalupe National Park as a sister unit of Big Bend (Texas) and Carlsbad Cavern (New Mexico) for tourism purposes. Soon Yarborough introduced a bill in the Senate and Douglas laced up his hiking boots to climb El Capitan, the dramatic 8,076-foot peak in the proposed Guadalupe national park that rises dramatically out of the Chihuahuan Desert.[33]

During his fourteen years in the Senate, Yarborough bemoaned Texas's unconscionable rank at the bottom in natural resource protection. And he was tireless in trying to turn that embarrassment around. In addition to Padre Island and Guadalupe, he wrote bills that established Fort Davis Historical Site in west Texas and the Alibates Flint Quarries National Monument in the Texas panhandle. Always at war with the irresponsible players in the oil, chemical, gas, and timber industries, Yarborough became a celebrated anti-pollution spokesperson. He was the sponsor of the 1962 amendment of the Bald Eagle Protection Act to permanently protect the golden eagle from demise, the 1966 and 1969 Endangered Species Acts, and the 1970 Water Quality Improvement Act. As his biographer Patrick Cox documented, "Yarborough cosponsored every major water, air, and solid-waste pollution measure passed by Congress from 1961 to 1970."[34] On the recommendation of Bill Douglas, the Kennedy family adopted the liberal Texan as one of their own. "Ralph Yarborough was a loyal friend and a tower of integrity," Ted Kennedy recalled. "He was a shining example to all of us who serve in public office. 'Discouraged' wasn't in his vocabulary. He taught us never to give up or give in and that, with a courageous attitude, victory was always possible next time or next year."[35]

CHAPTER 15

Campaigns to Save the Hudson River and Bodega Bay

Coretta Scott King (*right*) in New York City, marching with the anti-nuclear group, Women Strike for Peace. This demonstration took place across from the United Nations at Forty-Third Street and First Avenue. Left is Mrs. Dagmar Wilson of Washington, DC, the founder of the group.

I

President Kennedy faced the biggest challenge of his administration in October 1962 when a US spy plane returned from a flight over Cuba with photographic evidence of Soviet nuclear missile bases under construction. Ultimately, four were identified, in addition to a service air base. The threat was real; missiles shot from what Kennedy called "that imprisoned island" could reach American cities, including New York, Chicago, and Houston.

After consulting with a wide variety of advisers, Kennedy ordered a high-risk naval blockade of Cuba and invoked the Monroe Doctrine to demand that Nikita Khrushchev remove all Soviet missiles from the Western Hemisphere. War was averted when the Soviets, after tense negotiations, agreed on October 28 to dismantle the launchers in a secret deal. For his part, Kennedy simultaneously agreed to remove the United States' Jupiter missiles from Turkey (in a year's time) and to refrain from another Bay of Pigs–style invasion of Cuba. After a nerve-wracking thirteen days, Kennedy came out of the Cuban Missile Crisis with increased popularity, which helped Democratic candidates heading into the midterm elections.

Whenever Kennedy felt stressed in office, which was understandably often, he daydreamed about sailing the Atlantic. Throughout his career, many of his doodles were of sailboats. He was very much attached to the presidential yachts, the *Sequoia* and a ninety-three-foot motor craft that he renamed *Honey Fitz*, after his maternal grandfather. During Easter and Christmas, the boat went with him to Palm Beach, and in the fall, she was docked at Hammersmith Farm in Newport, Rhode Island. Seafaring allowed the president time to relax. Sometimes *Honey Fitz* was anchored along the Potomac so he could slip away for a twilight cruise to Mount Vernon as a welcome diversion from the White House.[1]

In the first wave of response following the missile crisis, most Americans were too relieved to bother with its environmental implications. With a lot of flag-waving, they celebrated the strength shown by their young president. The activist group Women Strike for Peace (WSP), however, had no compunction about protesting loud and clear in the wake of the near nuclear war.[2] They marched in front of the UN building in New York City by the hundreds and in other places across the country. The massive turnout was impossible to ignore.[3]

Women Strike for Peace was a genuine movement, not an organization. In the world of environmental activism, people of enormous egos, like Brower and Commoner, dominated nearly every corner. The women's group had no hierarchy, no officers, and none of the ego baggage of a personality-driven organization. The erstwhile faction started with a mother in the cobblestoned Georgetown section of Washington: Dagmar Wilson, who was horrified by the specter of nuclear war and the contamination of food and air by nuclear testing. In despair, she called some of her friends, they called their friends, and so on. The heat-lightning effect caused more than a hundred thousand women to participate in the group's pacifist marches and letter-writing campaign. Learning how many of her sisters truly cared about the impact of nuclear weapons on the health of children, Wilson called for a nationwide march on November 1, 1961. Ultimately, fifty thousand marched in forty cities. Both Jackie Kennedy and Nina Khrushcheva, the wife of the Soviet premier, sent letters of support to the group. President Kennedy watched the Washington rally from his White House window and acknowledged the movement's influence in a press conference.[4]

A few months later, fifty Women Strike for Peace activists traveled to Geneva, Switzerland, the site of ongoing disarmament talks with the Soviets, to make their anti–nuclear testing/anti-war views known to the delegates.[5] Among the protesters in Geneva working with Wilson was Coretta Scott King. On the eve of her departure for Switzerland, she told a reporter, "No sober minded person can ignore the possible ominous effects of a continuation of the arms race and atmospheric tests."[6]

In the wake of the Cuban Missile Crisis, Women Strike for Peace was so effective in turning the issue from averting nuclear war to the wholesale elimination of nuclear stockpiles that representatives from their ranks were summoned to testify before the House Committee on Un-American Activities. Some right-wing congressmen thought that Communists must be behind such a pacifist groundswell. Blanche Posner, a mother from Scarsdale, New York, used her time before the committee to speak about the lethal effects of strontium-90, the radioactive isotope created during hydrogen bomb explosions. Like the other mothers, many of whom brought their youngsters to the hearing, Posner was anything but un-American. Eventually, the committee investigation of Women Strike for Peace petered out in disgrace.[7] Ever since scientists had first raised concerns about strontium-90 in 1956 claiming that the isotope had penetrated America's milk supply, Posner had been concerned about children. She got involved with the scientists' Committee for Radiation Information at the Rockefeller Institute, attending a half-dozen of its forums. She desperately wanted anti–nuclear testing voices to be taken seriously by the Kennedy administration.

II

During that same autumn of 1962, in fact on the very September 27 publication date of *Silent Spring*, residents of the lower Hudson River woke up to a jarring announcement in that morning's edition of the *New York Times*, which reported that Consolidated Edison (Con Ed), the area's electric utility and the largest one in the country, was on the verge of building a mammoth hydroelectric plant at Storm King Mountain on the west bank of the Hudson just four and a half miles from West Point.[8] Storm King was treasured as the premier scenic wonder of the Hudson Highlands, but the company hadn't considered that or the cultural and historical sites that its enormous industrial plant would ruin. In off times, when Con Edison had excess power, Storm King would use it to pump river water into an enormous reservoir at the top of the small mountain. In times of peak usage, when metropolitan New York was known to have blackouts, water would be released from storage, dropping through turbines back into the river. The disruption to the water, which would be filtered and altered in temperature, would cause damage to the aquatic creatures within the river, and the desecration of Storm King itself would mar the landscape. The worst problem for the Hudson River ecosystem was starting operations at the same time as the announcement regarding Storm King: the Indian Point Nuclear Power Plant located twenty-eight miles downriver from Storm King. The facility drew large amounts of water directly from the river, which destroyed vast numbers of fish and wildlife, but just as damaging, it discharged extremely hot water right back into the river.

Indian Point was living proof that Con Edison was desecrating New York's historic and beautiful river.[9] The group Scenic Hudson stepped forward to challenge the private utility giant's plan to develop Storm King.

The battle over Storm King became the fiercest environmentalist confrontation of the 1960s. Carl Carmer, who had been a friend of

FDR and in 1939 had contributed the volume on the Hudson River in the *Rivers of America* series (for which Marjory Stoneman Douglas had written *River of Grass*), summoned the top environmental activists working in the region to his octagonal home in Irvington-on-Hudson, New York. The gathering included Leo Rothschild of the New York–New Jersey Trail Conference, Walter Boardman of the Nature Conservancy, and Harry Nees of the Sierra Club. Together with Robert Burnap and Virginia Guthrie of Scenic Hudson, they agreed on a strategy to block Con Edison's Storm King project by convincing the Federal Power Commission to disallow a permit on aesthetic, recreational, and educational grounds.[10] Frances Reese was another founder, an engine of eco-activism who became an expert on the plight of the striped bass in the river. She made the threat to *Morone saxatilis*'s survival a dominant theme in the case against Storm King.[11]

Whereas Con Ed said the plant was progress on the march, Carmer countered that "progress is made when the people preserve their inheritance of scenic, historic, and recreational values." He battled dragons by putting his writing skills to use. The broadside he produced warned, "There is ominous talk along the Hudson today, talk from self-appointed oracles who declare the march of cities inevitable. In only a few years, we are told, the City of New York may stride like a wild irresistible monster in mile-by-mile steps up the Great River of the Mountains. Soon the clamor of machines may be echoing from the Palisades—soon the noiseless deeps below the peaceful highlands may be sucked into the maw of the city."[12]

New York City undoubtedly had expanding electrical power needs. But lovers of the Hudson River Valley weren't concerned about that. Residents in towns such as Garrison on the east bank of the river, across from the Storm King Mountain site, saw the Con Ed plant as a sacrilegious attack on the storied highlands region made famous by the Hudson River School of landscape painters (1825–1870). Grand scenic vistas once described by Washington Irving and painted by Frederic Church would be marred by giant 345-kilovolt transmission lines conveying power from the village of Cornwall to the electrical grid downstate. For protoenvironmentalists, the battle wasn't just about a single utility plant but whether the Hudson as an ecosystem could be saved from industrial abuse. Con Ed owned many other coal-fired plants in New York, including an ugly steam plant along the East River in midtown Manhattan, and was therefore easy fodder for demonization. As *Fortune* magazine put it, Con Ed was "the company you love to hate."[13] On a hot summer night, however, when New York City was a veritable steam bath, residents wanted their air-conditioning to work. That's all many of them craved, and that made the battle to cancel plans for the Storm King plant more complicated.

Scenic Hudson attracted support from the "green" troubadour, Pete Seeger, a banjo player, whose song "Where Have All the Flowers Gone?" would become an anthem of the environmental movement. Seeger lived in Beacon, a city along the east bank of the Hudson. Determined to educate the public about the river, he commissioned the historically accurate Hudson River sloop *Clearwater* and launched it from Kingston. It plied the river, drawing attention to the effort to save the waterway from environmental devastation.[14]

Besides Seeger, the most effective early Hudson River activist was a sportswriter and fisherman named Robert H. Boyle. Born on August 21, 1928, in Brooklyn, Boyle had grown up in Manhattan. His heart, however, was most at home along the banks of the Hudson River fishing for perch and bass. After serving in the marine corps in the Atlantic Fleet during the Korean War, Boyle began his career at the United Press wire service. Throughout the fifties, he wrote for *Time*, *Life*, and a new magazine called *Sports Illustrated*. Fishing the Hudson from his home in Croton, New York, was a passion that led to Boyle educating the public about riverine conservation.

Boyle's interest in environmental protection showed early at *Sports Illustrated*; in September 1959, he wrote an article about a nationally known lepidopterist (butterfly expert), Vladimir Nabokov, the Russian-born novelist. His outing with Nabokov, the author of *Lolita*, was a long way removed from the World Series or the Army-Navy football game. In a very tangible way, Boyle succeeded in bringing environmentalism to many thousands of US sports fans. On the hike, Nabokov was looking for a particular butterfly in Oak Creek Canyon, "a sort of watch-pocket Grand Canyon," Boyle wrote, south of Flagstaff, Arizona. The novelist had exclaimed, "The thrill of gaining information about certain structural mysteries in these butterflies is perhaps more pleasurable than any literary achievement."[15] In a sense, that quote set the course of Boyle's own life, as he would use his estimable writing skills more and more to further his activist goals as a protoenvironmentalist and self-taught riverine ecologist.

Only three months later, Boyle wrote a *Sports Illustrated* article without a byline on Tule Lake in northern California, "quite possibly the largest refuge for migratory waterfowl in the entire world." Boyle, who lived in San Francisco at the time, had been disgusted by the plans of local officials to drain the Tule Lake wetlands system in order to irrigate more farmland. The Interior Department had used its power to avert that disaster. But Boyle remained incensed at the cavalier attitude of the Tule Lake water-level official in charge, whom he quoted: "Land worth $500 is too valuable to be dedicated to ducks."[16]

On another occasion, Boyle tagged along with Cus D'Amato, a top-tier boxing manager, to fish in upstate New York. His article about it recounted this story:

> D'Amato's fascination with fishing has led him into the larger world of nature. Not long ago, he told a friend, "I've been hearing about a guy I want to learn about. I've heard a lot of quotes, and after I heard his name with the fourth or fifth one, I said to myself that I gotta know this guy better."
>
> "Who's the guy?" the friend asked.
>
> "Thoreau," Cus said.[17]

Giving him a thumbs-up, Boyle delighted in D'Amato's desire to read *Walden, Cape Cod,* and *The Maine Woods*. It fit into Boyle's calling, which was to use his press pulpit at *Sports Illustrated* to bring street-smart or suburb-dumb Americans closer to the green-gladed countryside. The mid–Hudson River Valley wasn't far from the biggest city of them all, skyscraped New York, where too many people thought of a forested mountain, like Storm King, as nothing more than "the middle of nowhere."[18]

In late 1963, Boyle pitched a lengthy article about his beloved Hudson River to his editors at *Sports Illustrated*. Believing that the Wilderness Society and the Sierra Club were overly consumed with remote western land, Boyle sought to draw attention to the storybook

Hudson and the plight of polluted communities such as Beacon and Ossining. Once he got the assignment, he was off to the races.

Scenic Hudson had already raised a lot of funding for the cause. Many children of the rich—the families that had built great estates along the river—had ponied up: to them, the Hudson Highlands were imbued with cultural and spiritual significance. The nonprofit set up an office in the National Audubon Society building in Manhattan, on Fifth Avenue and Eighty-Eighth Street. Enter Robert Boyle to aid the cause, with binders of data proving that 90 percent of striped bass eggs in the Hudson were laid between Highland Falls and Dennings Point, where Storm King is located. Also working in favor of Scenic Hudson—if "favor" can be attached to atrocity—was the fact that by 1963, people noticed huge piles of dead fish, mostly striped bass, on a town dump near the Indian Point power plant. The best guess was that they'd been attracted by the warm water discharged by the facility and then become trapped. Workers had been directed to gather them before anybody noticed and truck them to a hidden dump. The sleuth Boyle heard about it, though.[19] In a *Sports Illustrated* article titled "A Stink of Dead Stripers," he began an offensive "intent on making the dead fish stench at Indian Point stick to Storm King."[20]

On the strength of Boyle's articles, Scenic Hudson received twenty-two thousand donations from forty-eight states. As the battle wound its way through the courts, Bill Douglas and Bobby Kennedy indicated that they would use their power to stop the project. The bitter dispute went on for seventeen more years before Con Ed canceled the plan and gave its share of Storm King to the state. On behalf of Scenic Hudson, Franny Reese signed the agreement in 1980 with her former nemesis, Con Ed.

The ripple effect of the Storm King fight gave birth to two durable environmental nonprofit mainstays, in addition to Scenic Hudson: Riverkeeper in 1966 and the National Resources Defense Council (NRDC) in 1970. Most impressive of all, the timeline from Scenic Hudson to Riverkeeper to NRDC led directly to the Clean Water Act of 1972. It moved the environmental movement in one other, even more radical way. In prioritizing the rights of animal species over human needs, through Reese's enlightened argument that the striped bass were intrinsically more important than the Con Edison power plant, the Storm King effort also laid a path to the Endangered Species Acts of 1966, 1969, and 1973.

When Boyle wrote "From a Mountaintop to 1,000 Fathoms Deep" for *Sports Illustrated* in 1964, he presented the Hudson as an ecological wonderland of sand sharks, yellow perch, white perch, sea sturgeon, pipefish, black bass, tomcod, butterfish, common jack, bullhead, pickerel, bluefish, menhaden, anchovies, American sole, summer flounder, smelt, sunfish, seahorses, trout—and striped bass—all mingling in "startling confraternity." Who but Boyle knew that the Hudson was such a natural hatchery for a diverse array of underwater life?[21]

Testifying before a Senate subcommittee about his *Sports Illustrated* articles, Boyle made clear that he *spoke* for the fish species of the world. "You do not build a power plant on top of a spawning bed of an irreplaceable fish resource," he fumed.

> I think if you want to put a monetary value on this resource you would very well find it would be worth many times that of the power plant and any revenue that the power plant can bring into a local community. The waters of Long Island Sound; the New York waters in which these Hudson bass feed, are probably the most heavily fished salt-water grounds in the world. . . . I would no more think of licensing a plant for that area than I would of slaughtering sheep in my living room. It's just not the place.[22]

Decades later, after the Storm King threat was over, Boyle worked tirelessly to teach New York City schoolchildren about the aquatic wonders in their own brackish backyard river.[23]

III

The 1962 midterm elections brought a major new conservationist voice into the Senate: Wisconsin Democrat Gaylord Nelson, who would cofound Earth Day in 1970. Every fall, Nelson went to the Wisconsin River to watch as columns of sandhill cranes congregate just behind the shack where Aldo Leopold had written much of *A*

Sand County Almanac. Nelson believed that conservationists needed to be active, making things happen, not waiting for disaster, and he showed it during his eighteen years in office. Politics was an emotional endeavor, he would say, and it was time for Wisconsinites to preserve the state's natural beauty in the tradition of native sons John Muir and Aldo Leopold, both of whom were heroes for him.

Once Kennedy and Udall, backed by the ORRRC report, fully embraced the establishment of national lakeshore parks in the Great Lakes, Nelson sprang into action. In northern Wisconsin, he hoped to establish the Apostle Island National Lakeshore, encompassing a chain of twenty-two Lake Superior islands off the Bayfield Peninsula. It was a vibrant habitat where more than eight hundred plant species thrived. Black bears roamed the islands, and the lakeshore was a nesting place for ring-billed gulls, double-crested cormorants, great blue herons, and cliff swallows.

Born in Clear Lake, Wisconsin, on June 4, 1916, Nelson, the son of a country doctor, spent his childhood exploring woods and waters, canoeing the nearby Saint Croix River, and fishing every day that he could. Encounters with wild animals during hikes, especially large ones such as wolves and moose, were treasured moments of his growing-up years. He grew determined to defend the wilderness in his home state. Nelson earned a law degree at the University of Wisconsin with the intention of entering politics. Drafted into the army during World War II, he was assigned to a segregated company, as one of four white officers in charge of two hundred Black enlisted men who saw action at Okinawa. After the war, Nelson moved to Madison to practice law. He resumed his dream of entering politics, hoping to make a difference in middle-class people's lives. After serving as a state senator for ten years and as governor for four, he was elected a US senator in 1962.

Living in the governor's mansion in Madison in the late 1950s, Nelson couldn't help noticing that trucks swept through the neighborhood, spraying DDT in billows that hung over the houses and everything in between. Seeing with his own eyes that DDT killed all kinds of insects and worms while adversely affecting birdlife, he spoke out against the chemical. The response was brutal; he was told in no uncertain terms to leave science to the scientists. Within a few years, though, he had a scientist on his side. "By the time I was elected to the U.S. Senate in 1963, I had read Rachel Carson's book, *Silent Spring.* Then I introduced the first legislation in the Senate to ban the use of DDT," he recalled.[24]

Nelson, in addition to shepherding the DDT ban on its slow route through Congress, gained a reputation as one of the Senate's most ardent environmentalists by originating legislation to create a national system of hiking trails, to include the 2,100-mile Appalachian Trail, which had never been under federal protection before. Throughout the Long Sixties, Nelson was instrumental in other key pieces of environmental legislation, including the Wilderness Act of 1964, and of course, Apostle Islands National Lakeshore of 1970. Unlike the other conservation senators of the era, he was more interested in modern ecology than old-fashioned national resource management. Under Nelson's Apostles National Lakeshore bill, commercial logging, mining, and quarries would be prohibited on all twenty-two islands in the proposed park.[25]

Another Democrat newly elected to the Senate in that 1962 midterm was George McGovern of South Dakota—for whom President Kennedy had campaigned during his western swing. The fact that Nelson and McGovern had won Senate seats—in Wisconsin and South Dakota, no less—spoke to the degree to which protoenvironmentalism had become mainstream. Concerned about the bloated Pentagon budget, McGovern soon sponsored legislation for the establishment of a National Economic Conservation Commission (NECC), aimed to train defense workers for what today could be called "green jobs." McGovern believed that blue-collar workers should be part of an environmental corps working on water pollution, air pollution, and toxic site cleanups.[26]

IV

In December 1962, the US Public Health Service assembled more than 1,400 activists at a National Conference on Air Pollution in Washington, DC. Among them were public health experts, engineers, physicians, scientists, legislators, industry big shots, and leaders of labor and civic organizations. The main topic was the role of the federal government in regulating air quality.[27] At a seminal panel, the burning of fossil fuels, together with the long-term effects of long-lasting chemicals and other pesticides on plants and trees, was deemed an urgent public health dilemma. Barry Commoner was one of the speakers, questioning whether science was growing beyond the ability of mankind to control when it entered the environment.[28]

The two top administration officials at the conference, Secretary of Health, Education, and Welfare Anthony J. Celebrezze and Surgeon General Luther Terry, emphasized that stationary sources of air pollution (e.g., factories) were under the control of state and local government. By contrast, emissions from moving sources, such as automobiles, could be addressed on a national level. Terry stated, "If we are going to clear the air, then it is particularly important that State and local air pollution control programs be extended in coverage and strengthened in-depth. We can assist through an expanded program at the Federal level."[29]

The December conference built on the Air Pollution Control Act of 1955, which had provided funds for government research on air pollution; it was an early challenge to state primacy. Seven years later, the scientific data was in. The panelists agreed that air pollution was causing chronic health problems. With that, though, the conference broke up and the Kennedy administration expressed yet again its disappointment that not nearly enough action was being taken, especially by Congress. After Kennedy sent a message to Capitol Hill early in 1962, specifically requesting movement on legislation to allow the federal government more leeway in regulating stationary sources of air pollution, nothing happened, except an extension of the Air Pollution Control Act: that is, more studies. The administration had high hopes for the December conference, but again, the conclusion was nebulous, except in recommending more studies. So it was that in the months that followed, there was much debate among Washington lawmakers over a replacement piece of legislation, the far stronger Clean Air Act, which among other things would establish a research and regulations program in the Public Health Service. Oil refineries and factories were targeted as the United States' worst polluters. In Los Angeles and New York City, there were calls for continued investigations

aimed at penalizing the sources of smog, but the irritating and potentially lethal effect of pollution was a national concern, with hazardous
compounds such as acetaldehyde, benzene, chloroform, phenols, and
selenium making Americans sick. The administration set 1963 as the
year to pass the Clean Air Act.

By Christmas of 1962, President Kennedy had established three
high-profile national seashores and a dozen national wildlife refuges.
Using his bully pulpit, he urged Americans to treat their natural resources with love, not act as despoilers. His administration was looking
at ways to regulate DDT and other pesticides. And it kept pushing the
Wilderness bill, which was still stuck in the House, as a necessity for the
well-being and preservation of democracy. Seven-year-old Scott Turner
wrote Kennedy that December bemoaning the vanishing wilderness
in the West: "We have no place to go when we want to go out in the
Canyon because there are [sic] going to build houses[.] So could you set
aside some land where we could play? Thank you four [sic] listening."[30]

Kennedy promised the boy that the Wilderness bill would soon be
passed by Congress and that young people needed to fight to protect
the environment. Because of Udall's promotional skills, national and
state park visits in the United States had surged to an all-time high.
And Udall, with help from many people, including Sharon Francis and
Wallace Stegner, was finishing *The Quiet Crisis*, scheduled to be published by Christmas of 1963. Summarizing 1962 for the *Washington
Post*, Udall wrote, "We are learning that the search of modern, urban
man is not for new ways to conquer nature, but for ways to save the
beauty of the out-of-doors so that, to use Robert Frost's words, man
can gain new insight from 'country things.'"[31]

And the New Frontier continued to fight for the preservation, or
the restoration, of the unswimmable Potomac River. To his credit,
President Kennedy had halted plans for highway construction that
would have marred the river's scenic qualities. Spurred by his brother
Bobby, JFK had also used his executive powers to further preserve the
C&O Canal between Washington and Great Falls, Maryland.[32]

On Bill Douglas's advice, Kennedy embraced Theodore Roosevelt's
notion of a national campaign promoting the virtues of physical fit-
ness. Hiking, jogging, canoeing, and bicycling were seen as ways to
stay fit. The rollout was that wilderness was shrinking and too many

Americans were staying indoors, suffering from what decades later would be called "nature deficit disorder." Inspired by a letter Theodore Roosevelt had written to the marines in 1903, Kennedy challenged members of the military and the White House staff to embark on fifty-mile hikes with a time limit of twelve hours. Plagued by his bad back, the president asked his brother Bobby to be his surrogate. While others in the administration demurred, Bobby unflinchingly took up the challenge one day in February, hiking from Georgetown to the presidential retreat at Camp David in temperatures that hovered around 25 degrees. According to a companion, Bobby "would have completed the fifty miles if he had to crawl" in order to fulfill Jack's national challenge. And that was close to what happened. "At Camp David," Robert Kennedy, Jr., recalled, "I waited anxiously for my father late into the night and saw him come in and collapse on the bed with my mother massaging his blistered and bleeding feet."[33]

V

Another vexing environmental battle was taking shape in 1962 just north of Point Reyes at Bodega Head, a rugged California seascape where cypresses were worked into knots by what the poet Robinson Jeffers called "the sailor wind."[34] The Pacific Gas and Electric Company (PG&E), backed by the Atomic Energy Commission, was making plans to build a nuclear power plant there. At 325,000 kilowatts of output, it would be the most potent nuclear reactor in the world. Local environmentalists were enraged; not only did the AEC seem to have a cavalier attitude toward potential water and air contamination, but the presence of a nuclear reactor would be detrimental to tidal sea life.[35] As proposed by the power company, the reactor would contain seventy-five tons of 2.5 percent enriched uranium, a highly radioactive fuel. County Administrator Neal Smith said in praise of the nuclear plant that the curves of the power lines would have "artistic" merit, just as redwood bark did. Bay Area environmentalists scoffed at such pablum. Public pressure mounted to stop the PG&E project at once.

Many of the same activists who had successfully lobbied for Point Reyes National Seashore spearheaded the anti–nuclear power crusade on the Bodega Bay tideland. David Pesonen, the Berkeley-based activist who had elicited the "Wilderness Letter" from Wallace Stegner in 1960, mobilized the grassroots effort through informative columns in the *Sebastopol Times*.[36] Due to his research, fishermen in Sonoma and Marin Counties worried about stock contaminated by thermal discharge into the Pacific, and the tourism industry complained that the nuclear plant would drive away visitors.

The showdown that ensued would have widespread ramifications for the burgeoning environmental movement. The Northern California Association to Preserve Bodega Head (NCAPBH) came up with all sorts of gambits to stop planning for the nuclear plant, ranging from petitions and letter-writing campaigns to lawsuits. The most innovative was the release of 1,500 helium balloons into the atmosphere, each emblazoned with "This balloon could represent a radioactive molecule of strontium 90 or iodine 131." On behalf of the NCAPBH, Pesonen commissioned a French expert on the subject to assess the safety of placing such a dangerous nuclear power plant so close to the San Andreas Fault. Meanwhile, Pesonen abruptly quit the Sierra Club, which he deemed too conflicted in its priorities to be effective. He preferred the boldly anti-nuclear and often undiplomatic NCAPBH.

The Sierra Club continued to fight the plant, but mainly on aesthetic grounds. "Doran Park beach [a two-mile stretch of beach and public campgrounds separating Bodega Bay and Bodega Harbor] would be effectively ruined as a recreation area by three rows of unsightly power lines, and Bodega Head itself would sport either a 400-foot smoke-belching stack or a huge sunken reactor," Philip Flint warned in the *Sierra Club Bulletin*.[37] The problem wasn't just that the Sierra Club was stuck in its ways, depending mostly on a certain type of national park advocacy. It was also hamstrung by many alternative energy environmentalists optimistic about nuclear power. Fossil fuels were dirty and required extraction from scenic landscapes and offshore fisheries. If nuclear could replace them in an ostensibly clean way, then many people in the Sierra Club and beyond considered that it might just be the energy panacea for the environmental movement.

Hundreds of thousands of Americans joined the protest against the plant at Bodega Head. Somewhat reluctantly, Udall got the Interior Department involved in the feud, which, as both Pesonen and the

Sierra Club could attest, was going to leave all concerned bloodied and bruised. He had the US Geological Survey experts study Bodega Head and its geology to provide a safety analysis (the 1962 equivalent of an environmental impact statement). Their conclusions were in line with those of the NCAPBH's geologist, who had declared that for a nuclear plant, "a worse foundation condition would be difficult to envision."[38] Udall, who was then among those who favored nuclear energy, nevertheless soon turned against PG&E. Building an atomic campus along the San Andreas Fault, he knew, was pure idiocy.

A long article by the influential Harold Gilliam, "Atom Versus Nature at Bodega," in the *San Francisco Chronicle* decried the arrogance of the proposed plant complex.[39] In many ways the "anti–nuclear power" movement of the Long Sixties was born during the Bodega Bay campaign, empowered by the new American ecological consciousness. Alfred Hitchcock happened to be filming *The Birds*, starring Rod Taylor and Tippi Hedren, at the scenic spot where PG&E wanted to build the Bodega Bay atomic compound; his use of the location served to verify its uniquely pristine beauty.

Finally, even the Atomic Energy Commission criticized PG&E for selecting a site in an earthquake zone. Under extreme duress from the public protest and battered in the media, the company canceled the facility in 1964.[40] It was a huge grassroots environmental victory with sweeping ramifications for other development versus conservation showdowns throughout California. When the Los Angeles Department of Water and Power began planning a nuclear power plant in Corral Canyon near Malibu, the idea was squashed by the same type of environmental coalition that had won the battle of Bodega Bay.[41]

The use of song and celebrity to win supporters to an ecological cause was one successful protest tactic developed in the Bodega Bay fight. The West Coast jazz revivalist Lucius "Lu" Watters performed older-style Dixieland at anti-nuclear festivals, and the folk singer Malvina Reynolds wrote "Take It Away," deriding the proposed Bodega atomic campus. Reynolds's other popular song, "Little Boxes," about suburban conformity in Levittown-like tracts, was recorded by her friend Pete Seeger in 1963 as a comment on how commercial developers tried to destroy open spaces.[42] By writing folk songs, Reynolds hoped to raise awareness of California's threatened forests, wildlife, shoreline, and enchanted deserts. "Take It Away" was long a clarion call of the anti-nuke movement:

> We've got to take over P.G.E.,
> It's become a dreadful pest.
> It's spreading atomic poison stuff
> Over all the Golden West.
> They're starting a plant at Bodega,
> A place that was wild and pure,
> They call it an atomic park,
> But it's an atomic sewer.
>
> TAKE IT AWAY, TAKE IT AWAY,
> There's a killer gang at the very top
> Of P.G. and E. today.
> We need that electric power
> To make our country run,
> But what's the use of electric juice
> When the people all are gone.[43]

Another singer-songwriter, Tom Lehrer, became famous in the 1950s for songs that deflated the arrogance, among other things, of the military-industrial complex President Eisenhower had ominously warned about in his farewell address of 1961. One of Lehrer's tunes targeted the rocket engineer Wernher von Braun, who had worked for Adolf Hitler before immigrating to the United States after World War II to design rocketry, originally for the army and later for the space program: "Once the rockets are up, who cares where they come down? 'That's not my department,' says Wernher von Braun."[44]

Lehrer, a fixture in northern California, wrote the 1965 song "Pollution," a favorite for Bay Area environmentalists. At protest rallies in Berkeley and Golden Gate Park, the Lehrer send-up against smog and dirty water was sung by folk musicians demanding a green future:

> Turn on your tap
> And get hot and cold running crud!

See the halibuts and the sturgeons
Being wiped out by detergeons.
Fish gotta swim and birds gotta fly.
But they don't last long if they try.

Pollution, Pollution,
You can use the latest toothpaste,
And then rinse your mouth
With industrial waste.[45]

Lyricists and social commentators such as Reynolds, Lehrer, and Seeger understood that buried deep in the American subconscious was a fear of radiation sickness and toxic chemicals. Marjorie Spock, one of the original anti-DDT activists, was starting to blame chronic respiratory illness on leaded gasoline. The consumer activist Ralph Nader was targeting corporations for using rivers as sewage canals. Martin Luther King, Jr., was rallying against toxic waste dumps and garbage incinerators in minority communities. Grievances and concerns about the lack of federal air and water pollution regulations turned environmentalism into a movement that politicians couldn't ignore. In Great Britain, activists were calling themselves "Anti-Uglies." In Germany, the first political group to use the name "Green Party" emerged, spawning similar entities in countries around the world and leading protoenvironmentalists in California to be called the "Greens." Battles were being waged and won. More and more, corporate polluters were perceived as the enemies of the Earth by millions of Americans, fed up with having their health damaged and their landscapes destroyed by growth-for-growth's-sake profiteers.

CHAPTER 16

The Tag Team of John F. Kennedy, Stewart Udall, and Rachel Carson

Albert Schweitzer (*left*) and Linus Pauling spending leisure time together in 1959 in French Equatorial Africa (today called Gabon). Both Nobel laureates were vehemently opposed to nuclear testing and sought seashore protection around the world.

I

For Rachel Carson, Barry Commoner, and others in the anti–nuclear testing movement, the year 1963 was a nightmare of predictions come true. Tons of radioactive material from the Nevada Test Site (the new name of the Nevada Proving Grounds) had been hurled into the stratosphere and were now drifting back to Earth in an invisible but steady drizzle. The Atomic Energy Commission had begun analyzing fallout samples for both strontium-89 and -90 in the late 1950s, and the disconcerting results were in. While medical professionals at Washington

University, in St. Louis, had been studying baby teeth, AEC chemists had been collecting samples of human tissue, and the verdict was bleak: people were getting cancer, leukemia, and other illnesses from nuclear debris. Bob Dylan's protest song "A Hard Rain's A-Gonna Fall" was a profound warning of planetary ruin at the end of a nuclear folly.

Not only was strontium-90 contaminating the environment; other, more esoteric bomb by-products, such as zirconium-95, ruthenium-103 and -106, cesium-137, barium-140, cerium-141 and -144, and the ubiquitous plutonium-239, had likewise been blasted into the stratosphere and were raining down in carcinogenic particles. Some US cities were being hit with more of the poisons than others. Even though Seattle was relatively near the Soviet test site at Novaya Zemlya, it received only one-tenth the amount of strontium-90 that fell on Westwood, New Jersey. The worst concentrations of radioactivity in the United States were detected in a band in the Midwest that was contaminating Missouri wheat and Iowa corn. According to the historian Richard L. Miller in *Under the Cloud: The Decades of Nuclear Testing*, "Plutonium was falling to earth in measurable amounts, and plutonium was one of the most carcinogenic substances known."[1] For the sake of planetary existence and sanity, the United States, the USSR, and Great Britain needed to negotiate a nuclear test ban treaty. Kennedy believed it. Khrushchev seemed to as well, but that didn't mean an agreement during the height of the Cold War would be easy.

In early 1963, the Animal Welfare Institute awarded the Schweitzer Medal to Rachel Carson. That recognition was satisfying to her. She had dedicated *Silent Spring* to Schweitzer, quoting his lines "Man has lost the capacity to foresee and to forestall. He will end by destroying the earth." She considered the Nobel laureate the "one truly great individual our modern times have produced."[2] Her four books exemplified his reverence-for-life philosophy, and she enjoyed nothing more than encounters with wild creatures. She was devoted to her sleek black cat, Moppet. "Whatever it may be, [observing animals] is something that takes us out of ourselves, that makes us aware of other life," she said in accepting the award.

From my own store of memories, I think of the sight of a small crab alone on a dark beach at night, a small and fragile being waiting at the edge of the roaring surf, yet so perfectly at home in its world. To me it seemed a symbol of life, and of the way life has adjusted to the forces of its physical environment. Or I think of a morning when I stood in a North Carolina marsh at sunrise, watching flock after flock of Canada geese rise from resting places at the edge of a lake and pass low overhead. In that orange light, their plumage was like brown velvet. Or I have found that deep awareness of life and its meaning in the eyes of a beloved cat.[3]

Of all the supportive letters Carson received, she cherished most the one from Schweitzer, dated March 16, 1963, and written in French. She displayed it on her Silver Spring writing desk in a picture frame. "Heartfelt thanks for your kindness towards me and my ideas," he said. "I am very touched by it. Yes, it is the most important issue for religion and philosophy; to make man truly human. I discovered this for myself, when I studied theology and philosophy in Strasbourg, in my youth. And I was pleased to discover that ethics that only take into account the relations of man with other men were fragmentary, and only those that take into account all creatures are complete, deep, alive, and full of energy . . . capable of producing a civilization truly human and moral."[4]

Compassion for animals was at the core of Carson's four remarkable books. Even as she described complex marine ecosystems in her sea trilogy, it was in tribute for the marine creatures that she spoke. "She had always been humane," her friend Ann Cottrell Free recalled. "She would pick up the stray cat, rescue the injured dog, return sea specimens to the sea after examination under the microscope. She did not look at the efforts of the humanitarian groups as unrelated to those of naturalists."[5] In the spirit of Schweitzer, Carson became a member of the nonprofit group Defenders of Wildlife, spoke on behalf of the New England Wild Flower Preservation Society, and wrote the introduction to Ruth Harrison's *Animal Machines: The New Factory Farming Industry*, an eye-opening book published in 1964 that lambasted the unethical treatment of poultry and livestock.[6] Harrison, a British animal welfare activist, exposed such grotesque indignities of postwar agriculture as battery cages for hens, solo crates for veal calves, and tether stalls for sows. Like Schweitzer and Harrison,

Carson firmly believed that an animal should have "pleasure in life while it lives," before it is slaughtered.[7]

Following Carson's acceptance of the Schweitzer Medal, she received a marvelous letter of philosophical musings from Thomas Merton, a renowned Trappist monk living in Kentucky. He was author of many books, including the best-selling *The Seven Storey Mountain*.[8] His biographer Monica Weis wrote that Merton's reading of *Silent Spring* "was an epiphanic event akin to other well-known and powerful moments of spiritual insight in his life."[9] Carson's book made him see a dark and intractable pattern in modern civilization set by rapid-fire technology, atomic fallout, species slaughter, thoughtless industrial growth, and new toxic materials. Merton told Carson that "the sickness" of humankind was "a very real and very dreadful hatred of life as such, of course, subconscious, buried under our pitiful and superficial optimism about ourselves and our affluent society."[10]

Carson admired Merton for his thought-provoking reverence for nature and his appreciation of the shared tenacity for elemental survival that all creatures possess. She considered as a kindred spirit any writer concerned with bayberry and juniper, waves and sea fog, the small paw prints of fox and the branch-stuffed nests of eagles. Another sage in that nature-loving category was Robert Frost, who died on January 29, 1963. "Rachel told me that Frost's death took a toll on her," Udall later recalled. "She had long admired his book *A Witness Tree*, which had been written at his 150-acre farm in Ripton, Vermont, in the early forties."[11] Carson and Frost had shared the belief that all of nature's moods—its temperatures, mists, sounds, silences, rain-drenched surface meanings, and mysterious undermeanings—were full of wonder and excitement.

Five months earlier, Kennedy had sent Udall and Frost to visit the Soviet Union as part of a Cold War cultural exchange. Though Soviet premier Khrushchev had upbraided Udall, saber-rattling about the Kremlin's cutting-edge nuclear missile technology, he was deferential to Frost.[12] "He was the great American poet of our time," Kennedy said of Frost. "His art and his life summed up the essential qualities of the New England he loved so much: the fresh delight in nature, the plainness of speech, the canny wisdom, and the deep, underlying insight into the human soul. His death impoverishes us all; but he has bequeathed his Nation a body of imperishable verse from which Americans will forever gain joy and understanding. He had promises to keep, and miles to go, and now he sleeps."[13]

Even though Frost rejected the label "conservation poet"—or any label besides "individualist," for that matter—Udall thought that the moniker was accurate. What Thoreau was to Walden Pond, Frost was to the wood lots, hardwood forests, and apple orchards of New England. "Frost was never so magical in appearance and thought as when he was in the woods," a Vermont friend and fellow poet, Wade Van Dore, wrote in the *Living Wilderness*. "His spirit then was almost chameleon. He could change color and shade into the tone of his surroundings as easily as that delightful small animal. Frost literally thawed in the woods, even though it might be mid-winter."[14] The last poem in Frost's last book, *In the Clearing*, published in 1962, was titled "In Winter in the Woods Alone," the final stanza of which is this:

> I see for Nature no defeat
> In one tree's overthrow
> Or for myself in my retreat
> For yet another blow.[15]

II

When the 88th Congress convened in January 1963, Henry M. "Scoop" Jackson became chairman of the Senate Committee on Interior and Insular Affairs (which would be renamed the Committee on Energy and Natural Resources in 1977). He was determined that the Wilderness bill being floor-managed by Frank Church would pass, and he was going to push for the designation of new national seashores. Three months later, the Senate did pass a slightly altered version of the Wilderness bill. Thinking that 1963 was the year to win passage in the House, the Kennedy administration applied extreme pressure on Wayne Aspinall to bend. Representative John Dingell, a Michigan Democrat and ardent conservationist, had lunches with Aspinall to try to soften him up.[16] Kennedy had made similar overtures to Aspinall since the Truman years, but convincing him to deem millions of acres of public land forever off limits to the road-building extraction industries would be difficult.[17]

Right off the bat, Jackson commissioned a two-year study on how best to establish North Cascades National Park in Washington State.[18] With Polly Dyer and Patrick Goldsworthy cheering him onward, the senator moved swiftly. Although Jackson appreciated the Forest Service's multiple-use policy, which stated that logging and recreation could be intermingled, Jackson wanted areas in parts of the North Cascades to remain roadless wilderness as a national park unit. From his Seattle office rooftop, he could see Mount Olympus to the west and Mount Rainier to the east, both national parks. Permanent protection of such North Cascades landmarks as Mount Baker and Glacier Peak was missing.

Jackson's Senate office disseminated an article by Harold Bradley in the *Sierra Club Bulletin* that summed up Scoop's thinking: American voters had "matured enough to seek commodities elsewhere and let the North Cascades masterpiece survive unspoiled for those who we should assume will surely be mature enough to cherish it."[19] With the wilderness lobby cheering him on, Jackson hoped that places such as Cougar Lakes and Alpine Lake in the North Cascades could be made off limits to development of any kind. In the fall of 1963, he encouraged a group called the North Cascades Study Team to hold public hearings in Wenatchee, Mount Vernon, and Seattle. The verdict Jackson wanted ultimately came in: both small-town and big-city residents wanted the new national park.

Another Democratic senator with an environmental agenda was Edmund Muskie of Maine. As of early 1963, the most adamant Clean Air Act supporter in Washington, DC, was the forty-nine-year-old Muskie, the chair of the Senate's first Subcommittee on Air and Water Pollution. Born in the textile mill town of Rumford, Maine, the son of a Polish father and a Polish American mother, on March 28, 1914, Muskie had a retentive mind for facts and worried from an early age about the health of mill workers. He was the valedictorian of his high school class and went on to college and law school in Maine, being admitted to the bar in 1939. After wartime service in the Navy, he returned to Waterville and became active in the Democratic Party, rising in state government to serve two terms as governor. In 1959, he became the first popularly elected Democrat to represent Maine in the Senate.

Muskie was respected in his largely Republican state in part because he was determined to draw business back home, however he could. Conservation wasn't at all high on his agenda when he first took his seat in the Senate, though the tourist industry interested him. At the start, he was relegated to an assignment considered disastrous for an ambitious Democrat: the Public Works Committee. In April 1963, he was named chairman of its new Subcommittee on Air and Water Pollution. "Air?" Muskie responded at the time. "What the hell do I care about air, coming from Maine?"[20] His "exile" to those assignments would turn out to be a godsend, giving him leverage over an issue—fighting pollution—that he soon believed to be as critical as war.[21]

As Muskie prepared his subcommittee to examine two major anti-pollution bills, he learned that the opposition to them didn't fall along party lines. On the contrary, he had a welcome ally in the ranking member on the Republican side, J. Caleb Boggs of Delaware. With bold support by them both, the Clean Air Act and the Clean Waters Restoration Act passed through committee and were awaiting approval by the full Senate. If the Clean Air Act became law, it would give the federal government interstate power to take action not only on mobile sources of air pollution but also on stationary sites—notably factories and power plants—that were up to then regulated by the states. Reading the anti–air pollution tea leaves was General Motors, which in December 1962 dumped the Ethyl Corporation from its portfolio because it was the primary manufacturer and fierce defender of lead gasoline (or more exactly, tetraethyl lead and similar additives).

The Clean Air Act became law in December 1963 without much delay, but the House took no action on the water pollution bill. Muskie thought that the way to get through to members of the House of either party who were holding up the bill was through their constituents. To prod them, he developed the first true modern environmental constituency by holding anti-pollution hearings in six cities in 1964. When Muskie's Subcommittee on Air and Water Pollution held hearings in Tampa, Florida, a crusading group called Women from Central Florida, led by Ann Belcher, brought such exhibits from Jacksonville as a dying plant and a white cotton sheet defiled beige after a day spent hanging outside on a clothesline. Soon the grassroots group was asking Muskie about emissions from Florida phosphate plants in Hillsboro and Polk Counties that were making plant workers sick.[22] Some of the hearings spawned rallies that inspired him to call the effort "our war" against "this menace to our health and welfare," adding that the war was "in its infancy."[23]

That was Muskie's way of admitting that, for the time being, the victories following from the Clean Air Act's passage were practically invisible: installing new framework and gathering more research. Factories would be little changed by the law, and the Big Four automakers (General Motors, Ford, Chrysler, and American Motors) were making no plans to abandon leaded gasoline, as Muskie had hoped they would. But the very fact that air pollution had become part of Americans' "mental furniture," as the environmentalist Bill McKibben wrote, along with urban sprawl and atomic fallout, was a hopeful sign.[24] Though Muskie didn't take on fossil fuel consumption in any meaningful way, he proved that a senator could gain national media attention by talking about clean air.

Once Muskie was anointed the anti-smog voice in the Senate, mail poured into his office from citizens concerned about significant levels of automobile emissions and industrial pollutants in their areas.

III

On April 3, 1963, *CBS Reports* aired *The Silent Spring of Rachel Carson*, a prime-time documentary hosted by Eric Sevareid and the investigative journalist Jay McMullen. To balance the program, CBS had invited Robert White-Stevens, a biochemist at American Cyanamid Company, to offer a rebuttal on air pollution. President Kennedy's science adviser, Jerome Wiesner, recommended that he tune in.[25] The show was sure to be controversial, but the drama started even before showtime. Just before airing, three of the show's five advertisers— Lehn & Fink, the manufacturers of Lysol, and two food producers, Standard Brands and Ralston Purina—pulled their commercials. Rachel Carson, they said, was an inappropriate salesperson for their product lines. CBS, known then as the platinum network, was undaunted.

As expected, Carson and White-Stevens had completely different views of what constituted the natural world. "Man's attitude toward nature is today critically important simply because we now have a fateful power to alter and to destroy nature," Carson explained. "But man is a part of nature, and his war against nature is inevitably a war against himself."[26]

Although Carson didn't look healthy, her even-keeled, fact-based delivery enabled her to shine. And the host, Eric Sevareid, made sure that the segment presented Carson as a gallant scientific reformer. No top-tier TV news personality was more dedicated to ecological preservation than the silver-haired Sevareid. His 1935 book *Canoeing with the Cree*, about his 2,250-mile trek north from Minneapolis into Canada, was beloved by nature lovers.[27] In Washington, he was a hiking companion of Bill Douglas and worked behind the scenes to get the Wilderness bill passed. Sevareid conducted the interview at Carson's Berwick Road house in Silver Spring, Maryland.

The *CBS Reports* episode, watched by more than 10 million viewers, was momentous for Carson, enabling her to reach working-class Americans who hadn't read either the *New Yorker* excerpts or *Silent Spring* itself. The program began with footage of pesticides being applied to trees and crops. On the show, Carson read the passage from *Silent Spring* that called pesticides "biocides" that enter the food chain. The United States, she made clear, had established a toxic and ultimately unlivable environment. "We have to remember that children born today are exposed to these chemicals from birth," she said. "Perhaps even before birth. Now what is going to happen to them in adult life as a result of that exposure? We simply don't know. Because we've never before had this kind of experience."[28]

Then White-Stevens, who wore a white lab coat to emphasize that he was speaking as a man of science, replied:

The major claims in Miss Rachel Carson's book, *Silent Spring*, are gross distortions of the actual facts, completely unsupported by scientific, experimental evidence and general practical experience in the field. Her suggestion that pesticides are in fact biocides, destroying all life, is obviously absurd in light of the fact that without selective biological activity these compounds would be completely useless. The real threat, then, to the survival of man is not chemical but biological—in the shape of hordes of insects that can denude our forests, sweep over our croplands, ravage our food supply, and leave in their wake a train of destitution and hunger, conveying to an undernourished population the major diseases and scourges of mankind.

If man were to faithfully follow the teachings of Miss Carson, we would return to the Dark Ages, and the insects and diseases and vermin would once again inherit the earth.[29]

White-Stevens's attempt to discredit Carson didn't work. By arguing that pesticides are benign, he came off as a myopic hack. Yet he was not a hack; he was for many years a professor at Cornell University and later at Rutgers University, where he headed its Bureau of Conservation and Environmental Science, specializing in pesticides and their effects on the environment.[30] Along the way, he wrote the standard textbook on agricultural pesticides. He was respected as a leading "environmentalist." Yet for the rest of his life he maintained that DDT was not a poison, even starting each semester's classes at Rutgers by putting a teaspoon of the chemical into his coffee and drinking it to make his point.[31] He argued that DDT was far superior to its alternative, mostly because there *was* no alternative. That being the case, he pointed to the fact that in tropical countries where DDT had been banned, cases of mosquito-borne malaria had surged and people had died, prompting the bans to be eased or lifted. It's easy to disparage White-Stevens in retrospect, but he represented a large sector of public opinion and the elite scientific community in the 1960s.[32]

Carson was facing down not only foolish fringe critics but also formidable career scientists who embraced DDT as a solution, not a problem. Also appearing on the documentary, weighing in on *Silent Spring*, were Orville Freeman, the secretary of agriculture; Luther Terry, the surgeon general; and George Larrick, the commissioner of the Food and Drug Administration. Freeman, determined to be honest when asked if the public was properly warned about the dangers of insecticides, snapped, "The answer I can say very quickly is 'no.'"[33]

Silent Spring and the *CBS Reports* show both had to teach the reluctant part of the population to think in terms of whole ecosystems, not merely the insect-human axis, and to look at ecology in the long term, as opposed to one season at a time.

The program helped Carson's cause by pointing out with disdain that the report on pesticides by the President's Science Advisory Committee had been delayed again—and yet again. The reason was the same as had occurred on the show: disagreement among the experts. Carson had the last word of the evening, eloquently explaining what the ecological term *balance of nature* meant and warning the audience about the "deadly products of atomic explosions" that the United States and the Soviet Union were releasing into the air, thereby making the future of human life precarious. She concluded, "I truly believe that we in this generation must come to terms with nature, and I think we're challenged as mankind has never been challenged before to prove our maturity and our mastery not of nature, but of ourselves."[34]

Television critics across the country gave CBS Reports advance ratings of five stars, some recommending it as the evening's "Best Bet."[35] With her steady, authoritative manner, Carson took on the chemical industry, and Americans heard her terrifying statement "We have put poisonous and biologically potent chemicals into the hands of persons largely or wholly ignorant of their potentialities for harm. . . . These chemicals are used with little or no advance investigation of their effect on soil, water, wildlife and man himself."[36]

The day after the documentary aired, Democratic senator Abraham Ribicoff of Connecticut demanded a congressional review of synthetic pesticides and other potentially harmful industrial and commercial substances. Ribicoff himself would soon chair a Senate subcommittee on the matter. "If it weren't for Rachel Carson, I never would have had these hearings. I was not aware of the extent [or] the importance of the problems she raised," he said later.[37]

The rest of the government's response to the documentary was mixed. The most that George Larrick, the FDA commissioner, could say was that *Silent Spring* had forced the agency "to take a new look at our responsibilities to the general public." Surgeon General Luther Terry described the harm to human health of long-term exposure to low levels of pesticides as an unanswered question. Freeman, not wanting to take a position, offered embarrassing doublespeak about the hazards of pesticides. Only Udall raved about Carson's performance. The department took the side of *Silent Spring* by committing to build a new laboratory at the Fish and Wildlife Service's Patuxent Wildlife Research Center in Prince George's County, Maryland, to study the effect of pesticides on wildlife.

Carson was invited to speak at the opening of the laboratory, but she declined owing to her deteriorating health, even though she lived a short half-hour's drive away. Udall filled in ably. Perhaps that was just as well. Everyone knew how Carson felt about the indiscriminate use of pesticides, especially DDT. The dedication of the Patuxent facility enabled Udall to state in no uncertain terms how he viewed them, and that became national news. After delivering extravagant praise of Carson and *Silent Spring*, he admitted to a case of resentment over the excitement surrounding NASA's moonshot venture. "It's not bad to be preoccupied with our own planet," he said, especially since, in terms of the effect of pesticides and pollution on nature, "it is already late evening on the conservation front."[38] He acknowledged that DDT had been beneficial in certain respects but said that its price was too high, naming the species that had been killed in large numbers and adding "Man, too, has been afflicted—in some cases fatally. These are the facts."[39]

The knowledge that Udall had rallied the Kennedy administration behind her work meant the world to Carson. "I am well aware that it took courage and forthrightness to make some of the statements contained in that speech," she wrote him. "During the years I worked on *Silent Spring*, there were times when I wondered whether the effort was worthwhile—whether the warnings would be heeded enough to change the situation in any way. Of course, I have been amazed and delighted by the many developments, but I can truthfully say that nothing has pleased me more than the tribute you paid."[40]

On May 6, 1963, Carson received an award from the Garden Club of America. Back in 1952, those horticulturally oriented women preservationists had awarded her the Frances K. Hutchinson Medal for her service in conservation. Eleven years later, in an endorsement of *Silent Spring*, they created a special citation. Carson traveled to

Philadelphia for the presentation ceremony. She accepted the recognition not as an individual but for the cause. "The need to conserve and protect the environment and all those essential interrelationships concerns all humanity," she said. "I am happy to feel that we are allies in conservation—in a fight to protect those resources on which our way of life is built."[41]

In 1963, the snarky way to mock ecologically minded people was to refer to the "cult of the balance of nature," a derogatory label replaced by "tree huggers" in the 1970s. Carson didn't invite such derision, never raising her voice or stooping to rabble-rousing. The environmentalist Bill McKibben described her later as the biologist who knocked "the shine off modernity." She was a classy gentlewoman who grew up in a more civil era, but who had a profound message for the rock 'n' roll–driven, groovy, swinging Sixties (and beyond).[42] Smoothing over the accusation that she was spreading radical ideas, Carson calmly defended her literary career. "We have heard inferences that the conservationists are only impractical dreamers or foolish sentimentalists," she told the Garden Club of America. "If I am dismayed by these statements, it is only because of regret that the speakers have somehow failed to understand the real issues of our world. I am sure history will record that the conservationists, far from being sentimentalists, were the tough-minded realists, facing the issues that *must* be faced today."[43]

On May 15, Kennedy had approved the long-awaited President's Science Advisory Committee (PSAC) report, titled *Use of Pesticides: A Report of the President's Science Advisory Committee*. PSAC (pronounced "pea-sack") had been established in the 1950s largely to engage physicists on issues related to space exploration and nuclear weaponry. It had been smart for Kennedy to turn to the blue-ribbon group to study DDT and other pesticides.[44] The original draft was even more hard-hitting than the final version. It had been softened at the request of Orville Freeman, who feared that if *Use of Pesticides* caused a panic, consumers would stop buying fish of all sorts. Even in its negotiated form, the report was clear on its overriding point, asserting that "elimination of the use of toxic pesticides should be the goal."[45] It was as if WARNING had been stamped on many of its pages. "Until the publication of 'Silent Spring' by Rachel Carson, people were generally unaware of the toxicity of pesticides," the report stated. "The Government should present this information to the public in a way that will make it aware of the dangers while recognizing the value of pesticides."[46]

By any objective analysis, PSAC confirmed that Carson had gotten it right: "Precisely because pesticide chemicals are designed to kill or metabolically upset some living target organism, they are potentially dangerous to other living organisms," the panel logically declared. For that reason, the report reached the seminal conclusion that "the hazards resulting from their use dictate rapid strengthening of interim measures until such time as we have realized a comprehensive program for controlling environmental pollution."[47] Zuoyue Wang, the author of a book about the committee, *In Sputnik's Shadow: The President's Science Advisory Committee and the Cold War*, called the report "a high point in the history of PSAC."[48]

The report unleashed a new wave of fury against Carson. Barry Commoner and Marjorie Spock had written about DDT's dangers, too, but they hadn't captured the public's imagination the way *Silent Spring* had after the *CBS Reports* documentary. Living in her modest house in Maryland, and living off of her book royalties and federal benefits from her time at Fish and Wildlife, Carson had completed her research without outside backing or institutional support. But she *did* have credibility as a respected nature writer and first-rate marine biologist. She represented scientific humility in the face of the unending mysteries of the natural world. After all, 80 million kilograms of DDT were sprayed across the United States in 1963 alone, without any firm idea of the long-term effects on untargeted species. Public opinion in her favor was bolstered further by the PSAC report.

The complaint that *Silent Spring* was an impassioned, biased polemic written solely to punish the synthetic chemical industry lingered. Put-downs by critics continued. Why didn't Carson write about how chemicals and technology were increasing human longevity? If she cared about world hunger, how could she not address the fact that pesticides were the linchpin of the modern agricultural revolution? Wasn't DDT essential in combating malaria in Africa? Big Chemical ignored the fact that Carson had claimed only that DDT was harmful to the environment; she hadn't called for it to be banned without further investigation. She had admitted that the pesticide had been "useful" in combatting malaria but noted that its effectiveness over time appeared to wane substantially as the insects developed resistance to it.[49] In that regard, she was rather conservative. Other scientists, writing in the *Lancet*, the British medical journal, went farther than Carson, stating that DDT, used for mosquito control, had a serious impact on reproductive health in wildlife.[50]

On June 4, 1963, Carson appeared before Abraham Ribicoff's Senate subcommittee to testify on her study of the toxicity of DDT and other synthetic pesticides. With the air of a patient teacher, she explained that such poisonous chemicals didn't simply vanish from the Earth's soil and water. She also volunteered ways to help minimize the hazards of pesticides. Two days later, she testified again, this time before the Senate Committee on Commerce. Going farther than before, she proposed the establishment of a Pesticide Commission, to be made up of independent experts from the fields of biology, genetics, and conservation who would help resolve disputes surrounding the use of the chemicals and make decisions in the public interest. Along with the recently issued PSAC report, the congressional hearings set the wheels of government action into motion. The coming years would see federal and state governments assuming a role in regulating the manufacturing and use of pesticides and other toxic substances, in parallel with the growth of public interest in protecting the environment. Rachel Carson had started a revolution in ecological thinking that gained foot soldiers daily, especially among college students, parents, scientists, and such powerful Washington lawmakers as Ribicoff, Muskie, Church, and Jackson.

Around the time of the PSAC report, Robert and Ethel Kennedy asked Carson to hold a seminar on ecology at Hickory Hill. Stewart Udall attended. Carson spoke about ocean conservation, DDT, and the need for more national wildlife refuges. Sitting at Carson's feet was nine-year-old Robert F. Kennedy, Jr., who would grow up to be an environmental lawyer and help found the Waterkeeper Alliance, a conservation nonprofit drawing together preservation groups from around the world working for the protection of specific bodies of water.[51] He said later that listening to Carson was "one of the most treasured moments" of his youth.[52]

Rachel Carson speaking before a Senate subcommittee studying pesticide spraying, on June 4, 1963.

During the course of that evening, the young Kennedy was shocked to learn from Carson that entire animal species such as the Carolina parrot and the passenger pigeon had become extinct. Full of youthful indignation, he began writing a book about environmental degradation and wildlife depletion with his father's full support. "Using embossed stationery that I'd received for Christmas, I wrote Uncle Jack requesting an appointment to air my concerns," he recalled. "He invited me for a private audience in the Oval Office. As a gift, I brought him a seven-inch spotted salamander I'd captured the previous afternoon and given the biblical name Shadrach."[53]

Bobby had poured tap water into a portable fishbowl and used that to transport the salamander to the White House. He didn't know that Hickory Hill had recently changed its water supply from well water to chlorinated town water. In the Oval Office, when he gave his uncle the salamander, it lay still in the bottom of the bowl. The president took a pen and poked at it, but it didn't move. "I think he might be dead," he said. Bobby wouldn't hear of it, insisting that Shadrach was just resting.

Kennedy walked outside to deposit Shadrach, who was indeed dead, in the White House fountain. Robert Jr. recalled, "I admitted to him that the salamander's lack of animation was striking." They talked about Carson and the onslaught of environmental degradation. The president made an appointment for little Bobby to discuss endangered species, such as the bald eagle and alligator, and the findings of *Silent Spring* with Udall.[54]

IV

Proud to have Carson part of the New Frontier, President Kennedy and Udall did their part, moving to save pristine shorelines from Maine to Florida as new National Park Service areas over the summer of 1963. Topping the list was Assateague, a thirty-seven-mile barrier island stretching from the eastern shore of Maryland to Virginia. It was a beautiful maritime environment located just three hours from the nation's capital and yet rich in shore life, from migratory birds to wild ponies. Much of its southern section had been protected by Franklin D. Roosevelt in 1943 as the Chincoteague National Wildlife Refuge; by 1960, developers hungered to turn the rest into a combination of Atlantic City boardwalk and tract houses, steps from the ocean.[55] Just as construction was beginning, the fierce Ash Wednesday great winter storm of 1962 obliterated a new road and most of the buildings on the island. Clearly, it was reckless to build along that part of coastal Maryland unless you liked storms and wanted beaches littered with rubble from shattered houses.

Recognizing that, the developers abandoned their more grandiose plans. Udall met with Maryland governor J. M. Tawes and others to jointly investigate how to proceed legally to establish Assateague National Seashore. Rogers C. B. Morton, the Republican congressman from eastern Maryland, said he favored the national seashore designation but he also believed that affected homeowners needed to have their say. (In truth, most were not yet homeowners but lot owners who had bought a half-acre parcel of land and were awaiting financial developments—such as the availability of insurance—before breaking ground on their beach houses.)[56] One of the lot owners was a man named Philip King, who might, one would think, be counted as a Udall ally; he had worked at the National Park Service from 1954 to 1961 as assistant to the director. As the president of the Ocean Beach Club, the group that represented the investors, he was awaiting Udall's visit to Assateague.[57] Homeowners on the Eastern Shore weren't happy with the NPS preservation plan. Having learned from his Prairie National Park debacle in Kansas, Udall took care to meet with locals during his trip to the area. As he intended to explain, his game plan called for the federal government to purchase the underdeveloped island from its many owners, create a system of dunes to halt future erosion, and create a national seashore to protect one of the most windy beach areas between Cape Cod and Cape Hatteras.

With reporters in tow, Udall arrived at Assateague to confront angry Worcester County citizens in a makeshift town hall forum. Before the event, photos had been taken of Udall admiring the wild ponies with two of his children and sinking his bare feet into the sand. Then came the meeting, when locals had their say. King scolded Udall unmercifully and was anything but a friend of the plan presented by his former NPS colleagues. He chastised Udall for orchestrating a federal land grab and impersonating a meteorologist. Others followed suit, with disingenuous diatribes about the Ash Wednesday storm having been one in a thousand years. Udall promised federal money for dune stabilization, which everybody in the area wanted. As King pointed out, though, that process would necessarily make beachside lots far less attractive—and less valuable. Udall held his own on seashore preservation, insisting that it was foolhardy to rebuild on an island so prone to hurricanes and winter storms.

"I don't see anything wrong with letting Americans build on their own land," King fumed to Udall. "If things had been done as you say, the Pilgrim Fathers would have had to go back to Holland. You want to keep the whole place a wilderness."

"If your attitude had prevailed, there would be no National Park System," Udall fired back.[58]

"You're from Arizona and you don't know a darned thing about beaches," King countered. "We own this land and we don't intend for you to take it from us; this isn't Russia or Czechoslovakia."[59]

What Udall had going for him was that the Ash Wednesday storm had lashed beaches from northern Florida to the south shore of Long Island, convincing record numbers of citizens that irresponsible development would be tempting fate. Eventually, the fact that houses on Assateague weren't safe took hold with most locals and even some island property owners. It took more than two years of congressional compromise and land purchasing, but on September 21, 1965, Congress authorized the creation of Assateague Island National Seashore.[60]

With the Assateague Island fight in the bag, Kennedy and Udall promoted a national seashore on Fire Island, a thirty-one-mile stretch of sand dunes—twenty-six miles of it undeveloped—northeast of New York City off Long Island's southern coast. To sell the idea, Udall started calling Fire Island "Central Park by the Sea." Developers— Udall was getting used to the charge—rejected the preservation effort as a federal land grab. The secretary believed wholeheartedly in the time-worn adage that every crisis affords opportunity, though. The same Ash Wednesday storm of 1962 that had slammed Assateague had also devastated Fire Island, leaving many houses on the five miles of developed land damaged or destroyed.

That, however, wasn't much of a game-changer for those who lived on the island. The houses had been built in a simple fashion so that they could be easily reconstructed. The storm, therefore, didn't afford Udall the crisis opportunity he coveted. One was at hand, though. The truly ominous news for Fire Island residents came in the form of a new plan proposed by New York's leading urban planner, Robert Moses, to build a parkway on the island. He'd already done the same on nearby Jones Beach, a barrier island just to the west. Those with houses on Fire Island took a stance on the proposal: they detested it. Ron Lee, the northeast regional director of the National Park Service, told the *New York Times* that the undeveloped part of Fire Island was "narrow and fragile at many places with no room for both a roadway and seashore preserve."[61] A battle ensued that ended with Moses's plans being scrapped. Defeating the parkway made residents fear other attempts and therefore support the designation of most of Fire Island as a national seashore.

The remaining challenge at Fire Island was the usual one for the NPS: how to combine the preservation of natural habitat with the encouragement of human visitors. But there was an added complication: regulating land use so that residents could stay without doing damage to the unique island. In the 1950s, "Our Vanishing Shoreline," which had motivated Senator Kennedy to save Cape Cod, had likewise recommended that Fire Island become a National Park Service unit.[62]

The Fire Island campaign enabled Udall to develop a philosophy for creating national parks in more high-end real estate markets, particularly in the East, where few doable opportunities remained. Even before the long-desired Wilderness Act was passed, he started conceiving how to pull off a second, eastern-oriented bill aimed at saving forest acreage in the White Mountains in New Hampshire, the Alleghenies in Pennsylvania, the North Woods of Maine, and the southern Appalachians in North Carolina.

With the Assateague and Fire Island campaigns progressing, Udall turned to the appointment of an Interior Department board to report on wildlife management in the national parks. It would also assess federal predator control policies, especially in the use of poisons on public lands. The resulting "Leopold Report," officially titled "Wildlife Management in the National Parks," was compiled by a board chaired by A. Starker Leopold, Aldo Leopold's son and a premier wildlife biologist. It concluded that "A reasonable illusion of primitive America could be recreated, using the utmost in skill, judgment, and ecologic sensitivity"[63] and it recommended that national parks be what the historian Dan Flores deemed "wilderness vignettes," presenting habitats as they were when "white eyes first fell" on them.[64]

Of pointed interest to wildlife devotees, even mammals with sharp fangs and claws, the seminal Leopold Report presented a searing denunciation of the federal predator control program. Udall was deeply startled by it. As the report detailed, agents from the Division of Predatory Animal and Rodent Control (PARC) had purposely poisoned the last remaining red wolves in Arkansas. Likewise, they had slaughtered Mexican wolves along the Rio Grande River, in Texas, Arizona, and Mexico, bringing them almost to the point of no return. The agency routinely used the chemical sodium fluoroacetate, known as Compound 1080, against coyotes, which was bad enough, but condors had almost become extinct from eating the carcasses of poisoned coyotes. PARC operatives in South Dakota poisoned a huge prairie dog town that was one of the last populations of black-footed ferrets on the Great Plains. Until the publication of the Leopold Report, environmentalists had focused largely on human-centered public health issues such as DDT poisoning, air and water pollution, toxic dumps, atomic power, abused watersheds, recreation, and renewable energy. After the Leopold Report, with Udall taking the lead, the federal government began to address the horrors committed against North American creatures such as the red wolf and black-footed ferret. Young Americans influenced by Rachel Carson, Jacques Cousteau, Sigurd Olson, and the nature documentaries of Walt Disney responded by calling for the Interior Department to ban PARC agents from killing species that are integral to American ecology. And that are beautiful creatures in their own right.[65]

The Kennedy administration embraced the Leopold Report, and it guided NPS policy for decades to come. Between the report and Scenic Hudson's insistence that even humble riverine species had a right to survive in modern America, an enormous change overtook the federal government. President Kennedy couldn't easily be near fur-bearing animals, due to allergies, but under his aegis, the tide began to turn in favor of the other Americans, those in the animal kingdom.

Another example of the Kennedy administration's responsiveness when presented with thorough scientific studies was JFK's signature on the Outdoor Recreation Act, which promoted the coordination and development of effective outdoor recreation programs as suggested in Laurance Rockefeller's ORRRC study.[66] As a result, the newly formed Bureau of Outdoor Recreation surveyed myriad river systems in the lower forty-eight states and Alaska to determine which of them needed federal environmental protection most urgently to preserve their aesthetic character.[67] With the bureau in place, a collaborative spirit brought together outdoor recreation programs in local, state, and federal agencies. Working with private landowners on conservation easements was promoted at Interior to a greater extent than before.

Udall was on a roll, with Carson as his conscience and new national seashores in the offing. But he wasn't always scoring accolades from the Sierra Club. In early 1963, David Brower made a last-minute entreaty for Udall to deactivate the newly completed Glen Canyon Dam, but to no avail. Instead, Udall looked forward to the time when Phoenix and Tucson would receive affordable power from Glen

Canyon. It was a shock, though, when he next announced plans to construct two dams on the Colorado River within the Grand Canyon region: Marble Canyon and Bridge Canyon. They'd be part of what he called the Pacific Southwest Water Plan, which would also import water from the Columbia River Basin and northern California for use in the desert Southwest.

It's true that Hoover Dam had already been in operation on the Colorado for decades, but it was downriver from Grand Canyon National Park. The very thought of damming the river just outside the park borders revolted environmentalists everywhere. "All hell was about to break loose," the historian Tom Turner wrote. "The Sierra Club and David Brower may have sacrificed Glen Canyon to save Dinosaur, but they weren't about to roll over in this invasion of a national park and a natural wonder known throughout the world. Brower took it as his personal duty to stop this outrage, and his board, this time, was in total agreement. He would organize, recruit, testify at hearings, publish books, and participate in the making of two films. He would stop the plan no matter what."[68]

CHAPTER 17

The Limited Nuclear Test Ban Treaty

Operation Teapot detonates its seventh bomb on a 400-foot tower over Frenchman Flat, Nevada, 1955. The nuclear explosion was one of many atmospheric tests conducted in the western United States. Though few people were concerned at the time, they produced radioactive fallout that spread throughout the country.

I

A reviewer for the *Birmingham News* criticized *Silent Spring* on the basis of its timing, as though the problems of the environment couldn't be urgent because other societal concerns were already terrifying. "It isn't enough to have the threat of atomic warfare, a population explosion, communist aggression, irreligion [*sic*], youth dereliction and other menaces," the reviewer wrote, "we must also face the prediction that chemical military warfare against insects is contaminating our air and sea and ground."[1]

Such despair was dominated by the first fear on the reviewer's list:

the overriding dread that if nuclear technology were unchecked, the world's nuclear powers would someday destroy the planet. The age of American invulnerability had ended after only four years with the development of a Soviet intercontinental missile program in 1949. As of October 1962, during the Cuban Missile Crisis, four nations had the bomb: the United States, the Soviet Union, the United Kingdom, and France. China was within two years of building one, and other countries were known to be trying to do so. The United States and the Soviet Union had already come close to deploying nuclear weapons in combat. After the missile crisis, banning atomic atmospheric and underwater testing became priorities for Kennedy, who hoped for a rapprochement with Soviet premier Nikita Khrushchev, one that might even put an end to the Cold War.[2] Preventing nuclear proliferation had long been a Kennedy objective. The first step, he believed, was banning nuclear testing, for all the urgent reasons that Schweitzer had detailed on Radio Oslo in 1958.

From 1951 to 1958, the Nevada Proving Grounds (renamed the Nevada Test Site in 1955) had been the locale of dozens of atmospheric tests of nuclear devices. Together with some seventy nuclear tests that the US military had conducted in the South Pacific after World War II, they had blanketed the earth with radioactive fallout. The short-lived moratorium on testing between 1958 and August 1961 had ended when the Soviets violated the tacit agreement. America followed suit, with 135 detonations over the next two years. The notion of a formal nuclear test ban treaty had been discussed since 1954, but it hadn't gained momentum with Cold War lawmakers in Washington, London, and Moscow. Nevertheless, Adlai Stevenson's promotion of a test ban, Martin Luther King, Jr.'s, War Resisters League oration, and Norman Cousins's SANE activism had incrementally moved the New Frontier policy needle that way. In the aftermath of the Cuban Missile Crisis of October 1962, Kennedy hoped once and for all to ban testing as a first step toward making the world a safer place.[3]

In late 1962, AEC leaders were infuriated that the Kennedy administration, paying attention to SANE's reports that radioactive fallout from Nevada in the 1950s was sickening residents within a hundred miles of the testing site, wanted the tests to stop. Even though the blasts gravely damaged desert ecosystems and caused cancer to spike among southwest residents, the AEC's media spin in the 1950s and early 1960s, by and large, remained decidedly pro-testing. Top reporters such as David Lawrence of the *Washington Post*, misled but gullible nonetheless, propagated falsehood after falsehood. "The truth is, there isn't the slightest proof of any kind that the 'fallout' as a result of tests in Nevada has even affected any human being anywhere outside the testing ground itself," Lawrence wrote.[4] The people affected most adversely by nuclear testing in the United States were Native Americans on reservations, rural "sagebrush" residents, and military personnel stationed in Nevada four to seven miles from the bombing range. But the fallout also spread long distances and no American was guaranteed to be unaffected.

Under JFK's aegis, US diplomats in Geneva struggled with their Soviet counterparts to produce an agreement. The specter of US officials conducting on-site inspections was an obstacle for the hypersecretive Kremlin. In the name of verification, the American negotiators insisted that seismographic stations be erected in the Soviet Union and that about twenty inspections overseen by the United Nations be conducted annually. To Khrushchev that term was unacceptable.

Even so, Khrushchev wanted a test ban deal with Kennedy. In November 1962, the Soviet ambassador to the United States, Anatoly Dobrynin, invited Norman Cousins, head of SANE, to speak to the premier in Moscow. Overtures had been made on Cousins's behalf by the Vatican, which wanted him to represent Pope John XXIII's desire for peace among the Cold War powers. Khrushchev appreciated SANE's dedication to ending nuclear testing and the year before had sent a letter to Cousins, saying that he hoped the grassroots movement would "bear real fruits."[5] Cousins's chairmanship of SANE impressed Kennedy, too. Under Cousins's initiative, Eleanor Roosevelt spoke at the May 1960 SANE rally in Madison Square Garden and newspapers published open letters against nuclear testing signed by Schweitzer, King, and the British philosopher Bertrand Russell, who had been jailed in London for his anti-nuclear activism.[6]

Cousins needed to inform the White House of his impending trip to Moscow. Cognizant of the Logan Act of 1799, which forbids US citizens to negotiate with foreign leaders without government approval, he sought clearance to converse with the Soviet leader. Kennedy offered more than just approval for Cousins to insert himself into international relations on behalf of the Vatican. In a meeting at the Oval Office, Kennedy said that Cousins could represent him, too, unofficially. "I'm not sure Khrushchev knows this, but I don't think there's any man in American politics more eager than I am to put Cold War animosities behind us and get down to the hard business of building friendly relations," Kennedy told Cousins.[7] That was the message he wanted Khrushchev to hear. Kennedy reasoned that Cousins would have instant credibility as a backdoor channel and unofficial negotiator. "My father was so charming," Cousins's daughter Dr. Candis Cousins Kearns reflected. "He could charm the birds out of the trees. First Albert Schweitzer, and then he got the Pope and Kennedy's full approval to negotiate with Khrushchev."[8]

In December 1962, Cousins flew to Moscow for the meeting with Khrushchev, speaking for almost three hours on behalf of the Vatican. Thinking he'd overstayed his appointment, Cousins stood up to leave—without delivering the message from the US president. He assumed it was just too late. "The chairman was reading my mind," he recalled of Khrushchev. "'Please sit down,' he said. 'How is President Kennedy?'" Cousins told him what Kennedy had said. They discussed nuclear testing for a few minutes. Before the end of the month, Kennedy received a letter from Khrushchev, the first official US-Soviet contact since the Cuban Missile Crisis. The president could barely believe his eyes. To get a nuclear test ban deal, the Soviet leader now begrudgingly agreed to allow two or three—not twenty—on-site inspections. Cousins had worked magic: this was a diplomatic breakthrough of epic proportions. In truth, Khrushchev had become aware that UN protocols provided new ways to inspect nuclear sites without the Soviets having to allow a posse of US atomic intelligence personnel to wander around their military installations. But Kennedy knew the Senate wouldn't ratify only three inspections.

Cousins briefed the Kennedy administration on Khrushchev's mental and physical shape. Whenever the subject of Cuba came up, Cousins reported, the Soviet premier's "eyes were in a deep stare."[9] Just after Christmas, Kennedy wrote Khrushchev back, insisting on eight annual inspections at the very least. "Khrushchev was livid because Kennedy's rejection of his offer caused him to look foolish," Cousins's biographer Allen Pietrobon surmised in the *Journal of Cold War Studies*. "He had gone out on a limb and extended a good deal of effort persuading his colleagues that the USSR could accept three inspections."[10] For the time being, the peace movement's hopes for an international test ban treaty faded.

Hoping to nudge the process in the right direction, Cousins decided to play the ace card that Khrushchev had given him in December. Actually, it may not have been all that much of an ace: in saying goodbye, the premier said something to the effect of, "call any time you'd like to chat." Possibly, he was just being polite. Cousins decided to play the card, though, letting the White House know that he could go back to Moscow anytime and speak with Khrushchev again if the president wanted him to. Kennedy picked up on the possibility, hoping to make up for the botched opening that came and went in December, and Cousins returned to Russia in April 1963 to speak with Khrushchev, who had not been insincere about chatting anytime. To humanize the mission, Cousins cleverly took along his two teenage daughters. The meeting, held at Khrushchev's Black Sea dacha, was far more serious than the first, lasting for eight hours. The premier was a marvelous host and made explicit his determination to achieve "peaceful coexistence" with the United States. But he was worried that the US-Soviet animosity had intensified dangerously since the end of World War II and that the time to act in behalf of world peace was now or never.

Upon returning to the United States, Cousins implored Kennedy to make a "determined fresh start" by delivering a speech friendly toward the Soviet Union. Cousins intuited that doing so in a very public setting could clinch a test ban treaty because the Kremlin would lose the suspicion that it was being toyed with. Taking Cousins's advice to heart, Kennedy had Ted Sorensen start writing an upcoming commencement address on the topic of peace. Feeling that his SANE advocacy was paying off, Cousins wrote a sixteen-page draft, too, and sent it to Ted Sorensen. His notes were a trove of facts and verbal olive branches. Drawing on some ideas of Cousins, Sorensen wrote one of the finest presidential speeches in US history, the American University "Peace Speech," of June 10, 1963.

That afternoon, to the surprise of the faculty and graduating students, Kennedy did more than wish them good luck in the future, he made a clarion call for new diplomatic negotiations with the Soviets, aimed at dramatically reducing nuclear tensions between the superpowers. There were no mentions of Soviet treachery, no references to their bragging about nuclear stockpiles, and no threats of mutual assured destruction.[11] In a manner rare in world history, the leader of a reigning superpower explored the topic of peace, not in a vague or hopelessly idealistic way but in a realistic tone. In the specific goal that Kennedy set, he echoed Hubert Humphrey's Senate resolution 148, which sought "a ban on all tests that contaminate the atmosphere or the oceans."[12] In the speech, Kennedy suggested "a treaty to outlaw nuclear tests." His words seemed novel, however, simply because of who he was. Stressing the idea of common humanity and a one-world ecosystem, Kennedy said, "We all inhabit this small planet" and "We all breathe the same air."[13]

The Kremlin took notice of the president's peace overture. To pave the way for its next move, it made sure the entire speech was published in the newspapers *Pravda* and *Izvestia*, which had a combined circulation of 10 million readers. The Soviets even allowed the BBC and the Voice of America to broadcast the speech throughout Russia without disruption. In keeping with the speech's promise, Undersecretary of State Averell Harriman, an experienced diplomat and former ambassador to the USSR, was chosen to resume negotiations on testing in Moscow. Great Britain also sent a representative. On July 25, 1963, after twelve days of talks, the three nations struck a deal: atmospheric, space, and underwater nuclear testing would be banned. The next day, in a prime-time television address, Kennedy announced the Limited Test Ban Treaty, claiming that the new status quo "is safer by far for the United States than an unlimited nuclear arms race."[14] Aiding JFK in the public square, Dr. King told the members of the Southern Christian Leadership Conference that activists in civil rights and disarmament were kindred spirits. He said that the SCLC wouldn't "accept the premise that our engagement in a struggle for racial justice in America removes us from the arena of concern over the Test Ban Treaty."[15]

After the Harriman mission, Kennedy convinced the Joint Chief of Staff to consider ways to go even further, on the basis that the cessation of tests was the logical starting point for future disarmament talks with Khrushchev.[16] First, though, the president needed sell the American public and their representatives on the test ban. To gather votes on Capitol Hill, Kennedy adopted a strategy that Women Strike for Peace had used: lobbying Congress through media channels aimed at families and women, in particular. The editors of *Cosmopolitan, Family Circle, McCall's, Parents Magazine,* and *Redbook* were persuaded to back Kennedy's historic ban in print. Leaving no lawmaker alone, WSP activists were highly effective for Kennedy on Capitol Hill. "I have said that control of arms is a mission that we undertake particularly for our children and our grandchildren, and that they have no lobby in Washington," JFK said in *Woman's Day.* "No one is better qualified to represent their interests than the mothers and grandmothers of America."[17] For the first time, a US president was working in sync with the anti-nuclear community, first by endorsing Cousins's use of backchannel methods in Russia and then by deploying Women Strike for Peace in Congress.

Ratified by the Senate on September 24 by a resounding vote of 80–19 (14 more than the two-thirds required to pass), the Limited Test Ban Treaty was instantly hailed for its promise of a practical and attainable reduction in nuclear fallout. The *New York Times* celebrated the "Atom Accord" as a major Cold War thaw. When word arrived that the editors of the *Bulletin of the Atomic Scientists* had responded to the treaty by moving the hand of the organization's Doomsday Clock back five minutes to twelve minutes before midnight, Kennedy said, "Well, let's get it to an hour before midnight."[18] Wisely, he had listened to SANE, WSP, and Barry Commoner's public science arguments about radioactive fallout.

Sorensen later stated that Kennedy considered the ban his greatest achievement.[19] Most presidential historians would concur, deeming JFK's multifaceted negotiation of the treaty a high point of the New Frontier. Ecologists universally welcomed the ban as a leap forward in making the world safer for people and animals. "The Nuclear Test Ban Treaty should be regarded," Commoner would observe in his bestselling 1971 book, *The Closing Circle,* "as the first victorious battle in the campaign to save the environment—and its human inhabitants—from the blind assaults of modern technology." He was also aware,

Dr. Linus Pauling, the American chemist and an adamant foe of nuclear testing, was awarded the Nobel Peace Prize in 1962. Here Pauling is holding up a sign as he joins picketers in front of the White House during a mass protest against the resumption of atmospheric nuclear testing. Dr. Pauling, a sixty-two-year-old native of Portland, Oregon, was also the recipient of the 1954 Nobel Prize for Chemistry.

though, that the Limited Test Ban Treaty was indeed limited, adding "It was only a small victory, for US and Soviet nuclear tests continue in underground vaults, and China and France, which are not bound by the Treaty, continue atmospheric testing."[20]

Predictably, in the wake of the Limited Test Ban Treaty, an uptick in underground testing occurred. As George Kistiakowsky, Eisenhower's second science adviser, and Herbert York, the first director of the Lawrence Livermore National Laboratory in Livermore, California, later complained, the ban made "uninhibited weapons development politically respectable."[21] From August 1963 to September 1992, there were 713 underground nuclear tests in Nevada. Ten more were conducted at other sites. Americans seemed to worry less about nuclear weapons proliferation, because there was no longer radioactive fallout compromising the safety of their children's milk . . . or so they were given to understand.

The Department of Defense wanted to know more about underground testing, mostly to learn the extent to which other nations could detect it. In conjunction with the AEC, it embarked on Project Vela, which selected Hattiesburg, Mississippi, as the site for detonations of two bombs one-third the size of the one dropped on Hiroshima. The explosions, in 1964 and 1966, occurred about 2,700 feet deep in a naturally occurring salt dome. Little thought was given to the impact environmentally; it was underground, after all. Even decades later, however, radiation was being detected at the surface in Hattiesburg. Local residents pointed to anomalies in cancer rates in the town. Once again, tragic proof was at hand of the habitual mistake of nuclear scientists: believing they knew more than they actually did about the power they unleashed.

In response to the new treaty, the AEC launched a public relations campaign aimed at minimizing the fear of radiation from underground nuclear detonations. Underground blasts at the Nevada Test Site increased as the AEC researched methods to reduce the amount of radioactivity they released into the atmosphere. One approach, according to the AEC's annual report the year after the Limited Test Ban Treaty, was "the development of thermonuclear explosives that derive only a small part of their energy from fission."[22] Underground testing became AEC's new bailiwick.

In addition, the AEC's Operation Plowshare, initiated in 1958, was still seeking ways to use nuclear explosions during peacetime, a strategy touted by the hydrogen bomb physicist, Edward Teller. Such detonations would presumably be legal according to the Test Ban Treaty, though it was certainly a gray area. Commoner listed some of the Plowshare proposals, including one near Seattle to use nuclear bombs to clear the path for a canal from the Columbia River to Puget Sound. None of the plans to use bombs in construction projects came to pass. The public was getting wise.

Women Strike for Peace was ecstatic that Kennedy had orchestrated the nuclear test ban treaty. Keeping their eyes on one goal, they had demonstrated how activists could make a difference by organizing for the greater good.[23] "When your destiny appears to be decided in ways opposite from the way you want it to go," Wilson told the *New York Times* in 1962, "you've got to get out there and concern yourself with it."[24] That same spirit carried millions of protoenvironmentalists into the fight locally and nationally during the third wave: volunteers who didn't, as Wilson put it, "want to go under without a shout."

Part III

The Environmentalism of Lyndon Johnson and Richard Nixon (1964–1973)

CHAPTER 18

JFK's Last Conservation Journey

Senator Gaylord Nelson of Wisconsin secured the "wild and scenic" designation for Wisconsin's Saint Croix, Namekagon, and Wolf Rivers as well as the federal preservation of the Appalachian Trail. In 1970, Nelson, after founding the first Earth Day, was able to realize his dream of bringing Wisconsin's twenty-two Apostle Islands into the National Park System.

I

In late September, just hours after the test ban treaty was ratified, four Democratic senators kept an appointment at Andrews Air Force Base with President Kennedy, who was in high spirits over his Cold War diplomatic negotiation with Khrushchev. Gaylord Nelson of Wisconsin, Joseph S. Clark, Jr., of Pennsylvania, and Hubert Humphrey and Eugene McCarthy, both of Minnesota, were to accompany Kennedy on the first leg of a whirlwind conservation tour through their home states. The second leg of the trip had Kennedy hopscotching across the Far West. Stewart Udall had planned the five-day, eleven-state trip as a more elaborate version of JFK's 1962 trip through the West. The theme of Kennedy's tour was that Americans had inherited a staggeringly beautiful country in need of proper stewardship. "It is an obligation," Udall wrote in the announcement of the journey, "to make sure that those who come after us, almost 300,000,000 by the year 2000, will benefit from the same rich inheritance."[1]

A key chapter in Udall's upcoming book *The Quiet Crisis* featured Gifford Pinchot, the first head of the US Forest Service under Theodore Roosevelt; his conservation legacy had remained the soul of the Service.[2] Even though the redoubtable Pinchot had died in 1946, Udall and Orville Freeman, the secretary of the USDA (which oversees the Forest Service), wanted Kennedy to begin his journey by visiting Pinchot's home, Grey Towers, in Milford, Pennsylvania; it was to be dedicated as a national historic site (the only one owned by the Forest Service). The Pinchot Institute for Conservation Studies was also officially opened that afternoon. Freeman, hoping to mend differences with Carson, requested that she be present at the event; she declined for health reasons.[3] Julius Duscha of the *Washington Post* wrote an appreciation of the administration's environmental record on the eve of the trip, concluding that "The Kennedy conservation record may be criticized as too little, but it has been in the public interest tradition of great conservationists like Theodore Roosevelt and Gifford Pinchot."[4]

The extensive tour was one way to rebut the idea that Kennedy's record on conservation was "too little." In less than three years, he had created three new national seashores (Cape Cod, Point Reyes, and Padre Island), which had been the first major additions to the National Park System in sixteen years. He was poised to establish four more areas—Canyonlands (Utah), Oregon Dunes (Oregon), Assateague (Virginia and Maryland), and Fire Island (New York)—soon. Beyond the parks that invited the public, Kennedy was strongly behind the Wilderness bill. Though it was still stalled in Aspinall's House committee, the president was supremely confident that Congress would pass the bill in 1964. By visiting national parks and federal recreation areas, reclamation and water resource projects, wildlife refuges and forestlands, he could showcase his New Frontier conservation agenda for western voters to applaud. Woven through Kennedy's speeches would be exhortations on curbing pollution and passing the Wilderness bill. Many of the states Kennedy planned to visit had voted for Richard Nixon in 1960, and the president hoped to put a couple in play for 1964. This was especially important because he was fearful of losing Deep South states because of his administration's determined leadership in combating Jim Crow segregation.[5]

Right according to plan, the press latched on to the symbolism of Kennedy's remarks at Grey Towers, in which he applauded Pinchot's scientific forestry principle of conservation. The president asked, "Have we ever thought why such a small proportion of our beaches should be available for public use, how it is that so many of our great cities have been developed without parks or playgrounds, why so many of our rivers are so polluted, why the air we breathe is so impure, or why the erosion of our land was permitted to run so large as it has in this State, and in Ohio, and all the way to the West Coast?"[6] Orville Freeman, who introduced Kennedy at the event, called his boss "the nation's No. 1 conservationist."[7]

After Milford, the president departed for Duluth, Minnesota, on Air Force One with Udall, Freeman, and the four senators in tow. Upon landing, the Kennedy party was greeted by more than fifty thousand fans hoping to get a glimpse of the president. He was then flown to Ashland, Wisconsin, enjoying en route a bird's-eye view of the Apostle Islands, which include shoreline and a chain of twenty-two wooded islands on Lake Superior in northernmost Wisconsin. The president's seatmate, Louis Hanson, a seasoned conservationist, was on the trip to speak about the scope and beauty of the Lake Superior island chain. The glittering bright blue lake captivated Kennedy, who said the Apostles' white sandy beaches and curving shorelines reminded him of Martha's Vineyard and Cape Cod. "His eyes lit up," Hanson recalled. "Here was the Massachusetts sailor seeing some of the best sailing water around." Suddenly two bald eagles soared from the Kakagon wetlands along the lake and glided near the helicopter; it was an unforgettable moment.[8]

That short flight over the Apostles made it a red-letter day for Gaylord Nelson, who had long been working to establish the bewitching island chain as a national recreation area or national lakeshore, with its exquisite beaches, sea caverns, and towering sandstone cliffs.

Senator Nelson's outstanding characteristic was easy to overlook at first. He was likable.[9] Unlike a great many politicians, he didn't exude a competitive spirit or the need to win whatever game was being played amid the interactions of Capitol Hill. Also, he liked people, never looking to belittle those with whom he was dealing. Nelson was, in other words, not a climber for the sake of climbing and so, when he nagged, cajoled, and badgered for the sake of the Apostle Islands, neither the president nor Udall could resent it. Nor, in the end, could they resist.

As the plans for a conservation tour took shape in late summer, Nelson had sent the White House staff conservation-themed quotes from Henry David Thoreau, Loren Eiseley, Joseph Wood Krutch, Wallace Stegner, and Nancy Newhall to insert into Kennedy's upcoming speeches. Once Kennedy announced the general plan for a western tour, Nelson wrote the president a five-page letter on environmental stewardship. Postwar technology, he lamented in his letter, had replaced open windows with air-conditioning, returnable bottles with plastic disposables, trains with diesel trucks—all at the expense of the natural world. "That the next decade or so is in fact our last chance can be documented with a mass of bone-chilling statistics— these statistics and what they mean will paint a picture with a compelling force understandable to everyone," he wrote Kennedy. "Rachel Carson's book on pesticides is a perfect example of the kind of impact that can be made with specifics. The situation is even worse in this country respecting water pollution, soil erosion, wildlife habitat destruction, vanishing open spaces, shortage of parks, etc."[10]

Contained in Nelson's letter were also helpful bullet points on how Kennedy could promote environmentalism as a Democratic Party priority in 1964. Though Nelson's prose was not in a literary sense on par with Stegner's "Wilderness Letter," there was something innocent and refreshing in it that impressed Kennedy.[11] Concerned that America's wilderness was disappearing, Nelson urged Kennedy to understand that conservationism was important to Americans of every stripe: "from ladies with a flower box in the window to the deer hunters with high powered rifles, the boaters who range from kids with flat-bottomed scows to the wealthy yachtsmen; family campers whose numbers are growing rapidly; bird watchers; skydivers;

wilderness crusaders; farmers; soil conservationists; fishermen; insect collectors; foresters; just plain Sunday drivers, etc."[12]

In response, Kennedy promised to make the Apostle Islands the second stop on his tour. Since 1930, the Wisconsin islands had been considered for national park status, and Kennedy's visit couldn't help but accelerate the pace of that designation.[13] The good news was that residents of the nearby towns of Ashland and Bayfield favored the plan, believing that a national park would bring tourism dollars to the economically distressed region. "If promptly developed, recreational activities and new national park, forest, and recreation areas can bolster your economy, and provide pleasure for millions of people in the days to come," Kennedy said in Ashland.[14]

The Mashkiiziibii—the Bad River Tribe of the Chippewa nation— wanted federal protection for the Kakagon Sloughs, a gorgeous unmarred wetland area flowing into Chequamegon Bay and Lake Superior, and Kennedy had been briefed about that, too.[15] The Sloughs, rivulets flowing through the wetlands, have long provided the Mashkiiziibii with delicious wild rice. The cleanliness of the water is, of course, crucial for growing the high-quality rice. (In 1973, the Kakagon Sloughs were accorded federal status as a national natural landmark and the Mashkiiziibii were given control over it, receiving many environmental awards for their stewardship in the years since.)

To Nelson's delight, Kennedy spoke eloquently in Ashland about the natural beauty of Lake Superior, the Apostle Islands, the Bayfield Peninsula, and the Bad River/Kakagon Sloughs area. "These islands are part of our American heritage," he said. "In fact, the entire northern Great Lakes area . . . is a central and significant part of the freshwater assets of this nation. We must act to preserve these assets."[16] The president added that Senator Nelson "has a "strong conviction, as I do, that every day that goes by that we do not make a real national effort to preserve our national conservation resources, is a day wasted."[17]

After the Apostle Island tour, Kennedy returned to Duluth to give an evening speech to delegates at the Northern Great Lakes Region Land and People Conference that Freeman had organized. In both this Duluth speech and the earlier one in Wisconsin, Kennedy linked economic development with conservation initiatives, realizing the region was a growing recreation area within a day's ground transportation of tens of millions of Americans.[18] Fully embracing the recommendations of "Our Vanishing Shoreline," his perennial favorite, Kennedy seemed to endorse Pictured Rocks (Michigan) and Apostle Islands (Wisconsin) as national lakeshores—to the rapt attention of Interior Department officials. The *Oshkosh Daily Northwestern* in Wisconsin ran the headline: "Upstate Area Deserves Park Site, Says Kennedy."[19] The Associated Press reported that the president was astounded that the Apostles could be bought for the "bargain price" of $500,000. Joining Kennedy for his announcement of support for national lakeshores on Lake Superior were hundreds of Mashkiiziibii and Gaamiskwaabikaang (Red Cliff band) Chippewa.[20]

The next day, September 25, after a quick stop at the University of North Dakota in Grand Forks, it was on to Billings, Montana, where Kennedy spoke at the Yellowstone County Fairgrounds. He suggested a new bill to make it simpler for state and federal governments to develop outdoor recreation areas, and he promoted the Youth Conservation Corps and congressional passage of the pending Wilderness bill.

By the time the tour reached Billings, it was clear that Kennedy wasn't preaching the gospel of wilderness and national park protection without also peppering his speeches with language that would ameliorate miners, lumbermen, and hydropower engineers. Arthur Edson, an Associated Press reporter based in Wyoming, observed, "Although Kennedy strongly believes in conservation, on this trip he is having a hard time getting much missionary zeal into it. He wanders from the text frequently, and occasionally doesn't even go near it." At the University of Wyoming Field House in Laramie, the president lauded new energy possibilities, such as advances in electrometallurgy and a process to derive gasoline from liquified coal, which would help western miners.[21]

At his big speeches in Duluth, Billings, and Laramie, Kennedy had initially been planning to avoid grandstanding about the Wilderness bill or the nuclear treaty. Ted Sorensen feared that if he mentioned the roadless bill, he'd be booed, and if he talked about ending aboveground testing in Nevada, he'd be heckled as being soft on communism. McGovern persuaded Kennedy to speak robustly on both major policy issues. "I told him to be boastful," he recalled. "That many westerners were fearful of strontium-90 and liked the idea of a permanent wilderness. I reminded Kennedy that FDR had won Utah four times in presidential races by championing the idea that national parks and outdoors recreation equaled jobs."[22]

After his Laramie speech, the third on a busy day, Kennedy spent the night at Jackson Lake Lodge in Grand Teton National Park. The Rockefeller family had funded the construction of the 385-room rustic architectural gem in 1955. Instead of hiking down Lakeshore Trail, the president sat in a rocking chair in the hotel building, read papers, and looked out of the giant picture windows at Jackson Lake and the pinnacles of the Tetons and Mount Moran. Though the hotel wasn't yet ten years old, the view from the lobby picture window was already storied. But even if JFK were inclined to hike, his bad back made it physically impossible for him to hit the trails.

On September 26, Kennedy visited the Hanford Engineer Works near Richland, Washington. Plutonium produced at Hanford had fueled the bomb used in the Trinity test of July 16, 1945, that had ushered in the atomic age. The 625-square-mile facility had never before been opened to visitors, but hordes of spectators were allowed on the premises to hear the president speak. He started by touting the treaty. "The Atomic age is a dreadful age," he said to a crowd of around thirty-seven thousand. "No one can say what the future will bring, no one can speak with certainty about whether we shall be able to control this deadly weapon, whether we shall be able to maintain our life and our peaceful relations with other countries."[23] The treaty, Kennedy said, was a step "on the long road to peace," one of grave importance to western residents cursed with the Nevada Test Site.

It pleased Kennedy to participate in the groundbreaking ceremonies for the construction of the new N reactor, the ninth on the nuclear campus, which would produce electricity for commercial and domestic customers. Keeping to his theme, Kennedy said that Hanford, "where so much has been done to build the military strength of the United States," would now, in the New Frontier, have "a chance to strike a blow for peace."[24] With the nuclear treaty signed, Kennedy hoped that atomic power could now be directed predominantly toward safe, clean energy purposes. He predicted that by 2000, nuclear energy would generate half of the nation's electricity. That optimistic forecast was not to be. According to the AEC's successor, the Department of Energy, by the year 2000 nuclear power represented less than 10 percent of the nation's electricity generation.[25] "Kennedy and I had high hopes that nuclear power from places like Hanford would get America off of fossil fuels that were poisoning our air with smog," Udall recalled. "We didn't, I'm afraid, properly take into account the dilemma of burying nuclear waste."[26]

For all Udall's professed interest in Native American culture, he didn't schedule a stop for JFK at any federal reservations. In fact, the administration generally turned a blind eye to living conditions on tribal lands in the American West. The Bureau of Indian Affairs, which was part of Udall's Interior Department, treated the citizens of the Navajo Nation, the largest Native American population in the United States, like subhumans. Uranium-mining companies ignored safety standards and left massive piles of contaminated tailings on Navajo land. Radioactive water was discharged into surface- and well-water reserves. Navajo tribal members suffered astonishingly high rates of cancer. The tap water on the reservations was undrinkable. At about that time, the National Indian Youth Council (NIYC) was launched in Albuquerque with representatives from twenty-one nations. Many of the founders, including Gloria Emerson and Herbert Blatchford (both Navajo), Clyde Warrior (Ponca), and Shirley Hill Witt (Mohawk), thought the New Frontier wasn't doing enough to help tribes protect sacred landscapes and develop sustainable non-mining-related economic incentives.[27]

After a quick flyover of the Oregon Dunes—still not a national park unit three years after Senator Neuberger's death—Kennedy and his party made their way to Mount Lassen Volcanic National Park in California.[28] While a park naturalist led Udall and the press corps on a hike around Manzanita Lake at dusk, Kennedy took a pass to relax at the park manager's residence.[29] Soaking up the Shasta County scenery, watching Steller's jays and chipmunks vie for cereal, the president suggested that the superintendent had "the best job in the world." He was enjoying nature for a few hours in a relaxed way. But the next morning, stepping out for a breath of fresh air before breakfast, he spotted a deer not far away. All by himself, in a very special moment, he made friends with the deer—which wasn't easy to do. Ducking back into the house, the president ransacked the kitchen for something to feed a deer. One of his aides then snapped a picture of the most powerful man in the world, standing in pink pajamas delightedly hand-feeding the deer cereal. The NPS was quick to point out that under most circumstances, feeding wild animals is verboten.[30]

Backtracking to the east, Kennedy flew to Utah to make a speech at the Mormon Tabernacle. Along the route from the airport to downtown Salt Lake City, well over 120,000 people awaited his car, to wave and hold up signs. That was one-third of the population. It seemed the whole city was smiling. Nothing quite like it had ever been seen in the rather quiet-living state. Kennedy ended his western tour with a major speech in Las Vegas dealing largely with water policy. In an effort to please local reclamationists, he praised Hoover Dam as the most significant infrastructure on the Colorado River, providing fresh water to Arizona, California, Nevada, and even Mexico. Kennedy was equally optimistic about the value of the Lake Mead–Hoover Dam complex. Speaking as the willing inheritor of Theodore Roosevelt's reclamation legacy, Kennedy said that audacious engineering was helping Nevadans survive droughts, wildfires, and water shortages.[31] Soon to come, as he said, would be the huge Glen Canyon Dam, with the creation of Lake Powell as a recreation hub 186 miles in length and soon to be famous for motorboating.

Earlier in the year, Kennedy had received a copy from David Brower of *The Place No One Knew: Glen Canyon on the Colorado*, an elegant large-format book the Sierra Club had just published. It featured the work of Eliot Porter, a New England photographer best known for his clean, vibrant images of Maine's Great Spruce Head Island. Having earned multiple degrees from Harvard and exhibiting widely, Porter was compared to Ansel Adams, though he shot in color. His photos of the natural world were so textured that they seemed three-dimensional.[32]

The aim of *The Place No One Knew* was to halt the construction of Glen Canyon Dam in order to preserve the gulches, rock walls, and remote canyons made by the Colorado and San Juan Rivers. It didn't accomplish its mission, of course; the Lake Powell reservoir started filling that year. With that, more of Glen Canyon disappeared by the day. On the upside, *The Place No One Knew* did trigger a federal review of all future reclamation projects and western rivers. And the book, in an indirect way, helped keep the passage of the Wilderness bill alive. The public's admiration of Porter's book proved that he could use landscape photography to make people aware of nature's beauty without compromising his artistic goals. Receiving praise from all quarters, he would be elected to the board of directors of the Sierra Club in 1965 and served until 1971.

Porter pointed out that over the decades, the Nevada wilderness had been largely overlooked by outdoor recreationists and tourists in favor of epic Grand Canyon scenery. Kennedy, the romantic realist, hoped to change that by announcing plans to save the "remaining un-spoiled shoreline" of blue Lake Tahoe in Nevada, calling the alpine lake "the gem of the Sierras." He also advocated a new Great Basin National Park in east-central Nevada to preserve groves of bristlecone pines, the oldest known nonclonal organisms in the world. The Great Basin National Park would also encompass Lehman Cave at the base of the 13,063-foot Wheeler Peak (owing to legal entanglements, the park wouldn't be established until 1986). "Much of the future of this State, in other words, rests on conservation," the president said in Las Vegas, "and this work must go forward in the 1960's."[33]

At that final rally of his conservation tour, Kennedy's rhetoric soared. "This is still a beautiful continent, but we want 'America the Beautiful' to be left for those who come after us," Kennedy declared. "Robert Frost, the late poet, once remarked, 'What makes a nation in the beginning is a good piece of geography.' Our greatness today rests in part on this good piece of geography that is the United States, but what is important is what the people of America do with it."[34]

Some anti-nuclear activists, encouraged by the Limited Test Ban Treaty, had hoped that Kennedy would use his Las Vegas speech to close the Nevada Test Site. It made sense as a follow-up of the historic treaty and the promotion of peaceful atomic energy in Hanford. But it was a pipe dream. With gallows humor, Nevadans still took pride in hosting atmospheric tests. You could get an "atomic hairdo" at the Flamingo Hotel along the Las Vegas Strip. Or dance as the pianist Ted Mossman played his boogie-woogie tune, "Atomic Bomb Bounce." Nevada citizens were not, by and large, concerned about nuclear testing. It was difficult for them to believe that nuclear testing in underground tunnels was bad for their health. At the time of JFK's visit, a public opinion poll showed that most Nevadans even wanted atmospheric testing to continue because they doubted that the Soviets would abide by a treaty banning it. Only a couple of leftist lawyers in Carson City, at most, were murmuring about legislation to compensate "atomic victims."[35]

For the most part, only the eccentric billionaire Howard Hughes, a conservative Las Vegas resident, was angry that the Nevada Test Site had continued underground atomic detonations after the signing of the Test Ban Treaty. Reverberations from the shock of the detonations routinely rattled the windows of Hughes's hotels, including those of his penthouse. For the rest of the decade, Hughes warred with AEC officials to get them to stop the testing. But the *Las Vegas Review-Journal* and *Reno Evening Gazette* ignored his pleas.[36]

From Las Vegas, Kennedy flew to Palm Springs, California, to rest at the home of Bing Crosby, see friends, and play golf. The Las Vegas speech aside, conservation nonprofits such as the Sierra Club and the Nature Conservancy were disappointed by Kennedy's western tour. Even the *New York Times* criticized him for "missing a great opportunity to focus attention on some really significant issues in the constant battle to conserve some of the natural beauty and wonder of America."[37] The list of complaints from environmental organizations was long. Why hadn't JFK visited the coastal redwood country in northern California, which Brower was pushing to become a national park? Why hadn't he visited Washington's North Cascades wilderness, where a grassroots effort championed by Patrick Goldsworthy and Polly Dyer was under way to establish it as a national park? Or visited the Grand Canyon National Park and Monument, which were being threatened by the Bridge Canyon and Marble Canyon Dams? There were plenty of wondrous places and preservation issues in the western states that were hanging in the balance and the president avoided them like the plague. The itinerary the White House planned made the trip a nice conservation tour and a stupendous soft-launch to the 1964 presidential campaign.

II

No sooner did President Kennedy return to the White House than Rachel Carson embarked on a journey to visit Muir Woods in California, with Marie Rodell as her travel companion. By then, Carson had to expend her energies carefully. The woman who couldn't muster the strength to go to the dedication of the Pinchot Institute in eastern Pennsylvania was making a cross-country trip to San Francisco. That was no double standard. It just showed how very much Carson wanted to see the big trees, as well as the Pacific Coast. With David Brower and his wife, Anne, as hosts, she and Rodell were driven through Marin County, north of the city, en route to the national monument. Whenever they stopped, Carson needed to use a wheelchair in order to conserve her strength.

The freshness of the Pacific air was intoxicating to her. She loved seeing the red-barked coastal trees' soothing, filtered light up close. Encountering windswept Bodega Bay and the sublime Point Reyes seascapes, havens for shorebirds, made her appreciate the Bay Area as an earthly paradise. "Once over the Golden Gate Bridge, one climbs up and up into those smooth, brown hills, so much of the road lined with eucalyptus trees," she wrote a friend.

> Then a long winding descent, one hairpin curve after another, into the canyon where the redwoods are . . . one sees great, burned-out stumps here and there, looking fresh enough to have resulted from a fire last year; yet the ranger said there had been no fire in Muir Woods for at least 150 years. . . . There was a marvelous freshness in the air, though I couldn't detect a distinctive odor. The under story of these woods is chiefly the California laurel—a huge tree.[38]

Studying the flora and fauna of the Bay Area was educational and enthralling to the Maryland-based biologist. Carson explored Marin County's redwood giants and the beach at Rodeo Cove, which as of 1972 became part of the Golden Gate National Recreation Area. She wanted to experience the vistas that Ansel Adams had been memorializing in photos since the 1930s. It fascinated her to learn that the legendary artist had refused to photograph the eucalyptus trees along the California coast because they weren't indigenous flora. "From Muir Woods we drove to the shore at Fort Cronkite, now part of the Golden Gate National Recreation Area," Brower recalled in 1995. "In the lagoon just inland were perhaps fifty brown pelicans having a hell of a good time, perhaps celebrating the beginning of their recovery with a pelican ballet, on that sunny day. I have to believe in magic, for what else could have led those pelicans to know that Rachel Carson would have preferred them to redwoods?"[39]

Brower may have been right, but he couldn't know why. When Carson was among the redwoods, she was uncomfortable precisely because she really and deeply preferred them to all else. Since earliest childhood, whenever she was in a truly natural setting—especially a forest or shoreline—she immersed herself in it. Though she appreciated that Muir Woods did not have the largest redwoods, she was enthralled by its completeness as an ecosystem: mature trees, young trees, fallen trees; a verdant understory dense with plants that had been around when dinosaurs roamed the planet; and a riparian corridor that was both nursery and lifeline for untold generations of wild creatures. "I longed to wander off, alone, into the heart of the woods, where I could really get the feeling of the place," she wrote to Dorothy Freeman. "Instead of being surrounded by people! And confined to a wheelchair."[40]

For the rest of his life, Brower associated California brown pelicans with Carson. *Silent Spring* had revealed that DDT thinned the shells of pelican eggs when the parents brooded. And he remembered how Carson had told a story of DDT-poisoned pelicans and worried about their fate. Sometimes, when dining at Sinbad's Restaurant on the San Francisco wharf, his favorite coffee and seafood hangout, Brower would count pelicans (his record for one day was 176) and ponder Albert Schweitzer's "reverence for life" philosophy or passages from *The Edge of the Sea*.[41] By the time of Carson's Bay Area visit, the pelican population in the eastern United States had crashed. In a calm but stern way, she told Brower that she feared their demise along the Pacific coast. And for good reason. DDT and other pesticides resulted in the fact that just one California brown pelican would successfully fledge at their primary breeding grounds in the 1970 season.

Another excursion took Carson and Rodell to Pacific Grove on the Monterey Peninsula to see the monarch butterflies congregate. Enraptured by Carmel-by-the-Sea, which the poet Robinson Jeffers had written about in such books as *Give Your Heart to the Hawks and Other Poems*, she drove by Tor House, the stone building that Jeffers had constructed by hand on the fog-shrouded Central Coast. Because Jeffers had been born in Allegheny, Pennsylvania, about twenty miles downriver from Springdale, her own hometown, Carson harbored a certain fondness for him. Jeffers's philosophy of "inhumanism," the belief that human beings are grotesquely self-centered and indifferent to the "astonishing beauty of things," made sense to her.

One afternoon, while in San Francisco, Carson spoke for an hour at the Kaiser Foundation symposium "Man Against Himself." *Silent Spring* was two years old, and the brutal attacks on the book were subsiding. A paperback edition was released, making the book more affordable for students, who made it a favorite. Ensconced in a chair onstage, Carson delivered her visionary address "The Pollution of the Environment" to an audience of 1,500 people. Instead of cataloging her illnesses, she simply said that arthritis had made her immobile. That evening, she called herself an "ecologist" for the first time. (She described ecology as a science focused on the interconnectedness of organisms and their ecosystems.) Though her discussion of the deleterious effects of pesticides on wildlife and her embrace of Barry Commoner's pioneering research on strontium-90 were predictable, her descriptions of how the by-products of atomic fission were being disposed of in the sea were jarring.[42] "We behave, not like people guided by scientific knowledge, but more like the proverbial bad housekeeper who sweeps dirt under the rug in the hope of getting it out of sight," she warned. "We dump wastes of all kinds into our streams, with the object of having them carried away from our shores. We discharge the smoke and fumes of a million smokestacks and burning rubbish heaps into the atmosphere in the hope that the ocean of air is somehow vast enough to contain them. Now, even the sea has become a dumping ground, not only for assorted rubbish, but for the poisonous garbage of the atomic age."[43]

Once back in Silver Spring, Maryland, Carson found an advance copy of *The Quiet Crisis* awaiting her. As Carson cracked it open, she read the handwritten inscription: "For Rachel—An educator-crusader for conservation whose stones have wide ripples!"[44] By the fall of 1963, the controversial writer of *Silent Spring* had become the darling of the American conservation movement. The National Wildlife Federation presented Carson with its first Conservationist of the Year Award. And she was the first woman to be awarded the Audubon Society's highest medal. "A prophet is not always honored in his own country," she said at the ceremony, "but the National Audubon Society *is* my country—I belong to it—and from its own strength and devotion to the cause of conservation I have drawn strength and inspiration."[45]

III

On October 3, 1963, Kennedy was in central Arkansas to pull the lever opening the Greers Ferry Dam on the Little Red River. Only the year before, he'd given his support to the preservation of two nearby waterways, the hundred-mile-long Current River and its thirty-mile-long tributary, Jacks Fork, in southern Missouri. The two were to form the core of an eighty-thousand-acre park, the Ozark National Scenic Riverways, in 1964.[46] Celebrating the opening of a dam on one river while pledging to protect two others in the vicinity was typical of the policy contradictions common in conservation. To a great degree, the trade-off worked, and the reservoir filled by the Little Red River became a lively resort, while the Current and Jacks Fork remained primitive.

Empowered by President Kennedy's conservation tour, the Limited Nuclear Test Ban Treaty, and the Ozark National Scenic Riverways, the administration pushed the Democratic leadership in Congress on the Wilderness bill. It concerned Kennedy that with his 1964 reelection campaign looming, that seminal piece of preservation legislation was languishing. Just before he departed for Texas in mid-November 1963 for a combination NASA space promotion and political fundraiser, he telephoned Representative Aspinall. They met for a friendly talk at the White House, but JFK's message was that the delays had to end. As an incentive, he offered his support for the Colorado Democrat's demand that new mining claims be allowed in national forest wilderness areas for a few more years. He wanted the bill to become law before the November 1964 presidential election, giving America's conservation voters a huge reason to reelect Kennedy-Johnson.[47] Aspinall remained cagey as the two parted.

The Greers Ferry Dam opening turned out to be Kennedy's last conservation-related activity, for the New Frontier came to an abrupt and terrible halt in Dallas on November 22, 1963. At 12:30 p.m. CST, three shots were fired at the presidential motorcade, wounding Kennedy. Immediately, the first couple's blue Lincoln convertible zoomed to nearby Parkland Hospital, but the president died within the hour.

Udall was informed of Kennedy's assassination in a telegram he received on a stopover in Honolulu, en route to Tokyo. The wire read, in part:

THE PRESIDENT, HIS LIMP BODY CRADLED IN THE ARMS OF HIS WIFE, WAS RUSHED TO PARKLAND HOSPITAL. THE GOVERNOR ALSO WAS TAKEN TO PARKLAND.

CLINT HILL, A SECRET SERVICE AGENT ASSIGNED TO MRS. KENNEDY, SAID "HE'S DEAD."[48]

Stuck at an airport more than 4,800 miles from Washington, knowing little and understanding less, Udall was numb. He immediately caught a plane and headed home to Virginia. On the way, he grabbed blank sheets of Office of the Secretary of the Interior paper and memorialized his feelings: "We are still in shock—why should one so young, so gallant—who had so much to give mankind, such a large role to play for peace—be struck down so insanely in full flight? Why does so much history end so cruelly? He had the mind of a modern president in a way none of his predecessors had. What he lacked in passion was compensated for by his great sense of self, his luminous mind could catch up and deal with the most complex problems of the age."[49]

On November 23, as Kennedy's casket arrived at the White House, Westinghouse radio broadcaster Sid Davis began reading Frost's poem "Stopping by Woods on a Snowy Evening" on the air but had to sign off, apparently overcome by emotion. Even in death, Kennedy still seemed to live, and indeed his visage and oratory would not die. The memory of his murder, the vacant horror, has haunted the country ever since.

When news of Kennedy's death reached *New York Times* editor John Oakes, he was speaking at a college campus. Matter-of-factly, he got on a pay phone and dictated his newspaper's editorial for the next day's edition. "The light of reason was momentarily extinguished with the crack of a rifle shot in Dallas yesterday," he declared. "But that light is, in reality, inextinguishable; and, with God's help, it will show the way to our country and our country's leaders as we mourn for John F. Kennedy in the darkening days ahead."[50]

Robert F. Kennedy became a white-water rafting and kayaking enthusiast after his brother's assassination in 1963. Here is Kennedy competing in the Hudson River's White Water Derby in 1967. RFK would take his entire family on rafting explorations down the Green, Colorado, and Snake Rivers in the American West.

CHAPTER 19

The Mississippi Fish Kill, the Clean Air Act, and American Beautification

San Francisco Freeway Protest, Golden Gate Park, 1963.

I

On November 3, 1963, Rachel Carson wrote Dorothy Freeman a note reminiscing about the "haunting music of geese" at the Mattamuskeet National Wildlife Refuge in North Carolina, gossiping that JFK's "very reddish brown" hair looked darker on TV than in person.[1] Nineteen days later, the leader she trusted with America's environmental future was dead. "For the 3½ days when we all lived so close to the tragedy I was numb and dazed—could think of nothing else," she wrote Freeman on November 27. "Yesterday I turned to writing my Audubon Medal acceptance because I must, but still my mind would not function. . . . Besides the feeling of personal sorrow and loss, which for me could scarcely be greater if it were a member of the family, there are the feelings of shock, dismay, and revulsion at the black aspects of our national life so strongly underlined—the bigotry, intolerance, and hatred preached by so many. The only ray of light I can see is that perhaps, because of his martyrdom, the noble ideals and aspirations that John Kennedy stood for will be understood as never before."[2]

Carson took solace from her association with Kennedy, a president who had held Big Chemical accountable after *Silent Spring*'s publication and orchestrated the Limited Nuclear Test Ban Treaty. JFK had stood by her and her research at a critical moment on the eve of her book's publication. Kennedy's instincts—on the Wilderness bill, national seashores, clean air, and clean water—had pushed the country in the right direction, an environmentally sustainable direction. Carson told Freeman that JFK had an "unusual grasp" and a surprisingly "deep concern" when it came to problems related to ecological conservation; she was uncertain whether Lyndon Johnson did.

Of the many places named after Kennedy in memoriam, two paid homage to the thirty-fifth president's love of the natural world. The Kennedy family received an avalanche of proposals as to where to build JFK's Presidential Library and Museum. The one they chose was in the southern part of Boston, a wide-open space overlooking the waters of Dorchester Bay. The architect, I. M. Pei, designed the smoked glass and white concrete edifice with angular walls to look like a "whitewashed baroque lighthouse" jutting outward to the slain president's beloved Atlantic Ocean.[3] In the 1970s, the landscape architect Carol Johnson was authorized to transform a former toxic landfill along the Charles River in Cambridge, Massachusetts, facing Harvard, the president's alma mater, into the lovely John F. Kennedy Memorial Park. Water cascaded down the sides of a granite fountain to depict the various New England rivers. Flowers now bloom on JFK's May 29 birthday.[4]

Even as Kennedy was laid to rest at Arlington National Cemetery, the Mississippi River was under ecological assault. Some 5 million fish were found floating in the lower river, belly up; indeed, fish had been dying by the millions in the Mississippi and also the Atchafalaya (a parallel river in lower Louisiana) for four years. Compounding the disaster, their decaying bodies had choked the intakes of regional power plants and contaminated public drinking water from Memphis, Tennessee, to Venice, Louisiana. Biologists from the Public Health Service conducted autopsies and found that the fish had been poisoned by the pesticide endrin. In addition, James Strain, the president of the State Board of Health in Louisiana, found endrin, heptachlor, DDE, and DDT in gulf shrimp beds. Fish that digested endrin became bloated with gas and fluid, making it extremely difficult for them to stay submerged. "When disturbed by the bow-wave of a ship, they come to the surface and remain there, thrashing helplessly," the *New York Times* reported.[5]

The company responsible for the presence of endrin in the rivers was the Velsicol Chemical Corporation, the Illinois-based pesticide maker that had sought to ban the publication of *Silent Spring*. Discharge from a Velsicol factory in Memphis that made endrin had caused the Mississippi River fish kill. Warnings went out across New Orleans and Baton Rouge for consumers to beware of poisoned fish and oysters in restaurants and grocery stores. All of the affected states struggled to grapple with the effects, but they were helpless to address the cause. In *Silent Spring*, Carson had nailed the consequences for birds and other wildlife of farms using synthetic pesticides on corn and cotton; now, clearly, fish and aquatic animals were being devastated, too, by runoff from local farms.[6] Humans consuming the fish were in danger; endrin and its chemical partners dieldrin and aldrin ravaged the human nervous system when consumed in significant quantities. Those three compounds were used widely on midwestern farms, especially those growing corn. Velsicol wasn't the only company under scrutiny; Union Carbide, Dow Chemical, Allied Chemical, Celanese, American Cyanamid, and others collectively used more than 3.7 trillion gallons of water annually in their manufacturing processes, with no regulation of the discharge.[7]

Abraham Ribicoff, Stewart Udall, John Dingell, and others decided to go after Velsicol for the damage to the Mississippi. Ribicoff proposed a fresh round of subcommittee hearings and drafted the

Clean Water Bill for the Senate. Udall ordered a departmental ruling forbidding pesticide use wherever there was a "reasonable doubt as to its environmental effects."[8]

What concerned Udall was that consumers liked modern conveniences such as aerosol deodorant, oven cleaners, food additives, gasoline, shampoo containers, Plexiglas, and plastic food wrap, so it was difficult to persuade them that chemical-based products were, in fact, a dire problem. "This highly innovative industry develops new ways to wage chemical warfare on water users so fast that pollution control authorities can barely keep track, let alone keep up," David Zwick and Marcy Benstock wrote in the 1971 book *Water Wasteland: Ralph Nader's Study Group Report on Water Pollution*. "No one can detect or treat many of the complex new compounds or know what their effects are in water; the rivers serve as giant test tubes for corporate chemists."[9]

The Mississippi fish kill made the first chapter of *Silent Spring* seem prophetic. Scores of organizations wanted to honor Carson for fighting the corporate chemists with good science. On December 3, still reeling from the Kennedy assassination, Carson received the Audubon Medal before an audience of five hundred in New York City. Accepting it, she described conservation as a cause with no end: "There is no point at which we will say, 'Our work is finished.'"[10] There was in her voice that evening a tone of resignation. Almost every hour she was awake, her activities were limited by deep fatigue and bouts of pain. A realist, she knew that *Silent Spring* would be her last book. "I keep thinking, if only I could have reached this point ten years ago! Now, when there is the opportunity to do so much, my body falters, and I know there is little time left," she told a friend.[11]

Three days later, on December 6, the American Academy of Arts and Letters inducted Carson as a select member (only the third woman out of fifty). The historian and sociologist Lewis Mumford said at the induction ceremony, "A scientist in the grand literary style of Galileo and Buffon, [Carson] has used her scientific insight and moral feeling to quicken our consciousness of living nature and alert us to the calamitous possibility that our shortsighted technological conquests might destroy the very source of our being."[12]

With the rest of the nation, Carson waited to see what sort of leader Lyndon Johnson would prove to be. For those against nuclear testing and in favor of the public health and ecology movements, it was an anxious time. LBJ was not known as a friend to environmental causes. The determined Michigan senator Philip Hart wasted no time in politely pestering the new president to establish the Pictured Rocks and Sleeping Bear National Lakeshores in his home state. The Sierra Club lobbied to move forward on the Canyonlands (Utah), North Cascades (Washington), and Redwood (California) national parks projects. Questions swirled among mainstream conservation leaders. There were no quick answers, but optimists were reassured by Johnson's evident love of the picturesque Texas Hill Country.

There was cause to believe that LBJ's attachment to the Hill Country was more than just public posturing. LBJ seemed happiest watching over his Pedernales River ranch in Stonewall, making sure that his 2,500 acres had enough luxuriant grass for his head of four hundred Herefords. That ranch, long his refuge from the hurly-burly of Austin and Washington politics, was his version of Hyannis Port. Though the region offered a relatively spare way of life, its beauty rooted residents in the land. The image Johnson cultivated was that of a western rancher showing respect for his property through proper land stewardship. A natural braggart, he would wax eloquent to reporters about his live-oak trees, meadows ablaze with wildflowers, and game galore. He didn't have to exaggerate; the spread along the Pedernales was gorgeous. From the White House, he would call his ranch foreman daily to check on weather conditions, grazing rotations, and maintenance-related issues. At dusk, especially in the spring, when the Pedernales and its gurgling creeks flowed with vigor, carpets of bluebonnets in bloom, the delicate smell of bluestem and sideoats grama grasses in the air, the setting was glorious. Lady Bird called the ranch "our heart's home."[13]

Democratic senators Frank Church and Ralph Yarborough wondered why LBJ had been opposed to the Padre Island National Seashore. Would he abandon other national seashore efforts in the pipeline? Only time would tell whether the new president was an instinctive preservationist like Jack Kennedy or a multiple-use conservationist like Wayne Aspinall. To which side of the Democratic Party did he tilt on public lands issues? Ansel Adams sent the new president a letter about the imperative of retaining Udall in the cabinet. "Seldom has any Secretary shown such devotion to the cause of conservation, and expanded such energy and intelligence in that domain," he wrote. "His accomplishments were all the more difficult because of hostility in Congress and the concerted efforts of pressure groups in the exploiting of timber, mining, and industrial world. He is a *great* Secretary of Interior, and it will be difficult indeed to find anyone to match him in these times."[14]

As Adams indicated, the third-wave conservation movement's immediate priority was keeping Udall at Interior. In his first three years, he had proven to be an honorable ally. Oakes, the *New York Times* editorialist, worked back channels through Bobby Kennedy—not a Johnson favorite, but a potent voice under the circumstances—to convince Johnson to retain Udall.[15] David Brower publicly praised Udall for defending *Silent Spring* early on and for injecting Interior with a shot of vitality. The Wilderness Society leaders Mardy Murie, Olaus's wife, and Stewart Brandborg also spoke up on the secretary's behalf. Mardy Murie had recently published the book *Two in the Far North* and hoped that Johnson would follow up Eisenhower's Arctic National Wildlife Refuge with a new Brooks Range National Park in Alaska.

In early December, Udall's book *The Quiet Crisis* was published by Holt, Rinehart & Winston, with an introduction by President Kennedy. Written with contributions and expert help from Wallace Stegner, Sharon Francis, Alvin Josephy, Jr., and other departmental special assistants, it was still the product of Udall's spirit. Every page burst with a blend of facts highlighting American conservation history and philosophical wisdom for a "land ethic of tomorrow." Udall warned that much of American technology, its commercial development, and its hyperindustrialization were ruining the natural world at an alarmingly rapid pace. He urged citizens to create "bright new chapters" in the conservation realm.

Most reviewers admired *The Quiet Crisis*, praising it for educating Americans about their nation's natural resource heritage and for profiling the conservation movement's founding fathers. The most visionary part of the book was Udall's promotion of a wild and scenic river system. "Generations to follow will judge us by our success in preserving in their natural state certain rivers having superior outdoor recreation values," he wrote. "The Allagash of Maine, the Suwannee of Georgia and Florida, the Rogue of Oregon, the Salmon of Idaho, the Buffalo of Arkansas, and the Ozark Mountain rivers in the State of Missouri are some of the waterways that should be kept as clean, wild rivers—a part of a rich outdoor heritage."[16]

Charles Stoddard, the director of the Bureau of Land Management, wrote Udall that the publication of *The Quiet Crisis* made his staff proud; the book served as both a memorial to JFK and a policy prescription for Johnson.[17] The president of Holt, Rinehart & Winston, Alfred Edwards, told Udall that *The Quiet Crisis* would take its place among the "lasting conservation classics." He said, "I only regret that our great mutual friend Robert Frost, who brought us together, could not be here to get his autographed copy from you."[18]

Of all the praise Udall received, a heartfelt note from Rachel Carson touched him the deepest. "I found in your book a direct and vital message," she wrote. "We all owe you a debt of gratitude for having written it. I know that I, for one, gained a broader perspective of the history of American conservation, and a deeper admiration and respect for those who struggled so long and faithfully to preserve even remnants of the continent it was. . . . When will the people fully understand and accept the obligation to the future—when will they behave as custodians and not owners of the earth?"[19]

One chapter in the book embraced nuclear power's promise, as Kennedy had done at Hanford, Washington. Udall called atomic energy development "the supreme conservation achievement of this century." He predicted that reactors would "'breed' energy from rocks," making dirty fossil fuels a thing of the past.[20] That was not a view that Rachel Carson or Barry Commoner shared. Sharon Francis had worried about that chapter, deriding it as "Stewart's hymn to a nuclear America."[21]

In its own way, *The Quiet Crisis* presented a united Udall-JFK front, and with the country in mourning over JFK, President Johnson favored continuity; he retained Udall. Bobby Kennedy was the only New Frontier cabinet official to leave during Johnson's first months in office. Understandably, RFK was shattered by his brother's death. To help heal, he took walks in the frosty Virginia woods. In the coming years, testing his mettle, he would climb mountains, raft down class 4 rapids, hike in the Mojave Desert, and swim in dangerous Atlantic currents. Perhaps because of the strength he drew from nature, he became a vigilant preservationist, reliably on the side of Defenders of Wildlife, the Wilderness Society, and the Sierra Club. Whatever conservation cause of the day Bill Douglas backed, Bobby Kennedy was on his side.

President Kennedy's death left Coretta Scott King spiritually numb. When he signed the Limited Nuclear Test Ban Treaty in 1963, she had, as her sister, Edythe Scott Bagley, observed, "the satisfaction of knowing she had participated in making history."[22] Just days before JFK was murdered, Mrs. King picketed with Dagmar Wilson at the United Nations in New York City, carrying a sign that read "Let's Make Our Earth a Nuclear-Free Zone."[23] Mrs. King didn't know it at the time, but her involvement in founding SANE and meeting with the Soviet delegation in Geneva had caught the attention of FBI director J. Edgar Hoover. The FBI began tracking her movements by tapping her telephone calls and reading her mail.[24] Though she wasn't subjected to the abuses by the FBI that her husband endured, she became a "subject of increasing interest" because of her anti-nuclear protests.

II

On December 8, 1963, Johnson met with Udall in the White House. There the two plotted to make the Wilderness bill into law in 1964. Udall explained his feud with Aspinall, and Johnson promised to lean hard on the stubborn Interior and Insular Affairs Committee chair. Going into the Oval Office, Udall understood, Johnson was imbued with a New Deal spirit but also had a chip on his shoulder about the New Frontier. Around that time, the bitter Johnson asked Hugh Sidey of *Time* magazine why the "Georgetown crowd" were so in awe of the Kennedys. Referring to Coco Chanel's classic perfume, Johnson was said to have asked, "How come when I say it, it comes out 'Horse Shit,' but when they say it, it comes out 'Chanel No. 5?'"[25] If Johnson felt competitive with JFK's legacy, perhaps it prodded him to surpass his predecessor in conservation policy. It was a good sign for Udall when Johnson said that he promised to have the same close relationship with him that FDR had had with Harold Ickes.

Udall soon learned that Johnson's mixture of ceaseless monologues, off-color jokes, chest-thumping, and looming close to one's face while wrapping his arm around one's shoulder was known as the "Johnson treatment." Ben Bradlee, the editor of the *Washington Post*, after a private meeting with the six-foot-three president, quipped, "You really felt that a St. Bernard had licked your face for an hour."[26] George Reedy, the White House press secretary, said that the treatment was "like standing under Niagara Falls."[27] The day after the White House meeting, Udall sent LBJ a memorandum listing Interior's top six initiatives for 1964: creating a Land and Water Conservation Fund; passing the Wilderness bill; developing a Youth Conservation Corps; creating Assateague (Maryland-Virginia) and Fire Island (New York) national seashores; establishing sea water conversion plants; and building a complex of dams on the lower Colorado River.[28] For the sake of the environment, Udall planned to collaborate closely with the Department of Health, Education, and Welfare (HEW) on air- and water-pollution-related issues.

On December 17, Congress paid tribute to the deceased Kennedy by passing the first Clean Air Act, a significant breakthrough which authorized the federal government to regulate stationary sources of air pollution. It also increased research on the negative health consequences of air pollution. The act, the first federal legislation to actually control air pollution as opposed to merely studying it, established the principle that where an air quality problem is interstate in character, the federal government can prosecute polluters. The US Public Health Service was charged with funding research on techniques of monitoring and controlling air pollution from stationary sources. The big shortcoming of the Clean Air Act, however, was that it didn't regulate air pollution generated by motor vehicles.[29]

Even though the Clean Air Act of 1963 became law when LBJ was president, the credit for its passage belonged to Kennedy, who had been outspoken about combating air pollution since the mid-1950s. The legislative victory on Capitol Hill largely belonged to Edmund Muskie, the first chair of the Senate Subcommittee on Air and Water

Pollution. To back up his rhetoric with data, he had his staff monitor such pollutants as ozone particulate matter, carbon monoxide, nitrogen oxides, sulfur dioxide, and lead. He went so far as to issue a policy statement about the "war" on air pollution in major cities. Following the passage of the 1963 act, his efforts increased, as he sought even stronger regulations.

The Clean Air Act of 1963 encouraged new anti-pollution advocacy groups. That winter, Carolyn Konheim, a young mother living on New York City's Upper West Side, noticed black soot all over her son's snowsuit. She teamed up with her neighbor Hazel Henderson, a pioneering environmental justice activist, to create Citizens for Clean Air. They aimed to demystify urban air pollution and explain to everyday citizens that their health was at stake. As Konheim and Henderson saw it, both Democrats and Republicans had been negligent in mandating emissions standards for industries and utilities. They wanted to mobilize the public to elect politicians who would prioritize clean air. When Republican congressman John Lindsay ran for mayor of New York in 1965, he adopted Konheim and Henderson's platform as his own. His campaign poster read: "Dirty air. Dirty lungs. Dirty laundry. He'll do something about air pollution."[30] With Citizens for Clean Air watching closely, Lindsay, who won the election, started his tenure by setting bold limits on sulfur emissions from fuel oil, and the sulfur dioxide content of New York City air was reduced dramatically. Lindsay also introduced an economic element to regulation, presaging the more modern system of pollution offsets. Within a couple of years, he asked Mrs. Konheim to join the staff of the city's Department of Air Resources, another Lindsay innovation.[31]

New York City wasn't alone; smog menaced dozens of other cities as well. Congress had amended the Air Pollution Control Act of 1955 to finance a study to be undertaken by the US surgeon general to investigate the health effects caused by exhaust from automobiles using leaded gasoline.[32] With around 80 million cars on American roads, air quality was getting worse by the day. The link between smog and car emissions was undeniable. Because the federal government wasn't authorized by law to control automobile emissions, California had set standards on its own behalf. With by far the most automobiles of any state, California was in a position to dictate to automakers, which it did. It mandated that all new cars sold in the state after 1966 be equipped with catalytic converters, devices that returned unburned gases to the combustion chamber. The phrase "California emissions" took on importance in the auto industry as a standard that had to be met through new engineering. Some companies built cars especially for the California market, some built all their cars with, in the vernacular, "California emissions." Consumer activists such as Ralph Nader started studying the public safety and health responsibilities of the Big Three automakers. Publicly, Detroit railed against any sort of regulation. But in actuality, they readied designs for cleaner cars of tomorrow.

Among the most common of the stationary sources of pollution were trash incinerators, or what Barry Commoner called "dioxin-producing factories."[33] Since the first "crematory" had been constructed in the United States in 1885, burning waste had seemed cost-effective and reliable. What was a better disinfectant than fire, which left ash

Automobile congestion plagued America in the 1960s. The Great Society made stamping out air pollution a major priority.

residue? But that disposal method was called out in the 1960s as triggering respiratory illness. Plants around the United States were shut down. Garbage burning became either illegal or tightly controlled because it emitted myriad types of pollution (especially problematic were gases that contained heavy metals and dioxins). Black soot from new incinerators contributed mightily to the curse of acid rain.[34]

III

At Johnson's first cabinet meeting, Udall relished the new president's storytelling about Texas Rangers and Alamo martyrs. "He is really in the saddle now," Udall wrote in his journal as Christmas 1963 neared. "Tough-minded, insistent on results, combines carrot and stick with rare skill of a great Majority Leader in the White House."[35] Still, he wondered whether LBJ was a genuine reformer such as FDR or a visionless power broker like Speaker of the House Sam Rayburn of Texas, who had, in Udall's words, "never climbed high enough up the hill to see the broader landscape."[36] In the coming years, one way Udall tried to ingratiate himself with LBJ was to invite him on sports outings. For example, he asked LBJ to join him, White House aide Bill Moyers, and Supreme Court justices Earl Warren and Arthur Goldberg on the fifty-yard line of Washington Redskins home football games. Even though LBJ declined the "boys' club" offer, he must have appreciated being asked.

The relationship between Udall and Johnson developed well in early 1964. At another cabinet meeting, Johnson praised a speech Udall had recently delivered in New York City, in which he had advised the crowd, "Know America First" before holidaying abroad to Europe, Mexico, or the Caribbean. To Johnson, "Know America First" was a smart slogan that would stoke the tourist economies of every state. When Udall suggested that the United States help developing nations in Asia and Africa establish their own national park systems, Johnson heartily agreed. With the weight of the United States' involvement in Vietnam on his shoulders, the president was optimistic that "environmental diplomacy" might bring goodwill to his Cold War foreign policy.[37] Pleased with Udall's western demeanor and conservation-driven work ethic, LBJ affectionately dubbed Udall "the Keeper of Natural Beauty."[38]

One of Udall's smartest moves had been hiring George Hartzog to replace Conrad Wirth as the head of the National Park Service, the change taking place in January 1964. Ever since their Missouri Ozarks river trip, Udall had seen Hartzog as an alter ego of sorts. Both men had grown up in a rural area, becoming horsemen at an early age. Whereas Udall had been a Mormon missionary for two years, Hartzog was a South Carolina minister by training. Both thought Wirth's Mission 66 effort to build more roads in national parks and jack up industrial tourism to be somewhat misguided. The gregarious Hartzog soon became a popular schmoozer on Capitol Hill with members of both parties. Stegner found Hartzog to be "one of the toughest, savviest, and most effective bureau chiefs who ever operated in that political alligator hole" of Washington, DC.[39] Both Udall and Hartzog also strove to racially integrate NPS. "The command came down from the White House from Lyndon Johnson," Udall recalled. "You know, diversify, integrate, give women opportunities, have opportunities for qualified blacks and help train them. That was part of Johnson's civil rights program."[40]

Lady Bird Johnson wasn't sure how best to assume the First Lady role suddenly thrust upon her by the event in Dallas. She didn't want to imitate Jacqueline Kennedy, but staying out of the public eye didn't appeal to her, either. Instead, she would feel her way into the role. Although she had lived in Washington since 1934, when she married Lyndon, then a congressional aide, it had never occurred to her to embrace urban beautification as a priority. Nor was the new "environmentalism" a draw. She had never cared about Douglas's C&O Canal campaign or Carson's findings about DDT. Gardening in her backyard, however, was a favorite hobby. Over the years, she had planted a weeping cherry, a pink dogwood, and a crab apple tree; she was so proud to have done so that she joked to her husband that her epitaph should be "She planted three trees!" Only during her East Texas childhood, when she had explored mysterious Caddo Lake, searching for wild violets and feeling "the crush of pine needles underfoot, the wind whispering overhead," had she felt like a naturalist.[41]

As Mrs. Johnson searched for a fresh policy initiative that could become hers, children's education seemed a compelling choice. She became the first honorary chair of Head Start, a part of LBJ's War on Poverty that prepared preschoolers in low-income families for the first-grade classroom. "Our hope was to rescue the next generation— there were thousands of children who became dropouts in the third and fourth grade," she said.[42] To promote the idea, she visited under-funded schools in New Jersey and Texas with film crews in tow. Over time, though, she realized that education was Lyndon's domain. She was more interested in the natural environments where she lived: the capital's Tidal Basin and the lazy-river parks along the Blanco, Guada-lupe, Rio Grande, and Pedernales in Texas. She was especially drawn to the beautification of America's roads and highways, so many of which were eyesores after decades of commercial abuse. It was a cause that spoke to her growing sense of conservation and historic preserva-tion as necessary themes in modern America.[43]

Those themes were under attack in the ongoing construction of the Cross Bronx Expressway, which activists fumed was hurting neighborhoods and parks in the densely populated New York City borough. In its time, the Cross Bronx Expressway was the costliest highway ever built in the United States.[44] Starting in 1948, the federal government had funded the project to provide drivers with a direct route from Westchester County across the Bronx to the George Wash-ington Bridge. Unfortunately, the route chosen cut directly through Crotona Park and displaced thousands of residents, mostly of color. If the chosen pathway had been a white middle-class stronghold, an alternative route most assuredly would have been found. The build-ing of the expressway was a high-profile example of environmental racism at work.[45]

The Cross Bronx was the dream of Robert Moses, whom Robert Caro, his biographer, called *The Power Broker* in the title of his book. Throughout the 1950s, Moses was the chairman of the city's Com-mittee on Slum Clearance. "Moses was notoriously fond of bulldozers, and ever anxious to clear away 'slums' and replace them with new buildings," the historian Kenneth T. Jackson recalled.[46] Moses was also a bigot who considered Blacks and Puerto Ricans dispensable. He was unable to sympathize with people living in squalor with poor san-itation services and schools. Treating poor people as deplorables, he behaved with sweeping and efficient authoritarianism, securing secret rights of way years in advance of using them. When a 1964 conference sponsored by the New York Academy of Sciences produced data about how ill workers were getting from asbestos when tearing down buildings to make way for the Cross Bronx, Moses scoffed that the findings were hocus-pocus science.[47] Caro reported, "A thick layer of gritty soot made the very air feel dirty. . . . Where once apartment buildings or private homes had stood were now hills of rubble, decorated with ripped-open bags of rotting garbage that had been flung atop them."[48]

If any highway being built in the mid-1960s could be considered anti-beautification, it was Moses's monstrous Cross Bronx Expressway. "If this trend continues, New York will wake up some fine day to find that we have sacrificed a major part of our parks to the speed mania," one New York resident, L. O. Rothchild, complained in the *New York Times*.[49] In addition to the destruction of neighborhoods and asbestos contamination, the highway funneled millions of cars and trucks through the borough every month, spewing exhaust.

The Cross Bronx Expressway problem, as Lady Bird saw in the early 1960s, was happening all across the country. Neighborhoods and rural ecosystems were being destroyed to make way for Eisenhower's Interstate Highway System and its offshoots, laying lines of poured cement from coast to coast without regard for historic or natural preservation. Though expressways benefited myriad aspects of modern life, she feared that too little thought was being given to the corresponding damage to civic life resulting from their destruction of green spaces and historic buildings.

By the time LBJ assumed the presidency, resistance to ugly freeways had erupted into a coast-to-coast grassroots movement supported by preservationists of all ages. An anti-freeway protest in San Francisco in May 1964 drew more than two hundred thousand people to Golden Gate Park. In Louisiana, the Vieux Carré Riverfront Expressway project stoked widespread dissent, referred to then as the "Second Battle of New Orleans."[50] Bostonians raged against projects such as the Southwest Expressway and the Inner Belt through Cambridge and Somerville. Lewis Mumford, the venerable sociologist, became the bane of the highway lobby, pleading with citizens to "forget the damned motor car and build the cities for lovers and friends."[51]

Joining the growing national movement against poorly planned highways, Lyndon and Lady Bird Johnson soon engaged executives of corporations, neighborhood associations, and government officials to join a new American beautification campaign. The First Lady had found her cause. "I began to think that in the White House I might now have the means to repay something on the debt I owed nature for the enrichment provided from my childhood onward," she wrote. "And since hometown in the next few years was still to be Washington, D.C., where better to start than in 'the nation's front yard?'"[52]

CHAPTER 20

The Great Society

RACHEL CARSON AND
HOWARD ZAHNISER'S LEGACIES

Following the death of Rachel Carson, a new generation of environmental lawyers was born. Here are Dr. Robert Riseborough (*left*) and attorney Victor J. Yannacone at one of the many DDT legal hearings they participated in.

I

On April 14, 1964, Rachel Carson died at her home in Silver Spring, Maryland. She was fifty-six. The cancer had attacked her liver, and nothing could be done to arrest its spread.[1] According to friends, she often compared dying with a free-flowing river running into the sea—the opposite of Frost's "West-Running Brook" and a peaceful vision that perhaps made her passing blissful. Until her last day, she communed with the sparrows and robins that congregated outside her bedroom window. As she requested, her remains were cremated. Her funeral service was held at National Cathedral in Washington, DC. Stewart Udall and Abraham Ribicoff were among the pallbearers. Her brother, Robert, buried some of her ashes beside her mother's grave in Springdale, Pennsylvania. Dorothy Freeman spread the rest along the ocean shore at Southport Island, Maine.[2] Carson's adopted son, Roger Christie, who was her grandnephew, would devote his adult life to keeping her flame alive.

Death didn't end the impact that Carson had worked so hard to achieve; *Silent Spring* had spawned a revolution. Although she rejected being painted as a crusader, her literary finesse had driven terms such as *ecology*, *interdependence*, and *balance of nature* into American parlance.[3] And like her beloved monarch butterflies, eels, and shorebirds, the powerful ideas in her landmark book migrated all around the world. Not only had she sounded the alarm about environmental degradation in *Silent Spring*, but her original, well-crafted ocean science observations in *Under the Sea-Wind*, *The Sea Around Us*, and *The Edge of the Sea* had helped launch global efforts in seashore preservation. Before Carson, the concept of balance of nature was seen by the public, if it was considered at all, as something distant, perhaps as the provenance of Indigenous people and bohemian seers such as Henry Beston and Henry David Thoreau. Her legacy lay in tying ecology and public health together and pushing them into the stream of postwar life.[4] "Henry David Thoreau is credited with making the nature essay a literary form," Carol B. Gartner, one of Carson's biographers, wrote. "Rachel Carson has done the same for the science book."[5]

Because Carson wasn't a professor, having spent much of her adult life working as a government bureaucrat, her gifts as a teacher have been given short shrift. But she taught the world that the effects of synthetic chemicals were insidious; that poisons enter the food chain, growing more dangerous as they become more concentrated; that DDT can and does change chromosomes; that pollution lingers long after its source has been removed; and that chemicals contaminate groundwater.[6] It was Carson, not Kennedy or Johnson, who taught Americans to wake up and cherish the earthly paradise that remains. But like Kennedy, she offered us a challenge. In the last chapter of *Silent Spring*, "The Other Road," she wrote these words:

We stand now where two roads diverge. But unlike the roads in Robert Frost's familiar poem, they are not equally fair. The road we have long been traveling is deceptively easy, a smooth super-highway on which we progress with great speed, but at its end lies disaster. The other fork of the road—the one "less traveled by"—offers our last, our only chance to reach a destination that assures the preservation of our earth.[7]

Carson's 1956 essay for *Woman's Home Companion*, "Help Your Child to Wonder," was reprinted in 1965, the year after her death, as a book, *The Sense of Wonder*, with added photography by Charles Pratt. In that work, Carson urged parents to teach their children to experience the "lasting pleasures of contact with the natural world," which are accessible "to anyone who will place himself under the influence of earth, sea and sky and their amazing life."[8]

With Udall pushing for the designation, the Rachel Carson National Wildlife Refuge along Maine's southern coast was established in 1966; her work for the Nature Conservancy was memorialized two years later, when the group preserved as the Rachel Carson Salt Pond Preserve a Maine tidal pool on the shores of Muscongus Bay, where she had studied cockweed, blue mussels, and green crabs when the tide receded. "What about the Nature Conservancy?" she famously asked. "It is the only group I know which is doing something practical about actually preserving areas."[9]

In 1980, President Jimmy Carter posthumously awarded Carson the Presidential Medal of Freedom. And in 2004, the National Audubon Society's Women in Conservation initiative, led by Allison Whipple Rockefeller, began presenting the annual Rachel Carson Award to women who dedicated themselves to global environmentalism; among the recipients were former secretary of the interior Sally Jewell; administrator of the EPA, Gina McCarthy; and the oceanographer Sylvia Earle.[10] Around the world, more than a dozen important awards are given in Carson's name. Two major, authoritative biographies—by Linda Lear and William Souder—have been published to widespread critical acclaim. In 1975, Carson's birthplace and home in Springdale, Pennsylvania—renamed the Rachel Carson Homestead—became a site on the National Register of Historic Places, and a nonprofit organization, the Rachel Carson Homestead Association, was established to manage the property. In 1991, the Department of Interior designated Carson's Silver Spring, Maryland, house as a National Historic Landmark. Before her death, Carson formed an organization to carry out her work. The Rachel Carson Council, opened in 1965, continues her legacy of environmental advocacy to this day.[11]

In early 1964, Carson had joined the board of Defenders of Wildlife. Its philosophy matched her own. "It is my belief that man will never be at peace with his own kind until he has recognized the Schweitzerian ethic that embraced decent consideration for all creatures—a true reverence for life," she wrote.[12] After her death, the group created the Rachel Carson Memorial Wildlife Education Fund, which enabled its magazine, *Defenders*, to publish educational supplements aimed at stopping the indiscriminate destruction of wildlife by the government's predator control programs.[13]

Just how correct Carson was in awakening the world to the peril of DDT became sadly evident in the ensuing years. In the Antarctic, where DDT has never been sprayed directly, penguins began showing traces of it in their bloodstreams. In Alaska, caribou absorbed and retained it in their soft tissue. The Food and Drug Administration tested and then seized over 25,000 pounds of coho salmon caught in Lake Michigan in 1969 after finding them to contain 19 parts per million of DDT and almost 0.3 parts per million of the insecticide dieldrin—concentrations considered perilous by the FDA and World Health Organization.[14]

The signs of Carson's revolution were appearing that spring of 1964 in many places: the quest for clean energy by some electric power companies; the growing awareness in the oil and automobile industries of the need to develop cleaner internal combustion engines; the cleanup of the noxious lower Mississippi River; the public understanding that insect management in agriculture shouldn't simultaneously ravage the rest of nature; and a widening recognition of the need to respect endangered species. There was also a movement afoot to renew recycling of metals, glass, and other consumer resources for the sake of the environment.

Curbs on pollution did not begin in the United States only in response to *Silent Spring*. Shortly after World War II, as noted, the US government passed the Federal Insecticide, Fungicide, and Rodenticide Act in 1947, next the Water Pollution Control Act in 1948, and then the Air Pollution Control Act in 1955. Each law was lacking in substantial ways. However, with the publication of *Silent Spring*, the pace of the environmental movement accelerated exponentially, especially on Capitol Hill. Just days before Carson died, Senator Abraham Ribicoff called for new subcommittee hearings over the Mississippi fish kill. Carson, who had learned of the news, told a friend she was "delighted" by Ribicoff's initiative.[15] Udall insisted that pernicious compounds embedded in the environment needed to be removed and chemicals such as DDT had to eventually be banned altogether.[16]

Senator Church defended the research reported in *Silent Spring* for the rest of his life. Fearing uncontrolled technological expansion, he railed against the federally sanctioned dispersal of 600 million pounds annually of pesticides, herbicides, fungicides, rodenticides, and fumigants into American landscapes. Gaylord Nelson introduced legislation to create a National Commission on Pesticides. "Through this massive, often unregulated use of highly toxic pesticides, every corner of the earth has been contaminated," he warned.[17]

Carson and *Silent Spring* launched what became known in environmental history as "the DDT wars." Biologist Charles Wurster had yet to read *Silent Spring* when he took a job as a researcher at Dartmouth College in Hanover, New Hampshire. During the evenings of April 15–18, 1963, DDT was sprayed there on the elm trees in the neighborhood surrounding the campus. Hoping to save the shade trees from the beetles that cause Dutch elm disease, local officials had conducted DDT spraying for several years.

One day, a student brought a very sick robin to the laboratory of Wurster, a postdoctoral researcher who was especially interested in the quite obvious decrease in bird populations in town. At his request, radio announcers and newspapers in New Hampshire and Vermont urged citizens to take any sick or dying birds they encountered to Wurster's laboratory. Within a few weeks, 151 dead birds, many of them American robins, were taken there. Dozens had been collected right on the Dartmouth campus by students. "Many of the birds from

Hanover exhibited tremors and convulsions before death, the typical symptoms of DDT poisoning," Wurster recalled. "DDT destabilizes nerves, causing them to fire spontaneously without control, so muscles twitch uncontrollably. We dissected the birds and they were analyzed for DDT content in a laboratory in Wisconsin."[18]

Wurster and three of his colleagues stayed on the case, using new computer technology to prove that food chain contamination by DDT had killed 70 percent of the robins in Hanover. They then wrote important articles for the peer-reviewed journals *Science* and *Ecology*. When Hanover city officials were confronted with the Dartmouth research, they agreed not to spray DDT anymore, switching to the far less dangerous insecticide methoxychlor. "We had spent two years stopping DDT in one town, but hundreds of towns were still using it," Wurster said later.[19]

In September 1965, Wurster moved to Long Island to become an assistant professor of biological sciences at the State University of New York at Stony Brook. He remained fully committed to his battle against DDT. Ultimately, his clients were birds—not only robins but also predatory and fish-eating birds such as the American bald eagle, osprey, peregrine falcon, and brown pelican. His goal was to save them from possible extinction in the wild by suing chemical companies to stop making certain products. He joined the Brookhaven (New York) Town Natural Resources Committee, a conservation-minded group of educators, scientists, and others. The committee members were inspired by *Silent Spring* and concerned about wildlife preservation, wetlands destruction, and pollution from duck farms and agricultural runoff. One member, Dennis Puleston, the director of the technical information division at Brookhaven National Laboratory, was aware of problems with the osprey population on eastern Long Island; for example, the shells of the birds' eggs became so thin that they cracked prematurely and the chicks inside died. Another member, Arthur Cooley, was a Cornell-trained science teacher at a local high school. A natural motivator, he taught his students to observe the marine ecosystem that surrounded Long Island. On field trips the students saw copious signs of pesticide poisoning exhibited by a range of creatures.

At a meeting of the Brookhaven Committee in April 1966, Wurster was tasked with writing a letter to the *Long Island Press* criticizing Suffolk County's Mosquito Control Commission, which had been spraying the north shore marshes with DDT since 1947. "If the decline in Long Island wildlife is to be checked, the use of DDT for mosquito control must be curtailed," he wrote the newspaper on May 6. Taking direct aim at the commissioner in charge, he stated, "It is alarming to think that the dissemination of such toxic materials is in the hands of a person who thinks they are harmless."[20]

When local leaders failed to respond, the group took the same approach that Marjorie Spock had taken seven years before: a major lawsuit. Spock's suit didn't win legally, but it did raise the issue of DDT in a public setting. More commonly, the preferred tools to fight threats such as the widespread administration of synthetic chemicals were political action and protests. The term *environmental law* did not yet exist. In lawsuits, courts were concerned primarily with tangible property losses; they considered the harm to wildlife caused by pollution or chemical spraying to be out of their jurisdiction.[21] Victor Yannacone, Jr., a lawyer in nearby Patchogue, was undeterred. "Sue the Bastards" became his motto.[22]

Yannacone's wife, Carol, had long enjoyed Yaphank Lake, a nineteen-acre body of water popular for recreation. The Suffolk County Mosquito Control Commission had started a DDT program to keep insects from breeding on the lake; soon dead fish were floating on its surface. When Carol Yannacone heard that the commission was planning to dump an even greater amount of DDT in Yaphank Lake in 1966, she asked her husband to use his legal skills to stop them.[23] "Rachel must have taken secret satisfaction," Udall wrote in an article about the *Silent Spring* author, "in the thought that the response to *Silent Spring* demonstrated that once the bird-watchers and wildlife lovers of the world united, they could prevail in many of the 'thousand little battles' for conservation across the land."[24]

Another person who was enamored by Carson was Sylvia Reade Earle, who dedicated herself to saving oceans. "Since I was a child, when I first heard about Rachel Carson and read her books, half the coral reefs in the world have disappeared or are in a state of terrible decline," she lamented in 2010. "Great loss of mangroves, sea grass and kelp forests around the world have declined and 90% of the fish we like to consume have disappeared. We have taken too much out of the sea not really respecting that there are limits to what we can extract and put into the ocean."[25]

Earle was born in Gibbstown, New Jersey, on August 30, 1935. When she was twelve, her parents moved to Clearwater, Florida, on the Gulf of Mexico. Rolling up her pant legs to wade into tidal pools, she was soon studying the diverse wildlife of the gulf, from manatees to sharks to turtles. An excellent student, Earle won a scholarship to Florida State University, followed by graduate work at Duke. Enamored of *The Sea Around Us*, she dreamed of diving deep to study coral reefs and kelp forests. The North Star in her career was the prospect of ocean exploration.[26]

In 1964, Earle was approaching thirty years old, struggling to balance the demands of raising her children with her desire to conduct underwater research around the world. That year, 1964, she left home for six weeks to join a National Science Foundation expedition to the Indian Ocean. As she recalled, "They didn't really think it through, that I would be the only woman with seventy men for six weeks at sea. . . . In fact I had a wonderful time. It was one of the best experiences for me as a scientist, seeing a different part of the planet that I had never imagined that I could explore before."[27]

In 1966, she received her PhD from Duke University. Her dissertation, *Phaeophyta of the Eastern Gulf of Mexico*, brought her attention from the oceanographic community. It was the first time that a marine scientist had made such a thorough study of that type of aquatic plant life: the division of algae characterized by brown pigmentation. Two years later, in 1968, she made a voyage to a hundred feet in the waters of the Bahamas in the submersible *Deep Diver*, discovering dunes on the seafloor. She was four months pregnant at the time. Her dream was to become an aquanaut, actually living underwater for days at a time, as astronauts did in space—and eventually she did. Whenever the opportunity for federal funding was appropriated for such a program, she was determined to be part of the team.

Earle built a sterling reputation as a scientist, with more than a hundred papers and several trade books to her credit. Still working in her mid-eighties, she became the nation's leading oceanographer and certainly the best known. "Because of the years I've had," she said, "observing as a witness to the nature of change on an unprecedented scale, I personally am driven to try to share the view and encourage people to connect to nature to understand that our lives are *totally* linked to the natural world."[28] Carson's sea trilogy inspired a generation of young people such as Earle to major in marine biology.

Because *Silent Spring* was so revolutionary, it has overshadowed Carson's classic sea trilogy.[29] Carson departed from the path of her earlier works to write it because a life-and-death crisis had come, one that needed the command of her pitch-perfect instincts as an environmental journalist. She admitted freely that she hadn't provided much original scientific research in her four books. But she had written them by conducting extensive firsthand observation along seashores and under the microscope. "I consider my contributions to scientific fact far less important than my attempts to awaken an emotional response to the world of nature," she said.[30]

Carl Safina, an oceanographer and author of numerous books on marine ecosystems, recalled that Carson's *Silent Spring* literally delivered a shock to his system when he was fourteen years old and growing up in Long Island. "I almost threw up," he said. "I got physically ill when I learned that ospreys and peregrine falcons weren't raising chicks because of what people were spraying on bugs at their farms and lawns. This was the first time I learned that humans could impact the environment with chemicals."[31]

Another environmentalist awakened by Carson at a young age was Al Gore, Jr., who would later become a US senator from Tennessee (1985–1993) and then vice president under Bill Clinton (1993–2001). During his senior year at Harvard in 1969, he enrolled in a class taught by Roger Revelle, Carson's oceanographer friend and authority on global warming, who had once been one of Udall's advisers. When Gore read *Silent Spring*, it "came as a cry in the wilderness, a deeply felt, thoroughly researched, and brilliantly written argument that changed the course of history." Later in life, he wrote the introduction to the 1994 edition of *Silent Spring*. There he speculated, "Without this book, the environmental movement might have been long delayed or never have developed at all."[32] Gore, who shared the Nobel Peace Prize in 2007 for educating the world about climate change, would credit Carson in numerous sold-out speeches as being the "shaft of light" who had opened his eyes to the interconnection of human beings and the natural world.

Individuals were motivated and groups were formed, but the federal government was slow to act. In May 1963, *Use of Pesticides: A Report of the President's Science Advisory Committee* had been greeted as a confirmation of *Silent Spring*'s conclusions. JFK directed his cabinet members to work on its recommendations.[33] In the fall of 1964, Udall, at last, instructed all agencies in the Interior Department that used pesticides on federal land to avoid those made with chlorinated hydrocarbons, notably endrin, dieldrin, chlordane, and DDT.

Progress on environmental issues raised by *Silent Spring* would accelerate. This made some critics resent the book even more. Blaming Carson for the "environmental hysteria" of the 1960s and early 1970s, one critic asserted that *Silent Spring* had been responsible for "the worst crime of the century."[34] It has been contended that bans on DDT in the developing world have caused millions of unnecessary deaths from mosquito-borne malaria. In 2007, *New York Times* columnist John Tierney attacked *Silent Spring* as a "hodgepodge of science and junk science."[35] Thomas Sowell, a conservative writer at the Hoover Institution, asserted, "There has not been a mass murderer executed in the last half-century who has been responsible for as many deaths of human beings as the sainted Rachel Carson."[36]

A book by two historians of science—Naomi Oreskes at Harvard and Erik M. Conway, at the Jet Propulsion Laboratory at Caltech—countered all charges from the American political far right. Their exposé *Merchants of Doubt: How a Handful of Scientists Obscured the Truth on Issues from Tobacco Smoke to Global Warming*, published in 2010, made a vigorous defense of *Silent Spring*. "Carson's argument was that any war on nature was one that we were bound to lose," Oreskes and Conway wrote. "Fish and birds were killed, while fast-evolving insects came back stronger than ever. Finally—and perhaps above all—it was a mistake to assume that the only harms that counted were *physical*. Even if DDT caused not one human death, humans would be affected: our world would be impoverished if spring came and no birds sang."[37]

II

On May 22, 1964, President Johnson delivered the finest and most historically significant speech of his career before eighty thousand people packed inside the University of Michigan football stadium in Ann Arbor. As he strode to the podium, with the Wolverines marching band playing "Hail to the Victors," Johnson delivered a powerful commencement address, galvanizing the nation. Scouring for a phrase in the vein of "New Deal" and "New Frontier," he settled on "the Great Society," first coined by White House speechwriter Dick Goodwin. The Great Society was to Johnson "a place where man can renew contact with nature," and so much more.[38]

The Civil Rights Act and the War on Poverty, Medicare and Medicaid legislation, immigration reform, assistance to public schools, voting rights, and the pending Wilderness Act—that long list made up only a part of the Great Society agenda. First and foremost, though, Johnson sought "an end to poverty and racial injustice."[39] And he didn't forget conservation, giving it a higher priority than any president since Theodore Roosevelt. The Great Society was a place where Americans could explore dense forests and swim in clean lakes. Essentially embracing Aldo Leopold's land ethic, Johnson called for community-driven conservation to take root in America's cities, villages, farms, and classrooms.

As Johnson pointed out, the very core characteristics of the nation were in danger: "The water we drink, the food we eat, the very air that we breathe, are threatened with pollution. Our parks are overcrowded, our seashores overburdened," he orated. "Green fields and dense forests are disappearing. A few years ago we were greatly concerned about the Ugly American. Today we must act to prevent an ugly America. For once the battle is lost, once our natural splendor is destroyed, it can never be recaptured. And once man can no longer walk with beauty or wonder at nature his spirit will wither and his sustenance be wasted."[40]

Conservation was raised to the level of a critical national issue that afternoon in Ann Arbor. To a greater degree than Kennedy, Johnson believed that that was where it belonged. "Johnson did care about conservation and beautification a lot," Goodwin recalled. "Not as much as Bobby Kennedy by a long shot. But as much as Jack, for sure. . . ."[41] Both presidents cared, but Johnson was determined to act more boldly. Instead of a crisis-by-crisis approach, Johnson wanted to sound a New Conservation alarm to ring throughout the federal, state, and local governments. Grassroots protoenvironmental organizations were encouraged by Johnson's promise to "establish working groups to prepare a series of White House conferences and meetings—on the cities, on natural beauty, on the quality of education, and on other emerging challenges."[42] In his Great Society, ecologists and biologists were elevated as indispensable first responders rushing to save nothing less than the future of the United States

Coretta Scott King was disappointed that Johnson's Great Society speech hadn't called for a nuclear-free world. She felt that the president missed an opportunity to connect nuclear disarmament to domestic programs, but she immediately embraced his notion of urban parks in economically disadvantaged and blighted neighborhoods. Her heroic efforts to ban nuclear weapons via SANE and WSP continued. She strategized regularly with Norman Cousins and Dagmar Wilson, as well as with fellow civil rights leaders Andrew Young and Jesse Jackson, who joined her anti-nuclear crusade.

Coinciding with the Great Society speech was Johnson's press to establish Fire Island National Seashore: a new oceanfront park within easy reach of millions of people in the New York metropolitan area. The bill was sent to the House in March 1964, but stalled when the Republicans filibustered civil rights legislation. On April 11, however, the House Interior Subcommittee finally approved the Fire Island National Seashore bill and sent it to the full committee.

Also passed by the Senate and destined for House approval in the fall of 1964 was the Ozark National Scenic Riverways bill. The troika of Johnson, Udall, and Frank Church hoped that the bill's passage was the forerunner to the present National Wild and Scenic Rivers system. Building on the recommendations of a joint

Agriculture and Interior study of the nation's waterways begun in 1963, Johnson's ardent support for wild-river measures soon elevated him to the exclusive "green" presidents' club along with Theodore and Franklin Roosevelt.[43] There is no doubt that in the broad issue of conservation, as well as in many others, Johnson would have been honored to be compared with FDR, his idol. Both were adamant about taking action in all directions at once, giving LBJ a hard reputation among aides and others for his unending, nagging, usually unpleasant impatience. FDR, of course, was never called "unpleasant," seemingly by anyone. But he exuded the impatience for government to *do something* that LBJ emulated. Fortunately for the environmental movement, that inclination on Johnson's part coincided with its coming-of-age in the mid-1960s.

Johnson had taken notice that anti-dam efforts were becoming a national rage with progressive conservationists. The Bureau of Reclamation was becoming a bogeyman for California, Arizona, and Colorado with liberal voters in ranching or utility businesses. The Bureau of Reclamation had initiated a plan in 1955 to build four dams on the fast-running Trinity River in California near the Oregon border. Starting upstream, it built the first two, the enormous Trinity Dam and smaller Lewiston Dam; they were completed in 1963. The next proposed site was on the Hoopa Valley Tribe Reservation. The Hoopa and nearby Yurok people depended on the Trinity for salmon, which was already being affected by the lower water levels and the inability of the fish to lay eggs upstream, past the dams. The Hoopa protested legally and morally, pointing out that without their river behaving like a river, they would starve. The Bureau of Reclamation canceled plans for the last two dams on the Trinity, but the environmental damage already done made the once-scenic river an ongoing disaster zone.[44] In consultation with Dr. Ira Gabrielson of the Wildlife Management Institute, Johnson realized that the days of big dams were ending and a new era of obligation to Indian nations and outdoor recreationists had arrived.

Just weeks after the Great Society speech, Johnson was on hand to personally transfer 271 acres from the Army to the state of New Jersey, doubling the size of Sandy Hook State Park, across Raritan Bay from New York City. Johnson used the symbolism of Sandy Hook, with the Manhattan skyline in the background, to speak eloquently about the need for open spaces near big cities. He pointed out that more than 8 million people lived within a twenty-five-mile radius of the newly expanded park. Poor people in the eastern United States couldn't be expected to take long road trips to Yellowstone and the Grand Canyon, he reasoned. The same was true to some extent for city dwellers all over the country. Therefore, urban wild lands were necessary for Americans of all stripes to enjoy without spending a fortune. It rankled him that at Sandy Hook State Park in 1963, before the federal land addition, more than 240,000 visitors had been turned away to avoid overcrowding. In the ceremony there, he expanded on his New Conservation message, mindful of such failures when he had made his aspirational Ann Arbor speech. He insisted:

> No nation anywhere at any time has had to face the multitude or magnitude of unprecedented problems that we must meet and master in maintaining the quality of life in metropolitan America. . . .
> We must clean up our air, clean up our rivers, clean up our streams, and open up the land for our people if we are to preserve the heritage and the healthiness of our American life.[45]

Among all the Great Society conservation initiatives in the summer of 1964, Johnson pressed hardest for Redwood National Park,[46] the last stands of redwoods, growing on a slender strip along the Pacific, a living link to the age of dinosaurs. The Forest Service predicted that by 1980, most redwood forests in private hands would be harvested by lumber companies such as Pacific Lumber Company and North Pacific Lumber Company. As it was, 95 percent of old-growth *Sequoia sempervirens* were gone. "Fractions of trees that were sprouting their leaves when Hannibal crossed the Alps now serve as durable shingles and siding, patio tables, and other amenities of our ephemeral culture," wrote Oliver Jensen, the editor of *American Heritage* magazine.[47] If federal protection wasn't enacted, *Sequoia sempervirens* would likely become extinct. Udall was anxious to help Johnson keep that tragedy from happening. "The most logical new proposal from our point of view would be the *endorsement by the*

President of a Redwoods National Park," he wrote Johnson on May 27, 1964. "'Save the Redwoods' is still the best understood and most honorable rallying cry of the conservation movement! . . . At the moment there is nothing I can think of that would do more to arouse interest and support for the President's 'save the countryside' program than this!"[48]

David Brower, who always had the right book under his arm and at the right time, presented the Johnsons with the big-format 1963 book *The Last Redwoods* by Philip Hyde and François Leydet, for which the interior secretary had written the foreword. The photographs of untouched stands of trees in the book were ethereal and full of foggy mystery. But there were also grim pictures of clear-cut redwood groves and gravely polluted California landscapes. The book was an effective lobbying tool for three conservation groups determined to establish a major Redwood National Park along the California coast north of Eureka: the Save the Redwoods League, the National Geographic Society, and the Sierra Club. The fact that President Johnson, in league with Laurance Rockefeller, was just as determined boded well for the preservation effort. And yet, even with all that support, it would become a melee, actually exposing that even well-intentioned groups could fall to squabbling over acreage allotments.

"*Sequoia sempervirens* is one of the glories of our continent, and the groves that still stand are one of the overpowering pageants of nature," Udall wrote in the foreword. "But we have learned, in managing the isolated enclaves of Redwoods now under state or federal protection, that no guardianship is sure unless a unit—a whole watershed—is placed under a single management plan that is ecologically sound."[49]

On June 25, 1964, Johnson was briefed on a new National Geographic Society study on how best to preserve the California coast redwoods. The report, *The Redwoods: A National Opportunity for Conservation and Alternatives for Action*, was publicly disseminated in the autumn of 1964. It stated emphatically that redwoods were an irreplaceable part of America's heritage.[50] That would seem to be an obvious fact. Many Americans, however, had an almost religious conviction that everything on Earth was there for mankind to exploit.

"Johnson often didn't read conservation reports," Udall recalled. "He was too busy with other work. But he latched onto the Redwoods.

He asked me if a Redwood National Park could be folded into the Wilderness bill. I told him it couldn't be. If he tried, the bill would blow up in his face. I couldn't get him to talk about Fire Island or North Cascades. Yet even when we met informally, he raised three issues with me: dirty water, Redwood National Park, and how Lady Bird was tracking with beautification."[51]

III

For three days in August 1964, Udall accompanied Lady Bird on a "Land and People" tour of Montana, Wyoming, and Utah. Along with the quick-thinking Liz Carpenter, Lady Bird's press secretary, they toured the Crow Indian Reservation and Custer Battlefield National Monument (now called the Little Bighorn Battlefield National Monument) in Montana, took a float trip down the Snake River, and explored the Wasatch Mountains in northern Utah. In the Utah ski town of Park City, Mrs. Johnson told gatherers, "The interesting thing about Park City is that you have developed the one resource that is least exhaustible, your natural beauty."[52] At the dedication of Flaming Gorge Dam on the Green River in Utah, approximately thirty-two miles downstream from the Wyoming border, she pleaded with the crowd, "Enjoy the beauty of your hills and protect it for your children."[53]

Udall had known Mrs. Johnson for years, but on the trip, working together for the first time, they developed a relaxed rapport. Udall recognized that the First Lady had "an instinctive feeling for the beauty of the country."[54] According to him, it was while she was floating down the Snake River in a rubber raft that she crystallized the idea of leading a national "beautification" effort. "Mrs. Johnson was extremely effective on her 'Land and People' tour of the West last week," Udall reported to LBJ. "Her interest in the out-of-doors and her concept of conservation have a national impact that is the finest kind of presidential politics."[55]

Like Lester Flatt and Earl Scruggs in bluegrass and Abbott and Costello in comedy, the Stew–Lady Bird tandem came to stand for something: the magnificence of national parks in a mid-1960s America torn asunder by Vietnam and civil rights marches. Virtually every

Lady Bird Johnson at Jackson Lake, in Wyoming, on the front porch of the recreation hall at the Rockefeller ranch, 1965.

National Park Service superintendent lobbied hard for the two to visit.[56] On a propeller plane flying across the Rockies, the First Lady chatted excitedly with Udall about new national parks such as Redwood and championing national river designations. "She just talked and talked about what the [Interior] department was trying to do, our programs; she was asking questions," Udall recalled. "She wanted to know everything."

The same summer as the "Land and People" tour, the Fire Island National Seashore campaign for congressional approval hit full stride, with President Johnson as the key ally. The goal was to turn the eastern end of the thirty-two-mile-long barrier island into a national park. The paragons of that conservation battle were many, but at the forefront was Maurice Barbash, a residential developer in a profession typically unmoved by grassroots land protection advocacy. But Barbash was known as the builder who loved trees because he had a penchant for clustering houses to leave room for them. A development of his on Fire Island, Dunewood, was like that, and he had kept one of the houses for himself.[57] To stop the highway that the state's chief planner, Robert Moses, was planning to build from one end of the island to the other, Barbash looked for a way to fight *for* something, rather than just *against* the highway.

In 1962, Barbash found it in a federal document that recommended, as one of many other things, making Fire Island a national seashore park. When he and a small group that included his brother-in-law made their push, few local officials or papers opposed the highway. That would change. Barbash spoke at Lake Placid, New York, where thirty-five national conservation and scientific societies discussed the Fire Island National Seashore proposal. As a result, the New York State Conservation Council endorsed it and the cause received widespread national attention. The *New York Times* and *Newsday* both called for a national seashore designation. Fire Island preservation was tracking.

In June 1964, the Senate Subcommittee on National Parks toured Fire Island. The delegation included several senators from western states: Alan Bible, a Democrat from Nevada; Leonard B. Jordan, a Republican from Idaho; Ernest Gruening, a Democrat from Alaska; and Milward Simpson, a Republican from Wyoming.[58] The trip was a success, because the lawmakers saw with their own eyes the gorgeous barrier island beaches, the wetland habitats, and a rare, primeval maritime holly forest recessed behind a double sand dune system. Late in July, the Fire Island National Seashore bill was unanimously approved by the House Interior Subcommittee. In August, after the House voted for the bill, at the urging of Senator Kenneth Keating, a Republican from New York, the Senate accepted the House version to avoid going into conference. On August 20, at the height of the summer tourist season, the bill passed Congress.

On September 11, 1964, Johnson signed the bill into law, handing out pens to Fire Island activists as souvenirs of the success of their David-versus-Goliath efforts.[59] The establishment of the Fire Island National Seashore was a huge triumph for Great Society conservation. Citizen environmentalists had said no to Robert Moses and won.

Johnson's vigorous support of the Fire Island National Seashore Act thrilled the Sierra Club, the Nature Conservancy, and other such nonprofits. Both of New York's Republican senators, Jacob Javits and

Keating, were taking the lead politically in promoting environmental protection laws in their state. While Democrats tended to be more interested in the new environmentalism, the movement didn't belong to one party or another.

By mid-1964, it was clear that Johnson was going to be a reliable conservation president, even a heroic one. He used to brag that he was "a creation of Congress." But as the historian Hal K. Rothman wrote, he was "truly a product of the Hill Country."[60] Indeed, LBJ spent a quarter of his presidency at his ranch in Stonewall on the Pedernales River, forcing the press corps to follow him there for big announcements. The president made seventy-four trips on Air Force One flying from Washington, DC, to Stonewall in just over four years.[61]

There was nothing citified about Johnson. At a Texas-style barbecue he threw for two hundred press correspondents, he simply ordered that a rostrum be placed on top of a hay bale so he could field reporters' questions—that's how country folk lived.[62] It was the mindset of a farmer, or a man who liked to get things done, or both. Being at the ranch, tending to cattle, organizing catfish fries, chatting with kinfolk, and captaining motorboat rides on the man-made Lake Travis rejuvenated Johnson's politics-weary soul, especially as the conflict in Vietnam heated up. The charming beauty of the Texas countryside and the swimmable Pedernales rooted him there. At heart, he was a country boy. In nearby Johnson City, he relished the small dairies, country stores, grain silos, and sawmills. By "small," Johnson meant a size of business that would not destroy the appearance, the health, and the quiet of the countryside.

As the Wilderness bill moved toward President Johnson's desk, the press compared him favorably to the United States' other cowboy president, Theodore Roosevelt. It was as if the White House in 1964 had become a frontier outpost, with Stetsons hanging on the Fish Room hat rack. LBJ had positioned himself as the Democratic leader in the roadless land protection movement in the American West, pulling leadership on that issue away from Hubert Humphrey. Projecting the president's rough-and-ready persona, the White House fed the media with reports of LBJ's outdoor skills from wood chopping to rabbit hunting to pecan gathering. "He's a crack shot, a fine horseman, a no-nonsense angler, and he knows every blade of the hill country of Texas," said *Argosy* magazine in October 1964.[63]

There was a basic truth to the tales. Ranching was undoubtedly at the core of Johnson's identity, and he gloried in western cowboy mythology. His press secretary, George Reedy, observed that the president's "self-painted portrait of a cattleman tending his herds, however, was difficult to accept with a straight face. He *did* know something about cattle, but he 'tended' them from a Lincoln Continental with a chest full of ice and a case of scotch and soda in the back seat."[64]

It was no secret that in many ways, Johnson strove to be the next FDR. The Great Society, in a sense, was the New Deal's bookend. Though that is true from a progressive policy perspective, he also wanted to emulate Theodore Roosevelt, who had once owned two ranches in North Dakota. LBJ sought to tap into his predecessor's mystique of carrying a big stick in foreign policy and saving wilderness areas from coast to coast. Johnson didn't think of himself as a southerner: he was a *westerner*. Like Theodore Roosevelt, he adhered to the code of Owen Wister's novel *The Virginian*. On his fifty-eighth birthday, in 1966, Johnson explained to the press gathered in Stonewall that "a good many" of his liberal programs, including those focused on education, poverty, and youth, had been advocated by Theodore Roosevelt: "I am a great admirer of the contributions he made to the Nation, as you can see reflected in our conservation program. . . . The President of that period and the President of today have a good many things in common—and we are getting some of them done now."[65]

After President Kennedy died in Dallas, President Johnson seized upon the Wilderness bill as his own initiative. He wholeheartedly agreed with the ORRRC's third report, "Wilderness and Preservation," which recommended immediate legislation to establish a National Wilderness Preservation System with Interior Department and USDA involvement. LBJ boasted that every American owned a stake in the national parks and national forests, which in total area came to six times the size of California.[66] When the Wilderness bill passed the Senate on April 9, 1963, by a 73–12 vote, Vice President Johnson had been pleased. As president, Johnson promised to twist Aspinall's arm and get the Wilderness bill signed into law before the 1964 presidential election. What he didn't know was that Aspinall's attitude had changed. But the president was a handshaker and he cultivated strong bonds with hook-and-bullet conservationists in the National Wildlife Federation, the Boone and Crockett Club, and the Izaak Walton League. Those conservation nonprofits mattered to him politically and needed to be on the side of the Wilderness bill.

It must be added that LBJ enjoyed listening to rustic outdoor tales told by frontier types and natural world raconteurs. Long before, the Texas folklorist J. Frank Dobie, a friend, had taught him the ancient art of fireside storytelling: always say that the bear and the trout weighed more than the scales showed. Senator Clinton Anderson, the chairman of the Senate Interior Committee, regaled Johnson with stories about hiking the New Mexico wilderness. Deep forest and streamside stories about bears spun by Bill Douglas easily captivated LBJ's attention, as they did that of his wide-eyed children. Lady Bird would say that her husband instinctively connected his childhood to the natural world; that is, to the quiet in the days before electricity.

"When I was growing up, the land itself was life," the president reflected,

> And when the day seemed particularly harsh and bitter, the land was always there just as nature had left it—wild, rugged, beautiful, and changing, always changing.
>
> And really, how do you measure the excitement and the happiness that comes to a boy from the old swimming hole in the happy days of yore, when I used to lean above it; the old sycamore, the baiting of a hook that is tossed into the stream to catch a wily fish, or looking at a graceful deer that leaps with hardly a quiver over a rock fence that was put down by some settler a hundred years or more ago?
>
> How do you really put a value on the view of the night that is caught in a boy's eyes while he is stretched out in the thick grass watching a million stars that we never see in these crowded cities, breathing the sounds of the night and the birds and the pure, fresh air while in his ears are the crickets and the wind?[67]

In conversations with Clinton Anderson and Frank Church, LBJ advised that the trick of marketing the new Wilderness classification was to make it seem as though it had come bottom up from the grass roots.

In the first half of 1964, Representative Aspinall held public hearings on the Wilderness bill in Denver, Las Vegas, and Olympia, Washington. It was an above-board exercise by Aspinall on behalf of the House Subcommittee on National Parks, Forests, and Public Lands to find out what western voters thought of the wilderness classification. It pleased him that the Wilderness Society commended him for fairness: "Excellent hearings, fairly and expeditiously conducted, once again demonstrated the great public support for wilderness preservation, and also revealed a new willingness on the part of former opponents to accept recently revised bills."[68] The hearings assured Aspinall that wilderness preservation had massive bipartisan public support. He had already had a change of heart, feeling since the Kennedy assassination that he wanted to make the Wilderness Act as a last favor to his late friend. When LBJ's congressional liaison came calling to put pressure on him regarding the Wilderness bill, Aspinall shooed him away: "You'll get it, the president will sign it. I just wish President Kennedy was here to sign it, but it will go to President Johnson."[69]

The last of Aspinall's hearings were held in Washington, DC, where Wilderness Society executive director Howard Zahniser testified. Over the years the two adversaries had become friends. Mutual respect cemented their closeness. That spring of 1964, the reclamationists and preservationists managed the art of compromise. After eight years of negotiation, they were at long last united, and an understanding of what constituted wilderness had been agreed. At the final House hearing on the bill, Zahniser testified, "It may seem presumptuous for men and women who live only 40, 50, 60, 70, or 80 years, to dare to undertake a program for perpetuity, but that surely is our challenge."[70]

Aspinall had finagled a concession from Zahniser to allow mineral exploration in the protected lands until 1984; out of desperation, Zahniser had concurred. The other key concession Aspinall won was a system of affirmative action under which Congress, not the White

House or federal land agencies, would recommend, debate, and vote for new wilderness designations. That laborious process made the acreage of wilderness designated after 1964 a slower affair because Johnson—or future environmentally inclined presidents—couldn't just say, "I so declare it," in a TR-like use of executive authority. Yet in the end, the final agreement benefited the preservation movement that Zahniser had led, for, as the historian Kevin R. Marsh explained forty-some years later, Aspinall's amendment had encouraged a "grass-roots citizen campaign" to protect more public lands under the Wilderness Act.[71]

On May 5, 1964, Zahniser died at fifty-eight of a heart attack at his home in Hyattsville, Maryland.[72] The emotional Aspinall mourned the Wilderness Society leader, his adversary, on the floor of Congress and promised that the bill his deceased friend had written would become law. On July 30, the House voted 374–1 for the bill.[73] Udall reported to Johnson that after "nearly six years of agonizing frustration," history was being made in the wilderness preservation world "due in part to the leadership efforts of Rep. Aspinall."[74] But Clinton Anderson gave the bulk of the credit where it belonged—to Zahniser—saying "He was steadfast in his devotion to the concept of wilderness as a fundamental part of American culture and tradition."[75]

Zahniser's passing shocked preservationists everywhere. Even his close wilderness allies Brower and Olson, who had plotted land boundaries with him, hadn't realized that he had only one lung and heart problems. Nobody, it seems, in the Senate or Congress knew just how terribly ill he had been. In one of his last essays, he reflected on how fragile Earth was, how important it was to save primeval landscapes. "The wilderness in America is still living," he wrote. "It is different to us, but it has not vanished. It no longer seems to contain mankind, as outer space does now seem to. We ourselves seem rather to contain the wilderness. The dear Mother who gave us origin, nurtured us—chastised us, too—lives with us now, lives, as we might say, through our care, but still lives."[76]

The Wilderness Act of 1964

President Lyndon Johnson congratulates Alice Zahniser after signing the Wilderness Act on September 3, 1964, at a White House ceremony.

I

Having been weaned on western cowboy movies and shoot-'em-up Zane Gray novels, Johnson embraced the American frontier ethos in the Space Age decade of Mercury, Gemini, and Apollo astronauts in a decidedly retro way, as a buckskin-clad flight back in time. Visitors to the White House often commented that Johnson's hands had a rocklike heaviness and solidity to them, noticeable even when he kept them clasped upon his knees. The minute disputes about canoe waters and alluvial gluts that Sigurd Olson was engaged in didn't interest him an iota. Neither did the transcendental thinking of Henry David Thoreau. The president viewed conservation from primarily a well-loved parcel of prairie perspective, though the ethereal beauty of an Ansel Adams photo of Yosemite touched him deeply, as great art does. By contrast, the web-of-life complexity embedded in Eugene Odum's textbook *The Fundamentals of Ecology* bored him. Quite simply, like Aldo Leopold, the president believed in proper citizen land stewardship and that vanishing wilderness demanded immediate preservation because it was "the very stuff that America is made of."[1]

At 10:30 a.m. on September 3, 1964, President Johnson signed the Zahniser-Aspinall compromise Wilderness bill into law in a Rose Garden ceremony. It was a prized moment. Though his face was seamed and weather-beaten, he seemed younger than his fifty-six years that day because of his ear-to-ear near-constant smile. Classifying what constituted *wilderness* had been a daunting endeavor since the 1950s, particularly because the postwar economy constantly needed more lumber, minerals, and other natural resources. But Johnson had been able to provide future generations with a gift of immeasurable value— pristine wilderness linked to the distant past of Lewis and Clark. Officially, the new law was "An Act to Establish a National Wilderness Preservation System for the Permanent Good of the Whole People, and for Other Purposes." But colloquially, it endures as the Wilderness Act. "It was a compromise bill," the historian Doug Scott explained in *The Enduring Wilderness*. "Aspinall voted for it—but the fundamental architecture as envisioned by Zahniser was intact."[2]

Zahniser's poetic definition of wilderness as "an area where the earth and its community of life are untrammeled by man, where man himself is a visitor who does not remain" survived the plethora of draft versions.[3] One aspect of the word "untrammeled" needed an asterisk, however: hunting and fishing were to be allowed in wilderness areas. Orphaned cougar cubs deep in the Bighorn Mountain wilderness probably wouldn't agree that humans had "left no trace."

Weeks before the Rose Garden ceremony, Hubert Humphrey, Johnson's running mate in the upcoming election, had written to the White House chief of staff, Lawrence O'Brien. He asked that LBJ honor Zahniser as the "father of the Wilderness bill." He also wanted credit for himself, writing, "I would like to add that yours truly, Hubert Humphrey, was the original sponsor, and turned it over to Clint Anderson here in the Senate when he was chairman of the Interior Committee. I have been the main cosponsor ever since the early 1950s. This bill was highly controversial in the beginning and I had to take most of the brickbats at the early stages. Like many other things, some of the controversy is dissipated in proper time and the bill passes. In the meantime, a few of us come out with the battle scars."[4] One more scar awaited Humphrey. While the senator wasn't at the signing ceremony, owing to a prior commitment, Johnson overlooked even mentioning his name.

The legislation's fiercest backers (minus Humphrey) elbowed to be in the official photographs that September 3. Front and center were Frank Church, who had been the bill's Senate floor manager; Clinton Anderson, who had shepherded it through the Senate; and John Saylor, the bill's strongest supporter in the House. Everybody present was in a good mood. Orville Freeman joked that he already missed the old Forest Service land designations "primitive areas," "canoe areas," "limited areas," "scenic areas," "wild areas," and "wilderness areas" (the distinction being based primarily on size).[5] As a courtesy gesture, the widows, Alice Zahniser and Margaret Murie, stood next to LBJ as he signed the bill, calling it an "important milestone" to raucous "hear, hear"s. The president, with visible warmth, gave the women the first signing pens.[6]

The Wilderness Act activated the legal mechanism to preserve large roadless acreage in perpetuity. But whereas the Antiquities Act of 1906 had allowed an executive decree for the establishment of national monuments, the Wilderness Act assigned to Congress power over wilderness designations and alterations of wilderness boundaries. In the power struggle between the White House and Congress, the Wilderness Act was a boon to the latter, as Aspinall had intended. Regardless, 9.1 million acres of roadless public lands had been preserved from bulldozers and chainsaws forever. As a bonus, the act ensured that more national forest acreage would be added in the future.

At the signing, Johnson's opening remarks were oddly lacking: "This is a very happy and historic occasion for all who love the great American outdoors, and that, needless to say, includes me." But that banality was all right, for those who were there could see the joy and reverence in the president's demeanor.[7] He went on to praise the 88th Congress as extraordinary conservation leaders for the ages. With the presidential election only two months away, Johnson promised that the wilderness classification would be followed by other ambitious New Conservation measures if he were elected. He said, "Action has been taken to keep our air pure and our water safe and our food free from pesticides; to protect our wildlife; to conserve our precious water resources. No single Congress in my memory has done so much to keep America as a good and wholesome and beautiful place to live."[8]

The poet Gary Snyder, in *The Practice of the Wild*, ably described wilderness areas by saying "These are the shrines saved from all the land that was once known and lived on by the original people, the little bits left as they were, the last little places where intrinsic nature totally wails, blooms, nests, glints away."[9] Brochures and signage prepared by the National Forest Service soon warned visitors entering a designated Wilderness that they weren't protected in case of accident, illness, or wildlife attack. Frontier mythology of the Jim Bridger variety was alive and well in twentieth-century America. The act promoted the kind of self-reliance and rugged individualism that thinkers from Ralph Waldo Emerson to Wallace Stegner had spoken to in their writings.

All fifty-four of the inaugural wilderness areas in 1964 were on Forest Service lands, largely because many of them were ready and relatively easy to designate. The smallest, Great Gulf in New Hampshire, near Mount Washington, just barely topped the minimum allowed acreage of five thousand. The biggest two were Bob Marshall Wilderness in Montana and Selway-Bitterroot in Idaho, which were almost tied at just shy of 1.1 million acres each. Others in the first group were Boundary Waters Canoe Area (Minnesota), Bridger (Wyoming), the Gila (New Mexico), Minarets (California), and Shining Rock (North Carolina). Only two of the fifty-four, Great Gulf and Shining Rock, were east of the Mississippi. The Wilderness Act mandated a ten-year review process, charging federal agencies to study and report to Congress on potential future wilderness areas.

The National Parks System was unhappy because the act gave the Forest Service most of the wilderness designations.[10] At the time, the NPS was struggling to handle the surging visitation to units. Yellowstone, for example, had to accommodate more than two million in 1964 alone. By contrast, the new wilderness areas in national forests, with a couple of exceptions, never topped one hundred thousand visitors a year. For that reason, the NPS was still promoting Conrad Wirth's Mission 66 infrastructure program, building amenities at national parks such as roads, cabins, telephone lines, and campsites.

For conservation nonprofits such as the Nature Conservancy, the National Wildlife Federation, and locally-focused groups such as the Montana Wilderness Association, the Wilderness Act was first and foremost a fundraising windfall. Nothing generates donor dollars quite like a major legislative victory. For the groups, though, the act also led to more work. The Sierra Club, led by Brower and his assistant, Michael McCloskey, and the new Wilderness Society began the arduous task of grappling with the wilderness laws. They also held public hearings, as mandated by the Wilderness Act, to resolve land rights disputes.[11] Clinton Anderson had once called wilderness protection a "national objective," and he used the Rose Garden event to praise "green" nonprofits.[12] "Those who roam the Wilderness years from now will marvel at its quiet and its charm," he said, "but they may never remember what organizations like the Sierra Club and The Wilderness Society did to preserve it." Anderson then added: "Those of us who do, will remain forever grateful."[13]

In 1995, when Orville Freeman was long retired from politics, he wrote to Bill Worf, a cofounder of Wilderness Watch, an organization that monitors waters in the National Wilderness Preservation System. Freeman was remembering that historic morning in September 1964 at the Rose Garden when the act was signed. He called it the "golden moment" of his entire eight years as secretary of agriculture. Yet the aura of triumph led to challenges. In great detail, Freeman documented how, from 1964 onward, special-interest groups tried to dismantle tenets of the Wilderness Act. "Everyone from ranchers to oil companies, to law enforcement agencies and the Defense Department asserted that we had to be 'reasonable or practical' in our interpretation and implementation of the bill," he wrote. "It was a singular challenge to attempt to write good, clear policy in the midst of such strong pressure. (Kind of like trying to get a fire lit in the middle of a Rocky Mountain blizzard, eh?)"[14]

Freeman was right to call the signing a "golden moment" that would loom large in US environmental history. The fact that the ORRRC's definition of wilderness was so widely embraced boded well for new classifications in coming decades. The *New York Times* editorialized that the Wilderness Act was a "landmark," and as time has gone by, historians have agreed. Roderick Frazier Nash, a professor of history at the University of California, Santa Barbara, in his seminal book *Wilderness and the American Mind*, published in 1967, deemed the law "one of the most remarkable intellectual revolutions in the history of human thought about the land."[15]

Hubert Humphrey said later in a speech in Chicago that "there is in every American, I think, something of the old Daniel Boone— who, when he could see the smoke from another chimney, felt himself too crowded and moved further out into the wilderness."[16] It was an intriguing thought: the frontier was not closed, so long as there was wilderness. Inadvertently, though, by evoking Boone, Humphrey drove home the point that the wilderness movement was led by well-to-do white men and aristocratic explorers. It was utterly lacking in socioeconomic, racial, ethnic, and geographic diversity and not much better as far as diversity by gender was concerned. Even when the Wilderness Act turned fifty years old in 2014, people of color were still underrepresented in the movement.

The positive press coverage that Lyndon Johnson received for signing the Wilderness Act reassured him that New Conservation was indeed a winning political issue, along with civil rights, education, and health care. It thrilled him that the act was politically bipartisan and had the backing of both the General Federation of Women's Clubs and the AFL-CIO. Johnson wasn't concerned about winning the conservationist vote in the general election. His Great Society conservation record towered over Barry Goldwater, his GOP challenger for the White House. Goldwater had opposed the Wilderness Act and favored dams wherever the Bureau of Reclamation wanted to build them in the Far West.[17] As it turned out, conservation issues didn't play a significant role in the 1964 presidential election.

News of the Wilderness Act's signing reverberated through local conservation groups as if it were the Fourth of July. Dreams of future wilderness areas exploded, after being pent up for years. Nowhere was the celebration more enthusiastic than in the Pacific Northwest. In Seattle, Portland, and Spokane, the documentary *The Wind in the Wilderness* aired, highlighting the newly designated Glacier Peak

Wilderness. The next ambition for the nonprofit group the Mountaineers was a national park designation in the Cascade range. At the time, it had just published an exhibit-format book, *The North Cascades*.[18] Momentum seemed to be on the side of North Cascades National Park, particularly if the Johnson-Humphrey ticket were to win that November.

Rachel Carson's editor at Houghton Mifflin, Paul Brooks, believed that September 3 was a seminal turning point. "Some of the groups who plan to push nature around, to strip-mine the southern Appalachians, to desecrate the remaining Indiana dunes, to kill the Wilderness bill, by delaying maneuvers in Congress, might do well to watch the bird-watchers, and listen," he wrote. "We are no longer in a minority; we have the votes and we intend to be heard."[19]

Perhaps somewhat naively, the Wilderness Act was ballyhooed as the fulfillment of Henry David Thoreau's dictum, "In Wildness is the preservation of the world"; of Theodore Roosevelt's splendid speech at the Grand Canyon, when he had said to leave the landscape "unmarred"; and, for that matter, of Lyndon Johnson's declaration that wilderness was "a source of America's greatness," "a part of America's soul."[20] It pleased Johnson immensely that the act would inspire a new generation of day hikers to join nonprofits such as the Sierra Club and the Wilderness Society, whose core membership was Democratic. Inasmuch as the Wilderness Act was indeed an opening act for the Great Society's New Conservation, it also highlighted the need for endangered species preservation, for, as the naturalist Lois Crisler had said, "Wilderness without animals is mere scenery."[21]

II

One could say that the Wilderness Act was the triumph of committed Minnesota outdoor recreationists, agitating for the preservation of the Boundary Waters. The flying wedge of Minnesota conservationists included six national figures: Sigurd Olson, Orville Freeman, Eric Sevareid, Olaus Murie, Ernest Oberholtzer, and Hubert Humphrey.[22] William O. Douglas, who was born in Minnesota, was an honorary member of the gang.

Olson, in particular, had been toiling his entire adult life for the preservation of the Boundary Waters. In the 1950s, he described the glories of the north country region, where "travel is still by pack and canoe over the ancient trails of the Indians and voyageurs."[23] Located along 150 miles of the US-Canada border in the northern third of Superior National Forest, the ecosystem constituted over 1,100 lakes and 1,500 miles of canoe routes. Bright blue lakes are not the only asset of the Boundary Waters Canoe Area (BWCA) as it was officially named. According to Paul Weiblen, an associate professor at the University of Minnesota, "The rocks of the BWCA are part of the earliest record of earth history on the North American continent. They provide evidence of the earliest weathering and erosion of rocks. They also record the first period of uplift which enlarged what is believed to be the primeval core of the continent, the vast Canadian Shield area."[24]

As with the other new wilderness areas, the process of creating the new entity engendered a tangle of local predilections and pressure to accept legal loopholes. "Wilderness should be sacred and quiet," Olson wrote, "just as the Indians felt in designating certain places as spirit and where no one talked."[25] Boundary Waters, unfortunately, has never been quiet even under its Wilderness designation. Olson and other conservationists managed to get the 800,000-acre tract across the finish line. But in doing so, it allowed legal exemptions to appease local economic boosters, who didn't want a newfangled wilderness in their backyard. They wanted fast-dollar commerce and a recreation area where motorboating was the tourist draw. In an uncomfortable compromise, the act provided a list of exemptions for Boundary Waters that included snowmobiling, motorboating, and logging.

In spite of powerful local developers opposing him in his hometown of Ely, Olson was unfairly criticized by national preservationists for embracing the motorboat clause. When Olson arrived at the Blackhawk Hotel in Davenport, Iowa, for a convention, he found terse telegrams and disquieting letters waiting for him. One read, "I always thought canoe men represented a breed of quiet strength and determination."

Insulted and rattled, Olson fired back: "For forty-two years I have fought threat after threat, any one of which could have destroyed the wilderness canoe country. We have not always achieved our objectives one hundred percent but have usually won substantially. In the process I hope I have learned some wisdom and strategy on how best to get things done. Never before, in all this time, has anyone ever felt my hand and convictions must be bolstered. My record can stand."[26]

The fight went on (and still does to this day), but in 1978, at the instigation of President Jimmy Carter, all the exemptions but the one for motorboating ended. Under the Wilderness designation, the Boundary Waters became more popular, not less. Soon BWCA took the honor of being the "most visited Wilderness Area in the United States," bringing in $100 million in economic activity for the region.[27]

Another area established in the first round by the act was the Bob Marshall Wilderness area in western Montana. Decades in the making, the million-acre Wilderness, considered the most primitive backcountry in the lower forty-eight, was first-class grizzly bear habitat. Bighorn sheep and mountain goats also ranged among its scattered boulders and cliff-face plateaus. Big river valleys extended deep into the national forests straddling the Continental Divide, from which the new wilderness area was carved out. Named in honor of Robert Marshall—an avid mountain climber and scientific forester who cofounded the Wilderness Society in 1935—"the Bob" is considered a Rocky Mountain nirvana by outdoor survivalists. While training as a forester, Marshall had worked at the Northern Rocky Mountain Forest and Range Experiment Station in Missoula. His book *Arctic Wilderness*, published posthumously in 1956, became a classic text on exploration. The Bob Marshall Wilderness Complex grew to include two other Wilderness areas, as well: Scapegoat and Great Bear. The three total 1.5 million acres, encompassing 1,100 miles of trails.[28]

One of the two Wilderness areas that Johnson established in the East was Shining Rock in the Appalachians of North Carolina, which contained an old-growth deciduous forest along the Shining Creek Trail; the yellow birch, sweet birch, and northern red oak there were among the most mature in the United States.[29] The area's other main feature is an outcropping of white quartz. Shining Rock is also famous for its numerous springs and waterfalls; it is the headwaters of the Pigeon River, a tributary of the Tennessee.[30]

In part as a favor to Ansel Adams, David Brower had worked hard to get the Minarets (today's Ansel Adams Wilderness) designation included in the first class of wilderness areas in 1964. The Minarets was a sacred landscape for experienced mountain climbers. This high Sierra tableau spanned 231,533 acres within the Inyo National Forest and Devils Postpile National Monument. It is hemmed in by Yosemite National Park in the northwest and what became the John Muir Wilderness in 1964 in the south. Although the greater part of the Minarets Wilderness is a valley of boulders and jagged peaks, there are exquisite stands of pine, fir, hemlock, and aspen. Most of the trees are traversed by the upper reaches of the San Joaquin River, crashing and foaming over slick rocks.

Because Adams loomed so large in the world of landscape photography, the art of Philip Hyde has taken a back seat in wilderness preservation history. But it was Hyde's exquisite art photographs, shot in color, that became the primary illustrations of the pioneering Sierra Club exhibit-format series. With his wide smile, all-seasons hiking boots, and crates full of cameras, Hyde was a photographer's photographer. He didn't get a wilderness area named for him, as Adams did, but he should have. Through the 1960s, the Sierra Club and others used Hyde's photographs for the campaigns to save wilderness areas in the High Sierra, Wind River Range, Canyonlands, North Cascades, Big Sur, Kings Canyon, Sequoia, Denali, Tongass, Monument Valley Navajo Tribal Park, and Oregon's Cascades. *American Photo* magazine named Hyde's photograph *Cathedral in the Desert, Glen Canyon, Utah, 1964* one of the top one hundred photographs of the twentieth century. "People are ever hurrying over the increasing highways that penetrate lovely country and either lacerate it or pass it by unseen," Hyde wrote. "A mind at peace may be found in any individual or people who have kept touch with what the land is saying and who lack the benefits of instant dissemination of the human troubles that make news. After reading Gandhi, I see that what we need now is a peaceful environmental revolution. The Earth will survive, but will man survive on the Earth?"[31]

The Wilderness Act culminated decades of Bill Douglas's preservation activism against excessive logging roads and clear-cuts. Not surprisingly, he embraced the Wilderness Act as a supreme achievement of the nascent environmental movement. Douglas hardly paused to enjoy the victory. Concerned that 81 percent of water consumption in the United States was for irrigation and that wilderness rivers were suffering severe ecological damage owing to impoundment, fertilizer runoff, dredging, pollution, and rerouting, Douglas next championed a Wild and Scenic Rivers bill. Unquestionably, his two-book series, *My Wilderness*, helped the bill gain traction on Capitol Hill. As he lamented while enjoying a high meadow in the Cascades, "An emptiness in life comes with the destruction of wilderness; that a fullness of life follows when one comes on immediate terms with woods and peaks and meadows."[32] Michael McCloskey said in his memoir, *In the Thick of It: My Life in the Sierra Club*, that no activist devoted more energy than Douglas to stopping a road along the pristine Minam River in eastern Oregon: a relatively small problem to some, but not to the justice. Douglas also devoted enormous time and effort to gain the Wilderness designation for the Cougar Lakes area in Washington State.[33]

In 1996, the historian William Cronon questioned the importance of wilderness in the opening chapter of his seminal book *Uncommon Ground: Rethinking the Human Place in Nature*, an essay he titled "The Trouble with Wilderness; or, Getting Back to the Wrong Nature."[34] On a purist level, he didn't think that "untrammeled" land, void of human influence, existed anymore. Wilderness areas, after all, welcomed human visitors. Though people were supposed to come and go "without leaving a trace," backcountry hikers followed man-made trails, left a scent, caught fish, often shot animals with rifles, and inadvertently brought human dominance in other ways into the so-called wilderness (with airplanes overhead and acid rain on the grass). Because humans had contaminated Earth, he rejected the idea that wilderness could really exist now.

On a more philosophical level, Cronon thought that the notion of visiting a wilderness area that is separate in every way from one's daily life contradicts the fact that every person is a part of nature. Guided by his careful reading of Aldo Leopold, he argued that a "wilderness area" should instead be intrinsic in every American's routine, however insignificant that wilderness might seem to be in comparison with a million-acre government preserve named after Bob Marshall or Ansel Adams. He felt that by compartmentalizing developed areas and wilderness ones, people are almost encouraged to sacrifice the former for the latter. Polluting or abusing populated places might be considered tolerable, Cronon argued, as long as other tracts are held pristine. He was quick to point out that he valued protected wild lands and was glad to see nature given every chance to thrive, but he suggested that the wilderness classification was itself misleading, giving people weekend satisfaction in a reserve called "nature," rather than a true sense of being, every day and in all seasons, part of nature.[35]

The strongest counterargument to Cronon's rethinking of the human place in nature came from the naturalist and writer David Quammen. Admitting that Cronon was right about how tricky the word *wilderness* was to define, he thought LBJ's classification was noble. "My own view is that the sensible way of defining wilderness is not in the stark dualistic terms that Cronon hears and denounces, but relatively, as a matter of degree," he wrote in 1999. "Wilderness can be anyplace where human impact upon the landscape is small, reminding us therefore that we are too."[36]

III

Just as momentous as the Wilderness Act was the Land and Water Conservation Fund (LWCF) Act, which was signed by Johnson in the same September 3 ceremony.[37] For the first time, the Interior and Agriculture Departments were given serious congressional appropriations to purchase wildlife habitat and treasured landscapes as the Nature Conservancy had been doing. This bank account for nature was funded with tax revenues received by the US government from sales of surplus land, along with a tax on motorboat fuel and a portion of national park admission fees. Starting in 1968, most of the LWCF funding came from offshore oil leases. As much as $900 million per year was designated to support parks, trails, open spaces, wildlife sanctuaries, and outdoor recreation sites. "The Land and Water Conservation bill," Johnson explained, "assures our growing population that we will begin, as of this day, to acquire on a pay-as-you-go basis the outdoor recreation lands that tomorrow's Americans will require."[38] (In 2020, the LWCF fund was made permanent as part of the Great American Outdoors Act.)

LWCF provided reliable money for land acquisitions for the NPS, the National Wilderness Preservation System, and the National Forest System. Sometimes buying real estate wasn't an easy proposition. When necessary, deals were struck under a life-tenancy agreement that allowed landowners to sell property to the federal government but stay put for a set amount of time, often twenty-five years. With the two major pieces of legislation signed on September 3, Johnson worked to designate a Redwood National Park, with the media and lawmakers following his lead. Of all the acrimonious negotiations that dotted the history of 1960s preservation, the fight for California's coastal redwood stands was the most fraught.

The Sierra Club book that Brower gave Johnson for Christmas in 1964, *The Last Redwoods*, with its introduction by Udall, juxtaposed photographs of giant old-growth trees with images of rivers swollen with debris and eroding hills terraced for highway construction. Timber companies, especially Pacific Lumber Company, lambasted Brower for demonizing their industry and becoming an arm of Udall's radical Interior Department. Brower responded that "the club's publishing mission is to take sides—to be unashamed of good propaganda."[39]

Unfortunately, after a century of redwood harvesting in California, few surviving tree stands were purchasable in 1964. Looking at a mottled map of the surviving groves still in private hands was disheartening, as not much of the famed redwood kingdom still existed. The best stands available for a park were 140 acres in the Redwood Creek area. They were owned by the Arcata Redwood Company. This parcel was home to what were then considered the world's tallest trees. *National Geographic* had ballyhooed the discovery of the "Mount Everest of All Living Things," a tree 367.8 feet tall on Redwood Creek, in a cover story.[40] The National Geographic Society then resolved to buy the tract. Photos of the magnificent tree ran in newspapers and magazines around the world. The problem was that Arcata wasn't willing to sell the redwood stands at a fair price.

Environmentalists could see that the Johnson administration would have to bring more money and pressure to bear. The Sierra Club launched a public relations blitz saying that the US government needed to acquire the remnant redwood stands along Redwood Creek before those heirlooms became lawn furniture or hot tubs. Echoing that sentiment, Bill Douglas admitted that a new national park effort at Redwood Creek would be "costly" but said that in the end, future generations would view saving the acreage as a greater gift than astronauts wandering around a lifeless lunar landscape kicking up moon dust.[41]

Ever since Udall had publicly promoted his support for a Redwood National Park in 1963, it had been controversial. The Save the Redwoods League lobbied fiercely for the parcel at Mill Creek near Crescent City; the Sierra Club thought Redwood Creek, approximately forty miles to the south near Orick, was a far better grove of the big trees. The National Geographic Society, led primarily by its close institutional ties with the league, ponied up $64,000 so that the Interior Department, which was short on money, could finish the survey.

The plan to locate the new national park on Mill Creek, as the league preferred, had a powerful ally in Laurance Rockefeller. Beloved by Lyndon and Lady Bird Johnson and just about everybody else in official Washington, LR was a longtime supporter of the Save the Redwoods League. His father, John D. Rockefeller, Jr., had helped preserve the Bull Creek redwoods in the early 1930s. Following in his father's footsteps, LR not only favored the Mill Creek option for "its astonishing beauty" but also argued with the league that it would trigger less of a war with the lumber lobby and its employees in Del Norte County than would the larger park on Redwood Creek that the Brower-Douglas alliance insisted upon. Both LBJ and LR favored the Mill Creek plan because it represented "the art of the possible."[42]

Entering the Redwood battle was Walter Reuther of the UAW, who wanted an even larger park than Douglas and Brower. "The Redwoods to me, have always been a true inspiration as I am sure they have been to all the fortunate Americans who have held an opportunity to see them," Reuther wrote Johnson in early 1966. "The preservation of a significant stand of these magnificent trees will be a truly monumental step, in the implementation of the 'Great Society.' Several months ago I asked our Recreation Department to gather the facts regarding size and location of a Redwood National Park. After careful consideration of the results of this study, the UAW supports the creation of a Redwood National Park of some 90,000 acres as proposed by Congressman Jeffery Cohelan of California in his bill H.R. 11723. This proposal contains virgin redwood forests of unequaled magnitude, ecological conditions most advantageous to redwood preservation, outstanding panoramic views, a long ocean-front beach, wildlife concentrations of major size, as well as a number of wilderness watersheds."[43]

President Johnson wrote Reuther back that, come hell or high water, his administration would establish Redwood National Park. The president summoned Sharon Francis, who was then his conservation staffer, and said, "I need you to tell me how much I have to spend to buy these redwoods." Scoop Jackson, Wayne Aspinall, and his budget experts all had different numbers, and the president wanted to know the exact dollar figure and what was doable. After talking with Udall and Ed Crafts at the Forest Service, Francis flew out for a whirlwind tour, taking Ansel Adams and a California natural resource officer with her. What she saw was a frenzy of logging—trees falling, trucks hauling. When she returned to the White House two days later, she told the president and First Lady that only a fragment of California's coastal redwood forest remained in private hands. "We should save every single redwood tree still standing this afternoon," she recommended.[44] Udall fumed that the timber companies were engaged in what he called "spite cutting."[45] In the end, Congress adjourned in 1966 with the plans for Redwood National Park dying in committee.[46]

Just when it looked as though Redwood was stillborn, however, Lady Bird Johnson entered the picture. Instead of arguing for Mill Creek versus Redwood Creek, she said that compromise was the path to conservation victory. At all costs, Udall, whom she dubbed "an expert salesperson," would have to cut through the boundary disputes and between the timber companies and the federal government for the sake of America's heritage. It was time for the Interior Department to negotiate. The Save the Redwoods League and the Sierra Club would have to do the same, for the same higher purpose. "Seeing the redwoods was one of the most profound experiences I'll ever have," she recalled. "Those cathedral-tall great trees were some of the oldest plant life on this planet. Their ancestors were here in the time of dinosaurs, and those very ones were probably little bitty saplings in

Lyndon and Lady Bird Johnson at the Texas Ranch in 1963. It was important to the first couple to be perceived as westerners, not southerners.

the time of Christ. It just levels you, gives you a sense of man's place here on this planet. The majesty of it all!"[47]

Bill Douglas also wanted California's redwoods to be rescued. But with too many cooks there already, he turned his formidable energy to the North Cascades campaign. Walt Woodward, a columnist for the *Seattle Times*, had written a series of seventeen articles in twenty-three days on the idea of a North Cascades National Park, which inspired Douglas to check out the proposed park boundaries on his own. The bumper-to-bumper traffic on State Route 20 headed toward the mountains had him cursing aloud. He counted twenty-seven automobiles queued ahead of him to get into an area being proposed for the park. Douglas hated that Yellowstone and the Great Smoky Mountains had become congested, and he worried that a similarly dismal fate awaited North Cascades. Of course, his car was the twenty-eighth in line, but that didn't stop him from belittling the others. "Potbellied men, smoking black cigars, who never could climb a hundred feet," he complained of the tourist swarm in a foreword to *The Wild Cascades: Forgotten Parkland*, the eleventh in the Sierra Club battle book series.[48]

Douglas's public squawks about too many tourists invading the Cascades didn't, in the end, preclude his support for the new national park, mainly because he despised the Forest Service's "multiple-use" doctrine and knew that Woodward's plan was the best option available. When *Sunset* magazine published a cover story titled "Our Wilderness Alps," intimating that indeed the North Cascades should become a national park, Douglas disseminated it far and wide.[49]

It wasn't the NPS, per se, that Douglas objected to, but Conrad Wirth's old Mission 66 infrastructure program, which had allocated vast sums to build roads, visitor centers, and other facilities by 1966. The iconoclastic former park ranger Edward Abbey spoke for Douglas when he wrote that the "onslaught of the automobile" and the "indolent masses" who drove them were turning parks into "national parking lots."[50]

In his 1965 book *A Wilderness Bill of Rights*, Douglas urged the Johnson administration to establish "a new set of procedural rules designed to preserve wilderness against destruction by other uses."[51] He offered pragmatic solutions for such controversial issues as the fencing of public land, mining claims in wilderness areas, desecration of wild and scenic rivers with dams, and the draining of wetlands. Arguing for the "natural beauties, scenic trails, trout streams, and the like in defiance of interstate highways," he railed against the "Great God of the Dollar."[52] Pitting extraction industries against the right of citizens to enjoy clean air and clean water, he declared wilderness "a living library" and "research laboratory for the ecologist." It was as if he thought trees and wildlife should have standing in the courts—an idea that he would indeed publicly support in 1972.

Every American, Douglas argued, has "the right to put one's face in clear, pure water, to discover the wonders of sphagnum moss, and to hear the song of whippoorwills at dawn in a forest where the wilderness bowl is unbroken."[53] During speeches throughout the 1960s at college campuses, the justice would quote the poet Ogden Nash in protest of ugly billboards along the federal highways: "I think that I shall never see / A billboard lovely as a tree. / Indeed, unless the billboards fall / I'll never see a tree at all."[54] Believing that "ferment" was essential to the college experience, he was a willing and popular speaker on campuses. His appeal to young people through speeches and articles helped to make environmental activism a form of rebellion for the students of the 1960s.[55]

Denouncing the "glamour" created by statistics about the popularity of man-made reservoirs such as Lake Mead and Lake Powell, Douglas painted a bleak picture of what happens when there are drastic fluctuations in water levels on associated rivers, such as the Savage, a tributary of the Potomac, and the Youghiogheny, a tributary of the Monongahela: "Drawdowns usually start by July first and as the water recedes the trunks often emerge and dreary debris appears in the form of bottles, oil drums, tin cans, auto tires, boards, driftwood, and paper stuck in slimy mud. These drawdowns produce eyesores, making some spots unapproachable and producing the ugliest of all scars on the face of the earth of which man is capable."[56]

While offering ways to regulate DDT and other insecticides in *A Wilderness Bill of Rights*, Douglas praised the poet Carl Sandburg for understanding what is important about the United States. He even urged teachers to assign Sandburg's poem "Wilderness" to schoolchildren:

> There is an eagle in me and a mockingbird . . . and the eagle flies among the Rocky Mountains of my dreams and fights among the Sierra crags of what I want . . . and the mockingbird warbles in the early forenoon before the dew is gone, warbles in the underbrush of my Chattanoogas of hope, gushes over the blue Ozark foothills of my wishes—And I got the eagle and the mockingbird from the wilderness.[57]

Roughly coinciding with the Wilderness Act, Douglas had written the important article "America's Vanishing Wilderness," for *Ladies' Home Journal*, in which he championed the establishment of preservation-driven "Committees of Correspondence"—an idea he lifted from the Revolutionary War era. Those committees had networked to keep the revolutionaries abreast of British maneuvers and to work for the common cause of American independence. Alaska senator Ernest Gruening entered the article into the *Congressional Record*. In the spirit of 1776's committees, Douglas used his Supreme Court office as a kind of clearinghouse for grassroots whistleblowers and activists to send in field reports of environmental desecration occurring in a myriad of geographical locales. Douglas became, in the words of the legal historian M. Margaret McKeown, a one-man environmental "reference library," providing nature-loving citizens tips on how to win local NIMBY (Not In My Backyard) fights, where to find data, how to get a *pro bono* lawyer, and what conservation nonprofits to contact for a particular battle.[58]

The bottom line was that if American citizens wanted to save a river or forest from ecological desecration, and wrote to Justice Douglas, he would sometimes write back with instructions on how to build a "green" coalition to defeat pro-development forces. "All I can do," he wrote to Marian Laurie from Brookfield, Illinois, "is urge that every group get thoroughly organized and extremely vocal to make sure that the basic values are not sacrificed as our population expands and our highways multiply."[59] When Meyer Lefkowitz wrote to the justice about sewage treatment problems in Florida, the justice suggested that he build a local environmental coalition to blame soap detergents with carbon molecules that "don't break down" for the water contamination crisis at hand.[60] Very few Washington power players would have bothered responding to citizen's inquiries, day in and day out. Douglas, however, considered such grassroots environmental organizing time well spent.

CHAPTER 22

Ending the
Bulldozing of America

President Johnson hands Lady Bird Johnson the pen after signing the Highway Beautification Act on October 22, 1965.

I

With its steep vertical cliffs, boulder-covered mesas, and seemingly endless maze of canyons and plateaus, Canyonlands was truly the Colorado Plateau at its awe-inspiring wildest. "As far as the eye can see, the Canyonlands country of southeastern Utah presents an array of visual wonders," Udall enthused when *Western Gateways* magazine interviewed him about the site. "Over millions of years the mighty lashes of wind and water gouged canyons, stripped rock layers until they bled in a fury of color, eroded the land into startling pinnacles and arches, and sliced away at the plateaued surface."[1] Twelve thousand years ago, Paleo-Indians traveled through and communed along the Green River, which courses through Canyonlands; by 1964, ancient petroglyphs found in the national park were deemed priceless antiquities by the Interior Department. Two of the earliest Paleo-Indian sites in North America, in fact, were discovered within the proposed park's boundary.

Ever since Stewart Udall laid eyes on the Canyonlands area from an airplane window in 1961, he'd wanted it to become a national park. For years, he had worked in tandem with Bates Wilson, the superintendent of nearby Arches National Park, to bring photographers and reporters to experience the scenic splendor of Utah's backcountry. Even though most San Juan County officials, prospectors, miners, and cattlemen, as well as the energetic Bureau of Reclamation commissioner, Floyd Dominy, opposed making Canyonlands a national park, Udall forged ahead. He commissioned faculty members at the University of Utah to do an economic study, which predicted ecotourism benefits for Moab, nestled along the roaring Colorado River, which had been a uranium-mining hub since the early 1950s and home to generations of mining and ranching families. Because the town was only thirty-two miles from the proposed northern entrance to Canyonlands, it would serve as the gateway to the national park. Motels, curio shops, and restaurants were already flourishing on Main Street because of the close proximity of Arches National Monument.

Working with Senator Frank Moss, a native son and Democrat, Udall promoted Utah as "America's National Park State" whenever the opportunity arose. He told Utah residents, "California is four or five times as big and they have four national parks? Why shouldn't Utah have five? Because you have the most scenic land." He said later, "That appeal worked with some people."[2] Actually, it was an idea that had occurred to several very savvy people.

Joseph Wood Krutch, the author of *Grand Canyon*, among other respected books of natural history, hoped that Canyonlands would become part of "a Golden Circle" of Utah parklands, with roads connecting the unit to Zion, Bryce Canyon, Capitol Reef, and Arches.[3] In the

1930s, Harold Ickes had sought such a wilderness network for Utah but had had to settle for a limited victory in the creation of the Capitol Reef National Monument and the expansion of Arches.[4] In 1964, though, with Johnson in the White House, a Democratic Senate and Congress, and Udall running Interior, preservationists were impatient to save the raw red-rock wilderness in Canyonlands.

Moab and the surrounding region were still benefiting economically from the postwar uranium boom, along with an increase in oil and gas exploration, and that bolstered the opposition to a park. Locals were making good money in the extraction industries and were not inclined to lock up a quarter million acres for hiking trails. In Salt Lake City, the state capital, battle lines were drawn between lawmakers hoping to protect the beauty and cultural riches of Canyonlands and those on the side of fast dollars and good-paying mechanized jobs.[5]

The preservationists, with their longer-term view of Utah's economy, teamed with the tourism industry to win the day. Just nine days after Lyndon Johnson signed the Wilderness Act into law, he signed another bill creating Canyonlands National Park. It was the first new national park since the Virgin Islands in 1957.[6] Udall called Canyonlands, thirty miles long and fifteen miles wide, a "pure park," three hundred million years in the making: Upheaval Dome, a sandstone crater split by a chalk-hued peak; Standing Rock Basin, a 4,720-acre desertscape of fins, benches, and monoliths; High Mesa, a 1,280-acre moonscape desert-bed with an "Island in the Sky"; and the Neck, a 640-acre tunnel between two surreal escarpments that was just wide enough for a Jeep to pass through. Douglas, Brower, and other public lands activists complained that the Canyonlands National Park warranted a million acres, not its meager 257,400 acres, a tenth of the acreage of Yellowstone National Park. In 2007, Udall recalled, "The good news was that Canyonlands was expanded in 1971 to its present 337,598 acres—not the million acres I wanted but a step in the Douglas-Brower direction."[7]

Once Canyonlands National Park was established, Udall pressed Johnson even harder to bring Congress to heel on the proposed Indiana Dunes National Lakeshore. Time was of the essence. The Bethlehem Steel Company, in need of a Midwest port, had just bulldozed more than two thousand acres there, shifting the sand on the site along the southern shore of Lake Michigan. Dunes are Earth's least stable landform, so the change in the shoreline's ecology was devastating. Udall urged the House Interior Committee to act promptly to save the remaining sand dunes on the lake's southern shore. "Among his modern predecessors only the redoubtable Harold L. Ickes equaled him in zeal and effectiveness in the cause of conservation," the *New York Times* observed at the time of the Indiana Dunes fight. "Secretary Udall's work in the field bestows great distinction upon the Johnson Administration."[8] For all of that, the Indiana Dunes remained in peril as the battle for national lakeshore status dragged on.

All of Johnson's landmark 1964 conservation achievements were passed in the middle of an election year, as the president faced the firebrand Republican nominee, Barry Goldwater, that November. Riding on Johnson's Great Society coattails, Democrats campaigning for office boasted that theirs was the party of cleaner air, prudent land stewardship, roadless wilderness, national parks, and purer water.

On September 7, 1964, an anti-Goldwater commercial called "Daisy" aired on behalf of LBJ's campaign. The ad showed a girl in a green meadow plucking the petals from a daisy. Suddenly, a militaristic male voice embarked on a countdown: "Ten, nine, eight, seven, six, five, four, three, two, one, zero." That was followed by a massive atomic bomb blast that made TVs light up with fireball imagery. President Johnson's voice then warned voters, "These are the stakes: to make a world in which all of God's children can live, or to go into the dark. We must either love each other. Or we must die." At the end, a stern announcer pleaded, "Vote for President Johnson on Nov. 3. The stakes are too high for you to stay home." The takeaway was that Goldwater was a madman, whereas LBJ would protect the world.[9]

Driving the dichotomy home, Johnson wrapped his arms around JFK's historic Limited Nuclear Test Ban Treaty just weeks before the election. Goldwater had always opposed it and threatened to withdraw from the agreement if elected. "This treaty has halted the steady, menacing increase of radioactive fallout," Johnson proclaimed on October 12. "The deadly products of atomic explosions were poisoning our soil and our food and the milk our children drank and the air we all breathe. Radioactive deposits were being formed in increasing quantity in the teeth and bones of young Americans. Radioactive poisons were beginning to threaten the safety of people throughout the world. They were a growing menace to the health of every unborn child."[10]

Throughout the fall, Mrs. Johnson campaigned on a whistle-stop tour across the country. Her train, the "Lady Bird Special," was decorated with red-white-and-blue bunting. She traveled 1,682 miles between October 6 and 9, made forty-nine stops, and delivered forty-seven speeches in eight southern states.[11] When campaigning on the West Coast, the First Lady would periodically talk about conservation, but in the South, she evoked NASA, technology, and civil rights. "Demanding and exhausting best describe those 'twenty-eight-hour days'—but that dimension was far exceeded by the exhilaration of seeing more of the land," she recalled of the grueling campaign. She said that her aspirations for the United States were rooted in her "longtime love of nature, reinforced by seeing the vastness of this land—its beauty and its blight." She continued, "Our eyes frequently met the majesty of America's splendor, but, sadly, all too often the evidence of neglect and abuse."[12]

On election day, LBJ drubbed Barry Goldwater: 486–52 in the Electoral College. It was a mandate for the Great Society, especially since the Democrats took thirty-six seats from Republicans in the House of Representatives and two in the Senate. With a strong Democratic majority in both houses of Congress, Johnson promised that his Great Society agenda would continue to be a priority. Indeed, during the first session of the 89th Congress alone, the Johnson administration would submit eighty-seven bills to Congress, and eighty-four of those—96 percent—would return to the Oval Office for the president's signature, akin to FDR's first hundred days in 1933 and even outstripping it in sheer volume. In the months after the election, LBJ was planning the kind of legislative sweep that would have made Roosevelt proud. Unfortunately, Lyndon's health became a problem; his gallbladder was removed and his irregular heartbeats were carefully monitored. Amid the joy of the landslide, Lady Bird worried that Christmas season that the heavy White House workload was killing her fifty-six-year-old husband.

Lee and Stewart Udall visited the Johnson ranch that December. Relaxed, yet excited about the conservation agenda for 1965, they

A snapshot of Joseph P. Kennedy, Sr., and Supreme Court Justice William O. Douglas taken by the pool at the Kennedy estate in Palm Beach, Florida, circa 1940. Douglas, a zealot for wilderness protection, was a frequent guest in the various Kennedy homes.

The ruined city of Hiroshima, Japan, in March 1946, seven months after the first use of an atomic bomb, on August 6, 1945. The detonation immediately killed an estimated 80,000 people. Radiation poisoning in the aftermath slaughtered another 62,000, raising the specter of the atomic age and, with it, environmental catastrophe.

Pool in a Brook, Pond Brook, a photograph taken in 1953 in New Hampshire by Eliot Porter (1901–1990). Pioneering new methods of color photography, as well as the "coffee-table" book, Porter brought intimate glimpses of the wilderness to millions of Americans and promoted the 1960s revival of Henry David Thoreau.

The US military tested the first hydrogen bomb (nicknamed Able) above the Bikini Atoll in the Marshall Islands on July 1, 1946, right after the Second World War. During the early Cold War years, the United States conducted twenty-three nuclear tests in the Pacific region, to the consternation of Rachel Carson. To this day, the Bikini Atoll remains toxic, with extremely high levels of strontium-90 in the well water.

Cathedral Peak and Lake by Ansel Adams, 1938. A native San Franciscan, Adams began selling his photographs of natural settings in the early 1920s. By the 1950s, his striking black-and-white images were ubiquitous, in magazines and books and on posters in dormitory rooms across the country. *Ansel Adams Archive © The Ansel Adams Publishing Rights Trust*

On August 7, 1961, President John F. Kennedy established the Cape Cod National Seashore. The new federal park was an integration of gentle ponds, tidal marshes, and towering dunes with the powerful Atlantic Ocean as backdrop.

Rachel Carson in the woods near her Maryland home, photographed by Alfred Eisenstaedt for *Life* magazine in 1962.

Two views of Sleeping Bear Dunes National Lakeshore by Susan Tusa in 2020. Sleeping Bear, in northern Michigan, was the last of three national lakeshores created between 1966 and 1970; all are in the Great Lakes region.

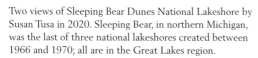

The Pulitzer Prize–winning poet Gary Snyder, born in 1930, drew inspiration for many of his works from Pacific Northwest ecosystems and Native American culture. At the beginning of his storied career in the 1950s, he was one of the Beat generation writers, and in the 1960s, bridged the emerging environmental and counterculture movements. This photograph was taken by Philip A. Harrington in 1969.

As a boy growing up in Alabama, E. O. Wilson (1929–2021) was interested in ants. After receiving his doctorate in biology from Harvard University in 1955, he became a global authority on ants. His studies led him to originate the field of sociobiology, which connected the interactions of insect and animal species with those of humans. He was a longtime proponent of biodiversity.

With the 106-mile Pedernales River in Texas as his inspiration, President Lyndon B. Johnson established the Wild and Scenic Rivers Act in 1968. The Pedernales was LBJ's natural respite from the hurly-burly of Washington politics.

Left to right: Olaus Murie, Howard Zahniser, and Adolph Murie on Cathedral Mountain in Alaska in 1961. All three were devoted to preserving the American wilderness during the postwar years, when that was a lonely calling. Their first major victory of the era was in 1960, with the establishment of the Arctic National Wildlife Refuge.

Vice President Richard Nixon attending the dedication of the visitor center at the Pinnacle Interpretive Shelter, Cumberland Gap National Historical Park, July 1959.

During most of her long life, Marjory Stoneman Douglas (1890–1998) worked hard to spread the understanding of the Everglades. Each victory was followed by yet more challenges, however, particularly the 1970 plan to build a massive airport right in the middle of her "River of Grass."

Robert F. Kennedy and William O. Douglas during a trip to the Soviet Union in 1955. In Uzbekistan, the grand mufti of Central Asia presented them with satin robes and colorful Uzbek hats—native finery that Douglas later called "one of my prized possessions." RFK learned how to fly-fish, hike, and canoe from his all-seasons hero, Justice Douglas.

Drakes Beach at Point Reyes, California, in a photograph taken by Philip Hyde for the Sierra Club book *Island in Time* (1962). Point Reyes became a national seashore on September 13, 1962, when President Kennedy signed the legislation to create the new federal park. The San Francisco Bay Area was a hotbed of environmental activism during the Long Sixties. *Estate of Philip Hyde*

Coast, North of Bodega Bay, California, by Ansel Adams, 1960. The nation's remaining dwindling public seashores were under pressure in the late 1950s, mainly from home builders, the tourist industry, and power companies. A grassroots citizens revolt in California stopped Pacific Gas and Electric (PG&E) from building a nuclear facility at Bodega Bay. *Ansel Adams Archive © The Ansel Adams Publishing Rights Trust*

In pajamas and slippers, President John F. Kennedy feeds a deer outside his cabin in California's Lassen Volcanic National Park on September 27, 1963. The snapshot was taken by an aide, Cecil Stoughton. After myriad attempts to nudge JFK close to nature during a tour of national parks in the West, he was finally won over in the dawn hours at Lassen, running around the cabin looking for more food to give to his new friend.

Philip Hyde had become a contributing photographer for the Sierra Club in 1950. Celebrated for his gorgeous color landscapes of the American West, he had a conservation portfolio that included the most acclaimed images ever taken of Glen Canyon, Redwood National Park, and the Navajo Wildlands. *South Rim, Winter* (1964) was from his series aimed at protecting the Grand Canyon from dams. *Estate of Philip Hyde*

In 1958, William O. Douglas led a protest march to save an undeveloped section of the Olympic Peninsula, to the west of Seattle, from a proposed highway. The 1950 publication of his first memoir—*Of Men and Mountains*—showed the public how intensely the "green justice" identified with Pacific Northwest wilderness areas. Douglas was unique as a member of the Supreme Court, serving thirty-six years—longer than any other.

Philip Hyde, *Cathedral in the Desert, Utah*, 1964. *American Photo* magazine named this one of the one hundred most important photographs of the twentieth century. In the same year, President Lyndon B. Johnson established Canyonlands National Park. *Estate of Philip Hyde*

Lady Bird Johnson and Secretary of the Interior Stewart Udall on a 1964 raft trip down the Snake River in Wyoming with the Teton Range in the background. Together, Stew and the First Lady helped spur Great Society legislation forward in the mid-1960s aimed at protecting unspoiled rivers and establishing new national parks, such as Redwood and North Cascades.

Lady Bird Johnson walking barefoot in the surf at Padre Island, off the Texas coast, in 1968. President Kennedy signed the bill protecting the pristine track of national seashore in 1962, but necessary state concessions took years to work out. The designation was finally official in 1968.

President Lyndon B. Johnson in the Texas Hill Country, where he was always at home on the range. On October 15, 1966, he authorized Guadalupe National Park in Texas, which opened to the public in 1972.

Aerial View of the North Cascades in Winter, by Stephen Matera, 2006. Though only three hours east of Seattle, the North Cascades are one of the most remote regions of the continental United States. When logging and hydroelectric interests encroached on the forest-rich region in the late 1950s, Pacific Northwest conservationists such as Polly Dyer and Henry M. Jackson moved to save the sublime wilderness as the North Cascades National Park.

Lyndon and Lady Bird Johnson in a field near their ranch on July 5, 1968. The Johnsons' love for every inch of their ranch lent them a connection to the land as deep as that of any first couple.

An aerial shot of the Santa Barbara oil spill, which began on January 28, 1969. The disaster destroyed life both in the ocean and on shore, testing the resolve of the newly installed Nixon administration.

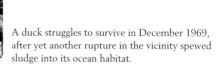

A duck struggles to survive in December 1969, after yet another rupture in the vicinity spewed sludge into its ocean habitat.

President Nixon cruising Lake Jackson, Wyoming, in August 1971 with Secretary of the Interior Rogers Morton (*left*). It was the first time Nixon had seen the Grand Tetons, a range that duly impressed him, even though, in his dark summer suit, he was freezing, with temperatures in the twenties on a summer morning in the mountains.

Pat and Richard Nixon walking near their beach house in San Clemente, California, on January 13, 1971. Both were native Californians; President Nixon, in particular, enjoyed romping in the waves. As an ocean conservationist, he was especially proud of establishing the Marine Mammal Protection Act (MMPA) and the National Oceanic and Atmospheric Administration (NOAA) in the early 1970s.

The Earth, as seen from Apollo 11, which was orbiting the moon in July 1969. Similar imagery was used as cover art on editions of *The Whole Earth Catalog*. It was an arresting, space-age sight that served as a reminder of how precious the planet truly is.

Copy of *The Whole Earth Catalog*, dated fall 1970. The brainchild of Stewart Brand, the oversized paperback was a cornucopia of alternative products and ideas, many of them geared toward sustainable living.

Denis Hayes was enlisted by Senator Gaylord Nelson of Wisconsin to be the original national coordinator of the first Earth Day. Taking time off from Stanford University graduate school, the twenty-five-year-old found pivotal support from the print media for the "teach-in." As a result, when April 22, 1970, arrived, Hayes was greeted by a crowd in New York City that stretched down Fifth Avenue for at least sixty blocks.

At the first Earth Day in New York City on April 22, 1970, people took the side of the planet, which, as the sign said, needed help.

Crowds jammed New York City on the first Earth Day. An estimated 20 million Americans responded to Senator Gaylord Nelson's call for an educational " teach-in" about the natural world. Later, in 1995, Nelson would be awarded the Presidential Medal of Freedom for founding Earth Day and fighting to outlaw DDT and other pesticides.

Robert Rauschenberg designed the first Earth Day poster to benefit the American Environment Foundation in Washington, DC. It features an image of a bald eagle surrounded by a photo montage of endangered animals; deforestation; and land, water, and air pollution. According to the Robert Rauschenberg Foundation, the artist's use of the American eagle "symbolically placed the United States at the center of a global problem."

Folk singer Pete Seeger at Hells Canyon along the Snake River in Oregon in 1972, preparing for a raft trip to protest the damming of the Snake. The composer of the song "Don't Ask What a River Is For" is seen sewing a SAVE HELLS CANYON bumper sticker onto a scarf, to use as the expedition flag. Seeger recorded a whole album of ecological songs in 1966, *God Bless the Grass*, with five songs contributed by Malvina Reynolds and liner notes by William O. Douglas.

The seventy-five-foot *Clearwater* sails the Hudson River past junk left on the riverbank. Pete Seeger originated the idea to build an authentic sloop in order to publicize the degradation of the Hudson; the sleek *Clearwater* was launched in 1969 and still sails.

David Brower (1912–2000) was the executive director of the Sierra Club (1952–1969) and Friends of the Earth (1969–1986). The *New York Times* called Brower the "most effective conservationist in the world." He was nominated three times for the Nobel Peace Prize for his work on various Earth stewardship causes. His Long Sixties campaigns resulted in establishing such national seashores as Cape Cod (1961), Point Reyes (1962), and Fire Island, New York (1964). He was an essential advocate in saving the Allagash Waterway in Maine (1967) and establishing Redwood National Park in California (1968).

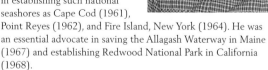

The poster for the Sierra Club's tenth wilderness conference in 1967, designed by Stanley Mouse and Alton Kelley. In the 1960s and early 1970s, Indigenous people were celebrated as heroic stewards of the land and water.

Sylvia Earle in a wetsuit along the shore of Alameda, California, photographed October 31, 2016, by Luisa Dörr. Earle, born in 1935, is a preeminent oceanographer and research scientist who has used her reputation to campaign for ocean preservation.

Robert Boyle, standing in the Hudson River with Storm King Mountain behind him, in a 2001 photograph by Mark Vergari. The Hudson, like most American rivers in populated areas, had been treated as a dump for more than a century. Boyle was a tireless citizen-activist, fighting to return it to its natural dignity.

Andy Warhol, *Bald Eagle*, 1983, from the Endangered Species portfolio, screen print on Lenox Museum Board. Warhol was a champion of President Nixon's historic 1973 law to prevent such American species as the polar bear, manatee, whooping crane, and American crocodile from vanishing.

Cesar Chavez (1927–1993), photographed by Arthur Schatz in 1969 for a cover story in *Time* magazine. As head of the United Farm Workers (UFW), Chavez became a progenitor of environmental justice, fighting to ban dangerous chemicals sprayed on agricultural produce.

Edward Abbey (1927–1989) circa 1979 at a lookout tower in Arizona's Tonto National Forest, where he once worked; the picture was taken by Buddy Mays. Abbey was the author of *Desert Solitaire: A Season in the Wilderness* (1968), a memoir of Utah's Arches National Monument. He was a ranger there in the 1950s.

On September 26, 1970, President Richard Nixon established the Apostle Islands National Lakeshore. These two wintertime views—*God's Window* and *Fire in the Sky*—were taken in 2014 by photographer Michael DeWitt. This national lakeshore lies off Wisconsin's Bayfield Peninsula and Lake Superior, the bluest and most pristine of the Great Lakes. Senator Gaylord Nelson lobbied for over a decade to preserve this archipelago of islands as part of the National Park Service (NPS).

Stewart Udall (1920–2010) of Arizona was secretary of the interior during the Kennedy and Johnson years. His book *The Quiet Crisis*, published in 1963, is considered a classic in conservation literature. When Udall died, the Department of Interior headquarters in Washington, DC, were named in his honor.

inspected cattle, walked along the Pedernales, ate peach preserves, baked bread, stargazed, and strategized. Over a meal, Stewart and Lady Bird agreed to transform Washington, DC, from a pathetically drab place to a city of gardens and greenery, akin to London or The Hague. To build on the work of Bill Douglas and the Kennedy family, the Johnsons agreed to a campaign to clean up the Potomac River, plant native trees, and preserve more green spaces around the Tidal Basin so the capital could become, in Udall's words, the "conservation showcase for the nation."[13]

Mrs. Johnson received widespread praise for adopting the beautification of Washington as her special cause in early 1965. Mail poured into the White House, offering free advice on everything from pruning trees to scrubbing sidewalks to clearing dump sites. Sharon Francis, Udall's assistant, was loaned to Lady Bird to help answer letters and coordinate the program in the White House with national conservation groups. The *New York Times* praised the Johnsons' beautification of the city as an endeavor long overdue.[14] *Washington Evening Star* columnist Mary McGrory, aware of the alliance of the First Lady and the interior secretary, said that Udall was metamorphosing into "the cosmetician of the American landscape."[15] To a president who was escalating US military engagement in Vietnam and fending off criticism in the segregationist South for the Civil Rights Act of 1964, the widespread praise on the environmental front was much appreciated. "I have been almost shocked . . . at the amount of response that we had to our natural beauty message," LBJ enthused to the press. ". . . The natural beauty theme has caught on."[16]

On January 4, 1965, President Johnson delivered his State of the Union address. With a massive TV audience watching (it was the first time the annual speech was given in the evening), he made a commitment early in the speech to reverse the nation's environmental woes in cities and towns. With passion in his voice, he called for the beautification of America. He drove forward a conservation agenda that was firmly rooted in the longtime tradition of Theodore Roosevelt and the newly minted one of Rachel Carson. "For over three centuries the beauty of America has sustained our spirit and has enlarged our vision," he said. "We must act now to protect this heritage. In a fruitful new partnership with the States and the cities the next decade should be a conservation milestone. We must make a massive effort to save the countryside and to establish—as a green legacy for tomorrow—more large and small parks, more seashores and open spaces than have been created during any other period in our national history."[17]

Building on the Wilderness Act and Kennedy's Cape Cod–Padre Island–Point Reyes trifecta of national seashores, Johnson proposed comprehensive wild-river legislation—the idea cherished by environmentalists, notably Gaylord Nelson, Bill Douglas, Sigurd Olson, and Frank and John Craighead, who were studying threats to grizzly populations in Yellowstone's riverine habitat.

Ever since Udall suggested a wild-river bill in *The Quiet Crisis*, proposals had poured into the Interior Department. As 1964 turned to 1965, the idea took hold with the conservation lobby on Capitol Hill. In the days following LBJ's address, the White House, the Interior Department, and the Forest Service received letters from ecologists suggesting waterways (or sections thereof) for what was being called a wild-river classification. Gaylord Nelson recommended seventy miles along both sides of the Saint Croix River, where a hydroelectric dam had been planned; he later convinced the Northern States Power Company to sell it to the government. Frank Church nominated the stretches of the Salmon River. Marjory Stoneman Douglas suggested Florida's Kissimmee, so rich in unusual flora and fauna. The Craigheads sought protected status for the Yellowstone River in Montana and Wyoming. Bill Douglas threw his clout behind the Buffalo in Arkansas and the Rogue in Oregon. Ed Muskie eyed the Allagash in northern Maine. Just about every conservation club, canoe outfitter, and dam fighter seemed to have a sparkling and swift or smooth and lazy river in mind. They were not necessarily famous rivers but ones known to those in the vicinity, those who feared that dams or commercial development could forever ruin their spellbinding essence.

It was a White House aide, Bill Moyers, who got the Wild Rivers bill rolling in earnest after LBJ's inaugural. Having consulted for Interior and Agriculture experts, Senators Nelson and Jackson, among others, he wrote President Johnson a synopsis delineating which waterways could quickly be designated. It included the Salmon and Clearwater (Idaho); the Rogue (Oregon); Saint Croix (Minnesota); the Rio Grande (New Mexico); and the Green (Wyoming). Moyers thought Johnson should try to save Florida's controversial Suwannee River—lionized by the Stephen Foster song—because the purchase of private land would be necessary; the others were Forest Service holdings. A slew of others were on a separate list of those that needed further study, including the Cacapon (West Virginia) and Wolf (Wisconsin).[18]

As the Wild-River bill became a White House priority, Johnson received immediate blowback from Congressman Aspinall. The reclamationist groused that the classification was "downright crazy"[19] and for the next three years he delayed hearings on it. Representative John Saylor, who had a list of sixty-six rivers he wanted included in the initial class, worked tirelessly to soften Aspinall on the new classification and eventually he convinced him to consider four rivers. To Saylor, a patient small-town politician, that was progress; at least Aspinall wasn't 100 percent opposed to the new classification scheme. Aspinall said he was willing to support it if in exchange he was granted appropriations for five new dams in Colorado.

A lawyer by training, Saylor had represented the southwestern Pennsylvania district surrounding Johnstown, known for its historic floods, since 1889. Of Mennonite heritage, the six-foot-four Saylor was nicknamed "Saint John" by environmentalists because of his river protection advocacy. He was a lifetime member of the Wilderness Society to boot. When Senator Church introduced the Wild-River bill in 1965, Saylor didn't like it conceptually. In the first place, he said, the language was vague, simply specifying that certain "glamour" rivers could not be dammed. In the second place, he rejected the idea that the only rivers worth saving were "wild"—that is, white-capped and ferocious. In response, he wrote his own version, the Wild and Scenic Rivers bill. It not only allowed that some very tame rivers also deserved protection but proposed a three-tiered classification system (wild, scenic, and recreational).

For a couple of years, the two bills circled each other without making any real progress. In 1967, Udall built on Saylor's trade-off, enticing Aspinall to move on the issue by promising in return federal help on water projects in Colorado. That promise, and the mounting pressure nationally to save wild rivers before it was too late, prompted Aspinall to submit his own bill, to the shock of conservationists. It had some strong features, including a ban on mining along all protected rivers, though it specified only four initially, and in his usual way, he insisted on a careful process—just short of obstructionist—for consideration of future additions. The House bill that moved forward in 1966 and 1967 was an amalgam of the Aspinall and Saylor versions, forged amid tense negotiations at every stage. In the end, it passed both houses of Congress in 1968 with only seven votes in total against. Johnson applauded the fact that the lawmakers offered protection not just for wild rivers but for "Wild and Scenic Rivers."[20] And he was ebullient, like a child on Christmas, that the bill had emanated directly from his 1965 State of the Union address.[21] Eight rivers, or segments, were specified in the bill as the inaugural wild and scenic rivers.

II

On January 20, 1965, just sixteen days after the State of the Union speech, Johnson delivered a short and poetic inaugural address. The novelist John Steinbeck had been tapped to offer ideas, though only a few of his lines made the final cut. After elaborating on the enduring appeal of the Bill of Rights and the American covenant, Johnson claimed that the essence of the United States was wrapped up in the nation's magnificent outdoors. "For this is what America is all about," he said. "It is the uncrossed desert and the unclimbed ridge. It is the star that is not reached and the harvest that is sleeping in the unplowed ground."[22] With the Wilderness Act to his credit, the president instinctively recognized that it was time to elevate New Conservation even higher on the list of Great Society national priorities.

The drumbeat in behalf of the environment continued the next month. On February 8, 1965, Johnson delivered a Special Message to the Congress on Conservation and Restoration of Natural Beauty to a joint session of Congress. The speech, written primarily by Richard Goodwin, confronted all of the United States' environmental concerns at once, showing the immense scope of the Great Society's ecological vision. Insisting on clean water, air pollution control laws, parks in urban areas, the establishment of more national seashores and lakeshores, national recreation areas, and, quite unexpectedly, scenic trails, Johnson swung to the fences. "Its concern is not with nature alone, but with the total relation between man and the world around him," he said of his creative New Conservation agenda. "Its object is not just man's welfare but the dignity of man's spirit."[23]

Having listened to his White House science team, Johnson warned Americans of a newly discovered danger: "Air pollution is no longer confined to isolated places. This generation has altered the composition of the atmosphere on a global scale through radioactive materials and a steady increase in carbon dioxide from the burning of fossil fuels."[24] Johnson was the first world leader to speak of the effects of air pollution on climate. Congress was reaching a consensus, shared by Johnson, that the Clean Air Act of 1963 didn't go far enough in addressing stationary sources of pollution; automobile emissions would also have to be seriously reduced. On that point, LBJ's "Conservation and Restoration of Natural Beauty" speech had struck a nerve. Previously, the issue had been left to the states, with California being the only one to set standards. LBJ obliterated the old prejudice against federal regulation of cars. It was a national problem because, after all, who didn't want clean air to breathe? The president pledged that the federal government would set a positive example in abating air pollution across the United States, with the Department of Health, Education, and Welfare leading the charge.

On issues related to the preservation of public lands and rivers, Johnson's speech seemed to have been lifted from Udall's playbook in *The Quiet Crisis*. The president promised that his administration would use the Land and Water Conservation Fund to establish new preserves, with a preference for ones close to large population centers. Taking the anti-dam stance that Bill Douglas had been promulgating for years, Johnson called for the federal preservation of free-flowing rivers in the Ozarks of Missouri. "Our stewardship will be judged by the foresight with which we carry out these programs," he concluded. "We must rescue our cities and countryside from blight."[25]

David Brower rejoiced over "Conservation and Restoration of Natural Beauty." In his mind, it meant that the North Cascades and Redwood National Parks had a fantastic chance of being established before the 1968 presidential election. Whether it was redwoods or overpopulation, wild and scenic rivers or sewage regulation, endangered species or eliminating smog, Johnson sounded determined to restore and repair the degraded American landscape. Letters of support poured into the White House from the members of the Sierra Club, the Nature Conservancy, the Garden Club of America, the National Parks Association, and other groups. Many senators, Church, Hart, and Douglas among them, congratulated him. The *Washington Post* deemed the February 9 message an "eloquent and moving visionary appeal for the ages."[26] Johnson was aware that he had gone further than any other president before him with his three speeches demanding action on the environment: his State of the Union address, his inaugural address, and the strongest, his Special Message to the Congress on Conservation and Restoration of Natural Beauty.

One popular passage of the speech was Johnson's call to build a cooperative public trails program for "the forgotten outdoorsmen of today."[27] If Lady Bird was personally interested in urban and roadside beautification, LBJ, the cowboy president, had a penchant for the National Trails Program. Influenced by the late Benton MacKaye, a founder of the Wilderness Society, Johnson called for the Bureau of Outdoor Recreation to study existing backcountry trails in the United States, report how well they served the public, and suggest federal legislation to create a comprehensive system. In the Senate, Henry Jackson and Gaylord Nelson advocated for a trail system. The bureau's fine report, *Trails for America: Report on the Nationwide Trails Study*, would appear in late 1966.[28] It recommended a network of national scenic trails, notably the Appalachian, Continental Divide, and Pacific Crest (on the Cascades and Sierras). Trails were more recreational than environmental, but they were part of the appreciation of nature crucial to promoting the movement in the 1960s.

On May 24–25, 1965, President Johnson held a White House conference on natural beauty (four years after JFK did the same for conservation). More than one hundred panelists in the program showcased a wide range of public and private visions. Once again, Laurance Rockefeller served as chairman of the sessions, which were held in the State Department auditorium. LBJ delivered a powerful special New Conservation message about how "uncontrolled waste products are menacing the world we live in, our enjoyment and our health."

But, hands down, the show stealer was Mrs. Johnson, who was at her very best in excoriating "civic ugliness and the decay of our cities and countryside."[29]

In response to the so-called Natural Beauty speech, in the spring, Ansel Adams visited the White House to discuss wilderness-related issues. LBJ and Adams got along like long-lost cousins. They both loved open spaces, wore Stetsons, and were void of Ivy League elitism. The president admired Adams's ethereal black-and-white photos of Point Reyes and Yosemite, which Udall had put on display at Interior. Adams felt the same about Johnson's leadership on conservation. As a result, Adams arranged that they'd collaborate on a book, composed of his pictures and Johnson's words, taken from his remarks as president. Nancy Newhall provided some material, as well. The result was 1965's *A More Beautiful America*, published that spring by the American Conservation Association. "Above all, we must maintain the chance for contact with beauty. When that chance dies, a light in all of us dies," Johnson wrote in the foreword.[30]

When the Canadian government in Ottawa announced that it was naming a mountain in the St. Elias Range to memorialize John F. Kennedy, his thirty-nine-year-old brother, having been elected as a senator from New York, pounced on the opportunity to be the first person to scale the 13,900-foot peak.[31] Douglas was unable to join him, but told Bobby to form a three-man team with James Whittaker of Seattle, the first American to climb Mount Everest, as the chief organizer and Barry Prather of Ellensburg, Washington, as the third climber.[32] The mountain, part of the Yukon Territory, jutted up dramatically from a level glacier plateau. It was an ascent for professional mountain climbers, not a novice, but RFK was determined to climb with the aim of placing a copy of his brother's Inaugural Address and a PT boat pin on the summit.

On March 24, 1965, RFK, Whittaker, and Prather, and a *National Geographic* team following, reached the top, dizzy with altitude sickness but triumphant. For Bobby, it had been an act of will and endurance. With temperatures at five degrees above zero, and wind whipping at eighteen miles per hour, it was considered ideal weather to make such an ascent. Besides an American and Canadian flag, RFK placed at the summit a dark rectangle with the Kennedy coat of arms. "President Johnson told me he thought the climb was a stunt," Udall recalled. "I knew better. It was a prayer and an act of self-determination."[33]

Wilderness, wild rivers, and national parks had a solid bipartisan constituency in both houses of Congress, but Johnson had to press hard on a more controversial issue: improving water quality. Clearly, Ed Muskie had struck a responsive chord in millions of Americans on clean air reforms, but water quality was without a powerful public face; LBJ seized the opening. It had become a national concern. The League of Women Voters, long aggressive in its approach to environmental conservation, sponsored highly responsible studies on water quality that Johnson read.[34] In 1966, it published the first, *The Big Water Fight: Trials and Tribulations in Citizen Action on Problems of Supply, Pollution, Floods, and Planning Across the U.S.A.*, and it made a film on the same topic to broaden awareness of water contamination in the United States. League members considered themselves citizen activists in the mold of Rachel Carson and Barry Commoner, trying to enhance environmental quality in America.[35] The head of the National Water Research Institute called the women in the League "the unsung heroines in the battle for clean water."[36] Another group that joined the fight was the Izaak Walton League, which sent experts to testify before Congress, breaking down technical information about water pollution.

Johnson was intellectually engaged in all aspects of water policy, from hydroelectric dams to protection of fish species. "[Johnson] always had a lot of insight on water problems and this grew out of the New Deal period and the dams that were being built in his own congressional district," Udall recalled. "He had an intimacy with water projects. He knew how they functioned and this, of course, was something that President Kennedy did not have."[37] Because so much Great Society legislation was being passed—and the Vietnam War and civil rights were claiming newspaper headlines—Johnson received less notice for his advocacy of what became the Federal Water Project Recreation Act of 1965. Sponsored by Aspinall in the House and Scoop Jackson in the Senate, it established a framework for national-local cooperation in opening reservoirs to water sports, including fishing.

More important, the new law required that all new federal water projects give priority to wildlife preservation and recreation. The act gave legislative viability to the notion that both of those objectives were benefits to American citizens.[38]

While LBJ was focusing on wild rivers and water-related recreation opportunities in the first half of 1965, Mrs. Johnson and Secretary Udall formed the Committee for a More Beautiful Capital, part of the campaign the pair had planned in December. The group included Elizabeth Rowe, the chair of the National Capital Planning Commission; a future mayor, Walter Washington, then with the National Capital Housing Authority; and Nathaniel Owings, the chair of the Temporary Commission on Pennsylvania Avenue. Deep-pocketed benefactors also joined Lady Bird's committee, among them the health activist Mary Lasker and Laurance Rockefeller, both of New York.[39] "That capable committee helped write a marvelous chapter in Washington's life simply by improving on what was already there," Mrs. Johnson later boasted.

Among the committee's accomplishments were the landscaping of hundreds of park sites and schools and playgrounds; the planting of nearly two million bulbs, 83,000 spring-flower plants, 50,000 shrubs, 25,000 trees, and 137,000 annuals; the landscaping of several major thoroughfares and freeways into the city; and innumerable clean-up, fix-up, and plant-up campaigns in the depressed parts of the district. To the public eye, what endures most from these years is drifts of yellow daffodils along the Potomac and in Rock Creek Park, and brilliant seasonal changes of color in the many triangles and squares.[40]

The First Lady's press secretary acknowledged that the effort was "a natural follow-up" to Jackie Kennedy's redecoration of the White House, yet it was meant from the beginning to have repercussions far beyond the nation's capital. Lady Bird's concept was to enhance her city in the hope of inspiring others to improve their hometowns with, in the examples she offered, "the planting of roadsides, the better design of public buildings, and the encouragement of open space in city and country."[41] Robert Boyle of *Sports Illustrated*, a committed Hudson River ecologist, joked that Lady Bird's efforts were "like putting lipstick on a corpse."[42] His assessment was too harsh. Contradictory to the Wilderness Act, as admirable as it was, Mrs. Johnson spearheaded the Great Society philosophy that environmentalism belonged in everybody's backyard, not merely at the end of a long road trip. There was something inherently noble about a denim-wearing First Lady patrolling Potomac River ravines marred by discarded tires, rusted buckets, smashed cinder blocks, and metal scraps, saying "This is unacceptable" to the reporters who were tagging along. Under Mrs. Johnson's direction, truckloads of rubbish were removed from the banks of the Potomac and Anacostia Rivers. She defended her beautification efforts against environmental critics who dismissed them as minor and local. That was the point: they were exactly that and needed to be repeated in a million minor and local ways across the country.

"Though the word beautification makes the concept sound merely cosmetic, it involves much more: clean water, clean roadsides, safe waste-disposal, and preservation of valued old landmarks as well as great parks, and wilderness areas," she explained. ". . . Beautification means our total concern for the physical and human quality we pass on to our children and the future."[43]

The First Lady soon turned attention to the needy neighborhoods of the nation's capital. She had help from San Francisco landscape architect Lawrence Halprin, who designed neighborhood gathering places and playgrounds. He also helped her committee close a landfill beside the Anacostia River and turn the reclaimed area into a popular riverside park. Lady Bird paid attention to unsightly businesses and successfully encouraged owners of parking lots and "fillin' stations," as she said with her gentle East Texas accent, to scrub clean their properties.

Mrs. Johnson spent much of 1965 pushing hard for pioneering legislation aimed at preserving the natural beauty of roadside America. Junkyards were the bane of her existence. Famously, she took aim at billboards that looked like giant pieces of litter on the landscape. Along the way, she made enemies of small businesses, such as the Wall Drug Store in Wall, South Dakota, which contended that it needed billboards to tell drivers which exit to take. In Montana, one billboard company erected a sign reading "Impeach Lady Bird." Undeterred, the First Lady pressed her campaign to limit roadside clutter, from both outdoor advertising and trash.[44]

While Lady Bird was warring against billboards, Bobby Kennedy became the personification of Wild Rivers. Over the Fourth of July weekend in 1965—along with Ethel and five of his nine children—he went kayaking and rafting along the Utah-Colorado border through white-water rapids of the Green River. Accompanying the Kennedys were mountain climber James Whittaker and his family and University of Denver ski instructor Willie Schaeffler. Their three-day, eighty-three-mile journey ended near Dinosaur National Monument, where RFK rode the School Boy rapid, plunging through the eons and the strata. He earned the "golden paddle" for his performance. Brower was enthralled that the Kennedys used their vacation to remind Americans that in the 1950s the entire Dinosaur National Monument ecosystem would have been forever marred by Bureau of Recreation dams if not for Sierra Club activists.

Bringing along a Time-Life journalist for the trip, RFK used the occasion to promote the Wild and Scenic River bill then circulating on Capitol Hill. Three of his closest Senate friends—Frank Church, Gaylord Nelson, and Walter Mondale—were doing the nuts-and-bolts legal work while RFK was presented as a modern-day Theodore Roosevelt in action whose colorful explorations brought attention to national parks and wilderness areas. Udall was jealous that he couldn't accompany the New York senator. "What spectacular fun we had," Ethel Kennedy recalled. "We camped at night along the river surrounded by 20,000-foot cliffs and mesas. Our time along the Green and Yampa was too short, though. We were, as a family, hooked on river conservation."[45]

Once Kennedy became New York's senator in January 1965, he advocated to save the Hudson River from industrial ruin; it would become his own beautification campaign. Consolidated Edison's proposal to build a $162 million hydroelectric power plant on Storm King Mountain was repugnant to RFK, who started kayaking on Hudson waters to raise ecological awareness. Frustrated that the decision rested in the hands of the Federal Power Commission, too restless to wait out what promised to be a decade-long legal battle, RFK pressed the Johnson administration to use its power to establish a riverway program in which Udall, at Interior, would work with the states of New York and New Jersey to develop a program to preserve and restore the Hudson River. This was a clever end run on Edison and ostensibly fit into the Great Society's push for Wild and Scenic River status. But Laurance Rockefeller, throughout, opposed RFK's attempt to shake hundreds of millions of dollars from Congress to clean the Hudson. LBJ asked White House aide Joe Califano to look into the fight brewing between Kennedy and Rockefeller and report back. "The Rockefeller plan (in which Laurance Rockefeller had a hand) is to preserve the Hudson *without* federal assistance," Califano wrote back. "It is running into tough opposition locally, Kennedy, Javits, and Representative [Charlie] Ottinger have proposed bills to clean up the Hudson *with* federal assistance."[46]

Working all the angles, consorting with anglers and outdoorsmen, RFK pushed for a Hudson Highlands National Scenic Riverway, which would include the segment of the Hudson River between Newburgh and the New York–New Jersey state boundary on its west bank and between the city of Beacon and the city of Yonkers on its east bank.[47]

On September 26, President Johnson signed a compromise bill that didn't get RFK the federal funding he sought for the Hudson but did involve the federal government in abating water pollution in the Hudson River basin. The act marked the beginning of major efforts to clean up the Hudson.[48]

Throughout the summer of 1965, while Lady Bird promoted beautification with fervor, LBJ continued to bear down on the pollution crisis. On October 2, he signed the Water Quality Act, which placed the initial burden for cleanliness of interstate waters on individual states, which were required to set standards and abide by them. It was a stepping-stone for the Clean Water Act of 1972.[49] Building on the Clean Air Act of 1963, the president signed an amendment to it, the Motor Vehicle Air Pollution Control Act, on October 10, setting the first national automobile emission standards and putting American carmakers on notice. The regulations would start with cars sold new in 1968. "Ours is a nation of affluence," Johnson said. "But the technology that has permitted our affluence spews out vast quantities of wastes and spent products that pollute our air, poison our waters, and even impair our ability to feed ourselves. At the same time, we have crowded together into dense metropolitan areas where concentration of waste intensifies the problem. Pollution now is one of the most pervasive problems of our society."[50]

A new agency, the Environmental Sciences Services Administration (ESSA), was created in the Commerce Department, with a new emphasis for a collection of long-standing agencies. The Coast and Geodetic Survey, the Weather Bureau, and the Central Radio Propagation Laboratory were transferred from the National Bureau of Standards to ESSA in the sweeping reorganization. ESSA's mandate was to engage in cutting-edge scientific research on the Earth's ecosystems, oceans, and atmosphere and in space. At heart, ESSA was tasked with keeping science free of interference from undue political or corporate influence.[51]

Entering the national debate on corporate responsibility to citizens for public safety and national air pollution reduction standards was the thirty-one-year-old Ralph Nader, whose 1965 book, *Unsafe at Any Speed: The Designed-In Dangers of the American Automobile*, attacked American automakers for everything from promoting the use of leaded gasoline to selling malfunctioning vehicles to building unnecessarily dangerous cars. Nader was born in 1934 in Winstead, Connecticut, to working-class Lebanese parents. When he was a junior at Princeton University, the national DDT battle was beginning, and he wrote a letter to the Princeton paper denouncing the destruction of songbirds by the DDT spraying on campus. "The DDT war was my beginning of being an environmental activist," he later recollected. "Robins were lying dead everywhere."[52]

After Princeton, Nader earned a degree from Harvard Law School. While hitchhiking across America, he visited Indian reservations in the Southwest, searching for ways to help First Peoples achieve equal rights and environmental justice.[53] On a visit to Yosemite, he befriended Ansel Adams and slept outside on the porch of the photographer's studio.[54] During the Kennedy years, he practiced law in Hartford, Connecticut. In 1964, Assistant Secretary of Labor Daniel Patrick Moynihan hired him as a consultant assigned to auto safety. With *Silent Spring* as an inspiration, Nader wrote *Unsafe at Any Speed* on his own time, studying case files from more than a hundred

Ralph Nader of Connecticut, a lawyer and critic of auto safety standards, stands on an overpass above Interstate 495, a beltway circling Washington, DC. An anti-pollution warrior, Nader was one of the principal architects of the Long Sixties environmental movement.

lawsuits then pending against General Motors for accidents involving its Chevrolet Corvair model. A master at gathering data, as well as a tireless researcher and memo writer, Nader concluded that US automobile companies were disregarding safety standards and putting style ahead of public health. The book's explosive publication made him the bane of Detroit's big three.

Unsafe at Any Speed became the most unlikely bestseller since *Silent Spring*. General Motors didn't take Nader's attack lightly. Just as the chemical companies had done with Carson, GM tried to discredit Nader's character. Operatives for GM, acting like the mob, tapped Nader's telephone and tried to compromise him by hiring prostitutes in a misguided attempt to trap him. The problem for GM was that Nader was squeaky clean and the Corvair was indeed poorly designed. At the time of publication, Nader was working as an unpaid consultant to Abraham Ribicoff, who had been such a staunch defender of *Silent Spring*. When Nader told Ribicoff that he was being followed by goons, the senator demanded that GM CEO James Roche testify before a Senate subcommittee. Under oath, Roche admitted that GM had hired a private detective agency to smear Nader. That revelation turned Nader into a consumer rights activist phenomenon.

Furthermore, Nader sued GM for invasion of privacy, settling the case in March 1966 for $425,000. Public esteem for Nader skyrocketed when he used the settlement money to found an activist organization, the Center for Study of Responsive Law. A year following the publication of *Unsafe at Any Speed*, Congress unanimously enacted the National Traffic and Motor Vehicle Safety Act. Speaker of the House John McCormack said the act's passage was owed to the "crusading spirit of one individual who believed he could do something: Ralph Nader."[55]

Unsafe at Any Speed wasn't an environmental book per se, but it did question whether automobile and petroleum companies were finding ways to reduce the public health effects of exhaust emissions. Nader was concerned that a conglomerate of federal agencies such as the Federal Energy Administration and the Federal Power Commission behaved like servants of the military-industrial-petrochemical order. His book was shocking in terms of the mass of data brought to bear on its narrow focus. And he was supported by protoenvironmentalists because he was trailblazing on how to hold reckless corporate polluters accountable.

Another 1965 publication was just as detailed but much broader in scope: the report by the President's Science Advisory Committee entitled *Restoring the Quality of Our Environment*. The 273-page book was the work of nearly a hundred scientists representing academia, government, and nonprofit institutions. Packed with evidence and forecasts regarding air, water, and soil pollution, starting with sources and continuing through long-term impact, the report was an omnibus of the darkest truths of assaults on the environment. Johnson didn't shirk its implications. After reading it, he even supplied a foreword to the publication.

At the end of the report was a twenty-three-page appendix, "Atmospheric Carbon Dioxide." That constituted the first official document, submitted to any government anywhere in the world, on the probable danger that rising CO_2 levels in the Earth's atmosphere posed: global warning and climate change.[56] The report referenced the cause as the burning of fossil fuels. The Earth's surface temperature could increase, on average, by five to seven degrees Fahrenheit, the report warned. That, in turn, could raise the level of the sea by ten feet:

Within a few short centuries, we are returning to the air a significant part of the carbon that was extracted by plants and buried in the sediments during half a billion years. Through his worldwide industrial civilization, Man is unwittingly conducting a vast geophysical experiment. Within a few generations he is burning the fossil fuels that slowly accumulated in the earth over the past 500 million years. By the year 2000 the increase in CO_2 will be close to 25%. This may be sufficient to produce measurable and perhaps marked changes in climate. The climate changes that may be produced by the increased CO_2 content could be deleterious from the point of view of human beings.[57]

Johnson's Great Society anti-pollution achievement and early climate change warning took a back seat to the all-consuming Vietnam War. Many of the key Democratic conservation-driven senators were slowly turning on LBJ over the divisive issue of the war. In the fall of 1965, Richard Goodwin, Johnson's speechwriter, resigned because of his moral outrage over it. Instead of cheering LBJ's New Conservation activism for, say, establishing Assateague Island National Seashore (Maryland) or the Delaware Water Gap National Recreation Area (New Jersey/Pennsylvania), let alone the epic Wilderness Act, antiwar liberals resented him for increasing the Vietnam draft to 231,000 new recruits each month. Antiwar protesters jeered him, shouting "Hey, hey, LBJ, how many kids did you kill today?" in front of the White House. Udall, in his private journals, questioned LBJ's sense of history, perplexed by his willingness to allow his Great Society domestic programs to be undermined by the unpopular war in Southeast Asia. "[Dean] Rusk and [Robert] McNamara have him in an alley with no exit," he lamented. "Bound, now, to bear the course, he turns to shallow appeals to patriotism. . . . It is rather pathetic really."[58]

Anybody can be valiant in a journal entry. Outwardly, Udall refrained from publicly criticizing Johnson's foreign policy. His reason, he later explained, was pragmatic. With his New Conservation team at Interior scoring environmental, beautification, and national lakeshore successes, he decided it was best to stay in his own lane. Just outside his office window, the evidence of Mrs. Johnson's beautification progress was plain as day. All around the Tidal Basin, trees and shrubs had been planted as part of the Committee for a More Beautiful Capital initiative. The Potomac River was cleaner than it had been in 1954, when Bill Douglas had made his first protest hike in the vicinity. Lee Udall took student groups to hike along the C&O Canal, advising them that the conservation of natural resources was a civic duty. So the interior secretary stayed focused and lobbied for "a real jackpot-cascade" of new national parks, lakeshores, and seashores in conjunction with the second session of the 89th Congress.[59]

Another reason that Udall kept his antiwar views to himself was that he'd been warned. His brother, Morris, serving in the House from Arizona, held the distinction of being the first member of Congress to oppose the war in Southeast Asia openly. At a face-to-face meeting with LBJ, Mo conveyed his disgust that 2,500 American servicemen had been killed in Vietnam in 1965. Johnson didn't appreciate the potshot criticism and made it abundantly clear that if his interior secretary denounced the war like his loudmouthed brother, he'd have an hour to clean his desk and leave.

In his loyalty to Johnson, Stewart was much like Lady Bird. Her biographer Jan Jarboe Russell observed, "In Vietnam, as in every other crisis of [Mrs.] Johnson's life, Lady Bird's role was to create a strange, even suffocating atmosphere of support around LBJ in which he could be assured that no matter how unreasonable his decision or petulant his behavior, his family and staff would stay forever loyal and submissive. She helped create a matrix of constant fidelity, which had the effect of restricting real debate about Vietnam or anything else."[60]

Reporting on the Vietnam War increasingly dominated press coverage of the Johnson administration. Some of LBJ's major contributions to civil rights still lay ahead, though, and the same was true of conservation. High on the president's list of environmental issues was "dirty water"—a phrase he'd shout at subordinates with disgust. Increasing river pollution, with its rank smell of chemicals on formerly pristine lakes, riled him to a state of belligerence. "Goddammit! No dirty water!" he'd shout at Udall and the HEW secretary, John Gardner.[61] Udall didn't have to be told. One afternoon, he took a group on a boat trip on the Cuyahoga River in northern Ohio. A press photographer snapped a picture of Udall holding his nose in disgust. "Phew," the secretary said, "one of the dirtiest rivers I've ever seen!"

When Senator Paul Douglas of Illinois, the former University of Chicago professor, prodded Johnson on the stalled Indiana Dunes National Seashore in late 1965, the president replied, "I'm on it," and then launched into a tirade about how the Chicago River was rusty with "pickle liquor" from steel mills. Reports that detergent foam was gushing out of faucets in cities along the Mississippi River also triggered a tirade. When it came to the Potomac River cleanup, he agreed with Bill Douglas: "Get the slime out," he'd bark at staff. He was appalled that the Merrimack River in New England was filthy brown, bubbling with nauseating gas; he wanted it so clean that children could swim in it. He also agreed with Bobby Kennedy that the Hudson River needed protection from polluters. Regularly, Johnson pointed his finger at midwestern senators about the horrible reports that the Missouri and Illinois Rivers flowed red with blood and offal from slaughterhouses. In speeches, he would quote Justice Oliver Wendell Holmes, saying "a river is more than an amenity, it is a treasure."[62]

Dr. Albert Schweitzer, the Nobel Peace Prize–winning philosopher, died on September 4, 1965, at his Lambaréné Hospital in Gabon. The legend of Schweitzer—the ninety-year-old physician who had renounced fame and fortune as a young, eminent theologian and musician to treat the medical and spiritual needs of lepers and the impoverished in Africa—had touched hearts the world over. The void he left was considerable. When LBJ learned the news at the ranch in Texas, he issued a statement: "The world has lost a truly universal figure. His message and his example, which have lightened the darkest years of the century, will continue to strengthen all those who strive to create a world living in peace and brotherhood."[63] A brokenhearted Norman Cousins wept for days upon the news. "His last walk was wonderful," Schweitzer's friend Walter Munz reported from Gabon. "Once more he passed through the orchard supported by his cane. He identified every tree he had planted and praised them for their sturdy growth, and for their beauty."[64] Only a rough wooden cross, hewn by the missionary himself, marked the grave in the damp red soil beside the Ogowe River. Bobby Kennedy put a telephone call through to Schweitzer's widow, Helene, expressing condolences. Lady Bird praised the great doctor for awakening people to be better earth stewards.

Just after Schweitzer was buried, on September 14, *CBS Reports* ran the most aggressively pro-environmental documentary in television history. Hosted by Charles Kuralt, "Bulldozed America" could have been a paid advertisement from the Sierra Club. With cameras capturing clear-cut redwoods, ugly tract housing, oceanside garbage dumps, and polluted rivers, *CBS Reports* jarred viewers into witnessing the grotesque mauling of the American Earth. Companies such as Weyerhaeuser and Arcata Redwood Lumber Company were villainized. With background music provided by the Weavers (Pete Seeger's old band), the Kuralt documentary led viewers to reject the damming of the Colorado River near the Grand Canyon National Park, support the Indiana Dunes National Seashore, and side with the conservationists trying to stop Con Edison from engineering the Storm King Mountain project on the Hudson. "You see, the American Dream was to level the wilderness, and I suppose the symbol of our power had become pretty much the bulldozers," Bill Douglas said in the documentary. "We're in the age of the machine, but the machine must not be our master. We need a new land ethic or else we're going to be consumed not by one great disaster, but by one thousand little brush fires all around the country that are too small to draw attention of anybody except the local people, and that will be lost fire by fire, battle by battle, until the whole of America is turned into a highway, into a junkyard."[65]

Lady Bird lived that very beautification credo to the nth degree. By early October, her anti-billboard campaign was coming to a head in the form of the Highway Beautification bill. Being in the thick of Capitol Hill intrigue made her squirm with discomfort. Kansas senator Bob Dole, a Republican, was not being particularly nice when he called the pending legislation "Lady Bird's bill."[66]

The first session of the 89th Congress was set to end on October 23. It had been productive in passing major pieces of LBJ's environmental legislation, notably the landmark laws on air and water quality. The president was in terrible pain most of the time in late September and early October, fighting a kidney stone, so staffers were surprised when he called an urgent meeting with his full cabinet, Democratic Party leaders, and his White House staff. The Oval Office was crowded with famous faces. "I thought, what in the world is going on?" Bob Hardesty, a speechwriter for the president, said. Later, he recalled how LBJ answered the question in his first sentence: "The Congress is about ready to adjourn and they haven't passed Lady Bird's Highway Beautification Act. And at the moment, we don't have the votes. Now, she wants that bill, and if she wants it, I want it, and by God, we are going to pass it!" With that, Johnson did what he did best: he talked his way through the Senate roster, asking those in the room after each name, "Who knows this person? How close are you? Who can make a call? Who's got chips to call in?"[67]

Johnson's army of influencers dispersed and did their best, but the vote was still uncertain. On October 7, the president was admitted to the hospital for an emergency operation to remove his gallbladder, as well as the kidney stone that had been plaguing him. That evening, members of the House were expected to attend a gala event in honor of the Congress, but leaders refused to adjourn until the Highway Beautification bill was settled one way or the other. As a result, on that memorable night, representatives stayed in session well past midnight, raucously debating it. Finally it passed.

About two weeks later, one day after Johnson returned from the hospital, on October 22, he signed the Highway Beautification Act into law at a ceremony in the East Room of the White House. The act authorized mandatory controls on outdoor advertisers and junkyards and gave financial aid to state governments for landscaping and scenic enrichment within and along highway right-of-ways. Outdoor advertising visible from within 660 feet of the right-of-ways was to be subject to control, except in industrial or commercial areas, as had been fought for by Mrs. Johnson. Junkyards within 1,000 feet of the right-of-ways, and visible from an interstate or primary highway, were now required to be screened or removed, except in industrial areas.[68]

President Johnson's heartfelt talk the afternoon of the signing was among the most relaxed and inspiring of his presidency. He made sure that his two daughters, Lynda and Luci, were present because he had decided to speak from the heart. He said:

Now, this bill does more than control advertising and junkyards along the billions of dollars of highways that the people have built with their money—public money, not private money. It does more than give us the tools just to landscape some of those highways.

This bill will bring the wonders of nature back into our daily lives.

This bill will enrich our spirits and restore a small measure of our national greatness.

Johnson reverted to storytelling that afternoon:

When I was growing up, the land itself was life. And when the day seemed particularly harsh and bitter, the land was always there just as nature had left it—wild, rugged, beautiful, and changing, always changing. . . .

Well, in recent years I think Americans have sadly neglected this part of America's national heritage. We have placed a wall of civilization between us and between the beauty of our land and of our countryside. In our eagerness to expand and to improve, we have relegated nature to a weekend role, and we have banished it from our daily lives.[69]

The beaming LBJ knew that it was a big New Conservation win for his administration.[70] The historian Robert Dallek was correct to write in *Flawed Giant: Lyndon Johnson and His Times, 1961–1973* that there was "no real priority" among LBJ's Great Society initiatives on civil rights, environment, education, and health care programs—he wanted them all. The Highway Beautification Act was perhaps the exception.[71] It was a gift to Lady Bird.

At one point, Johnson looked out at the top-tier conservationists in the East Room, Jackson and Rockefeller among them. His warm, soft-spoken, homespun Texas aura caught the audience off guard. For once, the braggadocious politician disappeared and he spoke warmly about the natural world as a father and husband. He said:

As I rode the George Washington Memorial Parkway back to the White House only yesterday afternoon, I saw nature at its purest. And I thought of the honor roll of names—a good many of you are sitting here in the front row today—that made this possible.

And as I thought of you who had helped and stood up against private greed for public good, I looked at those dogwoods that had turned red, and the maple trees that were scarlet and gold. In a pattern of brown and yellow, God's finery was at its finest. And not one single foot of it was marred by a single, unsightly, man-made construction or obstruction—no advertising signs, no old, dilapidated trucks, no junkyards. Well, doctors could prescribe no better medicine for me, and that is what I said to my surgeon as we drove along.[72]

Some observers wonder if Lyndon Johnson really cared about anything except power for its own sake. He seemed to let down his guard, though, on one of the best days in 1965: the signing of the Highway Beautification Act. Perhaps because he'd come home from the hospital only the day before, his mind was clear, his love of Lady Bird and the natural world evident in his storytelling.

CHAPTER 23

America's Natural Heritage

CAPE LOOKOUT, BIG BEND, THE GRAND CANYON

American environmental activist David Brower (1912-2000) (*center*), executive director of the Sierra Club, leads an ultimately successful protest against the proposed construction of dams on the Colorado River near the Grand Canyon National Park, Arizona, 1966.

I

When Lyndon Johnson ratcheted up the Great Society's commitment to preserve what he called "America's natural heritage" in his special message to Congress on February 23, 1966, he didn't need to search for a constituency; most Washington lawmakers were on his side. Point by point, the president offered specific conservation measures to fight air pollution, make cities healthy places, and save historic places, in addition to his extensive reform plan to clean the nation's waters. David Brower, Marjory Stoneman Douglas, Barry Commoner, and all those preservationists who had, as he once said, "stood up against private greed for public good," were enthralled. In fact, Johnson began the special message by quoting the epigraph of *Silent Spring*, Albert

Schweitzer's warning: "Man has lost the capacity to foresee and to forestall. He will end by destroying the earth."[1]

Americans in the twenty-first century should read the "Natural Heritage" speech to understand conservation leadership at its best. Only Theodore and Franklin Roosevelt equaled Johnson in prioritizing preservation and then driving appropriate measures into law around the power of his persuasive personality. Johnson rhapsodized about saving "uncharted forests, broad sparkling rivers, and prairies ripe for planting," using language that could have been lifted from Walt Whitman's *Leaves of Grass*. Then, sounding like John Steinbeck, the president insisted that California redwood groves be spared the lumberman's ax, for they were ambassadors from another time. Once again, he made it clear that national lakeshores needed to be established in Michigan, Wisconsin, and Indiana. Underneath the picturesque imagery, though, his special message was a warning in the style of *Silent Spring*. He said, "We see that we can corrupt and destroy our lands, our rivers, our forests and the atmosphere itself—all in the name of progress and necessity. Such a course leads to a barren America, bereft of its beauty, and shorn of its sustenance."[2]

Johnson's preservation theme was clear: the country was grappling with an unprecedented environmental crisis. Treasured landscapes had been compromised by synthetic pesticides, industrial debris, poor planning, and pollution of every type. He vowed to bend the will of Congress to pass bills aimed at arresting the unacceptable degradation of America's natural resources. Backed with fresh scientific data from the blockbuster report of the Environmental Pollution Panel of the President's Science Advisory Committee report, released in November 1965, he was well prepared to win Congress over. He described in heartbreaking detail the destruction of America's major rivers such as the Ohio, Mississippi, and Missouri by the discharge of treated and untreated sewage collected from a population of nearly 50 million citizens. Sounding like Bill Douglas on the loose, he lashed out at the chemical industries for poisoning rivers with by-products that wouldn't break down benignly in the water.

The good news was that the Water Pollution Control Act of 1965 had begun the process of stopping polluters by means of interstate water standards. It authorized a $3.4 billion grant program over a four-year period for the construction of sewage treatment facilities and removed the dollar limitation on grants to enable large cities to participate more fully and increased the grant percentage. In addition, the act led to a comprehensive investigation into the effects of pollution and sedimentation on fish and wildlife, outdoors recreation, water supply, and other beneficial water uses in estuaries and estuarine zones of the United States.[3]

What Johnson wanted from Congress in 1966 was appropriations to clean and preserve entire river basins from their sources to their mouths by having *one* federal regulatory agency uphold *one* standard. In his plan, the federal government would help low-income river communities receive funding for new sewage treatment plants. Going further, he called for the establishment of a National Water Commission to advise on the entire range of water resource issues. "The technology of water treatment must be improved," he stated. "We must find ways to allow more 're-use' of waste water at reasonable costs. We must remove or control nutrients that cause excessive growth of plant life in streams, lakes and estuaries. We must take steps to control the damage caused by waters that 'heat-up' after cooling generators and industrial engines."[4]

William Ruckelshaus, who in 1970 would become the first administrator of the Environmental Protection Agency (EPA) under Richard Nixon, credited Johnson's 1966 special message as the origin of the National Environmental Policy Act of 1970 and the Clean Water Act of 1972. "None of these environmental historians get it right," he groused later. "They all want to write about the Sierra Club and the Wilderness Society because it's romantic. But the story of the 1960s environmental movement should be centered on how Johnson and then Nixon launched an effective campaign for comprehensive sewage treatment in America. That's what has made the difference. The problem is that few academic historians want to become sewage treatment experts and the popular presses don't think it's newsworthy."[5]

What Ruckelshaus said was true; Johnson's determined efforts to curtail water pollution and reform sewage treatment protocols had been marginalized even at the time amid the excitement of the administration's Wilderness Act and Highway Beautification Act efforts. Johnson's public willingness to chastise corporations and citizens for treating the country's rivers as sewage drains for everything from poultry farm waste to hog manure to toxic chemicals didn't seize the public imagination.

Johnson's special message couldn't have been smarter or more realistic, yet it was largely ignored in the press, drowned out by scornful Vietnam War coverage; the number of US troops in South Vietnam would soon hit the 250,000 mark. Even as US planes began bombing Hanoi and Haiphong, the message made clear that LBJ wasn't finished with his expansionary New Conservation federal park initiatives. "It is possible to reclaim a river like the Potomac from the carelessness of man," he said. "But we cannot restore—once it is lost—the majesty of a forest whose trees soared upward 2,000 years ago."[6]

Building on Kennedy's Cape Cod feat, Johnson realized the same national park status for Cape Lookout, a pristine string of three largely uninhabited barrier islands situated off the North Carolina mainland and accessible only by boat.[7] Cape Lookout, which is a section of the southern Outer Banks, below Ocracoke Island, is possibly the finest stretch of white sand beach in the United States, a rich biosphere for conch, scallops, whelks, and sand dollars, among hundreds of other life-forms. Due to its proximity to the Gulf Stream, tons of Scotch bonnets (the North Carolina state shell) regularly washed up on the beaches. The Outer Banks had long been renowned for fishing, with charter boats taking anglers out from the mainland to catch wahoo, king mackerel, and tuna. Johnson predicted that the national seashore designation would enhance that recreational activity and that tourists from all over the world would be drawn to photograph the wild Banker horses that lived on Shackleford Banks, one of the islands making up the Cape Lookout complex. Johnson raved about the picture-postcard, black-and-white diamond-patterned vintage lighthouse on the Cape, as well as the supposedly sunken remains of Blackbeard the Pirate's circa 1700 ships.

At the White House signing ceremony for Cape Lookout National Seashore on March 10, Johnson, in good humor, caught the press by surprise. With Lady Bird at his side, he stood in the East Room, smiling at the small crowd. His wife seemed to understand the sincerity of what her husband said: "Unless we begin now to restore the environment in and around our cities, we will be condemning a large part of our population to an ugly, drab and mechanical fate."[8]

Johnson's oration that afternoon was significant because he used "the clear water and the warm sandy beaches"[9] of coastal North Carolina as a basis to explain what the New Conservation effort meant to him in personal terms and why the Great Society's "necklace of national sea and lakeshores" was the ultimate gift to future generations of Americans. Made just fifteen days following the Special Message on Preserving Natural Heritage, the remarks showed that Johnson's desire to create Parks for People wasn't a passing impulse. If there was a moment when Johnson rose to the rhetorical heights of Jack Kennedy or Martin Luther King, Jr., it was in his rambling "I See an America" speech at the Cape Lookout signing:

I see an America where city parks and plazas, as numerous as today's parking lots, bring rest and relaxation to shoppers and to office workers. If I were the mayor of any big city in America today, I would immediately put the best minds in my city to work to plan and program areas within reach of my population, and see if I could effect a coordination between the city, the State, and the Nation so that the people who make up the industrial genius of this empire will have a place to take their kids and to relax and to rejuvenate themselves for the mighty production that may be ahead in the following years.

I see an America where city streets are lined with trees and city courts are filled with flowers.

I live in a little community of 600. My daughter wrote an article last year and they paid her for it. She took the money and put four live oak trees around the plaza of that town. They were planted in memory of her grandmother, as I observed last week. And it made the town look like a different place. The bank planted some flowers in front of its doors. The merchants down the street got some shrubs and put them up there.

That can be and should be and is being done all over America. Because that is an assignment that doesn't have to come out of Washington. That is something that every lady who belongs to a garden club can make a contribution to. That is an assignment that every businessman can contribute to. That is a business getter. People will come to look at his lawn and his window and admire it and maybe make a purchase while they are there.

I see an America where our air is sweet to breathe and our rivers are clean to swim in.[10]

This marvelous "I see an America" oration drew only a smattering of publicity nationwide. As usual, most newspaper editors and TV producers were too distracted by Vietnam and Dr. King to consider much else. Few reporters understood that protecting America's treasured landscapes was truly dear to LBJ's heart. When the president said, "I do not want my children, or my grandchildren, or those who may come after me that may bear my name, to ever be able to point to blight and trash as their inheritance from me," he meant every word of it.[11] To read the speech today, it may be hard to realize how hopeful, even quaint, some of his imagery seemed in 1966.

Shortly after the establishment of Cape Lookout, the Fish and Wildlife Service funded the Rachel Carson National Wildlife Refuge in ten scattered units along fifty miles of southern Maine coastline between Kittery and Cape Elizabeth. Created in cooperation with the state of Maine, the refuge protected essential salt marshes and estuaries for migratory birds. The complex embodied Carson's sea trilogy and conservation activism by returning to nature the environment she most cherished, without much disturbance by humans. And closer to Cape Lookout, the Rachel Carson Estuarine Reserve—a cluster of islands and marshes totaling 2,315 acres near Beaufort, North Carolina, accessible only by boat—was preserved as a habitat for osprey, snowy egret, and Wilson's plovers.[12]

Following President Johnson's lead, Maine voters soon authorized a $1.5 million bond to protect the "wilderness character" of the Allagash and St. John Rivers. Those waterways drained in a storied region of wilderness in the Maine North Woods not far from Thoreau's noble Mount Katahdin. The lobbying of Bill Douglas, Ed Muskie, and Stewart Udall had paid off. The area needed rehabilitation, after logging actually altered the flow between the area's myriad waterways and lakes. The Johnson administration worked with Maine's legislature to designate the Allagash Wilderness Waterway in 1966. Later in July 1970, it became part of America's National Wild and Scenic River system.[13]

II

When the spring rains came to Washington that early April 1966, the First Lady, with Stewart Udall and George Hartzog, the National Park Service director, as companions, set off for Big Bend National Park in Texas, which covered more than eight hundred thousand acres along the green-tinted, slow-flowing Rio Grande. The towering Chisos peaks made it a very different ecosystem from her acreage in Stonewall near Austin. Pausing to look around, to soak in the rugged grandeur and majestic low mesas, Mrs. Johnson, in full booster mode, declared the desolate "big sky vistas" of west Texas her muse. Her "priceless days" in the fresh air of Big Bend country was like an anti-stress tonic.[14]

Everybody was in a playful mood. Mrs. Johnson had photos taken of seventy or so reporters as if *they* were the celebrity subjects of interest. Those in her entourage thought of the trip as a break from the hurly-burly of Washington politics. The national press corps, tagging along, were given a comic warning by Udall: "You are headed for the wide-open spaces. It is two hours to everything! Relax, take a tranquilizer, enjoy the landscape. It's bigger than all outdoors. It IS all outdoors! Get with the wilderness spirit!"[15] The airport nearest to the park was in Presidio, Texas, known to Mrs. Johnson only because, with its temperatures routinely about 100 degrees, it was often registered as the hottest place in the United States. There, one woman from Marfa, Texas, came up to her and said, "Nobody ever stops here. All they do is fly over us."[16] Mrs. Johnson recorded in her audio diary that she had greeted a herd of skittish antelopes grazing near the runway and seen a placard that read WELCOME LADY BIRD, UDALL AND YOU ALL. It was a festive arrival, made more so by the Sul Ross State college band loudly playing "The Yellow Rose of Texas."[17]

The two-hour drive from Presidio to Big Bend was through "Pancho Villa Country," territory also famous for its Comanche Indian lore. "The scenery was an ever-changing panorama of mountains and what looked like volcanic flows, swirls of tumbleweed, a few struggling bluebonnets, cactus and yucca, and finally great expanses of nothing at all except creosote bushes with little tiny yellow blossoms on them," she wrote. "It was a harsh, forbidding land, hostile to man, a land of arroyos and mountains barricaded by boulders and armored with plants that 'either stick or sting or stink,' as somebody has written."[18]

When Lady Bird arrived in Panther Junction, a visitor center in the park, she was greeted by a mariachi band and the Odessa Chuck Wagon barbecue outfit. Dressed like Elizabeth Taylor in the movie *Giant*, which had been filmed in the nearby town of Marfa, Mrs. Johnson praised the splendor of Big Bend and vowed to help the National Park Service protect other Texas ecosystems such as the lush Big Thicket and bayous, along the Louisiana border.

When JFK had visited Yosemite, in 1962, he had turned down the invitation to hike to Vernal Falls, remaining behind at the lodge in his navy blue suit. In contrast, Lady Bird, in blue jeans and checkered shirt, the local attire, wanted to explore every inch of Big Bend as a Girl Scout might. Udall was attentive to the First Lady's needs. Her cottage in the national park was redecorated and sanitized.[19] Instead of taking his usual anxious, lunging strides, he held back, not wanting to walk ahead of her in west Texas. To some, that shift in habit reeked of sucking up to the boss's wife. To others, it was a masterstroke for the preservationist movement, which at that very moment was advancing Wild and Scenic Rivers legislation through Congress. Acting like a schoolteacher, Udall spoke to Mrs. Johnson about the sandstone cliffs, piñon pines, and peregrine falcons. "Stew is a natural outdoorsman," the First Lady remarked the next day.[20]

Hiking up Lost Mine Trail, Mrs. Johnson observed with childlike enthusiasm the haughty chickadees, sky-dancing hawks, tiny purple wildflowers, and mighty oaks. "Stew and I set the pace and a naturalist came along beside us to tell us about the wildlife," Lady Bird recorded in her diary. ". . . On the way up, surrounded by the marvelous vistas, Stew and I both said at once, 'On a clear day you can see forever!' And you can. In one direction you see into Mexico. It's a wild, free land and it does something to you."[21]

No tour company could have planned a better trip to Big Bend than the one the NPS had unfurled for Lady Bird. Everything was perfectly choreographed. The El Paso Natural Gas Company paid for catering, so taxpayers weren't stuck with the bill. At dusk a full moon rose higher and higher. Wildlife experts lectured about mountain goats, rattlesnakes, and the strangeness of the landscape. Navajo blankets were spread out on the ground for stargazing. Old English ballads from the Appalachians and cowboy songs, such as "Little Joe the Wrangler," were sung hootenanny style. Thick T-bone steaks were on the menu, as were riveting frontier tales about the rancher Charles Goodnight and the Comanche leader Quanah Parker. It was conservation politics as fiesta. "Sometimes I think the Lord made up in this Western country for what he didn't give us in rainfall and in verdant vegetation with the glory of the sky," Mrs. Johnson reflected in her diary. "It was the most superb theater, fit subject for a symphony or a poem—but for me just an hour of delight that was almost tangible—of the heightened feeling of being alive. . . . What a night to remember!—sheer magic, and a day worth five ordinary days of living."[22]

At dinner, Mrs. Johnson wore a bright red dress that caught Hartzog's eye. Playing stage manager, he insisted that the First Lady stand next to a spectacular cholla tree in full bloom. The twilight lighting was perfect, and Hartzog clicked away on his no-frills camera. Bursting with excitement, he sent the roll of film to Marathon, Texas, to be developed overnight and then released to the press. Late that evening, a runner brought the pictures to Hartzog. "I don't think you are going to like these," he said.

"Why?" Hartzog asked the runner. "Didn't they come out?"

"Oh yeah," he said, "they came out real nice—in black and white!"[23]

After a good night's sleep, Lady Bird was served "cowboy" coffee, ham, sourdough biscuits, and pancakes for breakfast. April 3 was Palm Sunday, and the entourage was observant. The day would be full, but not quite as insouciant as the previous one. After chow, the First Lady embarked on a five-hour rubber-raft journey down the Rio Grande through the Mariscal Canyon, hemmed in by the Chisos Mountains. But it wasn't an unscripted expedition. Greyhound buses spat diesel fumes down dirt roads to the Tally put-in spot. A tape recording of coyotes was played to create an atmospheric effect. Secret Service officers in rafts were omnipresent on the ten-mile journey, with Lady Bird's raft taking the lead. Portable latrines were erected along the way. It all added up to an overkill of VIP treatment.[24] All 139 people who floated with the First Lady that afternoon down the Rio Grande River—more than the population of many a west Texas town—were treated as if dignitaries. Lady Bird's press secretary, Liz Carpenter, was barking orders like a Broadway stage manager. A ticked-off Secret Service agent was heard to say, "I keep this gun for just one reason."

"Why?" someone asked.

"To shoot the first son of a bitch that tries to rescue Liz Carpenter in case she should fall in," he snapped.[25]

The journey down the Rio Grande is one of the great outdoor adventures the National Park Service offers. In a raft, the traveler looks up to see enormous cliffs rising with giant pipe organs and cathedral spires like lording shadows. It was chilly on the river in the morning, only to turn broiling hot by lunchtime. "There were always cries of 'Man overboard!'" Mrs. Johnson wrote in her diary. "And I remember while shooting down the rapids I saw one man clinging to a rock and fighting a losing battle to maintain his grip against the swift rush of water."[26]

Once the raft trip ended at Rio Grande Village Camp Ground, margaritas and Carta Blanca beer were served for a warmhearted toast to US-Mexican relations. In her diary of the Big Bend trip, Mrs. Johnson raved about seeing a roadrunner in an ocotillo, green cottonwoods, Colima warblers, and "all kinds of cactus," along with Spanish dagger, agave, and lechuguilla.[27] Udall handed out faux certificates to the press members who had rafted, designating them "Original American Wet-bottoms." And he praised Lady Bird for her uncomplaining demeanor and overall frontier spirit. "You have had a wilderness experience," Udall told her. "I think you will look back five, ten, or twenty years from now and remember this as spectacular."[28]

III

While Lady Bird was in Texas, President Johnson signed an executive order establishing the President's Council and the Citizens' Advisory

Committee on Recreation and Natural Beauty, a development inspired by his wife. LBJ joked about why he didn't accompany Mrs. Johnson on her Rio Grande rafting excursion with Udall and Hartzog. "We have a wonderful Big Bend Park that Lady Bird has been visiting, but it is a long way for me to get to stroll in the woods," he chuckled. Then he squinted as if to make a profound point: "It is a long way for people to go who live in Washington and Baltimore and Philadelphia and New York and Los Angeles. We need some of those areas right where the people live. There is not a front yard in the country, there is not an apartment house in the land, there is not a public building anywhere, there is not a street, there is not a sidewalk, and there is not a road that cannot be improved and made more beautiful."[29]

LBJ knew that America's blighted cities in 1966 were in dire financial and environmental shape. Urban parks, often blacktopped, were beset by drug dealers and petty criminals. Garbage pickup in low-income zones was spotty at best. Reports that children in schools were falling sick from lead and asbestos contamination made front-page news. Deadly smog was especially thick in northeast cities that spring and summer. Determined to rectify the health hazard, Norman Cousins, still editor of the *Saturday Review* and still a leading anti-nuclear activist, chaired New York City's Air Pollution Task Force, which sought to eradicate chronic smog in the five boroughs. Cousins told the *New York Times* that day after day when he looked out the window of his apartment on Park Avenue at 35th Street, he saw the hideous towers of a power plant "spouting smoke and fumes into the air." Wedded to scientific data, he oversaw a pollution crisis report that was released that May and caught LBJ's attention. Reflecting on how his work with SANE from 1957 to 1963 contributed to President Kennedy's Limited Nuclear Test Ban Treaty, Cousins insisted that the emerging fight against smog, atomic radiation, and the Vietnam War were intertwined. "It's the same thing, a fight to preserve the human environment."[30]

Like Norman Cousins, LBJ was keenly sensitive to the problem of pollutants and dangerous chemicals in impoverished American neighborhoods. But even he was caught off guard by the revolutionary struggle that emerged in California's San Joaquin Valley in 1966, led by Mexican American agricultural leaders and a venerable reform group called the National Consumers League. The setting was a hearing in Visalia, California, the home base of the National Farm Workers Association (NFWA), a new union struggling to organize produce pickers. Three senators from the Subcommittee on Migratory Labor—Harrison Williams of New Jersey, Bobby Kennedy of New York, and George Murphy of California—were on a short tour to investigate conditions for California farmworkers. The star witness on March 16 was Cesar Chavez, the leader of NFWA (which was changed to the United Farm Workers in 1972). He spoke bluntly about the economic plight of laborers who were paid far below the standard minimum wage. Another witness, Elizabeth Coleman, was on the board of the National Consumers League, a group started by the Nobel Peace Prize winner Jane Addams that had been advocating better treatment of migrant workers since 1906. Citing the fact that agricultural work was the third most dangerous employment of any sector of the economy, Coleman listed exposure to pesticides as one of the two perils of a picker's work, elevating it to equal status with the danger of farm machinery.[31]

Significantly, the only other people at the Visalia hearing who mentioned agrichemicals were farm owners, who threatened that if wages for field-working staff increased, they would replace the weed pickers with herbicides. That, of course, would expose the remaining pickers to even more chemicals. Chavez was nearly overwhelmed by the challenge of organizing laborers in the face of such dire threats, but the workers' health was his main concern. For that reason, he sought to establish a rudimentary clinic when the NFWA organized its first strike. The reports from physicians and nurses at the clinic were alarming. "Union leaders and health care workers realized the pervasive and collective nature of the pesticide problem," wrote the historians Laura Pulido and Devon Peña in the journal *Race, Gender & Class*.[32] When Cesar Chavez and Dolores Huerta, the driving forces behind the NFWA, saw the ghastly reports, they added health and safety to their labor demands, and by that, they meant strict regulation of the use of agrochemicals.

Press reports circulated that Cesar Chavez had been greatly influenced by *Silent Spring*, which had awakened him to serious medical problems caused by pesticide sprays that were adversely affecting agricultural farmhands. The first time DDT was banned in the United States wasn't by the Environmental Protection Agency in 1972; it was by a NFWA contract with grape growers in 1966–1967. Both of Chavez's grape boycotts, beginning in the late 1960s, focused on pesticides as a health threat to consumers as well as to farmworkers. Chavez's longtime nurse (and later physician), Marion Moses, became one of the nation's foremost authorities on the effects of pesticides on agricultural workers. Chavez's last, and longest, tactical fast, in 1988, was to protest the pesticide poisoning of farmworkers and their children.[33]

The concept born in the Central Valley of California was that those directly affected by the use of chemicals by any controlling entity— whether an employer or a neighboring factory—had the right to object to exposure to those chemicals and even restrict it legally. The principle was morally sound: nobody should get sick while picking fruit and vegetables. At the time, in the mid-1960s, Chavez and Huerta put clauses insisting on environmental justice measures in the union contracts they submitted to farm owners. Throughout the arduous process of establishing the union and securing the contracts with farm owners, they made sure that the clauses remained. Chavez used the pesticide issue to bring environmental activists onto his side and as a show of power to the owners of big farms, who were used to monopolizing the power in the Central California agricultural zone.

Kennedy sided with Chavez's cause in the press. There wasn't anything about Cesar Chavez that RFK didn't grow to admire. When the Senate hearing ended, the union leader announced that he would be walking three hundred miles from Delano to Sacramento to protest unfair labor practices and the use of pesticides. A Roman Catholic, Chavez declared his march to be a *peregrinación* (pilgrimage) to uplift the spirits of farmworkers in their struggles to unionize.[34] He said, "Since this is both a religious pilgrimage and a plea for social change for the farmworker, long advocated by the social teachings of the Church, we hope that the people of God will respond to our call and join us for part of the walk just as they did with our Negro brothers in Selma."[35]

Singing hymns and holding placards that read ¡VIVA LA HUELGA! and VIVA LA CAUSA, the protesters marched during the Catholic holy season of Lent.[36] The march ended on Easter Sunday, when Chavez arrived in Sacramento as a true Gandhian martyr in spirit and in body—walking with blistered and bleeding feet to awaken the world to the plight of farmworkers and migrant field hands. Even though that Easter RFK was at home in McLean, Virginia, overseeing an egg hunt with his family and the Udalls, he read about Chavez's historic march walk with boundless appreciation. As Ethel Kennedy recalled, her husband considered the NFWA leader a "brother" in the crusade for human rights.[37]

IV

Because her highway beautification effort and the Big Bend trip were press successes, Mrs. Johnson headed to California in September 1966 to keep promoting the New Conservation. At the start of the trip, she visited Point Reyes National Seashore, where she hosted children involved in the Audubon Society and in the Sierra Club and also a constellation of labor union leaders. The event took place in front of the attractive beige cliffs of Drakes Beach, one of Ansel Adams's favorite subjects to photograph. She told the guests, "One of the dominant facts of modern times is that Americans are now more and more divorced from natural surroundings." She intimated that every American needed a place to recreate—a national or state park to love. "The growing needs of an urban America are quickening 'the tick of the conservation clock,'" she warned. That evening, she arrived at the San Francisco Opera to attend a performance and was greeted with a bomb threat. Police arrived from all corners. Protesters paraded in front of the venue with signs that read, LADY BIRD, BRING OUR TROOPS HOME NOW and BEAUTIFY AMERICA, BUT DEFOLIATE VIETNAM?

In her 1966 book *Styles of Radical Will*, Susan Sontag compared LBJ's Vietnam War with the genocide of Native Americans and wilderness conquering. "The white race is the cancer of human history," Sontag wrote, "it is the white race and it alone—its ideology and inventions—which eradicates autonomous civilization whenever it spread, which has upset the ecological balances of the planet which now threatens the very existence of life itself."[38] Usually calm and collected, Mrs. Johnson was rattled by "the aura of madness, sort of a mob spirit" in California, and the writing of Sontag.[39] She saw first-hand how the Vietnam War was tearing the country apart.

After San Francisco, Mrs. Johnson headed to Carmel to promote California's first "scenic highway." State Senator Fred Farr had asked her to dedicate the Big Sur Coast strip of Highway 1, which she did gladly. Starting in Carmel, her entourage visited Hurricane Point and Bixby Creek Bridge. Ansel Adams was there to greet and photograph the First Lady peering out into the blue Pacific. For a day, Mrs. John-son forgot about Vietnam and urban riots and enjoyed the symphony of waves crashing and seagulls crying.

Throughout 1966, the Johnson administration was engaged in one of the most controversial environmental battles of the decade: the Bu-reau of Reclamation plan to build dams on the Colorado River at both ends of Grand Canyon National Park. To the shock of environmental-ists then and now, Udall (who oversaw the Bureau) favored it. Perhaps he hoped to return to Arizona someday and run for either governor or senator and so felt obliged to take a pro-dam stance, which would be popular with the voters. Turning a cold shoulder to him, Brower, Adams, and other preservationists did not share his enthusiasm for the recreational opportunities afforded by the reservoir that would be created by each gargantuan dam, and Udall unexpectedly became the bane of the Sierra Club. He was assailed as a turncoat for support-ing the Colorado River Basin Project, as the twin dams were called. Around the time Mrs. Johnson was in Big Bend, Brower described himself to Udall "as a conservationist who has supported you strongly and wants to improve that support." Brower continued, however, that Udall could "hardly win a blacker mark . . . so far as I can see it, your conservation career is at stake and our parks and wilderness too."[40]

Most Indigenous tribes in Arizona opposed the double damming of the Colorado, but Udall never allowed them a public hearing. As assuredly as Martin Luther King, Jr., broke with Lyndon Johnson over Vietnam, Brower broke with Udall over the proposed Marble Canyon and Bridge Canyon Dams on the Colorado River. The Sierra Club had been wrong in 1955, he said, to allow the dam at Glen Canyon ("The place no one knew") to be constructed in exchange for protecting the sanctity of Dinosaur National Monument. Not this time.

In 1966, Brower refused to let the desecration of Grand Canyon

National Park happen on his watch at the Sierra Club. Like Thoreau in the Mexican-American War, he even pledged to go to jail before he would allow the dams to be built and flood the Grand Canyon, turning ecosystems adjacent to a sacred World Heritage Site into man-made reservoirs for pleasure boats and electricity for Nevada casinos. The general public, he believed, could be mobilized to save the Grand Canyon.

The Johnson administration, with Floyd Dominy in the commissioner's seat at the Bureau of Reclamation, saw the proposed dams as cash registers that would finance the Central Arizona Project, which would divert Colorado River water to provide Phoenix and Tucson with electric power and the environs with irrigation. Because Brower was a comfortable public speaker, at ease explaining ecological science and ready with comic turns of phrase, he was popular on the university circuit. Visiting campuses, he recruited volunteers for the Save the Grand Canyon cause. The charismatic Brower even convinced three recent MIT graduates—Alan Carlin, an economist with the RAND Corporation; Laurence Moss, a nuclear engineer; and Jeff Ingram, a mathematician—to lend their analytical skills to the Sierra Club to help sway public opinion.[41]

Throughout 1966, the battle over the Grand Canyon grew in ferocity week by week. Each side was dug in: environmentalists, in favor of sanctity for the Grand Canyon, and reclamationists, in favor of hydroelectric power salvation. Aspinall, chairing the House Interior Committee, and Udall, running the Interior Department, wanted the dams and had considerable clout on their side. The Sierra Club had moxy. Heading the resistance was Brower, who appeared on TV shows such as *The Tonight Show*, starring the enormously popular Johnny Carson, and CBS News to argue for leaving the Grand Canyon alone. He had an ally in John Oakes, the Sierra Club board member who was a senior editorial writer for the *New York Times*. A long editorial in the paper on January 17, 1966, stated, "If this dam is built, it will destroy irreplaceable scenic and archeological values. . . . It might therefore be supposed that the arguments for building it are exceptionally strong. In fact, they are not."[42] Joseph Wood Krutch also defended the need for the Grand Canyon to remain undiminished by bulldozers. "The canyon is at least two things besides spectacle," he wrote. "It is a biological unit and the most revealing single page of earth's history anywhere on the face of the globe."[43]

Leading politicians on the Western Slope of the Rockies—both Democrats and Republicans—insisted, to the contrary, that the massive project was necessary to make the desert bloom. They believed that the Grand Canyon dams would rival the Hoover Dam and the gargantuan Colorado–Big Thompson Project reservoir network in the generation of power. Brower testified at congressional hearings in 1965 and 1966, eloquently describing what would be lost if dams were built. A Republican senator accused him of being a patsy for California water interests, which competed for every drop they could get. They had no desire to see Arizona siphon off billions of gallons. But Brower never lost his cool. He became even more of a media superstar when *Life* magazine profiled him on May 27, 1966, in an article entitled "Knight Errant to Nature's Rescue," with gorgeous photographs of the Grand Canyon to illustrate the story. The writer, Hal Wingo, described Brower as unquestionably America's "No. 1 Working Conservationist."[44]

Enjoying the limelight, Brower hatched a controversial plan to rattle the public about the danger to the Grand Canyon posed by the proposed dams. He had been impressed with advertisements created by the edgy San Francisco agency Freeman & Gossage. In particular, he admired their charmingly written campaign for Whiskey Distillers of Ireland, which had run in the *New Yorker*. He also liked that it used his favorite typefaces, Centaur and Arrighi, the same ones he had long used in his exhibit-format books. Without waiting for clearance from the Sierra Club board, he drafted language for an advertisement in the form of "An Open Letter to Stewart Udall," and then took it to the Freeman & Gossage offices, located in an old firehouse in the Barbary Coast neighborhood of San Francisco.

After days of haggling, Brower and the advertising firm came up with two different statements. Brower's version was headlined "Who Can Save the Grand Canyon? You Can and Secretary Udall can too, if he will." At the bottom of this ploy was a Sierra Club coupon asking for a financial contribution. The Freeman-Gossage ad—the one that went into the *New York Times*—was far more cunning. The headline read, "Now Only You Can Save Grand Canyon From Being Flooded . . . For Profit." It provided seven coupons addressed to decision-makers such as President Johnson, Interior Secretary Udall, and Representative Aspinall. When the full-page ad appeared in the *New York Times* that June, it didn't just hit a nerve; it electrified the political discourse of the nation. "I never saw anything like it," Dan Dreyfus of the Bureau of Reclamation complained. "Letters were arriving in dump trucks. Ninety-five percent of them said we'd better keep our mitts off the Grand Canyon and a lot of them quoted the Sierra Club ads."[45]

When Udall saw the *New York Times* stunt, he was apoplectic, feeling it was a betrayal by Brower. The Interior Department, under his leadership, had fought in tandem with Brower. That very month the Sierra Club, with Udall's approval, was lobbying Congress to establish the North Cascades and Redwood National Parks. In retaliation, Stewart had his brother Mo direct an IRS agent to deliver a legal notice to the Sierra Club's San Francisco office stating that since the ad had sought to influence US government decisions, it had nullified the club's status as a nonprofit.

The IRS decision to withdraw the Sierra Club's tax-exempt status didn't intimidate Brower one iota. The IRS action became headline news across the country, reported as the federal tax bully Goliath picking on little David Brower, an earnest national park defender from Berkeley. The week the first *New York Times* ad ran, Sierra Club membership soared. Likewise, the Sierra Club book *Time and the River Flowing: Grand Canyon* (1964) had a surge in sales. Striking back at Stewart (and Mo), Brower placed a second ad in the *New York Times* on July 25, 1966:

**Dinosaur and Big Bend. Glacier and
Grand Teton, Kings Canyon,
Redwoods, Mammoth, Even Yellowstone
and Yosemite. And
The Wild Rivers, and Wilderness.**

How Can You Guarantee These, Mr. Udall,
IF GRAND CANYON IS DAMMED FOR PROFIT?

Once again, bushels of coupons and letters poured into Udall's Interior Department office. Seemingly overnight, Udall became the demon of counterculture activists, environmentalists, and every other type of outdoors-loving American. Feeling scapegoated by Brower, Udall pondered his next chess move. He thought he was media savvy, but Brower had beat him to the punch. Like a lightning bolt, a third Sierra Club ad ran in six newspapers. It was the most controversial of them all: SHOULD WE ALSO FLOOD THE SISTINE CHAPEL SO TOURISTS CAN GET NEARER THE CEILING?[46]

The Sierra Club challenged the IRS in court for rescinding its tax status and lost. The Sierra Club suffered a drop of donors because contributions were no longer tax-deductible, but it gained a sharp increase in membership. Brower observed, "People may not know whether or not they like the Sierra Club, but they know what they think about the IRS."[47]

While the self-aggrandizing Brower was a prickly person to deal with, his campaign to save the Grand Canyon was a masterpiece of environmental activism. Once he turned national attention to the dams, he then made clear that the Sierra Club was also opposed to logging the last redwoods in California; the construction of a power plant at Storm King Mountain along the Hudson River in New York; and a plan to build a new air cargo terminal in the San Francisco Bay Area. In November 1966, he testified before a House committee against the Grand Canyon dams. Mo Udall, who was on the committee, tried to paint Brower as a wilderness extremist, unable to compromise. Brower won the showdown:

MR. UDALL: What would the Sierra Club accept? If we have a low, low, low Bridge Canyon Dam, maybe 100 feet high, is that too much? Is there any point at which you compromise here?

MR. BROWER: Mr. Udall, you are not giving us anything that God didn't put there in the first place, and I think that is the thing we are not entitled to compromise. . . .

MR. UDALL: You say that you will continue the fight and try to defeat the bill unless it contains a provision setting aside that dam site once and for all in the Grand Canyon National Park.

MR. BROWER: We have no choice. There have to be groups who will hold for those things that are not replaceable. If we stop doing that we might as well stop being an organization, and conservation organizations might as well throw in the towel.

MR. UDALL: I know the strength and sincerity of your feelings, and respect them.[48]

Shortly after that televised exchange, both Udalls threw in the towel on damming the Colorado River. The bill authorizing the dams withered away without ever coming to a vote in Congress. The biggest influence on Stewart Udall's about-face on the dams, however, wasn't Brower; it was the fact that Senator Henry Jackson, the chairman of the Senate Interior Committee, opposed them because they were "the primary means of financing the importation of water into the Colorado River Basin from the Pacific Northwest."[49] He didn't like that part of the master plan. Nevertheless, Brower claimed the victory.

In September, before the proposal collapsed, Brower sent President Johnson a petition signed by California's residents in strong opposition to the Bridge Canyon (Halipai) and Marble Gorge Dams being built. It was the wording of the petition that was noteworthy: "As power plants ONLY we can see no reason why alternative means cannot be used, such as thermal or atomic plants to provide Arizona with its energy needs." That Brower was willing to consider nuclear energy to stop the dams must have surprised Johnson, who did read the petition.[50]

In the aftermath of the Grand Canyon showdown, the Bureau of Reclamation had to find an alternative way to electrify Arizona's growing metropolises. The solution chosen was the Navajo Generating Station, a coal-fired power plant near Page, Arizona. Philip Hyde was photographing the Navajo wild lands for the Sierra Club when that unfortunate turn of events happened. "Once more, the environmentalists buckled down the battle to save a last piece of the natural river, and once more—for the second time in a century—they were victorious," Donald Worster wrote in *Rivers of Empire: Water, Aridity, and the Growth of the American West*. "Once more, however, they lost something as well, for the energy to make the CAP [Central Arizona

Project] go would be derived instead from coal strip-mined on Hopi sacred lands at Black Mesa in northern Arizona and burned in the Navajo Generation Station near Page, polluting the crystalline desert air with ash and poison gas."[51]

In the end, Lyndon Johnson received scant credit for anything he did on the conservation front in 1966. He couldn't catch a break. That his administration ultimately stopped the dams at Marble Canyon and Bridge Canyon was buried under criticism by the *Washington Post* and the *New York Times* of the vengeance that the IRS had inflicted on the Sierra Club. The Great Society's urban renewal effort was painted by the New Left as white American gentrification that would only make cost-of-living prices soar in lower-income communities. Johnson's would-be allies in the conservation world castigated him for protecting America's natural heritage while simultaneously destroying the tropical ecosystems of Vietnam with napalm.

Photographer Philip Hyde was a leading Sierra Club activist for saving such treasured havens as the Canyonlands, Redwoods, and the Grand Canyon. Many of his original compositions are considered priceless conservation manifestos.

When, in September 1966, Johnson launched a Water for Peace program green-lighting the Army Corps of Engineers to build dams in so-called Third World nations, Bill Douglas protested. From Johnson's point of view, he was ensuring foreign populations would have water. From Douglas's, the president was wreaking unnecessary havoc on global ecosystems that perhaps no one fully understood. Johnson, raised in the 1910s in the arid state of Texas, where water was sometimes a luxury, was slow to recognize that environmentalism's constituency was no longer impressed by the impulse to generate electricity from hydroelectric dams. FDR had been dead for two decades and hydroelectric dams weren't in vogue. Compromise, for the environmentalists, would never be a viable option when it came to large-scale construction projects the Bureau of Reclamation or Army Corps of Engineers had up their sleeves.

CHAPTER 24

Defenders

HISTORICAL PRESERVATION, ENDANGERED SPECIES, AND BEDROLL SCIENTISTS

Endangered whooping crane eggs are transferred in portable incubators en route to their new home at the Rare and Endangered Species Division of the Patuxent Research Station in Laurel, Maryland.

I

Over the Fourth of July weekend in 1966, Bobby and Ethel Kennedy went on a four-day rafting journey down the Snake River in Idaho's Hells Canyon, alternating between inflatable rubber rafts and kayaks. Once again, the Hatch Brothers were the outfitters for the expedition. Just as JFK loved the oceans, RFK's heart turned to fast-flowing rivers, particularly ones like those in Idaho where he could run the rapids. What had become abundantly clear to his daughter, Kerry Kennedy, was that her father was happiest when coaxing rainbow trout into a landing net. "When Dad was river-running, he was in his element," she recalled. "All of my strongest memories of him, in fact, were in this kind of nature setting."[1]

Joining the Kennedy family on the Snake River journey through Hells Canyon were former NASA astronaut John Glenn and his wife, Annie; the singer Andy Williams and his wife, Claudine Longet; and the legendary photographer Harry Benson. At the time, RFK was working with Frank Church to keep rivers in Idaho undammed and to launch a pollution cleanup on the Hudson. "My father was crestfallen over how America's rivers had been abused," RFK Jr. recalled. "He wanted us kids to be protectors of wild rivers and wilderness. So the Snake trip had conservation meaning behind it."[2]

Unobtrusive, yet talkative, and a hearty companion, Benson was instantly adopted by the Kennedy family. "It was a photographer's dream," he recalled. "They let me go about my business as if invisible. My main goal was trying to keep my camera dry. Bobby would constantly try to help me cover my equipment when we were in fast waters when you had to hold on. The light in Idaho was so wonderful at dusk when we all sat by a river campsite and recounted the days' strenuous activities."[3]

For Kerry Kennedy, who would become an important human rights activist in the twenty-first century, her father's love of the Hells Canyon gorge and Rocky Mountain Bighorn sheep was contagious. "The Snake was so cold, yet Daddy kept looking for cliffs where we could all dive into the river. This was part of his regime for us kids to find ways to overcome fear. Daddy revered William O. Douglas, who had taught him how to camp properly. Few know this, but Daddy loved photographs. He must have taken hundreds with Douglas. Just seeing a hawk or a deer brought him so much joy. At campfires along the Snake, he would read aloud from *America's Best Loved Poems and Ballads*, things like 'The Cremation of Sam McGhee' and 'Casey at the Bat.' On other river trips it was Thoreau or Rudyard Kipling's 'If.'"[4]

While the Kennedys' adventure on the Snake River was making for incredible photographs and backcountry stories, just like the previous year's Green River run along the Utah-Colorado border, the more desk-bound Lyndon Johnson spent the summer working to sign a flood tide of seven conservation bills into law at an October 15, 1966, White House ceremony. Smiling broadly, thoroughly in his element, he praised the hard-bargaining 89th Congress and Stewart Udall for prioritizing public lands protection in the American public policy arena. "When our forefathers came here they found nature's masterpiece," he said in storytelling mode. "They found a beautiful, rich, varied, fertile land, a whole continent to farm and hunt on, and to explore."

Then Johnson turned serious. Because of our releases of untreated sewage and industrial waste, smog and water pollution, excessive clearcutting, and wetlands drainage, the country was in an acute ecological crisis. The time had arrived—in fact, was actually long past due—for hard-and-fast New Conservation measures to be enacted. LBJ's Great Society wanted to reconnect *all* Americans—urban and rural—to America's bountiful national heritage by significantly expanding the National Park System and recreation opportunities in existing national forests. To Johnson, his establishment of the Delaware Water Gap National Recreation Area, which soon attracted more than ten million visitors annually, was a crown jewel of this New Conservation grid.[5] "We are creating recreation areas where they will do the most good for the greatest number, for all of our people—near our cities, where most of our people live," he said. "We are putting national parks and seashores where a man and his family can get to them. The father that is the mechanic can load his five children in his car, and in an hour or 2 hours, or 3 hours, take them to a nearby playground."[6]

The first bill Johnson authorized was the stark Guadalupe National Park in west Texas. The park was configured around the massive limestone dome of El Capitan (8,085 feet), the highest peak in the state. Strange as it seemed, the entire Chihuahuan Desert setting had once been a reef flourishing beneath the waters of an ancient inland sea, the same vanished sea that had carved out the honeycomb of Carlsbad Cavern, forty miles to the northwest.[7]

At that same ceremony, Johnson established Pictured Rocks National Lakeshore, the first site to receive the coveted designation. Located in the Upper Peninsula of Michigan, Pictured Rocks had dazzling waterfalls that spilled over mineral sandstone cliffs and into Lake Superior.[8] When the sun shone on the fifty- to two-hundred-foot layered cliffs and groundwater trickled out of their cracks, mixing with the minerals in the rock, a kaleidoscope of gorgeous colors revealed itself: red and orange iron, blue and green copper, brown and black manganese, and white limonite. On an even larger scale, intricate turrets and spires had been sculpted from rock by centuries of unceasing waves and harsh weather.

Nobody was more elated by the establishment of Pictured Rocks than Philip Hart, Michigan's Democratic senator. Since the early 1960s, he had campaigned for the national lakeshore with the same vigor as his Minnesota canoeing friend, Sigurd Olson, had on behalf of the Boundary Waters. After World War II, Hart had begun to hope that the Upper Great Lakes region—the northern portions of Minnesota, Wisconsin, and Michigan—would grow beyond its reputation as a resource-rich frontier of interest chiefly to extraction industries. He saw it as the "North Woods," a scenic destination whose beauty and opportunity for outdoor recreation would draw tourists from Midwest cities such as Chicago, Milwaukee, and Detroit. Whenever Michigan's Upper Peninsula came up in conversation, Hart's expression became childlike, as if the cooling shade of the woods and bright blue waters of Lake Superior had been etched into his being. Johnson had entrusted Hart with shepherding the Voting Rights Act past the threat of a filibuster by southern legislators; the saving of Pictured Rocks was a reward of sorts. Hart was elated—but not satisfied. Immediately after the Upper Peninsula park was signed into law, he began maneuvering to make Michigan's Sleeping Dunes the second national lakeshore.[9]

To please the California conservationists who were present and as a courtesy to Lady Bird, who was standing nearby, President Johnson spoke at the signing ceremony about the bills enacted that day authorizing money for the enlargement of California's Point Reyes National Seashore. The site had an uptick in tourism since its designation in 1963. Unfortunately, it hadn't seen an equally massive increase in funding for land acquisition. With passage of the Land and Water Conservation Act (LWCA) in 1964, funding was now available for additional habitat purchases. "We increase the land in the Point Reyes National Seashore in California," he announced, laughing. "And if we don't stop Mrs. Johnson going out there, we will increase it some more, I am afraid."[10] There was a scheme afoot to save offshore Pacific waters just beyond the Marin County peninsula as a federal marine sanctuary.

The third area that Johnson's signature established that afternoon was the Big Horn Canyon National Recreation Area, situated between Billings, Montana, and Sheridan, Wyoming. It was the traditional hunting grounds of the Crow people and part of the newly protected land was on their reservation. The completion of Yellowtail Dam had flooded the seventy-one-mile-long Bighorn Canyon, which had soon become a premier recreational boating and fishing hotspot. The NRA transformed eastern Montana into a world-class trout fishing area.

The most unusual new unit, established in Fairfax County, Virginia, was the first national park for performing arts. Wolf Trap National Park, a 130-acre property, would host concerts and educational programs alike. The stage opened in July 1971, and it's considered to be LBJ's more rustic version of the John F. Kennedy Center for the Performing Arts, which was then under construction along the Potomac River. In the coming decades, such music greats as Bob Dylan, Aretha Franklin, and Bill Monroe would perform at Wolf Trap.

The National Historic Preservation Act of 1966, another project promoted by the First Lady, was an integral part of the signings that October. The act provided and expanded the National Register of sites, buildings, districts, and even objects. It established a matching federal grant program to states and the National Trust for Historic Preservation for planning, acquisition, and restoration of historic properties.[11] A few months earlier, Mrs. Johnson had written the foreword to *With Heritage So Rich: A Report of a Special Committee on Historic Preservation*, commissioned by the US Conference of Mayors with a grant from the Ford Foundation. The inspiration came during the 1965 White House Conference on Natural Beauty. The report questioned the sacrifice of historic buildings to the exigencies of the urban renewal movement sweeping through American cities and towns.

In startling fashion, *With Heritage So Rich* reported that almost half of the twelve thousand structures documented by the Historic American Buildings Survey of the National Park Service since the survey began in 1933 had been leveled. "This is a serious loss and it underlies the necessity of prompt action if we are not to shirk our duty to the future," Mrs. Johnson wrote. "We must preserve and we must preserve wisely."[12]

Lady Bird's interest was goaded by the failure of Eisenhower's

Interstate Highway System to preserve historic places such as the Storyville District in New Orleans and the fairgrounds in San Antonio. She was dismayed that homes in tree-rich cities such as Buffalo, Detroit, and Baltimore were being bulldozed to make way for highways. The famous Oregon House in Bucyrus, Ohio, built in 1829 as the first inn on the overland stage route, had recently been razed to build a restaurant. Boston had allowed its opera house to meet a similar fate. Lady Bird saw this as missing out on history tourism revenue. Since 1960, Colonial Williamsburg, in Virginia, for example, had become America's number one tourist attraction; the Rockefeller family had donated millions of dollars to the restoration effort. Why couldn't such refurbishment, she wondered, be replicated to create economic engines in Charleston and Savannah?

The National Historic Preservation Act insisted that the thoughtless tear-down attitude toward old buildings had to end. With a little white paint and reconditioning, old Victorian houses could take on a new shine. Such buildings as the Battle Creek Post Office in Michigan and the Masonic Temple in Cincinnati weren't "eyesores" to Mrs. Johnson but centerpieces of historic districts. Determining what "historic" was and wasn't, the 1966 act established a sensible process for architectural preservation. Vintage historic structures and monuments that would be affected by federal projects—or by work that was federally funded—had to be documented to standards issued by the secretary of the interior. The law required individual states to determine historic sites in their jurisdiction. All fifty were required to open historic preservation offices to complete an inventory of culturally important sites. The law also created the Advisory Council on Historic Preservation and the National Register of Historic Places, an official list not only of individual buildings and structures but also of districts, objects, and archaeological sites. All of these measures were implemented in the very nick of time, before the wrecking ball decimated the cultural atmosphere and old-style elegance out of America's urban neighborhoods.

The objectives outlined in *With Heritage So Rich* were visionary and implemented within coming years with considerable success.[13] Because of the act, Montana's Butte-Anaconda area, the world leader in copper production in the late nineteenth and early twentieth centuries, was designated a historic district and has been protected as a national heirloom. Other such districts—Virginia City, Nevada; Brooklyn Heights, New York; Annapolis, Maryland; and Eureka Springs, Arkansas, sprang up. Entire blocks, in some cases, were renovated with exacting care and plenty of bronze plaques. Landmarks such as Ohio's Fort Meigs site, a War of 1812 battlefield; Maryland's St. Mary's First City, a Catholic community; and Kansas's Medicine Lodge Peace Treaty site, where Plains Indians had agreed, in 1869, to be moved to federal reservations, were restored and eventually opened to the public.

II

Even more important to New Conservationists, Johnson reversed the prevailing opinion on the treatment of wildlife in the United States by signing the comprehensive Endangered Species Preservation Act (ESPA) of 1966.[14] It was Ralph Yarborough, the only southern senator to vote for the Civil Rights Act, who introduced the legislation, inserting in the *Congressional Record* accounts of extinct American species such as the Oahu thrush (1825), Steller's sea cow (1854), Labrador duck (1876), sea mink (1890), Texas grizzly (1890), plains wolf (1895), passenger pigeon (1914), heath hen (1932), and Laysan Island rail (1944).[15] The act's precursor was a policy recommendation in 1964 that had led the Interior Department to establish the Committee on Rare and Endangered Wildlife Species. The following year, LBJ, spurred by Udall, asked Congress to pass legislation to help protect such endangered species as the American alligator, peregrine falcon, bald eagle, and whooping crane from extinction. Once Udall's committee published its assessment, *Rare and Endangered Fish and Wildlife of the United States* (known as the "Red Book" because of the color of its cover), momentum grew for legislation on behalf of North American species.

The much-publicized ESPA act, which went into effect in 1967, allowed the Fish and Wildlife Service to spend up to $15 million per year to buy acreage specifically as a habitat for the listed species. Moreover, the Departments of Agriculture, Defense, and the Interior were mandated to preserve the habitats of endangered species on all of the land that they managed. Other federal agencies were encouraged, though not required, to protect struggling species. The act also mandated that federal land agencies preserve endangered species habitat in the public domain "insofar as is practicable and consistent with their primary purpose."[16]

After the ESPA act of 1966 was signed, other wildlife protection laws were passed, including ones to stop fishermen from clubbing sea otters, which were considered pests in California for eating red abalone, and from hunting whales by using tracking devices attached to their bodies. A longtime curator at Harvard University's Museum of Comparative Zoology, Harold Coolidge, Jr., a founding director of the World Wildlife Fund, sought to help other nations follow suit and develop their own Endangered Species Preservation Acts. President Jimmy Carter would later honor Coolidge for providing an "early warning" about the plight of endangered species.[17]

Under the act, the first list of threatened and endangered species was compiled from field reports and data gathered by top wildlife biologists from the fifty states. Before long, seventy-eight species were on the list: fourteen mammals, thirty-six birds, three reptiles, three amphibians, and twenty-two fish. Most controversial was the insistence that carnivores such as the timber wolf, the grizzly bear, and the Florida panther—all the bane of livestock farmers—be rescued.[18] Such protection of predators was new but indicated the understanding of ecology and the fact that predators were crucially important to the health of an ecosystem. Under the act, wild animals, including predators, were accorded rights. Most Americans welcomed the new rules. Environmental news was usually terrible—a smog incident or an oil spill—but the very idea that species in peril could avoid extinction brought a dollop of hope in the mid-1960s. For the first time, US government biologists were developing data on how the warming planet, dying oceans, recent species extinctions, and flood/drought displacements were affecting North American wildlife. By example, the act also taught citizens to respect and appreciate more common wildlife—raccoons, coyotes, deer, egrets, and herons—that they might see in their own backyards.

The Defenders of Wildlife, under the unbending leadership of Mary Hazell Harris, its executive director, ran an amazingly effective public awareness campaign to promote the Endangered Species Preservation Act of 1966. The Defenders rightfully felt that their activism had contributed to the victory in a major way. Even so, Harris refused to pull punches or curry favor with the Johnson administration. In 1966, she published a withering full-page editorial in *Defenders* magazine, lambasting LBJ's use of defoliants in Vietnam and allowing the Department of Defense to conduct chemical-biological field tests at the Dugway Proving Ground in Utah.[19] "Such action shows complete disregard not only for many innocent people, but for the animal and plant life of Vietnam and for the Earth," she wrote.[20] At the time the Endangered Species Preservation Act was enacted, the Pentagon admitted that over 2 million acres of Vietnam had been blanketed with defoliants, mainly 2,4D and 2,4,5-T.[21]

Testifying at a public hearing in Congress, Harris objected to "the use of poisons in controlling predatory mammals that are causing damage. It regards present Federal predator programs as inefficient, as far behind the times."[22]

Pointing out that it was the rare predator that attacked livestock, she called for an end to "indiscriminate operations against a whole species."[23] Picking up on that theme, the eminent Canadian wolf expert Douglas Pimlott suggested to Congress something previously unthinkable: that wolves be reintroduced into Yellowstone, which happened in 1995.[24]

The novelist and essayist Edward Abbey worked as a caretaker of the Defenders of Wildlife's seventy-thousand-acre Aravaipa Canyon refuge in southern Arizona after the Endangered Species Preservation Act took effect.[25] Known as the Whittell Forest & Wildlife Area, the refuge was a haven for mountain lions, javelina, a few black bears, and a pack of wolves, along with a herd of whitetail deer, to name the primary furbearers. "I was walking along Aravaipa Creek one afternoon when I noticed fresh mountain lion tracks ahead of me," Abbey wrote in the essay "Freedom and Wilderness," included in his

1977 nonfiction book *The Journey Home: Some Words in Defense of the American West.* "Big tracks, the biggest lion tracks I've seen anywhere. Now I've lived most of my life in the Southwest, but I am sorry to admit that I had never seen a mountain lion in the wild. Naturally I was eager to get a glimpse of this one."

After extrapolating the importance of endangered species from numerous angles, Abbey told his readers that he had indeed glimpsed a mountain lion. It had happened at dusk in Aravaipa Canyon. He had a five-second stare-off with the hundred-pound creature before retreating back down the desert trail. Detailing the affection he'd felt for the big cat, Abbey said the brief encounter had been one of the "shining" moments in his life. "I want my children to have the opportunity for that kind of experience," he wrote. "I want my friends to have it. And someday, possibly one of our children's children will discover how to get close enough to that mountain lion to shake paws with it, to embrace and caress it, maybe even teach it something, and to learn what the lion has to teach us."[26]

As a class of creatures, reptiles are generally stigmatized as menacing pests, snakes in particular; in the 1960s, few nonprofits had a stake in their survival. Some of these reptiles were in desperate straits and, because they needed protection, became beneficiaries of the Endangered Species Preservation Act. Archie Carr, the Florida-based herpetologist so admired by Rachel Carson and E. O. Wilson for keeping field observations in natural history vibrant, used the passage of the law as an occasion to teach that turtles, snakes, alligators, and lizards were essential to the health of ecosystems. Celebrated as the "man who saved sea turtles," he defended the right, the very necessity, of snakes—even venomous ones—to live unmolested in the natural world, writing in *Audubon*:

> I have heard little worrying over the future of snakes, and this to me is depressing. Snakes are not degenerate beings, punished with leglessness for ancient sins, as people once said. A snake is an elegant product of a hundred million years of natural selection. Its loss of legs was an evolutionary advance, a means of living successfully in unexploited ways. But, because those ways are secret, the decline of snakes in our changing world has gone on almost

unmonitored. Others have spoken for cranes and whales, and I
hasten to say these words in praise of snakes, whose silent spring
is also far along.[27]

Archie Carr's book *The Everglades* (1973) made a jarring bookend
to Marjory Stoneman Douglas's more wistful *The Everglades: River
of Grass*. Whereas Douglas had celebrated the Everglades, Carr felt
compelled to document the environmental degradation of the swamp
since President Truman had opened it as a national park.

"So being a naturalist, living in the woods, and having the peculiar background I have, I am especially susceptible to the disease
of bitterness over the ruin of Florida over the partly aimless, partly
avaricious ruin of unequaled natural riches of the most nearly tropical
state," he wrote in *A Naturalist in Florida* (1994). "But in my case I
decided simply, 'What the hell, you cry the blues and soon nobody
listens.' And that made me see that there was really no sense writing
another vanishing Eden book at all."[28]

By tagging loggerhead and green sea turtles, Carr effectively monitored their whereabouts in the ocean, taking notes on where they

Wildlife ecologist Archie Carr was passionate about conservation, especially sea turtles. To
honor him, the US Department of the Interior created the Archie Carr National Wildlife
Refuge along Florida's mid-Atlantic Coast.

buried clutches of eggs on beaches. The naturalist Edward Hoagland affectionately called Archie Carr and his threadbare ilk "bedroll scientists."[29] In 1991, the twenty-mile section of Florida coastline from Melbourne Beach to Wabasso Beach, along State Road A1A, became the Archie Carr National Wildlife Refuge, a sanctuary for loggerhead and green sea turtles, to honor the "turtle man of Florida."

Building on the traditions of Adolph Murie and the Craighead brothers, who pioneered in the use of electric means of monitoring wildlife, Great Society–era specialists used radio telemetry, tranquilizers, radio collars, and other technical aids to study species on the brink of extinction. The Aleutian Canada goose was a prime example. Ornithologists worried that *Branta canadensis leucopareia*, with its distinctive white neck band, was vanishing. In 1962, Bob "Sea Otter" Jones, the manager of the Aleutian Islands National Wildlife Refuge, took a boat to Buldir Island. With biological intuition, he carefully decided that if members of the goose species still existed, they would congregate there. Starting in the 1750s, Arctic and red foxes had been brought to the Aleutian Islands by fur traders who wanted their pelts. As the foxes had proliferated, they feasted on the geese. But far-flung Buldir, known for its high, rocky cliffs surrounded by fierce ocean breaks, was so windy and isolated that the fur merchants had never released foxes there.

Jones, whose coarse demeanor exuded the enduring fortitude of granite surrounded by glaciers, rather recklessly launched a dory and headed to the remote island. Taking to the fierce ocean in such a small vessel was a high-risk venture, but Jones couldn't resist. As Rachel Carson would have appreciated, Buldir provided a thriving ocean environment, one not yet controlled or mauled by the industrial order. As the boat neared shore, Jones saw a few sea otters frolicking in an underwater kelp forest—a rare sight. Then what seemed to Jones like a sky-shining miracle occurred. "The first four geese appeared almost at once, and when the circuit of the island was completed, fifty-six had been counted," he recalled.[30] That was the only record of nesting Aleutian Canada geese since 1938.[31] After taking photos, Jones returned to his base in Homer, Alaska, to determine a proper restoration strategy.

The next year, he returned to Buldir and procured eighteen

goslings and shipped them to the Fish and Wildlife Service's Patuxent Wildlife Research Center in Maryland to be bred. For the next twenty years, a reintroduction program was undertaken to repopulate Alaska with the Aleutian Canada goose. Trial and error ensued. Collaborations with Russia and Japan were inaugurated. And eventually, the Fish and Wildlife Service succeeded in building sustainable populations. By 2000, the small but plucky goose was no longer considered endangered or threatened, the two classifications developed for designated species. "Humans nearly drove the Aleutian Canada goose to extinction, and humans, through the Endangered Species Act, saved this magnificent bird," Bruce Babbitt, the secretary of the interior at the time, declared in 1999.[32]

The Aleutian Canada goose was just one of many avian species that survived with help from the Endangered Wildlife Research section at Patuxent. A dozen biologists based there from 1965 to 1980, under the leadership of Ray Erickson, studied species at risk, twenty in Hawaii alone. Patuxent was a kind of modern-day Noah's Ark, rehabilitating endangered species from decades of decline in the wild. Propagation programs were specific to each species. Bald eagles are a good example. As Rachel Carson had warned, the problem that bald eagles suffered was DDT accumulation in eggs, causing their tissues to become weak. Building on her conclusions in *Silent Spring*, the Patuxent researchers pioneered a protocol for incubating eggs in their laboratory. Before long, more than 120 eagle chicks had been hatched in captivity. Another Patuxent program for the eagles was anchored around building elevated platforms, known as "hacking towers," where the birds could nest. "We were extremely cautious when working with species whose numbers were so few," Erickson recalled. "Before initiating our work with whooping cranes, we used three species of sandhill cranes to develop safe rearing methods for the whoopers. When we began looking at black-footed ferrets, I made a trip to Russia to pick up four dozen polecats, a similar species, for our initial propagation studies."[33]

The statuesque whooping crane is typically five feet tall with blinding white plumage, raven black primary feathers, a red crown, and fierce yellow eyes. Due to unregulated hunting and habitat loss, the magnificent birds had been slaughtered to near extinction. In 1937, only twenty whoopers were documented in the wild; they wintered in the Aransas National Wildlife Refuge in Texas (which President Franklin Roosevelt had established as a safe haven). Once the Endangered Species Preservation Act became law, biologists at Patuxent initiated a captive breeding program from twelve eggs gathered in far-flung locations in a valiant effort to increase flock numbers. One of the eggs, incubated in Colorado, hatched a crane that biologists named Canus. As a fledgling, Canus was shipped to Patuxent to be the first whooping crane in LBJ's endangered species recovery project. Canus ended up siring a large quota of cranes that were later released into the wild. Because of the act, the number of whooping cranes reached fifty-seven by 1970 and six hundred by 2022.[34] Captive breeding at Patuxent worked. "The bird has become the emblematic endangered species, thanks in part to its fierce charisma," Jennifer Holland wrote in *National Geographic*. "Standing nearly five feet tall, it can spy a wolf—or a biologist—lurking in the reeds. It dances with springing leaps and flaps of its mighty wings to win a mate. Beak to the sky, it fills the air with whooping cries."[35]

Sadly, since the Great Depression, two of the worst destroyers of wildlife were the predator control units of the US Fish and Wildlife Service and the Bureau of Land Management. They were the two main targets of Mary Hazell Harris at Defenders of Wildlife. Congressman Saylor called the federal predator control agents "sinister and contemptible." At issue was the way the agencies, in deference to stockmen, killed coyotes and wolves and other species in wholesale operations. Among the poisons used by the agencies were sodium fluoroacetate, commonly called 1080; cyanide; arsenic, put into honey buckets; thallium, injected into bait carcasses; and strychnine, concealed in sugar-coated pills. "The whole public domain is filled with these poisons," Bill Douglas lamented. "They're poisoning wildlife— weasel, mink, fox, badger, coyotes—faster than animals are born. And carrion-eating birds—eagles, magpies, Canada jays, Clark's nutcrackers, woodpeckers—feed on poisoned bait and go away to die."[36] (Until 1972, the only restriction on selling such lethal poisons to consumers was that the labels had to be registered with the USDA.)

Another serious threat to wildlife populations was the explosion of suburbia in the postwar years. As middle-class and upper-class citizens

relocated from crowded cities to the suburbs, wildlife lost crucial habitats. Species were forced to flee from tract houses, golf courses, strip
malls, big-box stores, fast-food joints, overlit gas stations—and endless parking lots. Rainwater, instead of being absorbed into the soil,
fell on rooftops and pavement and rolled off, aggravating flood problems. Nonindigenous plants introduced to ornament the developed
landscape wreaked havoc on the natural order. Stormwater carrying
lawn fertilizer and weed killer was swept into ponds and creeks. Open
spaces and forests vanished as suburbia mushroomed.[37] Defenders of
Wildlife started a campaign to keep the habitats of endangered or
threatened species from being destroyed to build bedroom communities. (The suburban habitat issue would be directly taken on in the
1969 and 1973 amendments to the Endangered Species Preservation
Act of 1966.)

Udall thought it important for Fish and Wildlife to choose a charismatic mammal to serve as the public face of the new Endangered
Species Preservation Act. The winner would join the two birds that
had already drawn a great deal of publicity: the bald eagle and the
whooping crane. Smokey Bear, introduced in 1944, had worked wonders for the US Forestry Service, but Udall wanted a living mammal that families would find lovable, not another cartoon. Neither
the Indiana bat nor the Delmarva Peninsula fox squirrel, both on the
endangered list, would excite children, he decided. Instead, his choice
was the irresistibly cute Key deer (*Odocoileus virginianus clavium*) of
Florida. The deer, standing barely three feet tall, lived only in the
southernmost Keys, where they were being crowded out by humans.
John Gottschalk, the director of the Fish and Wildlife Service, said,
"Endangered Key deer cannot live in the midst of pizza parlors, and a
marsh blacktopped for an airport will not sustain marsh dwellers."[38]
The first habitat the service bought as part of the new law was more
than 2,300 acres that it set aside for the white-tailed dwarf deer. The
frail-looking deer still had a tough battle for survival; fifty years later,
less than one thousand existed, and they were still on the list of endangered species.

No previous New Conservation rollout captured the public imagination as did the Interior Department's work with endangered species. Suddenly, the wingspan of whooping cranes and the plight of

the osprey were in the news. Magazines published stories about condors, bighorn sheep, and alligators. Women's clubs organized boycotts against merchandise made of alligator skin and NBC News used the *Huntley-Brinkley Report* to denounce alligator poachers. Vultures, once the objects of scorn, were declared by Edward Abbey to be a charismatic species, essential to healthy ecosystems. The timber wolf, once maligned as a monster that frightened Little Red Riding Hood, was treated sympathetically by scores of newspaper outdoors writers in the LBJ years.

In Florida, the Endangered Species Preservation Act triggered a long-running "Save the Manatee" movement. In 1966 only three thousand remained in the state. Floridians had been trying to protect the charismatic sea cows since 1892, when it had become illegal to kill them. The reduction in seagrasses on which manatees feed, changes in the quality of the water in which they live, an increase in motorboat traffic, and human activity in general caused the population to dwindle. But once manatees were declared endangered, Floridians stepped up to help them recover. Manatees have huge tails and flippers that they use to dig up the roots of submerged vegetation. They are whiskered and adorable, and millions of Florida schoolchildren through the years have learned to treasure them as Florida's mascot. Eventually, state lawmakers in Tallahassee offered a manatee license plate, and Florida Power & Light embraced the marine mammal protection cause as a high priority. Adult manatees consume about a hundred pounds of vegetation daily, so maintaining the seagrass ecosystems along the Atlantic and Gulf Coasts was the first step for Florida's manatee enthusiasts. In 2017, however, manatees were downgraded from "endangered" to the more alarming "threatened." Regardless of the government status, massive die-offs continued to occur due to starvation from loss of aquatic vegetation to eat. This survival of manatee as a wild species had turned dire in 2022 due to climate change.[39]

Some of the endangered species on the 1966 list weren't well known. In the Great Plains, for example, the black-footed ferret, a mink-like predator with a black mask like that of a raccoon, was among the rarest mammals in the world. When John James Audubon painted it and introduced it to the scientific world in 1851, many other naturalists thought his illustration was a flight of fancy. At the time, black-footed ferrets were actually abundant and widely distributed throughout the Great Plains from Texas to Canada. They were seldom reported simply because they are fossorial and nocturnal, sleeping all but a few hours a day.[40]

In the twentieth century, the black-footed ferret became endangered because of diseases, prairie dog eradication programs, and the fact that the prairie habitat on which it depends had been plowed for crops. Prairie dogs are the ferret's main food, and so the mass slaughter of them by livestock ranchers and wheat farmers had the unintended consequence of bringing the ferrets to the brink of extinction. In fact, twice in the first dozen years after making the endangered species list in 1966, the species was declared extinct. But in 1980, in a stunning development, a group of black-footed ferrets was discovered in Wyoming. With very careful cultivation of the survivors, the population has been rising, and now about five hundred exist in the wild or in captivity. Without Johnson's ESPA, there would probably have been none alive in the twenty-first century.[41] The new law was integral to a trend gaining strength in the 1960s of people learning to treat all living creatures with respect. There was a long way to go, but even as early as 1966, in the first iteration of the Endangered Species Preservation Act, the US government prohibited furs from spotted cats such as jaguars and ocelots from being traded. Udall later observed, with the act in mind, "Some people think that environmental protection began with Earth Day in 1970, but a lot of action led up to that."[42]

ESPA wasn't operating in a bubble. The Canadian government joined forces with the United States to move "spare" whooping crane eggs from the breeding lands of the Northwest Territories to Patuxent Wildlife Research Center outside of Washington. Corporations proved that they weren't impervious to the need to help critters survive. Udall convinced oil companies to suspend coastal drilling when whooping cranes were in the neighborhood. The Northern States Power Company of Minneapolis protected bald eagle sites on the thirty thousand acres it owned along the Saint Croix River. Boise Cascades of International Falls, Minnesota, protected the eagles on an eight-hundred-thousand-acre spread and on its other lands with intermingling ownership. Weyerhaeuser Company of Tacoma, Washington, guarded eagle nesting in tree farms in the Pacific Northwest. The Red Lake Chippewa Tribe joined the ESPA effort by likewise protecting eagles on its half-million-acre reservation in northern Minnesota.

Tangentially connected to the 1966 law was an intensified interest in discerning just exactly what plants were indigenous to North America. In the early 1960s, a gregarious wild foods aficionado named Euell Gibbons had become a literary sensation with such books as *Stalking the Wild Asparagus*, which encouraged readers to find and eat foods that were natural or even wild. He shifted the concept of "health food"—a catchall term for unconventional diets—to "organic food."[43] Gibbons and others in the growing movement advocated simpler foods grown without chemicals. Cesar Chavez was an enthusiast, protesting the use of agrochemicals not only for the sake of farmworkers but also because he and his family consumed only fruits and vegetables grown without them.[44] Gibbons and his followers stressed a related theme, one that was counterintuitive to most Americans at the time: prioritizing indigenous plants. That was revolutionary.

The organic food movement became parallel to environmentalism, encouraging not only better practices in agriculture but also a preference for native edible species such as squash, pumpkin, grapes, elderberry, cranberry, and blueberry, which would be undeniably healthy for the surrounding plants and animals, as well as for consumers.[45]

III

Johnson had one more conservation win as 1966 came to a close. Although it wasn't front-page news in most regions, it was a remarkable accomplishment. At long last, on November 5, the Indiana Dunes National Lakeshore—situated at the southeast tip of Lake Michigan—was established (in 2019 it was elevated to a national park). Johnson had fulfilled Paul Douglas's dream by saving fifteen miles of beaches, sand dunes, marshes, bogs, ferns, prairies, rivers, oak savannas, and woodlands in Porter and LaPorte Counties. He honored the 8,721-acre setting as the culmination of a fifty-year "Save the Dunes" effort, one that was within striking range of ten million people who lived within one hundred miles of its radius, giving residents of Metro

Chicago, in particular, a place close by for a wilderness outing. "When I was young, I wanted to save the world," Senator Douglas said. "In my middle years I would have been content to save my country. Now I just want to save the dunes."[46]

Senator Douglas hugged Johnson as if the president had saved his life that November afternoon. In the epic struggle between preservationists and reckless polluters, the good guys won. Since 1952, when the Save the Dunes Council had been formed through the advocacy of Dorothy Buell, a retired English teacher, the site had been the focus of one of the great preservation battles of the Midwest. Jack Kennedy couldn't get it done; LBJ did. In the end, Indiana Dunes was a compromise that would shape northwest Indiana's identity forever. In managing to balance the needs of nature, industry, and community, the Johnson administration had a lot to be proud of. Save the Dunes continued as a nonprofit organization, collaborating with the Paul H. Douglas Center for Environmental Education in Gary to protect the biodiversity of the complex ecosystem.[47]

Udall considered himself privileged to be able to personally tell the ailing Carl Sandburg the news of the national lakeshore designation for Indiana Dunes. Ever since Robert Frost died, Udall had clung to Sandburg as a literary friend, and in 1964, he had encouraged Lady Bird to read *The War Years* (1939) and *The Prairie Years* (1926) of Sandburg's six-volume biography of Abraham Lincoln. "He was shaggy-haired (his white hair, almost in a Dutch bob), rugged, completely untrammeled in his conservation," Mrs. Johnson wrote about meeting Sandburg in her White House diary. "He had somewhat the same attitude as Lyndon, I thought. He hadn't heard about the rules."[48]

Even though Sandburg had moved to a mountain farm called Connemara in Flat Rock, North Carolina, with his wife, Lilian, he never stopped campaigning to save the Indiana Dunes. "I knew Carl was very sick because when I was at Connemara, he talked about *Honey and Salt* being his last book of poetry," Udall recalled. "So it was a joy to see his facial expression when I told him that the Indiana Dunes was poised to be saved. It was one of the greatest moments of my life."[49]

Sandburg died eight months later. To honor the "People's Poet,"

Udall worked to make Connemara a national historic site. Rare for an Interior Department head, Udall oversaw the entire NPS acquisition and renovation of the farm where Sandburg had spent the last twenty-two years of his life. On October 18, 1968, in record time for a new national park area, the Carl Sandburg Home National Historic Site was opened to the public. Udall was even able to classify the poet's pet goats as a "historic herd" in keeping with Sandburg's final wishes.

The Sandburg home moved so expeditiously through Congress because the National Park Service under George Hartzog's leadership was championing "living history" at the time. Following a proposal by Marion Clawson in the journal *Agricultural History* in 1966, the Johnson administration sought to re-create farm activities at certain sites to increase tourism. There was a new mountain farm, for example, in the Great Smoky Mountains National Park in Tennessee, and reconstructed period crops and livestock were added to the Booker T. Washington National Monument in Virginia. Under the Johnson administration, with Hartzog in charge, forty-one areas were selected for living history demonstrations, ranging from candle makers at the Saratoga National Historical Park (New York) to bread bakers at Hopewell Furnace National Historic Site (Pennsylvania). The Park Service added chickens, cows, and horses at certain units to increase the appeal for school groups. Maintaining the Sandburg goat herd fit the Great Society living history endeavor perfectly.[50]

CHAPTER 25

"Sue the Bastards!" and Environmental Justice

Whitney Young, Reverend Martin Luther King, Jr., Walter Reuther, and Pastor Eugene Carson Blake in the lobby of the Lincoln Memorial, August 23, 1963, during the March on Washington for Jobs and Freedom.

I

As an ecological thinker, Martin Luther King, Jr., understood that people of color were exposed to environmental hazards in the workplace and neighborhoods in far greater numbers than their white counterparts. Black children in Baltimore were getting sick from lead paint. Rats carrying hantaviruses roamed Harlem. Discarded tires in Chicano communities in Los Angeles were breeding grounds for disease-carrying mosquitoes. Toxic disposal sites were being located in poor, rural communities in Alabama and Louisiana. Dilapidated apartments in Chicago were infested with triatomine bugs that carried Chagas disease. In California alone, more than sixteen thousand pesticides had been registered, with the net effect being that Mexican-American agricultural workers were getting seriously ill.[1] Gases such as carbon monoxide (CO), sulfur dioxide (SO_2), nitrogen dioxide (NO_2), and ozone (O_3) were all being studied by multiple federal bureaus for their deleterious effects on the air quality near industrial sources. "The cities," King complained in 1967, "are gasping in polluted air and enduring contaminated water."[2]

To King, environmental justice, intertwined with social justice and human rights, was part of a "revolution of values," including civil rights, freedom from urban blight, and nuclear disarmament, "forcing America to face all its interrelated flaws."[3] Regularly, King preached that a deeper attachment to the natural world would be necessary for the American dream to be fulfilled. Drawing from Thoreau and Schweitzer, King emphasized the need to respect all of God's living creatures. "Although God is beyond nature he is also immanent in it," he wrote. "Probably many of us who have been so urbanized and modernized need at times to get back to the simple rural life and commune with nature. . . . We fail to find God because we are too conditioned to seeing man-made skyscrapers, electric lights, aeroplanes, and subways."[4] There wasn't anything mystical about his views. Martin and Coretta Scott King had four children and like all parents, they wanted their children to grow up with safe water to drink, healthy air to breathe, and birdsongs to enjoy. Joining forces with the Kings on environmental justice issues were the National Farm Workers Association (NFWA), led by Cesar Chavez, and Walter Reuther of the United Auto Workers.

Between 1966 and 1968, with Democrats controlling the presidency and both chambers of Congress, forty separate proposals relating to environmental policy and protection were introduced.[5] The 90th and 91st Congresses passed such enlightened public safety legislation as the National Traffic and Motor Vehicle Safety Act, the Fair Packaging and Labeling Act, the Federal Hazardous Substances Act, the Wholesome Meat Act, the Natural Gas Pipeline Safety Act, the Flammable Fabrics Act Amendment, the Child Protection Act, and the Hazardous Materials Transportation Act.[6] None of those was entirely an environmental law, but all had the same underpinning in the

Great Society's conviction that regulation is a federal responsibility. Helping to bring public consciousness to the blizzard of laws was the National Wildlife Federation, which in 1966 produced its first documentary film, *At War with Waste*.[7]

Regardless of the quantity of the Great Society's forward-thinking legislation, new environmental protection laws, to the dismay of Barry Commoner and Norman Cousins, tended to be focused on the regulation of end results rather than prevention. Although LBJ collected a large wall full of signing pens that are now displayed at his presidential library and museum in Austin, there was no overriding office— like the later Environmental Protection Agency (EPA)—to monitor and enforce compliance in the late 1960s. Everything of consequence pertaining to environmental health was done piecemeal and without proper federal coordination. Interagency cooperation was scant. Dangerous environmental practices could survive in the cracks of inadequate state laws and also of state and local governments unable or unwilling to enact or enforce regulations.

And for all of LBJ's voting and civil rights accomplishments in 1964 and 1965, Black and Latino neighborhoods were still receiving the brunt of industrial pollution run amok. Far too frequently, pollution attacked citizens who lived in what King called "islands of poverty": the poor quarters of cities, the bottomlands of rural villages, the smog-thick zones where environmental quality was sacrificed in the interest of corporate profit.[8]

The city of Fort Myers, Florida, for example, purchased a large block in the historically Black community of Dunbar in 1962 under the ruse that the acreage would be used for a development called "Home-a-rama." The city instead buried waste from its water treatment facility in the site's ponds and surrounding scrubland. Unbeknownst to Dunbar residents, more than twenty-five thousand cubic yards of sludge from the water treatment plant was poured into pits in the tract. Not until 2007 did residents discover the residue, which contained elevated levels of arsenic that were toxic and had very possibly made generations of citizens ill. According to Harriet A. Washington in *A Terrible Thing to Waste: Environmental Racism and Its Assault on the American Mind*, "The site bore no identifying signs, was unfenced and was surrounded by African American families whose numbers soon grew explosively in a building boom."[9] Fort Myers wasn't unique. Most American cities in the 1960s protected white neighborhoods from the levels of toxic exposure that communities of color were forced to endure. Poor parts of Atlanta, Detroit, and Cleveland didn't even have municipal sewage treatment plants.

Such environmental racism was vividly illustrated on May 8, 1967, when an eleven-year-old African American boy named Victor George drowned in a garbage-laden pond at the Holmes Road Land Fill in Houston, Texas.[10] That the tragedy occurred in Houston was no surprise; the Bayou City was notorious for irresponsibly disposing waste in Mexican American and African American wards. "In place of NIMBY (Not in My Back Yard) politics, Houston practiced a PIBBY (Place in Blacks' Back Yard) policy," the esteemed sociologist Robert D. Bullard observed. "Government and private industry targeted Houston's black neighborhoods for landfills, incinerators, garbage dumps, and garbage transfer stations."[11] The bayous, or "bi-yohs" as locals called them, were treated as end points for raw sewage. Houston's African American communities, the largest in the South, still had open sewers and unpaved roads when Johnson was president. Only after the Clean Water Act of 1972 was passed was Houston compelled to regulate its untreated sewage and garbage that, as the esteemed sociologist Stephen L. Klineberg noted, "was discharged daily for decades into the bayous."[12] Houston had no serious zoning laws to protect residents—notably Black and Latino ones—from the dumping of toxic industrial effluents and plain old rubbish in their wards.

Holmes Road was in the poor Black neighborhood of Sunnyside. Victor George had gone to the landfill, which was next to a park, to see if any objects had been left there. Seeing a piece of Styrofoam in a hole full of water, he tried to retrieve it, fell into the hole, and drowned. Victor George's death shocked students at Texas Southern University, a Historically Black College. They staged a sit-in at the dump entrance and demanded its immediate closure. Houston police arrested dozens of the young protesters, which only increased the students' zeal to achieve social and environmental justice. Rallies to free the TSU students erupted throughout the surrounding Third Ward, and police responded by blockading all roads leading into and out of the university and ordering the campus closed. After dusk, police invaded, inciting the students to throw bottles, shoes, and rocks at them. Believing that the barrage of flying debris came from a male dormitory, police surrounded the building and exchanged gunfire with students inside. The police fired some four thousand rounds into the dorm and then stormed the building. Nearly five hundred pajama-clad students were arrested, and police paraded them out of the building with their hands behind their necks. Once outside, they were ordered to lie face-down on the pavement.

Remarkably, no students were killed in the TSU incident, which became known as the Houston riot. However, one officer, rookie Louis Kuba, lost his life to gunfire. TSU students Douglas Waller, Floyd Nichols, Charles Freeman, John Parker, and Trazawell Franklin, Jr., were arrested and charged with Kuba's murder. (Waller and Nichols had not even been on campus during the deadly incident. Waller, in fact, was already in jail on felonies related to the landfill protests.)

The charges against the TSU Five were dismissed after three years on the basis of wrongful arrest. Ultimately, authorities concluded that Officer Kuba had been killed by a ricocheting police bullet. But the root causes of both the initial protests and the police's disproportionate response remained: environmental racism, which allowed municipalities and industry to dump waste in Black and Latino neighborhoods without fear of pushback.

By 1967, grim public health reports on economically poor communities were widespread. In Detroit, the United Auto Workers, boldly led by Walter Reuther, entered the fray to support poor Michigan communities being adversely affected by pollution. The UAW sponsored the Down River Anti-Pollution League to protect the towns of Wyandotte, Lincoln Park, and River Rouge from the ravages of dirty air and poisoned water. To launch the effort, the UAW held public forums, which documented a host of effects from asthma to eye irritation and from peeling paint to blighted playgrounds. Social workers were brought in to help communities deal with environmental psychology.[13]

II

Scoop Jackson perceived that despite progress in many directions, the federal government's environmental protection efforts lacked cohesion. He came to believe that if his Senate staff drew up a comprehensive National Environmental Policy bill, LBJ would sign it into law.[14] Unlike Muskie or Nelson, Jackson was a Cold War foreign policy hawk who supported Johnson's elongated Vietnam War strategy. On environmental issues, however, he remained the most effective preservationist on Capitol Hill. After becoming chairman of the Committee on Interior and Insular Affairs in 1963, he had a hand in every key conservation law passed during the Johnson presidency and beyond. His chief ally was John Saylor, the Pennsylvania Republican, who was still the ranking member of the same committee in the House.

Jackson and Saylor plotted conservation strategies weekly. One of their goals in 1967 was to establish two giant national parks, North Cascades and Redwood, before the 1968 presidential election. LBJ was on board for them. But besides the new federal parks, the bipartisan Jackson-Saylor team explored how to streamline national environmental policy. Jackson asked his top aide, William Van Ness, Jr., to develop a framework for regulating air and water pollution that would establish a holistic federal approach to environmental stewardship for decades to come.

Born to a blue-collar Montana family in 1938, Van Ness spent most of his childhood in Washington's Puget Sound region and graduated from the University of Washington School of Law, where he earned praise as articles editor on the *Washington Law Review*. Van Ness was intending to pursue a doctorate at Yale Law School when Senator Jackson's office approached him, saying it needed an energetic lawyer to serve as the Interior Committee's special counsel. After interviewing with Jackson, whom he later described as "a hell of a nice man, with an open mind, and full of common sense," Van Ness decided to move his family to Washington, DC. As a favor of Jackson, Bill Douglas personally swore Van Ness into the Bar Association of the District of Columbia as a precursor to his new job.[15]

For nearly two centuries, the federal government had embarked on engineering projects with one priority: accommodating the nation's expansion, mainly for the sake of economic growth. The environment was a relatively new consideration, one that didn't appeal to the Army Corps of Engineers, the Bureau of Reclamation, or the new Department of Transportation, established in 1967. Adding to the confusion was the fact that departments and bureaus didn't communicate their high-ticket construction plans to one another; every bureau was a fiefdom unto itself. Jackson and Van Ness saw the helter-skelter pattern all over the country but nowhere as ominously as in Florida's Everglades ecosystem. The "river of grass," in Marjory Stoneman Douglas's words, was threatened by two arms of the federal government; frustration was audible in Van Ness's voice even fifty years later, when he recalled the situation. "In Florida, we had the Everglades National Park, which was a big area, very popular to visit. There was also a plan by the airline industry to build a super jetport *in* the Everglades, covering part of the Everglades National Park, because it was above water, it was flat, it would be cheap to do. . . . And at the same time, the State of Florida was trying to dry up a lot of land around the Everglades, building dikes and canals, because this was very valuable agricultural land. So you had different Federal agencies—Transportation pushing the airport, and Corps of Engineers pushing this water project and trying to dry up the Everglades and the Interior Department wanting to protect the park."[16]

As Van Ness discovered, the agencies were untethered, and because of that, the Everglades biosphere was in grave peril. He said, "They didn't talk to anybody, they weren't required to put a document together that detailed what their projected plans were, no requirement that they consult with any of the other agencies. That was the heart of the problem. We saw this repeatedly involving land use."[17]

Van Ness and Senator Jackson conceived a law that would open to scrutiny—and coordination—any government project that would affect the environment. Jackson secured the services of Lynton Keith Caldwell, a professor at Indiana University, with a long-term commitment to environmental issues; officially, he would serve as a consultant to the Interior Department. In practical terms, Caldwell and Van Ness spent most of their time in 1967 and 1968 pulling together the National Environmental Policy bill. Its overweening goal was to coordinate and enforce planning, cooperation, and transparency where

America's natural resources were concerned. The term they invented was *environmental impact*.[18]

Caldwell, who retired from academia in 1986, recalled that there was little resistance to the bill. Disappointingly, however, nonprofits such as the National Audubon Society and National Wildlife Federation ignored the pending legislation. In a 2003 interview, he said, "This may surprise you, but environmental organizations were not in on it all. As I recall, the only one who attended [our] hearing was the Sierra Club. Others were invited but simply didn't appear. When we were trying to develop the strategies to use we approached different organizations and visited quite a few. The problem was, once again, there was no sense of a unified approach at that time. Many of them wanted to follow their own agendas, pursue their own fights."[19]

Even with bipartisan support, the bill needed to be handled carefully in its journey through Congress. Saylor turned to Representative John Dingell to do a lot of the heavy lifting; Jackson recruited Frank Church, whose love of western rivers was legendary. Van Ness and Caldwell were wary of two groups that made the path complicated. Caldwell said, "Opposition came from those scientists who objected to Rachel Carson's findings (in *Silent Spring*) on the toxicity of some chemicals. And concepts were different then. Perhaps the greatest opposition came from 'conservationists' who were very opposed to the 'preservationists.' The conservationists, who held a more utilitarian, 'wise use' concept of resources, were afraid the nation would get into worse shape if greater restrictions were imposed."[20]

Privately, LBJ was encouraging Jackson to move swiftly on the National Environmental Policy bill. Despite having signed historic Great Society anti-pollution legislation, made the Wilderness Act a reality, and established five national seashores and lakeshores, the president was even less well regarded among environmental activists in 1967 than before as his relentless prosecution of the Vietnam War continued to escalate. It was up to two forward-thinking people from the state of Washington—Jackson and Van Ness—to iron out the legal construct of the National Environmental Policy bill.

III

Always on the march, Bill Douglas continued to promote the federal government's constitutional right to protect natural resources such as Pacific Northwest salmon runs. He addressed it in his 1967 opinion in *Udall v. Federal Power Commission*, which determined whether to allow construction of the High Mountain Sheep Dam on the Snake River (in Hells Canyon). "That determination can be made only after an exploration of all issues related to the 'public interest,'" Douglas wrote, "including future power demand and supply, alternate sources of power, the public interest in preserving reaches of wild rivers and wilderness areas, the preservation of anadromous fish for commercial and recreational purposes and the protection of wildlife."[21]

The environment was elevated to the Supreme Court with Douglas's advocacy in *Udall v. FPC*. "Douglas's intervention signaled the judiciary's heightened concern for environmental considerations," the historian Paul Sabin noted, "its skepticism toward agency planning, and its openness to the lawyers who sought to represent environmental claims in court."[22] Around the time Douglas wrote that opinion, he took Senator Robert F. Kennedy's children salmon fishing in the Puget Sound area.

By 1967, Douglas, who had distanced himself from the Johnson administration on Vietnam, had become a "green" institution to himself and the self-appointed watchdog of Washington State's public lands. Warring with extraction companies was his idea of golf or tennis—a fun sport. When the Kennecott Copper Company announced plans to extract copper from an area near Image Lake in the Glacier Peak Wilderness Area, he objected loudly. Glorious from any point of view, the serrated crests and needle peaks were in danger of being forever desecrated by a mining operation. The Forest Service had instituted the Glacier Peak preserve, at about a half million acres, in 1960. It was not far from the proposed North Cascades National Park; in the same mountain range, in fact. Leading an environmental protest near the proposed Kennecott mine site that year, Douglas vehemently fought to keep the Suiattle Valley sacred. "We must not turn everything into dollars, let's save something for the spiritual side of life," he told the 150 people who attended the camp-in to save the Cascades.[23] Scoop

Jackson used his influence to stop the mine, at least temporarily, in late 1967. At about the same time, LBJ, feeling the heat from Douglas and Jackson, urged Congress to establish the North Cascades National Park on approximately 250,000 acres east of Seattle. That, however, was only half the acreage Douglas and the North Cascades Conservation Council wanted.

Douglas's disenchantment with LBJ continued to grow, with the Vietnam War as the justice's primary beef. But Douglas was also alienated by the White House's refusal to declare nuclear power detrimental to the environment. By 1967, fourteen nuclear power stations were generating electricity in the United States, and more were planned or under construction.[24] Douglas understood the energy benefits of nuclear power but nevertheless believed that without an acceptable way to dispose of the waste, the plants would put citizens at high risk of exposure to radiation.

A more personal complaint that Douglas had with LBJ was the president's approach to enjoying the natural world. To Douglas, the immaculate khaki suits Johnson wore while overseeing his cattle ranch were vulgar; the fashion statement smacked of a drugstore cowboy. Douglas visited the LBJ Ranch and was appalled by the intrusions of mechanized contraptions on nature outings. Not realizing, perhaps, that the Buddhist-inspired Douglas thought of rivers as divine entities, Johnson tried to impress him by driving his floating 1962 Amphicar, the only mass-produced civilian amphibious automobile, into the Pedernales River.[25] On another occasion, the president invited Douglas to go boating on the Colorado River reservoir, Lake Lyndon B. Johnson (commonly called Lake LBJ), which was opened in 1965, to honor the president. "We went out in a speedboat going sixty miles an hour while Secret Service men nervously rubbed suntan lotion on his exposed skin," Douglas recalled. "He would direct the boat around one buoy and head for another, all the while screaming for more speed. Finally, we slowed down and I left the speedster for the patrol boat, giving my place to one of his favorite blondes. As I rejoined Lady Bird I asked, 'What is Lyndon doing when he's not going sixty miles an hour in a speedboat?' 'Going six hundred miles an hour in an airplane,' was her answer."[26]

By contrast, Douglas thought that Senator Robert F. Kennedy, who had learned the art of fly-fishing in Washington State as his pupil, had it right. During the Johnson years, RFK embraced river ecology as his top conservation policy concern. Douglas took a "paternal pride," as his wife, Cathy, put it, in the fact that little Bobby had become a wilderness activist.[27] On Memorial Day 1967, RFK participated in the Hudson River's White Water Derby, earning third place in his class. RFK used the kayaking weekend to raise public awareness about both the Wild and Scenic River bill and the saving of Storm King in the public battle with Con Edison.[28] A month later, working with the mountaineer Jim Whittaker, RFK organized a white-water rafting trip down the Colorado River through Grand Canyon National Park for his family and friends.[29]

Udall helped RFK set up a family visit to Utah's Rainbow Bridge National Monument before the raft outing.[30] "The Kennedys had hiked a short trail from the Lake Powell docks in Rainbow Bridge Canyon to the national monument," Udall recalled. "I wanted him to see that if the dams near the Grand Canyon were built, it wouldn't devastate the national park as Brower had preached."[31] His tactic backfired; RFK turned against the Marble Canyon and Bridge Canyon Dams and wasn't impressed by Lake Powell, probably because Douglas had soured him on man-made reservoirs.

Senator Robert F. Kennedy and his family and friends bounce along peacefully during a part of a four-day rubber raft trip down the Grand Canyon on the Colorado River.

"The Grand Canyon outing was one of our last family river trips," daughter Kerry Kennedy realized. "I was seven years old. Going down the Colorado we heard this screeching from up a cliff. We looked with binoculars and it was a little dog in distress. Daddy left the boat and climbed the cliff, which seemed like the Empire State Building. He snatched the little dog and brought her down to us. He handed her to me as a gift. 'You take care of her, Kerry,' he said. I burst with excitement. We called her Rocky, the 'rockhound.' Cut to a few days later, a couple claimed that Rocky was their dog. They took him back from me. I was on the plane crying when daddy came back with a covered cardboard box in his hands. It was Rocky, my pet. He bought the dog from the couple. She lived with us at Hickory Hill and gave birth to puppies."[32]

Stewart and Lee Udall completed their own Grand Canyon rafting trip shortly after the Kennedy excursion. That was no coincidence. With mounting opposition from western development interests to his reversal regarding the Grand Canyon area dams, he desired the same kind of warmhearted press that Kennedy's trip had received. Mo Udall led his own trip on the Colorado that summer, but he still favored the two Bureau of Reclamation dams near the Grand Canyon and hoped to convince the nine members of the House Interior Committee who went with him on the trip to agree.[33]

IV

By 1967, Charles Wurster of the State University of New York at Stony Brook remained frustrated that half a decade after *Silent Spring* had exposed the hazards of DDT, the pesticide was still legal in all fifty states. Four years earlier, he had helped prove that DDT was responsible for bird deaths on the Dartmouth campus, leading the town of Hanover to ban its use.

One evening, he received a phone call from an attorney, Victor Yannacone, Jr., of Long Island. He was handling his wife's lawsuit against the Mosquito Control Commission (MCC) for contaminating Yaphank Lake with DDT, causing a fish kill. "Vic wanted to know if I knew anything about the effects of DDT, would I support his lawsuit, and did I know any other scientists who would do likewise," Wurster recalled. "I said yes, yes, and yes."[34]

Yannacone later recalled the atmosphere of the time: "Everyone had given up hope. . . . Rachel Carson had been Stalinized out of science and 'better living through chemistry' was the watchword of industry."[35] On June 6, 1966, he filed the first case in the "DDT wars" in the Suffolk County Supreme Court in Riverhead, Long Island, before Judge D. Ormonde Ritchie. At the time, the practice of "environmental law" was embryonic. It was Yannacone, in fact, who coined the phrase, and first *Newsday* and then the *New York Times* ran with it.[36]

Anti-DDT strategy sessions were held with the Yannacones, Wurster, and a group that included George Woodwell, the senior ecologist at Brookhaven National Laboratory; Robert Smolker, a colleague of Wurster in the biological sciences department at Stony Brook; and Arthur Cooley, a high school biology teacher and naturalist. The war room for the Suffolk County DDT litigation was the second floor of an old house in the town of Patchogue where Yannacone maintained a law office. It emerged as the nursery of the new environmental law movement, which adopted Yannacone's battle cry, "Sue the bastards!" On August 14, 1966, just a short while after Yannacone filed papers, Judge Ritchie issued a temporary injunction enjoining the MCC from further use of DDT.

The day after the court injunction was issued, a fumigation truck rumbled down Yannacone's street in Patchogue and sprayed his lawn with DDT. "Vic rushed out with a paper towel, which he immediately brought to me in Stony Brook for analysis," Wurster recalled. "Yes, it was DDT! Vic charged off to Riverhead, filing a motion claiming contempt of court. The judge was irate, and the Commission lost more credibility. They claimed they were just emptying their trucks. I guess they didn't know what else to do with it, and the driver didn't realize he had picked the wrong yard on the wrong street."[37]

The Suffolk trial was held in late November 1966, presided over by Judge Jack Stanislaw, who had to look up the word *ecology* in a dictionary before entering the Riverhead courtroom. The Yannacone-Wurster group enlisted top scientists as witnesses, submitted an appendix to their suit containing a hundred academic papers documenting DDT's deadly effect on wildlife, and presented evidence that the herbicide had concentrated in food chains, poisoned osprey and eagles, and destroyed salt marshes in Suffolk County. By contrast, the MCC had no biologists speak on its behalf and stuck to the defense that the anti-DDT scientists had only circumstantial evidence that the pesticide was killing birds and fish. Stanislaw extended the temporary injunction through late 1967. Eventually, the courts ruled against Yannacone, but by then, the case had inspired change. The hiatus alone forced MCC to look beyond DDT; it adopted instead the insecticide Sevin. "We didn't know it then, but DDT was never to be used again on Long Island," Wurster recalled. "In 1967 it was prohibited by the Suffolk County Board of Supervisors and by the Town of Huntington, and by 1971 DDT was banned in New York State."[38]

Yannacone's class action suit of late 1966 was a watershed moment in US environmental history. Though he and his fellow plaintiffs ultimately lost, their strategy of filing lawsuits against government at all levels proved to be effective, especially when all else failed. At the time, Yannacone said, such suits were considered "just this side of bomb throwing," something you didn't do "in polite circles."[39] The Suffolk County suit changed the calculus and encouraged other citizen groups to sue in the name of environmental protection and public health.

Inspired by Yannacone's efforts, the National Audubon Society established a new defense fund focused on bringing anti-pesticide and anti-insecticide cases to court. On September 30, 1967, in a presentation at the society's convention in Atlantic City, Yannacone explained that his environmental strategy was modeled on the civil rights movement and in particular on the Legal Defense Fund of the National Association for the Advancement of Colored People. "The time has come for you who are committed to the preservation of our environment to . . . enter the courtroom to protect our natural resources," he said. ". . . Experience has shown that litigation seems to be the only civilized way to secure immediate consideration of such basic human rights. Litigation seems to be the only way to focus the attention of our legislators on the basic problems of human existence. The major social changes which have made the United States of America a finer place in which to live have all had their roots in fundamental constitutional litigation."[40]

The social questions that Yannacone posed in Atlantic City were profound:

What can you do when a municipality decides that the highest and best use of a mighty river is the city sewer? What can you do when timber and paper companies cut down entire forests of Redwoods and other exotic species in order to "reforest" the area with fast-growing pulpwood trees?

It is time to assert your basic rights as citizens. Rights guaranteed by the Constitution and derived from Magna Carta. It is time to establish once and for all time that our natural resources are held in trust by each generation for the benefit, use and enjoyment of the next. Today, while there is still time, you must knock on the door of courthouses throughout this nation and seek equitable protection for the environment. You must assert the fundamental doctrine of equity jurisprudence—a doctrine as old as the Talmud or the New Testament or the Roman Law—a doctrine as old as civilization.[41]

Yannacone's hard-hitting Atlantic City speech declared war on corporate polluters but his "Sue the bastards!" approach proved to be too radical for the National Audubon Society. Its general counsel, the Wall Street lawyer Donald C. Hays, worried that wealthy donors would revolt. Disappointed by Audubon's lack of backbone, the Yannacones realized that they would have to create an organization themselves to take environmental issues to court.[42]

So it was that on October 6, 1967, the pugnacious "Long Island Ten"—the Yannacones, Charles Wurster, H. Lewis Batts, Jr., Robert Burnap, Dennis Puleston, Robert Smolker, Anthony Taormina, George Woodwell, and Arthur Cooley—met at Brookhaven National Laboratory to incorporate the Environmental Defense Fund (EDF) as a not-for-profit public benefit membership corporation in the state of New York.[43] Those ten committed activists were determined to finish the work that *Silent Spring* had started. From its humble start, EDF would grow into one of the largest and most important environmental advocacy groups in the United States, with an eventual membership of 1.5 million.

In autumn 1967, the EDF flexed its legal muscle in western Michigan, where Fremont, Muskegon, Lansing, and other towns were using DDT and dieldrin to combat invasive Japanese beetles. Nobody there seemed to remember the Great Mississippi Fish Kill of 1963 or any of the other lessons that had so concerned Rachel Carson. EDF sued the Michigan Department of Agriculture on behalf of plaintiffs in nine affected municipalities. "Our complaint not only detailed the harmful effects of dieldrin and DDT to wildlife but also offered harmless yet effective alternative pest control strategies," Wurster wrote. "We had learned that *battles are easier to win when alternatives are offered*."[44] The court handed EDF a defeat, dismissing the case, but once again, the legal pressure alone brought results: the Michigan Department of Agriculture and various municipalities together decided to find alternatives to DDT. And the cooperative Extension Service of Michigan State University withheld its statewide reconciliations of DDT for its projects.[45]

The EDF's next target was Wisconsin, a state that had become a hotbed of environmental activism in the Kennedy and Johnson years. Partnering with local environmentalists, EDF began collecting data to prove that DDT was contaminating the state's meat, eggs, milk, and other dairy products, leading to a nearly six-month-long hearing in the State Assembly that culminated in a statewide ban. EDF was helped financially in Wisconsin by the National Audubon Society's Rachel Carson Fund.

The EDF's well-publicized and ultimately effective work in New York, Michigan, and Wisconsin during its first year led to a surge in environmental legal activism across the United States, just as the phrase "environmental law" became part of America's lexicon as a result of the Suffolk DDT case. "Sue the Bastards" was a rallying cry on college campuses. Prior to 1967, there had been no environmental law movement or any significant support for environmental litigation until the conclusion of the DDT Wars and the subsequent enactment of the National Environmental Policy Act in 1970. The original two-volume edition of *Environmental Rights and Remedies* (1972) became the first environmental law treatise.

The Natural Resources Defense Council, formed in 1970, the Sierra Club Legal Defense Fund, established in 1971, and other "green" groups were soon established to join the fight against polluters in the courts. "A court of equity is the *only* place to take effective action against polluters," said Yannacone. "Only in a courtroom can a scientist present his evidence, free from harassment by politicians. And only in a courtroom can bureaucratic hogwash be tested in the crucible of cross-examination."[46] Wurster put the mission in even more aggressive terms: "Conservationists are a nice, placid, quiet, law-abiding group of citizens, some of the best we've got, but they have a way of talking to themselves in a closed ecosystem. They are legally weak, scientifically naive and politically impotent. They lack an offense. EDF isn't content to do things in the usual, slow way with limited accomplishment. We want to do more faster, even if we have to crack a few skulls."[47]

Though Yannacone believed that litigation was the most effective way to protect general human and civil rights, Ralph Nader believed that launching consumer protection attacks on defective products and manufacturing pollution excesses was the most viable way to effect change. He was gearing up to expand beyond his initial focus on automotive safety and take on Fortune 500 corporations. He argued that the free enterprise system would survive only if corporations were held responsible for public health and environmental stewardship. "The question is not whether we can build a car that won't pollute the air," he told the *New York Times Magazine* in 1967, "the question is whether we can overcome the resistance of the auto industry and the oil industry to get it built."[48] By 1968, Nader built the ad hoc organization Nader's Raiders into a national youth force for consumer protection.

Out in Central California's agricultural valleys, Cesar Chavez was building an impressive national following. His NFWA had scored a major victory in 1966, securing the first genuine collective bargaining agreement for farmworkers in the continental United States. When Chavez called for a boycott of table grapes from a certain California grower, millions of American consumers complied. Around the same time, the California Department of Public Health's Bureau of Occupational Health published a pamphlet titled "Occupational Disease in California Attributed to Pesticides and Other Agricultural Chemicals," which documented 1,347 cases of human sickness caused by pesticides and insecticides in California the previous year, mainly among the state's farmworkers.[49] Building on the NFWA's newfound strength, Chavez sought contracts that required employers to provide toilets, hand-washing facilities, and clothing protective against pesticide exposure and to refrain from spraying pesticides while workers were in the field.

Martin Luther King, Jr., in a 1966 telegram, told Chavez that "our separate struggles are really one."[50] The pesticide issue soon became an important part of the NFWA's protests and boycotts, transforming Chavez into a national leader for environmental justice. He tried to negotiate contracts that required agribusiness companies to test farmworkers for pesticide exposure on a regular basis. "The real problem is that the workers don't even know how dangerous this stuff is," he told Steven V. Roberts of the *New York Times*. "They call it dust or spray, but we call it poison. It's a subtle death, like quicksand. You don't know what's happening until it's too late."[51]

By that point, it was clear to leaders such as Nader, Chavez, Wurster, and Yannacone that the federal agencies set up to protect health and the environment, especially the Department of Health, Education, and Welfare (now Health and Human Services), had failed in their missions. Thousands of industrial chemicals commonly found in consumer products had never been tested; and then, it was done only after a chemical was discovered to be harmful.[52] The standard of scientific proof provided by the chemical companies was insanely low. By the time they pulled a chemical from the market, people had already died unnecessarily. Years before the term *toxic dump* was in the public parlance, places like the Stringfellow Acid Pits (to be designated a Superfund hazardous waste site in 1980) near Santa Monica, California, a vast cauldron of heavy metals, solvents, DDT, and hydrochloric acid, were contaminating local residents.[53]

Another environmental concern was that the Atomic Energy Commission (AEC) was routinely testing nuclear weapons underground in various locations, with nearly two hundred tests in Nevada alone. Another took place at Amchitka Island in Alaska's Aleutian chain. Udall was drawn into the issue of nuclear proliferation when the village council of Point Hope, Alaska, wrote and pleaded for his help in preventing Project Chariot—a proposed AEC detonation to construct an artificial port—from contaminating the caribou and game birds upon which the people depended for sustenance. Udall announced that he wanted to review AEC's scientific studies and then advise them if he saw unpreventable adverse effects from the detonation. The AEC, however, canceled Project Chariot before Udall had further comment.

All of America's underground nuclear testing took place between August 1963 and December 1967. According to the lawyers pursuing a case against the AEC, radioactive gases had seeped to the surface as a result of all those detonations, and 33 percent of them had vented radioactivity into the air.[54] Wasn't that prohibited by the Limited Nuclear Test Ban Treaty? During one Nevada test of nuclear space rockets, in fact, levels of air contaminants along Route 95 between Reno and Las Vegas temporarily rose to two hundred thousand times their normal level. Explosions weren't the only source of poisons. Around Rocky Flats near Denver, Colorado, where warheads destined to be used in Vietnam were manufactured, public-minded scientists proved that serious amounts of plutonium had escaped the plant. "'Sue the bastards!' was spreading out of Long Island to the Southwest," recalled Stewart Udall. "I didn't pay enough attention to this because I was preoccupied getting Redwood and North Cascades National Parks created. Going after AEC seemed counterproductive."[55]

V

Bill and Cathy Douglas spent the Supreme Court's 1967 summer recess at their forest cabin in Goose Prairie, Washington. Together they hiked the nearby forests and rode horses stabled at the Double K Mountain Ranch. As Douglas told Bobby and Ethel Kennedy, it was blissful until the Air Force began testing supersonic aircraft over the eastern slopes of the Cascade Range, producing shattering sonic booms that disturbed local residents and frightened the local horses.[56] "In that pristine quiet, the sound was really magnified," Cathy Douglas recalled. "Bill kept saying there had to be a way for parts of America to be protected as solitude areas."[57]

Indignant, protective of his right to peace and quiet, and worried about the effect on area residents, Douglas fired off a sharply worded letter to Defense Secretary Robert S. McNamara, the mastermind behind the US bombing in Vietnam, requesting that the test flights avoid the Goose Prairie area. "The sonic boom is having disastrous consequences in the valley," he complained. "We are accustomed to having pictures knocked off our walls. Our neighbor lost many rocks from his chimney. One day last week a horse wrangler riding on American Ridge was nearly killed being thrown violently into a tree. The next day Mrs. Douglas and I were up Thunder Creek, and happily not in a part of the trail that had a precipitous drop-off. A sonic boom sent our horses about two feet in the air, and three feet sideways." There was no need, he argued, to disturb the central Washington backcountry when the "nearby Pacific Ocean should offer plenty of space for maneuvering. . . . I write you because local objections go unnoticed. Your Air Force people out this way are very callous."[58]

Twelve days later, the unimpressed McNamara advised Douglas that the sonic booms would continue. Jittery horses and loose bricks were simply the price Americans had to pay for defeating communism in Southeast Asia. Furious at McNamara's casual dismissal, Douglas went for the jugular in his response on July 28: "The victims of this callous federal program for which you claim responsibility are so helpless that I am arranging to put your letter in the public file so that their attorneys will have the benefit of your confession of responsibility. I assure you, Mr. Secretary, that your 'villagers' here are not as voiceless and impotent as your 'villagers' in Vietnam."[59]

Going over McNamara's head, determined to "ban the boom," Douglas wrote to President Johnson on August 12, mocking the defense secretary and predicting that locals would file lawsuits.[60]

A Supreme Court justice threatening a president with lawsuits was no laughing matter. Johnson tried to placate Douglas but in the end was obliged to support McNamara's assertion that ending supersonic tests in the Pacific Northwest airspace would be "impractical" and potentially even "perilous" for the US war effort.[61] More successful was Senator Church of Idaho, who used one of Douglas's arguments about avalanche risk to convince the Pentagon to suspend wintertime supersonic flights over the Sun Valley ski resorts. But that wasn't the broad win that Douglas sought.[62]

LBJ's refusal to intervene soured Douglas on LBJ's New Conservation. Bolstering his efforts, Douglas joined forces with David Brower of the Sierra Club to protest noise pollution in America. Describing high-decibel aircraft clamor as "an increasing scourge of urban life and country life," he blamed it for causing sleep deprivation, hearing loss, and environmental damage. He railed against airports such as New York City's John F. Kennedy International Airport, where air traffic, he said, was forcing nearby schools to institute "jet pauses" that deprived students of an hour of education every day. Estimating that 40 million Americans were exposed to potentially hazardous noise, he encouraged people to sue the companies and military operations responsible, and he called on Congress to pass a Noise Control Bill to establish limits and standards.

On the subject of sonic booms, not only Douglas but a growing number of Americans were disgusted by the federal government's massive funding to develop a supersonic passenger plane by 1970. President Kennedy had initiated the project in 1963, and eventually it was assigned to Boeing. The cost to the government topped $1 billion, but it was regarded, even by Kennedy, as a challenge similar to that of the NASA moonshot, but for the airline industry. Simultaneously, a partnership of British Airways and the French aeronautics firm Aérospatiale also sought to put a supersonic transport (SST) into the air. A competition between the Americans and the Europeans ensued.

Douglas deemed supersonic passenger planes unnecessary as well as harmful. Senator William Proxmire of Wisconsin, a Democrat, agreed and held hearings on the environmental effects of the Boeing SST. Building on Sierra Club fact sheets, a new group called Citizens League Against the Sonic Boom emphasized the inefficient fuel consumption of such planes. It also charged that SSTs would damage the ozone shield and indirectly cause an uptick in skin cancer. An anti-SST coalition of sixteen environmental groups was established, led by the Sierra Club. One Brower-inspired newspaper ad described the new plane in blunt terms:

BREAKS WINDOWS, CRACKS WALLS,

STAMPEDES CATTLE, AND WILL

HASTEN THE END OF THE

AMERICAN WILDERNESS.

That effort helped kill the American SST effort, though it was a taxpayer revolt that ultimately led to its demise.[63]

Another project, described by some as an Army Corps of Engineers boondoggle, came to Douglas's attention in 1967. When residents of the area surrounding the Red River in east-central Kentucky realized that construction work was about to begin on a dam there, they hastily organized a Cumberland County chapter of the Sierra Club to protest it. Such a dam would flood the paradisiacal Red River Gorge, with its unusual profusion of native plants, natural arches, waterfalls, and birds in every season. The dam had been authorized in 1962, but many locals didn't see any real need for it—except as an excuse to spend millions of dollars of government money in the notoriously impoverished eastern Appalachians. Among the members of the increasingly desperate group were Carroll Tichenor, a horse breeder; Oscar H. Geralds, Jr., an attorney; and Wendell Berry, the agrarian poet.[64] When the group had nowhere left to turn, it sent a letter to Douglas, the patron saint of free-flowing rivers.[65]

Douglas didn't respond to the letter, though. And so the Kentuckians sent him a pretty postcard of Red River Gorge, figuring that if he just saw the place, he wouldn't ignore the preservation cause. That proved to be the lucky charm. Douglas wrote a postcard in return, offering to lead a hike along the Red River, with one proviso: the group would have to pay his airfare. That was his way of ensuring that the protesters were as committed as they expected him to be. In truth, paying alimony to three ex-wives, he was always hard pressed for money.

On November 18, 1967, Bill and Cathy Douglas led a hike along the Red River Gorge (which was in the boundary of the Daniel Boone National Forest) for a lively contingent of around six hundred conservation-minded citizens from Boy Scouts to reporters from the New York Times, Time, and CBS.[66] The Douglas protest met a counterprotest of Kentuckians carrying not only placards that read "Sierra

Club Go Home" and "DAM the Gorge," but loaded guns. The unde-terred Douglas proclaimed the quiet, lonely dells and tree-rich ravines "one of the great untamed beauty spots in the country which we must do all we can to save."[67] That wasn't hyperbole; Douglas was right that the gorgeous Kentucky landscape deserved to be a national park. Wendell Berry, in *The Unforeseen Wilderness: Kentucky's Red River Gorge*, would write eloquently about the "poplars and hemlocks of a startling girth and height" that brought relief from the trash found along the riverbanks of the Red River.[68]

"This was the first public hike I'd been on," Cathy Douglas Stone recalled. "Everybody was clamoring around. But when I laid eyes on the gorge, I knew we were in the right to raise our activist voices. This landscape was very much worth saving. Nobody who visits it could say otherwise. I felt proud that we were saving just a geographically beautiful place."[69]

After the hike, in a speech on November 19 at the Phoenix Hotel in Lexington, Douglas denounced the Army Corps of Engineers as "public enemy number one," an agency "obsessed with building dams" that tore apart all the interconnected webs of river valley life. In front of an enthusiastic audience, he explained why he was constantly pro-testing: "You only win a few of these battles. The tide is running very strong against the little areas of wilderness we have left. You must be willful if there are to be any alcoves of solitude left."[70] The Bill and Cathy Douglas ramble in Kentucky led to a new study of the dam project, a different site for construction, and an overall delay of about a decade. After further protests, the idea of flooding the Red River Gorge was scrapped.

Soon after the Kentucky protest, Douglas made an appointment to see LBJ about stopping the construction of the Oakley Dam near Decatur, Illinois, part of the $120 million Sangamon River Project. The dam had the bipartisan support of three senators: Charles Percy, a Republican from Illinois; Everett Dirksen, a Republican from Illinois; and Paul Douglas, the Democrat from Illinois.[71] The justice was dis-turbed that the proposed reservoir would flood the historic village of Allerton Park. Holding "Wild Bill's" memo, LBJ asked the justice his thoughts about North Vietnam and the United States entering peace negotiations. Not mincing words, Douglas told the president that he opposed everything about his administration's Vietnam policy, that the only recourse was to call the war a mistake and withdraw US troops. The war was costing taxpayers $2.5 billion a month, had killed thousands of young Americans, had ripped society to shreds, and had damaged US relationships with NATO allies around the world. "When I was finished," Douglas recalled, "he crumpled the memo I had given him without reading it and put it in his pocket."

Douglas's fundamental disagreement with his old friend Johnson was that the Texas politician thought every person had a price. Because Johnson believed in raw politics, he viewed the new breed of environmentalists with suspicion; Douglas thought of them as profiles in courage. How do you make a deal with activists who want to punish Fortune 200 companies or the Army Corps of Engineers without any thought of making a personal profit? By contrast, Senator Ed Muskie was using the issue of smog to advance his political stature, and that made sense to Johnson. The Environmental Defense Fund lawyers, who were paid to harness the power of the courtroom, also fit into the president's worldview. To LBJ, the Sierra Club under Brower was a fundraising gambit; that was understandable. He preferred anything or anyone with whom it was possible to negotiate. That wasn't Bill Douglas. By 1967, their friendship had deteriorated, neither man really caring.

On Christmas Day 1967, at Ebenezer Baptist Church in Atlanta, Martin Luther King, Jr., delivered a prophetic sermon laden with environmental justice overtones. Echoing Albert Schweitzer, he linked the natural world to the promise of Christ not fully realized. Interconnectedness, he knew, was what ecology was all about. Just as from the 1950s onward he had protested for nuclear disarmament, he had likewise joined the vanguard of the new ecological consciousness sweeping the land. "It really boils down to this: that all life is interrelated," he preached to the holiday congregation. "We are all caught in an inescapable network of mutuality, tied into a single garment of destiny. Whatever affects one directly, affects all indirectly. We are made to live together because of the interrelated structure of reality."[72] That, of course, was an eloquent definition of ecology.

CHAPTER 26

The Unraveling of America, 1968

Whittier High School student Richard M. Nixon was enamored by the majesty of the Grand Canyon. Running for the White House in 1968, Nixon promoted himself as a Theodore Roosevelt–style conservationist.

I

Lyndon Johnson's stubbornness was on display in early 1968 when Stewart Udall organized a CBS television special broadcast from Ford's Theatre, where John Wilkes Booth assassinated Abraham Lincoln in 1865. Udall considered the renovated Washington landmark a crown jewel of the Johnson administration's National Historic Preservation Act of 1966. To celebrate the reopening, Udall organized a gala on January 30, hosted by actors Helen Hayes and Henry Fonda. It was a tailor-made opportunity for the administration to garner credit for its beautification and historic preservation work. But many of the entertainers, including Robert Ryan, Harry Belafonte, and the folk singer Odetta, opposed Johnson's Vietnam policy. At the last moment, the petulant LBJ boycotted the gala, sniping at the peace activists in the cast to a White House aide, "Why shall I honor *them* with my presence?"[1]

There was a much bigger reason for Johnson's foul mood. Hours before the Ford's Theatre event, on the Vietnamese lunar New Year, the North Vietnamese Army and the National Front for the Liberation of South Vietnam had begun the Tet Offensive, a massive assault on South Vietnam. Eighty thousand troops were bombarding dozens of provincial capitals and attacked the US Embassy in Saigon, making a mockery of the administration's claim that the United States was easily winning the war. Udall thought LBJ's decision to skip the Ford's Theatre event was self-savaging. That night, he wrote in his diary, "What [Johnson] desperately needs right now, and wholly lacks, is a little of Lincoln's 'with malice toward none' spirit. A good dose of it might even re-elect him—but he's a carrier of grudges who can't forget a thing."[2]

The CBS special received critical raves, and within a few weeks, the theater reopened to the public with a play based on Stephen Vincent Benét's narrative poem *John Brown's Body*.[3] That success, despite Udall's hopes, was of little help to Johnson. The Vietnam War, mistaken and unwinnable, had undermined the administration to the point that top advisers were preparing to quit. That February and March, the wounded Johnson regularly lashed out at subordinates. "I was being forced over the edge by rioting blacks," he later told the historian Doris Kearns Goodwin, "demonstrating students, marching welfare mothers, squawking professors, and hysterical reporters."[4]

Frank Church, Gaylord Nelson, George McGovern, and other anti-war senators considered President Johnson to be an albatross around the neck of the Democratic Party. The fiercest critic of all was Senator Eugene McCarthy, a Benedictine-trained graduate of Saint John's University in Collegeville, Minnesota, who broke with his party's lack of moral leadership over Vietnam and had already announced his presidential candidacy, aimed at taking down LBJ. Poetic with a mordant sense of humor and baked in contrarianism, McCarthy had served nine years in the House of Representatives and ten years in the Senate, as part of a new farm-labor movement. Declaring that new Clean Air laws represented "an issue that's better than motherhood," McCarthy earned an A-plus record from conservation groups keeping score.[5]

Rumors swirled around Washington that Robert F. Kennedy, following McCarthy's move, might likewise challenge LBJ for the party leadership. If Kennedy won the Democratic nomination, he planned to tap Udall to be his running mate.[6] "My father adored Stewart," RFK Jr. recalled. "They were beyond close. I sat in on a meeting at Hickory Hill that my dad had with Frank Mankiewicz and Ted Sorensen. My dad wanted Stewart—it was a visceral thing for him. He loved Udall because he was the physical embodiment of Woody Guthrie's song 'This Land Is Your Land.' And he needed a Westerner on the ticket with him."[7]

Udall was ideally suited to be RFK's vice presidential choice. With fancy footwork, he had managed to stay out of the Vietnam morass, thereby securing his standing with anti-war doves. And as the progenitor of both Kennedy's New Frontier and Johnson's New Conservation environmental agendas, he was a hit on college campuses. Indeed, he was nationally respected as the southwestern conservation hero who had established Canyonlands National Park, fought to get the Wilderness Act passed, drove the Indiana Dunes National Lakeshore and Fire Island National Seashore into existence, and, at the last minute, after a change of heart, saved the Grand Canyon by relinquishing the twin megadams on the Colorado River.

With the Ford's Theatre event behind him, Udall accelerated the administration's efforts to win passage before the November election of four "Forest Bills" being hotly debated in Congress: Redwood National Park, North Cascades National Park, the Wild and Scenic Rivers System, and the National Trails System. It was the same old story. All the bills easily passed in the Senate. To accomplish the ambitious feat of passing in the House, however, would require the support of Representative Aspinall, whose advocacy on behalf of the timber, mining, and oil and gas exploration industries was cemented into his public persona. As Brower quipped, preservationists saw "dream after dream dashed on the stony continents of Wayne Aspinall."[8]

Of the four bills, Redwood National Park was the most troubled, as well as the most urgent. In his 1967 State of the Union address, LBJ had pleaded with the legislative branch to save the coastal redwoods north of San Francisco. Weeks later, in a special message to Congress, he repeated his call to action: "This is a 'last chance' conservation opportunity. If we do not act promptly, we may lose for all time the magnificent redwoods of Northern California."[9] LBJ's main foe was California governor Ronald Reagan, a Republican, who had said, "A tree's a tree—how many more do you need to look at?"[10]

Refusing to grapple directly with Reagan, LBJ turned to Laurance Rockefeller, whose family philanthropy in support of the Save the Redwoods League was well known in Sacramento political circles. With his connections to the Republican Party and the business community, Rockefeller met with Reagan on Johnson's behalf and conducted a shuttle diplomacy between the warring principals. Most of the lumber companies supported, if anything, a compact park at Mill Creek, a site that the Save the Redwoods League and some other conservation nonprofits firmly believed constituted the most valuable forestland. The Sierra Club rejected that position, insisting on an extensive park at Redwood Creek.[11] Governor Reagan came out against pulling acreage anywhere from private interests for a national park. In addition, he was perhaps the only governor ever to demand compensation from the Interior Department in the form of an exchange for state parkland that would become a national park.[12]

The Sierra Club was in full cry, pressing its case to save the redwoods in the midst of the political melee. Taking a multimedia approach, David Brower followed up the publication of *The Last Redwoods* by distributing an educational film called *Zero Hour in the Redwoods*; co-producing an Oscar-winning short documentary, *The Redwoods*; and taking out expensive full-page "battle ads" in the *New York Times*.[13] The year before, Congress had begun in earnest to craft the Redwood National Park bill. The sticking points were, first, the size; second, the location; and, third, Reagan's scheme to force an exchange of state parkland for other federally-owned tracts. (Of course, cost was an issue, too, but it cut across the other three.) A cadre of politicians in both houses of Congress considered the "Reagan exchange" to be the most incendiary item on the list. They thought the California governor should have been grateful to have a new national park in his home state—one that was bound to attract legions of tourists to Humboldt and Mendocino Counties.

As a step forward in the heated process, characterized by one compromise after another, Senators Scoop Jackson and Thomas Kuchel cosponsored a Redwood National Park bill that specified a 64,000-acre park. That was 21,000 acres larger than Laurance Rockefeller's proposal, which Johnson and Udall, as well as the Save the Redwoods League, had already accepted. But it was 33,000 acres smaller than the park the Sierra Club wanted. The bill specified Mill Creek as the site. A hard push by a group of senators for an amendment precluding any kind of exchange of lands failed, and the bill passed the Senate by a wide margin.

In the spring of 1968, the same fights emerged in the House hearings. Representative Aspinall, who had delayed action on a Redwood National Park for five years, was a lightning rod for the squabbles. That fact exasperated David Brower and his followers. When Aspinall attended hearings in Eureka, a gateway city of the proposed park, he sided with the angry citizens who claimed that timber industry jobs would be severely reduced if 64,000 acres of redwood groves were saved as a jumbo federal park. Brower feared that with the November 1968 presidential election on the horizon, all hope for Redwood National Park might evaporate, especially if the Republican Richard Nixon won the White House. The possibility of a new president may or may not have troubled Aspinall. But it certainly made him cognizant that his control over the high-profile Redwood National Park bill could slip away if Nixon won. Ultimately, he allowed his Interior Committee to authorize a bill. That alone was reason for New Conservationism to cheer after so much time in deadlock. The bill called for a park consisting of a Redwood Creek unit and a Mill Creek unit. It didn't provide for Governor Reagan's land exchange. And then there was the shocker: the negotiated size was only 28,400 acres. The Save the Redwoods League deemed it insulting.

The punishing outcry over the reduced acreage in the bill surprised Aspinall. In fact, he received more angry mail for the half-loaf Redwood National Park proposal than for any other initiative in his twenty-four years in Congress. The syndicated columnist Drew

Pearson mocked "molasses-moving Aspinall" for seeing dollar signs instead of God's majesty when it came to redwood groves. On Capitol Hill, seventy-three congressmen signed a petition protesting the pathetically small acreage, most of which consisted of nothing more than the three state parks already marked for inclusion.[14]

The wily Aspinall had a plan, though, and the even more wily LBJ appreciated his cunning. In the 1950s, Johnson had exploited the same arcane, complex, and even strange rules during his years in the House and Senate. He wanted an expansive Redwood National Park badly but couldn't help admiring the way Aspinall had wielded power over the Interior Committee by slowing the pending legislation to a standstill; once that occurred, the Coloradan had shaped a bill that suited his overall plans. Other major conservation bills were also maneuvering through the Interior Committee in 1968, and that, too, worked to Aspinall's advantage. It was left to LBJ to pressure Aspinall on Redwood National Park through a revolving door of intermediaries. "He let us know all the time," Aspinall recalled of the atmosphere surrounding the bill.[15]

Aspinall proved his power-broking abilities by forcing first his committee and then the whole House to vote for his small-park/no-exchange bill by tacitly threatening to delay a vote on a Redwood National Park indefinitely—a fate that, members knew, might doom it to oblivion. Aware that he was playing for high stakes, with other deals and compromises awaiting passage of the Redwood National Park bill, Aspinall promised his colleagues that the size could expand "in conference," when the House and Senate would reconcile their versions. Against every inclination and with Brower fuming in the background, the conservation-driven representatives John Saylor of Pennsylvania and Mo Udall of Arizona voted "yes" on the Aspinall version. The final vote was 389–15.

Afterward, the conference managers kept the two-part location, allowed the exchange, and, as Aspinall had predicted, expanded the size, ultimately to 58,000 acres. No one was completely happy with the final bill. But the grief, at least, was distributed quite equally among the many nonprofits, timber companies, government officials, and citizens taking a strong interest in the fate of the giant trees. For LBJ, the main thing was that the Redwood NP would be established before election day. After the bill passed with voice votes in both houses, former NPS director Horace Albright wrote to Udall, "You are clearly at the top now, and I don't see how any successor can reach the pedestal on which you are standing."[16]

On March 6, 1968, President Johnson, for the first time in American history, delivered a message to Congress devoted solely to Indian affairs and to the people: their plight and goals, education, programs, and plans to improve their living conditions. Over the decades, other presidents had been asked to address Native American equality concerns—only LBJ did. In an unprecedented move, he established by executive order a National Council of Indian Opportunity to review federal programs for Native Americans. "I propose a new goal for our Indian programs," Johnson said. "A goal that ended the old debate about termination of Indians and stressed self-determination."[17]

Matching money to his rhetoric, LBJ recommended that Congress appropriate a half billion dollars for programs aimed at Native Americans—about 10 percent more than in fiscal year 1968. In addition to establishing three federal fish hatcheries on Indigenous lands, LBJ sought to improve education and provide health care benefits on reservations—environmentalism wasn't in the mix. Instead, LBJ boasted that under his leadership almost 130 manufacturing plants began operations on reservations, providing employment for four thousand people. These included the General Dynamics Corporation missile parts plant at Fort Defiance, Arizona, and the Fairchild Semiconductor Corporation transistor-assembly facility at Ship Rock, New Mexico (but on the Navajo reservation). Johnson also sent US government biologists to work on the restoration of wildlife habitats in six states with Indian reservations to help hunting and fishing opportunities.

On March 8, hoping to shine a spotlight on his environmental policy achievements in order to showcase them in a presidential election year, Johnson delivered the remarkable Special Message to the Congress on Conservation, also known as the "To Renew a Nation" speech. Brimming over with quotes from Theodore Roosevelt, the president minced no words about how concerned he was about industry and technology disembodying the American landscape. "From the great smoke stacks of industry and from the exhausts of motors and machines, 130 million tons of soot, carbon and grime settle over the people and shroud the Nation's cities each year," he warned. "From towns, factories, and stockyards, wastes pollute our rivers and streams, endangering the waters we drink and use. The debris of civilization litters the landscapes and spoils the beaches." It was as if Johnson painted the same dystopian landscapes in words that Rachel Carson had used in her opening fable in *Silent Spring*. It's not unreasonable to claim that the voice of Carson had melded into that of LBJ. Both were visionary conservation leaders who espoused an American heritage of not only raw wilderness, clean shorelines, and unbroken forests but, as the president put it, "safe environment for the crowded city."[18]

It's heartbreaking to read Johnson's special message. LBJ got it completely right about environmental stewardship but was all wrong about Vietnam, which destroyed his presidency. Whether it was asking Congress for $2 billion for community waste treatment plants or passing a Safe Water Drinking Act, banning surface mining in the Appalachians or cleaning all US waterways so the discouraging sign POLLUTED WATER, NO SWIMMING would not be posted anymore, LBJ was an exemplary conservation leader. It wasn't just that he talked a "green" game. In a noble way he backed his rhetoric with decisive legislative action. In his "To Renew a Nation" special message, he called for seven new national wilderness areas, to be created from existing national forests: Mount Baldy (Arizona), Desolation (Washington), Pine Mountain (Arizona), Ventana (California), and Siccane (Arizona). "We are now surveying," he exclaimed, "unspoiled and primitive areas in Arkansas, Oklahoma, Georgia, and Florida as further possible additions to the wilderness system." In coming years he hoped an Eastern Wilderness Act could be passed to save uncultivated lands from Maine to Florida.

But Johnson couldn't win over Democratic liberals to praise his New Conservation achievements because saying anything praiseworthy about the Tet-beleaguered president was akin to being pro–Vietnam War. In 1964, LBJ had considered Senator Eugene McCarthy of Minnesota as his vice presidential running mate but had instead chosen Hubert Humphrey, an AFL-CIO favorite. On Great Society domestic issues, including conservation measures, McCarthy was a blue-dog Democrat. But he was fiercely opposed to the Vietnam War on moral principles, and thus, after he launched his presidential campaign, he became the high-profile leader of a movement to dump Johnson. On March 12, 1968, he came within a few hundred votes of defeating Johnson in New Hampshire, the first primary of the campaign season. Determined anti-war college students knocked on doors for McCarthy in snowy cities such as Hanover and Nashua. Many hippies cut their long hair, shaved, and, as the slogan went, got "Clean for Gene."

White House aides would later recall how fatigued and tormented Johnson had been after New Hampshire. McCarthy became the bane of his existence.[19] Vietnam had boxed him in politically, and there was no escape. On March 31, in quiet tones, he told a prime-time evening television audience that he wasn't going to seek reelection. Udall watched the abdication, which he'd been given a heads-up about, at his McLean home along the Potomac and gasped in frustration. For a president to yield and admit folly wasn't an easy thing to do. Johnson had the look of a ruined adventurer: drained, run down, out of cards to play and clearly beaten. "There wasn't an iota of his signature swagger visible," Udall recalled. "I spoke to Bobby Kennedy shortly after, and he told me he, [too,] was caught by surprise."[20]

Driving to West Virginia days later, Udall stopped at a cliff overlooking the confluence of the Potomac and Shenandoah Rivers. The undulation of blue ranges in the distance held him transfixed, and the fresh air was an elixir. He resolved then to continue avoiding the dove-or-hawk politics of the war. Instead, in the months leading up to the presidential election, he'd focus on getting the quartet of important forest bills—Redwood National Park, North Cascades National Park, National Wild and Scenic Rivers System, and National Trails System—through Congress and signed into law by his lame-duck boss. From time to time, he shared moments of good fellowship with LBJ, only to be treated days later with a studied and inexplicable detachment. Perplexed, his pride wounded, Udall took such flashes of coldness as personal affronts. In conversation with former LBJ speechwriter Richard Goodwin, who was working for Bobby Kennedy, he resolved not to let the "Johnson treatment" hinder progress on his agenda in the six months he had left at Interior. In a letter to the president and First

Lady the day after the announcement, Udall wrote, "Your leadership has given the nation a new outlook on conservation—a new concept of land stewardship for the future. I believe you have elevated the aspirations of the American people to new horizons—and history will surely honor both of you for your actions."[21]

Just as Udall was starting to plot a big conservation finale before November, a catastrophe rocked the country: on an ill-omened April 4, Martin Luther King, Jr., was assassinated on the balcony of the Lorraine Motel in Memphis. In 1964, at the age of thirty-five, King had won the Nobel Peace Prize, the youngest person ever to do so. Perhaps the greatest moral leader America every produced, he was in Memphis to help striking sanitation workers when a white supremacist shot him. An overlooked aspect of King's career is his tireless activism for underserved families; he sought more urban parks, cleaner communities, mold-free public housing, preventive health care, and outdoor recreation facilities. He had championed the banning of lead paint in Atlanta, Memphis, and Chicago. In 2004, Udall said, "King was an environmental justice leader before the term was even used."[22]

Sickened by the rot and refuse in degraded slums, King had defended poor people whose health was damaged by pollution and subpar sanitation. "Certainly Martin should be considered a leader in bringing public awareness to environmental degradation in Black communities," Georgia congressman and civil rights leader John Lewis recalled in 2010. "Blacks in the back of the bus were sitting where the gas fumes engulfed them. That's where you smelled the gas. There was no air-conditioning in buses when Martin was alive. Black passengers got sick and dizzy. You can consider the Montgomery Bus Boycott, which King led, as environmental justice act to a certain degree."[23]

Unlike the wealthy and upper middle classes, poor people, King would say, couldn't just flee big cities for a long weekend on the Snake of Idaho or Allagash of Maine. "White flight, gentrification, the decay of urban neighborhoods, and interstate highway construction, which flattened minority communities," Andrew Young recalled, "were routinely addressed by the Southern Christian Leadership Conference."[24] King's City on a Hill was a community where children weren't poisoned by acrid smoke and asbestos. Representative Ron Dellums, a

California Democrat who represented Berkeley and Oakland, and who would cofound the Congressional Black Caucus in 1971, went so far as to nominate Brower for the Nobel Peace Prize because the activist was a Martin Luther King–styled "publicist for a revitalized environment," "protector of an endangered planet," and "public conscience to the delicate balance of life on earth."[25]

Just four days after her husband's death, Coretta Scott King flew to the accursed city of Memphis, sick at heart, and led a people's protest herself. Three weeks later, she orated against the Vietnam War in New York's Central Park, standing in for her slain husband.[26] Over the summer of 1968, she fled with her four children to New Hampshire to give them all a chance to heal. Craving the solitude of nature, the grief-stricken family hiked the shores of Lake Winnipesaukee near Wolfeboro. During those weeks, Mrs. King spoke often with Norman Cousins about keeping the anti-nuclear movement alive.

II

Roughly coinciding with King's assassination was Mrs. Johnson's "Crossing the Trails of Texas" tour. A small retinue of reporters accompanied her. While she was in Texas, neighborhoods in northwest Washington were burning as a result of rioting in the aftermath of King's death. Around 115 American cities had erupted in spasms of violence following the killing of King. In Texas, Mrs. Johnson was visibly troubled by the sudden unraveling of America: Secret Service personnel were on high alert owing to serious threats on her life.

The high point of Mrs. Johnson's visit was San Antonio's Hemisfair, a world's fair that had just opened. Most impressive was the extension of the San Antonio River by a quarter mile to connect with the fairgrounds. The city's Conservation Society had rehabilitated twenty historic buildings, funded in part by the National Preservation Act of 1966. "Will Rogers once remarked that the two towns in America with the most personality were New Orleans and San Antonio," Lady Bird reflected of her Hemisfair experience. "Today both towns enjoy a booming tourist trade. At various points in their past, they refused to let their personalities be devoured by the onslaught of so-called progress—the metro dollars increased as a result. Commerce capitalized on the natural gift of waterfronts and natural heritage of many bloodlines."[27]

Accompanied by George Hartzog, Lady Bird also enthused about history-tourism while visiting the Battle of Goliad site, where Texas settlers had attacked the Mexican Army in 1835. And she spoke affectionately of the Big Thicket bayous and forests near her childhood home. Texas senator Ralph Yarborough referred to the East Texas cypress wilderness as the "biological crossroads of America."[28] Mrs. Johnson praised Yarborough for introducing legislation in 1966 to protect Big Thicket, as well as the ecologically rich Big Cypress tract in Florida.[29] A national park at Big Thicket would be within a hundred miles of 3 million people, most of them in Houston; such proximity fit well with the administration's Parks for People agenda.

Five months before his wife's tour, LBJ had signed a bill establishing the National Park Foundation (NPF), a nonprofit that allowed the National Park Service to obtain tax-deductible donations.[30] In coming years, corporations such as the Walt Disney Company and Union Pacific Railroad would become annual sponsors. "Here is another adventure in partnership between government and concerned private citizens—designed to serve the common good," LBJ proclaimed.[31] While Mrs. Johnson was in Texas, she met with some of the first big-dollar donors to NPF. She crafted her message to appeal to them, recommending that the IRS and Texas tax laws reward the preservation of a handful of Houston's landmark buildings on the recent model of Old San Juan in Puerto Rico. Though some urban planners were swooning over recently planned satellite cities such as Reston, Virginia, and Columbia, Maryland, Mrs. Johnson preferred the restoration and greening of urban downtowns.[32]

Lady Bird swooned over the new Padre Island National Seashore at the southern tip of Texas's Gulf Coast. She had been invited to dedicate it. Political life was strange: her husband had opposed the seashore park in 1961, but, seven years later, the Johnsons claimed credit for its existence. Rolling up her pants to walk along the sandy beach, enjoying the tides, Lady Bird felt her anxiety dissipate under the cloudless blue sky.[33] Wearing a beach hat to block the sun at the ceremony, Mrs. Johnson noted that her husband called Padre Island a pearl in the "necklace of national seashores" and praised the weather-worn beauty of driftwood on the shifting Gulf dunes. At her side were Udall and Hartzog. "Legends of early Indians, of shipwrecked Spanish galleons, are part of Padre Island, and I hope, Mr. Hartzog, there will be occasions when some gifted story-teller could bring them to life as part of the regular program here," she said. "I've been to so many national parks, and that is one of the great things they do. They weave in the history of the island, the history of man and nature—the whole ecology—sitting around the campfire or in the visitors center."[34] To her point, though "National Parks" conjures up romantic images of wide-open, wild places, in truth, nearly two-thirds of the United States' 423 national park units are historical or cultural sites.

There was no easy banter between Mrs. Johnson and Stewart Udall on the Air Force One flight back to Washington. Intuitively, Lady Bird knew that Stew's allegiance was to Bobby Kennedy. "I felt there was a withdrawal of his enthusiasm for the Johnsons," she wrote in her diary. "I would not be surprised if he got out of the Cabinet. But there is still a real dedication in him for all the work of conservation and I think he gives credit to Lyndon for all his effectiveness in that field. He talked about the things that might be done in the next months to nail down further the conservation program. I was very, very tired. And it was, somehow, a sad conversation that left much unsaid."[35]

Udall was indeed bonded to RFK, though when the New York senator announced his candidacy for the Democratic nomination for president on March 16, 1968, Udall felt that he couldn't publicly endorse him. He was, of course, entirely supportive.

The core of RFK's following were blue-collar workers, poor whites, and people of color. Cesar Chavez began a fast on February 15, 1968, both to promote the NFWA and to unite warring factions within the organization. RFK, with ties to the Latinx community, cheered him on from New York and Washington. By week two of the hunger strike, reports emerged that Chavez was dying. His nurse, Marion Moses, urged him to take vitamins and sip some soup and grapefruit juice. Moses had grown up in West Virginia. She earned a BS degree in nursing at Georgetown University in 1957. After bouncing around a few hospitals, she became head nurse in the medical and surgical unit of Kaiser Foundation Hospital in San Francisco. In Berkeley one afternoon, in 1966, she saw a sign on a campus bulletin board about Chavez's farmworkers' union needing medical workers. Immediately, she drove to their headquarters in Delano, California, and for the next five years worked *pro bono* at the very humble health clinic on the NFWA's Forty Acres retreat. In addition, she became Chavez's personal health care provider.

Over time, Moses became an authority on the chemicals used in agriculture and the ways they were making migrant pickers seriously ill.[36] Few people, in fact, knew more about the toxicity of pesticides than Moses. In coming years, she would write such important books as *Harvest of Sorrow: Farm Workers and Pesticides* (1992) and *Designer Poisons: How to Protect Your Health and Home from Toxic Pesticides* (1995).[37]

RFK sent Chavez a telegram urging him to break his fast, insisting that his death wouldn't benefit the cause of the farmworkers' union. Chavez asked Kennedy to visit him. By the time RFK arrived in Delano, Chavez had lost thirty-five pounds. After twenty-five days—one more than Gandhi's longest fast in India—he was ready to end his fast. Just holding his head up had become difficult. With Moses overseeing the historic moment, RFK handed Chavez a piece of Mexican bread and declared him "one of the heroic figures of our time."[38]

When Chavez had started connecting labor rights and environmental health, the National Audubon Society, the National Wildlife Federation, and the Sierra Club withheld their support, deeming Chavez too radical a figure. They didn't back the NFWA's grape boycotts or get involved in its agenda on environmental health. There was a heavy taint of racism in their rejection of him. When Bill Douglas led hikes to protest dams and Dagmar Wilson organized anti-nuclear demonstrations, they were embraced as environmental leaders by the establishment. When Chavez observed that low-income people and communities of color were disproportionately affected by environmental dangers, virtually all of the blue-chip conservation groups refrained from backing his action.

A promoter of organic farming, Chavez would spend the rest of his life crusading against pesticides, promoting clean air and water, and insisting that produce should be grown without harming workers. At his home in Keene, California, he grew an organic garden including grapes and pears. Having adopted the French system of intensive

Robert F. Kennedy broke the twenty-five day fast of Cesar Chavez on March 10, 1968. Together, they were early voices on behalf of environmental justice.

farming, he dug deep beds and mastered composting techniques. "His yields were phenomenal," his son Paul Chavez recalled. "His fruits and vegetables were world class, and he used less water. My dad was thinking about both environmental justice and sustainable agriculture. He was an all-around big environmentalist."[39] By the time Chavez died in 1993, he was still far ahead of his time, trying to awaken the American public to risks of growing food without respect for the overall ecosystem and public health concerns.[40]

On June 5, 1968, the forty-two-year-old Bobby Kennedy won the California and South Dakota primaries in his quest for the presidency. It looked as though he was going to surge past McCarthy and Humphrey. Then, after delivering a victory speech at the Ambassador Hotel in Los Angeles, he was shot by a Palestinian terrorist. A busboy named Juan Romero held RFK's head and placed rosary beads in his hand. Kennedy asked him, "Is everybody okay?" and Romero replied, "Everything's going to be Ok."[41] The entire tragedy was broadcast live on television. Ethel Kennedy, three months pregnant, rushed to her husband's side and prayed. The journalist Pete Hamill recalled that Kennedy had "a kind of sweet accepting smile on his face."[42]

Twenty-six hours later, Robert Kennedy was dead. Seeing him buried next to his brother in an evening service at Arlington National Cemetery was too hard for Udall to bear. The cord that had bound his political and personal life together had been irrecoverably cut. "After Bobby died, I couldn't work for a few weeks," he said later. "I eventually realized that he would have wanted me to forge forward with the conservation bills like Redwood and North Cascades and the wild and scenic river systems. He had gotten involved with the whole Scenic Hudson cause. I continued to push those forward as my way to honor Bobby, my friend, my hero, and to help the Johnsons secure their own conservation legacies."[43]

Robert Kennedy, Jr., was fourteen years old when his father was murdered. Until then he had planned to be a veterinarian. In the summer of 1968, he went on safari in Africa with Lem Billings (JFK's closest friend) and determined that wildlife conservation was his true calling. In the fall he enrolled in Millbrook School, a boarding school for boys in Duchess County, New York, attracted to it because of its ornithology program. To heal from losing his father, he and a friend took up falconry and captured raptors with snares. "We flew wild red-tails, falcons, and goshawks and pioneered many of the game-hawking techniques still used by American falconers," RFK Jr. recalled. "We talked about hawks every spare moment at meals, between classes, and after chapel."

Falconry led RFK to become an environmental lawyer determined to save hawks and kestrels from DDT poisoning and habitat destruction. Four ecosystems, the Shawangunks, the Catskills, the Adirondacks, and the Hudson River, had found a militant, lifelong environmental protector in him.[44] Once he earned his law degree from Pace University, he crusaded to ban polychlorinated biphenyls (PCBs) from being released by industries along the Hudson River and elsewhere. PCBs were used in the 1960s and later by power companies as dielectric (electrically nonconductive) fluids in transformers and capacitors. They were carcinogenic to humans and wildlife. Much like DDT, they were persistent in the food chain and remained in the environment indefinitely.[45]

III

In early August 1968 the Republican Party, as anticipated, nominated Richard Nixon for president. Since losing to Kennedy in 1960, he had struggled and scrapped his way back to the top of the GOP heap. During the campaign, he styled himself as a Theodore Roosevelt reclamationist and wildlife conservationist. With subtlety and skill, he courted Izaak Walton League members in the Midwest, Conservation Fund members in the Northeast, and even Ralph Nader's young consumer activist coalition. Far more than Governor Reagan, Nixon championed the National Park Service. "The grandeur of California's quaint redwoods, like the immense horizon of California's ocean, for my father growing up was a part of California's mystique of limitless opportunity for a young man," his daughter Tricia Nixon Cox recalled in 2020. "Like his love of swimming in the Pacific Ocean, he never tired of visiting California's redwoods forests."[46]

Odd as it might seem, in 1968 Nixon admired the thrust of Nader's consumer advocacy work. Not that he would have said it in public. Nader had artfully broadened his focus from his initial critique of auto safety and was pressuring American corporations on many fronts, including pollutants and their effects on a workforce and its surrounding community. As the 1968 presidential election heated up, Nader, in a sobering *New Republic* article, blamed West Virginia and Ohio coal-mining corporations for causing black lung disease.[47] And these mining companies responded by attacking Nader as a radical opposed to capitalism. The United States had the world's biggest coal reserves and a longtime problem with miners contracting respiratory diseases.[48] Nader's fact-based exposés against Big Coal contributed to both houses of Congress passing the monumental Occupational Safety and Health Act of 1971.[49]

Nixon's daughter, Tricia, who was twenty-two in 1968, believed that environmental protection was an urgent concern of the sixties generation, just as Nader contended. "My future husband, Edward Cox, took time off from his summer work with Ralph Nader to attend the '68 convention," she recalled. "He attended my father's presentations to large groups of delegates and came away impressed with my father's frank talk about the need for federal environmental regulations.

Edward thought it brave politically since he knew the core philosophy of the delegates was instinctively against more federal regulations."[50]

Nixon's key adviser on environmental issues at the time was the razor-sharp conservation-driven attorney John D. Ehrlichman, whose law firm had been involved with Puget Sound's ecological issues for years. An only child, Ehrlichman was born on March 20, 1925, in Tacoma, Washington, but moved to Santa Monica, California, with his family after the Great Depression hit. Surfing in the Pacific and hiking mountain trails of the San Gabriels were his favorite teenage pastimes. A dedicated Boy Scout, he won the Distinguished Eagle Scout award and at eighteen joined the army air corps to serve in World War II, even though his father had died in 1940 while flying for the Royal Canadian Air Force. As a chancy lead B-17 navigator in the Eighth Air Force, he won the Distinguished Flying Cross. After the war, he attended UCLA on the GI Bill and graduated in 1948. Drawn to a career in land and water use law, he attended Stanford University Law School, graduating in 1951.

Moving to Seattle, Ehrlichman and his uncle Ben became perhaps the top land and water use lawyers in the Pacific Northwest. As a partner in Hullin, Ehrlichman, Roberts & Hodge from 1952 to 1968, his legal expertise in urban land use and zoning was legendary. In the mid-1960s, he garnered local press by blocking the development of an aluminum plant on Guemes Island, in a case that eventually went to the Supreme Court. The Ehrlichmans lived in Hunts Point, on the east side of Lake Washington. Alongside the plainer citizenry he regularly swam, boated, waterskied and fished in their backyard. Nobody worked harder to clean up Lake Washington from pollution than the bookworm Ehrlichman. In the late 1960s, he represented homeowners in Port Susan in suing Snohomish County over its approval of housing subdivisions on property owned by the Atlantic Richfield Company, which was remiss in procuring the permits needed to construct a refinery on the premise.[51]

Ehrlichman was also an angler extraordinaire. Whenever the opportunity arose, he went fishing all around King County, Puget Sound, and south to the Columbia River when salmon ran. "He and mom took us five children camping often as they could in the Cascades, the

Olympics and the Canadian Rockies," his daughter Jan Ehrlichman recalled. "Dad taught all of us to fish on these camping trips."[52]

In 1960, Ehrlichman worked on Nixon's presidential campaign. Even though Nixon lost, it was the beginning of a fast friendship between the two politically driven men. The level-headed Ehrlichman was also close to Henry Jackson and shared the senator's desire to protect salmon runs, have safe drinking water, and promote sustainable development. When an effort was needed to clean up gorgeous Lake Washington, adjacent to downtown Seattle, Ehrlichman took the lead, studying grassroots movements in the Great Lakes and Chesapeake Bay areas to mobilize citizens.

In 1962, Ehrlichman invited Nixon to boat around Lake Washington to fundraise and strategize for his campaign for California governor. Nixon was delighted to discover that Ehrlichman had become a conservation folk hero in Seattle for blocking the aluminum plant on Guemes Island. During their boat outings, Ehrlichman explained to Nixon the fundamentals of the post–*Silent Spring* landscape of environmental policy in America.[53]

The environment wasn't Nixon's "thing," as Ehrlichman put it, but the Lake Washington adventure would prove to be consequential in US history. In 1968, Nixon chose Ehrlichman, considered by the Democratic triumvirate of senators Henry Jackson, Ed Muskie, and Gaylord Nelson to be a "covert green," to be his chief domestic adviser and point man on all environmental quality policy concerns.[54]

Ehrlichman was instrumental in selecting Russell Train as Nixon's undersecretary of the interior. Train, a lawyer by education, had served in several government posts, but his true calling was the protection of global wildlife. Over the years, he was among the founders of several nonprofit groups dedicated to that pursuit, including the American branch of the World Wildlife Fund. He regularly went on African safaris with family and friends to study lions and rhinos. He was a refined, bighearted gentleman, at work or away from it. Believing that the Endangered Species Preservation Act of 1966 was too weak, he prepared reports for Ehrlichman that made the case for species and habitat protection. John McPhee interviewed Train for *Encounters with the Archdruid*. "Thank God for Dave Brower," Train said. "He makes it so easy for the rest of us to be reasonable." When Brower heard that, he responded, "Thank God for Russell Train. He makes it so easy for anyone to appear outrageous."[55]

As a California politician, Nixon knew in 1968 that the Sierra Club had a growing membership and media clout. And he had many longtime friends in the Save the Redwoods League. At times he took a genuine interest in marine mammal protection. Nevertheless, he had a special loathing of Bill Douglas, who, he fumed, had somehow received a free pass from the press for his blatant conservation-related conflicts of interest. According to Nixon, it was as if Douglas flaunted his lack of judicial restraint and the *New York Times* cheered him on because he made colorful copy as the "green" justice.[56]

By 1968, the Sierra Club was surging in influence and militancy. Senator Jackson even allowed club lawyers access to his Senate office to work on the North Cascades National Park campaign. Venturing into urban preservation, Brower published *Central Park Country: A Tune Within Us*, which included nature-themed poems by Marianne Moore.[57] Moore had her own heavy-duty bone to pick: she blamed LBJ for the bombing campaign that was destroying Vietnam's ecosystems.[58] The *Sierra Club Bulletin*, once so loyal to Jack Kennedy, eviscerated Johnson for his Vietnam policies, especially the spraying of defoliants such as 2,4-D and 2,3,5-T. With an unquenchable thirst for global peace, Brower called the Southeast Asia war "technological colonialism" that poisoned US soldiers as well as Vietnamese people, supposedly in the name of freedom.[59]

From August 26 to 29, 1968, the Democratic Party gathered at the International Amphitheatre in Chicago, Illinois, for a convention that was, as the novelist Norman Mailer later described it, "martial, dramatic, bloody, vainglorious, riotous, noble, tragic, corrupt, vicious, vomitous, appalling, cataclysmic."[60] Disheartened by the assassinations of King and Kennedy and horrified by the party's unwillingness to end the Vietnam War, protesters stormed Chicago. To thwart them, barbed wire ringed the Amphitheatre and six thousand men from the Illinois National Guard, along with eleven thousand city police officers, turned the city into a battleground. Todd Gitlin, one of the leaders of Students for a Democratic Society (SDS), an anti-war organization, paraphrased a lyric from the hippie anthem "If You're Going to San Francisco": "If you're going to Chicago, be sure to wear some armor in your hair."[61]

In the Amphitheatre, away from the chaos outside orchestrated by Mayor Richard Daley's police, the Democratic Party chose Hubert Humphrey as its nominee with Edmund Muskie as his running mate. That was a dream team from the perspective of conservation, but the candidates' refusal to commit to an immediate end to the Vietnam War caused many liberal Democrats to sit the election out.

There was an environmental plank in the 1968 party platform, but it was a laundry list of lame generalities. Throwaway lines such as the need to "work toward abating the visual pollution that plagues our land" and "focus on the outdoor recreation needs of those who live in congested metropolitan areas" hardly inspired youth activism.[62] But perhaps nothing related to ecology could have fast-tracked in that year of the Vietnam War, Black Power, women's rights, and the assassinations of MLK and RFK.

In 1968, the Democratic Party, headed by Humphrey, was slow to understand that a year after the Environmental Defense Fund had been founded, conservation had a new "Sue the bastards!" aggressiveness about it. The folksy days of Carl Sandburg lobbying to preserve the Indiana Dunes by telling homespun yarns had given way to urgent plans of eco-action with no room for compromise against philistines. Just as dunes shifted with the winds, the idea of what constituted conservation was changing as it came to embrace more ways that human activity posed a threat to the earth. A growing number of young people considered themselves environmentalists. Many were impatient to the point of desperation. Environmentalism was no longer a mere issue; it was fast becoming a political identity and a way of life.

In 1968, the Ecological Society of America asked the three presidential candidates, Humphrey, Nixon, and George Wallace (running for the American Independent Party), to put into writing their views on what constituted proper White House environmental leadership. Nixon ignored the request; Wallace wrote a few lines of pablum. Humphrey, by contrast, provided an adept three-page response calling for a prototype of the Environmental Protection Agency. "We need not only more ecologists," he said, "but a new breed of professional ecologists who are prepared to act as broad-ranging 'environmental specialists' in ecology, planning, political science, sociology, engineering, and other disciplines which relate to the totality of our environment."[63]

That summer saw the formation of the American Indian Movement (AIM) in Minneapolis. Founded by Dennis Banks, Russell Means, and Clyde Bellecourt, AIM was ready to challenge government policies that hurt their tribes. Armed and militant, AIM said that the federal government had robbed Indigenous people of their lands, exploited sacred religious sites such as New Mexico's Blue Lake (Taos-Pueblo) and Monument Valley (Navajo), and treated Native Americans like third-class citizens. Another valid point concerned the New Conservation movement. When land was saved in the form of a park or preserve, it was never placed under the management of its original stewards. Native American nations contended that they knew from centuries of experience how to best preserve their lands and waters but were ignored by the Interior Department.

In trying to break through to unleash environmentalism, nature writers, it might be said, were also becoming more aggressive. At the front of that pack in 1968 was Edward Abbey, whose book *Desert Solitaire: A Season in the Wilderness* galvanized environmentalists with a call to action to save the Southwest wilderness from hyper-industrialization. Abbey anchored his nonfiction book on his two seasons as a ranger in Utah's Arches National Monument during the late Eisenhower era. The result was a perfectly rendered hybrid of transcendental joy, coyote humor, in-your-face wrath, field science data, philosophical righteousness, and moral clarity. After a rave review in the *New York Times* that January, the memoir became a weapon of resistance, a polemic against despoilers of nature.[64]

Raised in the Alleghenies of western Pennsylvania, Abbey, entirely self-assured, had the scraggly beard of an Old West prospector and the iconoclastic poise of a Beat Generation bohemian. In *Desert Solitaire*, he called the 76,000-acre Arches National Monument, which would be upgraded to national park status in 1971, "the most beautiful place on earth."[65] He described the region, with its burnt orange cliffs, corroded monoliths, and natural bridges, in lively, eloquent prose. "Everything is lovely and wild, with a virginal sweetness," he wrote. "The arches themselves, strange, impressive, grotesque, form but a small and inessential part of the general beauty of this country."[66]

Abbey's detailed journals and notes from his time in the unfenced Utah backcountry formed the basis of *Desert Solitaire*. When he was on his rambles as a park ranger, he felt intoxicated, as if time were suspended. Awed by the eternal beauty all around him, mirthful and full of delight, he melted into the landscape, living in rustic simplicity and natural fellowship with the desert's wildlife and learning the ways of desert ecology. Inspired by Walt Whitman's dictum "Resist much, obey little," Abbey became a fierce watchdog of Arches and surrounding Utah terrain held sacred by the Hopi, Navajo, Ute, and Pueblo of Zuni. Patrolling in a Park Service pickup, often in uniform, he came to revile the bulldozers, hydroelectric dams, paved roads, and industrial tourism that defined Southwest development in the 1960s. He channeled that revulsion into ferocious, and at times anarchistic, prose.

In *Desert Solitaire*, Abbey denounced large-scale uranium mining in Utah's salmon pink tableland, and he reminded a cynical and distrustful public that the mission of the National Park Service was to preserve treasured landscapes in an "unimpaired" fashion. "Wilderness preservation, like a hundred other good causes, will be forgotten under the overwhelming pressure of a struggle for mere survival and sanity in a completely urbanized, completely industrialized, ever more crowded environment," he warned. "For my own part I would rather take my chances in a thermonuclear war than live in such a world."[67]

Abbey's *Desert Solitaire* was beloved by nature lovers, inasmuch as it mainly preached to the choir. *The Population Bomb* by Paul Ehrlich, on the other hand, was a Cassandra sensation. A biology professor at Stanford who specialized in butterflies, Ehrlich had already documented the bay checkerspot butterfly (*Euphydryas editha bayensis*), which was facing annihilation on the West Coast, largely because of changes in its microclimate. In 1967, Brower heard Ehrlich lecture at the Commonwealth Club of California and was unnerved. He persuaded Ehrlich to turn the lecture into a book, the subject being the scientific and behavioral challenges pertaining to overpopulation. Brower also introduced him to an editor at Ballantine. Within three weeks, Ehrlich had a draft of *The Population Bomb* ready for publication.

Ehrlich's primary concern was that ever-growing numbers of

humans couldn't be fed without causing major planetary environmental damage. Ever since he had read William Vogt's 1949 book *Road to Survival*, he had fretted about the havoc that *Homo sapiens* were wreaking on the world's ecosystems. *The Population Bomb* warned that the natural world had a finite carrying capacity and that since World War II, it had been breached. "Basically, there are only two kinds of solutions to the population problem," Ehrlich wrote. "One is a 'birth-rate solution' in which we find ways to lower the birth rate. The other is a 'death rate solution' in which ways to raise the death rate—war, famine, pestilence—find us."[68] He saw the United States as the major culprit, because with only 7 percent of the world's population, the country used 35 percent of its natural resources.

Selling more than 2 million copies, *The Population Bomb* triggered massive popular interest in the idea of zero population growth.[69] Ehrlich appeared as a guest on NBC's *Tonight Show* and was a highly effective spokesperson on the causes of overpopulation.[70] On the book's fortieth anniversary, he wrote, "It introduced millions of people to the fundamental issue of the Earth's finite capacity to sustain human civilization."[71] More than any other US senator, Frank Church was alarmed by the book's grim data. Speaking in southeastern Idaho, he warned, "Every 24 hours a city the size of Salt Lake is being added to the population of the world." That "swarming human horde," as he put it, would lead to global ecological ruin.[72]

The other writer who put environmental ethics front and center before the American reading public was N. Scott Momaday, a Kiowa, whose novel *House Made of Dawn* (1968) won a Pulitzer Prize for Fiction.[73] Thanks to Momaday, at long last, Native American environmental heroes like Chief Seattle, Standing Bear, Black Elk, Lame Deer, and Hyemeyohsts were put on the pedestal with Thoreau for their wisdom on the interrelationship with man and nature.[74] His novel is credited with sparking the Native American Renaissance. Superb writers such as Leslie Marmon Silko and Louise Erdrich have ascribed Momaday with inspiring them to focus on Native American topics in their literature. In his essay "An American Land Ethic," Momaday wrote that his Kiowa ancestors knew to regard the earth, sky, and waters as sacred treasures. This fine essay was included in *Ecotactics: The Sierra Club Handbook for Environmental Activists* (1970).[75]

One other book related to the environment was published in 1968 that captured the public imagination in a futuristic way, just as the sixty-nine-year-old Henry Beston, author of *The Outermost House* (1928), died at his Chimney Farm in Maine. Twenty-nine-year-old Stewart Brand published the inaugural *Whole Earth Catalog* in a print run of only one thousand copies. In an oversized, magazine-like format, it offered alternative ways to think about planetary living, linking techno-utopianism, environmentalism, and the counterculture. Brand stitched together such disparate subjects as Buckminster Fuller geodesic domes, NASA, LSD, the Jefferson Airplane, and organic farming in a surprisingly coherent way. In the mid-1960s, he had been one of the novelist Ken Kesey's Merry Pranksters and had participated in the parties about which Tom Wolfe had written in his book *The Electric Kool-Aid Acid Test*. Amazed by satellite photos of lonely Earth floating in the dark, Brand imagined a world in which computer technology, space exploration, and green thinking would work in cosmic unity.[76]

CHAPTER 27

Lyndon Johnson

CHAMPION OF WILD RIVERS AND NATIONAL SCENIC TRAILS

(OCTOBER 2, 1968)

Avenue of the Giants, Humboldt Redwoods State Park, California, 1964. This photograph, from Philip Hyde's Sierra Club Exhibit Format Series book *The Last Redwoods*, was central to the campaign to establish Redwood National Park in 1968.

I

Four years of horse-trading on Capitol Hill had paid off with the passage of four so-called "Forest Bills" establishing the Wild and Scenic Rivers and National Scenic Trails designations, plus two new national parks. Those New Conservation achievements showed that thoughtless growth and the reckless exploitation of natural resources could be checked and demonstrated that the Wilderness lobby was still alive and well in Washington. On October 2, 1968, with the presidential election only a month away, Lyndon Johnson convened conservation leaders at the White House for the grand signing ceremony.[1]

In his opening words, LBJ greeted Chief Justice Earl Warren, who was present; Justice William O. Douglas hadn't been invited because of his Vietnam War dissent. "In the past 50 years, we have learned—all too slowly, I think—to prize and to protect God's precious gifts," Johnson said at the East Room event. "Because we have, our own children and grandchildren will come to know and come to love the great forests and the wild rivers that we have protected and left to them."[2]

One could see in the signing ceremony a dichotomy that was central to Johnson's persona as a conservation president. A more media-savvy president would easily have signed into creation the two national parks—Washington's North Cascades and California's Redwood—in a natural setting. Instead, a ruddy-faced Johnson, dressed in his usual dark gray suit, conducted the signing in a businesslike way. There weren't even beautiful Ansel Adams, Philip Hyde, or Eliot Porter photographs on posterboard easels to elevate the historic meeting above Kiwanis Club status. The Forest Bills ceremony, as unfurled by the White House, might have been an IRS conference on tax reform or the budget deficit. It was Dullsville, USA. But the White House's mundane rollout contrasted sharply with his rhetorical eloquence about what the the natural world meant to LBJ, who spoke affectingly at the signing about the grandeur of wild America.

Throughout his career, Johnson got things done by sticking close to the US Capitol or the White House when he was in Washington. At 1600 Pennsylvania, he remained not just indoors but what one might call *very* indoors, surrounded by piles of papers, billows of cigarette smoke, competing TV broadcasts, and ringing telephones. Though it's hard to generalize, most white-collar people who established their careers in the first two-thirds of the twentieth century lived with the same dividing line between work-and-office time and personal-and-outdoors time. Udall was the representative of a more modern style. He duly attended the signing ceremony on October 2 but then flew off the next day to Wisconsin's idyllic backcountry, where he made public remarks about the New Conservation laws in the fresh forest-scented air, dressed appropriately for a day spent white-water rafting. For Johnson, such relaxed "in nature" moments were reserved for his time at his Texas ranch.

Environmentalists found it easier to relate to someone such as Stewart Udall than LBJ. Most of them were like "Stew" in that they were always on the lookout for the thrill and rush of outdoors adventure. Bill Douglas and David Brower were archetypes of the boots-on hikers whose identity in youth lay in embracing the windblown challenge of nature. The wilderness gang had fetishized snow-clad volcanoes such as Washington's Mount Baker and fast rivers like West Virginia's Gauley. Johnson was at a personal disadvantage with the main thrust of 1960s-style preservationism because that kind of hardy adventuring wasn't the way he rolled. Frank Church and Scoop Jackson might smile patiently when LBJ said that he enjoyed nature most during a picnic with Lady Bird in San Antonio. And they surely rolled their eyes when he described the beauty of the pastoral Texas Hill Country as seen while riding in an aqua convertible that could be driven across the Pedernales. Unlike Kennedy, who was an Atlantic sailor, LBJ globbed Coppertone suntan lotion on his sun-drenched face to motorboat around Lake Lyndon B. Johnson as if he were actor Slim Pickens in the movie *Dr. Strangelove*. "There was a disconnect between Lyndon Johnson and environmentalists," George McGovern recalled. "We didn't trust him or give him proper credit for the Wilderness Act or the Wild and Scenic Rivers System because of the Vietnam War."[3]

Even though Johnson wasn't kinsmen of those who associated untamed nature with strenuous adventure, he forged ahead on conservation legislation to protect wild places because it was part of America's frontier heritage. If some people hiked the remote 8,751-foot Guadalupe Peak in west Texas, that was fine by LBJ. But if others were just as exhilarated by tossing a football in the Great Swamp National Wildlife Refuge in Morris County, New Jersey, about twenty-five miles west of Manhattan's Times Square and just off the Garden State Parkway, that was fine, too. In Johnson's view, public lands preservation and environmentalism had to be accessible for the psychological good of American families and the ability for them to make day outings. "The wonder of nature is the treasure of America," he said.

"What we have in woods and forest, valley and stream, in the gorges and the mountains and the hills, we must not destroy. The precious legacy of preservation of beauty will be our gift to posterity."[4]

Even though LBJ was set apart by the Browerites (who saw every issue from the top of El Capitan) or RFK adventure junkies (who enjoyed shooting rapids in Class 4 whitewater rivers), he gave the Silver Spring revolutionaries one of its best single days ever by signing the four bills into law that second of October 1968. What once again became vividly clear was that LBJ was modeling his New Conservation agenda after Theodore Roosevelt. Johnson had said to reporters that they were America's two cowboy presidents who understood the western frontier. "Whenever I pictured Teddy Roosevelt," he told historian Doris Kearns Goodwin, "I saw him running or riding, always moving, his fists clenched, his eyes glaring, speaking out against the interests on behalf of the people."[5]

The most TR-like of Lyndon Johnson's October quartet was the Wild and Scenic Rivers Act, an idea that Laurance Rockefeller's Outdoor Recreation Resources Review Commission (ORRRC) had first promoted in 1962 and Stewart Udall had trial ballooned in *The Quiet Crisis*. A linear variation of the National Wilderness Preservation System established in 1964, the Wild and Scenic Rivers Act provided three classifications for protected rivers: recreational (swimming, fishing, etc.), which might have limited development along the shoreline; scenic, which were undeveloped but accessible by road; and, most precious of all, wild. Those stretches of wild rivers were "generally inaccessible except by trail, with watersheds or shorelines essentially primitive and waters unpolluted," the new law stated. "These represent vestiges of primitive America."[6] Under the WSR law, a river could earn one of those designations by Congressional acts incorporating the river into the national system or by the secretary of the interior's designation after state legislation ratification and a request from a governor.[7]

Since the days of George Washington, the federal government had been involved in many aspects of river infrastructure, including managing transportation, dredging, and later flood control. But the Wild and Scenic Rivers Act, building on the "national rivers" designation that had been used to protect the Ozark rivers under Kennedy, added two new concerns to the national oversight: aesthetics and ecology.[8] Maintaining the beauty and health of American rivers that were still pristine was the Great Society's innovative contribution to the philosophy that had informed the Wilderness Act of 1964.

The House of Representatives Committee on Interior and Insular Affairs, chaired by Wayne Aspinall, hammered out sixteen iterations of a wild-river bill during the Johnson years. The House, in the end, voted 265–7 in favor of the final version. Disregarding fierce protests from the Army Corps of Engineers, the Bureau of Reclamation, and the Federal Power Commission, the Senate—led by Frank Church, Gaylord Nelson, and Walter Mondale—passed the legislation more easily. Upon Johnson's signature, specially selected free-flowing streams were at long last saved for "their natural scenic, scientific, esthetic, and recreational value," which lawmakers affirmed outweighed their "value for water development and control purposes."[9] Call it the triumph of Bill Douglas over the clank of twentieth-century machines. The law principally banned dams and other major waterways protected from the designated stretches of remarkable rivers but had little effect on private land.

In a warm, generous mood in the East Room, Johnson recalled that after the Udall family's rafting trip down the Colorado in 1967, his interior secretary had returned to Washington recommending that every family should get to know at least one stupendous river. In this spirit, Johnson spoke in a highly personal way of the Pedernales:

> I played on it as a child. I roamed it as a college student and I visited it frequently as President. But my wife has some more specific plans for me to go back and walk it with her—both sides, I think.
>
> I'm signing an act today which preserves sections of selected rivers that possess outstanding conservation values.
>
> An unspoiled river is a very rare thing in this Nation today. Their flow and vitality have been harnessed by dams and too often they have been turned into open sewers by communities and by industries. It makes us all very fearful that all rivers will go this way unless somebody acts now to try to balance our river development.[10]

Because LBJ would be returning to his Texas ranch—and the Pedernales—once his term ended, the Wild and Scenic Rivers Act would be a ringing validation that he was a great conservation president. America's 2.9 million miles of rivers, he rightfully believed, were the irreplaceable veins and arteries of the United States for future grandchildren to enjoy.[11] The stretches of eight free-flowing inaugural rivers that were first to receive the WSR designation totaled 789 miles. They were the Middle Fork of Idaho's Clearwater with its two tributaries, the Lochsa and the Selway, for 185 miles; the Eleven Point for 44 miles (Missouri); the Middle Fork of the Feather for 154 miles (California); the Rio Grande for 53 miles and its tributary the Red River for 4 miles (New Mexico); the Rogue for 85 miles (Oregon); the Middle Fork of the Salmon for 104 miles (Idaho); the Wolf for 25 miles (Wisconsin); and the Saint Croix and its tributary, the Namekagon, for 200 miles (Wisconsin and Minnesota).[12]

From that day forth, each wild and scenic river was to be administered by a federal agency: the National Park Service, the US Forest Service, the Fish and Wildlife Service, or the Bureau of Land Management. With the gleam of accomplishment in his eyes, LBJ boasted that a hundred other rivers would soon become part of the system in the 1970s.[13] It wasn't hyperbole. By 1988, dozens of rivers (plus tributaries) had been incorporated into the National Wild and Scenic Rivers System. That very year, with Ronald Reagan as president, Oregon passed a phenomenal river bill that added another thirty-three fast-flowing rivers and tributaries with strict management into the LBJ system.[14]

LBJ's inaugural class of WSRs all captured the unbridled spirit of American waterways that roar, surge, pound, ramble, and weave through all fifty states. The Clearwater River (Middle Fork) in northeast Idaho was the premium steelhead trout domain, while the Lochsa tributary was populated by cutthroats. California's Middle Feather included one of the premier wild trout fisheries on the West Coast. In his 1974 book *Sierra Whitewater: A Paddler's Guide to the Rivers of California's Sierra Nevada*, Charles Martin correctly deemed the Feather River the most beautiful in all of California with its clear water bordered by white gravel banks and thick green forests. The segment of the Rio Grande saved by the act in north-central New Mexico was one of the best whitewater runs in the Southwest. Its waters are thick with brown and rainbow trout. The Rogue River in southwest Oregon represented one of the classic wild-river rafting trips in North America. The lower portion of the river passed through the Siskiyou Mountains, prime habitat of big salmon (up to forty pounds) and steelhead (fifteen pounds). Zane Grey, the popular writer of western novels, had owned a cabin along the Rogue and had urged that that part of Oregon should be designated a national park.

Rivers, Johnson believed, from the far-flung waterways of Alaska, Idaho, and Wyoming to those coursing through the rural countryside of Vermont and Maine, were the lifeblood of America. At the signing, he told Udall he felt frustrated that the designation of the sapphire-colored Guadalupe River in Texas had been sabotaged by the Kerr County, Texas, Chamber of Commerce.[15] Two sections of free-flowing eastern rivers, West Virginia's Cacapon and Shenandoah, had been omitted at the last minute because of opposition by landowners and the all-powerful coal industry.[16]

Making clear that the October 2 act itself was just the opening salvo, President Johnson named twenty-seven other waterways for study and possible inclusion in the National Wild and Scenic Rivers System.[17] He then recommended that people read John Graves's 1959 book *Goodbye to a River*, which described how a dam had ruined the Blanco River in Texas. "President Johnson believed that the National Wild and Scenic Rivers System would be the notable piece of conservation legislation of his presidency," Liz Carpenter recalled. "Once he returned to Texas, to the banks of the Pedernales, he would talk about all of Texas's rivers. If Kennedy had the seashore and Lady Bird wildflowers, then Lyndon thought of himself as the River King. But [Gaylord] Nelson and [Frank] Church were the driving forces of making it happen on Capitol Hill."[18]

Working against the Cold War–era obsession of channelizing, canalizing, dredging, filling, or damming rivers, the outdoorsmen Frank and John Craighead (twin brothers) had contemplated a wild and scenic rivers designation in the fifties. LBJ's act, the jubilant Frank Craighead believed, was as innovative as the National Park Service Act of 1916.[19] The magazine *National Parks* made the most obvious comparison: "The legislation did for wild rivers what the Wilderness Act did for wild landscapes."[20]

Ansel Adams, David Brower, and Margaret Murie, among other committed activists, sent Johnson their congratulations. For white-water river outfitters in such towns as Salmon, Idaho, and Grants Pass, Oregon, the act represented their industry's coming of age. Riverine conservation nonprofits immediately lobbied for more designations from the exciting list of rivers to be studied: sections of the Missouri once traveled by Lewis and Clark; the Chattooga of the Carolinas and Georgia, soon to be made famous by the James Dickey novel *Deliverance*; and the Allagash Wilderness Waterway, about which Henry David Thoreau had written in *The Maine Woods*.[21] All the major women's magazines, including *Redbook*, *American Home*, and *Good Housekeeping*, had gotten behind Johnson's rhapsodic embrace of living rivers.[22]

Udall was the Johnson administration's point man for calming the fear of Washington lawmakers and western reclamationists worried that the act would inhibit economic development west of the Mississippi River. Preaching the doctrine of free-flowing rivers, he looked congressional supporters of dams in the eyes and told them, "We're going to balance things out."[23] After the White House ceremony, he reflected, "For too long, we have regarded our rivers as something to carry away our waste."[24]

A self-serving reason the signing was celebratory for Udall was because it buried the grudge Brower and Douglas harbored toward him over his Grand Canyon dams advocacy in the mid-1960s. Together with Gaylord Nelson, Udall had helped convince the Northern States Power Company to cede to the US government seventy miles along both sides of the Saint Croix River where a power dam had been planned by the states of Wisconsin and Minnesota.[25]

Nelson had pushed the wild and scenic rivers legislation forward like a steam engine without brakes. Recruiting fellow river enthusiasts in the Senate to the Grand Cause was his impassioned calling. In 1963, when he wanted to establish the Saint Croix as a national river, he knew he had to unite with Walter Mondale, then the attorney general of Minnesota, because the river runs along the border between Minnesota and Wisconsin. In 1965, Nelson delivered a keynote speech to the Minnesota Conservation and Water Pollution Control Commissions, a speech known since as the "Bill of Rights for American Rivers." He said, "Call the roll of the great American rivers of the past, and you will have a list of the pollution problems of today." He then named cherished waterways: the Androscoggin in Maine; the Connecticut, the Hudson, the Delaware, the Ohio, the Mississippi, the Missouri, and the Minnesota. "The story in each case is the same: they died for their country," he said. "They died in the name of economic development."[26]

Just as the room turned somber, Nelson gave the New Conservationists a shot of political optimism, telling them that he would make sure the entire Saint Croix and its tributary, the Namekagon, became national scenic waterways. At the time of this appeal, he was drafting the "forever wild" bill with his thoughtful Senate colleague Walter Mondale of neighboring Minnesota.

When LBJ signed the Wild and Scenic Rivers Act, it included 152 miles of the Saint Croix, 98 miles of the Namekagon, and 24 miles of the Wolf River in Menominee County, Wisconsin. Thus, because of the Nelson-Mondale alliance, three of the nine rivers designated by LBJ were in Wisconsin. Nelson prided himself that Wisconsin held the honor of having the only rivers east of the Mississippi classified in the WSR Act, though he quipped, "They weren't very far east of it."[27]

To celebrate LBJ's signing, Nelson took Udall on a Wolf River raft trip. Wisconsin was a swing state and Nelson was up for reelection that November. As it turned out, Udall was his top out-of-state endorser.[28] At an evening event in Green Bay, the legendary Packers coach Vince Lombardi called Nelson "the nation's number one conservationist." He added, "I would vote for him if he were a Republican or Democrat."[29] In Wisconsin, endorsements didn't get any more powerful than that.

Frank Church, the Senate sponsor of the Wild and Scenic Rivers Act, was thrilled that it provided Idaho with a significant part in the new system, assuring future generations that long stretches of the state's rugged mountain rivers would remain unspoiled and free flowing. Church told the Idaho Recreation and Park Association that he hoped the law's enactment on October 2 was only the first step in saving Idaho's wild rivers. His Senate office was working hard to add the main stem of the Salmon River (also known as the "River of No Return"), as well as parts of the Saint Joe, Bruneau, Moyie, and Priest Rivers, to the National Wild and Scenic Rivers System. Beyond that, Church said, he wanted a statewide effort to stop industries from polluting the Snake River. In a 1969 speech, he hoped that Idaho would someday be known as the Wild Rivers State instead of the Gem State. In that same speech, he lamented that seventy thousand dams had been built in the United States.[30]

Others besides Nelson and Church deserved credit for safeguarding wild rivers. Two Democratic senators—Mike Mansfield of Montana and Clinton P. Anderson of New Mexico—clustered around Johnson at the signing. The House of Representatives' hero of the conservation moment was John Saylor, who deemed the Wild and Scenic Rivers Act a high-water mark of his fifteen years in office, along with the passage of the Wilderness Act, the defeat of the Echo Park and Grand Canyon Dams, the preservation of the Ozark riverways, and the saving of Idaho's Hells Canyon.[31] Rowing, paddling, rafting, and drifting down America's rivers such as the Rogue, the Green, the San Juan, and the Rio Grande was suddenly the new rage in the outdoor recreation world. "A nation without flowing rivers," Edward Abbey wrote in *One Life at a Time, Please*, "would be a nation without hope."[32]

In 1998, Bruce Babbitt, Bill Clinton's first-rate interior secretary, praised the Wild and Scenic Rivers Act. "It changed the way we view rivers everywhere," he said. "It is not enough to call this law an act of Congress. It was more than that. It was a turning point—the end of one era and the beginning of another. It also set into motion a swirl of ideas. It opened a window and exposed a new way of seeing the landscape. It showed us that rivers are more than scenic resources—they are ecological life sources that mirror the health of the land around them."[33] By the time the act celebrated its fiftieth anniversary in 2018, the system had protected more than 13,000 miles of 289 major rivers (and a total of 495 rivers, forks, and tributaries) in forty states and Puerto Rico.

II

Senator Nelson was deeply involved with another law that Johnson signed on October 2—the National Trails System Act. In 1965, he had written two bills that had been folded into the rivers and trails acts that Johnson embraced. Henry Jackson was a staunch ally in promoting the National Trails System Act (NTSA), which would provide citizens with an intimate relationship with the natural world and transform them into committed conservationists.[34] The Jackson-Nelson team got NTSA rolling. The act was an outgrowth of Nelson's advocacy for the Bureau of Outdoor Recreation's 1966 *Trails for America: Report on the Nationwide Trail Study*, which had recommended three types of trails: national scenic trails, park and forest trails, and metropolitan area trails.[35]

Encouraged by both Freeman and Udall to move forward, Johnson had heartily embraced the premise. The latter two categories had shifted as the bill had progressed through Congress, becoming focused, like the first, on purpose rather than locale. Two key components of the NTS were immediately created by the resulting legislation: the multistate Appalachian and Pacific Crest Trails, both more than two thousand miles long and in danger of losing their wilderness characteristics. With the passage of the national trails law, they were protected. The National Trails System Act fulfilled the dream that the Wilderness Society's cofounder Benton MacKaye had voiced in 1938: an American trail network that would allow citizens to backpack long distances along both coasts.[36] There was already a fine system of trail huts and hostels in New Hampshire, where they were operated by the Appalachian Trail Club, in many cases under permit from the Forest Service. Such well-built rest structures would be emulated along the new federal designations. The National Trails System, according to the new law, consisted of three types of trails: national scenic trails, national recreation trails (the majority), and connecting and side trails. As the historian Samuel P. Hays pointed out, "Many of these trails were revivals of those established for firefighting by the Civilian Conservation Corps during the 1930s."[37]

The unsung bureaucratic hero of the National Trails Act was the US Forest Service under the leadership of Edward Cliff. "We really

RFK and Henry M. "Scoop" Jackson had become good friends in the 1950s. Even though Jackson supported the Vietnam War, RFK admired the Washington senator's tenacity on driving the Forest Bills through the Senate.

provide more recreation than the Park System does," he pointed out in an oral history for the Johnson Presidential Library, "but we have larger areas."[38] The Pacific Coast Trail, for example, would be administered by the Forest Service. It extended 2,350 miles from the Mexico-California border northward, generally, along the mountain ranges of the West Coast states to the Canada-Washington border near Lake Ross. On October 2 of that year, more than 100,000 miles of the trails were already established and administered by the Forest Service on the National Forest and National Grasslands units. The development of these trails had begun with the establishment of the National Forest System in 1905. What Johnson's Forest Service excelled at, however, was the cobbling of cooperation agreements with ranchers, farmers, woodland owners, timber operators, and other rural residents. This was no small feat, and the Great Society pulled it off.

The National Trails System Act of 1968 made wise use of existing cooperative agreements among land trusts, local and state governments, and private landowners. Wooden mileage and directional guideposts would be erected along the clay-rutted route to help hikers, who would now be able to walk the Appalachian and Pacific Crest Trails without a compass, map, or directional confusion. Johnson made it clear that states were encouraged to develop their own trail systems, and many did. For example, Michigan opened routes connecting the three Great Lakes. Bicycle trails were added.[39] Putting in his two cents that October, Bill Douglas advocated a Potomac Trail from Washington, DC, to northwest Pennsylvania. And Vermont senator George Aiken began lobbying for a North Country Trail from the Green Mountains of Vermont to North Dakota.[40]

Among the many national historic trails eventually designated were westward migration routes, such as the Santa Fe National Historic Trail, the Long Walk National Historic Trail through New England to Canada, and the Oregon National Historic Trail. Among the most popular with tourists was the Lewis and Clark National Historic Trail, established in 1978, which in the 1950s had been a dream of the late Senator Richard Neuberger. In the 1980s, to honor Native American history and the terrible part played by the forced displacement marches, the Trail of Tears and Nez Perce National Historic Trails were formed. Thirty more trails have been created over the years. Hiking trails, Nelson said, "represent perhaps the most economical form of public investment in outdoor recreation. There ought to be a place to hike within an hour's reach of every American."[41]

III

The historian Jared Farmer rightfully called Redwood National Park, which Johnson also signed into law on October 2, 1968, nothing more than "an ungainly hybrid, a strip park whose boundaries encompassed three spectacular state parks [to remain under the state ownership and management] unrelinquished by California."[42] It existed, though, in a heavily logged northern California landscape, and that was itself an achievement. Through the four years of the bill's laborious path through Congress, LBJ had maintained his personal investment and interest in protecting old-growth California redwood trees.

At the signing ceremony, Johnson declared the ancient giants the enduring symbols of American beauty. Saving them was as essential for his New Conservation legacy as protecting bald eagles, whooping cranes, and wildflowers. The new national park totaled 58,000 acres (28,000 of which were the three California state parks) and 10,900 acres considered virgin-growth timber, saved, in the nick of time, in the waning days of LBJ's presidency.[43] Sadly, though, all around Redwood National Park, logging continued as Douglas predicted. The Save the Redwoods League tried to kick in an additional $15 million to $20 million to purchase the stands of trees along Skunk Cabbage Creek, only to be ignored or rebuffed by lumber companies.

LBJ predicted that the Redwood National Park designation would be seen as the capstone of his New Conservation agenda:[44]

The redwoods will stand because the men and women of vision and courage made their stand—refusing to suffer any further exploitation of our national wealth, any greater damage to our environment, or any larger debasement of that quality and beauty without which life itself is quite barren. Yes, the redwoods will stand. . . . They will declare for all to hear, when other great conservation battles are being fought: We stand because a nation found its greatest profit in preserving for its heritage its greatest resource, and that is the beauty and the splendor of its land.[45]

LBJ wasn't a student of scientific forestry as FDR had been. He didn't read the Yale School of Forestry silvicultural journal or plant evergreens along the Hudson River to sell as Christmas trees. But he knew in his heart that *Sequoia sempervirens*, the coast redwood, has no equal in America. A single redwood tree was more noble than the grandest country estate of any Vanderbilt. Always bragging about how big things were in Texas, he had nothing but patriotic admiration for the world's tallest trees, found exclusively within forty miles of the Pacific Ocean from south of Monterey Bay all the way north across the Oregon border. But by concentrating on the trees, he missed an opportunity to link the permanent protection of the Eel, Navarro, Russian, Mattole, Klamath, Mad, and Smith Rivers to the conservation of California's redwoods. Johnson thought of the redwoods as stand-alone attractions, but in fact, they were part of intricate ecosystems. What LBJ accomplished in northern California wasn't preservationism writ large but astute timber politics chess playing. Although he deserved credit for establishing Redwood National Park (something is better than nothing), he missed the moment to forever save the coastal redwoods ecosystem, not merely the groves of big trees.

The point wasn't lost on Bill Douglas. When Brower later wrote to Douglas about participating in a Sierra Club salute to Lady Bird Johnson, the justice agreed under the ironclad condition that he wouldn't have to praise her husband. "I take with a grain of salt the so-called conservation achievements of LBJ," the hardheaded Douglas wrote Brower.

> If we had a fighting conservationist in the White House, we would certainly have more than 58,000 acres in the Redwood National Park. If we had had such a person in the White House we would not have lost the Little Tennessee River to the TVA. I don't think any credit is coming to him for the Red River Gorge or Allerton Park in Illinois, both of which I had something to do with. Everything that was done was in spite of LBJ's attitudes. I can go through the whole list that way. Anyway, it doesn't detract from the salute to the First Lady. But I hope any publicity which is released will play down the achievements of LBJ as a conservationist, because the guy, in my view, is a complete phony on that score. He is the one who has given the Corps of Engineers more promotion, more backing, more encouragement than any other person in our history.[46]

Despite Redwood National Park's shortcomings, the Sierra Club president, Mike McCloskey, proclaimed the second of October a great victory over the "dirty-tricks opposition" of the most powerful combination of financial interests that had ever stood against a proposed federal park.[47] Brower, however, was more jaded. Like Douglas, he considered the new Redwood National Park a glass-half-empty disappointment; he was convinced that LBJ should have fought harder for more coastal redwoods and not been cowed by Aspinall's bluffs and the timber lobby's intimidation tactics.

Douglas, in cahoots with Brower, complained loudly to people who mattered in Washington, DC, that in getting the Redwood National Park bill passed, LBJ had accommodated the Arcata Redwood Company by shrinking the overall acreage and leaving out the ancient grove at Mill Creek. He fumed that, ecologically speaking, the boundaries had been gerrymandered to make no sense for long-term preservation. And he worried that the deal would cost more than triple the $92 million that Congress had authorized to purchase the land. Where would the rest of the money come from?

At Mill Creek, Douglas and Brower pointed out, two wilderness areas were now legally sandwiched between logging zones. This undermined the sacredness and sanctity of the protected big tree groves. Deriding LBJ as a practitioner of half-loaf conservation, Douglas insisted that a president with the spine of Theodore Roosevelt would have stood toe to toe with the Senate plan of 64,000 acres instead of worrying about Aspinall's obstruction. The suspicions of both Douglas and Brower about Johnson's "false-front" Redwood Park had merit. No sooner had LBJ established the national park than three logging companies—Arcata, the Simpson Lumber Company, and Georgia-Pacific—started retribution clear-cutting the old-growth trees in the Mill Creek and Redwood Creek areas.[48]

Looking for public relations windfalls, the National Park Service named an ancient stand of giant trees in the Redwood National Park "Lady Bird Johnson Grove." That November, Lady Bird visited the new park. At a dedication ceremony the following year, Presidents Nixon and Johnson stood at her side; Governor Reagan was there, too. Though many prime trees were still in the hands of timber companies, the grove was the perfect way to honor Lady Bird's conservation achievements. After Reverend Billy Graham praised the redwood grove as a "great cathedral," Mrs. Johnson made her remarks.[49] "Conservation," she said, "is indeed a bipartisan business because all of us have the same stake in this magnificent continent."[50]

IV

The fourth conservation bill that LBJ signed on October 2 established the 555,000-acre North Cascades National Park, which, combined with other parks, preserved a total of 1.2 million acres in northwest Washington State.[51] The grassroots movement to protect that part of the country went back to the days of Franklin D. Roosevelt and had been ratcheted up in 1951 when Douglas's book *Of Men and Mountains* was published and activist Polly Dyer dug in her heels. Seven years later, Kerouac's novel *The Dharma Bums* again called attention to the mountain expanse.[52] The fight to preserve the strikingly beautiful region from a multiple-use fate had become a primary mission of the Sierra Club and the Mountaineers.

Stewart Udall had decided in 1961 that North Cascades needed to be a national park. To let the US Forest Service continue to control its virgin forests and the largest glacier system in the lower forty-eight states would have meant losing the jewels of the Cascades. Now, near the end of LBJ's presidency, it was finally part of the national park family. At the signing, park promoters such as Patrick Goldsworthy of the North Cascades Conservation Council and Brock Evans of the Sierra Club were toasted for their perseverance. "The North Cascades National Park and its adjoining acres in what have been called the 'American Alps' is next door to the Pacific Northwest's most populous communities," LBJ pointed out.[53] It mattered to the president that the new federal park was less than three hours by car from Seattle, allowing for weekend recreation. "It modified the national park concept in a way that recognized wilderness as the most significant resource," the historian David Louter wrote in *Windshield Wilderness: Cars, Roads, and Nature in Washington's National Parks*, "yet still held the door open for America's motoring masses."[54]

There was general feeling in the East Room that Johnson was destined to be remembered as Mr. Conservation in history. Melville Bell Grosvenor, the chairman of the National Geographic Society, lauded Johnson for the preservation of wild places in the public domain: "You have done more serious conservation, the preserving of the natural beauty and scenic wonders of the United States, than any other president in history."[55]

As election day approached, LBJ said that nothing he had accomplished in the Great Society had given him "a greater sense of reward" than his conservation work with Lady Bird and Stewart Udall.[56] Compelled to grudgingly endorse Hubert Humphrey for president, he could at last brag that he had signed more than three hundred environmental measures in five years, a feat, he said, that would have "staggered the imagination" of Theodore Roosevelt and Franklin D. Roosevelt.[57]

Newspaper headlines in the first week of October referenced frustrations in Vietnam, racial unrest, urban riots, and a bitter political scandal involving Johnson's friend Abe Fortas. In spite of these troubles, Johnson maintained an outwardly sunny disposition, believing his status had been elevated in history by signing the four Forest Bills. An editorial in the Scripps-Howard newspapers observed, "It may not have been a very good week for Lyndon B. Johnson on some fronts, but on the conservation front it was spectacular."[58]

CHAPTER 28

Taking Stock of New Conservation Wins

Richard Nixon lived in San Clemente, California, with his wife, Pat, along the Pacific Ocean. He is responsible for creating National Oceanic and Atmospheric Administration (1970) and the Marine Mammal Protection Act (1972).

I

When Richard Nixon defeated Hubert Humphrey on November 5, 1968, the New Conservation movement shepherded by Kennedy and Johnson seemed to be over. LBJ had only a couple of lame-duck months left. Udall, however, wasn't ready to call it quits. While writing *The Quiet Crisis*, he had learned how Theodore Roosevelt had used the Antiquities Act of 1906 to establish Mount Olympus

National Monument, the core of today's Olympic National Park, two days before his term ended in 1909. That executive power tool was subsequently used by other presidents who wanted to instantly protect threatened places of extraordinary scenic value. For instance, Herbert Hoover set aside 4 million acres in national monument areas as he exited the White House in early 1933, including Death Valley (California), Great Sand Dunes (Colorado), and White Sands (New Mexico).[*][1]

Just before Election Day, LBJ had met with a delegation of Floridians, gathered in the Fish Room at the White House, to establish Biscayne National Monument. The new monument was in the northernmost part of the Florida Keys (4,200 acres of uplands and over 12,000 acres of submerged reefs).[2] This was the world's third largest coral reef. The uninhabited low-lying islands, barrier reefs, mahogany groves, dense mangrove thickets, and tidal marshes were American heirlooms of the first order. Furthermore, endangered American crocodiles and manatees lived in Biscayne's aqua-blue waters.[3] The national monument provided shelter for deep-sea fish as well as the smaller shore life that Rachel Carson wrote about in *The Edge of the Sea*.

Throughout 1968, Johnson asked that the legal work to establish Biscayne Bay be hashed out. That he succeeded before leaving the White House via an executive order was impressive. As he explained,

> "It will give our people almost 200,000 acres of islands and their adjoining bay and ocean waters, and they are all brimming with tropical plant and animal life. On these islands grow trees that were unknown anywhere else—and Presidents ahead of me [Herbert Hoover and Franklin D. Roosevelt] used to go there for their retreat. I have seen their pictures in the club rooms of the old days. These are the last remnants of a vast forest which once covered much of Florida. In these waters are rare tropical animals which now will be assured a haven from destruction.[4]

---

* All three were eventually upgraded to national parks.

Biscayne National Monument became so popular that it was up-graded by Congress to national park status in 1980 during the Jimmy Carter administration. The "no fishing zones" the National Park Service managed, however, became extremely controversial. With Miami erecting condo towers, high-rise hotels, and a $3 million marina, Biscayne NM was a gallant victory for LBJ's New Conservation determination to protect the upper Florida Keys as a diving and snorkeling paradise. Although this area has been near a well-traveled sector of the Florida coast since the beginning of European occupation, and in earlier times by Native Americans, it remains relatively undisturbed. "Relics of the boisterous era of Spanish galleons, loaded to the gunwales with gold bullion, and English sea rover searching the seas for the loot lie rotting in these reefs," Udall wrote of Biscayne. "Several ship hulks, wrecked on these treacherous shoals, are within the monument boundaries."[5]

LBJ took great pride that Biscayne National Monument was located near Miami, only about twenty miles away by car. It fit in perfectly with his "Parks for People" initiative. Once, as a senator, he had stayed at the private Coco Lobo Club in Biscayne Bay, where he had stuffed himself with stone crabs, drunk fruity cocktails, and smoked cigars while he looked out over the water. The fact that working-class Floridians could now vacation in this ecosystem was deeply significant to Johnson. Disregarding Miami developers clamoring to build an industrial seaport on a coral reef, complete with a four-hundred-foot channel dredged to it from across the bay, Johnson firmly sided with the preservationists.[6]

Johnson told the crowd assembled at the White House signing:

You know in our early days when Theodore Roosevelt and some of our other presidents were so conservation minded and they were trying to have playgrounds and national parks for our country; they were located in the West. The Grand Canyon and Yellowstone—if you could afford a round trip ticket or if you had a month to go by jalopy out there with your family—why you could get to see some of the glories of nature.

But the Redwoods and Assateague and Fire Island and Biscayne are all going to be in short distances from population centers, where you can take Molly and the babies on a Sunday afternoon and get back to nature. They are not off in far off remote locations.[7]

Instead of celebrating Biscayne NM as a stand-alone victory, Udall pushed his boss to create even more national monuments. Shortly after the 1968 election, he sent the president a memorandum proposing the preservation of 7.5 million acres of public land, much of it in Alaska; he explained that thanks to the Antiquities Act, this could be done swiftly through a raft of executive orders. To Udall, it was urgent: deposits of oil had been discovered at Prudhoe Bay, and drilling in the Arctic couldn't be far behind. Udall thought the preservation of this Beaufort Sea wilderness, the wildest and most remote in the United States, was more important than commercial uses. Subzero in the winter, but abundant with wildlife, it was home to fox, bears, caribou, moose, and birds by the millions. He wanted the Interior Department to control 3.5 million acres of the Brooks Range as the Arctic Circle National Monument; he also wished to enlarge Mount McKinley and Katmai National Monuments and two other sites in Alaska, Cape Newenham National Wildlife Refuge by 265,000 acres and Clarence Rhode National Wildlife Range by 1 million acres.

Udall wanted to get it while the gettin' was still good, before Nixonites flocked to the nation's capital. In Utah, he plotted to add 215,000 acres of canyonlands to Capitol Reef National Monument and double the acreage of Arches National Monument, the area that Edward Abbey had written about in *Desert Solitaire*.

Stewart and Mo Udall had made the creation of Sonoran Desert National Monument in southeastern Arizona their personal crusade. They hoped to convince LBJ to combine a newly preserved tract including the existing Organ Pipe Cactus National Monument along with Cabeza Prieta National Wildlife Refuge and most of the US Air Force's Barry M. Goldwater Range. The main ecological reason was that the desert mountains were essential habitats of the Sonoran pronghorn antelope.[8] Udall also sought to establish Marble Canyon National Monument, just north of the Grand Canyon, along a stretch of the Colorado River that he had once rafted. Among other things, that would allow Udall, the "father" of Canyonlands National Park, to end his eight-year tenure at Interior with a major victory—one that would have pleased Harold Ickes—in the realm of Colorado Plateau preservation. In all, Udall compiled a list of seven new monuments or additions to existing ones.

Udall was confident that LBJ would green-light his whole list. According to the historian Thomas G. Smith, "All targeted areas were in the unappropriated public domain, so there would be no budgetary concerns."[9] However, out of nowhere, Udall faced a roadblock when DeVier Pierson, the White House special counsel, warned Johnson that although there wasn't a legal impediment to going forward with Udall's list, there could be political fallout from Congress. Pierson, an Oklahoma attorney who had been advising Johnson since 1967, said that using the Antiquities Act in that way and failing to consult Congress formally would be like poking a grizzly bear.[10] Such executive branch bravado instinctively went against LBJ's well-honed congressional instincts.

Pierson believed that Aspinall, in particular, would probably feel double-crossed by Johnson if a slate of executive orders suddenly rained national monuments. Aspinall, who was still chairing the House Interior Committee, believed strongly in congressional prerogative. Pierson reminded LBJ that his New Conservation accomplishments, such as the Wilderness Act, the Highway Beautification Act, and many others, were consequential and also irreversible, having been signed into law after full congressional review and approval. Blindsiding Congress by using the Antiquities Act to claim 7.5 million acres could damage Johnson's bipartisan legacy in conservation. Even more worrisome, because president-elect Nixon had chosen Alaska governor Walter Hickel to be his interior secretary, for Johnson to make a land grab of the Brooks Range, in Hickel's home state, could be construed as an affront to the incoming Nixon administration. A transition feud would inevitably disrupt the orderly transfer of office, Pierson argued, pitting Johnson-Udall against Nixon-Hickel.

Hospitalized with the flu, Johnson did not want a flap with Nixon and remained undecided. "My heart fell at this," Udall recalled. "My sixth sense told me the big package was in trouble. But there was nothing left to do but hope—and bide our time."[11]

Johnson met with Udall and Pierson in the White House in early

January to discuss the proposal. The president's initial response was one of tepid procrastination. A week later, Udall claimed that he had consulted with key players in Congress and had received almost unanimous support for the new monuments. The next day, he sent LBJ a memo quoting Aspinall as being reluctantly on board—but on board. When asked specifically about Aspinall by LBJ, Udall said that the congressman's response to the package had been "Well, you know I don't favor action dealing with public lands by Executive Order but knowing you and President Johnson I am not surprised. But there's no point in my opposing you now."[12]

Secretary of Defense Clark Clifford, who had also attended the January meeting about the monuments, had eyed Udall dubiously and cautioned that the acreage should be debated in Congress, as North Cascades and Redwood had been. Udall countered, "Mr. President, I want you to go down in glory. . . . I don't think I have ever given you bad advice on any major issue."[13]

With time running out, Udall forwarded LBJ an upbeat memo about how Theodore Roosevelt, Calvin Coolidge, and Herbert Hoover had made similar Antiquities Act proclamations on their way out of the White House. Even Eisenhower, Udall pointed out, had established the Chesapeake and Ohio Canal National Historical Monument and protected the Arctic National Wildlife Refuge at the end of his White House tenure. If anything, last-minute executive decrees on behalf of America's natural heritage were a presidential tradition. If Johnson were to announce the new monuments as a Christmas gift to the American people on television, Udall predicted, the response would be enthusiastic. Concerned about inflaming the Nixon-Hickel team, LBJ again stalled, telling Udall for a second time that he'd think his proposal over.[14] Udall carped in his journal, "Johnson, like a great oak with umbrella limbs, is so domineering and so self-centered he destroys the very people he wants to help most."[15]

Working all the angles, Udall sent Johnson a personally inscribed copy of his new book, *1976: Agenda for Tomorrow*; the president graciously wrote a thank-you note.[16] The American public soon received an even better gift, but it was courtesy of NASA. On Christmas Eve 1968, Colonel Frank Borman and his Apollo 8 crew, on their fourth orbit around the moon, suddenly saw Earth rise beyond the lunar

President Lyndon Johnson welcomed a delegation of Native Americans at the White House alongside Secretary of the Interior Stewart Udall. He was proud that the Great Society improved living conditions on reservations.

horizon. "Oh my God," Bill Anders exclaimed. "Look at that picture over there! There's the Earth comin' up. Wow, is that pretty!"[17] The photos snapped by the Apollo 8 astronauts were the first widely available images of the planet in its wholeness taken by human hands. The radiant Earth, alive with clouds and oceans, illuminated against the black of eternity, instantly became an icon, inspiring a whole wave of planetary thinking and ecological awareness. As the biologist Lewis Thomas noted, whereas the moon was "dead as an old bone," the blue-green earth was "the only exuberant thing in the cosmos."[18]

In January, LBJ, wanting a seamless transition, was still wavering on Udall's national monument package. Time was running out. Suspecting that Udall didn't really have Aspinall's support, Johnson telephoned the congressman to review revisions in the package. According to what Udall had reported, Aspinall would be accommodating.

"Mr. President, Secretary Udall has never spoken to me about this particular matter," Aspinall said.

"What?" Johnson snapped, blood pounding in his ears.

"That's exactly right," Aspinall said.[19]

Understandably furious, Johnson canceled his scheduled January 18 meeting with his interior secretary. In another personal call to Aspinall, full of apologies, LBJ asked if the national monument proposal would be acceptable if the acreage were cut by 95 percent. Meanwhile, Udall only made matters worse by foolishly releasing to United Press International the list of new national monuments without permission from the White House. Upon seeing the leak, Johnson telephoned Udall to say, "That's a hell of a way to run a department," ordered him to rescind the press release, and hung up. Moments later, Pierson telephoned Udall and accused him of rank insubordination. Later, Pierson called the conversation "the stormiest session I have had with a cabinet officer." Udall told Johnson, "You have my resignation right now."[20]

With only forty-eight hours left before Nixon's inauguration, Johnson did not accept Udall's resignation.[21] But his unhappiness with the secretary increased when he read in the *Washington Post* that Udall planned to rename the NFL-used District of Columbia Stadium, built on National Park Service land (the Anacostia Park), after Robert F. Kennedy. With Ted Kennedy and Frank Church backing his action, Udall went ahead with a quick dash of a pen; legally, he didn't need the president's permission. If he had wanted to, he could have renamed the stadium for LBJ; it was his prerogative. But he chose RFK, instead, disgusted by LBJ's jealousy of his recently assassinated friend.[22] "The President still hasn't buried Bobby Kennedy," he sniped.[23]

On Sunday, January 19, 1969, the day before Nixon's inauguration, Udall took the Brandeis-Douglas cure for stress: he hiked along the C&O Canal. The temperature was in the low twenties, so he bundled up. As he walked, he took stock of his eight years as secretary of the interior. When the Cape Cod preservation battle had been in the news back in 1961, Ansel Adams had prophesied that "within the span of the present administration we will win or lose the fight for the wilderness and the ideals of the National Parks."[24]

Reflecting on Adams's declaration, Udall mentally inventoried the national seashores and lakeshores that Presidents Kennedy and Johnson had established with his steady guidance: Cape Cod National Seashore in 1961; Point Reyes National Seashore and Padre Island National Seashore in 1962; Fire Island National Seashore in 1964; Assateague National Seashore in 1965; Cape Lookout National Seashore, Indiana

Dunes National Lakeshore, and Pictured Rocks National Lakeshore in 1966. As director of the National Park Service since 1964, George Hartzog had presided over the greatest expansion of parks in the nation's history: new kinds of parks (Buffalo and Ozarks) had been authorized; large new traditional parks (Redwood and North Cascades) had been established; and millions of acres of pristine lands had been set aside to preserve the cultural and natural heritage of Utah and Alaska, our last frontier.[25]

The list was long. Udall, in fact, had helped establish a record sixty-four National Park Service areas, more than any of his predecessors had. The runner-up, Harold Ickes, who had served from 1933 to early 1946, five more years than Udall, had helped establish fifty-one national park units.[26] When it came to new US Fish and Wildlife refuges, the Kennedy and Johnson administrations had together contributed 115 to the system. Udall was proud that the two Democratic presidents he had served had seized the public desire to save America's natural heritage from the maw of the industrial-chemical technological order.

To Udall's mind, the New Conservation movement had begun in earnest with the Dinosaur National Monument victory in 1956, which had enabled Zahniser and Brower to elevate public lands preservation as a national priority. When Robert F. Kennedy went whitewater rafting with his family down the Green River, for Udall, it was validation that "the saving of Dinosaur was an epic achievement for the ages."[27] In addition, postwar American affluence had made out-of-the-way parks such as Dinosaur more popular than ever. The combination of extra pocket money and leisure time enabled nonprofits such as the Nature Conservancy and the Sierra Club to flourish with new members and donors. Another result of the postwar affluence was the emergence of an "activist" upper middle class—college educated, idealistic, concerned, and financially well off. The nation had never before had such a "mass elite" of college-educated people disposed to ecological thinking. Knowledgeable about the teachings of Aldo Leopold, Eugene Odum, and Rachel Carson, these postwar activists were sophisticated, resourceful, politically potent, and dedicated to protecting the natural world in the aftermath of Nevada nuclear testing and the alarms about DDT.

During Udall's years at Interior, a lot had changed, including the nomenclature; the term *conservation* became antiquated, though LBJ clung to it as an homage to Theodore Roosevelt. *Environmentalism* was the new byword of those ecologically aware Americans who sought to protect not just one place or species but the entire system of life on Earth.

Another even more serious change had been in the public's fickleness about nuclear energy. During Udall's time at Interior, he believed that nuclear power was the energy of tomorrow. It's true that he had been against the nuclear reactor planned for Bodega Bay in California after citizen groups had published a report stating that the reactor might trigger an earthquake. But on the whole, Udall was bullish about nuclear energy's potential, heartened by the prediction that by 1980, it could provide about 37 percent of the United States' electricity.[28] Most green activists, however, grew increasingly suspicious about its safety.

One reason Udall was successful at Interior was his uncanny ability to persuade CBS News broadcasters such as Walter Cronkite, Charles Kuralt, and Eric Sevareid to back LBJ's New Conservation campaigns. A segment on the nightly news promoting the saving of the North Cascades or the need to stop raw sewage from being dumped in the Missouri River was worth millions of dollars in publicity. Well-written books such as Edward Abbey's *Desert Solitaire* and Paul Ehrlich's *The Population Bomb* mattered, too. Udall understood that the media served the function of a honeybee, pollinating the news cycle with stories about strontium-90, leaded gasoline, and radioactive fallout. The public understandably reacted with fear and misgiving.

In the 1960s, some environmentalists even became media stars. When Rachel Carson appeared on *CBS Reports*, it was a prime-time news event. Cesar Chavez was profiled on the July 4, 1969, cover of *Time*. Ralph Nader was featured on the cover of *Newsweek* in 1968; a year later, Barry Commoner was on the cover of *Time*, called the "Paul Revere of Ecology." Paul Ehrlich regularly appeared on *The Tonight Show* warning about overpopulation and limited natural resources. Robert F. Kennedy became the public face of river exploration.

With the exception of Rachel Carson and Eugene Odum, perhaps no American worked harder to promulgate the concept of ecology in the public square than Barry Commoner.[29] As the publisher of the journal *Nuclear Information*, he wielded a broadening influence, and the changes he made in the title of his publication reflected that. In the middle of the 1960s, it became *Scientist and Citizen*. By the end of the decade, the periodical was renamed *Environment*.[30] The outgoing Udall believed that Commoner was right; pollution was not a result of too many people but rather of commonplace chemical products, fossil fuels, and industrial processes that marred, perhaps permanently, the natural world. All of Commoner's scientific learning, from attending Harvard to spraying DDT in the Pacific during World War II to organizing the "Baby Tooth Survey" at Washington University to writing the book *Science and Survival*, had taught him that technology wasn't a god.[31] "The environmental crisis is a grim challenge," he wrote in *Field & Stream*. "It also is a lofty opportunity. From it we may yet learn that the proper use of science is not to conquer nature, but to live within its scope."[32]

To Commoner, the planet's ecological survival didn't mean the shelving of technology. Not at all. But it did demand that technology develop scientific protocols to make innovations compatible with the natural world. Whether it was the megawattage of power plants, the megatonnage of nuclear bombs, or the mega-growth of the global population, clearly the Earth's habitat was deteriorating at an astonishing rate. Commoner warned, "Human beings have broken out of the circle of life, driven not by biological need, but by the social organization which they have devised to 'conquer' nature: means of gaining wealth that are governed by requirements conflicting with those which govern nature. The end result is the environmental crisis, a crisis of survival."[33]

The academic credentials of Commoner, Ehrlich, and Odum ensured that environmentalists could not be written off as faddists. And they weren't alone; attention to environmental studies was growing at such institutions as the University of California at Santa Barbara, the University of Georgia, the University of Wisconsin, and the University of Michigan, as well as at Stanford and Harvard Universities. More and more during the 1960s, those universities had laboratories, funding, computers, data banks, and interchanges of information dedicated to the expansion of the understanding of Earth science.

In the end, Udall thought, he and Johnson had made a potent team.

He appreciated the fact that the president had always been receptive to his ideas. In an interview two years later, he recalled, "[If] you could come up with a good idea and say, 'This is good for the land and good for the people,' he bought it." He dismissed the falling-out in January 1969 as having been due to budgetary matters and said that otherwise, he couldn't think of a major policy disagreement between them.[34]

Another blessing that Udall counted was that President-elect Nixon had chosen the Ivy League–educated Russell Train, who believed "in the environment first, Nixon second."[35] Tapping Train, an ex-director of the World Wildlife Fund, to chair a task force on Natural Resources and the Environment,[36] was fortuitous. Much like Bill Douglas, Train prided himself on being able to identify flora and fauna the world over. "The Bible has the Golden Rule," Train recalled of his conservation education. "The Golden Rule says, 'Thou shalt do unto others what you would have them do unto you. To my way of thinking, those others include the whole community of this Earth, all the living things—and inanimate as well—and we damage that extraordinary structure at our peril."[37]

Russell "Russ" Train was born on June 4, 1920, in Jamestown, Rhode Island, on Narragansett Bay, where his parents rented a summer place. His father had been a rear admiral in the US Navy during World War I. The highlights of Russell's childhood were his outdoors excursions along the forty-mile-long Bouquet River in the high Adirondacks backcountry. His father had taught him to follow Theodore Roosevelt's "strenuous life" philosophy of learning how to survive in a wilderness environment. "I grew up a Republican," Train recalled in a 2004 interview with *Mother Jones*. "You know, you sort of inherit these things. My great-grandfather was a Republican member of Congress during the Civil War. I certainly have always felt the Republican Party stood for conservation values."[38]

As a Princeton University undergraduate, Train joined the Army Reserve Officers Training Corps (ROTC), and when World War II erupted, he went on active duty, seeing combat in the Okinawa campaign in the Pacific theater and rising to the rank of major. After the war, he earned a JD from Columbia University Law School. From 1948 to 1958, he held top positions in the federal government connected to finance: legal adviser for the Congressional Joint Committee on Taxation, where he became an authority on tax law; chief counsel to the then majority adviser to the House Ways and Means Committee; and assistant to the secretary of the treasury and head of the department's legal advisory staff.

In 1954, he married Aileen Bowdoin, and they had four children together. Even though the family home was in Washington, DC, the Trains frequently went on safari to Africa to study wildlife. In his memoir, Train recounted his life-changing encounter with mountain gorillas, deeming it a "heart-stopping," momentous event that had turned him into an endangered species zealot. With Aileen as his travel partner, he explored Kenya, Rwanda, the Belgian Congo, and Uganda. Train wrote that he had been enthralled by "the cold early mornings, the heat of midday spent in the shade of an acacia, the incessant calling of the doves, the nights full of stars, the occasional roar of the distant lions, and the call of a hyena." There was a paradoxical duality about Train; he was both President Eisenhower's US Tax Court administrator and founder and head of the African Wildlife Leadership Foundation, which helped developing nations in Africa create wildlife parks and reserves.

In 1965, Train resigned his position on the tax court and became president of the Conservation Foundation. Founded in 1961, the nonprofit sought innovative ways to bring environmental issues to the forefront in all Washington, DC, bureaucracies. Also serving on the Conservation Foundation's advisory board was Lynton Keith Caldwell, an astute political scientist at Indiana University who regularly consulted with Train about environmental policy issues. In 1968, Train was appointed to the bipartisan seven-person National Water Commission. When Nixon won the presidential election that year, he asked Train to chair a task force on natural resource and environmental issues. This cabal of conservationists recommended that Nixon establish a White House office on environmental policy, an idea that culminated in the passage of the National Environmental Policy Act (NEPA). Train's excellent work on the "green" task force prompted Nixon to appoint him first as undersecretary of the interior (1969–1970), then as chairman of the newly formed Council on Environmental Quality (1970–1973), and as the administrator of the Environmental Protection Agency (1973–1977).

Russell Train was the leading activist on endangered species in the Republican Party during the 1960s and 1970s. His fingerprints are all over President Nixon's two epic wildlife-protection pieces of legislation, the Endangered Species Act of 1969 and 1973. In 1994, Train was elected chairman emeritus of the World Wildlife Fund Council.

Even with all the social change of the tumultuous 1960s, Udall still considered himself essentially an old-breed frontiersman like Bill Douglas and an endangered species activist in sync with Russell Train.[39] Enlarging the National Park System had always been a priority during his eight years as interior secretary and he looked forward to more units being added during the Nixon years. High on Udall's list were Voyageurs National Park (Minnesota) and Kauai National Park (Hawaii). He also hoped that Congress would convert large-acreage national monuments such as Glacier Bay (Alaska) and Death Valley (California) into full-fledged national parks, with vast tracts within their boundaries designated as wilderness.[40]

Back from his C&O Canal hike, his mind clear of worry, Udall wrote Lyndon and Lady Bird a warm letter. "I'm painfully sorry—that our last two days ended in discord and disarray," he lamented. "For my part, I choose to regard it as remarkable that during five years and two months of collaboration we had only one, sad quarrel. And even then, the fault lay in poor communication, not rancor. In any event, I want both of you to know that I have wiped the slate clean days ago. As far as I am concerned, our relationship now is what it always was during those warm, rich, unforgettable years. Robert Frost, in his last years, used to sum up life as he saw it in his late eighties by closing his letters with the salutation, 'Friendship is all.'"

The betrayed Lyndon Johnson disregarded that olive branch and never spoke to Udall again.

After more than five years of carefully calibrating his relationship with Lyndon and Lady Bird Johnson, Udall was leaving government out of favor. Then, on the morning of January 20, the day of Nixon's inauguration, LBJ signed executive orders enlarging the Arches, Capitol Reef, and Katmai National Monuments and creating a new Marble Canyon National Monument near the Grand Canyon. That was, in Udall's estimation, a ridiculously small package and a stinging personal rebuff. Arizona's Sonoran Desert, his special request, had been scuttled. Most humiliating, Johnson hadn't even consulted with him about the executive orders that had added 384,000 acres (in contrast to Udall's 7.5 million) to the National Park System.[41]

Not lost on those outside of the LBJ-Udall scuffle was the fact that Marble Canyon had, at long last, been saved via an executive order—this was the dream of Joseph Wood Krutch fulfilled. It encompassed the section of the Colorado River canyon in Arizona from Lee's Ferry to the confluence of the Little Colorado River—the beginning of the Grand Canyon. The site had once been eyed for a hydroelectric power project, but instead of the Marble Canyon dam, the United States had a new national monument by that name, one that would be incorporated into the Grand Canyon National Park in 1975.

In the coming years, Udall would boast that his national monument wish list had a kinetic life of its own. It continued to circulate inside Interior throughout the presidencies of Nixon and Ford. That alone gave the remaining sites a certain priority in the ongoing discussion of public land protection. In fact, Udall's master plan for Alaska would be realized to a great extent in 1980, when President Carter, with encouragement from Mo Udall in the House and Frank Church in the Senate, used his executive authority to designate 56 million acres of Alaskan wilderness as federally protected national monuments.

History has shown that, counter to the grumbles of Douglas and

Brower, Lyndon B. Johnson was one of the great American conservationist presidents, one who did not receive enough credit. The reasons for his neglect and devaluation are many. First, nobody will ever loom larger than Theodore Roosevelt in the conservation realm. Not only did the twenty-ninth president save 234 million acres of wild America, but he created the US Forest Service and Federal Bird Reservations, the progenitor of the US Fish and Wildlife Service. A Harvard-trained naturalist, he had written the catalog "The Summer Birds of the Adirondacks in Franklin County, N. Y." and a book called *The Deer Family*. And his bending of the Antiquities Act to save the Grand Canyon is epic stuff. Likewise, Theodore's distant cousin Franklin used the Civilian Conservation Corps to plant 3 billion trees while establishing eight national parks and eight hundred state parks. The Roosevelts' conservation legacies were hard acts for any president to follow. Johnson's presidency was defined by the Great Society, with its groundbreaking legislation on civil rights, voting rights, health care (the creation of Medicare and Medicaid), and early childhood education (Head Start). Those major domestic policy accomplishments tended to overshadow his New Conservation accomplishments.

The legacy of Johnson's predecessor, John F. Kennedy, also made it tricky to evaluate LBJ's record on a stand-alone basis. With his youth, charisma, and energy, JFK seemed to flood the nation's capital with conservation ideas. On his watch many future units of the National Park System were first marked as potential acquisitions: for example, Fire Island, Indiana Dunes, Pictured Rocks, Canyonlands, Cape Lookout, Guadalupe, North Cascades, and Redwood. It was Kennedy who embraced Rachel Carson and her prescient warning about the dangers of chemical pollution, atomic fallout, and DDT in *Silent Spring*. Udall, Johnson's very effective interior secretary, was, of course, a holdover from the Kennedy administration. Thus, many of Johnson's accomplishments, such as the creation of new national parks, national trails, historic preservation, and urban recreation areas, could be seen as merely executing Kennedy's New Frontier agenda. But Johnson was the leader who established the era's most important milestones in conservation, public health, and environmental protection. In addition, some of the biggest conservation accomplishments of the early

Nixon administration, especially the creation of the Environmental Protection Agency, were realized because the groundwork had been laid during the Johnson administration.

Another factor that hindered LBJ's green legacy was his press office. Acclaim for the president's New Conservation agenda often seemed to flow to the First Lady, to Udall, or to any number of others, including the legislators who steered the bills through Congress. To Johnson's detriment, most of his major conservation bills were signed at dull, routine ceremonies at the White House. Standing at a podium in the Rose Garden or East Room wasn't the same as going down the turbulent Rio Grande on an inflatable raft in Big Bend's Mariscal Canyon.

LBJ's press secretaries didn't make important conservation bill signings such as September 3, 1964, and October 2, 1968, seem memorable or even noteworthy. Why couldn't LBJ have signed the Wild and Scenic Rivers Act next to Idaho's fast-flowing Selway River? Couldn't he have signed the legislation creating national scenic trails on the Appalachian Trail or Blue Ridge Mountains picnic area? Why weren't there photographs of Johnson feeding orphaned bald eagle chicks in Ohio or petting a manatee in Florida? The optics of the New Conservation were awful. LBJ's press secretaries, Pierre Salinger, George Reedy, Bill Moyers, and George Christian, didn't stage events to capture the drama of the efforts to preserve the American Earth.

As a result, no fewer than four books have been written about Lady Bird Johnson as the American beautifier, whereas LBJ's contributions to conservation have been largely neglected. This is a grave oversight, given that none of the national parks, lakeshores, or seashores created during his term would have made it to signing without his savvy management of legislation as it wended its way through the House and the Senate. Where Johnson excelled, thanks in part to his speechwriters Douglass Cater and Richard Goodwin, was in delivering speeches on New Conservation that were utterly remarkable in explaining the need to protect natural beauty in America. Yet Johnson's delivery was stilted and couldn't compare with JFK's relaxed and assured delivery. Even so, to read LBJ's speeches on conservation, which contain refreshingly personal passages, is to hear from one who truly knew what was at stake and cared about protecting Earth from misuse.

It's not too late to honor President Johnson. Dwight D. Eisenhower's name is on the National System of Interstate and Defense Highways, better known as the Interstate Highway System, in tribute. Likewise, the Wild and Scenic River System should be christened the Lyndon B. Johnson Wild and Scenic River System. Johnson's advocacy of unharnessed rivers, in particular, after he read John Graves's *Goodbye to a River*, was a radical departure from longtime federal programs that dammed rivers to produce hydroelectric power. His Wild and Scenic Rivers Act was a visionary law passed purely on behalf of the aesthetics of rivers. The act was on the side of trout, turtles, salmon, and salamanders—not irrigation. Throughout his New Conservation program, Johnson protected our natural resources with a Schweitzerian "reverence for life" philosophy. His determination to get the Endangered Species Act and the Wild and Scenic Rivers Act done before leaving the White House proved his firm commitment to natural beauty writ large.

Because the environmental intelligentsia of America didn't like Lyndon Johnson, rightfully blaming him for the mistaken Vietnam War, he never won awards from the Sierra Club or the National Audubon Society. "The men of ideas think little of me," he confided to the biographer Doris Kearns Goodwin a few years later, "they despise me."[42]

Perhaps if Johnson had dubbed his management of natural resources "New Environmentalism," instead of clinging to "conservation," he'd be remembered now as one of the great public land and river protectors in US history. Lady Bird Johnson later claimed that it was Lyndon's "White House boys" that wouldn't allow her to use the word *environment*. They insisted, she said, that it be "Lady Bird's Beautification Project." The First Lady believed that only Lyndon could have built a congressional coalition around a series of environmental laws that made America a better place. Fortunately, Lyndon and Lady Bird, working in tandem, made conservation mainstream in the 1960s and prepared the ground for 1970s environmentalism; once that took hold, as Nixon soon found out, there was no turning back.

CHAPTER 29

Santa Barbara, the Cuyahoga River, and the National Environmental Policy Act

President Nixon examines the Santa Barbara sand in March 1969, two months after an oil spill devastated the California beach community.

I

When Lyndon and Lady Bird Johnson returned to Stonewall, Texas, following Nixon's inauguration on January 20, 1969, working the ranch on the Pedernales River slowly healed wounds left by more than thirty years on the Washington battlefield. At sixty, Johnson—overwrought, smoking heavily, and harassed by angina pain—threw himself into overseeing his spread. With the Secret Service hovering around, he devised a cutting-edge irrigation system, ordered pipe laid in the Pedernales shallows, built a poultry coop, planted strains of drought-resistant grasses, watched over his herd of Herefords, and attended the local cattle auctions. At Lyndon B. Johnson State Park, just across the river, he often greeted visitors, escorting them to watch a slide show about his love of the Hill Country, a region that arced from Austin two hundred miles southwest to the Mexican border. The LBJ Ranch was in the heart of the Edwards Aquifer system of freshwater springs. "He's become a goddamn farmer," a friend groused. "I want to talk Democratic politics, he only talks hog prices."[1]

Leaving current politics to others, he started writing *Vantage Point: Perspectives of the Presidency, 1963–1969* with the help of assistants, and raised funds for his presidential library, planned for the University of Texas at Austin. Lady Bird volunteered to promote the Texas Highway Department's ongoing effort to landscape older roadsides and guarantee that new ones retained elements of floral beauty. She used the Lyndon B. Johnson State Park to hold fundraisers for the program and barbecue picnics to celebrate its construction engineers, foremen and crews, and botanists. "I thought I ought to meet some of the men who actually do the job along our highways, to thank them for preserving one of the wildflowers' last citadels," she wrote. "I have learned well that when springtime brings forth wildflowers along roadsides, it is because a maintenance foreman had the foresight to hold off mowing until after the plants had gone to seed the year before." She also paid tribute to the workmen who spread Texas Indian paintbrush or bluebonnet seeds "on bare spots where there were no flowers."[2]

Once out of the White House, the Johnsons worked hand in glove with the National Park Service to maintain their Texas home exactly as it had been when LBJ had been president (he had spent 490 days, about a quarter of his White House tenure, at the ranch). Scores of tourists came by, hoping to glimpse or photograph the famous couple among the rolling hills covered with cedar, Spanish oak, bald cypress, and cottonwoods. The NPS soon operated two units, the ranch in Stonewall and Johnson's boyhood home in nearby Johnson City. The ranch house acreage included Junction School, where the thirty-sixth president had long before gone to nursery school, a reconstruction of his rustic birthplace, his grandfather's house, an open hangar housing the jet that the US government had assigned to the president in retirement, a runway for it, and a display barn.[3] The site in Johnson City soon acquired other vintage structures situated along an oval walking path. The Show Barn Complex educated visitors about the significance of the ongoing Hereford cattle operations. "This is my country," Johnson would say, "the Hill Country of Texas."[4]

Back in Washington, Richard Nixon was sorting out the chaotic legacy left by Johnson, notably the bitter division caused by the Vietnam War. Pundits mistakenly assumed that environmental concerns would be deprioritized by Nixon, but they underestimated his political cunning. Because the Democratic Party controlled both houses of Congress, the Big Five environmental senators—Church, Muskie, Hart, Jackson, and Nelson—were pressing the new president for a stringent new set of federal anti-pollution laws. To their pleasant surprise, Nixon was amenable; the White House had hired a crackerjack team of GOP conservationists, all lawyers who took environmental issues seriously, in particular, the power quartet of John Ehrlichman, Russell Train, John Whitaker, and William Ruckelshaus. Depending on the group he was addressing, Nixon's attitude on the importance of conservation would, in the words of biographer Evan Thomas, "wax and wane."[5]

A key to understanding President Nixon's political attitude toward the environment in 1969 can be found in a story told by Train. At a dinner in New York just before Nixon's inauguration on January 20, 1969, Train was granted a golden opportunity to converse one-on-one with the incoming president for five to ten minutes. After mulling it over for two hours, Train decided to use his face time with Nixon to discuss the pollution crisis and endangered species. As he told the story:

> I talked to him about the importance of the environment, a concern which involved every geographic region of the country, which involved all kinds of people and interests, and which could be used to help unify the nation and bring people together. He nodded his head and indicated that he understood. He said, "that sounds pretty good. But, what about the poor and the blacks living in

the inner cities?" Of course, he had instantly put his finger on an extremely important aspect of what I had been saying, one which was very often overlooked, and is still overlooked. So, I discussed the relationship between poverty and the environment with him, about lead paint, about the fact that people in the cities suffered more from air pollution than others.

I tell this story to underline the fact that from the beginning, Nixon had a keen appreciation of the political importance of the environment. I think it clearly influenced the first three years of the Nixon Administration. It certainly was a great help to us in achieving our agenda at the Council of Environmental Quality.[6]

On January 28, 1969, at 10:45 a.m. PST, crude oil from Union Oil Company's drilling platform, located in the Pacific Ocean less than six miles off the coast of Santa Barbara, began leaking uncontrollably, forming a bubbling oil slick, in some places an inch thick, that drifted to shore with each lapping wave. Nixon had been president for only eight days. Within hours, exquisite California beaches were thick with black sludge. Union Oil had been founded in Santa Paula, California, in 1890 and in 1969 was the eleventh largest US oil company with global operations.[7] Overnight, the petroleum giant joined Dow Chemical Company, the manufacturer of Agent Orange, as one of the most loathed corporations in the United States. The oil company had been awarded a waiver by the US Geological Survey that had allowed it to construct a protective casing around the drilling hole that was sixty-one feet short of the federal minimum requirement. "The public reaction was triggered not so much by any particular regulatory failure at Santa Barbara," Train recalled in his memoir *Politics, Pollution, and Pandas: An Environmental Memoir*, "as by the stark images, which served as a sort of visual metaphor for perceived environmental outrages throughout the country."[8]

A few days later, a Coast Guard contingent reported that thirty-five square miles were blanketed in 3 million gallons of oil within twenty-four hours after the blowout.[9] Around 3,600 seabirds perished from suffocation and exposure to the spill. The *Los Angeles Times* ran a pitiful photo of a human hand holding a convulsing, oil-encrusted duck that had washed up onto the sand, covered in oil and barely alive. When a bird touched oil, its feathers, which are meant to insulate it, clumped, exposing patches of bare skin to the frigid water. More than 80 million Americans watched such oil-soaked birds on TV struggle to take flight. An emergency wildlife medical center was set up at Carpinteria State Beach. Each rescued bird was doused in a solution called Polycomplex A-11, which dissolved the petroleum. The bird was then washed in a tub of fresh water, dried, and put in a warm place to raise its body temperature before pneumonia set in. "I am amazed at the publicity for the loss of a few birds," the Union Oil president, Fred Hartley, scoffed after the spill, only to get hammered by the national media for his callousness.[10]

Santa Barbara, traditionally a Republican community, united in collective disgust over the stench of dirty crude oil that infiltrated citizens' breathing air, regardless of party affiliation. In town hall meetings, gatherings, rallies, and demonstrations, locals complained that tar balls were as ubiquitous as seaweed on the beaches. Young volunteers—from Big Sur, Menlo Park, Berkeley, North Beach, Marin County, and Los Angeles—rushed to Santa Barbara to help the creatures in need.[11] The explosion had been so seismic that it had cracked the ocean floor in five places. Hay straws and pitchforks were used by volunteers to clean the beach. Nationally, the environmental disaster caused an uproar heard well into the twenty-first century. The fact that a serene California coastline had been ruined by the unchecked power of Union Oil mobilized citizens into a protest frenzy.[12] John Whitaker—a White House adviser on environmental issues to Nixon who had earned a PhD in geology from Johns Hopkins University—deemed the oil spill "comparable to tossing a match into a gasoline tank; it exploded into the environmental revolution, and the press fanned the flames to keep it burning brightly."[13]

Santa Barbara immediately became a battle cry against all types of industrial pollution, particularly leaks from offshore drilling rigs on the Pacific coast. Though an able crew managed to stopper the top of the well, the highly pressurized gas and oil continued leaking into the water through faults and fractures in the upper layer of the ocean floor.[14] Three years after the Santa Barbara spill, the California State Lands Commission ordered a moratorium on all new offshore drilling in state waters, even on existing leases. President Nixon had just purchased an oceanfront estate in San Clemente, California, on the Pacific Ocean. The ghastly scenes of Santa Barbara covered in oil sickened him, too. Nevertheless, he wasn't prepared to shut down drilling operations along the Pacific coast willy-nilly, for fear of alienating the powerful oil and gas lobby. Instead, the administration proposed investigating what had gone wrong on the Union Oil platform and adopting tighter regulations for future oil drilling in all federal waters off the coast of California. Even though the spill wasn't Nixon's fault per se, many Californians blamed him—the price of being president when an environmental catastrophe unfolds. The oil spill in Santa Barbara was the worst in the United States until 1989, when the *Exxon Valdez* discharged 11 million gallons of crude oil off the coast of Alaska.

Working against Nixon was the fact that the fifty-year-old ex-governor of Alaska, Walter Hickel, was his interior secretary. Born on August 18, 1919, in Ellinwood, Kansas, Hickel grew up poor and never forgot the indignities. A survivor of Dust Bowl ravages, he sympathized with hard-luck loggers and fisherfolk eking out a blue-collar living in his adopted home of Alaska. Straight shooting, clumsy with words, dyslexic, and prone to telling cornball jokes, Hickel was the Yogi Berra of American politics in the 1960s and 1970s. Hickelisms such as "You can't just let nature run wild" and "A tree looking at a tree really doesn't do anything" made him the boob and bane of college-educated environmentalists. Just weeks before his Senate confirmation hearing, in full dissent mode, Hickel said he was opposed to "conservation for conservation's sake." Nevertheless, he survived that inquisition and was confirmed as interior secretary by a solid 73–16 margin.[15] Although Hickel believed that the Alaskan wilderness—its craggy pinnacles, salmon runs, blue coastal ice sheets, and caribou-stomped tundra—constituted an earthly paradise, his rubber-stamp support of Big Oil had caused Democrats to pounce on him during his tense Senate confirmation hearing earlier that month.[16]

Senators Walter Mondale and George McGovern, in particular, had grilled the homespun businessman as if he were a buffoon unfit to be a dogcatcher, let alone run the Interior Department. Liberals had chosen Hickel as a target of political opportunity, and they ridiculed him before the TV cameras as if he were a duck in a shooting gallery.[17]

With uncharacteristic malice, David Brower testified at the hearings that Hickel was a Petroleum Club toady with no environmental record, unconcerned with the welfare of migrating caribou herds in the Arctic National Wildlife Refuge or the millions of migrating birds in the Yukon Delta. What Brower misrepresented was that Hickel, at heart, was a Theodore Roosevelt hunting-fishing conservationist who was open-minded about balancing economic growth with proper land custodianship.

On February 2, having been interior secretary for only a few days, Hickel flew to Santa Barbara for an on-site inspection of the damage caused by the ghastly oil slick. As Nixon's on-the-ground inspector, he investigated the crime scene, learning that Union Oil had tried to install a directional well when a blowout had occurred (owing to inadequate casing on the drill).[18] Trying to contain the spill, Union Oil, at Hickel's direction, transported logs on barges commandeered from Long Beach, along with a plastic "sea curtain," to the site.[19] Federal, state, and local authorities worked together to seal off the oil and gas bubbling from the Pacific floor. Even prisoners from California penitentiaries pitched in on the beach cleanup.

Back in Washington, the White House domestic adviser, John Ehrlichman, blamed LBJ for the contaminated coast of Santa Barbara.[20] His rationale was that in 1968, the Johnson administration had auctioned off nearly six hundred square miles of channel leases in California, raising $603 million. Even the New York Times deemed that sale "the prelude to the trouble."[21] And Ehrlichman was correct to charge that federal drilling regulations hadn't been properly examined for fifteen years.[22]

On his inspection tour of Santa Barbara, the rough-and-tumble Hickel listened to citizens' grievances with focused empathy. But he reiterated that the Nixon team had inherited the lease and permanently revoking would require going through a legal process. Yet, he was visibly disturbed that the Union Oil Company, partnering with Texaco, Gulf, and Mobil, was still operating two additional platforms in two-hundred-foot-deep water and still producing thirty thousand barrels of oil a day. Because the Interior Department hadn't developed any process for oil spill containment, Hickel was forced to improvise. Surveying the oil-smeared beaches, he asked Edward Weinberg, a lawyer from the solicitor general's office, what he could do to stop Big Oil from continuing to drill with their four other Santa Barbara wells. The answer was "Nothing."

"What do you mean, nothing?" the infuriated Hickel shot back. "Give me some authority to at least suspend operations."[23]

Weinberg insisted that that would be unlawful. Such an order had never been issued by an interior secretary. As the lawyer rattled off facts about federal drilling leases, Hickel scoffed at him. "Quit all exploration and production in this lease tract," he ordered authorities along the beach. "That is an order."

"Mr. Secretary," Weinberg pleaded, "you can't do that."

"I just did," Hickel said.

Hickel personally told Union Oil's Fred Hartley that it was a direct order from President Nixon. With that, Hartley complied, telling his superintendents on the scene to suspend operations at the other wells.[24] Soon thereafter, Hickel, in a clever public relations gambit, announced an "Ecological Preserve": a 21,000-acre, two-mile-wide buffer zone surrounding the oil-scarred beaches.[25]

In Santa Barbara, Hickel barked orders like a three-star general on D-Day, confident that President Nixon would back up his action. The White House wanted him to "downplay [the spill] a little bit, and so I flew out there, . . . Oh, my God," Hickel recalled, "and I called the White House right after that. I said, 'This is a disaster.' I said, 'You've got to face it. It's a disaster,' and it was a disaster. And the people in Santa Barbara, when I went up to see them, they were really angry that the United States wasn't really moving that fast."[26]

Over the following days, more than 3.3 million gallons of oil rushed out from drilling-induced fissures on the ocean floor.[27] A pissed-off Hickel, seeing waves thick with crude oil congealing on the shore, convinced Nixon to embrace the Santa Barbara oil spill as a galvanizing moment in US environmental regulatory policy. Thanks largely to Hickel, the administration did not cover up for Union Oil; it used the crisis to champion oceanic conservation, shoreline protection, and accountability on the part of Big Oil.[28] In an unexpected move, Hickel proclaimed that environmental protection was now the top Nixon administration priority.

On February 11, Nixon, the first native Californian president, issued a statement directing his science advisers to recommend ways to "most rapidly assist" in restoring the beaches and water around Santa Barbara. Furthermore, he promised to prevent future environmental tragedies. When he toured the beaches on March 21, he was visibly shaken by the scope of the destruction, as seen from a helicopter. "What is involved is something so much bigger than Santa Barbara," Nixon told the press. Once on shore, protesters shouted, "Get oil out! Get oil out!" at the president. Speaking louder, Nixon said, "What is involved is the use of our resources of the sea and land in a more effective way, and with more concern preserving the beauty and the natural resources that are important to any kind of society that we want for the future. I don't think we have paid enough attention to this. . . . We are going to do a better job than we have done in the past."[29] Santa Barbara, he told the crowd, had "touched the consciences of the American people."[30]

A no-nonsense Nixon asked the oil companies to clean up the damage zone with the least amount of White House executive direction possible. If groups like the Environmental Defense Fund went to court armed with irrefutable scientific evidence against Union Oil, that was all right by the two-sided Nixon, who tried to be pro-environment while simultaneously trying not to become the bogeyman of the entire fossil fuel industry. Get Oil Out (GOO), a grassroots effort led largely by Santa Barbara women seeking to end offshore drilling in California, was dismissed by the Nixon administration, however, as left-wing radical overkill. Nixon couldn't tolerate anarchist Abbie Hoffman and hippie Allen Ginsberg demanding that the United States wean itself off its gasoline addiction.[31] "I remember how pissed off he was," White House adviser Dwight Chapin recalled. "We went out and walked the beach and he was disgruntled by the damage. I don't remember what action he took but I remember his attitude was on the side of the serious conservationists like Ehrlichman, Train, and Scoop Jackson."[32]

A year after the spill, a group of local University of California at Santa Barbara activists, mobilized by environmental studies professor Roderick Frazier Nash, authorized the "Santa Barbara Declaration of Environmental Rights." It called for an "ecological consciousness" that recognizes people as being part of a community of living things sharing a fragile earth and rallied individuals to government, to take responsibility for its preservation.[33] Nash had published *Wilderness in the American Mind* in 1967 and was now, in the wake of the oil spill, melding wilderness, ecology, ocean preservation, and public health into an iron fist of higher education protest.[34] A year after the Santa Barbara spill, the US Geological Survey reported that only eight barrels of oil were leaking daily from Block 402, a quarter of which was recoverable. Nevertheless, traumatized Santa Barbarans continued to complain that there was a "chronic" amount of residual oil on the surface of the drilling area in the channel.

The Santa Barbara oil spill stoked a national debate over whether California and other states should use nuclear power for their growing energy needs. Even though Brower and Nader viewed coal as an ecological pariah, they nevertheless were largely opposed to nuclear energy because of the public health hazards of radioactive waste disposal. So were millions of other concerned citizens. Such nuclear waste was persistent and would remain radioactive for two hundred thousand years. But Nixon and Ehrlichman both thought that nuclear power could make the United States energy self-sufficient by freeing it from its dependency on foreign oil. Furthermore, Nixon, in the coming years, would promote Project Independence: the building of a thousand nuclear plants by the year 2000. He considered nuclear energy (what Eisenhower had called "Atoms for Peace") the best way to wean America off of coal burning, diesel trucks, and Middle Eastern oil. "What literally may become the 'hottest' conservation fight in the history of the U.S. has begun," Robert Boyle wrote in *Sports Illustrated* in early 1969. "The fight is over nuclear power plants and the damage they can inflict on the natural environment. The opponents are the Atomic Energy Commission and utilities versus aroused fishermen, sailors, swimmers, homeowners, and a growing number of scientists."[35]

What made nuclear power such a Day-Glo controversy in California was that construction had commenced on Pacific Gas and Electric's enormous Diablo Canyon Power Plant near Avila Beach, two hours north of Santa Barbara along the coast. "The radiation from the waste produced by a city's nuclear power plant," Barry Commoner warned, "would be sufficient to kill 100 times the city's population."[36]

Because the electricity-generating plant was being built only a mile from the eighty-mile-long Shoreline Fault, the post–Santa Barbara oil spill worry of many environmentalists was what would happen to the Golden Bear State if the PG&E nuclear campus was struck by a 7.1 magnitude earthquake. It was the Bodega Bay debate of 1962 redux, and President Nixon was on the side of PG&E.

Proponents of nuclear power such as Nixon and Ehrlichman tried selling the modern energy source as safe, clean, quiet, and odorless. But that wasn't how students of Rachel Carson and other environmentalists saw it. After six years of contentious hearings, referendums, and litigation, PG&E won, and Californians were forced to take a "fingers crossed" approach to generating electricity at the Diablo Canyon Power Plant for its ever-increasing population.[37]

II

After the Santa Barbara spill, news outlets began covering ecological disasters and alternative energy with accelerated interest. When, on June 22, 1969, the eighty-five-mile-long Cuyahoga River, which bisected Cleveland and fed into Lake Erie, caught fire, northeastern Ohio found itself in the crosshairs of the snowballing national debate on the environment. Suffering through racial violence, high crime rates, and a shrinking economy, Cleveland was still a great manufacturing metropolis. But its waterways were seriously polluted. The Cuyahoga had long been treated as a sewage canal for industrial debris by Cleveland-area companies. The river had caught fire at least a dozen times before, most dramatically in 1952. Adding insult to injury, Lake Erie appeared to be dying. Of the sixty-two public beaches along the lake, only three were deemed swimmable in the summer of 1969.

Every day, Buffalo, Toledo, Detroit, and other cities joined Cleveland to dump a combined total of 1.5 billion gallons of untreated or poorly treated waste, including nitrate and phosphorous discharge, into the lake. Belly-up fish and nonbiodegradable rubbish were ubiquitous along the shoreline. Excessive algae growth in the lake threatened the health of the body that provided water for 12 million people in the northern states. The pollution of Lake Erie had negatively impacted the lake's essential $12.9 billion tourist industry and world-class fisheries.

Ridiculed as being the "Mistake on the Lake," Cleveland became the emblem of all that was wrong in urban America. Out of the miasma, luckily, emerged one of the great hands-on crusading mayors in US history. Carl Stokes, the first African American elected as mayor of a major US city, wasn't an environmentalist per se, but he had been raised by a single mother in a public housing complex on the eastern side of Cleveland, where slum dwellers saw, smelled, and even tasted the polluted waters of their manufacturing-based city. Long before the term had street credence, Stokes was promoting *sustainable living* for the general welfare of all citizens regardless of their social or economic status. Following the policy ideas of Mayor John Lindsay of New York City, Stokes drew up a sprawling agenda that included decreasing air and water pollution, of which 40 percent emanated from Ohio.

But with all the abandoned houses and factory blight when Stokes took office in November 1967, he had his work cut out for him. Early in his administration he pressed for a $100 million bond to fund the cleanup of the Lake Erie shoreline. To Stokes, clean water and breathable air were central to environmental justice, public health, and future downtown redevelopment. "Stokes understood," the historians David and Richard Stradling wrote, "that the environmental crisis existed well beyond the river, even beyond the soot fall from the industrial smoke plumes. Stokes realized that the environmental crisis was inseparable from the broader decay of his city."[38] Under his leadership, Cleveland passed the $100 million bond for environmental cleanup, but that alone couldn't save the Cuyahoga from the decades of pollution produced by the Standard Oil refineries, Republic Steel, and Harshaw Chemical.

The ostensible cause of the Cuyahoga River fire was a spark that had fallen from a train crossing a railway bridge onto the river below, igniting industrial debris and waste floating on the surface. Flames—in many places four or five stories high—had erupted as if an oil refinery had blown up. Within thirty minutes, firefighters had extinguished the blaze. The Cleveland *Plain Dealer* ran an editorial two days after the fire headlined "Cleveland: Where the River Burns." As catastrophes go, the Cuyahoga River blaze was a minor one. Nobody was injured, and only a few railroad trestles were damaged. There was no footage or photos of the industrial accident. Grappling with factory closures and decaying ghettos, the Cuyahoga fire was low-grade news even to most Clevelanders. Because the Cuyahoga had caught fire many times before, locals processed the Sunday morning mishap as "another sad chapter in the long story of a terribly polluted river."[39]

Such indifference, however, soon changed when *Time* magazine, in an unbylined article, made the dirty Cuyahoga the poster child of poisoned American waterways. In muckraking fashion, the August 1 issue cataloged the city's environmental atrocities. As if quoting from William S. Burroughs's *Naked Lunch*, it assailed Cleveland's water pollution problem in jarring fashion: "No Visible Life. Some river! Chocolate-brown, oily, bubbling with subsurface gases, it oozes rather than flows."[40] Most waterways were in trouble. The Missouri River was dotted with grease balls the size of cantaloupes that reckless Nebraska meatpacking factories emitted. The Mississippi River near St. Louis was toxic from synthetic chemical discharges. *Time*'s list of "severely polluted" waterways, included the Chattahoochee, the Hudson, the Potomac, the Monongahela, and the Milwaukee.

At its core the *Time* article—which received so much attention that *Time* began a weekly "Environment" section in the magazine—made readers wonder why state and federal courts dealt with water pollution cases via nuisance law. In the public outcry from the story were the seeds for the Nixon administration to create an Environmental Protection Agency and pass a tough Clean Water Act. Ehrlichman convinced Nixon that the one-two punch of Santa Barbara and Cuyahoga meant that environmental protection was the moonshot of the moment. Yet *Time* misled readers, for the story was accompanied by a harrowing photo of a boat almost consumed in flames and streams of water from bridge-bound firefighters trying to extinguish the inferno. What *Time* failed to tell its readers, in an act of fake-news deceit, was that it had printed a stock photo of a Cuyahoga accident that had been taken seventeen years earlier.[41]

Nixon had been handed the ecological mess, and history, he knew, would judge him on how well he responded to the dirty air and water crisis. Just eight months earlier, President Johnson had joyously

Cleveland reporter Richard Ellers lifting his hand out of the Cuyahoga River's thick, oily sludge. In the summer of 1969, the Cleveland river caught fire from industrial pollution and raw sewage.

established the Wild and Scenic Rivers classification, a late-inning Great Society victory in sharp contrast to the dirty-water headlines with which Nixon had to contend. Through no fault of his own, Santa Barbara and the Cuyahoga handed the new Republican president an environmental crisis atmosphere. Pressure was mounting for him to embrace the National Environmental Policy bill that Senator Henry Jackson's able staff was working on.

The White House carefully monitored a grand jury investigation of the causes of the Cuyahoga fire in the fall of 1969. The Jones & Laughlin Steel Corporation was found liable for releasing substantial amounts of cyanide into the river. A successful cleanup effort ensued to rid the river of the heavy black gunk in it. In the nonprofit world, environmental coalitions were formed to stop the eutrophication of Lake Erie caused by the ravages of raw sewage, agricultural runoff, and chemical contamination. *Time* called Lake Erie "a gigantic cesspool," owing to agricultural runoff (largely manure and synthetic chemicals) into its water from southwest Ontario and northwest Ohio. Mayor Stokes promised to make cleaning the water body the centerpiece of urban renewal.

Once Lake Erie was declared dead in 1970, Nixon took it upon himself to reverse the verdict. Cleveland may have been the butt of late-night comics, but Nixon knew that winning Ohio's electoral votes would be essential in 1972. In 1971, the children's book *The Lorax*, by Dr. Seuss (Theodor Geisel), was published in which the mustachioed title character, modeled after Gifford Pinchot, is "in search of some water that isn't so smeary. I hear things are just as bad up in Lake Erie."[42] (After 1986, inspired by successful Lake Erie restoration efforts and prompted by a letter from students, Geisel revised the subsequent editions.)[43]

As Dr. Seuss realized, the good news was that Lake Erie hadn't actually died. In 1998, President Bill Clinton honored the Cuyahoga by designating it an American Heritage River, in league with the Hudson, Mississippi, Potomac, and Rio Grande, saying "Once so polluted it caught on fire, the 100-mile-long Cuyahoga became a stark symbol of the plight of America's rivers and a rallying point for passage of the Clean Water Act, one of the nation's landmark environmental laws."[44] In 2019, American Rivers named the Cuyahoga "River of the Year" in tribute to "50 years of environmental resurgence."[45] Yet record-setting algae blooms and associated dead zones (oxygen-depleted areas where algae had died and decomposed) continued to wreak havoc on the tourist and fishing industries there.

Made furious by the Cuyahoga River fire, Bill Douglas began looking into the culprits behind Lake Erie's destruction via his national network of environmental whistleblowers. At the age of seventy, the justice was grappling with the radical concept of rivers and lakes as plaintiffs because they "spoke for the ecological unit of life that is part of it."[46] One of Douglas's aides recalled him bursting into his Supreme Court chamber one afternoon, moaning "Lake Erie is dying!" with the wail of an Old Testament prophet.[47] His clerks assumed that saving the lake was going to be the "green" justice's next environmental crusade now that North Cascades was a national park.

Other players in the Silent Spring Revolution likewise seized upon Lake Erie as the symbol of industrial disease. Barry Commoner used Lake Erie's ruin in lectures as an emblem of America's deterioration writ large.[48] And Nader's Raiders investigated why the federal government had allowed Cuyahoga and Erie to be subjected to sewer overflows, invasive species, algae blooms, and mercury pollution. Out of Ralph Nader's effort grew a four-hundred-page book, *Water Wasteland* by David Zwick and Marcy Benstock, which used as its frontispiece the misidentified *Time* photograph.[49]

In later years, Douglas, Commoner, and Nader all pointed to the Cuyahoga fire as the prime catalyst for the cascade of water pollution legislation such as the Clean Water Act of 1972 and the Great Lakes Water Quality Agreement of 1972 passed into law by Congress during the Nixon years. The crisis had the effect of stoking such a swirl of ecological concern that Americans put their collective shoulders to the wheel to guarantee that Lake Erie didn't die from, as the novelist Kurt Vonnegut put it, "Clorox bottles and excrement."[50]

What concerned Ehrlichman, Train, Hickel, and Whitaker was that Nixon in the fall of 1969 was losing the environmental issue to the Democrats. Nixon was advocating for the construction of an Alaskan pipeline and had punted on the issue of banning DDT. Furthermore, Nixon supported the supersonic transport (SST) because the United States had to stay militarily ahead of the Soviet Union, and the Johnson administration had already pumped billions of taxpayer money into the aerospace project. It also mattered to Nixon that the two Democratic senators from Washington State didn't want SST canceled. The coalition against SST—including economists Milton Friedman and John Kenneth Galbraith—was fierce. Even some Boeing executives were against the costly program, which the Senate voted against funding in early 1971.[51]

In this competitive atmosphere, Nixon signed his first truly important environmental bill, the Endangered Species Act of 1969. No one fought harder for this act than Russell Train, who became enamored with endangered species during savanna safaris in Africa. Increasing the enforcement of the 1966 Johnson-Udall act (also known as the Red Book), Nixon and Train now banned the importation of wildlife threatened anywhere in the world and expanded the list of protected species to include certain mollusks, crustaceans, mammals, fish, and amphibians. Nixon had a minimal role in the passage of this Democratic bill but seized credit; the White House pushed photos of the signing ceremony far and wide. This became Nixon's environmental policy pattern in coming years: let Democrats do all the hard legislative hammer-and-tong cobbling while he took media credit for himself at high-profile signing ceremonies.

III

A month after *Time*'s Cuyahoga River exposé, Senator Gaylord Nelson was flying to Seattle, having just inspected the Santa Barbara disaster zone. As in Cleveland, untreated municipal sewage and industrial wastes around Seattle were causing eutrophication in Puget Sound and Lake Washington. As a lawyer, Ehrlichman had earned local-hero status in Seattle a few years earlier for stopping a proposed aluminum plant on Guemes Island in the San Juan chain. On the airplane ride, Nelson read newspaper reports of anti–Vietnam War teach-ins being held on a dozen college campuses that fall of 1969. Working on a speech to deliver at the Washington Environmental Council in Seattle that upcoming September 20, aimed at banning DDT and Great Lakes preservation, Nelson, in a flash of inspiration, decided to spearhead a national environmental teach-in. "I am convinced that the same concern the youth of this nation took in changing this nation's priorities on the war in Vietnam and civil rights," he said in Seattle, "can be shown for the problem of the environment."[52]

The Associated Press ran a story about Nelson's environmental teach-in idea, and the rest is history. To plan the awakening, the University of Michigan was chosen as the December venue with a later teach-in slated at the University of Toronto. Walter Reuther's UAW, Nelson knew, would fund the Ann Arbor and Toronto summits. Without Reuther there would have been no Earth Day.

Indeed, from the Ann Arbor proceedings of December 1969 would spring the first Earth Day, scheduled to be held on April 22, 1970. The environmental woes that needed to be faced head-on were vast. Eco-problems were everywhere. Just off Staten Island, for example, the Arthur Kill blob (a raw sewage wave) near Fresh Kills Landfill had wiped out virtually all fish species in the water there. Boston Harbor stank so badly that people on wharves and docks vomited. California's

Santa Monica Bay was so poisoned that dolphins and whales had disappeared. Something large scale had to be done.[53] The Nixon administration *had* to act. It was especially important to Nelson and Reuther to make sure that the multitude of Earth Day teach-ins would be bipartisan. In that spirit, they made Representative Pete McCloskey, a California Republican, co-chair.[54]

From his Texas ranch, Johnson applauded the Nixon administration for expanding on his Great Society environmental accomplishments in 1969. To Johnson's delight, instead of trying to undo his Endangered Species Preservation Act of 1966 or Water Quality Act of 1965 and Clean Air Act of 1963 in his first White House term, Nixon sought to strengthen them. Because of such courtesies, Johnson made a point of not publicly criticizing Nixon for his delayed actions on Santa Barbara, Cuyahoga, or anything else. For that matter, Johnson even sympathized with the White House whenever Nixon took press heat over Vietnam policy, saying "I once told Nixon that the presidency is like being a jackass caught in a hailstorm. You've got to just stand there and take it."[55]

Not long after the Cuyahoga fire, NASA launched Apollo 11, the first manned mission to the moon. At Nixon's invitation, the Johnsons attended the liftoff at Cape Kennedy in Florida on July 16, 1969. Neil Armstrong and Edwin "Buzz" Aldrin's landing the *Eagle* on the lunar surface was an epic moment in world history. What Nixon intuitively understood was the way NASA's space exploits and environmental protection intersected in profound ways. "Now that we have seen our planet as it appears from outer space, all men of all nations can appreciate it more clearly than ever before as their common home, small and round and one with a thin and precious atmosphere on which we all depend and with no artificial boundaries to divide our energies or keep us apart," Nixon wrote in *Fortune* magazine. "Man has applied a great deal of his energy in the past to exploring his planet. Now we must make a similar commitment of effort to restoring that planet. The unexpected consequences of our technology have often worked to damage our environment, now we must turn that same technology to the work of its restoration and preservation."[56]

Some pundits, however, ridiculing Nixon's lavish embrace of technology as Pollyannaish, criticized NASA's Apollo program on environmental grounds. *Esquire* magazine, for example, asked American luminaries what Armstrong's first words on the lunar surface should be. Bill Douglas answered, "I pledge that we the people of the earth will not litter, pollute and despoil the moon as we have our own planet."[57] Frank Church, who had turned anti-dam, complained, "Today, we asphalt Americans, our eyes smarting from pollution, are rocketing men to the moon" while wallowing "in our own litter."[58]

The microbiologist René Dubos of Rockefeller University, who in 1969 won a Pulitzer Prize for *So Human an Animal: How We Are Shaped by Surroundings and Events*, about the interconnection between environmental degradation and psychological human health, predicted that baby boomers would be "remembered as the generation that put a man on the moon while standing knee-deep in garbage."[59]

The Pulitzer Prize–winning novelist Norman Mailer, who had long denounced plastic, turned his 1970 book about NASA's Apollo program, *Of a Fire on the Moon*, into a scathing diatribe about the proliferation of synthetic products. He mocked the Apollo 11 capsule's interior, "crammed with the bank of instruments eighteen inches over each astronaut's head as he lay in his plastic suit on a plastic couch—lay indeed in a Teflon coated Beta-cloth (laid on Kapton, laid on next to Mylan, next to Dacron, next to neoprene-coated nylon) spacesuit on his Armalon couch—plastic, that triumphed reason over nature."[60]

Mailer's and Dubos's warnings about technology run amok proved prescient on September 10, when a nuclear device twice as big as that dropped on Hiroshima was detonated in an underground site in Rulison, Colorado, with very limited media outrage. Cliffs crumbled, and the earth shook violently. A local resident told *Look* magazine that it "was like a train rushing up the canyon."[61] A couple dozen student activists from the University of Colorado protested the detonation near the Atomic Energy Commission site; the Boulderites were flabbergasted that the brutal blast had occurred without any provision having been made for their safety. One activist later recalled that "The ground went crazy," as if a magnitude 9 earthquake had struck.[62] Roadblocks had been erected far from the detonation site to corral a gaggle of reporters, FBI agents, scientists, and lawmakers. The test was part of Project Plowshare, which existed to develop peaceful uses for nuclear explosions. At Rulison, the AEC partnered with the Austral Oil Company of Houston. The hope was that the underground explosion would free natural gas hidden 8,400 feet below the ground surface.

That it did. However, the resulting radioactivity caused the gas to be radioactive and therefore utterly unusable by consumers. Hoping to avoid serious lawsuits, David Miller, the project's spokesperson, described twenty-eight damage claims filed by residents nearby as "small stuff."[63] Public protests by Coloradans, in a spasm of NIMBY outrage, led to the test site being closed. Project Rulison was a failed effort, but it matters in US energy history because it reconfirmed that natural gas production could be enhanced by cracking tight formations of underground rock. In 1976, a historical marker was erected on a gravel road in Garfield County along Route 338, commemorating the nuclear experiment. The sign should be revised to read "The Birthplace of Modern Fracking."

IV

The changing of the guard from Johnson to Nixon in 1969 coincided with a similar shake-up at the Sierra Club. In May, between the traumas of the Santa Barbara oil spill and the Cuyahoga River fire, Brower was dismissed as the club's president for overspending and failing to adhere to board policies.[64] The dynamics of the club had changed markedly in the late 1960s as its power to influence Washington lawmakers had soared.

By the time Nixon was inaugurated, even Ansel Adams had turned against Brower's aggressive stance. Other preservationists, however, stuck with Brower. His departure marked the end of an era (1955–1969) during which the Sierra Club had scored major preservation victories against the Bureau of Reclamation, the Army Corps of Engineers, the Forest Service, and commercial developers. In his resignation letter, Brower warned activists against allowing conservation to be used as a "cosmetic" for the modern world: "We have to develop, and soon, a deeper devotion to conservation as an ethic and conscience in everything we do. . . . We cannot go on fiddling while the earth's wild places burn in the fires of our undisciplined technology."[65]

In September 1969, Brower held a press conference and announced the establishment of three new organizations: Friends of the Earth, the John Muir Institute for Environmental Studies, and the League of Conservation Voters. When asked why three, not one, Brower replied, "The earth needs a number of organizations to fight the disease that now threatens the planet: 'Cirrhosis of the environment.'"[66] From 1969 to 1979, Brower was most indelibly identified with Friends of the Earth, which had offices in Washington, DC, and Berkeley, California. Name the environmental concern—pollution, nuclear technology, the industrial contamination of rivers, genetic engineering, urban runoff, deforestation, pesticides, climate change—and Brower had a well-informed opinion.[67]

On the literary scene, Brower was the protagonist of John McPhee's *Encounters with the Archdruid*, a well-reviewed 1971 nonfiction book tracking Brower's battles with corporations, reclamationists, the government, and multiple-use agencies.[68] Because McPhee, a *New Yorker* staff writer, wrote graceful prose, Brower emerged as the long shadow of John Muir in the age of Nixon. Brower was the bridge from the old-style "wilderness mountains thinking" to environmental justice. "Always prescient, he focused on biodiversity issues, ad hoc campaigns, and kept a positive slush fund to give activists cash for their causes," Steve Chapple, a coauthor with Brower of *Let the Mountains Talk, Let the Rivers Run: A Call to Those Who Would Save the Earth*, recalled. "Dave would hold court at Enrico's in North Beach and fund eco-activists with pocket cash."[69]

Bill Douglas was furious about Brower's dismissal from the Sierra Club. When the "green" justice was asked to speak at a club event in New York City later that year, he agreed, primarily to express his fierce displeasure with the shabby way the board had treated Brower. With the new Sierra Club executive director, Mike McCloskey, twitching in the front row, Douglas scolded attendees about "going soft" now that Brower was gone. Understandably, McCloskey "felt let down, even insulted" because he had helped Douglas with preservationist battles in Oregon and Washington.[70]

Yet soon McCloskey would be in Douglas's good graces for convincing the Sierra Club to sue Walt Disney Productions and the Forest Service to prevent their building an enormous winter resort in a wildlife reserve in the Sierra Nevada. McCloskey won a preliminary injunction against the project around the time of Apollo 11.[71] The case *Sierra Club v. Morton* eventually reached the Supreme Court and liberalized the rules of standing (who could decide who had the right to bring suit). In a famous dissent, Douglas concluded that suits could be filed on behalf of forests, oceans, wildlife, and more.[72]

At issue was Mineral King Valley in California's southern Sierra, a wilderness kingdom of ancient conifers, granite peaks, and black bear dens. The Forest Service, Douglas's all-seasons foe, announced that Disney was going to build a $35 million ski resort in the remote wilderness. If approved, there would be a dozen ski lifts, a non-denominational church, an ice-skating rink, grocery stores, diners, a conference center, two large hotels, a heliport, a massive underground facility to house resort services, and a Country Bear Jamboree attraction. Disney anticipated 2.5 million visitors within its first year of operations. The fact that public protest and court actions caused Disney to not build the complex was considered a win by most environmentalists.

For a passing moment in the fall of 1969, global warming due to fossil fuel excess was bandied about the White House. Back when Lyndon Johnson was president, his science adviser from 1964 to 1969, the chemist Donald Hornig, had warned the annual gathering of America's utilities that increases in carbon dioxide emissions were "triggering catastrophic effects on the planet."[73]

The Hornig warning lingered on in some government agencies. Yet in 1969, with the Vietnam War tearing the nation apart, alternative energy—solar, wind, nuclear, and hydrogen fuel cells—couldn't net much press attention. Not only were American cars gas guzzlers, but oil companies ran TV commercials daily. Even after Santa Barbara, combating CO_2 fossil fuel emissions didn't lend itself to popular public policy discourse. Quite simply, the issue of climate change was far too complicated for most lawmakers even to contemplate. Stewart Udall, in his 1968 book, *1976: Agenda for Tomorrow*, though, had expressed his global warming concern in no uncertain terms: "Our technology is now pervasive enough to produce changes in world climate, even as inadvertent side effects, which could melt enough of the world's glaciers to cause the inundation of all coastal cities."[74] Nonetheless, Udall admitted that the US government, mired in Vietnam, had failed to educate the public about the accelerating climate situation.[75] If, however,

Gaylord Nelson had been president instead of merely a "green" visionary in the US Senate, perhaps the twenty-first-century climate crisis
would have been solved: in 1969, he called for a ban on pesticides,
outlawing ocean waste dumping, *and* the end of the internal combustion engine.[76]

Nixon's White House domestic affairs adviser, Daniel Patrick
Moynihan, was one of those in government who at least tried to turn
the ship. On September 17, 1969, Moynihan wrote Ehrlichman a
startling memorandum. Considered the White House intellectual,
Moynihan was known as the book-learned freethinker of the Nixon
administration. The president had given Moynihan a wide berth.
Moynihan had learned from his alarmed friend Robert White, the
head of the US Weather Bureau and brother of the journalist Theodore White, that the planet's atmosphere was steadily warming in
response to carbon dioxide emissions. It was only a matter of time,
Moynihan feared, until global warming became an existential planetary threat. His memo read:

FOR JOHN EHRLICHMAN

As with so many of the more interesting environmental questions, we really
don't have very satisfactory measurements of the carbon dioxide problem. On
the other hand, this very clearly is a problem, and, perhaps most particularly, is
one that can seize the imagination of persons normally indifferent to projects
of apocalyptic change.

The process is a simple one. Carbon dioxide in the atmosphere has the effect of a pane of glass in a greenhouse. The CO_2 content is normally in a stable
cycle, but recently man has begun to introduce instability through the burning
of fossil fuels. At the turn of the century several persons raised the question
whether this would change the temperature of the atmosphere. Over the
years the hypothesis has been refined, and more evidence has come along to
support it. It is now pretty clearly agreed that the CO_2 content will rise 25% by
2000. This could increase the average temperature near the earth's surface
by 7 degrees Fahrenheit. This in turn could raise the level of the sea by 10 feet.
Goodbye New York. Goodbye Washington, for that matter. We have no data
on Seattle.

It is entirely possible that there will be countervailing effects. For example, an increase of dust in the atmosphere would tend to lower temperatures and might offset the CO_2 effect. Similarly, it is possible to conceive fairly mammoth man-made efforts to countervail the CO_2 rise. (e.g., stop burning fossil fuels.)[77]

Moynihan went on to suggest that global warming was "a subject that the Administration ought to get involved with. It is a natural for NATO." He noted that one government official who "knows a great deal about this" was Hubert Heffner, the deputy director of Nixon's Office of Science and Technology. It was Heffner who responded to Moynihan's urgent memo on Ehrlichman's behalf: "The more I get into this, the more I find two classes of doom-sayers with, of course, the silent majority in between. One group says we will turn into snow-tripping mastodons because of the atmospheric dust and the other says we will have to grow gills to survive the increased ocean level due to temperature rise from CO_2."[78]

Heffner dutifully asked the Environmental Science Services Administration to examine the issue of global warming carefully. Robert White, the administrator of the agency, did study climate change, coming out strongly for the reduction of carbon emissions from automobiles, trucks, and factories. For the time being, though, all that the so-called silent majority of Americans heard from Nixon on global warming was silence. There were too many other environmental issues to prioritize: getting lead out of gasoline, punishing the coal industry, promoting the SST, and addressing the grievances of Native American leaders about reservations being federally sponsored toxic waste dumping grounds.

Having spent two years perfecting the legal language in his Senate office, Scoop Jackson had introduced what became the National Environmental Policy Act (NEPA) to colleagues in February 1969. Jackson's notion of the Environmental Impact Statement (EIS), a fundamental principle of NEPA, was debated in Senate hearings on the bill. The role of an EIS, as envisioned by Jackson, was for federal agencies to provide an analytic review of pros and cons for major federal building or extraction that would have a significant impact on the surrounding environment. The EIS was a government form that outlined the impact of a proposed project on its environment. All projects had to have a Plan B alternative, including scrapping the project if its environmental impact was deemed unacceptable. The EIS requirement became Section 102 of NEPA. "Nobody seemed to pay much attention to it [at the time]," Scoop Jackson's Senate aide Bill Van Ness later recalled. "I wanted the EIS to be short enough to be easily read and understood by cabinet officers and other federal decision makers."[79]

That July, the Senate passed S. 1075 and referred the bill to the House. Representative John Dingell of Michigan had a similar bill to the one Jackson was shopping. That September, the House passed Dingell's bill, H.R. 12549, leading to the two bills going to a joint Senate-House committee to hammer out their minor differences. That was accomplished in early December 1969, when both the House and Senate passed the final version of the act the week before Christmas recess. The amount of grunt work Dingell did was extraordinary. All that remained to be seen was whether Nixon would sign the far-reaching NEPA into law or veto it.[80]

Once Jackson and Muskie had agreed on a final version of NEPA, it fell to Train to convince Nixon to sign it. Back when Train had headed the Conservation Foundation he had served as a pioneer in shopping the environmental impact statement, which was the heart of NEPA. What Train told Nixon was that if he vetoed the bill, it would be met with a congressional override and gift the Democrats with the "environmental issue" for years to come. "It didn't take much persuasion," Train recalled, "as long as Muskie didn't get to claim it as his accomplishment."[81]

On January 1, 1970, President Nixon was in a festive mood, excited to watch Texas play Notre Dame in the Cotton Bowl on TV in his ten-room San Clemente mansion, purchased shortly after the Santa Barbara spill. On a bluff overlooking the Pacific Ocean, La Casa Pacifica had a sandy beach and a swimming pool surrounded by cypress trees. Unexpectedly, the president asked the correspondents who had followed him to southern California to convene in the library of his Western White House. There he signed the farthest-reaching environmental protection legislation of all time, aimed at cleaning up and detoxifying American landscapes. The National Environmental Policy Act was the handiwork of pro-conservation Democrats and Republican worker bee Representative John Saylor of Pennsylvania. Nixon had opposed the bill throughout most of 1969. However, as an astute politician, he recognized that after Santa Barbara and Cuyahoga, the legislation matched the concerned mood of the country. With Ehrlichman urging boldness, Nixon cunningly framed NEPA as his own bill to demonstrate that he cared as much about the environment as Lyndon and Lady Bird Johnson combined. "It is particularly fitting that my first official act of the new decade is to approve the National Environmental Policy Act," Nixon said in San Clemente. "The 1970's absolutely must be the years when America pays its debt to the past by reclaiming the purity of the air, its waters, and our living environment. It is literally now or never."[82]

Considering that environmental legislation such as the Wilderness Act had taken eight arduous years to hammer out, the passage of NEPA had come together rather quickly, written and passed by both houses of Congress in a scant year. NEPA, a brilliantly articulated, concise, visionary, bendable statute, would be the foundation of US federal sustainability efforts well into the twenty-first century.[83] The first clause of the law established "the responsibilities of each generation as trustee for the environment for succeeding generations." The second clause, as Douglas had called for since the Kennedy presidency, amounted to a Bill of Rights that assured all Americans "safe, healthful, productive, and esthetically and culturally pleasing surroundings."

As Jackson had engineered it, federal agencies were mandated by NEPA to issue Environmental Impact Statements to go along with proposals for "major Federal actions significantly affecting the quality of the human environment." In essence, NEPA was, at last, a federal attempt to end future environmental disasters such as the Donora smog incident, the Santa Barbara oil spill, and the Cuyahoga River fire. It wasn't just a pipe dream about encouraging "productive and enjoyable harmony between man and his environment";[84] it was an actionable legal blueprint, as the historian Jack E. Davis put it, "for resuscitating natural systems."[85] NEPA also charged all agencies—Interior, Agriculture, Transportation, Defense, Commerce, and the rest—to coordinate any "decision making which may have an impact on man's environment" and to include in every recommendation scientific data on environmental impact.

There was an important third section of NEPA that established the Council on Environmental Quality (CEQ) in the executive branch (along with the obligation of the president to present an annual report on the current state of the environment). The CEQ became the federal environmental watchdog apparatus that would inform future environmental laws. As a result of NEPA's stringent enforcement, in fact, the AEC's nuclear licensing process was ended, and the Outer Continental Shelf oil drilling and Trans-Alaska Pipeline were shelved until proper EISs could be secured. Critics of NEPA, including the Chamber of Commerce and by the late 1970s the Department of Energy, complained that the law stunted economic growth. In counterstatements, Jackson and Dingell insisted that NEPA provided for a safer environment for American businesses to flourish in. Overall, NEPA called balls and strikes like an umpire in baseball. It established rules for regulations to control the use of natural resources, waste removal, and management of recreational opportunities. As historian J. Brooks Flippen put it, NEPA was "no paper tiger" but a juggernaut of highly qualified environmental investigations operating in the heart of the federal bureaucracy.[86]

The role of the CEQ, however, was somewhat diminished a year later, in December 1970, when the new Environmental Protection Agency (EPA) was created as a result of NEPA's being anchored more firmly. Most important, NEPA served as the catalyst of a Wild West boom in environmental law. Without NEPA, there never would have been the Federal Water Pollution Control Act (amended in 1972), the Toxic Substances Control Act (1976), the Resource Conservation and Recovery Act (1976), the Safe Drinking Water Act (1974), and many more noble attempts by the federal government to take the environmental protection of American citizens seriously.[87]

Given the magnitude of NEPA, Nixon was coy about the rollout. His signing of the legislation seemed purposefully subdued: on a holiday . . . in a beach house . . . just before a football game . . . with few reporters present. All of that translated into NEPA's receiving only modest publicity. Perhaps with his reelection in 1972 on his mind, Nixon didn't want to rub NEPA into the faces of the oil, gas, construction, real estate, and chemical industries. Some scholars, however, see the New Year's Day signing as an attempt to procure a windfall of press because there weren't any other big breaking news events on that football-driven hangover holiday. Either way, with the stroke of Nixon's pen, the United States no longer consisted merely of people. NEPA sought legal protection for "the air, the aquatic, including marine, estuarine, and fresh water, and the terrestrial environment, including, but not limited to, the forest, dryland, wetland, rangeland, urban, suburban, and rural environment."[88]

With a dose of hyperbole, NEPA was instantly dubbed the Magna Carta of environmental law. After centuries when it had been assumed that the natural world existed for Americans to exploit and plunder, NEPA declared that the official policy of the US government was to "create and maintain conditions under which man and nature can exist in productive harmony." Overnight all major companies and commercial developers needed to hire environmental lawyers to assess all new-product introductions or shovel-ready infrastructure projects.[89]

Furthermore, the Nixon administration encouraged citizens to weigh in on such proposals in hearings and through a public comment process. NEPA promised a meaningful say to anyone affected by a development project: individuals, small-business owners, faith-based organizations, tourism groups, conservation nonprofits. And it required the federal government to consider alternatives or mitigation for projects that threatened lasting public health or ecological harm to communities. By working with the Democrats and sharing leadership equitability with Henry M. Jackson and John Dingell, Nixon had enacted an expansive federal law that towered over the more tightly focused legislation that LBJ had signed. And before long NEPA had seized the attention of world governments. As Sharon Buccino noted in the *New York Times* on NEPA's fiftieth anniversary, "Small wonder that some 160 other countries have emulated this model with laws of their own, seeking to replicate a process that has proved its worth time and again."[90]

Magnanimously, Nixon credited Jackson and Dingell for inspiring NEPA while relishing the way he had stolen the Democratic Party's thunder on the environmental policy front from Muskie. He had earned the right to gloat. But, in the end, NEPA could also be credited to myriad true-blue environmentalists who had put their shoulders to the wheel to protect the American earth, water, and sky from 1945 to 1970. One could argue that Carson and Brower were part of the spirit of NEPA, as were the Environmental Defense Fund, Citizens for

Clean Air, Defenders of Wildlife, the National Park Foundation, and the Nature Conservancy. Middle-class Americans in the Izaak Walton League who wanted more hunting and fishing opportunities had been part of the drive for NEPA, as had Carl Stokes invoking environmental justice in Cleveland and Cesar Chavez marching against pesticide use in the agricultural belt of central California. So had Victor Yannacone, Charlie Wurster, Marjorie Spock, and Barry Commoner in their brave condemnations of DDT. On the left, the folk singers Pete Seeger, Joan Baez, and Malvina Reynolds had injected respect for ecology into the youth counterculture movement of the sixties at festivals and protest rallies. In Republican circles, wealthy philanthropists such as Laurance Rockefeller and Mary Lasker, who had used their vast financial resources and Eastern Establishment connections to recommend that the environmental impact of all new building projects be considered, had also led us to NEPA. The nature-driven prose of Henry Beston, Marjorie Stoneman Douglas, Wallace Stegner, and Robert Frost were instilled into NEPA, too.

What baffled the press was why Nixon had emerged as the Rough Rider of environmental activism. There was little about his 1968 campaign that pointed to his caring so deeply about pollution and natural resource depletion. Nixon aide John Whitaker had the most pragmatic answer. "Ehrlichman sold him on the environment," he believed. "He made Nixon see it was politically dangerous if he didn't get on board. He brought pollsters to say, 'this thing is catching fire!'"[91] Another factor was Nixon's disdain for Muskie, who was trying to be Captain Clean Air and ride a "green" wave into the White House port in 1972. Luckily for history, Ehrlichman's notes were archived at the Hoover Institution at Stanford University. On October 23, 1969, he wrote, "need a bold stroke" to "pull the rug from under Muskie."[92]

There were other caring environmentalists in Nixon's life besides Ehrlichman, Train, and Whitaker. His wife, Pat, was deeply distraught over the Santa Barbara oil spill and urged her husband to pursue conservation as if he were the reincarnation of Theodore Roosevelt. The president's brother Ed, who lived in Seattle, was a serious environmentalist close to Scoop Jackson, who promoted national park expansion for the Golden Gate National Recreation Area in San Francisco

Bay and an upgrade of the Endangered Species Act. White House speechwriter William Safire was onto something when he observed that "intellectually Nixon's heart was on the right while his head was with FDR, slightly left of center."[93]

Then there was Christopher DeMuth, a twenty-three-year-old White House aide whom Nixon personally urged to help him protect oceans and marine mammals. The deal was that DeMuth, who would later mature into a leader of the Natural Resources Defense Council, should keep his paper trail away from Maurice Stans, the secretary of commerce. "If Maury Stans is involved, he'll bring in all his business friends and nothing will happen. So, stay away from Stans," Nixon instructed DeMuth in paternal fashion. "I'll protect you. We'll let him know when we're done."[94]

It was the presidencies of John F. Kennedy and Lyndon Johnson that had elevated into the public sphere in a serious fashion from 1961 to 1969 the third-wave conservation that birthed NEPA. Nixon was the inheritor of new laws that stipulated that economic growth was proper only if it upgraded Americans' "quality of life." Those two Democratic presidents had opened the legal floodgates and mobilized concerned citizens, giving the country NEPA. Politically, the blueprint to combat the disease of overconsumption had been drawn during the New Frontier and Great Society years, with Senators Jackson, Church, Muskie, Hart, and Nelson doing yeoman legislative work with crackerjack staffs that at long last improved the overall outlook for environmental protection in the United States. Because all Americans want a sustainable future, NEPA became the popular tool for arresting environmental degradation from the 1970s into the twenty-first century.[95] The three "E" words—ecology, environment, and ecosystem—had caught fire in 1970s America. Conservation had almost overnight become as antiquated as the term "I Like Ike."

CHAPTER 30

Generation Earth Day, 1970–1971

President Richard M. Nixon holds up Senate Bill 1075 after signing it in San Clemente, California, on New Year's Day 1970. The bill, known as the National Environmental Policy Act, set standards that made "environmental impact" a factor in the US economy.

I

As a southern California native, President Nixon had witnessed agricultural abuses and environmental degradation firsthand from Los Angeles smog to the Santa Barbara oil spill, wildfires in the San Gabriels, to San Joaquin valley droughts, and High Sierra habitat loss. It really wasn't hard for John Ehrlichman to convince his boss that the 1970s were indeed the decade of the environment. After signing NEPA, in fact, Nixon told reporters that he had recently driven all over Orange County with his Florida friend Charles "Bebe" Rebozo, to show him the ocean-view sights. To his chagrin, he had been startled to see cookie-cutter suburban subdivisions blanketing the hillside like anthills all around Dana Point and Laguna Beach. Both Nixon and Rebozo agreed that within a decade, ticky-tacky strip malls would erase Orange County's citrus grove virtues and lazy-day charm. What Nixon loved about San Clemente was its original historic Spanish-style structures that recalled a long-ago California vibe, walking the sandy beachfront and relaxing at oceanside restaurants, with patios facing the Pacific Ocean.

On January 22, 1970, aware that ecology mania was in the American breeze, Nixon delivered his first State of the Union address, one-third of which was devoted to environmental policy. The question he posed was, he believed, the great question of the seventies:

Shall we surrender to our surroundings, or shall we make our peace with nature and begin to make reparations for the damage we have done to our air, to our land, and to our water?

Restoring nature to its natural state is a cause beyond party and beyond factions. It has become a common cause of all the people of this country. It is a cause of particular concern to young Americans because they, more than we, will reap the grim consequences of our failure to act on programs which are needed now if we are to prevent disaster later.[1]

President Nixon's State of the Union address demonstrated how quickly policy positions considered radical circa 1960 had become national concerns with mobilized citizens' support a decade later. The program he was promoting was the most costly and comprehensive in US history. Among other things, he wanted to expand the Solid Waste Disposal Act of 1965 and requested $10 billion from Congress for municipal waste treatment plants throughout the United States to make rivers blue and lakes swimmable again. "We can no longer afford to consider air and water common property, free to be abused by anyone without regard to the consequences," the president said. "Instead, we should begin now to treat them as scarce resources, which we are no more free to contaminate than we are free to throw garbage into our neighbor's yard."[2]

Nixon, a genius political negotiator, was being preemptive in his address. No longer would the regulation of industry emissions be monitored by states, which competed so intensely for factories and manufacturing jobs that regulating companies to protect public health was beyond them. Quite simply, Nixon wasn't going to let anti–Vietnam War liberals such as Church, Muskie, and Nelson seize the high ground on his White House watch. CBS News reporter Roger Mudd asked Muskie, "Has the president preempted the environmental issue from the Democrats?" It was a smart question. CBS News correspondent Dan Rather flat-out said, "What it boils down to is that the president has caught the Democrats on the environment—and he's walked away with their clothes."[3]

Stories circulated over how a teenage Nixon was captivated by the jaw-droppingly gorgeous Grand Canyon when he visited the national park on a high school trip. As president, Nixon enjoyed being the face of Yellowstone, Big Bend, and Zion. As part of his Legacy of Parks initiative, he ordered federal agencies to unload surplus real estate properties, many owned by the Department of Defense, which were then converted into federal parks in which working-class Americans could enjoy myriad recreational opportunities.[4] Furthermore, the extraordinary "Earthrise" photograph taken by an Apollo 8 astronaut, Bill Anders, on Christmas Eve 1968 was adopted by the Nixon administration as a visual representation of the new environmentalism. That same month, coinciding with Nixon's State of the Union address, CBS News declared 1970 the Year of the Environment and started using "Earthrise" as the bumper image on its nightly news broadcast.[5]

For the White House environmental team, led by Ehrlichman, the Rubicon between commercial developers and preservationists came to a head in south Florida. Nixon was forced to decide which side he was on. Plans for a huge new jetport in Big Cypress near the Everglades, the one that septuagenarian Marjory Stoneman Douglas brought to a stop, had broken ground years before Nixon entered the White House. At the request of Joseph Browder of the National Audubon Society, Mrs. Douglas helped found Friends of the Everglades in 1969.

Wearing a floppy hat and wielding a cane, she journeyed to half a dozen Florida counties to deliver speeches opposing the jetport. Promoted by the Dade County Port Authority, the jetport, if built, would be the largest airport in Florida. At the time of Nixon's speech, the jetport site had been purchased, brush clearing had begun, and the first concrete runway had been completed, covering the delicate wetlands habitat on which it was built.[6] The construction did not put an end to the resistance by the Interior Department to cease and desist. Florida governor Claude Roy Kirk, a Republican, was in favor of the jet facility. So was the head of the Florida Department of Transportation, whose comment on the controversy was: "I call the Everglades a swamp. My children can't play in it."[7]

Floridians with access to the Nixon White House, notably Nathaniel Reed of the US Fish and Wildlife Service, told the president that the mammoth jetport would contaminate the main body of the Everglades through its feeder waterways, laying waste to the supposedly protected national park.

Ehrlichman likewise told Nixon the jetport had "boondoggle" written all over it. Then, in a strange twist, Nixon's valet, Manuel Sanchez, weighed in on the side of the Friends of the Everglades. Nixon, in conference with Russell Train, was laughing off Reed's arguments about the biodiversity value of Florida's swamps, saying that at the Key Biscayne golf club, he'd asked time and again that swaths of mangrove trees be cut down so that golfers could see the ocean as they played. That was when Sanchez walked in with coffee for the president. Nixon, still jawing with Train, said, "But the environmentalists all raised hell." He continued, "Russ, nothing lives in those mangroves except gooney birds." Then, turning to Sanchez, he asked, "That's right, isn't it, Manuel?"

"No, Mr. President, that is not right," Sanchez said. "You know when we are down at Key Biscayne, and I have my day off, I go fishing. I know all the fish I catch when they are very small they grow up in those mangroves and that is where they are protected and if you took those mangroves out, we wouldn't have any more fish."

In Train's recollection, Nixon just stared at Sanchez for a moment, undoubtedly shocked to hear his obedient valet so bluntly disagree with him—but also, perhaps, impressed by the point Sanchez had made. Eventually, Nixon joined the Everglades preservation coalition, and in September 1969, almost exactly a year after ground had been broken on the jetport project in a ribbon-cutting ceremony, the president ordered the construction work to cease. Its concrete runway survived as a training site for pilots tasked with protecting wildlife.[8] Further influenced by Bebe Rebozo, who lived in Key Biscayne, Nixon, speaking from Camp David, described his banning of the jetport in 1970 as "an outstanding victory for conservation."[9]

Many Florida Republicans were incensed. Environmentalists, however, especially Mrs. Douglas, were delighted. Nixon's defense of the Everglades even pleased the *New York Times* and *Miami Herald* editorial boards.[10] Evidence was accumulating that because of NEPA regulating industry at the federal level, there would be, as William Ruckelshaus put it, "No place for the polluters to hide."[11]

It was well known among rare book dealers that Ehrlichman had one of the finest US presidential history collections. Believing that the environmental crusade was what William James called "the moral equivalent of war," Ehrlichman convinced Nixon that if he united with Henry Jackson, a pro–Vietnam War friend of the White House, a true bipartisan environmental awakening might take hold in the early 1970s, with Nixon leading the charge. The notion that NEPA assured that economic and environmental quality were compatible was contagious, and Nixon had caught the bug.[12] "I'm an old lady," Mrs. Douglas joked about her alliance with Nixon. "I've got white hair, I've been around here forever, and no one can afford to be rude to me."[13]

If Nixon's signing of NEPA and his Everglades protection were the first big-time indications of the administration's serious commitment to environmental protection, the State of the Union address on January 22, with its plea that "Restoring nature to its natural state is a cause beyond party and beyond factions," was seen by surprised Democrats as the Great Reconfirmation that Nixon indeed harbored a genuine TR streak.[14] Ed Muskie, however, derided the address as environmental grandstanding and posturing for accolades.

When Nixon delivered a Special Message to the Congress on Environmental Quality on February 10, he listed fourteen executive orders and twenty-three legislative proposals to combat pollution and provide parkland. "The time has come when we can wait no longer to repair the damage already done," he told Congress. But many on Capitol Hill still doubted his resolve to follow through on his progressive thirty-seven-point program.[15] In a matter of months, though, his White House would send Congress the outline of his visionary Environmental Protection Agency (EPA). It was to encompass a number of previously existing government offices—but not too many. "What we needed—and what the public wanted," Russell Train later explained, "was an organization with a clearly defined mission: to be the sharp, cutting edge of environmental policy in the government."[16]

II

Gaylord Nelson, the Wisconsin senator, became the media darling in the early months of 1970 as Earth Day approached.[17] Somewhat mild-mannered as a speaker, he could nevertheless talk with authority about PCBs in mother's milk, PCBs in Michigan cows, and poisons leaking from rusty drums on Milwaukee docksides. Back in 1963, when he had convinced President Kennedy to visit the Apostle Islands, the nation wasn't ready for an Earth Day "youthquake" movement.[18] In 1970, though, with NEPA a reality, US citizens were demanding federal action. Nelson recruited Pete McCloskey, the ecology-minded Republican congressman from California, to serve as his environmental teach-in cochair. Together they appointed a twenty-five-year-old Stanford University graduate student, Denis Hayes, to direct the national teach-in. It was Hayes who hired the advertiser Julian Koenig, celebrated for his "Think small" Volkswagen campaign, who branded the teach-in Earth Day (the other serious contenders were Green Day and Ecology Day).[19]

Born in Wisconsin on August 29, 1944, Hayes grew up in Camas, Washington, a depressed paper mill town on the Columbia River. To escape the drip-drop rainy dreariness, he explored the "pristinely beautiful" world of thick forests, swift rivers, and forlorn foothills rolling away into the Cascades range of Washington State.[20] In 1964, Hayes had graduated from Stanford University with a degree in history; he was also class president.[21] An anti–Vietnam War dove and early solar power advocate, one deeply determined to help launch a global ecological awareness crusade, he was ideally suited to be a young leader in the ecoconsciousness-raising movement that was afoot on college campuses. Working with the UAW to print a "green" newsletter, he rented a dingy office in Dupont Circle in Washington, DC, with some seed money from Walter Reuther. On his office wall was a photo of Mickey Mouse ears drawn on the heads of David and Julie Nixon Eisenhower. Hayes's office vibrated with a Woodstock feel (a lot of long-haired men, peace symbols, rock music, and leather sandals).[22] With intense blue eyes and a long Lincolnesque face, Hayes proved a master at working with the media and recruiting young people to save the planet.

At first, Hayes, McCloskey, Reuther, and Nelson planned Earth Day as a single event. With astonishing speed, however, scores of colleges and universities began organizing Earth Day protests against galloping commercial development, overpopulation, and the deterioration of the planet. Schools joined in with special programs. In quick order, the UPI ran a Hayes-driven story, "America's Youth Rallies to the War on Pollution," on February 22, followed by a two-part CBS News special, *The Environmental Crusade*. The *New York Times* reported on March 2 that Earth Day was galvanizing into being a "wide environmental protest."[23] Nonprofits that had been campaigning in the same spirit for years or even decades shared their values with young people through the Earth Day teach-in. Momentum was on the side of the environmentalists as never before. Ehrlichman, who had met Hayes once at Stanford University, invited him to the White House to discuss Earth Day and anti-pollution laws with Nixon. But Hayes refused, determined not to be co-opted.[24]

Opponents of Earth Day condemned the planned event as an unpatriotic deflection from the war in Vietnam. Barry Goldwater pointed out that April 22, 1970, was the hundredth anniversary of the birth of Vladimir Lenin and warned Americans that Earth Day was a Communist plot. Nelson had a masterful comeback for the Red baiters: "A person many consider the world's first environmentalist, Saint Francis of Assisi, was born on April 22. So was Queen Isabella. More importantly, so was my Aunt Tillie."[25]

Earth Day 1970 mobilized a vast political alignment that was both shocking for the time and welcomed by practically everyone.

Democrats and Republicans, rich and poor, city dwellers and rural farmers, millionaire tycoons and labor leaders, along with students of all ages—20 million Americans in all (10 percent of the US population)—took to streets, parks, and auditoriums to demonstrate for a healthy, sustainable environment. Hayes's vision was, why not let young people be the ombudsmen for the planet? "It was a gamble, but it worked," Nelson would later say.[26]

Anyone of a certain age remembered decades later where they were on that historic first Earth Day. Stewart Udall lectured at the New School in New York, pointing out that the United States represented only 6 percent of the world's population but was consuming "35 to 50 percent of the world's nonrenewable resources."[27] His syndicated column, "Udall on the Environment," was carried in newspapers across America that spring. Unhampered by government restraints, he sided openly with Earth Day activists he admired, such as Gaylord Nelson, Pete McCloskey, and Ansel Adams. By turning against the Colorado River dams that would have marred Grand Canyon National Park, Udall felt liberated in 1970—willing and able to put the Corps of Engineers and the Bureau of Reclamation on the defensive in the American Southwest, if need be. At public forums centered around Earth Day, he apologized for the Glen Canyon Dam blunder and spoke about wild and scenic rivers. It was clear he idolized Brower, who was a genius in raising public awareness about environmental degradation.

That Earth Day, Udall realized that Rachel Carson had launched an "eco-activist revolution" in 1962 by trying to halt the spraying of DDT on living creatures and plants.[28] It wasn't simply because she had written authoritatively and had amassed facts against the chemical industries. What mattered was the way she had unsheathed the word *chemical*, as if it were a death knell to human existence. Phrases such as chemical control, chemical war, chemical assault, and chemical destruction were all Carsonisms.[29] The irony was that in 1970, DDT was still being sprayed from airplanes in the United States to kill insects.

All three major television networks covered Earth Day as if it were the Super Bowl. NBC's *Today Show* focused on ecology for the entire week of April 20 to 24. ABC News aired three prime-time environmental specials during the same time frame. Walter Cronkite of CBS

News aired a week of Earth Day programming with the intensity of the 1960 Kennedy-Nixon debates. "Every night on color television we saw yellow sludge flowing into the blue rivers," William Ruckelshaus recalled, "every day as we drove to work, we saw smudges against the barely visible blue sky."[30] In retrospect, Ruckelshaus thought that the proper Earth Day slogan should have been "Americans are fed up!"[31]

Even though Nixon refused to appear on TV shows to celebrate Earth Day, suspicious that they were liberal traps, he urged his White House environmental team of Hickel, Train, Ruckelshaus, Whitaker, and Ehrlichman to "Hit Carson, Cavett, and the think-type programs."[32]

John Lindsay, the New York City mayor, who was hoping to close the Fresh Kills Landfill on Staten Island, shut Fifth Avenue down for a massive environmental rally centered around a mall of educational booths and ecology exhibits.[33] In Philadelphia, local universities organized a huge event featuring Senator Ed Muskie, who declared that "an environmental revolution" had arrived and slammed Nixon for spending "twenty times as much money on Vietnam than to combat water pollution."[34]

Former vice president Hubert Humphrey delivered a major speech at an Indiana high school, while Henry Jackson addressed students at the University of Washington.[35] Senator Edward Kennedy gave a major environmental policy address at Yale University. The Sierra Club published *Ecotactics* (with Michael McCloskey writing a searing foreword).[36] The indomitable Ralph Nader seemed to be everywhere at once on Earth Day, as if he had body doubles. Paul Ehrlich, whose *The Population Bomb* had sold two million copies, was the hot ticket in Ames, Iowa, and Toledo, Ohio. Cesar Chavez attended Mass with his family in the Central Valley of California and planted fruit trees. (Furthermore, he was suing to force California grape growers to sign an agreement that July after a five-year strike.) At the Washington Monument, near the White House, a crowd of ten thousand people gathered to hear Pete Seeger and Phil Ochs sing songs such as "Where Have All the Flowers Gone?" Troops of Girl Scouts from Connecticut came to the nation's capital to pull discarded tires and garbage out of the Potomac River. Everybody, it seemed, rolled up their sleeves or sharpened their pencils to participate in that Earth Day teach-in.

Bill Douglas planned to celebrate Earth Day by hiking along the C&O Canal. When he invited his friend Scoop Jackson, fresh from the NEPA victory, to lace up for the hike, the Washington senator burst out laughing. Making environmental legislation was his forte, not endurance treks in the muddy Maryland countryside, he said. Plus Jackson was slated to speak in Seattle that day. Once back at the Supreme Court from his Earth Day hike, Douglas petitioned President Nixon for the C&O Canal to be upgraded to national park status. Behind the scenes, Jackson joined Douglas's cause as a favor to the septuagenarian justice. With consummate skill, Jackson, who still backed the Vietnam War and SST, steered a bill to establish the Chesapeake & Ohio Canal National Historical Park through the Senate and then helped procure congressional approval. In a note to Douglas around Christmas 1970, Jackson wrote, "I am pleased to report that at long last we were able to get your C&O Bill approved and on the way to the White House." (After President Nixon signed the bill into law in 1971, the *Washington Post* playfully declared that the new national park had been "walked into existence" by Douglas.[37])

Congress had adjourned for Earth Day so all of its members could participate in teach-in events. The senator of the moment, Gaylord Nelson, had assisted Democratic policy makers far and wide in writing environmental speeches. In dramatic fashion, governors Nelson Rockefeller of New York and William Cahill of New Jersey signed laws establishing state environmental agencies. The National Education Association boasted that 10 million children had taken part in Earth Day teach-ins. But there were also a few incidents of violence. In San Francisco, "environmental vigilantes" dumped oil into a Standard Oil reflecting pool to protest the Santa Barbara oil spill and others that were certain to happen if offshore drilling were continued.[38]

Proud that Earth Day organizers had rejected financial contributions from Mobile, Standard Oil of New Jersey, and other corporations, Representative John Saylor spoke to students at Penn State University on Earth Day, offering commonsense advice. "Boys and girls, I don't buy this antigrowth idea," he said. "We need growth to pay for the environmental improvements that have to be made. You can't build a sewage treatment plant unless people have money to pay for it."[39] That Saylor came off as a fuddy-duddy wasn't important to the ecology-minded students. They celebrated the sixty-two-year-old Saylor for his promotion of the Wilderness Act, the defeat of the Grand Canyon dams, the preservation of the Current and Jack Forks Rivers, and the saving of Hells Canyon in Idaho. And no Republican of his postwar generation had fought harder on behalf of Pennsylvania's state parks and forest reserves than Saylor.

Senator Frank Church delivered a brilliant jeremiad in Idaho on Earth Day against uncritical acceptance of technology and a national mania for economic growth. "Our rivers are clogged with filth," he scolded. "Smog is the 'air apparent' in our large cities. Our eyes are assaulted by unrelieved urban ugliness. Much of our land is hideously defaced."[40] Church had become the conscience of the Democratic Party in the Cold War, the most honest lawmaker in the Senate. At the time of Earth Day, he coauthored the most pronounced legislative effort to end the Vietnam War: the Church-Cooper Amendment of 1970. If one lawmaker was the soul of Earth Day, besides Gaylord Nelson, it was Church.

Lyndon and Lady Bird Johnson applauded the Earth Day volunteerism from their Texas ranch. According to Liz Carpenter, Mrs. Johnson's all-purpose assistant, they viewed the event as if it were the capstone of the Great Society's New Conservation triumphs.[41] Nixon, she believed, was merely trying to keep pace with Johnson's impressive environmental record of having organized nine environmental task forces and signed over three hundred environmental measures into law from 1963 to 1969.[42] Sadly, his astonishing environmental policy feats weren't mentioned in the Earth Day press coverage: LBJ was ancient news.

President Nixon was deeply suspicious of Earth Day. Nevertheless, he allowed Hickel's Interior Department to participate in the teach-in extravaganza. With press in tow, Hickel canoed the Chena River in Fairbanks and spoke at the University of Alaska. Vice President Spiro Agnew, who considered Earth Day a scam, thought Hickel had made a fool of himself. Nixon took the safe middle course on Earth Day; he simply planted a tree on the White House South Lawn. That historic April 22, Nixon refused to deliver a speech. In a photograph that went viral, Nixon stood in his wrinkle-free suit while the First Lady, Pat Nixon, in high heels, did all the soil shoveling.[43]

The reason Nixon didn't savor the nationwide teach-ins was a galloping paranoia. Earth Day, he feared, was a trap by liberal academics and left-wing activists. The president, in fact, asked the FBI director, still J. Edgar Hoover, to spy on high-profile Earth Day teach-in events held at colleges and universities. Fearful that the whole environmental youthquake had been plotted by antiwar radicals in Berkeley, Menlo Park, Ann Arbor, Madison, and Ithaca, the president sulked the afternoon away.[44] Representative McCloskey later recalled:

> I was friends with John Ehrlichman at that time, who was an environmental lawyer, incidentally, before he went to jail for Watergate.
>
> And he called me after Earth Day—he was laughing as hard as I ever heard, and he said, "Pete, I've got this report from [FBI director] J. Edgar Hoover to deliver to the president tomorrow," because the president was so paranoid that Earth Day was going to be a bunch of anti-war kids gathered that he had put them under surveillance by the FBI.
>
> He read me part of the report: "There's a bunch of girls with flowers in their hair, and they're wearing only three garments, no bras. And it was very benign. They were a little drunk, [there was] a little pot, maybe a little love out under the bushes, but these girls sat in the grass patting their dogs, and it was a very benign affair."[45]

Walter Reuther spent Earth Day at the University of Michigan and Oakland University near Detroit, where his daughter Elisabeth was studying liberal arts. With heartfelt conviction, he spoke about democratic values, defending society's underdogs, and the importance of protecting the environment. "Father electrified the audience," Elisabeth Reuther Dickmeyer remembered, "making clear sense that to preserve and honor our planet is our only option, as the alternative is existential. I was never more proud of him than I was on that day."[46]

Bizarrely, both Reuther and his wife, May, died only days later when their chartered Learjet 23 mysteriously burst into flames upon arriving in Pellston, Michigan, near UAW's environmental campus at Black Lake. The UAW leader had previously survived two assassination attempts and had barely escaped with his life in a previous airport mishap. Now he was dead like the Kennedys, King, and Malcolm X before him. Speculation was rampant that he had been murdered for his social justice activism. That has never been proven. What is known is that the Learjet 23 had a crucial part, the altimeter, placed upside down.

Because May Reuther also perished in the crash, there wasn't anybody to wear the UAW environmental flag with the verve of Walter Reuther (though Douglas Fraser, who would later become UAW president, tried). The loss of Reuther was devastating to the Silent Spring revolution. Almost single-handedly he had kept labor in the eco-mix.

From a practical, long-term perspective the most important development that emanated from the Earth Day hullabaloo was the creation of the Natural Resources Defense Council (NRDC) that June. Growing out of the Storm King showdown on the Hudson River, NRDC, under the leadership of John H. Adams, started suing government agencies that succumbed to "regulatory capture" by corporations. Adams, unlike Brower, wasn't interested in becoming famous. But he wanted NRDC to be bigger than EDF. In that quest, he hired the best scientists, economists, reporters, fundraisers, and administrators he could find with the goal of developing an NAACP-style legal defense fund for the environment. During the 1969–1970 congressional session, a mind-boggling eight thousand environmental bills were introduced. NRDC, a pioneer of environmental activism, would grow into a powerhouse of 1.2 million members by the time Barack Obama became president.[47]

History has honored Gaylord Nelson mightily for the Earth Day idea. Nobody paid him a greater tribute than Bill Clinton did when he presented the Wisconsin senator the Presidential Medal of Freedom in 1995, noting that he was the fountainhead of all the acts that had grown out of the grand teach-in event: the Environmental Protection Agency, the Clean Air Act, the Clean Water Act, the Safe Drinking Water Act, and many more.[48]

Earth Day turned out to be a cultural phenomenon in 1970 and 1971, a "supercause" that inspired a wave of hit songs with environmental themes that attracted young people to the movement. Marvin Gaye's "Mercy, Mercy Me (the Ecology)," a track on his classic *What's Going On* album, was a shamanlike prayer for humanity to stop the destruction of Earth. In his smooth tenor voice, he lamented, "Oil wasted on the ocean and upon our seas, fish full of mercury / Ah oh mercy, mercy me / Ah things ain't what they used to be, no no / Radiation under ground and in the sky / Animals and birds who live nearby are dying / Oh mercy, mercy me / Ah things ain't what they used to be / What about this overcrowded land / How much more abuse from man can she stand?"[49]

The popular rock band Three Dog Night had a hit with "Out in the Country," a paean to the green pastures and forests that rank commercialism was destroying. The lyrics were written by Paul Williams: "Before the breathin' air is gone / Before the sun is just a bright spot in the night-time / Out where the rivers like to run / I stand alone and take back somethin' worth rememberin'."[50]

The most profound Earth Day song of 1970, the one whose lyrics were regularly quoted in outdoors magazines and eco-conferences, was the Canadian singer-songwriter Joni Mitchell's "Big Yellow Taxi," a cut from the album *Ladies of the Canyon*. "They paved paradise / And put up a parking lot" became a green power mantra. It turned out that Mitchell's classic song was inspired by a trip to Hawaii where she looked out of her hotel room window and saw lush green tropical forests on a mountainside slope beside a huge asphalt parking lot. The notion of replacing Paradise with something as visually repulsive as a parking lot was an epiphany to Mitchell. Her line on human nature— "Don't it always seem to go / that you don't know what you've got 'til it's gone"—was adopted as a potent slogan for environmental awareness.[51]

Around the time of Earth Day, a vegetarian food movement had taken hold of the counterculture. There was a huge increase in organic farming. The following year, Frances Moore Lappé's *Diet for a Small Planet* was published, selling over 3 million copies. Connected to this uptick in vegetarianism was the Farm Animal Rights Movement, based on ethical considerations. Overall, Earth Day had produced an avalanche of books about the environment and animal rights. Brower earned a reputation as a great talent scout for ecology themed titles. "By any measure," historian Adam Rome wrote, "the rise of what conservationists began to call 'eco-publishing' was breathtaking."[52]

The greatest legacy of Earth Day was hard to quantify: it changed people's lives. High school and college students sought to start a career in environmental studies. Brent Blackwelder, for example, riveted by "green" activists on their soapbox, volunteered for Earth Day. Instead of becoming a math teacher as he had planned, he turned eco-activist. In 1973, he helped found American Rivers, a big-time nonprofit working both to increase the number of the Wild and Scenic River units and to halt the unnecessary damming of rivers. In 1994, Blackwelder was hired as president of Friends of the Earth, Brower's world-beat nonprofit organization. Many of the top environmental activists of the twenty-first century—such as Bill McKibben and John Kerry—were baptized into the environmental movement thanks to the first Earth Day extravaganza.[53]

III

The bipartisan, cross-generational harmony of Earth Day was short-lived. Eight days after Earth Day, on May 4, members of the National Guard killed four unarmed students at Kent State University in Ohio and wounded nine others, merely for their protesting the Nixon administration's military incursion into Cambodia. Not only was it the only time that students had been killed in an antiwar rally in US history, but any goodwill Nixon had garnered from his Earth Day olive branch to the youth of America went down the drain. How could Nixon truly care about the environment while US troops were spraying one-eighth of Vietnam with chemical defoliants? By spraying napalm to kill vegetation in Vietnam, the administration was committing ecocide.[54]

Discombobulated by Kent State and growing antiwar youth protests at colleges and universities, Nixon suddenly appeared in a spontaneous 5:00 a.m. visit to an antiwar protest being organized at the Lincoln Memorial. Hoping to connect with the young people, he spoke about foreign policy, football, and the environment. "I just wanted to be sure that all of them realized that ending the war, and cleaning up the streets and the air and water, was not going to solve spiritual hunger, which all of us have," Nixon wrote in his diary.[55]

Distraught by the Kent State shooting, Walter Hickel wrote a letter critical of Nixon's Vietnam War policy that was leaked to the press.

The line that irked Nixon most was "I believe this administration finds itself today embracing a philosophy which appears to lack an appropriate concern for the attitude of a great mass of Americans—our young people."[56] Having established NEPA and planted a tree on the White House grounds for Earth Day, Nixon was outraged at his interior secretary's insubordination and insult. If there had been a guillotine near Lafayette Park, Train later joked, Nixon would have marched "Wally the Rube" to it under armed arrest.

Instead, that summer, Nixon had Ehrlichman ax Hickel out of all administration happenings. He was persona non grata in the White House. That July, to really punish Hickel, Nixon placed the new National Oceanic and Atmospheric Administration (whose acronym, NOAA, was pronounced "Noah") in the Commerce Department under Secretary of Commerce Maurice Stans instead of the Interior Department, where it belonged. NOAA soon had a $270 million budget and twelve thousand employees.[57]

Getting into the president's face about the Vietnam War guaranteed a one-way ticket out of Washington for Hickel. Insubordination of any kind infuriated "the boss," whose law-and-order stance wasn't just for show. Anyone not sympathetic to the National Guard at Kent State, for example, wasn't part of his Silent Majority constituency. When push came to shove, Nixon aligned with the Chamber of Commerce over NEPA. He once told Ehrlichman that in a "flat choice between smoke and jobs" he was on the side of jobs."[58] Ehrlichman recalled that when presented with a report on environmental issues, the president usually sat quietly, barely listening, and giving it his time only because he knew that land and water cleanup had to be done. The exception was when new national park units—such as Voyageurs (Minnesota) and Apostle Islands (Wisconsin)—were brought into the conversation. Nixon loved establishing new parks.

Nevertheless, Nixon allowed Russell Train free range to wax philosophical on such issues as pollution-free cars, Lake Erie cleanup, and shaming Procter & Gamble to cut phosphate emissions. And if it hadn't been for Ehrlichman, there was no telling if NEPA and EPA would exist. As a land lawyer in Seattle, Ehrlichman had fought developers in the Puget Sound area who constructed housing tracts, stores, or factories without following zoning laws. According to a former law partner, Egil "Bud" Krogh, Ehrlichman, since his days at UCLA, was a "passionate supporter of preserving the natural environment in the Northwest. He was a camper, a hiker, a fisherman. He was one of the first effective environmental lawyers out here."[59]

After Nixon signed the National Environmental Policy Act and the first Earth Day was past, the Council on Environmental Quality (CEQ), the advisory team approved by Nixon to help develop and clarify NEPA's regulations and produce guidelines, chaired by Russell Train, set up shop on Jackson Park just off Lafayette Square. Train began the implementation and enforcement of NEPA laws. Behind his desk was a marvelous painting of an orange orangutan and other wildlife art from his extensive travels to Africa. Warmhearted and avuncular, Train became the administration's point man on endangered species. As CEQ head, he roused the ire of the US Chamber of Commerce, which wanted the Republican James Watt, a former lobbyist for big business, to lead the CEQ, as well as Goldwater conservatives. Train also got an earful from the Sierra Club, which carped that the Nixon administration wasn't forcing compliance with NEPA in regard to Environmental Impact Statements (EIS).[60] "I was getting it from both sides," Train recalled. "That meant we were being effective."[61]

On July 9, 1970, President Nixon, in sync with Congress, established the Environmental Protection Agency, a new federal agency tasked with overseeing US environmental policy: consolidating all main US government programs aimed at regulating pollution, clear-cut timbering of forests, dam building, strip mining, and oil extraction within delicate ecosystems. It would have departmental status. The Seattle zoning master, Ehrlichman, was at the top of his game in establishing EPA protocol and streamlining more than eight federal agencies that dealt with pollution. Once again, the White House collaborated with Scoop Jackson in the Senate and John Dingell in the House to establish the EPA. After a mandatory period during which Congress could have blocked the presidential reorganization, the EPA was approved by both houses. The plan was for it to begin operating in December, after the November midterm election.

On the same night that the EPA was formally organized, July 9, 1970, Nixon established the National Oceanic and Atmospheric Administration (NOAA)—which officially opened for business on

John Ehrlichman, Nixon's domestic policy adviser from 1969 to 1973, did more to protect America's troubled environment than any other administration official.

October 2 and was housed in the Department of Commerce. It was responsible for analyzing atmospheric functions (including weather and climate), improving the conditions of the oceans by enforcing the sustainable use of resources and coastal and marine ecosystems, and supplying environmental information to the public. Another of NOAA's mandates, making coastal surveys, had been initiated by Thomas Jefferson. NOAA managed eight major agencies, one of which oversaw the US Weather Bureau. "We already have the scientific, technological and administrative resources to make an effective, unified approach possible," Nixon said of NOAA. His directive specified, especially in relation to the ocean, that "We must understand the nature of these resources, and assure their development without either contaminating the marine environment or upsetting its balance."[62]

That summer of 1970, nobody was more excited about the creation of NOAA than Sylvia Earle, the PhD marine scientist at the

Mote Marine Laboratory & Aquarium in Sarasota, Florida. In a high-profile competition, she was selected for the Tektite project, which involved an installation fifty feet below the surface of the sea in the US Virgin Islands, where scientists could live and conduct research. Even though Earle had logged more than a thousand research hours underwater, she was rejected in the first wave of recruits owing to gender bias. But she persisted.

In 1970, coinciding with NOAA's official founding on October 3, 1970, Earle was chosen to lead the first all-female team of aquanauts in Tektite II. That venture, sponsored jointly by the US Navy, the Department of the Interior, and NASA, enabled teams of marine scientists to live for weeks at a time on the ocean floor and study the ecosystem. In all, ten teams would descend in Tektite II and conduct nine different research projects. Mission 6 was the first and only all-female roster. The pressure at forty-seven feet down is nearly ten times that on the surface, powerful enough to crush a soda or soup can. Once a diver gets to thirty feet, his or her lungs retract to half their regular dimension. So being an aquanaut means one has to be ready to have a "face off with mortality" every time one deep dives.[63] Because the habitat was pressurized, that would be a concern only if the system failed.

Though the aquanauts of Mission 6 did not get the widespread attention the Apollo astronauts did, the *New York Times* did run continuing coverage, going so far as to explain the types of breathing and diving apparatuses used by the female scientists when they exited the facility in diving gear to explore.[64] Everything was going perfectly for them when a significant earthquake hit the Virgin Islands. A wave of fear spread through the government agencies sponsoring Tektite II. Earle radioed her bosses to reassure them: "We got up at once and checked the habitat carefully to be sure no damage had been done to the controls and communications systems. Then we went back to bed."[65]

When it came to marine science, Earle had the right stuff. Once she surfaced, bubbling with excitement over her marine experience, she became a star oceanographer, the Jacques Cousteau of her generation. After the Tektite II mission was complete, she traveled on scientific missions to the Galápagos Islands, the waters of Panama,

China, and the Bahamas, and the Indian Ocean. For the remainder of that decade, in fact, she regularly broke marine submersion records. Whether it was investigating the World War II battleship graveyard in the Caroline Islands of the South Pacific, tracking sperm whales from Hawaii to New Zealand, or becoming the first human to walk on the seafloor untethered at a lower depth than any human being before or since, she transcended her hero Rachel Carson's knowledge of oceanic life.

Twenty years after Nixon established NOAA, President George H. W. Bush appointed Earle to serve as the agency's chief scientist. In addition to her other duties, she became a global ambassador for ocean protection. In 1998, *Time* magazine chose her as its first Hero for the Planet. "Earthlings take for granted that the world is blue, embraced by an ocean that harbors most of the life on the planet, contains 97 percent of the water, drives climate and weather, stabilizes temperature, generates most of the oxygen in the atmosphere, absorbs much of the carbon dioxide, and otherwise tends to hold the planet steady," she wrote in 2012, "a friendly place in a universe of inhospitable options."[66]

IV

In the summer of 1970, long-haired counterculture youths from the San Francisco Bay area descended on Yosemite National Park. On any given day, the park recorded around 15,000 overnight and 18,000 transient visitors. More than 2 million people a year loved the park (aka Yosemite City) to death, creating a crisis of law enforcement, garbage pickup, and traffic control. Everything became difficult to manage.[67] John Muir's glorious natural wonderland was being ruined by air pollution and litter, which were ubiquitous. Two years earlier, the NPS had discontinued the Firefall spectacle that President Kennedy had witnessed in 1962 because the crowds had become too large and unruly.[68] But that didn't slow the surge of summer visitors.

Because the $4 daily admission fee to Yosemite was reasonable, vans full of hippies and college students made the park a party destination. Young revelers poured into the Yosemite campgrounds along the Merced River and near the base of Yosemite Falls wearing tie-dyed T-shirts and sandals; marijuana smokers and stereo blasters turned Yosemite into a woodsy Haight-Ashbury. Complaints poured into the park superintendent's office about lewd behavior, public intoxication, panhandling, and other degenerate behaviors.[69]

Overwhelmed Yosemite rangers complained about visitors illegally parking in pristine meadows, along rivers, on roadsides, and in open fields. Tow trucks were called in from Merced and Fresno to remove illegally parked cars and campers that were causing gridlock. Field rangers received a flood of complaints about hippies dancing nude, smoking bongs, beating on drums, and allowing their dogs to run unleashed.

As the Fourth of July approached, the park superintendent, Robert L. Arnberger, asked rangers to tighten up law enforcement. Rumors of a spontaneous Grateful Dead rock concert extravaganza—a Woodstock West—and illegal fireworks serving as main attractions rattled police.[70] One conservative park goer from Houston, Texas, L. E. Curran, was at the park that Fourth of July and complained that "greasy-headed Spock spooks" had invaded Yosemite from the Bay Area like a gang of anarchists. Furious, he suggested "using tranquilizer guns and pellets on them."[71]

What happened next is known as the Stoneman Meadow Riots. On July 4, rangers accompanied by wranglers, maintenance employees, and even wildlife specialists stormed the open space with batons and megaphones, trying to scare away the youths. CBS News caught it on film. "Before my very eyes we watched these children stampeded, several being clubbed, and two thrown to the ground, handcuffed, and led off for jail," John Fisher, a physician and former Florida state senator, told President Nixon in an outraged open letter published in the *Sacramento Bee*. Deeply distraught, he told Nixon (and thereby the public) in detail how a ranger had physically mauled his own twenty-year-old daughter.[72]

The Yosemite rangers pushed the riotous partygoers from Stoneman Meadow to Camp 14 as a containment strategy. A number of tourists, particularly families with small children, were frightened by the mayhem and fled Yosemite, not wanting to be part of a law-and-order smackdown. Trouble had overtaken paradise. The *New*

York Times lamented that "the angst of our age arrived in the national parks" and stayed for the rest of the summer.[73]

A few of the young people threw rocks at the NPS rangers, calling them "park pigs." Having stood up to the authorities, the revelers went wild, igniting a bonfire in the road, rolling a police car over, and scuffling with members of law enforcement. Feeling shorthanded, the NPS called for police backup from the nearby communities of Madera, Merced, and Fresno. US marshals arrived. After two days of disturbances, law enforcement got Yosemite in hand, arresting 174 people, including 41 minors, mainly for alcohol- and drug-related matters. But there were also charges of assaulting a federal officer. Nobody died, but there were injuries.[74]

The battle for Yosemite continued that summer. NPS employees, for the most part, blamed the young people for being reckless and arrogant. The *Berkeley Barb* called for a "10,000 person freak army" to descend on Yosemite on Memorial Day weekend. That weekend, some rangers at the entrance to Yosemite refused to let hippie types enter the premises, a clearly illegal measure. Nixon's Interior Department had a public relations issue on its hands. Taxpayers were used to thinking of park rangers as kindhearted naturalists, not police carrying pistols and handcuffs. In 1971, the Stoneman Meadow Riots triggered the creation by the NPS chief, George Hartzog, of improved law enforcement training. Rangers in Yosemite and elsewhere were told to toughen policing. At Yosemite, the superintendent's office adopted a zero-tolerance policy in an effort to end the criminal activity in the park.

That August, just weeks after the Stoneman Meadow Riots, *Time* covered "The Rise of Anti-Ecology."[75] The article pointed out that the oil and gas lobby loathed the Earth Day agenda. It also took to task the white upper-middle-class dominance of the environmental movement. The Brower-versus-Wayburn feud at the Sierra Club was nothing more than two elite white men in a power struggle. Many African American civil rights groups were put off by the outdoorsy groups concerned about Yosemite and the Grand Canyon but not about systemic racism and poverty in crime-ridden cities such as Oakland, California, and Newark, New Jersey.

Truth be told, the Nixon White House did care about poverty, racism, crime, environmental justice—certainly Ehrlichman and Moynihan did. Hoping to score points with the green lobby for establishing the EPA that summer, Nixon invited leaders from the Nature Conservancy, the Wilderness Society, and other "green" nonprofits to a White House discussion on the environment. The president started the conversation by stating, "All politics is a fad. Your fad is going right now. Get what you can, and here's what I can get for you." After Nixon spoke about his plans to expand the National Park Service, it was Sierra Club president Phillip Berry's turn to talk. Bitter over Cambodia, full of pique and indignation, Berry rudely snapped, "I don't believe what you believe in." For five minutes Nixon and Berry squabbled with malice in their voices. Fed up with being disrespected, Nixon abruptly left the room.[76]

When the EPA opened its doors on December 4, 1970, its first director was the thirty-eight-year-old William Ruckelshaus. Born on July 24, 1932, to a family of Indianapolis lawyers, he graduated from Princeton University, followed by a law degree in 1960 at Harvard Law School. Returning to Indiana, he was appointed deputy state attorney general and was assigned to the State Board of Health, working as counsel to the Indiana Stream Pollution Control Board. It was in that capacity that he gained his first major environmental experience.[77] In the coming years, the moderate Republican would win a seat in the Indiana House of Representatives. But he lost a US Senate run to Birch Bayh, a Democrat, in 1968. Everybody in Indiana knew that Ruckelshaus was a genius administrator but only a flawed retail politician. From the GOP talent grapevine, Nixon learned that Ruckelshaus would be the platinum-plated lawyer to take on the daunting job of being EPA's first administrator.

Ruckelshaus's core mission at EPA was to pull together, like a bureaucratic herder, the myriad efforts of the various agencies overseeing the United States' environmental interests and federal agendas. Nuclear waste, DDT, corporate polluters, insect plagues, dirty water, wildfires, public health, and economic development were all placed in his wheelhouse. As the founding administrator of EPA, he established the agency's mission and outlined the pecking order of environmental priorities. The agency's objective, made clear in Title I of NEPA, was to encourage productive and enjoyable harmony between US citizens and their environment, and to preserve the natural world for future generations. The key mechanism for initiating an EPA recommendation for any future project would be the submission of an Environmental Impact Statement (EIS). After a scientific review of the EIS of, say, a dam or an interstate highway, the EPA would decide on the effects on the natural world of the proposed program and suggest any modifications necessary to reduce potential environmental damage. The EIS was the handiwork of the political scientist Lynton Keith Caldwell, lawyer Bill Van Ness, and Senator Jackson.[78]

The initial issues that Ruckelshaus tackled were organizing protocols for the new agency: finding solutions for the United States' most polluted cities, such as Gary, Indiana, and Detroit, Michigan; curbing abuses of corporations disregarding laws; enforcing emissions standards for six pollutants and automobiles; enforcing air quality standards, which were applicable to all fifty states; and banning DDT as a biocide. Because America's environmental crisis respected no local or state boundaries, the EPA under Ruckelshaus was determined to play an activist role. With *Silent Spring* and *The Quiet Crisis* as catalysts, Ruckelshaus sent a broad message from day one that the EPA would set standards and exercise bold leadership. When hiring people at EPA, he valued their judgment and sense of humor as well as their résumé. In order to establish an honest and collegial workplace, he pledged that the EPA would be a "fishbowl" of transparency.

Even though Nixon admired captains of industry, he was willing to burst open and shake down the multimillionaire bad guys if they tried to circumvent EPA enforcement. "I've had an awful lot of jobs in my lifetime, and in moving from one to another, have had the opportunity to think about what makes them worthwhile," he reflected in a 1993 oral history. "I've concluded there are four important criteria: interest, excitement, challenge, and fulfillment. I've never worked anywhere where I could find all four to quite the same extent as the EPA. . . . At EPA, you work for a cause that is beyond self-interest and larger than the goods people normally pursue. You're not there for the money; you're there for something beyond yourself."[79]

Five days after starting at EPA, Ruckelshaus gave the keynote address at the Second International Clean Air Congress. That was followed by a scolding address on December 11 to the mayors of

Cleveland, Detroit, and Atlanta, giving each city six months—that was it—to fully comply with the Nixon administration's water pollution standards or face legal proceedings. The aggressiveness of Ruckelshaus, who served as first administrator until April 1973, was something to behold. Soon the EPA published a water analysis that documented that 85 percent of US water pollution problems were instigated by point sources of pollution (such as sewage treatment plants or industrial discharges) and only 15 percent by nonpoint source pollution (runoff from farms and city streets and sewer overflows). Never kicking the can or obfuscating the truth, Ruckelshaus was an old-fashioned midwesterner who believed in the ironclad virtues of constitutional law. The duty of the EPA, he would say, was informing an engaged populace about polluters.[80]

As A. James Barnes, one of his assistants, recalled years later, "Bill used to describe the process of getting the new agency up and running as being like trying to perform an appendectomy on yourself while running a 100-yard dash."[81] Ruckelshaus's new EPA was the last bastion of trust in environmental action on a significant national scale. Companies couldn't be trusted, and neither could states or other, even smaller, governmental entities. To Ruckelshaus, EPA had to be the last resort, without cynicism or cronyism, to prioritize healthy air, wildlife abundance, and uncontaminated waterways. As a Harvard-trained lawyer, he knew that for the EPA to work, enforcement would have to have tiger-sharp teeth.[82]

V

President Nixon had done a remarkable job of living up to his environmental State of the Union address. When NOAA opened its doors on October 2, it was clear to Scoop Jackson that the president had cleverly let the Democrats do all the work while he seized credit for the results. Yet as the midterm election of 1970 neared, Nixon was outflanked by Gaylord Nelson and Denis Hayes, who had organized a movement to unseat the "Dirty Dozen" (the twelve members of Congress with the worst environmental policy records). That November 4, the movement removed seven of the anti-greens, thereby demonstrating that the Silent Spring revolution still had political clout in America. "The Dirty Dozen campaign grew out of Earth Day," Denis Hayes recalled. "It was our clearest victory yet."[83] Shortly after election day, as expected, Nixon replaced Hickel as interior secretary with the affable Rogers Morton.

On paper, Morton seemed like a wise choice. A so-called Brooks Brothers Republican, Morton had been elected to the House of Representatives from Maryland in 1962, marketing himself as a Chesapeake Bay conservationist. Stewart Udall was his close friend. In 1969, Nixon chose the loyal Morton to serve as chairman of the Republican National Committee. The dismissal of Hickel allowed Morton, who was easily confirmed, to become the only secretary of the interior from the East Coast in the twentieth century. "The questionable aspect of the coming appointment is Mr. Morton's almost total lack of involvement in the great environmental movement of the past few years," the New York Times editorialized. "For three terms he sat on the House Committee on the Interior without making a noticeable impression on conservationists—or on the environment, for that matter—and then shifted to another committee."[84]

The Times was right to be suspicious of Morton. Whereas Hickel was known for directness, openness, and frontier independence, Morton was a GOP dandy in league with the oil and gas world. The Interior Department that Morton had inherited was vastly diminished compared to the Udall years. Three new agencies—EPA, CEQ, and NOAA—had usurped some of the power of Interior (which remained responsible for the management of federal public lands). Refusing to speak out on environmental issues, as Udall and Hickel had done, Morton was, in the end, a kinder, gentler version of Eisenhower's first interior secretary, Douglas McKay.

While Udall celebrated national parks, and Hickel was a hunting-fishing conservationist, Morton favored employees at Interior working with the Bureau of Mines, the Office of Coal Reserves, and the Office of Oil and Gas. Morton was one of the cabal of Nixon administration figures in 1971 that sought to convince Republicans that the environment was an electoral loser. His greatest accomplishment at Interior was overseeing the construction of the Trans-Alaska Pipeline System and promoting offshore drilling in the mid-Atlantic coast states of

New York, New Jersey, Maryland, and Virginia (which failed). "Rogers was dapper, warm, and nice to be around," Train recalled. "He called himself a conservationist, but he was really an energy man."[85]

For two years, various tribes demanded that the Nixon administration return certain natural wonders owned by the US government because they were sacred religious places. As an olive branch of sorts, after the midterm election, Nixon agreed to return Blue Lake and 48,000 acres of land in northern New Mexico to the Taos Pueblo Indians. This tribe had been lobbying to get Blue Lake—located high on Wheeler Peak in the Sangre de Cristo Mountains—returned from the Forest Service since the 1950s, when Severino Martinez (governor of Taos Pueblo) declared that the sacred spot belonged to the Taos Pueblo tribe: "Blue Lake is the most important of all our shrines because it is part of our life, it is our Indian church, we go there for good reason, like any other people would go to their denomination and like a shrine in Italy where the capital of the Roman Catholics worship is different: people go visit and give their humble words to God in any language that they speak. It is the same principle at the Blue Lake, we go over there and talk to our Great Spirit in our own language and talk to Nature and what is going to grow, and ask God Almighty, like anyone else would do."[86]

While the Kennedy and Johnson administrations punted on the Blue Lake issue, Nixon, to the surprise of the press, sided with the Taos Pueblo. On December 15, the president signed into effect Public Law 91-550, approved in bipartisan fashion by the US Congress. "This is a bill that represents justice, because in 1906 an injustice was done in which land involved in this bill, 48,000 acres, was taken from the Indians involved, the Taos Pueblo Indians," Nixon said. "The Congress of the United States now returns that land to whom it belongs."[87] This edict represented a dramatic change in federal Indian policy, inaugurating a new era of self-determination and the end of the trustee relationship between the US government and the Native American tribes. Blue Lake was no longer under the jurisdiction of the Carson National Forest but the autonomous government of the Taos Pueblo.

The Blue Lake win meant that First People might be able to reclaim natural wonders under the auspices that these sacred landscapes were part of their religious, cultural, and legal rights. In mid-August

1971, around one thousand people celebrated the return of their ancient Blue Lake ground in Taos Pueblo by holding prayer sessions, holy dances, buffalo meat feasts, hot pepper stew, and commemorative addresses. It was a cause of widespread jubilation in Indian country nationwide.[88] One Taos Pueblo man gestured to the nearby San Juan River and said, "The water in this river comes from Blue Lake, our ancestors came out of Blue Lake, long ago. Blue Lake nourishes everything. It is the source of our wisdom, of our life. . . . Do you understand?"[89]

On December 31, 1970, Nixon signed the Clean Air Act Amendments into law, eleven months after his State of the Union Address, when he had said, "We still think of air as free. But clean air is not free." Taking advantage of the fact that Ralph Nader was feuding with Ed Muskie, the chair of the Senate Subcommittee on Air and Water Pollution, for arguing that the culprit of environmental degradation was one of individual consumption rather than Big Industry pollution, Nixon found his opening. While Muskie worked on clean air legislation, Nixon took the high road, saying that he was willing for the federal government to be responsible for target cleanup dates and strict regulations. NEPA, EPA, and the Clean Air Act of 1970 meant, as the historian Robert Gottlieb put it, "a new environmental presence in Washington and providing the first block in the construction of a new policy framework."[90] The Clean Air Act soon reaped impressive results. When Nixon signed the amendments into law, US vehicular transportation emitted 180,000 tons of lead into the atmosphere. Twenty years later, emissions had decreased by 99 percent to 1,600 tons.[91]

The Clean Air Act of 1970 included standard-setting mechanisms, an increased federal role in the development of control technologies, and impressive target dates for eradicating or vastly reducing emissions or discharges, especially more visible or treatable forms of pollution such as sulfur dioxide emissions. Surpassing its namesake of 1963 by a long shot, the act set tough standards on stationary and motor vehicle emissions and gave the EPA the authority to oversee a comprehensive program to enforce them.

What gave the act fangs was it provided provisions for citizen suits. It also established four innovative regulatory programs: the National

Ambient Air Quality Standards (with the EPA designating the acceptable levels of carbon monoxide, nitrogen dioxide, sulfur dioxide, particulate matter, particulate matter hydrocarbons, and photochemical antioxidants); State Implementation Plans (SIP); New Source Performance Standards (NSPS); and National Emissions Standards for Hazardous Air Pollutants (NESHAP).

While the 1970 Clean Air Act was often referred to as the "Muskie Act," it was, in truth, the "Ruckelshaus Plan." With great fortitude, Ruckelshaus sought to reduce air pollution by 90 percent under the act's provisions. Because he was empowered to enforce the Clean Air Act, the toughest piece of environmental legislation ever passed by Congress, the Sierra Club turned to him to be the cleanup cop. The club took out a full-page ad in the *Los Angeles Times* begging Ruckelshaus to "Hang Tough" and not be bullied by companies that dumped poisonous waste into the atmosphere.[92]

Once Nixon signed the Clean Air Act, he sought public credit for himself, which was fair enough. There was one ironclad White House rule that Ehrlichman implemented once again: at no point should Muskie be allowed to "hog the cameras" and take credit for EPA or the Clean Air Act.[93] The Maine senator was banned by Nixon from entering 1600 Pennsylvania Avenue. Muskie, whose staff had worked hard on the Clean Air Act of 1970, wasn't even invited to the White House signing ceremony in early 1971. It pleased the sadistic Nixon to stiff the Maine senator he lived to loathe.

After three years of protesting, Bill Douglas finally got the noise-control regulation he craved from the Nixon administration. In 1970, Congress enacted the Occupational Safety and Health Act (OSHA), which, among other things, governed noise in all factories owned by companies engaged in interstate commerce. The loudest continuous noise a worker could be exposed to during an eight-hour day was ninety decibels. But that was only a partial victory for Douglas. His ire now was at the more than 1 million motorcycles, 700,000 pickup trucks, 600,000 four-wheel-drive cars, 80,000 snowmobiles, and 50,000 dune buggies "roaring and churning" their way onto BLM lands alone.

"These motorized visitors are mostly marauders who are in the wild areas not because they love nature but because these unpatrolled areas give them a place to rampage beyond the reach of law," he wrote. "Vandalism, misdemeanors, and major crimes are spawned by the adventurers, and the silence of the woods is shattered and the wilderness pillaged."[94]

Besides fighting noise pollution, Douglas continued to rally against utility companies in Texas, with a few exceptions, because they exercised the power of eminent domain. It was yet another crusade. Pipelines, railroad trestles, and transmission lines were constructed in Texas by "modern Ahabs" without any environmental consideration whatsoever. In *Farewell to Texas: A Vanishing Wilderness*, Douglas pilloried the Army Corps of Engineers for drowning rich bottomlands and erasing "free-flowing rivers rich in archeology, history, beauty, and adventure." In equally biting prose, he eviscerated stockmen for overgrazing, lumber barons who pillage acreage, and vandals who "poison historic trees in order to sabotage giant oaks and old-growth areas in Dallas and Houston." Then there were Texas poachers who shot bald eagles, red-headed woodpeckers, and alligators for kicks with no legal repercussions. Douglas called for yet another amendment to the Endangered Species Act, one that would protect animals' habitats and marine mammals.[95]

For both Brower and Douglas, unnecessary dams in Texas, Colorado, and elsewhere remained the devil spawn of the industrial-military technological complex in the age of Nixon. To Brower and Douglas, free-flowing rivers remained the "ultimate metaphor of existence." Recruiting young people for Friends of the Earth at college campuses, Brower declared, "I hate all dams, large and small." Once, when asked why environmentalists were opposed to all hydroelectric dam proposals, he had a quick response: "If you are against something, you are for something. If you are against a dam, you are for a river."[96] And being "for" a river meant keeping streams alive with trout, salmon, and other aquatic life.

Looking back on Nixon's first three years in the White House, Russell Train pointed out that Nixon, as he sought reelection in 1972, could genuinely "point to the passage into law of major legislative proposals of his administration, including: air quality legislation, strengthened water quality and pesticide control legislation, new authorities to control ocean dumping."[97] Train pointed out that federal funding for environmental endeavors had increased fourfold from LBJ to Nixon and continued to grow in 1971. "In water quality alone, federal funding had grown fifteen-fold." What Train had said was true: Nixon had earned bragging rights as an environmental president. Instead of grappling with individual bureaucratic fiefdoms, the heads of CEQ and EPA afforded the future president the ability to streamline environmental policy.

While Nixon basked in the aura of his NEPA, EPA, NOAA, and Clean Water Act accomplishments of 1970, Black congressmen homed in on urban renewal, environmental justice, systemic racism, and urban pollution issues with the creation of the Congressional Black Caucus (CBC) in early 1971.[98] This group of thirteen African American representatives, all Democrats—Shirley Chisholm of New York, William Clay, Sr., of Missouri, George Collins of Illinois, John Conyers, Jr., of Michigan, Ron Dellums of California, Charles Diggs, Jr., of Michigan, Walter Fauntroy of the District of Columbia, Augustus Hawkins of California, Ralph Metcalfe of Illinois, Parren Mitchell of Maryland, Robert Nix, Sr., of Pennsylvania, Charles Rangel of New York, and Louis Stokes of Ohio—began studying how chemical dumps, dirty water, and poisoned air were disproportionately affecting people of color. It was the largest class of African Americans ever to serve simultaneously in congressional history.

Just as the CBC was gelling as a formal organization, President Nixon gifted it a publicity windfall by refusing to meet with its members. Furious at the snub, the CBC boycotted Nixon's January 22, 1971, State of the Union address.[99] Backed into a corner and vilified in the press for refusing to meet the African American leaders, Nixon capitulated. On March 25 he invited the thirteen CBC lawmakers to the White House. During the meeting, the CBC presented Nixon with sixty-one urgent recommendations to promote affirmative action, eradicate racism, and improve living conditions in Black neighborhoods.[100]

The most vocal environmental justice leader of the CBC was Representative Ron Dellums, whose California 7th District included residents in Berkeley and Oakland. Dellums understood the lineage from civil rights to the anti-nuclear efforts of SANE and Women for Peace, which had segued into the anti–Vietnam War demonstrations on college campuses across the United States. As the John the Baptist

Fifteen members of the Congressional Black Caucus pose on the steps of the US Capitol in 1977. *Front row, from left to right*: Barbara Jordan of Texas, Robert Nix, Sr., of Pennsylvania, Ralph Metcalfe of Illinois, Cardiss Collins of Illinois, Parren Mitchell of Maryland, Gus Hawkins of California, Shirley Chisholm of New York. *Middle row, left to right*: John Conyers, Jr., of Michigan, Charles Rangel of New York, Harold Ford, Sr., of Tennessee, Yvonne Brathwaite Burke of California, Walter Fauntroy of the District of Columbia. *Back row, left to right*: Ronald Dellums of California, Louis Stokes of Ohio, and Charles C. Diggs, Jr., of Michigan.

figure of the burgeoning environmental justice movement, Dellums knew that the separation of race and environment was an artificial construct. The shared foes, the evil demons that needed to be slain, were corporate greed, government indifference, and White privilege. To Dellums, an unabashed socialist, the philistines who knowingly exploited and destroyed the natural environment were the same people who abused racial and ethnic minorities and had imposed the military draft on young men to fight in an immoral war in Southeast Asia that was slaughtering citizens with B-52 bombers, Agent Orange, and other lethal technological equipment.

More than the other CBC leaders, Dellums thought that David Brower of Friends of the Earth was a planetary leader who not only sought a cleaner San Francisco Bay but wanted poor people to have equal access to recreational activities. When Brower garnered media attention for stopping the Grand Canyon dams and advocating for

North Cascade National Park in the late 1960s, he was, to Dellums, a fellow progressive traveler in the noble fight against racial injustice and equitable distribution of wealth. In coming years, Dellums would write the Nobel Committee in Oslo that his Berkeley constituent David Brower was deserving of its high honor. "I nominate David R. Brower, the world's leading voice to stop the war against the planet earth, for the Nobel Peace Prize," Dellums wrote, full of constituent pride. He considered Brower a "personal hero."[101]

When Nixon announced on April 30, 1971, the deployment of US ground troops to Cambodia, Dellums and Brower shouted foul. Friends of the Earth led the charge. Brower organized a group of top conservation leaders to call on Congress to demand a quick withdrawal from Cambodia and the ending of the Vietnam War. Dellums circulated the official demand to offices on Capitol Hill. In a May 14 "Open Letter to President Nixon," the Browerites condemned the Southeast Asian war and "an ecological disaster that ultimately destroys both the land and the people it purports to protect." The letter denounced chemical defoliation, saturation bombing, and the systematic devastation of agriculture. Signatories inscribed that US foreign policy had to be tethered to an ecological worldview that was planetary in nature. "A planet *is* at stake," they warned, "and to save it we must begin by giving up the policy of destruction that leads with relentless logic to a My Lai—and to a widening of war in the interest of 'shortening it.'"[102]

On Earth Day 1971, Barry Commoner served as ecobuddy to CBS anchorman Walter Cronkite, commenting that the incineration of plastics caused widespread air pollution. Commoner was educating the public about the way discarded plastics had tragically turned America's rivers and lakes into ghoulish trash heaps. To Commoner the very survival of the human race depended on saving the "ecological fabric" of Earth from unnatural synthetic products, fossil fuels, and nuclear radiation.[103]

That fall of 1971, Barry Commoner's profound *The Closing Circle: Nature, Man, and Technology* was published to enthusiastic reviews. Arguing that humans were the only species capable of producing toxic materials absent from nature, he blamed chemical manufacturers for injecting the planet's first known "intrusion into an ecosystem of a substance wholly foreign to it." According to Commoner, whose celebrity had increased since his *Time* magazine cover story, the Earth had no defense mechanism to break down "literally indestructible" plastics. With scientific exactness, he wrote how plastics were being dumped in the ocean, thereby decimating marine life. To Commoner, ecological survival didn't mean the shelving of technology; it did demand that creators of technology develop scientific protocols for new innovations that were copacetic to the natural world that technological power altered. People would have to be cured of what Wendell Berry called "the disease of overconsumption."[104]

On the second Earth Day, Nixon hitched his wagon to the nonprofit Keep America Beautiful, which was airing TV spots showing the "Crying Indian," aka Iron Eyes Cody, shedding a tear over the stinking debris and garbage strewn across the United States' treasured landscapes. Even though Iron Eyes Cody had been born Espera Oscar de Corti in 1904 in rural southwestern Louisiana to parents who had emigrated from Sicily just two years prior to his birth, the Hollywood actor became the ubiquitous "Native American" symbol of combating pollution. The hoodwinked public didn't know that Iron Eyes Cody had no Indian blood. The famous one-minute PSA showcased the actor, black braided and buckskinned, a single feather in his bonnet, paddling down a litter-strewn river in a birch-bark canoe enveloped by a scrim of industrial smog and mounds of disgusting litter. Having appeared in more than a hundred western movies, a favorite sidekick of the actors John Wayne and Errol Flynn, Iron Eyes Cody had an impressive Hollywood résumé. On TV series such as *Bonanza* and *Gunsmoke*, he was credited as "Indian Chief" or "Indian." Though Sicilian, he came to embody the Hollywood norm of what a Native American male figure *should* look like: dark skinned with a sun-wrinkled face, long hair, a suspicious gaze, and overwhelmed with ecological concern.

At one point in the famous Keep America Beautiful public service announcement, Iron Eyes Cody dramatically hauled his canoe ashore onto a desecrated riverbank. Smog had yellowed the Los Angeles sky. Suddenly, in the TV ad, an automobile driver flings a bag of garbage out the window. It breaks wide open like a piñata. An array of soiled fast-food wrappers and household debris falls on the old Indian's moccasins. The camera pans to his mournful face as a slow-moving pear-shaped tear rolls down his face.

Iron Eyes Cody became a TV advertising icon in the Keep America Beautiful campaign against litter. Unbeknownst to most viewers, his real name was Espera "Oscar" DeCorti, a Sicilian American from Louisiana.

The TV-watching public around Earth Day 1971 wondered if he was really crying. Or was it fake tears? The fact that his long black braids were a wig and his dark complexion was created with makeup wasn't known at the time. On the TV spot, a basso profundo voice says, "Some people have a deep, abiding respect for the natural beauty that was once this country, and some people don't."[105]

"The tear was real," DeCorti later insisted. "I would look up at the sun, and when I looked back down the tears would stream down. Then I would wipe away the tear from one cheek. When I'd turn to the camera, what it would capture was the other tear."[106]

The ubiquitous TV spot appeared with the words GET INVOLVED NOW. POLLUTION HURTS ALL OF US posted as it ended. For millions of Americans, the "Crying Indian" became the ultimate symbol of environmental idealism in the Nixon years.[107] The spot won film prizes and is considered one of the most effective PSAs of all time. Suddenly a mega-celebrity of sorts, Iron Eyes Cody was instantly recognizable in airports and grocery stores. Fans crowded him for photos and autographs. He was honored with a Hollywood Walk of Fame star in 1971. It's been estimated that his "Crying Indian" face was seen on so many billboards, posters, and magazine ads that he became the most recognizable "Native American" of the twentieth century. Upon closer scrutiny, however, the fact that the long black braids were a wig and his dark complexion was created with makeup wasn't the only fraudulent part of the PSA in which a voice says, "Some people have a deep abiding respect for the natural beauty that was once this country, and some people don't."[108] Those deceptive TV spots and posters weren't promoting the "green environmental movement" message. They were on behalf of Nixon's Keep America Beautiful Inc.

Just as the Iron Eyes Cody commercial aired, Oregon enacted a container deposit act as a litter control measure. This was the first of such legislation in the United States. Bottles constituted 40 percent of nationwide litter. In fact, beer and soda companies were responsible for 173 million bottles and 263 million cans annually in Oregon. After 1971, the state required a five-cent deposit on all bottles. When they were returned, customers got their money back. Coca-Cola didn't like the whole bottle deposit measure, seeing it as hurting sales. The Atlanta-based soda company fought back by manufacturing more cans.[109]

Just how determined Nixon was to be perceived as an environmentally concerned president was evident on a chilly summer morning in 1971. In order to promote his "Legacy of Parks" initiative, Nixon vacationed with his wife, Pat, in the Grand Teton National Park. Soon Nixon would ask Congress to increase the HUD budget by $15 million to $20 million to acquire more open spaces for outdoor recreation purposes.[110] To Nixon's credit, in 1971 he introduced "Legacy of Parks" by converting closed federal lands to public spaces for outdoor recreation. The effort transformed over eighty thousand acres of land into 642 parks, something that afforded access, as Nixon put it, to the "casual tourist and avid outdoorsman" alike.[111]

One morning in the Tetons, Nixon emerged from his Lake Jackson cabin to take a boat ride to be photographed with the snow-covered Grand Teton mountains as a backdrop. Laurance Rockefeller and Russell Train had convinced the president to see for himself the prettiest summits in the National Park System. Once Nixon had arrived in Jackson Hole, he had determined that the signature 13,770-foot peaks would be the ideal backdrop for an outdoors photo op. After conferring with Secretary of the Interior Rogers Morton, who had accompanied him on the trip, he agreed to get up early to be captured in the golden light. The problem was that the morning of August 19 that Nixon chose saw temperatures dramatically drop. Because Nixon had no proper outdoors wardrobe, Morton's deputy at Interior, Nathaniel

Reed, was tasked with finding him winter clothing: a coat, flannel shirt, gloves, scarf, and hat.

Once those were procured, Morton and Reed went to the Colter Bay Marina with the garments. The Secret Service took the clothing articles and told Reed to lie down on the boat's bottom with the gear. Eventually, Nixon arrived at the dock wearing a summer suit, dark tie, and white dress shirt with no undershirt. Determined not to look like a sissy, he posed for photos in his suit. The photographers clicked away. Then it was anchors aweigh as two press boats followed on each side. Accompanied by his twenty-three-year-old daughter, Julie, Nixon ordered Reed to stay hidden on the bottom until they were well away from the cameras; at that time, he would covertly put on the winter gear. As Nixon and Morton stood looking at the Tetons, Nixon loudly exclaimed, "It is beautiful!"

And in an irritated whisper, he said to Reed, "Nathaniel, it is cold as hell." With the photographers still clicking away, Nixon knew he should have dressed warmly, and he was angry at Reed for not making him do so. Four or five minutes went by with Nixon waving and appearing to bask in nature's grandeur when he snapped, "I am freezing my ass off. Get me the hell out of here." Suddenly a Secret Service agent got onto a microphone and firmly announced, "The president has a very important telephone call. The photo session is now over."[112]

Though the Lake Jackson episode was comical, it went a long way to demonstrate how determined Nixon was to be perceived as a nature lover. To Ehrlichman, the event, part of Nixon's "Legacy of Parks" initiative, was a visual way to remind voters that Nixon was a Theodore Roosevelt conservationist at heart. By the end of 1971, Nixon led Ed Muskie, his likely rival, by around 12 percent in the polls. Frustrated that Nixon was stealing the environmental issues, Muskie resorted to snarkily quipping at fundraisers, "It appears that this administration has undergone an environmental metamorphosis, emerging from the cocoon not as a butterfly but as a moth."[113]

CHAPTER 31

Nixon's Environmental Activism of 1972

THE GREAT LAKES PROTECTION, THE DDT BAN, AND THE STOCKHOLM CONFERENCE

William Ruckelshaus was the first administrator of the Environmental Protection Agency and gave it tremendous authority. Under his leadership, DDT was banned from usage in the United States in 1972. An Indianapolis lawyer, he was unyielding on implementing NEPA's environmental impact statement laws.

I

In early 1972, President Nixon summoned his EPA administrator, William Ruckelshaus, for a private one-on-one Oval Office chat. Not realizing that Nixon thought 96 percent of the civil service was out to destroy him, Ruckelshaus mistakenly assumed they would strategize about the pending DDT ban and getting the lead out of gasoline. After a few minutes of niceties, however, Nixon cut to the chase: he wanted the EPA to stop being so cozy with the green lobby. Nixon's high-powered friends from General Motors, Ethyl Gasoline Corporation, and the Chamber of Commerce had warned the president that the EPA had turned anti-business. "You better watch out for those crazy enviros, Bill!" the president said. "They're a bunch of commie pinko queers!"[1]

Nixon thought the "fad" of environmental politics had reached its zenith in early 1972. As White House adviser H. R. Haldeman had written in his diary, "On Domestic Policy P is deeply troubled that we are being sucked in too much on welfare, envir., and consumerism." The president's persistent cynicism about the environmental movement was in blinding contrast to his administration's precedent-setting accomplishments in the field. Nixon, for example, would approve certain EPA regulations, then turn up the air conditioner in San Clemente, Washington, DC, and Camp David so as to make rooms chilly enough that he had to build a fire to stay warm. Overall, he tolerated Democratic politicians such as Scoop Jackson and John Dingell because they were engaged in conservation politics. But he disdained grassroots environmentalists such as George McGovern and Gaylord Nelson, whom he claimed fundraised by exaggerating the death of the Great Lakes. To Nixon, these senators were left-wing grandstanders who preferred porpoises to people. Yet when Nixon learned how Neil McElroy, the CEO of Procter and Gamble, wouldn't praise his administration's environmental record in the Great Lakes for fear of offending Ed Muskie and the Democratic majority over phosphates, the president turned vindictive. Haldeman reported that the "President is thoroughly disgusted and told me to have [John] Whitaker screw P and G."[2]

The big pending environmental legislation of 1972 was the Clean Water bill (which *Congressional Quarterly* called "the most comprehensive and expensive environmental legislation in the nation's history"). That was no small matter for President Nixon. Poll after poll revealed that people wanted US waters to be fishable and swimmable and to have zero water pollution discharge by the mid-1980s. Furthermore, the sentiment was widespread that companies should be prohibited from discharging toxic amounts of pollutants into US waterways. Yet the expensive $24 billion Clean Water bill was a pill Nixon wasn't keen to swallow; he saw it as a Democratic trap. For one thing, it was Nixon's nemesis Ed Muskie's pet legislation. In early 1972, Nixon mistakenly thought that Muskie, "Mr. Air Pollution Control," would be the Democratic nominee for president. Why give the audacious Muskie a Clean Water Act feather in his cap, Nixon reasoned, when the ambitious Maine senator vilified his administration in the press weekly as not implementing NEPA properly nor policing industrial discharges?

Another Nixon concern was that the Clean Water bill, which had passed the Senate in November 1971, reeked of liberal overkill. Powerful corporations and taxpayers, Nixon worried, would feel the brunt of the $24 billion bill. Certain big businesses had spearheaded a counterattack against the draconian federal water pollution regulations that year. Chemical industries and agribusiness were vital fundraising constituencies for the GOP, and Nixon knew he needed their financial help to win reelection in 1972. Extraction industries, in particular, loathed Muskie's hyperregulatory bill.

Backing the Muskie bill was Brower's Friends of the Earth, who partnered with twenty-five other public interest groups pushing Washington lawmakers to pass the legislation and attacked Nixon for overstating the price tag. The House publicly worked to slow the bill down.[3] Antiregulatory rhetoric, in fact, had become doctrine with conservative Republicans. Dutifully, Ruckelshaus had informed Nixon that Brower had a point: about half of US streams and rivers, 70 percent of lakes, ponds, and reservoirs, and 90 percent of surveyed ocean and coastal areas were violating EPA water-quality standards. Moreover, EPA data showed that the water impairment of America varied by source. Why punish all companies, Nixon shot back, because of a few bad pollution actors? He preferred the idea that individual states should decide on water quality standards to enact. However, it wasn't a plausible or sensible policy. What good did it do for Louisiana to usher in strict water pollution laws if Illinois was going to treat the shared Mississippi River as an industrial sewage canal? Fearing the political risk of antagonizing Big Agriculture, the oil and gas industries, and port communities by supporting the Clean Water bill, Nixon threatened to veto the legislation if the price tag remained astronomical. No matter what, he promised that the federal government would punish illegal water polluters on a case-by-case basis.

When it came to the Clean Water bill, Nixon was putting pragmatic politics that presidential election year ahead of environmental principles. His green team of Ruckelshaus, Ehrlichman, Whitaker, and Train pleaded with him to sign the bill. Young voters' influence had skyrocketed in 1971 after the passage of the Twenty-sixth Amendment, which had lowered the voting age from twenty-one to eighteen, instantly creating 11 million new voters, many of them interested in ecology.[4] Furthermore, the fact that the *Los Angeles Times* had hired three full-time environmental reporters led White House advisers to believe that land and water stewardship remained a top voter concern in California. Though California, with its 45 electoral votes, mattered mightily, Nixon calculated that by establishing the Golden Gate National Recreation Area in the San Francisco Bay area that fall and lobbying for the EPA to require the sale of unleaded low-octane gasoline at most of America's filling stations by 1974, he would receive the "green" reelection support of conservation-concerned Californians.[5]

Nixon was probably mistaken not to co-opt Muskie's farsighted Clean Water bill as he had done with the Clean Air Act of 1970. Step by step, since 1969, Nixon had grown into being an extraordinary environmental president. If he signed the Clean Water bill into law just before the November 4 election, he could honestly boast that his environmental record equaled those of Franklin Roosevelt and Lyndon Johnson. Once a modified version of the Clean Water bill passed the House in March 1972, a group of Republican senators and congressmen, joining Ruckelshaus and Train, begged him to take the high ground, to forget about Muskie and grab the glory of a Clean Water Act in an election year. The unbudgeable and fiscally prudent Nixon held firm that $24 billion was an unseemly blackmail attempt by Democratic liberals to punish corporations. Instead, he preferred to go after major corporate environmental polluters legally, to make an example of them. In Ohio, for example, the administration could be environmentally vigilant in holding Lake Erie polluters accountable. Before long, Nixon's attorney general, John Mitchell, filed federal lawsuits against twelve companies in northeastern Ohio for discharging dangerous quantities of cyanide into waterways near Cleveland.

Truth be told, in early 1972, Nixon wasn't worried that his veto of the Clean Water bill would tarnish his enviro-certifies. His historic visit to China from February 21 to 28, complete with photos of him dining with Chou En-lai and walking along the Great Wall, was international news. Nixon, with the help of National Security Advisor Henry Kissinger, had pulled off a major geopolitical realignment coup. Even the *New York Times* and *Washington Post* credited Nixon with the opening of China. On top of that, he signed an Anti–Ballistic Missile Treaty (ABM) with the Soviet Union that May, traveling to Moscow to negotiate with Soviet premier Leonid Brezhnev. It was the first time a US president had visited the Soviet Union since World War II. Nixon's China and Russia foreign policy gambits both were TV broadcast winners and reaped huge public approval for Nixon. The voters saw him in the spring of 1972 as a diplomat writ large, a global statesman, even though his expansion of the Vietnam War into Cambodia and Laos continued to stoke widespread protests throughout the United States.

For President Nixon, the opportunity to score a daily-double win in both US foreign policy and clean water environmental activism occurred on April 15, when he signed a joint agreement to start cleaning up the Great Lakes. The pact, as a prelude to Earth Week, was known as the Great Lakes Water Quality Agreement and cost the US taxpayers $3 billion over five years (the Canadians spent only a seventh of that amount). At the Ottawa signing ceremony Nixon sounded like a committed environmental leader determined to save the Great Lakes. While there were still issues related to the phosphate problem, no one could honestly say his administration wasn't trying to stop excessive industrial discharge from entering Lake Michigan and Lake Erie. While Ottawa didn't garner much press attention, it was an impressive moment in the realm of environmental diplomacy.[6] The Great Lakes were, as the *New York Times* put it, "America's ecological problem children," and Nixon was giving them the funding love they so desperately needed.[7]

II

Nixon was disappointed that his Great Lakes Water Quality Agreement received minimal press attention. To his mind, he was pioneering in the realm of environmental diplomacy. A disinterested Nixon treated Earth Week 1972, from April 17 to 22, with general indifference. Yet there were other environmental issues that snatched his attention that spring. Early in the year, Buffalo Creek in West Virginia and the Buffalo River in Arkansas garnered headlines of very different kinds. "It was strange that both Buffalo waterways coincided within days of each other," Ruckelshaus recalled. "The Buffalo River was a triumph of proper conservation stewardship, while Buffalo Creek demonstrated how brutal coal companies were when it came to clean waterways and workplace safety for miners. At heart, both Buffalo events, regardless of differences, were about dams."[8]

The Buffalo Creek flood occurred at 8:00 a.m. on Saturday, February 26, 1972, when the Pittston Coal Company's retaining dam wall, built to hold a huge coal-waste refuse pile, exploded open along the border between West Virginia and Kentucky. Most chilling of all, just days prior, government inspectors had said that the structure, which dammed the stream in Middle Fork Hollow in the mountains of West Virginia, was in fine shape. So the collapse took residents by utter surprise. Traveling at a speed of thirty miles per hour, the coal slurry rolled like a tsunami from the hill and devastated sixteen towns, leaving 125 people dead and thousands homeless. Most of those who died were women and children, crying for help, unable to struggle out of the heavy black water choked with swirling debris and splintered homes. The flood unleashed around 132 million gallons of black wastewater accumulated over fifteen years. The wall of water crested over thirty feet high upon a string of coal towns situated along the narrow, crowded valley of Buffalo Creek. "There have been many coal-mining disasters," Gerald M. Stern wrote in *The Buffalo Creek Disaster*. "But the Buffalo Creek disaster is unique. This time it was not the strong, working male coal miners who died in the mines. That has become so commonplace in coal mining as to be expected. No, this time it was the miners' defenseless wives and children, caught, unprepared for death, in their beds one Saturday morning."[9]

Pittston Coal Company, in its legal findings, used the alibi that the grave disaster had been an act of God. But the company had been grossly negligent, according to the ACLU report *Disaster on Buffalo Creek: A Citizens' Report on Criminal Negligence in a West Virginia Mining Community*.[10] The dam that broke was owned by the Buffalo Mining Company, a West Virginia corporation whose sole stockholder was Pittston Coal Company. From its Park Avenue offices, Pittston concocted the "act of God" defense to avoid class action legal suits. The tactic backfired. The citizens of West Virginia were livid. Robert O. Weedfall, the state's climatologist, responded immediately to the bogus defense: "'Act of God' is a legal term. There are other legal terms—terms like 'involuntary manslaughter' because of stupidity and criminal negligence."[11]

Many of the four thousand survivors of the Buffalo Creek disaster, those who had been able to extract themselves from the sludge and debris, had been exposed to toxic substances. Scientific studies of the effects of coal slurry on human cell tissues had recently revealed that black tar caused cancer to proliferate. As one medical report stated, "Chronic exposure to the metals found in coal slurry can damage virtually every part of the body. Health problems caused by these metals include intestinal lesions, neuropathy, kidney and liver failure, cancer, high blood pressure, brittle bones, miscarriages and birth defects among others."[12] Here was an Appalachian disaster in which a wealthy company had killed Americans and destroyed the infrastructure of an entire swathe of West Virginia. The term *environmental justice* wasn't yet in the American lexicon, but it was clear that fourteen Appalachian mining communities had been victimized by reckless and uncaring mining operations. Reports of a surviving "miracle baby" and corpses being found in ruins dominated radio and TV broadcasts for more than a week.

Even before the Buffalo Creek disaster, Pittston had had an abysmal record of ignoring government regulations. In the same year, federal inspectors found more than five thousand safety violations in Pittston's mines for which the Bureau of Mines assessed the company $1.3 million in fines. Though the state of West Virginia sued the company for $100 million for disaster and relief damages, Governor Arch Moore, a Republican, ended up settling for a minuscule $1 million.

The ACLU stepped in and conducted a devastating independent study of the catastrophe. Its final report documented gross negligence on the part of Pittston. "FIRST, the flood was not an act of God," the report began. "This terrible tragedy was caused by men, acting through corporations and governments."[13]

The Buffalo Creek tragedy, West Virginia's worst man-made disaster, was a painful story to follow closely. There were frenzied media reports about missing citizens, toxic water, and a makeshift morgue set up at the South Man Elementary School in the town of Man. The state police discovered bodies floating in the Guyandotte River, twenty-four miles downstream. Many of the surviving coal miners, desperate for work, held no grudges against the company but wanted financial compensation for the flood. Coal-gas explosions and black lung disease were curses of the trade. President Nixon was in Asia that February when the Buffalo Creek disaster unfolded but promised Governor Moore federal relief aid for West Virginia families.

With the Buffalo Creek disaster and Nixon's trip to China dominating media attention, the establishment of America's first "national river," the Buffalo River in Arkansas, became only feel-good back-page news. After years of intense lobbying by Dr. Neil Compton, painter Thomas Hart Benton, Bill Douglas, and scores of grassroots activists, the pristine 135 miles of the pale jade–colored lower Buffalo River was saved from dams and degradation. On March 1, 1972, President Nixon signed the bill creating the Buffalo National River, ending the recurring designs by the US Army Corps of Engineers to construct a dam on Arkansas's unblemished river. Interior Secretary Rogers Morton had chosen that date for Nixon because it was a hundred years to the day that President Ulysses S. Grant had signed the bill to establish Yellowstone National Park in 1872, America's first. At long last, the Buffalo National River became the fifth National Park Service area in Arkansas and arguably the most significant preservationist victory west of the Mississippi since the Everglades had opened as a national park in 1947.

That March, the Ozark Society celebrated the federal national river designation with picnics and boat trips along the protected Buffalo. Out-of-state tourists came to experience the towering limestone bluffs, two-hundred-foot-high Hemmed-in-Hollow Falls, the highest waterfall between the Appalachians and the Rockies, and other free-flowing rivers. Governor Orval Faubus, the unlikely segregationist turned conservationist, was honored by the Ozark Society for fighting on behalf of the Buffalo throughout the 1960s. But to Dr. Compton, the real hero was Senator J. William Fulbright, a Democrat from Arkansas. "The Ozark Society and its many thousands of friends extend to you our sincere appreciation to the senator who made it all possible for your efforts on behalf of the newly created Buffalo National River," Compton wrote Fulbright. "To you goes the distinction of being our first public official to recognize its value and to use the influence of your office to its ultimate establishment."[14]

What was so special about the Buffalo River preservation was that Arkansans themselves had awakened to the stunning beauty of their state. Rural citizens in the Ozark Mountains had fought to keep the Buffalo undammed and unpolluted for its entire length. Though the Buffalo National River was administered by the National Park Service, the US Forest Service also protected a million acres of the Ozark National Forest where the river began. Vast acres of Arkansas wilderness forests, including Ouachita National Forest, St. Francis National Forest, and Barkshed Recreation Area, were preserved. Ozark communities now understood what ecotourism, as promoted by Laurance Rockefeller, was all about: rural communities making money via a prosperous outdoors recreation industry instead of the extraction of national resources. Instantly, the Buffalo National River became one of the most popular attractions in Arkansas, averaging eight hundred thousand visitors a year.[15] A few years after the historic designation, the slogan on Arkansas license plates changed from "Land of Opportunity" to "The Natural State."[16]

There were many southern Democrats, usually those sympathetic to George Wallace, who thought the Nixon administration was a "bunch of ignoramuses" when it came to environmentalism. In their minds, Nixon should have made Canada pay more to clean up the Great Lakes and dammed the Buffalo River—and, more important, sided with the mine owners' "act of God" claim in West Virginia. What galled these southern Democrats most, though, was NEPA. As the Democratic representative from Alabama, Robert Jones saw it, the Tennessee-Tombigbee Waterway was a $386 million project for a 253-mile canal from Mobile to the Tennessee border, approved by Congress and endorsed by the Nixon White House. Yet because Ruckelshaus didn't like that, the project used NEPA to stop the building via a court order. The EPA's rationale for the stoppage: the Corps of Engineers didn't submit a detailed environmental impact statement. If the Tennessee-Tombigbee Waterway Project could be stopped, Jones carped, then Ruckelshaus could stop any major federal infrastructure project in the Deep South.[17]

III

In early 1972, even though the NEPA and the EPA had come about in 1970, DDT had still not been banned. The tenth anniversary of *Silent Spring*'s publication was near, and there were scores of peer-reviewed articles filled with data demonstrating that DDT contamination was wiping out wildlife. Brown pelicans in California laid eggs that shattered and produced near zero chicks. There was no doubt about it: the sight of dead and dying birds turned the public against DDT. The bald eagle population was still in sharp decline because of DDT poisoning. No creature was more cherished in the American imagination than that enduring symbol of national greatness. Yet, as Rachel Carson had warned, DDT and other chemicals used in landscaping and on crops made the bird's extinction a real possibility. The bald eagle eats mainly fish. DDT carried by runoff in waterways poisoned fish and, in turn, some predators that ate them. In that way, eagles were doubly threatened: by ingesting DDT in their food and by fish stocks shrinking to levels too low to sustain the eagle populations as DDT caused salmon and walleye in North America to die off. Other bird populations whose diets contained toxic levels of DDT were also dwindling: the peregrine falcon, the California condor, the brown pelican, and the osprey were all in trouble. By 1972, a state was lucky if it had even one breeding pair of osprey around.

From the publication of *Silent Spring* to the DDT wars on Long Island to the founding of the Natural Resources Defense Council in 1970, the grassroots effort to ban DDT never ceased. Environmental groups doggedly petitioned the Department of Agriculture, which pared down the list of allowed uses of DDT but stopped well above zero. When the EPA opened for business in December 1970, it was given most of the responsibility for regulating pesticides, including DDT. The environmental groups pinned their hopes on Ruckelshaus, the administrator, but instead of implementing an outright ban, he ordered a temporary one and launched yet another study of that most studied chemical. When the results showed that the pesticide wasn't harmful enough to justify its discontinuation, environmentalists cried foul, pointing out that the researchers had been heavily drawn from the USDA and its corporate allies, parties known to favor DDT. Ruckelshaus rescinded the ban.

With that, Stewart Udall was inflamed, coauthoring an editorial that ran in papers all over the country. "Ruckelshaus' honeymoon is over," he stated. "The capitulation to the agrochemical industry came as a profound shock to environmental leaders in Washington."[18] Ruckelshaus had a different view. As he said in a 1971 speech to the National Audubon Society, of which he was a longtime member, he was personally in favor of a ban, but it wasn't that simple. He explained, "I was compelled by the facts to temper my emotions . . . because the best scientific evidence available did not warrant such a precipitate action. However, we in the EPA have streamlined our administrative procedures so we can now suspend the registration of DDT and the other persistent pesticides at any time during the period of review."[19]

Blessed with a keen intelligence, methodical mind, and droll sense of humor, Ruckelshaus was the conscientious public servant extraordinaire. And he had the street smarts to establish EPA's formative protocol. In 1972, as the agency's first administrator, he institutionalized the centralization of the environmental regulatory system as the smartest way to protect public health and the environment in perpetuity. His EPA was famous for forcing factory owners to provide detailed evidence about any materials they dumped into US waterways.[20] "My impression in those days was that pollution was essentially a problem caused by competition among the states for the location of industry within their borders," he said in a 2005 interview.

When we began to enforce pollution laws, they were pretty broad in modern terms and only addressed flagrant pollution. I mean, there were a lot of cities without any sewage treatment and there were industries discharging absolutely untreated material into the waterways, killing fish. But whenever we pushed a major company very hard, there was always the threat they would move to the South where the governors said, in effect, "Come on down here, we don't care, we need your business, we need jobs." My impression was, if you simply centralized all of this oversight and enforcement activity, you could bring such states and governors in line because there wouldn't be any place for them to run and hide.[21]

Agribusiness forced the EPA to hold hearings on the results of its DDT study, hoping to secure the use of the nondegradable pesticide once and for all. The hearing was chaired by the pro-DDT judge Edmund Sweeney. The outcome was predictable, but the methodical and nonflashy Ruckelshaus wasn't; he declined to attend even one session of the seven-month-long inquiry. Behind the scenes, he was working to lay the legal and technical groundwork that would make a DDT ban effective. Environmentalists wanted him to act yesterday: Ruckelshaus wanted the ban to last until tomorrow.

On June 14, 1972, the two paths to the same goal finally came together, with Ruckelshaus brushing aside the opposition, shaking off the threat of lawsuits, and declaring an immediate, outright ban on nearly all uses of DDT, as well as related pesticides that he feared would be detrimentally substituted for the more notorious chemical. Still allowed, however, were exports of DDT for use by the World Health Organization and US foreign aid program to control malaria; use in public health emergencies in the United States, such as to prevent malaria outbreak after a flood; and use—strictly regulated—on three vegetables for lack of anything else to reliably control their pests. (Later, all uses in the United States were banned in the 1980s.) Environmentalists, notably the World Wildlife Fund board of directors, continue to fight for a ban on DDT outside the United States, where its only use in the twenty-first century, sanctioned by the World Health Organization, is for disease control.

At the end of 1972, DDT was banned in the United States after ten years of legislative battles. Ruckelshaus was immediately targeted by chemical industry lobbyists. He dismissed their attacks as "bad science." Research showed that DDT was just as Rachel Carson had said: an "elixir of death." Full of fury, Nixon told Ruckelshaus, "I completely disagree with this decision." Cognizant that farmers were a huge voting bloc, the president threw a tantrum at Whitaker: "I want plenty of efforts to get it reversed."[22] Ruckelshaus's courage was rewarded when bird populations began to recover in sync with the ban. Thirty-five years later, the bald eagle was populous in its old haunts. In 2007, in a momentous conclusion to a frightening wildlife-in-peril episode in US history, it was taken off the endangered species list.

As demonstrated by the Great Lakes Water Quality Agreement, the Nixon administration understood that the environmental movement had turned global in 1972. At the historic June 5–16 United Nations–sponsored conference in Stockholm, where the theme was attacking global environmental problems, Ruckelshaus announced the US ban on DDT. Out of the 114 nations that sent delegates to Stockholm was born Earthwatch, a sophisticated global network for monitoring environmental problems such as atmospheric pollution and ocean degradation. The Nixon administration tapped Christian A. Herter, Jr., an environmental special assistant to Secretary of State William P. Rogers, to be the US representative at the UN convocation.[23] The Soviet Union and other Warsaw Pact countries boycotted the UN event because East Germany wasn't asked to participate; thus, the United States happily played the leading conservation stewardship role in Stockholm. The UN conference, in the end, galvanized other nations to adopt their own national versions of the United States' NEPA. In the spirit of Stockholm, the United Kingdom sought to ban sulfur and nitrogen dioxides emitted from coal-fired power plants because it returned to Earth as acid rain in Germany and Sweden, killing lakes and forests.

The Stockholm Declaration, as it became known, listed twenty-six global principles, including the protection of world wildlife and prevention of oceanic pollution. The UN delegates approved such US proposals as a ten-year moratorium on whale hunting. Also, the United Nations launched a massive education initiative about proper natural resource management protocol. A highlight of the conference was the powerful speech Indian prime minister Indira Gandhi gave linking ecological stewardship with the alleviation of poverty. Maurice Strong, secretary general of the United Nations Conference on the Human Environment, claimed that Stockholm launched "a new liberation movement—liberation from man's thralldom to the new destructive forces which he himself has created."[24]

Even though the Stockholm meeting worked out well for the Nixon administration, Ruckelshaus caught quite a bit of flak at the conference from what he called "a wild group, a mix of anti-war people and environmentalists." When the EPA administrator tried to speak in Stockholm, he was heckled for the United States' indiscriminate bombing and use of herbicides in Vietnam. Only when anthropologist Margaret Mead intervened and pleaded for the audience to listen so Ruckelshaus could get a sentence across without being booed did the audience calm down. Overall, the Nixon administration came out way ahead; Stockholm was a big win for the White House. On June 20, Nixon told the press, "I am proud that the United States is taking the leading role in international environmental relations."[25]

At the time of the Stockholm conference, Senator George McGovern of South Dakota looked like a shoo-in to be the Democratic nominee to run against Richard Nixon. After early victories in New Hampshire and Illinois, Ed Muskie's campaign collapsed; the Maine senator finished fourth in Wisconsin and Florida. Ralph Nader had damaged Muskie's reputation as an environmentalist when his Center for the Study of Responsive Law suggested he should be stripped of his title "Mr. Pollution Control" because he was lazy about punishing corporations for ignoring regulatory laws.[26]

If environmentalism became a key voting issue in 1972, some liberal pundits believed, the benefit would go to a Democrat over Nixon. But that was wishful thinking. Once McGovern procured the nomination, Nixon's reelection was essentially a foregone conclusion. And no matter what McGovern said, Nixon's environmental record was strong, and the White House had astutely claimed credit for every powerful bill passed by the Democratic Congress regarding anti-pollution policy from 1969 to 1972. McGovern hammered on Nixon for hedging on the Clean Water bill of 1972, which the president threatened to veto if Congress and the Senate dared to pass it without shaving billions of dollars off the price tag. But Nixon brushed the criticism aside.[27] McGovern's overall attack that Nixon didn't "care" about the habitat of piping plovers or white pelicans rang hollow, even to serious Auduboners.

When it came to the June 7 California Democratic primary, McGovern narrowly defeated Hubert Humphrey by 5 percentage points. It was determined by surveys that McGovern had won the state because the environmental vote had gone his way. Two-thirds of Californians favored his position on a ballot measure calling for super-ambitious pollution controls. Humphrey, the senator who had sponsored the Wilderness bill in the Senate back in 1956, had been "outgreened" by McGovern.[28]

Throughout the summer of 1972, while the Clean Water bill was being debated, Nixon stood arm in arm with Senator Mark Hatfield, a Republican from Oregon, and Representative John Dingell to state that saving endangered species was a top US government priority.[29] As Democrats and Republicans both held their political conventions in Miami over the summer, a bipartisan agreement was also forged to protect marine mammals from mass slaughter. A fifteen-year moratorium on the slaughter of whales, seals, and dolphins was called for by Friends of the Earth. Film footage of the clubbing of baby seals and the massacring of dolphin pods shown on the evening TV news had stirred public outrage. Senator Philip Hart sought to move the debate on the Marine Mammal Protection bill from the Senate Commerce Committee to the Interior Committee. What frustrated the Michigan Democrat was that his Senate colleague Humphrey was holding up the bill with his demand for a tuna industry exemption. The *New York Times* backed Hart's position with a strongly worded op-ed stating "Interior has a strong conservationist tradition while the oceanic administration in Commerce is the former Bureau of Commercial Fisheries with a new name but the same old exploitative, pro-industry attitudes."[30]

As Nixon knew, the protection of marine mammals had been energized by television shows such as *The Undersea World of Jacques Cousteau* and the prime-time TV series *Flipper*, which had taught Americans to appreciate how deeply intelligent dolphins were. On community bulletin boards in grocery stores and on college campuses across the United States, flyers appeared protesting the slaughter of dolphins by tuna industry nets asking "Would you kill Flipper for a tuna fish sandwich?" There were also grim stories about ocean acidity tied directly to the amount of carbon in the atmosphere and river flows that directly harmed manatees and dolphins. Marine biologists were calling for the restoration of seagrass meadows and kelp forests to help seals and otters survive. Nixon, reading public sentiment, promised to sign the Marine Mammal Protection bill that fall. The bill protected these marine creatures and population stocks declining to the point where they stopped being significant factors of the ecosystem where they lived.

As the November election neared, the League of Conservation Voters rated US senators on the environment. Only Gaylord Nelson of Wisconsin had a perfect 100 score. George McGovern did well with 74.[31] Within the increasingly liberal Democratic Party, being an environmentalist was a prerequisite to winning elections. That September in 1972, in Colorado, Representative Wayne Aspinall, the chairman of the House Interior Committee, was defeated by the League of Conservation–backed Alan Merson. The age of pro-dam reclamation Democrats was over.[32]

On the Republican side, Governor Russell Peterson of Delaware made protecting beaches from coastal pollution his brand issue. He had stopped the Shell Oil Company from constructing a refinery on 5,800 acres near Smyrna that it had acquired for that very reason. The act also blocked a consortium of the fourteen largest oil corporations in America from erecting smokestack factories in Delaware Bay to feed existing and future refineries in the Northeast. Peterson bemoaned the fact that in the seven beachfront counties of New Jersey, more than 20 percent of the 233,563 acres of salt marsh had been drained since Eisenhower had become president.[33] Peterson considered himself a Nixon Republican.

Once again there were no US presidential debates in 1972, leaving the four Kennedy-Nixon debates of 1960 an aberration in US history at that time. With McGovern trailing Nixon in all the major public polls, environmentalists pinned their hopes on the House and the Senate to pass the visionary Clean Water bill. In early October, just weeks before the election, the Senate voted in favor of the bill 74–0 and the House followed with a 366–11 vote. Ed Muskie had triumphed. Leading Republicans pleaded with Nixon not to veto the popular measure. Unswayed, Nixon, late on October 17, vetoed the bill, saying he planned to attack pollution in "a way that does not ignore other very real threats to the quality of life, such as inflation and increasingly onerous high taxes."[34]

The Nixon administration was divided on the Clean Water bill: the Office of Management and Budget, the Council of Economic Advisers, and the Department of the Treasury were for a veto, while the Environmental Protection Agency, the Council on Environmental Quality, and the departments of State and Interior were against it.[35] October 18 was the showdown day. Would the House and the Senate have the will—and the votes—to override a Nixon veto? Would the United States begin to resuscitate rivers poisoned with oil, clogged with grease blobs and gook, or contaminated with human waste and phosphates? Would Nixon really let his loathing of Muskie prohibit him from seizing historical credit for the Federal Water Pollution Control Act Amendments? On the initial vote before the Senate, Muskie, who had introduced the legislation a year earlier in 1971, turned eloquent:

> Today, the rivers of this country serve as little more than sewers to the seas. Wastes from cities and towns, from farms and forests, from mining and manufacturing, foul the streams, poison the estuaries, threaten the life of the ocean depths. The danger to health, the environmental damage, the economic loss can be anywhere.
>
> Just a ten-minute walk from this Chamber, the Potomac River is a hazard to health. The Georgetown Gap in the District of Columbia's sewer pipelines allows 15 million gallons of raw sewage to pour into the river every day.[36]

Republican senator Howard Baker of Tennessee said with equal passion that the $24.6 billion cost was worth it. "Study after study, public opinion poll after public opinion poll have revealed that this nation's economy can absorb the costs of cleaning up pollution without inflation or without a loss in economic productivity," he noted. "As I have talked with thousands of Tennesseans, I have found that the kind of natural environment we bequeath to our children and grandchildren is of paramount importance. If we cannot swim in our lakes and rivers, if we cannot breathe the air God has given us, what other comforts can life offer us?"[37]

Nixon wasn't listening. The Senate and House overrode his veto of the Federal Water Pollution Control Act Amendments of 1972 in the wee hours of the morning. Nixon continued to insist that the bill was "budget-wrecking." McGovern seized on Nixon's anti-environmental veto as proof that the Nixon administration's record was "hypocritical platitudes coupled with spineless inaction."[38] Brushing McGovern off as a lightweight, Nixon insisted that the cost of the act was "staggering."

Within the Republican Party, Nixon swatted away his detractors as if pissants. Pete McCloskey ran against Nixon as an anti–Vietnam War and pro-environment candidate in 1972 only to find near zero traction. Yet he had boldly stopped the AEC and Pacific Gas and Electric from constructing high-power pipes smack-dab in the city center of Woodside, California.[39] On all environmental matters, he was a GOP maverick who kept Nixon on the defense. As a leader of Earth Day, he was popular with Hollywood stars, and he enjoyed charging Nixon with crimes against the environment.[40]

A few days later, after his veto of the Federal Water Pollution Control Act Amendments, Nixon, as if atoning for sins, used the executive authority provided under the Antiquities Act to establish two new national monuments. The first was Fossil Butte in southwest Wyoming, one of the biggest deposits of freshwater fish fossils in the world. Back in the dinosaur days, that area of Wyoming had been a subtropical lake ecosystem. Placid water and fine sediment had conspired over time to make for prime conditions for preserving fossils. Protecting Fossil Lake and portions of the Green River ecosystem would enable scientists to study the fossils of fish, alligators, bats, turtles, and small horses.[41]

The second, Hohokam Pima in Arizona's Sonoran Desert, was a sacred Native American archaeological site. Within the monument's boundary were more than sixty midden mounds (ancient trash heaps), a central plaza, two oval-shaped areas thought to be dance floors, and an intact irrigation network inhabited by Hohokam peoples from about AD 300 to 1200. The ancient village, located within the Gila River Indian Reservation near Sacaton, Arizona, was thought to once have had two thousand human inhabitants.[42]

On October 21, 1972, Nixon enthusiastically signed the Marine Mammal Protection Act (MMPA). That law not only prohibited the killing of marine mammals except under rare, well-specified conditions, it also set standards for the protection of the habitats of whales, seals, otters, walruses, and other marine mammals from importation, exportation, hunting, capture, or any other form of harassment, but it did so with a publicity blitz. Strange as it might have sounded to young antiwar protesters, Nixon had a soft spot in his heart for whales, which he loved watching from his backyard in San Clemente. The mass destruction of these magnificent mammals by the Russians and Japanese sickened him. In promoting the Marine Mammal Protection bill throughout 1972, Nixon, working with lawyer Lee Talbot, strong-armed the Department of Defense to stop killing whales and other marine mammals in American waters and ban the import of blubber. The military insisted that they needed sperm whale oil to fuel ships. Nixon investigated the conundrum and discovered that the Dupont Corporation could now easily (though more expensively) produce artificial whale oil. At Nixon's urging, the Pentagon capitulated. Whale oil was banned. With that last hurdle resolved, Nixon signed the Marine Mammal Protection Act with his wife, Pat, cheering him on.[43]

Nixon's MMPA reflected the more scientific approach to protecting marine access. In addition, the very existence of MMPA showed that every environmental law didn't have to accrue benefits to mankind, such as laws related to resource management or recreation. MMPA was unlike most previous environmental laws in that it was an act of caring for nonhuman species. And six days after signing the MMPA, Nixon signed the Coastal Zone Management Act, which mandated that coastal states develop management plans to offset the negative impact of humans on coastal areas.[44]

There were all sorts of environmental initiatives emanating from the White House as election day neared. On October 21, Nixon signed the Federal Environmental Pesticide Control Act of 1972. It was a clever have-it-both-ways law. On the one hand, the EPA was given jurisdiction to ban certain pesticides; on the other, farmers would be paid for their financial losses incurred during a transitional time period. Flashing two V's for victory, Nixon, closing in on McGovern, declared the act to be a win for both agriculture and the environment.[45]

Just days before the election, Nixon also signed the Golden Gate National Recreation Area into existence. Golden Gate NRA stretched south from Point Reyes through much of western Marin County to the headlands and Fort Baker; then over to the Presidio and down the oceanfront of San Francisco; encompassed Alcatraz, Crissy Field, Fort Mason, and the Maritime Park on the city's northern wharfs; and continued to San Mateo.[46] The White House had worked closely with the cosponsors of the act, California representatives William S. Maillard, a Republican, and Phillip Burton, a Democrat, to fund the $120 million effort for the acquisition of Alcatraz Island and Fort Mason from the US Army. Other exquisite ecological and historically important landscapes in the San Francisco Bay were likewise safeguarded for inclusion in Nixon's new national recreation area (NRA).

With Golden Gate NRA, Nixon had established one of the biggest urban parks in the world, something Kennedy and Johnson had been unable to do. Though the original impetus for the new NRA had come from Burton, who, as chairman of the Subcommittee on National Parks, had galvanized the Bay Area in the late 1960s and 1970s to support the urban federal park, Nixon deserved credit for embracing this major Sierra Club initiative. The true grassroots leadership for Golden Gate NRA was Edgar Wayburn, a lifelong leader whom Burton called "my guru."[47]

Polar opposites on almost all political issues, Burton and Nixon shared a belief that the Bay Area's recreational opportunities should be available to anybody regardless of their socioeconomic status. Burton was the national champion of urban parks as "parks for people, where the people are." Sierra Club president Wayburn called Burton "a big engine . . . for the rest of us to lay track."[48] Nixon, who distrusted Burton, Wayburn, and the Sierra Club, nevertheless supported them on the Golden Gate NRA. It was, he believed, good for California. "We did not know how to say thank you to Nixon," Brower later admitted. "[He] had great promise and did great things."

The other NRA that Nixon established that late October was Gateway in New York City and Monmouth County, New Jersey. Gateway NRA included more than twenty-six thousand acres of land, and there were national park areas all around. Jamaica Bay National Wildlife Refuge, part of Gateway National Recreation Area, held the distinction of being the only wildlife refuge in the entire National Park System; it was home to more than 330 species of birds. In grand collaborative fashion, the National Park Service and the Nature Conservancy were determined to improve the ecological health of coastal areas surrounding New York City. Old forts and airports were brought into the urban park to use as history tourism sights. At Sandy Hook, seals often hauled out on the beach for tourists to snap photographs. Salt marshes, dunes, brackish ponds, and woodlands were saved at Gateway NRA from becoming environmental sacrifice zones.

Gateway National Recreation Area was a monument to thoughtful urban planning. More than 20 million people lived within two hours' travel time of Gateway units such as Breezy Point in Queens; Floyd Bennett Field; Great Kills Park; Miller Field and Fort Wadsworth on Staten Island, along with the sandy beach running between Great Kills and Fort Wadsworth; Hoffman and Swinburne Islands off Staten Island; and Sandy Hook, New Jersey. Most of the land and beach property was owned by New York City, New York State, or the federal government. But all the properties were controlled by the Department of the Interior. If one were to choose a spearhead who could claim to be the "conceptualizer" of Gateway, it would be Representative William Fitts Ryan, a Democrat from New York who sadly died of throat cancer just six weeks before the NRA became a reality. But Nixon, too, deserves accolades for embracing the largeness of the federal park vision.

Both Golden Gate and Gateway were important conservation accomplishments for which Nixon never received proper credit. They were the two conservation trophies he was most personally proud of because they had brought "parks to the people where the people are."[49] Following the success of Golden Gate and Gateway, the National Park Service soon followed with other urban parks, such as Cuyahoga Valley (near Cleveland), Chattahoochee River (near Atlanta), and Santa Monica Mountains (near Los Angeles). In 1972, the US Congress, following Nixon's lead, appropriated $200 million from the Land and Water Conservation Fund to back initiatives with the goal of conserving wild areas around big cities.

Even though President Nixon and the Sierra Club weren't natural allies, they collaborated again in 1972 on the establishment of the Cumberland Island National Seashore, the southernmost two barrier islands on Georgia's Atlantic coast. Situated between Jekyll Island on the north and Florida's Amelia Island on the south, these salt marshes, mud flats, and tidal creeks were magical in every way. The problem the National Park Service faced was that descendants of Andrew Carnegie had sold three thousand acres of the island to the real estate mogul Charles Fraser, who had helped commercially develop Hilton Head Island, South Carolina. A civil disagreement ensued between Fraser (in favor of beachfront houses) and Brower (in favor of protecting wilderness).

The *New Yorker*'s John McPhee had the wherewithal to travel around Cumberland Island with Fraser and Brower, scribbling away in his notebook, documenting their divergent views. Much as McPhee had when he traveled the North Cascades and Grand Canyon with Brower, he used the Cumberland Island squabble, as delineated in *Encounters with the Archdruid* (1971), to great effect.[50] In what Brower thought of as a compromise, he proposed that 90 percent of Cumberland be left undeveloped, with the remaining 10 percent allotted as environmentally sound houses and businesses. Once again, Brower unleashed an inventive preservation campaign, winning over public support for the national seashore designation. Groups like the Georgia Conservancy joined forces with the National Park Foundation, which ended up buying the disputed land outright from Fraser. With the landownership issue resolved, Nixon, on October 23, 1972, established Cumberland Island National Seashore (totaling 36,415 acres). In coming years, the northern part of the island became a federally designated Wilderness area.

Because Cumberland National Seashore encompassed twenty-three different ecological communities, Nixon agreed that only three hundred tourists at a time be allowed on Cumberland Island. Georgia governor Jimmy Carter, a moderate Democrat, played a major conservation role regarding the saving of Cumberland, as did ecologist Eugene Odum. Sea turtles, dolphins, and whales thrived in the protected waters, and Archie Carr praised Nixon for picking up the signing pen. John F. Kennedy, Jr., to honor his father's environmental legacy as the "National Seashore President," got married to Carolyn Bessette on Cumberland Island in 1996.[51]

As predicted, on November 7, 1972, Nixon won by a landslide, one of the biggest in US history (60.7 percent to McGovern's 35.5). In fact, he won every state except Massachusetts. But a deep dive into that election's polling data shows that the environment was still a hot issue. The Republicans lost two Senate seats, while the Democrats still controlled the House. Politicians with an ardent environmental stance had by and large won election. And there was other good news for environmentalists. In New York, a $1.5 billion environmental bond issue was passed. Similarly, North Carolina passed a visionary "environmental bill of rights." Of the fifty-seven gubernatorial and congressional candidates endorsed by the League of Conservation Voters, a remarkable forty-three had won.[52]

As 1972 ground to a close, with Congress adjourning for the holidays, the prevailing assumption was that 1973 would become the year of energy with the environment taking a back seat. The secret negotiations with the North Vietnamese were at a crossroads and pointed toward the end of the United States' longest war. Nixon was on top of the world. In a number of touching and sentimental ways, Nixon commiserated with his friend Lyndon Johnson, exiled in Texas, crippled by terrible health. Despite his serious heart problems and persistent cough, the Texan continued chain-smoking cigarettes. Michael DeBakey, a pioneer of heart bypass surgery, declined to operate on him because two of his arteries were too severely damaged. When Johnson complained of debilitating stomach cramps, a physician diagnosed him with diverticulosis (small pouches growing on the large intestine).

Refusing to talk with the national press, except on one occasion with Walter Cronkite of CBS News, which didn't go well, LBJ made just one major speech in 1972. It was an impassioned speech on civil rights and was delivered at the LBJ Presidential Library in Austin. Dressed in a western-style suit with two large breast pockets, a yellow shirt, and a burnt orange Texas Longhorns tie, the ex-president's remarks on racial injustice were heartfelt and inspirational. But that single event drained Johnson to the point of near collapse. And Watergate woes were festering around Nixon, whose 1973 would turn out to be vastly more problematic than he could have imagined when celebrating the New Year in oceanside San Clemente.

Last Leaves on the Tree

Nothing is a grander symmetry than this: Water, drinkable; our air, breathable; birds, built and blurred on a breeze; trees having hugs sighs into the heavens; our children, giggling and gilded in grass. Earnest for the first time, we must earn this turned Earth back.

—Amanda Gorman, "Earthrise" delivered at Joe Biden's Presidential Inauguration (January 20, 2021)

President Richard Nixon, pictured in 1971, stopped a jetport from being built in Florida that would have destroyed the biological integrity of the Everglades National Park. The areas of conservation policy that personally interested Nixon most were seashore and marine mammal protection.

I

On January 22, 1973, Lady Bird was shopping in Austin when she received an urgent call from the Secret Service: her sixty-four-year-old husband had died of a heart attack. She was shattered. Her only consolation was that Lyndon would be buried in the Ranch's family cemetery, alongside his parents and grandparents. A few days later, under freezing rain, with President Nixon attending, Johnson's remains were committed to the earth, surrounded by live Spanish oaks whose moss "hung like mourners" over the burial site along the Pedernales.[1] America's thirty-sixth president had returned to the Texas soil he had long embraced as his all-seasons essence. "My deepest attitudes and beliefs were shaped by a closeness to the land," he had once said, "and it was only natural for me to think about preserving it."[2]

After the burial, Mrs. Johnson moved to Austin and busied herself with urban beautification efforts. She marveled that the city, with its cedar groves, freshwater holes, and deer aplenty, was an "environmental vineyard" where one could still find solitude among the colorful meadows of Indian paintbrush and bluebonnets. A few years before, she'd been asked to chair the Town Lake Beautification Committee, a group trying to establish a hike-and-bike trail and plant crepe myrtles and other flowering trees around Austin's Colorado River reservoir. With the support of the mayor's office, she'd planted native shrubs throughout Zilker Metropolitan Park and filled the area's open spaces with redbud, pear, and peach trees. "The riverfront is one of the 'trademarks' of Austin," she boasted after her beautification effort took hold. "When the city is on television, you are shown scenes of the capitol building, the University of Texas Tower, and a portion of Town Lake."[3]

Mrs. Johnson loved Luke 12:27: "Consider the lilies, how they grow: they neither toil nor spin, yet I tell you, even Solomon in all his glory was not clothed like one of these." As she approached her seventieth birthday in 1982, she manifested that beauty for Austin by creating the Wildflower Research Center at the University of Texas at Austin,[4] dedicated to inspiring the conservation of native plants. She called the center "my last hurrah" and by 1995 had seen it outgrow its downtown location and move ten miles southwest to a forty-two-acre Hill Country site, where its transcendently serene lily ponds and garden paths wound through beds supporting 650 native plant species. Two years later, the sanctuary was renamed the Lady Bird Johnson Wildflower Center, and to this day its educational programs teach children and adults about gardening, sustainable land use, water conservation, and the dangers of the chemical insecticides Rachel Carson warned about in *Silent Spring.*

"It is not just one organization, one location," Mrs. Johnson's daughter Luci Baines said of the center in 2007. "It is a philosophy that will endure long after my mother is not here, and I think there is no legacy she would treasure more than to have helped people recognize the value in preserving and promoting our native land."[5]

Two days prior to LBJ's death, President Nixon had delivered his second inaugural address without a peep about the environment.[6] With thick frost and a few patches of dry, cold snow on the ground in Washington, Nixon, a supremely transactional politician, had focused on "peace in our time." In the coming months, he would privately carp that environmentalism was a fad driven by liberalism, but he was also proud of his sterling record as an old-fashioned conservationist. Relaxed with his family, he sometimes lamented how the bison had nearly been hunted to extinction in North America and the passenger pigeon actually vanished as a species. At Camp David, he would hand-feed his pet deer, Nibbles, and he'd long been known (and ridiculed) as an obsessive dog lover. Fascinated with pandas since receiving Hsing-Hsing and Ling-Ling from China in 1972, he was proud to be an international spokesperson for wildlife conservation. He considered Stockholm a genuine accomplishment of his foreign policy legacy. And in early 1973 he was ready to up the ante by making the Endangered Species Act of 1969 even more inclusive of wildlife habitats.[7]

Impressed by the amount of favorable mail he had received for the Marine Mammal Protection Act from whale and dolphin enthusiasts, Nixon approached Russell Train, the chairman of the Council on Environmental Quality, to prod both houses of Congress to draft visionary legislation that would protect vanishing North American species and their habitats. On the Democratic side, Representative John Dingell of Michigan was the workhorse of what became the Endangered Species Act of 1973. Working closely with Dingell was the stellar Nixon team of Undersecretary of the Interior E. U. Curtis Bohlen, a bold environmentalist and former State Department official; Clark Bavin, who headed the Fish and Wildlife Service's law enforcement division; Lee Talbot, a Council on Environmental Quality conservation expert; and of course, Russell Train, Talbot's boss. Together they cobbled together the bill that would become the landmark Endangered Species Act of 1973, an expansion of the more limited 1966 and 1969 acts, aiming to preserve animals at risk of extinction by protecting their remaining numbers and protecting their habitats.[8] "The World Wildlife Fund fed me some useful data," Train recalled. "Lee Talbot did the heavy lifting, and I was open to learning about every creature in peril."[9]

As part of his efforts, Train convinced Nixon that the bald eagle should serve as the act's poster child. The media coverage of the congressional hearings focused on charismatic creatures such as gray wolves and polar bears. But unlike the 1966 and 1969 laws, the 1973 bill wasn't just focused on charismatic megafauna (large birds and mammals). Remarkably, Nixon's legislation gave protection to every type of endangered species—regardless of size or habitat range. Even plants would be protected.[10] "The result," Talbot recalled, "was one of the strongest pieces of legislation ever submitted to Congress."[11]

In the spirit of Nixon's Legacy of Parks program of 1971—which transferred 80,000 federal acres to state and local governments for outdoor recreational purposes—the president now wanted Train's team to repurpose parts of Camp Pendleton (near the Western White House in San Clemente) for oceanfront citizen enjoyment. "Nixon used to talk about how Camp Pendleton was a wildlife nirvana," Train recalled. "Natural resources hadn't been exhausted even though the camp was ringed by freeways, drainage facilities, canals, and the like."[12] During Nixon's second term, US Army soldiers and marines found themselves protecting endangered seabirds, counting wild horses, constructing quail guzzlers, fastening radio collars on deer, stocking ponds with fish, and arresting game poachers.[13] Nixon's Legacy of Parks project, in the end, resulted in an astonishing 642 new parks by 1976.[14]

Just as Nixon's earlier initiatives had co-opted the conservation issue from his Democratic rivals such as Ted Kennedy and Ed Muskie, perhaps he hoped that his conservation bona fides could help blunt the impact of the growing Watergate scandal. After all, majorities among both the American people and Capitol Hill lawmakers supported wildlife protection laws, so it could only help Nixon to publicly show his support for osprey, grizzly bears, manatees, and other beloved endangered creatures.

In late April 1973, during the throes of Watergate, Ruckelshaus left the EPA to serve as acting FBI director; two months later, he was named deputy attorney general. Having moved to the second post at the Department of Justice, he navigated the Watergate affair gingerly but was forced to resign not quite three months later during the so-called Saturday Night Massacre, when he refused to obey Nixon's order to fire the Watergate special prosecutor, Archibald Cox.

Nixon's past environmental victories did not sit well with much of the corporate world in 1973. Companies were livid about the Environmental Impact Statement (EIS) requirements. Pacific Gas and Electric, Exxon, and Texaco were still angry at Nixon for the federal overreaction to the Santa Barbara oil spill of 1969 and his signing the National Environmental Policy Act (NEPA). CEOs from extraction companies in the Rocky Mountain states now criticized Nixon for endorsing Earth Day. Members of conservative think tanks funded by Coors wondered why groups such as the Environmental Defense Fund (EDF) and the Natural Resources Defense Council (NRDC) had tax-free status. If LBJ and Udall could strip the Sierra Club of its nonprofit gravitas for having taken out advertisements in the *New York Times*, they wondered, couldn't Nixon do the same with the Scenic Hudson and Get the Oil Out crowd?

The Izaak Walton League successfully sued the US Forest Service to halt clear cutting in national forests. Their victory began a turning away from "get out and cut" first policies on public lands. Tired of being demonized as super-polluters, corporate leaders began a full-fledged counteroffensive, attacking environmentalists as "mystics," "sentimental bird-watchers," "kooks," "hippies," and "alligator freaks"[15] blocking the march of capitalism. Eventually, those anti-environmentalists settled on "tree huggers" in the 1980s.

II

Except for supporting the Endangered Species Act, Nixon felt that his days of environmental policy heavy lifting were over. Having defeated McGovern by one of the biggest landslides in US history, he saw 1973 as the year to end the Vietnam War via a peace accord, shrink government costs, and negotiate a nuclear arms control agreement with the Soviet Union. The administration's environmental thought leader, Ruckelshaus, no longer at EPA, couldn't talk with authority anymore about how pesticides were decimating freshwater mussels and how the ivory-billed woodpecker had vanished. Instead, Nixon relied more on pro-business types such as the right-wing White House adviser H. R. Haldeman.

With the able Russell Train taking over EPA and John Whitaker appointed to undersecretary of the interior, the only environmentalist within easy range of the Oval Office remained John Ehrlichman, who was up to his eyeballs in Watergate woes. Replacing Ehrlichman, Ruckelshaus, and Train whispering into Nixon's ears were Secretary of Agriculture Earl Butz and the new secretary of commerce Pete Peterson. Neither cabinet secretary thought the environmental crisis worthy of any more White House attention. The change of strategy became apparent on February 15, 1973, when Nixon delivered his third and final environmental address to Congress. "I can report to the Congress," he said, "that we are well on the way to winning the war against environmental degradation—well on the way to making our peace with nature."[16]

For Frank Church and Gaylord Nelson it was as if Nixon had declared the environmental crisis over with the administration having saved the day. Fairly enough, the core of Nixon's message that February was that there had to be a balance between economic growth and environmental protection, or else inflation and overtaxation would burden the middle class. Nevertheless, on Nixon's watch, the United States became the global leader in protecting planetary ecological awareness. When the International Union for Conservation of Nature drafted the Convention on International Trade in Endangered Species of Wild Fauna and Flora, the fingerprints and ideas of Train were all over the document.

Geopolitics in the Middle East, however, would soon throw a monkey wrench into the Sierra Club world and deal the environmental movement a devastating blow. On October 6, 1973, Egyptian and Syrian forces launched coordinated attacks on Sinai and the Golan Heights, both of which had been occupied by Israel after the 1967 Six-Day War. With the Soviets arming Egypt and Syria, Nixon soon began airlifting weapons and military supplies to Israel. In response, the member nations of the Organization of Arab Petroleum Exporting Countries (shorthanded for OPEC) proclaimed an oil embargo that targeted the United States and other nations backing Israel. Between the embargo's announcement on October 17, 1973, and its end in March 1974, the price of oil quadrupled, leading to shortages and sending gas prices soaring.

Oil and gas companies blamed US environmentalists for the "oil shock," arguing that the country could produce all the fossil fuels it needed for energy independence were it not for onerous EPA and Clean Air Act regulations. Seasoned environmentalists such as David Brower, Barry Commoner, and Ralph Nader were vilified for their campaigns to limit emissions from cars and coal plants, desulfurize fossil fuels, and ban leaded gasoline. Soon the pendulum had swung decisively in favor of those demanding that Congress stop bemoaning the Santa Barbara mishap and instead start offshore drilling on behalf of US national security and energy interests.

Even though Americans' penchant for big cars, fast driving, and freedom of choice received a fair share of the blame, the EPA became a scapegoat for consumers ravaged by inflation and high prices at the pump. Some GOP conservatives denounced Nixon for signing the Clean Air Act of 1970, which had prevented more than 100 million tons of US coal from being mined. Others scoffed at Nixon's EPA for having supported the construction of mass transit systems to provide a more energy-efficient, lower-pollution alternative to cars. Public servants such as Gaylord Nelson, Frank Church, William Ruckelshaus, and John Ehrlichman were all savaged by what Train called the brutal "shotgun blasts" of the oil and gas lobby. When Congress circumvented NEPA to approve the Trans-Alaska Pipeline on January 3, 1974, the writing was on the wall. The energy crisis had put the

Silent Spring revolutionaries on the defensive for the first time since JFK had assumed the presidency in 1961.

Some pragmatic environmentalists felt that the energy crisis obliged them to compromise with companies such as US Steel and Mobil. Norman Cousins, then serving as the chairman of Mayor Lindsay's New York City task force on environmental pollution, feared that the energy shortage would doom the ban on heavily polluting fuels such as coal. Speaking that December at the annual meeting of Keep America Beautiful, a public service organization favored by Nixon and 100 companies, trade associations, and labor unions, Cousins retreated from his usual environment-first advocacy to address the sudden energy shortages: "Because of the energy crisis we are going to have to use coal, but no one talks about switching to low-sulphur coal. I would feel much better if they also were going to use electrostatic precipitate cleaners and other effective means of employing modern technology in the combustion of coal so as not to sacrifice the environment to meet the energy crisis."[17]

After Ruckelshaus stepped down as EPA administrator to run the FBI in April 1973, the job of enforcing NEPA fell to his successor, Russell Train. "Some have been saying that all we have to do to warm things up is burn the Clean Air Act," Train complained. "Some tell us that environmental effort is responsible for the energy crisis. Some tell us that a little pollution never hurt anybody. What these people say is simply not true, and they know it."[18] Luckily for Train, Ruckelshaus had left EPA in fine bureaucratic shape with pragmatic mission statements, top-notch scientists, and sensible outlines for long-range environmental policy—all of which remain relevant more than half a century later.

Working closely with Undersecretary of the Interior E. U. Curtis Bohlen, Nixon made it clear in early 1973 that new tools were needed to save vanishing species and protect the environment. Revisions of the Endangered Species bill were numerous in the coming months, with John Dingell the chief puppeteer on Capitol Hill. Within administration circles, Russell Train worked diligently with congressmen to gain approval of the catalytic converter—a device that converts toxic gases and pollutants from internal combustion engine exhaust "into less toxic pollutants by catalyzing a redox reaction." This helped the Clean Air Act of 1970 achieve the automobile emissions reduction that it promised. Train also initiated "green diplomacy" with Soviet ambassador Anatoly Dobrynin, thereby setting a more hopeful global environmental standard.

With his White House besieged by Watergate woes and the tricky Arab oil embargo, Nixon happily embraced the Endangered Species bill that fall as the glorious capstone to his conservation legacy. Contrary to what later critics would allege, it was among the least controversial bills he ever signed.[19] On December 19, 1973, after a year of congressional negotiations, the Senate approved the final measure 92–0, followed a day later by the House, 355–4. On December 28, Nixon signed the act into law, putting teeth into the effort to forever protect endangered species. From his home in San Clemente he proudly stated, "Nothing is more priceless and more worthy of preservation than the rich array of animal life with which our country has been blessed."[20]

The Endangered Species Act of 1973 was largely noncontroversial. As the *Washington Post* editorialized, the vast majority of Americans clearly supported a serious commitment to protecting vanishing wildlife (even the National Rifle Association was for it).[21] Bestselling books by Faith McNulty, such as *The Whooping Crane: The Bird That Defies Extinction* (1966) and *Must They Die? The Strange Case of the Prairie Dog and the Black-Footed Ferret* (1971), had helped awaken Americans to the plight of rare North American creatures nearing extinction.[22]

The Marine Mammal Protection Act had brought loads of positive TV news coverage of Nixon's concern for polar bears and manatees. With EPA protections in place and DDT banned, eagles, osprey, pelicans, condors, and other avian species began recovering their numbers. Furthermore, Nixon tasked Train with leading the environmental publicity campaign in 1974 as a showcase of concern for endangered creatures. Nixon's law has been credited with inspiring the ban on the trade in elephant tusks and efforts to redesign fishing nets to protect endangered turtles in the Gulf of Mexico.[23] Both LBJ and Nixon deserve accolades for their visionary concern for wildlife and for shaping lasting and effective protections. "The 1966 act had taken an initial step in creating a more mechanical approach to the problem

of vanishing species by requiring the Fish and Wildlife Service to begin maintaining a list of native species endangered with extinction
while the 1969 act extended that official listing to foreign species as
well," the historian Mark V. Barrow wrote in *Nature's Ghosts*. "But
it was not until the 1973 legislation, with its broad notion of 'taking'
and harsh penalties for those who harassed or harmed any listed species, that the federal government placed itself at the center of a far-
reaching program to identify, monitor, and rescue species that seemed
to be struggling to survive."[24]

Because of the energy crisis, the Vietnam War, and the Watergate
scandal, the Endangered Species Act of 1973 didn't attract the amount
of public attention that the Nixon team had hoped. The species that
netted the most love was the whooping crane. Under a new EPA recovery plan, America's tallest bird became the beneficiary of the captive breeding program for cranes conducted in Patuxent, Maryland,
and Baraboo, Wisconsin.[25]

Florida, a state Nixon loved dearly, was the great beneficiary of the
Endangered Species Act of 1973. Everglades National Park devised an
American crocodile sanctuary that kept nesting areas of these prehistoric creatures protected from harassment of any kind. Florida Power
and Light managed their populations of baby crocodiles in cooling
canals. State herpetologists studied them on northern Key Largo, and
the US Fish and Wildlife Service established the Crocodile National
Wildlife Refuge in 1980. Endangered status led to habitat conservation, thereby saving the *Crocodylus acutus* from extinction in North
America. By the time Joe Biden was president, the crocodile population was greater (and over a larger range) than had been historically
documented, even appearing in waterways around Miami.

Likewise, Florida panthers found new post-1973 protected status
in the Everglades and the Big Cypress National Preserve. State wildlife
researchers determined that panthers also had a genetic problem due
to breeding, so they imported some Arizona mountain lions to diversify the gene pool. The state led research and monitoring on both state
and federal lands and designated a large area of cypress as Fakahatchee
Strand Preserve State Park. In 1989, the US Fish and Wildlife Service
created the Florida Panther National Wildlife Refuge. The panther
rebounded and now ranges farther north in Florida. Automobiles are

still a major threat, but when Interstate 75 was constructed, wildlife underpasses were smartly added.[26] The NHL hockey team in Jacksonville, Florida, was named the Panthers in 1993 as a tribute to the *Puma concolor coryi*.

In 2020, the Richard M. Nixon Presidential Library and Museum opened a permanent outdoor exhibit celebrating the Endangered Species Act of 1973. There were wonderful sculptures of bald eagles, grizzly bears, and Sierra Nevada red fox. Interactive exhibits provided children with tools to understand how a species gets unlisted and what constitutes a critical habitat. Looked at from the 2020s (and beyond), it's clear that the Endangered Species Act of 1973 was one of Nixon's most cherished accomplishments. "With climate change and natural area loss pushing more and more species to the brink, now is the time to lift up proactive, collaborative, and innovative efforts to save America's wildlife," Secretary of the Interior Deb Haaland said in September 2021. "The Endangered Species Act has been incredibly effective at preventing species from going extinct and has also inspired action to conserve at-risk species and their habitat before they need to be listed as endangered or threatened. We will continue to ensure that states, Tribes, private landowners, and federal agencies have the tools they need to conserve America's biodiversity and natural heritage."[27]

III

With the Endangered Species Act under his belt and Watergate weighing him down, Nixon began 1974 by signing the Emergency Highway Energy Conservation Act, which established a national fifty-five-miles-per-hour speed limit, and in midsummer, he signed the Federal Energy Administration Act, the first coordinated energy policy in the history of the United States. Environmentalists greeted those actions with enthusiasm, but they were, in fact, rare high points in a chaotic year.

What Scoop Jackson called "impeachment politics" was not good for the environmental movement.[28] With the Watergate scandal raging, the White House was under siege and every serious figure in national politics was steering well clear of the president. In February, the House of Representatives voted to proceed with impeachment. Soon after, a grand jury named Nixon as an unindicted coconspirator, and Special Prosecutor Archibald Cox subpoenaed the White House to turn over sixty-four secret tapes. Two days after Nixon signed the Federal Energy Administration Act, the House Judiciary Committee held its first impeachment hearings.

In a televised broadcast on August 8, 1974, Nixon announced to the nation that he would resign, effective the next day. As promised, at 10:00 a.m., he departed from the South Lawn of the White House, flying aboard Marine One for the final time. At Andrews Air Force Base, he boarded Air Force One, renamed *Spirit of '76*, and flew to El Toro Marine Corps Air Station in California.

On September 8, 1974, Nixon accepted a pardon from President Gerald Ford "for all offenses against the United States which he has committed or may have committed or taken part in during the period from January 20, 1969 through August 9, 1974."[29] Though Nixon would never confess to any wrongdoing, many in the public saw his acceptance of a pardon as a tacit admission of guilt.

Nixon was lucky to get a pardon. On January 1, 1975, exactly five years since NEPA had become the law of the land, during the Watergate trials, John Ehrlichman was convicted of conspiracy, obstruction of justice, perjury, and other charges. Having been convicted of a felony, he was disbarred and spent eighteen months in prison in Arizona. Once out of jail, in 1978, he lived in Santa Fe, New Mexico, as a writer of four novels and of 1982's *Witness to Power* memoir. Out of all of Ehrlichman's myriad accomplishments during the Nixon years, the one he was most proud of was leading the effort to deed Blue Lake—considered sacred waters by Indigenous people—back to the Taos Pueblo. "In the second half of his life, New Mexico's San Juan River became one of his favorite places," his son, Peter Ehrlichman, recalled. "When in Seattle, he loved Issaquah Creek, near Bellevue."[30] Upon his death in 1999, Leonard Garment wrote a tribute in the *New York Times* about his old boss in the Nixon years: "Ehrlichman served his sentence. He went on to write good books. He moved in with the Indians in New Mexico, grew a beard, married a beautiful woman, and gave learned, funny lectures. But his Watergate image remained."[31]

Nixon, from 1969 to 1974, truly sought to return the Republican

Party to the conservation tradition of Theodore Roosevelt when it came to natural resource management and environmental leadership. Where he suffered was that the press was always suspicious of his motives while never giving him proper coverage of his leadership regarding Golden Gate and Gateway National Recreation Areas, clean air, endangered species, and anti-pollution laws. After Ronald Reagan, an anti-environmentalist, won the 1980 election, Nixon's "greenism" streak was largely drained from the bloodstream of the GOP. And Nixon himself hadn't written much about his environmental accomplishments in his own *RN* memoir.[32] When Nixon traveled to Alaska, he marveled at the unparalleled beauty of the landscapes. "Now I get Hickel," he said. "If this were my home, I'd be an environmentalist too."[33]

In the wake of Nixon's White House resignation and the advent of Reaganism, it was left to Russell Train to remind historians of the thirty-seventh president's undeniable record of accomplishments. As president of the World Wildlife Fund from 1978 to 1985 and its chairman from 1985 to 1994, Train defended Nixon's environmental record every chance he got. His humanity was so great that he felt genuine pain for Nixon's suffering in exile. Especially after President George H. W. Bush awarded Train the Presidential Medal of Freedom, he told stories about how Nixon had boldly endorsed Endangered Species out of some hidden soft spot in his heart for creatures. In addition to historic wildlife protections, Nixon had unexpectedly backed policies punishing toxic waste and sewage treatment pirates. Once while touring a treatment plant in Chicago with Train, Nixon had been offered a glass of freshly decontaminated water. "I never drink before lunch," he had demurred and later lamented that whereas LBJ would be remembered for redwoods, he would go down in history as "the sewage treatment president."[34]

Train died in 2012, having established the annual J. Paul Getty Award for Conservation, the "Nobel Prize for conservation," under the sponsorship of the World Wildlife Fund (WWF), whose membership had reached 1.3 million and had an annual operating budget of $100 million.[35]

William Ruckelshaus, having resigned as deputy attorney general rather than carry out President Nixon's illegal order to fire the independent special Watergate prosecutor, returned to Washington, DC, in 1983 to become President Reagan's second EPA administrator, serving until 1985. He was a legend at EPA for having consolidated fifteen federal agencies with environmental briefs into a single organization with 8,800 employees. "I was able to keep the Nixon-era environmental agenda alive in the Reagan years," he recalled, "but only to a low-burner degree. Times had changed. Reagan loved the outdoors, liked the idea of roadless wilderness, but worried the Endangered Species Act was a trick to stunt real estate developers, dam building, electrical companies."[36] And Reagan nixed the US government's solar power agenda, including taking down the solar panels that President Carter had installed on the White House roof.

Having stabilized EPA, Ruckelshaus returned to Seattle to become the head of Browning-Ferris Industries, a waste management company that sought to break organized crime's stranglehold on big-city garbage removal. He turned to business management and the practice of law, proud that at EPA he had been instrumental in establishing clean air and clean water as essential parts of every federal decision. He did so by steering a steady course toward his goals during the Nixon and Reagan years, without allowing himself to be distracted from the core principles of honor and patriotism with which he had been raised in Indiana. In 2004, Ruckelshaus helped President George W. Bush write the timely report *An Ocean Blueprint for the 21st Century*.[37]

When William Ruckelshaus died at eighty-seven in 2019, he was credited by the *New York Times* with developing EPA's organizational structure in its formative years. Among his accomplishments were taking enforcement action against horribly polluted cities and corporate polluters; setting health-based standards for air pollutants, including automobile emissions; requiring states to submit regular air quality plans; and banning the general use of DDT.[38] It's still not too late to name the EPA headquarters on Pennsylvania Avenue in Washington, DC, after Ruckelshaus.

Scoop Jackson died in his hometown of Everett in 1983 at age seventy-one. His legend looms large in US Senate history and at the University of Washington in Seattle, where his voluminous public and private papers are housed. The nation honored his sterling conservation career with the 103,000-acre Henry Jackson Wilderness in Washington State, straddling the North Cascade Mountain Range.

There is minimal evidence that Nixon was overtly proud of what he accomplished as a "green president." For the most part, he preferred discussing his Cold War foreign policy. In 1991, though, after a speech in New York, the former president bumped into EPA administrator William Reilly, who had served as a senior staff member on Nixon's Council on Environmental Quality. "I know you," Nixon greeted Reilly with a huge smile, hand extended. "You're at EPA, and I founded EPA. I'm an environmentalist, too."[39]

IV

Throughout 1973 and 1974, Bill Douglas kept charging that the FBI, the CIA, and the White House were out to destroy him because of his environmental and anti-war activities. Convinced that his telephones were tapped—even the phone booth in front of Whistlin' Jack's Outpost & Lodge near his cabin in Goose Prairie, Washington—he'd gone so far as to hire former Supreme Court colleague Abe Fortas to sue Pacific Northwest Bell on his behalf for federal invasion of privacy—a short-lived effort.

Douglas had undoubtedly been surveilled during his judicial career, just probably not at a pay phone. In 1970, then House minority leader (and later president) Gerald Ford had led an effort to impeach Douglas for brazen partisanship, and the bad blood between the justice and other government officials was a matter of very public record. In one contemporary editorial cartoon, a US bomber marked DEFENSE DEPARTMENT was shown emptying its bomb bays over a cabin and a billboard reading "Welcome to Goose Prairie, Washington, Home of Justice Douglas."

As happens with octogenarian Supreme Court justices, the wolves on both the left and the right began clamoring for him to retire. On December 31, 1974, while vacationing in the Bahamas with his wife, Cathy, Douglas suffered a massive stroke that paralyzed his left leg. Friends urged him to step down from the Court. He refused, but his health woes accumulated. At last, on November 12, 1975, after thirty-six years on the bench—the longest tenure up to that time—Douglas reluctantly hung up his robes. Accolades poured in from all corners. In his honor, in 1984, Washington's Cougar Lakes area was renamed the William O. Douglas Wilderness (169,081 drop-dead gorgeous acres not too far from Yakima). Furthermore, the National Park Service dedicated the Chesapeake & Ohio Canal National Historical Park to Douglas for "his immense efforts in preserving and protecting the natural and historical resources" of the United States. The *New York Times* described the event as the honoring of "a prophet."[40]

At the ceremony, though enfeebled and in a wheelchair, Douglas was still a fighter, his steel blue eyes missing very little. Eight Supreme Court justices were in attendance, with Chief Justice Warren E. Burger speaking to honor his friend. Then it was Wild Bill's turn. With his wife, Cathy, and Ted Kennedy at his side, he reminisced that Louis Brandeis had hiked the C&O Canal annually. It bothered Douglas that Supreme Court associates had been afraid to join him along the trail as Udall and the Kennedys had. But there was still time. "I promise to get well, and be able to walk it again," he said.[41] Then he praised the type of outdoors-loving Americans he most admired: "all those who have no portfolio but who have two strong legs and like to hike, and to listen to the pileated woodpecker."

The seemingly indomitable Douglas died in 1980 at age eighty-one at Walter Reed Hospital in Bethesda, Maryland. Survived by his fourth wife, Cathy, and two children with his first wife, Mildred and William Jr., he was buried at Arlington National Cemetery, just 150 yards from the graves of Jack and Bobby Kennedy. Along with Rachel Carson, he led a remarkable generation of environmental activists, wholly varied in style, who had awakened post–World War II America to a profound understanding that its seashores, mountains, deserts, lakes, rivers, wildlife, and flora were more valuable than all the gold and oil in the world.

The year Douglas died, Barry Commoner ran for president on the Citizens Party platform, mustering a meager 233,052 votes. Soon thereafter, he moved his Center for the Biology of Natural Systems from St. Louis to Queens College in New York City. Keenly aware of the disparities in the realm of public health, Commoner grasped the concept of eco-racism before it was labeled as such. Privileged suburban whites enjoyed greenbelts and well-maintained parks, but America's urban ghettos, he complained, were plagued with industrial waste, hazardous emissions, and peeling lead paint. In poor Black communities in rural Alabama and Mississippi, raw sewage was routinely pumped into wastewater lagoons. "Through its history, the black community can be a powerful ally in the fight against environmental degradation," Commoner wrote in *The Closing Circle*. "Blacks need the environmental movement, and the movement needs the blacks."[42] Ahead of his peers as usual, Commoner remained committed to ending environmental racism until he died in 2012, worried that the effects of climate change would disproportionately impact the world's poor.

The 1980 election spelled doom for the top Democratic Party environmental warhorses in the Senate. Gaylord Nelson was defeated for reelection in Wisconsin but continued to be a national leader on the environment. In 2002, he appeared on *To Tell the Truth* as a contestant with his founding Earth Day as his calling card. The United Nations established an Environmental Leadership Medal in his name. On December 8, 2004, President George W. Bush signed legislation designating 80 percent of the landmass of Wisconsin's Apostles National Lakeshore as the federally protected Gaylord Nelson Wilderness. He died in 2005 at his home in Bethesda, Maryland, at the age of eighty-nine.

The other blow in the 1980 election came with the defeat of Frank Church in Idaho by a mere four thousand votes. "Losing Nelson, Church, and the others was a catastrophe," Denis Hayes recalled. "Reagan people such as James Watt, Anne Gorsuch, and Bob Burford and the whole Coors-affiliated anti-environmental zealots tried to undo what Kennedy, Johnson, and Nixon had accomplished. The extraction industry folk were in charge. Up until 1980, environmentalism was bipartisan. That all changed with Reagan. The pragmatic, problem-solving, wilderness-loving wing of the Republican Party disappeared overnight. It became fashionable to attack serious environmentalists as socialist fools."[43]

Nevertheless, even while running for reelection in 1980, Church succeeded in establishing Idaho's River of No Return Wilderness. That had been his dream since LBJ enacted the Wilderness Act in 1964. Combining the old Idaho Primitive Area, the Salmon River Break Primitive Area, and other protected areas, at 2.4 million acres it

remains the largest Wilderness in the United States outside Alaska. To
honor his long environmental career, the area was renamed the Frank
Church–River of No Return Wilderness Area in 1984.[44]

Ansel Adams continued taking his ethereal photographs of windblown trees, mountain vistas, and redwoods standing tall in the fog,
and in the late 1970s, art museums worldwide began acquiring his
works for high sums of money. His conservation advocacy remained
relentless, and though the Vietnam War had kept him at arm's length
from Nixon, he became personal friends with both Gerald Ford and
Jimmy Carter, whom he photographed in November 1979 for the
National Portrait Gallery. With his trademark ardor, Adams continued to fight on behalf of California's parks and beaches from his
home studio in Carmel Highlands on the Big Sur coastline. Following his death in 1984, the Minarets Wilderness in California's High
Sierra was expanded and renamed for him. His archive was acquired
by the University of Arizona in Tucson, where it is housed along with
Stewart Udall's personal papers.

From 1980 until his death on April 22, 1994—coincidentally on
the twenty-fourth anniversary of the first Earth Day—Richard Nixon
lived in Bergen County, New Jersey, with his wife, Pat, and dogs. Because he was so proud of his foreign policy accomplishments, he seldom sought to counteract his Watergate disgrace by pointing to his
remarkable NEPA and EPA accomplishments. He always harbored a
suspicion that environmentalism was a left-wing attempt to "system
destroy" capitalism.[45]

In California, however, where coastal real estate was expensive,
Nixon had learned that rich Republicans wanted their communities to
be clear of pollution and litter. As a San Clemente and later New Jersey homeowner, he appreciated that NIMBY sentiment of his friends
such as Charles "Bebe" Rebozo in Florida. "Given that environmental
issues were foundation to a healthy nation, Nixon considered them inherently conservative," his young research assistant Monica Crowley,
who published two books of her own reminiscences, *Nixon off the Record* (1996) and *Nixon in Winter* (1998), recalled. "He was therefore
frustrated and disappointed when the Republicans later ended the political and policy high-ground on those issues, and allowed the left to
hijack and redirect them for more radical policy purposes. During the

Reagan, Bush, and Clinton years, Nixon gave careful thought to ways in which the GOP could re-embrace a responsible environmental agenda and reclaim the mantle of thoughtful conservancy."[46] Nixon is buried next to his wife, Pat, in Yorba Linda, California, on the lovely grounds of the Richard Nixon Presidential Library and Museum.

When Lady Bird Johnson died on July 11, 2007, and was buried next to her husband in the family gravesite on the Ranch, the cabal of Silent Spring revolutionaries were largely gone. Posthumously, she was bestowed the Rachel Carson Award by the National Audubon Society for "women who have made outstanding contributions to the conservation and environmental movement."[47] Two weeks after her death, the Austin City Council passed a resolution to rename the Town Lake reservoir Lady Bird Lake.

Stewart Udall remained on the national scene well into the twenty-first century. With his chiseled face, weathered skin, dark dancing eyes, and combed-back gray hair, he looked like an aged amalgam of Lyndon Johnson and Geronimo, a sun-drenched westerner "with a distant look in his eyes."[48] For a few years, Udall taught at the Yale School of Forestry and wrote a syndicated column on the environment, inspired by his admiration for Bernard DeVoto's old *Harper's* essays. Although his friendship with the Kennedy family never buckled, most of his political work at the time was directed toward helping his brother, Mo, win the 1976 Democratic presidential nomination, a fight he ultimately lost to Jimmy Carter. During Carter's presidency, Udall and his wife, Lee, moved from Virginia to Phoenix, Arizona, where he practiced law, served on the board of the Environmental Defense Fund and the National Wildlife Federation, and became a hot commodity on the college lecture circuit.

Udall spent more than a decade as a personal injury lawyer, notably for victims of radiation poisoning from surface and underground nuclear detonations at the government's Nevada Test Site. Beginning in the late 1970s, he led suits on behalf of "downwinders" in Nevada, Utah, and Arizona who'd developed leukemia, thyroid cancer, and other illnesses after exposure to nuclear fallout, as well as miners, many of them Navajo, poisoned by radon gas while mining uranium ore to power the expanding US nuclear arsenal. "Had it been so inclined, the AEC could have required the installation of low-cost ventilation systems to protect uranium miners," he fumed. "AEC safety experts could have designed such a program in a matter of weeks. . . . the AEC's decision to put the flow of ore ahead of human health was a reckless act that sacrificed the lives of hundreds of miners."[49]

Lamenting that Marysvale, Utah, was a "lung cancer laboratory," Udall apologized to southwesterners for having allowed such desecration of human health to take place while he was a US congressman. "We bought into their new Edens—and into the thesis that the atomic scientists were wizards who would transform life on earth," he wrote.[50]

Just as Carson had been relentless in her *Silent Spring* investigation of DDT, Udall became a crusader exposing "the big lies of the bomb testers."[51] He traveled to every radiation-affected community in the American West, and, inspired by the $950,000 compensation awarded to the Marshall Islands in the 1980s for atomic tests, sought an even larger settlement for his 1,200 clients who'd been unwilling "human guinea pigs"[52] in the government's experiments. Beyond refusing to take a retainer for his work, he also organized fundraisers to keep his crusade alive, including an evening at the Santa Fe Indian School featuring Pete Seeger and Edward Abbey to benefit Navajo downwinders.[53]

Udall was crestfallen when his radiation case failed to win in US District Court.[54] After losing again in the appeals court, he grew depressed. But in 1990, President George H. W. Bush signed the Radiation Exposure Compensation Act (RECA), which read, in part, "The Congress recognizes that the lives and the health of uranium miners and of innocent individuals who lived downwind from the Nevada tests were involuntarily subjected to increased risk of injury and disease to serve the national security interests of the United States. . . . The Congress apologizes on behalf of the Nation to the individuals . . . and their families for the hardships they have endured."[55] Udall felt vindicated not only by the apology but by the accompanying $100 million in damages awarded to the Southwest's radiation-sick survivors.

"The compensation was far too little," he said. "But it lifted my spirits. The arrogance of the US government after World War II regarding the Nevada tests was revealed for future generations to ponder. I thought of Jack Kennedy a lot in the 1980s. His American University speech, followed by the Test Ban Treaty with the Soviet Union and Britain, was the beginning of making amends for Hiroshima and Nagasaki and Nevada. I thanked Bush for signing RECA. The circle was closed, and I could sleep with a more clear conscience."[56]

Long a prolific writer, with books such as *America's Natural Treasures: National Nature Monuments and Seashores* (1971) and *To the Inland Empire: Coronado and Our Spanish Legacy* (1987) to his credit, Udall turned to writing *The Myths of August: A Personal Exploration of Our Tragic Cold War Affair with the Atom* (1998), a history of government lies about and the human toll of US atomic testing. By then, President Bill Clinton had signed an executive order and established an interagency working group to promote and advance environmental justice principles across the United States. "When Clinton signed the environmental justice executive order in 1994," Udall recalled, "I had a feeling that all we'd done since *Silent Spring* wasn't for naught."[57] Throughout Clinton's presidency, Udall was an outspoken voice on the need for the EPA to expand its mission to protect both human health and the environment.

After the turn of the millennium, by then a longtime resident of Santa Fe, New Mexico, Udall, in his eighties, was still able to hike up the ten miles (and four thousand feet in elevation) from the floor of the Grand Canyon to the South Rim. In 2008, he celebrated as his son Tom was elected US senator from New Mexico. But time finally caught up to him. On January 31, 2010, at age ninety, he died. Two years earlier, he'd written a letter with Lee to his grandchildren: "Go well, do well, my children. Cherish sunsets, wild creatures and wild places. Have a love affair with the wonder and beauty of the earth."[58]

Stewart Udall was the last surviving member of John F. Kennedy's original cabinet. A few months later, President Barack Obama renamed the Department of the Interior headquarters in Udall's honor. Even conservative Republicans in Congress had voted in favor of the name change. Later that summer, after memorial services in Santa Fe, Phoenix, and Washington, Udall's ashes were scattered at national parks in the American West and on a desert vista in St. John's, Arizona, his childhood home. The fact that Congress had passed the Radiation Exposure Compensation Act allowed him to die with a clear conscience. By 2022, when its last payments were distributed, the families of the Southwest downwinders and uranium miners Udall had represented had received some $2.4 billion in compensation.

Considered the most successful interior secretary in American history, Udall had spearheaded national seashore and lakeshore campaigns, racially integrated his agency by hiring Robert Stanton and other minority staff members, wrote the influential *The Quiet Crisis*, and elevated the stature of Rachel Carson, Carl Sandburg, and Wallace Stegner into the conservation pantheon. He was a wilderness warrior and metaphorized into a wild river maven. His environmental reach across the land, at once broad and deep, personified what was so inspiring about the environmental movement during the Kennedy, Johnson, and Nixon years. Without educating the public about the sins of wilderness desecration, as Udall did, it's doubtful that NEPA and the EPA would ever have been established. By letting the slick-rock canyons of the Southwest speak for themselves, protecting the last vestiges of Pacific Northwest forests, and embracing the aesthetic beauty of wild rivers such as the Snake and the Saint Croix, Udall encouraged Americans to serve the Earth just as Jack Kennedy had asked them to serve their country. It had been Rachel Carson's revolution, but Udall, the last leaf on the tree, emerged in the four decades following her death as the environmental justice steward of her long-ago dreams back in Springdale, Pennsylvania, along the banks of the Allegheny River.

Acknowledgments

It is time for us to kiss the earth.

—Robinson Jeffers, "Return" from *The Selected Poetry of
Robinson Jeffers* (1938)

Numerous first-rate US environmental and presidential history books have fueled my *Silent Spring Revolution* endeavor. The Weyerhaeuser Environmental Books series at the University of Washington Press, under the general editorship of William Cronon, is a phenomenal academic enterprise from which I profited. Many of these first-rate Weyerhaeuser titles can easily be discovered in my notes and in the selected bibliography. Rachel Carson has attracted a slew of fine biographers. Linda Lear, William Souder, Arlene Quaratiello, Robert K. Musil, Paul Brooks, Carol B. Gartner, Nancy Koehn, and Mark Hamilton Lytle blazed my trail. Diana Post and Cliff Hall allowed me the privilege of lingering for an afternoon at Carson's home on Berwick Road in Silver Spring, Maryland, where she wrote *Silent Spring* from 1958 to 1962 on a desk that is still there. Thanks, too, to Carson's nephew Roger Christie and his wife, Wendy Sisson, who caretake Carson's Southport, Maine, cottage with consummate attention.

The Rachel Carson Papers, housed at Yale University's Beinecke Rare Book and Manuscript Library, provide scholars access to manuscripts, notebooks, letters, newspaper clippings, photos, and other material relating to Carson's research and publications. The majority of the collection's material was a bequest to Yale University in 1965, with later additions gifted by Houghton Mifflin Company in 1968, Marie Rodell (Carson's literary agent) in 1973, and Lois Darling in 1983.

Also of importance is the Linda Lear Center for Special Collections and Archives at Connecticut College, which holds the Lear/Carson Collection. This depository consists of archival materials Lear gathered for her biography *Rachel Carson: Witness for Nature* (1994) and the edited anthology *Lost Woods* (1999) as well as personal papers given to Lear by Carson's colleagues and friends. Additionally, the Dorothy Freeman Collection, consisting of letters between Carson and Freeman, is housed at the Ladd Library at Bates College in Lewiston, Maine. This openhearted correspondence has been ably edited by Martha Freeman into *Always, Rachel: The Letters of Rachel Carson and Dorothy Freeman, 1952–1964* (1994).

The Rachel Carson Council in Washington, DC, as envisioned by Rachel Carson, was founded in 1965, after her death, to carry on her environmental work. Directed by Dr. Robert K. Musil, the RCC advocates for climate justice on Capitol Hill and is backed by a national network of more than sixty colleges, with thousands of activists. More than thirty outstanding young environmental leaders serve as RCC fellows, receiving financial support and mentoring as they work from campuses nationwide on campaigns for fossil fuel divestment, renewable energy, food insecurity, and environmental justice. Musil—deeply knowledgeable about all aspects of Carson's life and the anti-nuclear movement—graciously proofread a final draft of this manuscript.

The John F. Kennedy Presidential Library and Museum in Boston, Massachusetts, holds a treasure trove of New Frontier conservation materials related to JFK (including his establishment of National Seashores and his 1962–63 conservation tours to the American West). Boxes of correspondence in the White House Subject Files were particularly helpful, especially the "Natural Resources" boxes 641 through 651. For JFK on the 1960 campaign trail, box 1031 (conservation speeches) proved vital. At the library, I received assistance at various times from Alan Price, Tom Putnam, Stacey Chandler, and Karen Alder Anderson. The JFK Library Oral Histories are pure historical gold. My longtime John F. Kennedy Foundation friends Rachel Day Flor and Elizabeth Murphy have supported this book from its conception, always pointing me in the right direction.

Multigenerational thanks go to members of the Kennedy family, who have informed this book in myriad ways. For decades, I've met with Ethel Kennedy at her homes in Palm Beach, Florida, and Hyannis Port, Massachusetts, to informally discuss the New Frontier–era political events. When August rolls around, I sometimes stay at Kerry Kennedy's lovely Cape Cod home with my family to sail, relax, and seize relief from the unrelenting heat in Austin, Texas. Every year, I work with the Robert F. Kennedy Human Rights Foundation at the request of President Kerry Kennedy and Executive Director Lynn Delaney.

Before his death in 2009, I toured Cape Cod with Senator Ted Kennedy to discuss his family's longtime interest in protecting treasured seashores. To my astonishment, he had memorized passages of Thoreau's "Walking." His wife, Victoria Reggie Kennedy, now US ambassador to

Austria, has done a wonderful job creating the Edward M. Kennedy Institute for the United States Senate. In recent years, I have sailed with Ted Jr. and Kiki Kennedy, Patrick and Amy Kennedy, Chris and Sheila Kennedy, and Joe and Lauren Kennedy in Hyannis Port. Jean Kennedy Smith, JFK's brilliant sister, was generous, recalling Henry Beston, the Outer Cape dunes, and the Atlantic milieu growing up. Before she died in 2020, she allowed me to crash at her New York City home so that I wouldn't waste money on a hotel room. Her son Stephen Kennedy Smith was my coauthor for *JFK: A Vision for America* to celebrate our thirty-fifth president's one hundredth birthday. Stephen's hospitality and kindness toward me are boundless. US Ambassador Caroline Kennedy Schlossberg, from her post in Australia, has likewise extended numerous courtesies to me.

Residing in Austin, Texas, has simplified my ability to study the extraordinary lives of Lyndon and Lady Bird Johnson. My family regularly hikes along the Pedernales River at the Reimers Ranch in Bee Cave, Texas, where thoughts of Mr. Johnson's propulsive ambition are often present in my mind. Liz Carpenter, who passed away in 2010, and her delightful daughter, Christy Carpenter (and husband Harvey Alan Levin), proved to be storehouses of Johnson family lore. Thomas Smith's biography *Stewart L. Udall: Steward of the Land* (2017) was indispensable to understanding the minutiae of Johnon's conservation world. Mark Winkleman has become the intellectual guru of the Austin book scene and has introduced me to many LBJ-relevant people over the years. I treasure him.

Often at dusk I jog in Austin's Zilker Park, Lady Bird's exquisite urban beautification triumph. Lynda Johnson Robb and Luci Baines Johnson, the president's delightful daughters, have enlightened me about their family's passion for the natural world at numerous dinners over the years. Luci and her husband, Ian Turpin, work diligently to help St. Edward's University in Austin become a higher-learning school of academic excellence. At the Lyndon Baines Johnson Presidential Library and Museum, the White House Center Files and first-rate oral history transcripts were very useful. In the LBJ–Lady Bird world, special thanks go to Mark Updegrove, Larry Temple, Claudia Anderson, Joe Califano, Chris Banks, Jennifer Cuddleback, Kathryn Hillhouse, Mark Lawrence, and chief archivist Jenna De Graffenried. Sharon Francis, a personal assistant to both Lady Bird Johnson and Stewart Udall, shared her unpublished memoir and hosted the Brinkley family at her Charlestown, New Hampshire, farm.

In central Texas, about fifty miles west of Austin, is the Lyndon B.

Johnson National Historical Park. During LBJ's presidency, the Ranch was known by the press as the "Texas White House." The federal park was authorized on December 2, 1969, by President Richard Nixon. Thanks to superintendent Justin Bates for running this bucolic "living history" site so well. I urge visitors to Austin to visit the LBJ Ranch (worth the drive) and the University of Texas's Lady Bird Johnson Wildflower Center.

I was lucky to interview many key Nixon-era figures profiled in this book. The indomitable Russell Train and kindhearted Nathaniel Reed were generous with their trench-warfare wisdom; both are now deceased. I greatly profited from J. Brooks Flippen's excellent *Nixon and the Environment* (2000) and authoritative *Conservative Conservationist: Russell E. Train and the Emergence of American Environmentalism* (2006).

Former EPA Administrator William Ruckelshaus—the eagle-eyed Indiana lawyer who oversaw the enactment of the Clean Air Act of 1970, and of the DDT ban of 1972—was a constant source of encouragement. Ruckelshaus consolidated a disparate set of environmental programs into a single federal agency: the Environmental Protection Agency (EPA). Fact: Ruckelshaus was one of the most able public servants in American history. In 2008, he asked me to conduct an oral history for the University of Washington–Seattle and Washington State University–Pullman (joint endeavor) of his entire life after enthusiastically reading my book *The Wilderness Warrior*. The loving hospitality of Bill and his wife, Jill, at their home in Medina, Washington, is forever ingrained in me. It was through Bill that I was introduced to hiking in Seattle's forest-rich Discovery Park, a Nixon-era urban playground. His death in 2019 devastated me. I wish he had been alive to proofread *Silent Spring Revolution*.

I was adopted as an auxiliary member of the Seattle conservation community's intelligentsia by Tom and Sonya Campion. Joel Connelly, the great environmental journalist of the Pacific Northwest, educated me about the North Cascades National Park battle and six Washington wilderness fights. And Doug Scott, a leading expert on the Wilderness Act, regaled me with background stories about the Long Sixties conservation zeitgeist. The University Archives at the University of Washington–Seattle houses the voluminous papers of both Henry M. Jackson and the North Cascades Conservation Council. And the William O. Douglas collection at the Yakima Valley Museum in Yakima, Washington, was likewise helpful. John Concillo, a Seattle film director, collaborated with me on *Liberty & Wilderness*, a documentary on Douglas's environmental activism. When in the Pacific Northwest I marvel that the threatened Chinook salmon (*Oncorhynchus tshawytscha*) swim through every major geographic zone of the Puget Sound area from urban Seattle/Tacoma/ Everett to agriculture-dominated rivers to lightly populated forests of the Olympic Peninsula. We must keep Washington State's streams flowing freely.

On the Richard Nixon front, I'm indebted to Ed Cox and Tricia Nixon Cox of New York City for helping me understand our thirty-seventh president's environmental policy motivations. Two of John Ehrlichman's children, Jan and Peter, couldn't have been kinder in sharing information about their father's deep connection to nature. Make no mistake about it: John Ehrlichman was one of the most significant environmentalists of the twentieth century, irrespective of his involvement with the Watergate nightmare.

At the Richard M. Nixon Presidential Library and Foundation, I was assisted by Jim Byron, whose knowledge of White House history is immense. Special thanks also to Frank Gannon, Mike Ellzey, Irwin Gellman, Jason Schultz, Ray Price, Ryan Pettigrew, and Dwight Chapin, all part of Nixonland. Professor Luke Nichter of Chapman University, who proofread chapters, knows more about Nixon than anybody alive. The H. R. Haldeman Diaries housed at the presidential library in Yorba Linda, California, are deeply illuminating on Nixon's shifting moods about environmental politics. The published version of *The Haldeman Diaries: Inside the Nixon White House* (New York: G. P. Putnam's Sons, 1994), edited by Stephen E. Ambrose, unfortunately focuses primarily on Watergate and Vietnam. Luckily, for environmental history, the unexpurgated Haldeman diaries are now available to scholars in Yorba Linda. The Atomic Museum in Las Vegas, Nevada, helped me better comprehend the magnitude of the Nevada nuclear tests of the 1950s and 1960s.

Words cannot express how much my friendships with Paul Chavez (Cesar Chavez's son) and Marc Grossman (farm worker activist extraordinaire) have meant to this book. They run the Cesar Chavez Foundation in Keene, California. Their noble mission is to expand on Cesar and Helen Chavez's environmental justice work. Likewise, two civil rights icons—the late John Lewis and the indispensable Andy Young—helped me understand the linkage between Martin Luther King, Jr., and the environmental justice crusade as personified by the Chavez movement. Elisabeth Reuther (daughter of the UAW leader Walter Reuther) and Representative Debbie Dingell (D-Michigan, wife of the late John Dingell) helped me better grasp Michigan politics in the 1950s, 1960s, and 1970s. Wayne State University in Detroit is the home of the vast Walter P. Reuther Library and Archives of Labor and Urban Affairs (including the Cesar Chavez and Walter P. Reuther papers). Today the UAW is known as the International Union United Automobile, Aerospace, and Agricultural Implement Workers of America.

The Gaylord Nelson Institute for Environmental Studies at the University of Wisconsin–Madison houses a plethora of documentary evidence about its namesake, the visionary founder of Earth Day. The Frank Church Institute at Boise State University is the well-organized home of the Idaho conservationist's personal and public papers. Thanks in Boise to Gary V. Weinske, Bob Kustra, Marlene Tromp, Monica Church, and Peter Fenn. At the Minnesota Historical Center in Saint Paul, the voluminous Hubert Humphrey and Orville Freeman papers were invaluable.

When the Interior Department building was named after Stewart Udall on June 8, 2010, I was asked by then secretary of interior Ken Salazar (now US ambassador to Mexico) to deliver a celebratory eulogy at the historic ceremony. It was a great honor that enabled me to meet scores of Udall's friends and relatives. His son, former US senator Tom Udall of New Mexico, has kept his father's environmental legacy alive in the twenty-first century. Tom is now the US ambassador to New Zealand and Samoa. Stewart Udall's papers are housed at the University of Arizona Library, Special Collections, alongside the Edward Abbey and Ansel Adams collections.

As the presidential historian at the New-York Historical Society in New York City, I'm grateful to work with CEO and president Louise Mirrer. Thanks to other friends at NYHS, including Agnes Hsu-Tang, Dale Gregory, Alex Kassl, John Monsky, Susan F. Peck, Andrew H. Tisch, Russell Pennoyer, Robert Caro, Miner Warner, Nancy Newcomb, and the formidable Pam B. Schafler. This post offered me the opportunity to have informal talks with many surviving members of the New Frontier and Great Society. In Oyster Bay, New York, I benefited from conversations with Heather Johnson (executive director of Friends of the Bay) and Marie Salerno (founding president and CEO of National Parks of New York Harbor Conservancy and the Federal Hall National Memorial).

For years now, I've been on the advisory board of the James Madison Council at the Library of Congress, my favorite federal institution. For this project, the Manuscript Division was very gracious in facilitating my access to the essential William O. Douglas and Barry Commoner papers collections. Special thanks to my friends Sue Siegal and David M. Rubenstein for always backing my action.

My intrepid friend Tim Palmer is the dean of American Rivers history. All of his books are golden. I was honored that he took time out from wilderness wanderings and rafting adventures to proofread various chapter drafts, saving me from the indignity of error. Because of the length of the book, I have refrained from asking anyone to proofread more than a few chapters. Other colleagues in the world of books who came to my aid include Walter Isaacson, Patricia Limerick, M. Margaret McKeown, Jane Mayer, Sara Dant, Char Miller, Doris Kearns Goodwin, Carl Hiaasen, William Souder, Michael Eric Dyson, and Beth Laski.

Conversations with Denis Hayes, John Dean, Pete Seeger, and George McGovern about Earth Day 1970 were extremely helpful. Adam Rome's *The Genius of Earth Day* (2014) is a masterpiece of historical scholarship. In addition, Allen Pietrobon, professor of history at Trinity University in Washington, DC, helped me better understand the remarkable antinuclear SANE career of Norman Cousins. For Florida history, I benefit from the Friends of the Everglades and the Marjory Stoneman Douglas Biscayne Nature Center. James Kushlan—former director of the Patuxent Wildlife Research Center and author of *Biscayne National Park* (2017)—carefully proofread chapters and answered my queries. Nobody, in my opinion, knows more about South Florida's ecological history than James Kushlan. The University of Miami library is the researcher-friendly depository of the Marjory Stoneman Douglas Papers, which include books, manuscripts, correspondences, photographs, diaries, newspaper articles, videos, and other primary materials of her indispensable life. She lived to 108 years old and kept a treasure trove of Floridiana documents to the very end.

On September 17, 2014, I was master of ceremonies for the 50th Anniversary of the Wilderness Act celebration, held at the Mayflower Renaissance Hotel in Washington, DC. The event featured powerful addresses by Senator Harry Reid (D-Nevada), Secretary of Interior Sally Jewell, and John Podesta, who was then counselor to President Barack Obama, leading the White House's climate and conservation efforts. The John P. Saylor Wilderness Leadership Award was presented to Senator Martin Heinrich (D-New Mexico). Also that night, the Howard C. Zahniser Lifetime Achievement Award went to Tim Mahoney for his forty-year career defending the National Wilderness Preservation System. All of these leaders helped me better understand environmental politics from a Washington insider perspective.

I'm on the advisory board of the Barack Obama Presidency Oral History Project at Columbia University in New York. Owing to my interest in environmental history, I was tapped as the lead interviewer of Gina

McCarthy, Christy Goldfuss, and other key climate change players of the Obama years. These long-form interviews provided me the opportunity to understand how NEPA, the Clean Air Act, and the EPA still anchor current environmental policy thinking. At Columbia, thanks to my advisory board colleagues, Lee C. Bollinger, Peter Bearman, Karida Brown, Jelani Cobb, Derek Chollet, Robert Dallek, Farah Jasmine Griffin, David Hollinger, Ira Katznelson, Kenneth Mack, Hannah Hankins, Helen Milner, Alondra Nelson, Michele Norris, Vicki Lynn Ruiz, Theda Skocpol, and Keith Wailoo.

The entire manuscript was enhanced by knowing Marilynn and John Hill, Helene Galen and Jamie Kabler, Bob and Oatsie Shrum, Chip and Jane Wiser, Geoff and Aileen Cowan, Brian and Victoria Lamb, Karen and Ben Cooper, John Avlon and Margaret Hoover, Andie Tucher, Steve Scully, Jeff Justice, Anderson Cooper, Andrea Lewis, Jim Irsay, Wolf Blitzer, Caryn Musil, Susan Swain, Douglas Bradburn, Matt Hannafin, Tom Stallings (Rice University), Rini Marcus, Ed O'Keefe, David Morton, Meena Bose, Shelly Austin, Hal Haddon, Melani Walton, Luke Metzger (Environment Texas), Michael Adams (son of Ansel Adams), Kati Anderson (the Walden Woods Project), Ben Riley, Harry Dennis, Adam Aron, Randy and Laurie Hatzenbuhler, David Hyde (son of Philip Hyde), Bob Utley, and Melody Webb. Also, both Victor John Yannacone, Jr., (cofounder of the Environmental Defense Fund) and Tom Turner (David Brower biographer) thoughtfully vetted parts of the manuscript. And my everyday inspiration, Terry Tempest Williams, shined over the project as if he were the reincarnation of Rachel Carson. My pal David Friend, the creative editor of Vanity Fair, paid me the high honor of editing my preface (nobody is better). As a contributing editor of Vanity Fair, working with Radhika Jones, I must say, is sublime.

Special thanks to the following historians and environmentalists who assisted me in answering various questions: Kabir Sehgal, Ariana Piper, Jack Loeffler, Robert Kennedy, Jr., Rick Ardinger, George Tobia, Mark Harvey, Deborah Dejah, Scott Einberger, William Cronon, Dick Beahrs, Mark Berejka, Sandy Bihn, James Bruggers, John de Graaf, Jay Udall, Lynn Udall, Denis Udall, Rob Bradley, Kenneth Brower, Jamie Rappaport Clark, John Cornely, Dan Chu, William Alsup, Cathy Douglas Stone, Karen Daubert, Lori Ehrlich, Tina Flourney, Ed Forgotson, Orly Jaffe, James M. Johnson, Michael J. Kellett, Michael Kern, John Kerry, Thomas C. Kierman, Patti Kenner, John Knox, Karen Bates Kress, Howard Labanara, Glenn Paulson, Susan Livingston, J. Michael McCloskey, Ralph Nader, Bob Latta, Maribeth Oakes, Peter O'Neill, Gillian Sorensen, Adam Frankel, Walter Robb, Shinee-Erkh Picon, David Raskin, Katrina vanden Heuvel, Steve McPherson, Joan Burroughs, Tim Richardson, Simon Roosevelt, Tweed Roosevelt, Theodore Roosevelt IV, Winthrop Roosevelt, Jill Kastner, Irston Barnes, Lesley Kane Saynal, Shannon Smith, D. Burr Udall, Jay Udall, Thomas Strickland, William Shafroth, Jamie Williams, Nikki Bufta, Brian Deese, John Leshy, Harry Dennis, Bill Howley, and John Suiter.

In the spring of 2022, the Eugene McCarthy Center at St. John's University in Collegeville, Minnesota, invited me to lecture around Earth Day. The center's then president, Jim Mullen, and his wife, Mari, are leaders of the "Return to Civility" movement in America and are dear friends. At St. John's, I benefited from time spent with Professor Derek R. Larson, author of the excellent *Keeping Oregon Green: Livability, Stewardship, and the Challenge of Growth, 1960–1980.*

The Society of Environmental Journalists (SEJ) provided me the opportunity to deliver two different keynote addresses with variations on Rachel Carson's impact. Thanks to such SEJ leaders as Sadie Bab, Sara Schonhardt, Michael Kodus, Kathiann Paskus, Tony Banboza, Sam Eaton, Rico Moore, Donovan Quintero, Meaghan Parker, Luke Ryan, Jay Letto, Megan Jeanelto, Christine Bruggers, Joseph A. Davis, Adam Geenon, Cindy MacDonald, and Beth Parke for keeping me apprised of modern-day public lands and natural resource battlegrounds. The talented writer Cynthia Barnett of the University of Florida tapped me to deliver keynote addresses for SEJ in Pittsburgh and Houston. George A. Cevasco and Richard P. Hammond's *Modern American Environmentalists: A Biographical Encyclopedia* provided very useful information in a crunch.

Three students assisted me with research at various stages of this book. Jackson Moffatt, a law student at Lewis & Clark in Portland, Oregon, worked with me in Austin during the COVID pandemic. My wife, Anne, and I adopted Jackson for six months as a member of our extended family. Jackson is a phenomenal deep-sea diver and is currently interning with the General Counsel of the NOAA's Oceans and Coasts section. Joshua Paul, a high schooler in Greenwich, Connecticut, who was interested in environmental law, helped me investigate California air pollution laws, Native American environmental lawsuits, and much more. Deeply interested in public affairs, he is now a student at Dartmouth University.

Sage Ranaldo, a recent graduate from the University of Toronto, also moved to Austin during the summer of 2021 and assisted me in myriad ways. Sage is a radio DJ, old-style scholar, and new-style IT hand.

His parents (Lee Ranaldo of the band Sonic Youth and his wife, artist Leah Singer) are family friends. They have raised Sage into an all-around scholar and gentleman.

Anna Carlton helped me in countless ways. Our family benefits from her generous spirit, academic insight, and marvelous sense of humor. And her parents, Don and Suzanne Carlton, are responsible for making the Dolph Briscoe Center of American History at the University of Texas–Austin such a pleasant place to research. Deborah McGill of North Carolina, at one crucial stage, helped me proofread chapters for accuracy.

Our close family friends Emma Juniper and her father, Christopher Juniper, taught me much about the counterculture 1960s and the environmental sustainability movement in America. At ICM, my agent, Sloan Harris, navigates me through the sharp, twisted turns of book publishing in our time of corporate consolidation.

Today, many of the conservation nonprofits from the 1950s, 1960s, and 1970s, which I write about, have evolved to work on climate change and environmental justice challenges of the twenty-first century. Their leadership is continuing to make a difference. I salute Michael Brune (executive director of the Sierra Club), Jennifer Morris (CEO of the Nature Conservancy), Janis Searles Jones (CEO of the Ocean Conservancy), Jamie Williams (president of the Wilderness Society) Melyssa Watson (executive director of the Wilderness Society), Collin D'Mara (CEO of the National Wildlife Federation), and Tom Kiernan (president and CEO of American Rivers).

Likewise, Mark Madison, historian at US Fish and Wildlife Service's National Conservation Training Center in Shepherdstown, West Virginia, shared backstories with me about Edward Abbey, E. O. Wilson, and Howard Zahniser. In 2020, Madison acquired Rachel Carson's personal possessions from her Silver Spring home, including the Underwood typewriter she used when writing about Chincoteague National Wildlife Refuge. Mentioning Abbey reminds me that Doug "Hayduke" Peacock is still wandering the Rockies, keeping track of grizzlies and plotting how to punish despoilers of our natural resources.

My friend Alison Whipple Rockefeller is the founder of the annual Rachel Carson Award for women in conservation. She has taken the global lead in educating the public about the enormous role women have played in environmental history—and she is a marvelous poet and public lands activist to boot. I've learned so much about Laurance Rockefeller, Sylvia Earle, and Rachel Carson from her.

At the University of California–Berkeley, I was invited to lecture on Rachel Carson. Special thanks to such Bay Area friends as Nate Brostrom, Carolyn Merchant, Jon Jarvis, Bernadette Powell, John Knox, Keith Gilless, and Alice Waters for making my stay so memorable. The Sierra Club Collection at the Bancroft Library, University of California–Berkeley was an incalculable treasure trove of Long Sixties conservation history.

Rice University has been my academic home for fifteen years. Recently retired President David Leebron, along with his wife, Y. Ping Sun, always supported my needs as an intrepid research scholar. Rice's dean of humanities, Kathleen Canning, is a rock star and buddy. Other colleagues at the university who have been supportive of this effort include Allen Matusow, Doug Miller, Jeff Falk, Tom Stallings, Philip Bedient, Lora Wildenthal, Carl Caldwell, Stephen Klineberg, Marcie Newton, Theresa Cisneros, Jim Blackburn, and Caleb McDaniel. The new president of Rice University, Reginald DesRoches, provided me with a sabbatical from teaching in 2022 that allowed me to meet my publishing deadline.

HarperCollins has been my publisher now for eleven books. I treasure my friendship with CEO Brian Murray and publisher Jonathan Burnham more than they know. Early on, I envisioned *Silent Spring Revolution* as an environmental history of the JFK–Rachel Carson–Stewart Udall years, planning to end the narrative in 1964, after Carson died and LBJ signed the Wilderness Act that fall. But Jonathan Jao, my esteemed editor, thought I should go for the whole enchilada—the US environmental movement from Hiroshima to the Endangered Species Act of 1973; he has my enduring gratitude. Helping Jao at Harper was his super assistant David Howe, who juggled myriad tasks with professional grace. And Harper's exemplary production team of designer Elina Cohen and production manager Diana Meunier supported my effort to write a prestigious book in the age of political quickies and celebrity confessions. They're pros. My appreciation for publicists Kate D'Esmond and Tom Hopke is boundless. Susan White, one of the nation's premier photo researchers, helped me locate indelible images that bring my historical characters to life.

My longtime friend Julie M. Fenster helped me organize, cut repetition, and provide tactical guidance throughout the writing of the manuscript. Julie's editorial skills and literary judgments are of the Day-Glo kind. I appreciated her noble battlefield assistance in every way, shape, and form.

My assistant for much of this project was Erika Holmes, a mother of three children in Austin. There were a hundred ways she made this book better. She is an environmentalist with a conscientious disposition for scholarly work, and makes every day sparkle with her warmth and charm. Her husband, Garrick Bell, a career intelligence and law enforcement professional, was kind to share Erika with me at my home office.

My ninety-four-year-old father, Ed Brinkley, remains my Rock of Gibraltar. When I was a boy, he took me hiking in the Appalachians and Rockies, and I haven't stopped hitting the trail since. His personal Walden Pond is Cook Forest State Park, along the Clarion River in western Pennsylvania. Mine is the 315-mile-long scenic Hudson River as it flows downward from its Tear of the Clouds source and empties into New York City's grand harbor. My sister, Leslie Brinkley, a longtime reporter for San Francisco's ABC News affiliate KGO, kept me apprised of all sorts of Bay Area environment stories while I labored on this book.

None of this historical journey back to the Long Sixties would have been possible without my loving wife, Anne. We're partners in every sense of the word. Together, we have raised three remarkable children—Benton Grace, Johnny Cleland, and Cassady Anne—who have all matured into naturalists at heart. Whether rafting down the Deschutes River in Oregon, exploring Walden Pond in Concord, or following Lady Bird Johnson's footsteps in Big Bend, they made writing *Silent Spring Revolution* an adventure. Our family has stuck by David Brower's sage advice to environmentalists: "Have a good time saving the world. Otherwise, you're just going to depress yourself."

National Wildlife Refuges

Kennedy Administration

Wapanocca National Wildlife Refuge	Arkansas	February 1, 1961
Washita National Wildlife Refuge	Oklahoma	April 15, 1961
Ottawa National Wildlife Refuge	Ohio	July 28, 1961
Wyandotte National Wildlife Refuge	Michigan	August 3, 1961
Moody National Wildlife Refuge	Texas	November 9, 1961
Harris Neck National Wildlife Refuge	Georgia	May 25, 1962
Delevan National Wildlife Refuge	California	September 12, 1962
Cross Creeks National Wildlife Refuge	Tennessee	November 9, 1962
Eastern Neck National Wildlife Refuge	Maryland	December 27, 1962
Anahuac National Wildlife Refuge	Texas	February 27, 1963
John Heinz National Wildlife Refuge at Tinicum	Pennsylvania	March 18, 1963
Alamosa National Wildlife Refuge	Colorado	July 25, 1963
Pahranagat National Wildlife Refuge	Nevada	August 6, 1963

Prime Hook National Wildlife Refuge	Delaware	August 8, 1963
Merritt Island National Wildlife Refuge	Florida	August 28, 1963
Lake Woodruff National Wildlife Refuge	Florida	November 18, 1963

Johnson Administration

Choctaw National Wildlife Refuge	Alabama	January 27, 1964
Lee Metcalf National Wildlife Refuge	Montana	April 10, 1964
Toppenish National Wildlife Refuge	Washington	April 27, 1964
Pee Dee National Wildlife Refuge	North Carolina	May 13, 1964
William L. Finley National Wildlife Refuge	Oregon	June 17, 1964
Clarence Cannon National Wildlife Refuge	Missouri	August 11, 1964
Cedar Island National Wildlife Refuge	North Carolina	August 18, 1964
Cibola National Wildlife Refuge	Arizona/ California	August 21, 1964
Kootenai National Wildlife Refuge	Idaho	August 31, 1964
Eufaula National Wildlife Refuge	Georgia/ Alabama	September 1, 1964
Hatchie National Wildlife Refuge	Tennessee	November 16, 1964
Cedar Point National Wildlife Refuge	Ohio	December 18, 1964
Ankeny National Wildlife Refuge	Oregon	January 18, 1965
Conboy Lake National Wildlife Refuge	Washington	April 14, 1965
Grays Lake National Wildlife Refuge	Idaho	June 17, 1965
Browns Park National Wildlife Refuge	Colorado	July 13, 1965

Sherburne National Wildlife Refuge	Minnesota	September 8, 1965
Baskett Slough National Wildlife Refuge	Oregon	October 22, 1965
Seedskadee National Wildlife Refuge	Wyoming	November 30, 1965
Ridgefield National Wildlife Refuge	Washington	January 27, 1966
Las Vegas National Wildlife Refuge	New Mexico	April 25, 1966
Maxwell National Wildlife Refuge	New Mexico	April 26, 1966
Flint Hills National Wildlife Refuge	Kansas	September 1, 1966
Muscatatuck National Wildlife Refuge	Indiana	October 6, 1966
Brazoria National Wildlife Refuge	Texas	October 17, 1966
Rachel Carson National Wildlife Refuge	Maine	December 21, 1966
San Luis National Wildlife Refuge	California	February 2, 1967
Arapaho National Wildlife Refuge	Colorado	September 26, 1967
UL Bend National Wildlife Refuge	Montana	October 30, 1967
Target Rock National Wildlife Refuge	New York	December 15, 1967
St. Vincent National Wildlife Refuge	Florida	February 12, 1968
Bear Lake National Wildlife Refuge	Idaho	May 9, 1968
Hobe Sound National Wildlife Refuge	Florida	September 23, 1968
Seatuck National Wildlife Refuge	New York	September 26, 1968
Grulla National Wildlife Refuge	New Mexico/ Texas	November 7, 1968
San Bernard National Wildlife Refuge	Texas	November 7, 1968
Amagansett National Wildlife Refuge	New York	December 16, 1968

Oyster Bay National Wildlife Refuge	New York	December 18, 1968
Buck Island National Wildlife Refuge	Virgin Islands	January 8, 1969
Fisherman Island National Wildlife Refuge	Virginia	January 17, 1969

Nixon Administration

Mason Neck National Wildlife Refuge	Virginia	February 1, 1969
Fisherman Island National Wildlife Refuge	Virginia	January 17, 1969
Buck Island National Wildlife Refuge	Virgin Islands	January 8, 1969
Umatilla National Wildlife Refuge	Oregon	July 3, 1969
Umatilla National Wildlife Refuge	Washington	July 3, 1969
Wassaw National Wildlife Refuge	Georgia	October 20, 1969
Nomans Land Island National Wildlife Refuge	Massachusetts	April 29, 1970
Ninigret National Wildlife Refuge	Rhode Island	August 12, 1970
Sachuest Point National Wildlife Refuge	Rhode Island	November 3, 1970
Sequoyah National Wildlife Refuge	Oklahoma	December 11, 1970
Wallops Island National Wildlife Refuge	Virginia	March 11, 1971
Conscience Point National Wildlife Refuge	New York	July 20, 1971
St. Johns National Wildlife Refuge	Florida	August 16, 1971
Julia Butler Hansen Refuge for the Columbian White-Tail Deer	Washington	December 17, 1971
Lewis and Clark National Wildlife Refuge	Oregon	April 19, 1972

Plum Tree National Wildlife Refuge	Virginia	April 24, 1972
Wapack National Wildlife Refuge	New Hampshire	May 17, 1972
Attwater Prairie Chicken National Wildlife Refuge	Texas	July 1, 1972
Seal Island National Wildlife Refuge	Maine	July 24, 1972
Thatcher Island National Wildlife Refuge	Massachusetts	July 25, 1972
Pearl Harbor National Wildlife Refuge	Hawaii	October 17, 1972
Meredosia National Wildlife Refuge	Illinois	October 25, 1972
Hanalei National Wildlife Refuge	Hawaii	November 30, 1972
Great Dismal Swamp National Wildlife Refuge	North Carolina	February 22, 1973
Great Dismal Swamp National Wildlife Refuge	Virginia	February 22, 1973
Pond Island National Wildlife Refuge	Maine	March 9, 1973
Humboldt Bay National Wildlife Refuge	California	April 4, 1973
Huleia National Wildlife Refuge	Hawaii	April 25, 1973
Nantucket National Wildlife Refuge	Massachusetts	May 1, 1973
Swan River National Wildlife Refuge	Montana	May 14, 1973
Salinas National Wildlife Refuge	California	June 27, 1973
Occoquan Bay National Wildlife Refuge	Virginia	June 29, 1973
Rose Atoll National Wildlife Refuge	American Samoa	August 24, 1973
Franklin National Wildlife Refuge	Maine	September 19, 1973
Supawna Meadows National Wildlife Refuge	New Jersey	September 23, 1973

Block Island National Wildlife Refuge	Rhode Island	November 1, 1973
Nansemond National Wildlife Refuge	Virginia	December 20, 1973
Sevilleta National Wildlife Refuge	New Mexico	December 28, 1973
Hopper Mountain National Wildlife Refuge	California	February 6, 1974
Billy Frank Jr. Nisqually National Wildlife Refuge	Washington	February 21, 1974
Cabo Rojo National Wildlife Refuge	Puerto Rico	May 20, 1974
Oxbow National Wildlife Refuge	Massachusetts	May 24, 1974
Baker Island National Wildlife Refuge	Baker Island	June 2, 1974
Howland Island National Wildlife Refuge	Howland Island	June 27, 1974
Seal Beach National Wildlife Refuge	California	July 5, 1974
Petit Manan National Wildlife Refuge	Maine	July 9, 1974
Egmont Key National Wildlife Refuge	Florida	July 10, 1974
Jarvis Island National Wildlife Refuge	Jarvis Island	July 27, 1974
Trustom Pond National Wildlife Refuge	Rhode Island	August 15, 1974
San Pablo Bay National Wildlife Refuge	California	December 18, 1974

National Parks

Kennedy Administration

Russel Cave National Monument, AL	May 11, 1961, proclaimed
Cape Cod National Shoreline, MA	August 7, 1961 authorized, established June 1, 1966
Fort Davis National Historic Site, TX	September 8, 1961, authorized; established July 4, 1963
Fort Smith National Historic Site, AR	September 13, 1961, authorized
Piscataway Park, MD	October 4, 1961, authorized
Buck Island Reef National Monument, VI	December 28, 1961, proclaimed
Lincoln Boyhood National Memorial, IN	February 19, 1962, authorized
Hamilton Grange National Memorial, NY	April 27, 1962, authorized
Theodore Roosevelt Birthplace National Historic Site, NY	July 25, 1962, authorized
Sagamore Hill National Historic Site, NY	July 25, 1962, authorized
Frederick Douglass National Historic Site, DC	September 5, 1962, authorized as Frederick Douglass Home; redesignated February 12, 1988

Point Reyes National Shoreline, CA	September 13, 1962, authorized; established October 20, 1972
Padre Island National Shoreline, TX	September 28, 1962, authorized; established April 6, 1968

Johnson Administration

Ozark National Scenic Riverways, MO	August 27, 1964, authorized; established June 10, 1972
Fort Bowie National Historic Site, AZ	August 30, 1964, authorized; established July 29, 1972
Fort Larned National Historic Site, KS	August 31, 1964, authorized; established October 14, 1966
Saint-Gaudens National Historic Site, NH	August 31, 1964, authorized; established May 30, 1977
Allegheny Portage Railroad National Historic Site, PA	August 31, 1964, authorized
Johnstown Flood National Memorial, PA	August 31, 1964, authorized
John Muir National Historic Site, CA	August 31, 1964, authorized
Fire Island National Shoreline, NY	September 11, 1964, authorized
Canyonlands National Park, UT	September 12, 1964, established
Ice Age National Scientific Reserve, WI	October 13, 1964 (affiliated area)
Bighorn Canyon National Recreation Area, WY-MT	Administered under cooperative agreement with Bureau of Reclamation DOI, December 31, 1964; established October 15, 1966
Arbuckle National Recreation, OK	February 1, 1965; absorbed by Chickasaw National Recreation Area March 17, 1976
Curecanti National Recreation Area, CO	Administered under cooperative agreement with Bureau of Reclamation DOI, February 11, 1965

Lake Meredith National Recreation Area, TX	March 15, 1965, cooperative agreement with Bureau of Reclamation empowers the National Park Service to administer area then called the Sanford National Recreation Area; name changed to Lake Meredith Recreation Area October 16, 1972; redesignated a national recreation area November 28, 1990
Nez Perce National Historical Park, ID	May 15, 1965, authorized
Agate Fossil Beds National Monument, NE	June 5, 1965, authorized
Pecos National Historical Park, NM	June 28, 1965, authorized; redesignated June 27, 1990
Herbert Hoover National Historical Park, IA	August 12, 1965, authorized
Alibates Flint Quarries National Monument, TX	August 21, 1965, authorized as Alibates Flint Quarries and Texas Panhandle Pueblo Culture National Monument; redesignated November 10, 1978
Hubbell Trading Post National Historic Site, AZ	August 28, 1965, authorized
Delaware Water Gap National Recreation Area, PA-NJ	September 1, 1965, authorized
Assateague Island National Seashore, MD-VA	September 21, 1965, authorized
Pennsylvania Avenue National Historical Site, DC	September 30, 1965
Roger Williams National Memorial, RI	October 22, 1965, authorized
Whiskeytown-Shasta-Trinity National Recreation Area, CA	November 8, 1965, authorized; established October 21, 1972
Amistad National Recreation Area, TX	November 11, 1965, cooperative agreement with US Section International Boundary and Water Commission, US and Mexico

Cape Lookout National Shoreline, NC	March 10, 1966, authorized
Fort Union Trading Post National Historical Site, ND-MT	June 20, 1966, authorized
Chamizal National Memorial, TX	June 30, 1966, authorized; established February 4, 1974
George Rogers Clark National Historical Park, IN	July 23, 1966, authorized
San Juan Island National Historical Park, WA	September 9, 1966, authorized
Guadalupe Mountains National Park, TX	October 15, 1966, authorized; established September 30, 1972
Pictured Rocks National Lakeshore, MI	October 15, 1966, authorized
Wolf Trap National Park for the Performing Arts, VA	October 15, 1966, authorized; renamed August 2002
Theodore Roosevelt Inaugural National Historic Site, NY	November 2, 1966, authorized
Indiana Dunes National Lakeshore, IN	November 5, 1966, authorized
John F. Kennedy National Historic Site, MA	May 26, 1967, authorized
Eisenhower National Historic Site, PA	November 27, 1967, designated
Saugus Iron Works National Historic Site, MA	April 5, 1968, authorized
North Cascades National Park, WA	October 2, 1968, established
Lake Chelan National Recreation Area, WA	October 2, 1968, established
Ross Lake National Recreation Area, WA	October 2, 1968, established
Redwood National Park, CA	October 2, 1968, established
Appalachian National Scenic Trail, ME-NH-VT-MA-CT-NY-NJ-PA-MD-VA-WV-TN-NC-GA	October 2, 1968, established

Carl Sandburg Home National Historic Site, NC	October 17, 1968, authorized; established October 27, 1972
Biscayne National Park, FL	October 18, 1968, as national monument; redesignated national park June 28, 1980
Mar-A-Lago National Historic Site, FL	January 16, 1969 (deauthorized in 1980)
Marble Canyon National Monument, AZ	January 20, 1969; absorbed by Grand Canyon National Park January 3, 1975

Nixon Administration

Florissant Fossil Beds National Monument, CO	August 20, 1969, authorized
Saint Croix National Scenic Riverway, WI-MN	September 4, 1969, under National Park Service; authorized October 2, 1968
Wolf National Scenic Riverway, WI	September 4, 1969, under National Park Service; later removed from system
William Howard Taft National Historic Site, OH	December 2, 1969, authorized
Lyndon B. Johnson National Historical Park, TX	December 2, 1969, authorized as national historic site; redesignated national historical park December 28, 1980
Theodore Roosevelt Island, DC	January 1, 1970; authorized May 21, 1932; transferred off of Public Buildings and Public Parks of the National Capital August 10, 1933, but not counted as separate area until January 1, 1970, in National Parks and Landmarks
Apostle Islands National Lakeshore, WI	September 26, 1970, established
Andersonville National Historic Site, GA	October 16, 1970, authorized
Fort Point National Historic Site, CA	October 16, 1970, authorized

Sleeping Bear Dunes National Lakeshore, MI	October 21, 1970, authorized; established October 21, 1977
Gulf Islands National Seashore, FL-MS	January 8, 1971, authorized
Voyageurs National Park, MN	January 8, 1971, authorized; established April 8, 1975
Lincoln Home National Historic Site, IL	August 18, 1971, authorized
Buffalo National River, AR	March 1, 1972, authorized
John F. Kennedy Center for Performing Arts, DC	Authorized as Natural Cultural Center September 2, 1958; name changed January 23, 1964; non-performing arts fns transferred from Smithsonian Inst to NPS June 16, 1972; transferred back to Kennedy Center for Performing Arts Board of Directors October 16, 1994
Puukohola Heiau National Historic Site, HI	August 17, 1972, authorized
John D. Rockefeller, Jr. Memorial Parkway, WY	August 25, 1972, authorized
Grant-Kohrs Ranch National Historic Site, MT	August 25, 1972, authorized
Longfellow National Historic Site, MA	October 9, 1972, authorized
Hohokam-Pima National Monument, AZ	October 21, 1972, authorized
Thaddeus Kosciuszko National Memorial, PA	October 21, 1972, authorized
Cumberland Island National Seashore, GA	October 23, 1972, established
Fossil Butte National Monument, WY	October 23, 1972, established
Lower Saint Croix National Scenic Riverway, MN-WI	October 25, 1972, authorized; combined with Saint Croix National Scenic Riverway February 1993

Benjamin Franklin National Monument, PA	October 25, 1972, designated; owned and administered by the Franklin Institute
Gateway National Recreation Area, NJ-NY	October 27, 1972, established
Golden Gate National Recreation Area, CA	October 27, 1972, established
Big South Fork National River and Recreation Area, KY-TN	March 7, 1974, authorized; transferred to National Park Service management October 22, 1976
Boston National Historical Park, MA	October 1, 1974, authorized
Big Cypress National Preserve, FL	October 11, 1974, authorized
Big Thicket National Preserve, TX	October 11, 1974, authorized
John Day Fossil Beds National Monument, OR	October 26, 1974, authorized
Knife River Indian Villages National Historic Site, ND	October 26, 1974, authorized
Martin Van Buren National Historic Site, NY	October 26, 1974, authorized
Springfield Armory National Historic Site, MA	October 26, 1974, authorized
Tuskegee Institute National Historic Site, AL	October 26, 1974, authorized
Clara Barton National Historic Site, MD	October 26, 1974, authorized
Sewall-Belmont House National Historic Site, DC	October 26, 1974, authorized
Cuyahoga Valley National Park, OH	December 27, 1974, authorized; established June 26, 1975, as a national recreation area; established as a national park in November 2000

Protection for Animals
Initial Endangered Species List:
1966–1967

This preliminary list consisted of 331 species divided into three categories of concern: 130 species either rare or endangered; 74 species at the edge of their range (and thereby at risk); and 127 species of "undetermined" status.

Mammals

Indiana bat: *Myotis sodalis*
Delmarva Peninsula fox squirrel: *Sciurus niger cinereus*
Timber wolf: *Canis lupus lycaon*
Red wolf: *Canis niger*
San Joaquin kit fox: *Vulpes macrotis mutica*
Grizzly bear: *Ursus horribilis*
Black-footed ferret: *Mustela nigripes*
Florida panther: *Felis concolor coryi*
Caribbean monk seal: *Monachus tropicalis*
Guadalupe fur seal: *Arctocephalus philippi townsendi*
Florida manatee: *Trichechus manatus latirostris*
Key deer: *Odocoileus virginianus clavium*
Columbian white-tailed deer: *Odocoileus virginianus leucurus*
Sonoran pronghorn: *Antilocapra americana sonoriensis*

Birds

Hawaiian dark-rumped petrel: *Pterodroma phaeopygia sandwichensis*
Hawaiian goose (nene): *Branta sandvicensis*

Aleutian Canada goose: *Branta canadensis leucopareia*

Tule white-fronted goose: *Anser albifrons gambelli*

Laysan duck: *Anas laysanensis*

Hawaiian duck (koloa): *Anas wyvilliana*

Mexican duck: *Anas diazi*

California condor: *Gymnogyps californianus*

Florida Everglade snail kite: *Rostrhamus sociabilis plumbeus*

Hawaiian hawk (ii): *Buteo solitarius*

Southern bald eagle: *Haliaeetus leucocephalus*

Attwater's greater prairie-chicken: *Tympanuchus cupido attwateri*

Masked bobwhite: *Colinus virginianus ridgwayi*

Whooping crane: *Grus americana*

Yuma clapper rail: *Rallus longirostris yumanensis*

Hawaiian common gallinule: *Gallinula chloropus sandvicensis*

Eskimo curlew: *Numenius borealis*

Puerto Rican parrot: *Amazona vittata*

American ivory-billed woodpecker: *Campephilus principalis*

Hawaiian crow (alala): *Corvus hawaiiensis*

Small Kauai thrush (puaiohi): *Phaeornia pulmeri*

Nihoa millerbird: *Acrocephalus kingi*

Kauai oo (oo aa): *Moho braccatus*

Crested honeycreeper (akohekohe): *Palmeria dolei*

Akiapolaau: *Hemignathus wilsoni*

Kauai akialoa: *Hemignathus procerus*

Kauai nukupuu: *Hemignathus lucidus hanapepe*

Laysan finchbill (Laysan finch): *Psittirostra cantans*

Nihoa finchbill (Nihoa finch): *Psittirostra cantans ultima*

Ou: *Psittirostra psittacea*

Palila: *Psittirostra bailleui*

Maui parrotbill: *Pseudonestor xanthophyrys*

Bachman's warbler: *Vermivora bachmanii*

Kirtland's warbler: *Dendroica kirtlandii*

Dusky seaside sparrow: *Ammospiza nigrescens*

Cape sable sparrow: *Ammospiza mirabilis*

Reptiles and Amphibians

American alligator: *Alligator mississippiensis*
Blunt-nosed leopard lizard: *Crotaphytus wislizenii silus*
San Francisco garter snake: *Thamnophis sirtalis tetrataenia*
Santa Cruz long-toed salamander: *Ambystoma macrodactylum croceum*
Texas blind salamander: *Typhlomolge rathbuni*
Black toad (Inyo County toad): *Bufo exsul*

Fish

Shortnose sturgeon: *Acipenser brevirostrum*
Longjaw cisco: *Coregonus alpenae*
Paiute cutthroat trout: *Salmo clarki seleniris*
Greenback cuttthroat trout: *Salmo clarki stomias*
Montana Westslope cutthroat trout: *Salmo clarki*
Gila trout: *Oncorhynchus gilae*
Arizona trout (Apache trout): *Oncorhynchus apache*
Desert dace: *Eremichthys acros*
Humpback chub: *Gila cypha*
Little Colorado spinedace: *Lepidomeda vittata*
Moapa dace: *Moapa coriacea*
Colorado River squawfish: *Ptychocheilus lucius*
Cui-ui: *Chasmistes cujus*
Devils Hole pupfish: *Cyprinodon diabolis*
Comanche Springs pupfish: *Cyprinodon elegans*
Owens River pupfish: *Cyprinodon radiosus*
Pahrump killifish: *Empetrichythys latos*
Big Bend gambusia: *Gambusia gaigei*
Clear Creek gambusia: *Gambusia heterochir*
Gila topminnow: *Poeciliopsis occidentalis*
Maryland darter: *Etheostoma sellare*
Blue pike: *Stizostedion vitreum glaucum*

Notes

Chapter 1: The Ebb and Flow of John F. Kennedy

1. John F. Kennedy, "Remarks at the America's Cup Dinner Given by the Australian Ambassador, September 14, 1962," John F. Kennedy Presidential Library and Museum, https://www.jfklibrary.org/Research/Research-Aids/JFK-Speeches/Americas-Cup-Dinner_19620914.aspx.

2. Anthony J. Czarnecki, "When the Kennedy Family Lived in Westchester County," *Westchester Historian* 93, no. 2 (Spring 2017): 39.

3. "A Plea for a Raise by Jack Kennedy; Dedicated to my [*sic*] Mr. J. P. Kennedy," letter from John F. Kennedy to Joseph Kennedy, Sr., February 1932, in John F. Kennedy, *The Letters of John F. Kennedy*, ed. Martin W. Sandler (New York: Bloomsbury Press, 2013), 6.

4. Barbara A. Perry, *Rose Kennedy: The Life and Times of a Political Matriarch* (New York: W. W. Norton, 2013), 11–16.

5. "Rose Kennedy Dies," *Life*, March 1, 1995.

6. W. Barksdale Maynard, *Walden Pond: A History* (New York: Oxford University Press, 2004), pp. 265–66.

7. Author interview with Edward Kennedy, July 3, 2007.

8. Henry David Thoreau, *Walden; or, Life in the Woods, and On the Duty of Civil Disobedience* (New York: Rinehart, 1948).

9. Robert F. Kennedy, Jr., *American Values: Lessons I Learned from My Family* (New York: HarperCollins, 2018), 47.

10. Jean Kennedy Smith, *The Nine of Us: Growing Up Kennedy* (New York: HarperCollins, 2016), 103.

11. James W. Graham, *Victura: The Kennedys, a Sailboat, and the Sea* (Lebanon, NH: ForeEdge, 2014), 17–18, 29–31, 49–51.

12. Henry David Thoreau, *Cape Cod* (Boston: Houghton Mifflin, 1864), reprinted in *The Writings of Henry David Thoreau*, vol. 5 (Boston: Houghton Mifflin, 1906), 272–73.

13. Henry Beston, *The Outermost House: A Year in the Life on the Great Beach of Cape Cod* (Garden City, NY: Doubleday, Doran, 1928).

14. Henry Beston, *Especially Maine; The Natural World of Henry Beston from Cape Cod to the St. Lawrence* (Brattleboro, VT: Stephen Greene Press, 1970), 24.

15. Author interview with Kerry Kennedy (Rose's granddaughter), August 17, 2015.

16. Ibid.

17. Beston, *The Outermost House*, 43.

18. "Henry Beston, 79, Author, Is Dead," *New York Times*, April 17, 1968, 47, https://www.nytimes.com/1968/04/17/archives/henry-beston-79-author-is -dead-his-outermost-house-told-of-life.html.

19. Robert Frost, "West-Running Brook," in *West-Running Brook* (New York: Henry Holt, 1928), 35.

20. Author interview with Edward Kennedy, July 3, 2007; Thoreau, *Walden*, 263.

21. Douglas Brinkley, *Rightful Heritage: Franklin D. Roosevelt and the Land of America* (New York: HarperCollins, 2016), 135.

22. Ibid., 171–75.

23. Robert Dallek, *An Unfinished Life: John F. Kennedy, 1917–1963* (New York, UK: Oxford University Press, 2011), 38.

24. Brian J. Murphy, "The Making of JFK," *America in WWII*, August 2008, 29.

25. Quoted in Graham, *Victura*, 157.

26. Kennedy, *American Values*, 31.

27. Author interview with Jean Kennedy Smith, May 15, 2008.

28. Vardis Fisher, *Mountain Man* (New York: William Morrow, 1965), preface.

29. Bruce Allen Murphy, *Wild Bill: The Legend and Life of William O. Douglas* (New York: Random House, 2003), 20.

30. William O. Douglas, *Of Men and Mountains* (New York: Harper & Brothers, 1950), 329.

31. Ibid., 292.

32. John D. Foster, "Bureau of Land Management Primitive Areas—Are They Counterfeit Wilderness?," *Natural Resources Journal*.

33. James F. Simon, *Independent Journey: The Life of William O. Douglas* (Berkeley: University of California, 1980), 7.

34. Daniela Deane, "Where the Kennedys and Gores Played," *Washington Post*, February 14, 2004.

35. William O. Douglas, *The Court Years, 1939–1975: The Autobiography of William O. Douglas* (New York: Random House, 1980), 301.

36. Yi-Fu Tuan, *Topophlia: A Study of Perceptions* (New York: Columbia University Press, 1974).

37. William O. Douglas, *My Wilderness: East to Katahdin* (Garden City, NY: Doubleday, 1961), 290.

38. Noah Feldman, *Scorpions: The Battles and Triumphs of FDR's Great Supreme Court Justices* (New York: Twelve Publishing, 2010), 171.

39. Author interview with William Haskell Alsup (Douglas's Supreme Court clerk, 1971 to 1973), July 6, 2021.

40. Victor John Yannacone, Jr., to author, February 25, 2021.

41. Michael O'Brien, *John F. Kennedy: A Biography* (New York: St. Martin's Press, 2005), 78–79.

42. John F. Kennedy to Lem Billings, May 15, 1936, John F. Kennedy Presidential Library, Boston, Massachusetts, JFKL.

43. Peter S. Britell, "Kennedy at Harvard," *Harvard Crimson*, November 4, 1960. See also John Clarke, "Selling J.F.K.'s Boat," *New Yorker*, May 19, 2015.

44. Quoted in Nigel Hamilton, *JFK: Reckless Youth* (New York: Random House, 1992), 513.

Chapter 2: Harry Truman

1. Quoted in Mark Fiege, *The Republic of Nature* (Seattle: University of Washington Press, 2012), 310.

2. Terrence R. Fehner and F. G. Gosling, *Battlefield of the Cold War*, vol. 1, *The Nevada Test Site, Atmospheric Nuclear Weapons Testing, 1951–1963* (Washington, DC: US Department of Energy, 2006), 26–28.

3. Richard Rhodes, *The Making of the Atomic Bomb* (New York: Simon & Schuster, 1986), 676.

4. Richard G. Hewlett and Oscar E. Anderson, Jr., *The New World 1939–46: A History of the United States Atomic Energy Commission*, vol. 1 (University Park, PA: Penn State University, 1962), 379.

5. A. Constandina Titus, *Bombs in the Backyard: Atomic Testing and American Politics* (Reno: University of Nevada Press, 1986), 16.

6. Ibid., 14.

7. Rhodes, *The Making of the Atomic Bomb*, 736.

8. Quoted in Bruce Allen Murphy, *Wild Bill: The Legend and Life of William O. Douglas* (New York: Random House, 2003), 608.

9. Quoted in Paul Boyer, *By the Bomb's Early Light: American Thought and Culture at the Dawn of the Atomic Age* (New York: Pantheon, 1985), 5–7.

10. Norman Cousins, *Modern Man Is Obsolete* (New York: Viking, 1946), 3–24.

11. Norman Cousins, *Albert Schweitzer's Mission: Healing and Peace* (New York: W. W. Norton, 1985); Norman Cousins, *Dr. Schweitzer of Lambaréné* (New York: Harper & Brothers, 1960).

12. Norman Cousins, "Modern Man Is Obsolete," *Saturday Review of Literature*, August 18, 1945, 5–7, https://rachelcarsoncouncil.org/wp-content/uploads/2018/01/RCC.Cousins.-SatRev.ModernMan.pdf.

13. E. B. White, "Notes and Comment," *New Yorker*, August 18, 1945, 13.

14. John F. Kennedy, "Remarks of John F. Kennedy, United War Fund Appeal, Boston, Massachusetts, October 8, 1945," John F. Kennedy Presidential Library and Museum, https://www.jfklibrary.org/archives/other-resources /john-f-kennedy-speeches/boston-ma-19451008.

15. Ibid.

16. David M. Blades and Joseph M. Siracusa, *A History of U.S. Nuclear Testing and Its Influence on Nuclear Thought, 1945–1963* (Lanham, Maryland: Rowman & Littlefield, 2014), 165.

17. Quoted in Titus, *Bombs in the Backyard*, 42.

18. Stafford Warren, "Conclusions: Tests Proved Irresistible Spread of Radioactivity," *Life*, August 11, 1947, 88.

19. "Nuclear Testing Legacy Is 'Cruelest' Environmental Injustice, Warns Rights Expert," UN News, July 16, 2020, https://news.un.org/en/story /2020/07/1068481.

20. Barton C. Hacker, *Elements of Controversy* (Berkeley: University of California Press, 1994), 6–9.

21. T. H. Watkins, *Righteous Pilgrim: the Life and Times of Harold Ickes*, vol. 2 (New York: Henry Holt, 1990), 118.

22. Ibid.

23. Harold Ickes, "Should Congress Vest Ownership of the Tidelands in the States?," *Congressional Digest*, October 1, 1943, 255.

24. "Harold L. Ickes Dead at 77; Colorful Figure in New Deal," *New York Times*, February 4, 1952, 1, 18, https://www.nytimes.com/1952/02/04/archives /harold-l-ickes-dead-at-77-colorful-figure-in-new-deal-selfstyled.html.

25. Coral Davenport and Lisa Friedman, "Interior Dept. Report on Drilling Is Mostly Silent on Climate Change," *New York Times*, November 27, 2021, A19, https://www.nytimes.com/2021/11/26/climate/climate-change-drilling -public-lands.html.

26. Drew Pearson, "The Washington Merry-Go-Round," syndicated column, November 16, 1949.

27. Michael Grunwald, "Harry Truman, South Florida, and the Changing Political Geography of American Conservation," in *The Environmental Legacy of Harry S. Truman*, ed. Karl Boyd Brooks (Kirksville, MO: Truman State University Press, 2009), 75.

28. Daniel Raimi, *The Fracking Debate: The Risks, Benefits, and Uncertainties of the Shale Revolution* (New York: Columbia University Press, 2018), 14.

29. Wendell Berry, *Our Only World: Ten Essays* (Berkeley, CA: Counterpoint Press, 2015), 24.

30. Quoted in Mark Harvey, "Sound Politics: Wilderness, Recreation, and Motors in the Boundary Waters, 1945–1964," *Minnesota History*, Fall

2002, 130, http://collections.mnhs.org/MNHistoryMagazine/articles/58
/v58i03p130-145.pdf.

31. David A. Dalton, *The Natural World of Lewis and Clark* (Columbia: University of Missouri Press, 2008), 198.

32. James K. Meissner, *Land Management: The Forest Service's and BLM's Organizational and Structural Responsibilities* (Washington, DC: United States General Accounting Office, 1999), 14.

33. Peggy and Edgar Wayburn, "Conservation in 1960: The Summing Up," *Sierra Club Bulletin*, January 1961, 12.

34. William O. Douglas, *The Autobiography of William O. Douglas*, vol. 2, *The Court Years, 1939–1975* (New York: Random House, 1980), 302.

35. Author interview with Edward Kennedy, July 3, 2007.

36. Author interview with Edward Kennedy, July 3, 2007.

37. Stewart L. Udall, *The Myths of August: A Personal Exploration of Our Tragic Cold War Affair with the Atom* (New York: Pantheon, 1994), 187.

38. Jordan Fisher Smith, "Remembering the Craigheads, Pioneers of Wildlife Biology," *New Yorker*, October 11, 2016; Doug Peacock to author, February 17, 2021.

39. Marjory Stoneman Douglas, *Voice of the River* (Sarasota, FL: Pineapple Press, 1990), 38.

40. Marjory Stoneman Douglas, *The Everglades: River of Grass* (New York: Rinehart, 1947), 1–3.

41. Ibid. Also see Jack E. Davis, *An Everglades Providence: Marjory Stoneman Douglas and the American Environmental Century* (Athens: University of Georgia Press, 2009), 360.

42. Author interview with Jim Kushlan, March 8, 2022.

43. Harry S. Truman, "Address on Conservation at the Dedication of Everglades National Park," December 6, 1947, Harry S. Truman Library and Museum, https://www.trumanlibrary.gov/library/public-papers/231/address-conservation-dedication-everglades-national-park.

44. Quoted in Davis, *An Everglades Providence*, 604–5.

45. Davis, *An Everglades Providence*, 408.

46. Michael C. Reis, "Researching Environmental History During the Truman Era in the National Archives," in *The Environmental Legacy of Harry S. Truman*, ed. Karl Boyd Brooks (Kirksville, MO: Truman State University Press, 2009), 119.

47. Scott Hamilton Dewey, *Don't Breathe the Air: Air Pollution and U.S. Environmental Politics, 1945–1970* (College Station: Texas A&M University Press, 2000), 116.

48. *Air Pollution*, Report by the New York Academy of Medicine Committee on Public Health, New York Academy of Medicine, April 27, 1964.

49. Daniel Costa and Terry Gordon, "Mary Amdur," *Toxicological Sciences* 56, no. 1 (July 2000), 5.

50. Ibid., 87–88.

51. Berton Roueché, "Annals of Medicine: The Fog," *New Yorker*, September 30, 1950, 33–51. Also, author interview with Edward Kennedy, July 3, 2007.

52. Author interview with Jean Kennedy Smith, May 15, 2008.

53. Charles O. Jones, *Clean Air: The Policies and Politics of Pollution Control* (Pittsburgh, PA: University of Pittsburgh Press, 1978), 27.

54. Udall, *The Myths of August*, 218.

55. Fehner and Gosling, *Battlefield of the Cold War*, 10.

56. Thomas G. Alexander, *Utah: The Right Place* (Salt Lake City, UT: Gibbs Smith, 1995), 367.

57. Quoted in Udall, *The Myths of August*, 203.

58. Emily Cook, Brigham Young University, "Bullock v. United States: Radioactive Sheep," Intermountain Histories, accessed May 19, 2022, https://www .intermountainhistories.org/items/show/163.

59. Titus, *Bombs in the Backyard*, 63.

60. John M. Cosco, *Echo Park: Struggle for Preservation* (Boulder, CO: Johnson Books, 1995), 25.

61. John F. Stewart, "William O. Douglas Oral History Interview," November 9, 1967, John F. Kennedy Presidential Library and Museum, Washington, DC, https://www.jfklibrary.org/asset-viewer/archives/JFKOH/Douglas%2C%20 William%20O/JFKOH-WOD-01/JFKOH-WOD-01, 6.

62. Richard J. Lazarus, *The Making of Environmental Law* (Chicago: University of Chicago Press, 2004), 52.

63. James F. Simon, *Independent Journey: The Life of William O. Douglas* (New York: Harper & Row, 1980), 139–40.

64. Author interview with Jean Kennedy Smith, May 2, 2017.

Chapter 3: Rachel Carson and the Shore of the Sea

1. Gina Hebert, "Rachel Carson Statue to Be Dedicated in Woods Hole, July 14," Marine Biological Laboratory, University of Chicago, June 12, 2013, http://www.mbl.edu/blog/rachel-carson-statue-to-be-dedicated-in-woods -hole-july-14/. Also see William Souder, *On a Farther Shore: The Life and Legacy of Rachel Carson* (New York: Crown Publishers, 2012), 43.

2. Rachel Carson to Dorothy Thompson, August 25, 1929, quoted in Jim Hain, "Rachel Carson and Woods Hole," Woods Hole Museum, http://woodshole museum.org/oldpages/sprtsl/v27n2-RCarson.pdf, 3.

3. Rachel Carson, *The Edge of the Sea* (Boston: Houghton Mifflin, 1955), xiii.

4. Rachel Carson, *Rachel Carson: Silent Spring & Other Writings on the Environment* (New York: Library of America, 2018), 424. First published in *Scripps College Bulletin*, June 1962.

5. Linda Lear, *Rachel Carson: Witness for Nature* (New York: Houghton Mifflin Harcourt, 2009), 7–18.

6. Ibid., 7–8.

7. Lear, *Rachel Carson*, 1–80; Arlene R. Quaratiello, *Rachel Carson: A Biography* (Amherst, NY: Prometheus Books, 2004), 1–5.

8. Rachel Carson, "My Favorite Recreation," *St. Nicholas Magazine*, July 1922.

9. Souder, *On a Farther Shore*, 26.

10. Rachel Carson, "Who I Am and Why I Came to PCW," Beinecke Library, Yale University, New Haven, CT.

11. "The Gentle Storm Center," *Life*, October 12, 1962, 105.

12. Souder, *On a Farther Shore*, 34–36.

13. Quaratiello, *Rachel Carson*, 25.

14. Rachel Carson, *Lost Woods: The Discovered Writing of Rachel Carson*, ed. Linda Lear (Boston: Beacon Press, 1998), 14–15.

15. Quoted in Paul Brooks, *The House of Life: Rachel Carson at Work* (Boston: Houghton Mifflin, 1972), 20.

16. Hain, "Rachel Carson and Woods Hole."

17. Rachel Carson, "Undersea," *Atlantic Monthly*, September 1937, 55–67. Carson originally wrote "Undersea" as a US Bureau of Fisheries brochure.

18. Linda Lear to author, March 9, 2021.

19. Albert Schweitzer, *Reverence for Life; The Words of Albert Schweitzer*, ed. Harold E. Robles (Anna Maria, FL: Maurice Bassett, 2017), 13.

20. Albert Schweitzer Center, *Animals, Nature and Albert Schweitzer* (Washington, DC: Flying Fox Press, 1982), 62.

21. James Brabazon, *Albert Schweitzer: A Biography* (Syracuse, NY: Syracuse University Press, 2000), 282–83.

22. Albert Schweitzer, *Civilization and Ethics: The Philosophy of Civilization, Part II*, trans. John Nash (London: A & C Black, 1923), 264.

23. Carol B. Gartner, *Rachel Carson (Literature & Life)* (New York: Frederick Ungar, 1983), 130.

24. Souder, *On a Farther Shore*, 44.

25. Quoted in Brooks, *The House of Life*, 5.

26. Mark Hamilton Lytle, *The Gentle Subversive: Rachel Carson, Silent Spring and the Rise of the Environmental Movement* (New York: Oxford University Press, 2007), 14.

27. Quoted in Brooks, *The House of Life*, 5.

28. Rachel Carson, *Under the Sea-Wind: A Naturalist's Picture of Ocean Life* (New York: New American Library, 1941).

29. Author interview with Jean Kennedy Smith, January 10, 2019. Also see Jean Kennedy Smith, *The Nine of Us: Growing Up Kennedy* (New York: Harper-Collins, 2016), 65–66.

30. "A Dramatic Picture of Ocean Life; UNDER THE SEA-WIND," *New York Times*, November 23, 1941, BR10, https://www.nytimes.com/1941/11/23/archives/a-dramatic-picture-of-ocean-life-under-the-seawind-by-rachel-l.html.

31. Quaratiello, *Rachel Carson*, 31.

32. Ann Cottrell Free, *Since Silent Spring: Our Debt to Albert Schweitzer and Rachel Carson* (Washington, DC: Flying Fox Press, 2007), 2.

33. Rachel Carson to Harold Lunch, July 15, 1945, Rachel Carson Papers, Beinecke Library, Yale University.

34. Steven Johnson, "The Secrets of Longevity," *New York Times Magazine*, May 2, 2021, 20–21.

35. Charles F. Wurster, *DDT Wars: Rescuing our National Bird, Preventing Cancer, and Creating the Environmental Defense Fund* (New York: Oxford University Press, 2015), 6.

36. Robert K. Musil, *Rachel Carson and Her Sisters: Extraordinary Women Who Have Shaped America's Environment* (New Brunswick, NJ: Rutgers University Press, 2014), 125; Linda Lear, *Rachel Carson: Witness for Nature* (New York: Henry Holt, 1992), 237.

37. Barry Commoner, *Scientific Statesmanship in Air Pollution Control* (Washington, DC: Public Health Service, 1964), 4.

38. Benjamin Ross and Steven Amter, *The Polluters* (New York: Oxford University Press, 2010), 6.

39. Rachel Carson, "Chincoteague: A National Wildlife Refuge," *Conservation in Action*, no. 1 (Washington, DC: Fish and Wildlife Service, US Department of the Interior), June 1947; Rachel Carson, "Parker River: A National Wildlife Refuge," *Conservation in Action*, no. 2 (Washington, DC: Fish and Wildlife Service, US Department of the Interior), June 1947; Rachel Carson, "Mattamuskeet: A National Wildlife Refuge," *Conservation in Action*, no. 4 (Washington, DC: Fish and Wildlife Service, US Department of the Interior), July 1947; Vanez T. Wilson and Rachel Carson, "Bear River: A National Wildlife Refuge," *Conservation in Action*, no. 8 (Washington, DC: Fish and Wildlife Service, US Department of the Interior), June 1950. All available at http://digitalcommons.unl.edu/do/search/?q=author_lname%3A%22Carson%22%20author_fname%3A%22Rachel%22&start=0&context=52045&facet=.

40. Carson, "Chincoteague: A National Wildlife Refuge."

41. Souder, *On a Farther Shore*, 123.

42. Rachel Carson, "Guarding Our Wildlife Resources," *Conservation in Action*, no. 5 (Shepherdstown, WV: NCTC), 1948.

43. Nancy Koehn, *Forged in Crisis: The Power of Courageous Leadership in Turbulent Times* (New York: Simon & Schuster, 2017), 399.

44. Rachel Carson to William Beebe, August 26, 1949, Rachel Carson Papers, Beinecke Library, Yale University.

45. American Medical Association, Council on Foods and Nutrition, "Health Hazards of Pesticides," *Journal of the American Medical Association* 137 (August 18, 1948): 1604.

46. Mark Z. Jacobson, *Air Pollution and Global Warming: History, Science, and Global Warming* (New York: Cambridge University Press, 2012), 176–80.

47. Martin V. Melosi, *Fresh Kills: A History of Consuming and Discarding New York City* (New York: Columbia University Press, 2020), 17.

48. Ted Steinberg, *Gotham Unbound: The Ecological History of Greater New York* (New York: Simon & Schuster, 2010), 258.

49. Curt Meine, *Aldo Leopold: His Life and Work* (Madison: University of Wisconsin Press, 2010), 524–25.

50. Aldo Leopold, *A Sand County Almanac* (Oxford, UK: Oxford University Press, 1949), 211, xxi.

51. William O. Douglas, *Of Men and Mountains* (New York: Harper, 1950), 4.

52. William O. Douglas, *Go East, Young Man: The Early Years* (New York: Random House, 1974), 203.

53. Author interview with Edward Kennedy, July 3, 2007.

54. Austin H. Clark, "From the Beginning of the World," *Saturday Review of Literature*, July 7, 1951, 13.

55. Quoted in Lear, *Rachel Carson*, 202.

56. Dorothea Cruger, "Object of Her Affection Is the Ocean," *Washington Post*, July 4, 1951, B3.

57. Rachel Carson, *The Sea Around Us* (New York: Oxford University Press, 1951), 15.

58. Ibid., 8–9.

59. Ibid.

60. Ibid., 120.

61. Quoted in Frank Graham, *Since Silent Spring* (New York: Houghton Mifflin, 1970), 8–9.

62. Rachel Carson to Mrs. Frank H. Griffen, Providence Garden Club, July 1952, *Garden Club of America Bulletin* 2 (1952): 78–80.

63. Rachel Carson to Marie Rodell, September 1952, Rachel Carson Papers, Beinecke Library, Yale University.

64. Rachel Carson, "Autobiographical Sketch," *New York Herald Tribune Book Review*, October 7, 1951, 14.

65. Cruger, "Object of Her Affection Is the Ocean."

66. Rachel Carson, "The Dark Green Waters," review of Gilbert C. Klingel, *The Bay*, *New York Times Book Review*, October 14, 1951, 20, https://archive.ny times.com/www.nytimes.com/books/97/10/05/reviews/carson-bay.html.

Chapter 4: William O. Douglas and the Protoenvironmentalists

1. Thomas J. Whalen, *Kennedy Versus Lodge: The 1952 Massachusetts Senate Race* (Boston: Northeastern University Press, 2000), 3–41.

2. Author interview with Edward Kennedy, July 3, 2007.

3. Quoted in Thurston Clarke, *JFK's Last Hundred Days: The Transformation of a Man and the Emergence of a Great President* (New York: Penguin, 2013), xii–xiii.

4. Ted Sorensen, *Counselor: A Life at the Edge of History* (New York: Harper-Collins, 2008), 52.

5. Author Interview with Gillian Sorensen, October 6, 2021.

6. Terrence R. Fehner and F. G. Gosling, *Battlefield of the Cold War*, vol. 1, *The Nevada Test Site, Atmospheric Nuclear Weapons Testing, 1951–1963* (Washington, DC: US Department of Energy, 2006), 206.

7. Joseph F. Pilat, Robert E. Pendley, and Charles K. Ebinger, eds., *Atoms for Peace: An Analysis After Thirty Years* (Boulder, CO: Westview Press, 1985).

8. Dwight D. Eisenhower, "Address by Mr. Dwight D. Eisenhower, President of the United States of America, to the 470th Plenary Meeting of the United Nations General Assembly," New York City, December 8, 1953, https://www.iaea.org/about/history/atoms-for-peace-speech.

9. Thomas G. Smith, "John F. Kennedy, Stewart Udall, and New Frontier Conservation," *Pacific Historical Review* 41, no. 3 (August 1972): 333.

10. Jacques Cousteau, *The Silent World* (New York: Harper & Brothers, 1953).

11. Author interview with Jean Kennedy Smith, May 2, 2015.

12. Quoted in Sylvia Earle, introduction to Rachel Carson, *The Sea Around Us* (New York: Oxford University Press, 2010), xii.

13. "Senator Kennedy Goes A-Courting," *Life*, cover, July 20, 1953.

14. Jackie Kennedy poem, quoted in Fredrik Logevall, *JFK: Coming of Age in the American Century, 1917–1956* (New York: Random House, 2020), 566–67.

15. Eugene Odum, *Fundamentals of Ecology* (Philadelphia: Saunders, 1953).

16. Betty Jean Craige, *Eugene Odum: Ecosystem Ecologist & Environmentalist* (Athens: University of Georgia Press, 2001), 85.

17. Arthur George Tansley, "The Use and Abuse of Vegetational Terms and Concepts," *Ecology* 16 (1935).

18. Sue Hubbell, introduction to Rachel Carson, *The Edge of the Sea* (New York: Houghton Mifflin Harcourt, 1998), xvi.

19. Betty J. Craige, *Ecosystem Ecologist and Environmentalist* (Athens: University of Georgia Press, 2002), 46–47.

20. Ibid., ix–x.

21. Jack E. Davis, *An Everglades Providence: Marjory Stoneman Douglas and the American Environmental Century* (Athens: University of Georgia Press, 2009), 413.

22. Ari L. Goldman, "Eugene P. Odum Dies at 88; Founded Modern Ecology," *New York Times*, August 14, 2002, A21, https://www.nytimes.com /2002/08/14/us/eugene-p-odum-dies-at-88-founded-modern-ecology.html.

23. "Eugene Odum," in *Modern American Environmentalists: A Biographical Encyclopedia*, ed. George A. Cevasco and Richard P. Harmond (Baltimore: Johns Hopkins University Press, 2009), 357.

24. Quoted in ibid., xiii.

25. Paul Brooks, *The House of Life: Rachel Carson at Work* (Boston: Houghton Mifflin, 1972), 258–59.

26. Robert K. Musil, "Rachel Carson's Cottage at the Edge of the Sea," Rachel Carson Council, Summer 2015, https://rachelcarsoncouncil.org /about-rcc/about-rachel-carson/rachel-carsons-cottage-at-the-edge-of-the -sea/.

27. Linda Lear, *Rachel Carson: Witness for Nature* (New York: Henry Holt, 1992), 260–61.

28. Henry Beston, "Miss Carson's First," review of *Under the Sea-Wind*, *Freeman*, November 3, 1952, 100.

29. Author interview with Judge Margaret McKeown, March 19, 2021.

30. William Blake, "Auguries of Innocence," 1863, Pickering Manuscript, Morgan Library & Museum, 14–15.

31. William O. Douglas, *My Wilderness: East to Katahdin* (Garden City, NY: Doubleday, 1961), 290.

32. Jack Kerouac, *On the Road* (New York: Penguin, 2003), 119.

33. Daniel Yergin, *The Prize: The Epic Quest for Oil, Money, and Power* (New York: Simon & Schuster, 1991), 409.

34. Author interview with Edward Kennedy, August 8, 2008.

35. Michael Frome, *The Forest Service* (Boulder, CO: Westview, 1984), 37.

36. William O. Douglas, *Farewell to Texas* (New York: McGraw-Hill, 1967), viii.

37. Bruce Allen Murphy, *Wild Bill: The Legend of William O. Douglas* (New York: Random House, 2003), 249–60.

38. (Hagerstown, MD) *Morning Herald*, May 25, 1953, 2.

39. John F. Stewart, "William O. Douglas Oral History Interview," John F. Kennedy Presidential Library and Museum, November 9, 1967, https://www .jfklibrary.org/sites/default/files/archives/JFKOH/Douglas%2C%20William %20O/JFKOH-WOD-01/JFKOH-WOD-01-TR.pdf, 37; author interview with Jean Kennedy Smith.

40. "Potomac Parkway," *Washington Post*, January 3, 1954, B4.

41. Shirley A. Briggs, *Washington: City in the Woods* (Washington, DC: Audubon Society, 1954).

42. "10 Canal Hikers Still Survive on 5th Day," *Washington Post*, March 25, 1954, 14.

43. "Associate Justice William O. Douglas," Chesapeake & Ohio Canal National Historical Park, National Park Service, https://www.nps.gov/choh/learn/his toryculture/associatejusticewilliamodouglas.htm.

44. William O. Douglas, letter to the editor, *Washington Post*, January 19, 1954, 14.

45. Adam W. Burnett, "Olaus Johan Murie," in *Modern American Environmentalists: A Biographical Encyclopedia*, ed. George A. Cevasco and Richard P. Harmond (Baltimore: Johns Hopkins University Press, 2009), 331–33.

46. David Backes, *A Wilderness Within: The Life of Sigurd Olson* (Minneapolis: University of Minnesota Press, 1997), 48–49.

47. "10 Canal Hikers Still Survive on 5th Day," *Washington Post*.

48. "Douglas Finishes His 189-Mile Hike," *New York Times*, March 28, 1954, 43, https://timesmachine.nytimes.com/timesmachine/1954/03/28/83323922 .html?pageNumber=43.

49. Andrew Schotz, "A Hike for All Time," (Hagerstown, MD) *Herald-Mail*, March 14, 2004, http://articles.herald-mail.com/2004-03-14/news /25024668_1_editorial-page-editor-hike-estabrook.

50. Murphy, *Wild Bill*, 333–34.

51. Barry Mackintosh, *Chesapeake and Ohio Canal: The Making of a Park* (Washington, DC: National Park Service, US Department of the Interior, 1991), https://www.nps.gov/parkhistory/online_books/choh/admin_history/his tory4.htm, 73.

52. Douglas, *My Wilderness*, 197.

53. Ibid., 195.

54. David Backes, *A Wilderness Within: The Life of Sigurd F. Olson* (Minneapolis: University of Minnesota Press, 1997), 227.

55. Sigurd Olson, *The Singing Wilderness* (New York: Alfred A. Knopf, 1956), 160–61.

56. Backes, *A Wilderness Within*, 252.

57. Sigurd Olson, *Lonely Land* (St. Paul: University of Minnesota Press, 1961).

58. Olson, *The Singing Wilderness*, 77.

59. Author interview with Robert F. Kennedy, Jr., June 7, 2018.

60. David A. Farenthold, "Potomac River's Health Rebounds," *Washington Post*, September 8, 2010.

61. Author interview with Ethel Kennedy, February 6, 2019.

62. Smith, "John Kennedy, Stewart Udall, and New Frontier Conservationism," 329–62.

63. John B. Oakes, "Conservation: Record of Congress," *New York Times*, September 5, 1954, X14, https://www.nytimes.com/1954/09/05/archives/conservation-record-of-congress.html.

64. William O. Douglas, *The Autobiography of William O. Douglas*, vol. 2, *The Court Years, 1939–1975* (New York: Random House, 1980), 309.

65. Robert F. Kennedy, Jr., *American Values: Lessons I Learned from My Family* (New York: HarperCollins, 2018), 76.

66. William F. Libby, "An Open Letter to Dr. Schweitzer," *Saturday Review*, May 25, 1957, 37.

67. Michael Egan, *Barry Commoner and the Science of Survival: The Remaking of American Environmentalism* (Cambridge, MA: MIT Press, 2009), 54–55.

68. Robert K. Musil, *Rachel Carson and Her Sisters: Extraordinary Women Who Have Shaped America's Environment* (New Brunswick, NJ: Rutgers University Press, 2014), 126. Also see Musil, "Rachel Carson and Nuclear War," Rachel Carson Council, https://rachelcarsoncouncil.org/rachel-carson-nuclear-war.

69. Eugene J. Rosi, "Mass and Attentive Opinion of Nuclear Weapons Tests and Fallout, 1954–1963," *Public Opinion Quarterly* 29, no. 2 (1965), 288–93.

70. Laurence S. Wittner, *Resisting the Bomb: A History of the World Nuclear Disarmament Movement* (Standford: Standford University Press, 1997), frontispiece.

71. Egan, *Barry Commoner and the Science of Survival*, 64.

72. Barry Commoner, *The Closing Circle: Nature, Man and Technology* (New York: Alfred A. Knopf, 1971), 65–66.

73. Stephen W. Royce, "Q&A," *Los Angeles Times*, November 19, 1953, 43.

74. "Women Protest Smog," *Los Angeles Times*, October 21, 1954.

75. "The Smog Battle: Two Doctors Give Their Medical View," *Los Angeles Times*, November 16, 1953.

76. A. J. Haagen-Smit, "Smog Control—Is It Just Around the Corner?," *Engineering and Science* 26, no. 2 (November 1962): 10.

77. Douglas Smith, "Fifty Years of Clearing the Skies," Caltech, April 25, 2013, https://www.caltech.edu/about/news/fifty-years-clearing-skies-39248.

78. Quoted in David Stradling, *The Nature of New York: An Environmental History of the Empire State* (Ithaca, NY: Cornell University Press, 2010), 182.

79. Elsie Robinson, "Smog Everywhere," (White Plains, NY) *Journal News*, March 8, 1949.

80. Clayton D. Forswall and Kathryn E. Higgins, "Clean Air Act Implementation in Houston: An Historical Perspective, 1970–2005," Rice University Environmental and Energy Systems Institute, Shell Center for Sustainability, February 2005, https://scholarship.rice.edu/bitstream/handle/1911/107669/SIP_2.pdf?sequence=1&isAllowed=y.

81. "London Smog Is So Thick Fires Burn Undetected," *New York Times*, December 19, 1953, 17, https://www.nytimes.com/1953/12/19/archives/london-smog-is-so-thick-fires-burn-undetected.html.

82. Edith Evans Asbury, "Smog Is Really Smaze; Rain May Rout It Tonight," *New York Times*, November 21, 1953, 1, 30, https://www.nytimes.com/1953/11/21/archives/smog-is-really-smaze-rain-may-rout-it-tonight-fourday-concentration.html.

83. Roy Popkin, "Two 'Killer Smogs' the Headlines Missed," *EPA Journal*, December 1986, 27.

84. John F. Kennedy, "Our American Cities and Their Second-Class Citizens," September 11, 1957, John F. Kennedy Presidential Library and Museum, https://www.jfklibrary.org/archives/other-resources/john-f-kennedy-speeches/new-york-ny-us-conference-of-mayors-19570911.

85. Steven Rosenberg, "Power Struggle," *Boston Globe*, November 6, 2008.

86. Special thanks to Susan Livingston of Marblehead, Massachusetts, for educating me about the sad and sordid history of the Salem Harbor Power Plant.

87. "A Cleaner Energy Future for Salem," Conservation Law Foundation, 2019.

88. Author interview with Edward Kennedy, August 8, 2008.

89. See Zoe Ackerman et al., "Blast Zone: Natural Gas and the Atlantic Coast Pipeline: Causes, Consequences and Civic Action," Rachel Carson Council, September 2017, https://rachelcarsoncouncil.org/wp-content/uploads/2017/09/blast-zone-final.pdf.

90. *Virginia's Toxic Coal Ash Problem* (Virginia Conservation Network, 2015), 7.

91. "Statement of Senator Kennedy Before the Select Committee on Water Resources, Boston, Massachusetts, December 8, 1959," John F. Kennedy Presidential Library and Museum, https://www.jfklibrary.org/assetviewer/archives/JFKCAMP1960/1031/JFKCAMP1960-1031-017.

Chapter 5: Wilderness Politics, Dinosaur National Monument, and the Nature Conservancy

1. Quoted in Elmo Richardson, "The Interior Secretary as Conservation Villain: The Notorious Case of Douglas 'Giveaway' McKay," *Pacific Historical Review* 41, no. 3 (August 1972): 333–45.

2. Linda Lear, *Rachel Carson: Witness for Nature* (New York: Houghton Mifflin Harcourt, 2009), 257.

3. Jack E. Davis, *An Everglades Providence: Marjory Stoneman Douglas and the American Environmental Century* (Athens: University of Georgia Press, 2009), 409.

4. Frank E. Smith, *The Politics of Conservation* (New York: Pantheon, 1966), 289.

5. "The Old Car Peddler," *Time*, August 23, 1954, 21.

6. "Fremont National Forest/White King and Lucky Lass Uranium Mines (USDA), Lakeview, OR, Cleanup Activities," United States Environmental Protection Agency, https://cumulis.epa.gov/supercpad/SiteProfiles/index.cfm?fuseaction=second.Cleanup&id=1001508#bkground.

7. Robert Gottlieb, *Forcing the Spring: The Transformation of the American Environmental Movement* (Washington, DC: Island Press, 1993), 250–51.

8. Doug Brugge and Rob Gable, "The History of Uranium Mining and the Navajo People," *American Journal of Public Health* 92, no. 9 (September 2002): 1416.

9. A. Constandina Titus, *Bombs in the Backyard: Atomic Testing and American Politics* (Reno: University of Nevada Press, 1986), 94.

10. Samuel Matthews, "Nevada Learns to Live with the Atom," *National Geographic* 103, no. 6 (June 1953): 839–50.

11. Bernard DeVoto, "The West: A Plundered Province," *Harper's*, August 1934, 355–64, reprinted as "The Plundered Province," in Bernard DeVoto, *Forays and Rebuttals* (Boston: Little, Brown, 1936), 46–65.

12. Wallace Stegner, *The Uneasy Chair: A Biography of Bernard DeVoto* (New York: Doubleday, 1988), 386.

13. Douglas Brinkley and Patricia Limerick, eds., *The Western Paradox: A Conservation Reader* (New Haven, CT: Yale University Press, 2000).

14. Quoted in Bernard DeVoto, *DeVoto's West: History, Conservation, and the Public Good*, ed. Edward K. Muller (Athens: Swallow Press/Ohio University Press, 2005), xxix.

15. Tom Turner, *David Brower: The Making of the Environmental Movement* (Berkeley: University of California Press, 2015), 15.

16. David Brower, "How to Kill a Wilderness" (not published at the time but later reprinted), in David Brower, *For Earth's Sake: The Life and Times of David Brower* (Salt Lake City, UT: Peregrine Smith Books, 1990), 125–28.

17. Kenneth Brower, *Hetch Hetchy: Undoing a Great American Mistake* (Berkeley, CA: Heyday, 2013), 1–2.

18. Ibid., 2.

19. Turner, *David Brower*, 67.

20. W. L. Rusho, "Bumpy Road for Glen Canyon Dam," *Bureau of Reclamation: History Essay for the Centennial Symposium* 2 (Denver: US Department of Interior, 2008), 531.

21. Lear, *Rachel Carson*, 180.

22. Mark Harvey, *A Symbol of the West: Echo Park and the American Conservation Movement* (Seattle: University of Washington Press, 1994), xv.

23. US Congress, Senate, Committee on Irrigation and Reclamation, Hearings, Colorado River Storage Project, 84th Congress, 1st Session, February 28, March 1–5, 1955, 679–96.

24. Brower, *Hetch Hetchy*, 2.

25. David Brower, "Grand Canyon Battle Ads," in Ernest Braun et al., *Grand Canyon of the Living Colorado* (San Francisco: Sierra Club and Ballantine Books, 1970).

26. Shirley Loui, "David Brower," in *Modern American Environmentalists: A Biographical Encyclopedia*, ed. George A. Cevasco and Richard P. Harmond (Baltimore, MD: Johns Hopkins University Press, 2009), 56–60.

27. Jackson Benson, *Wallace Stegner: His Life and Works* (New York: Penguin, 1997).

28. Patianne Delgrosso Stabile, "Wallace Earle Stegner," in *Modern American Environmentalists: A Biographical Encyclopedia*, ed. George A. Cevasco and Richard P. Harmond (Baltimore, MD: Johns Hopkins University Press, 2009), 487.

29. Wallace Stegner, *This Is Dinosaur: Echo Park Country and Its Magic Rivers* (New York: Alfred A. Knopf, 1955).

30. Edward Abbey, *Desert Solitaire: A Season in the Wilderness* (New York: McGraw-Hill, 1968), 135.

31. Rachel White Scheuering, *Shapers of the Great Debate on Conservation* (Westport, CT: Greenwood, 2004), 80.

32. Quoted in Kathy Mengak, *Reshaping Our National Parks and Their Guardians: The Legacy of George B. Hartzog, Jr.* (Albuquerque: University of New Mexico Press, 2012), 128.

33. David Brower, "Scenic Resources for the Future," *Sierra Club Bulletin* 41, no. 10 (December 1956).

34. Hal K. Rothman, *Saving the Planet: The American Response to the Environment in the Twentieth Century* (Chicago: Ivan R. Dee, 2000), 96.

35. Mark Harvey, *Wilderness Forever: Howard Zahniser and the Path to the Wilderness Act* (Seattle: University of Washington Press, 2005), 6.

36. Tim Palmer, *America's Great Forest Trails* (New York: Rizzoli, 2021), 68.

37. Douglas Brinkley, "Thoreau's Wilderness Legacy, Beyond the Shores of Walden Pond," *New York Times*, July 7, 2017, 12, https://www.nytimes.com/2017/07/07/books/review/douglas-brinkley-thoreaus-wilderness-legacy-walden-pond.html.

38. Howard Zahniser, "In the Month of May," in *The Wilderness Writings of Howard Zahniser*, ed. Mark Harvey (Seattle: University of Washington Press, 2014), 14–16.

39. Roderick Frazier Nash, *Wilderness and the American Mind* (New Haven, CT: Yale University Press, 2001), 221.

40. Howard Zahniser, "Wilderness Forever," in *Voices for the Wilderness*, ed. William Schwartz (New York: Ballantine, 1969), 100.

41. Michael D. Nichols, "Howard Zahniser," in *Modern American Environmentalists: A Biographical Encyclopedia*, ed. George A. Cevasco and Richard P. Harmond (Baltimore, MD: Johns Hopkins University Press, 2009), 535.

42. Harvey, *Wilderness Forever*, 106.

43. Kenneth L. Smith, *Buffalo River Handbook* (Little Rock: The Ozark Society Foundation, University of Arkansas Press, 2004), 109.

44. Steven Solomon, *Water: The Epic Sruggle for Wealth, Power, and Civilization* (New York: HarperCollins, 2010), 348–49.

45. Roderick Frazier Nash, *Wilderness and the American Mind* (New Haven, CT: Yale University Press, 2001), 320–21.

46. "Ikes Fete Saylor," *Scranton Times Tribune*, August 16, 1970, 40.

47. Bob Kelleher, "Forty Years On: Sigurd Olson and the Wilderness Act," Minnesota Public Radio, September 3, 2004, http://news.minnesota.publicradio .org/features/2004/09/03_kelleherb_wilderness/. For Humphrey's growing interest in wilderness conservation, see Kevin Proescholdt, Rip Rapson, and Miron L. Heinselman, *Troubled Waters: The Fight for the Boundary Waters Canoe Area* (St. Cloud, MN: North Star Press of St. Cloud, 1995), 2–10.

48. David Backes, *A Wilderness Within: The Life of Sigurd F. Olson* (Minneapolis: University of Minnesota Press, 1997), 265–66.

49. Doug Scott, *The Enduring Wilderness* (Golden, CO: Fulcrum, 2004), 19.

50. *Congressional Record*, June 7, 1956, 84th Congress, 2nd Session, 9772-83. Rep. John Saylor introduced the House companion bill as H.R. 11703.

51. Hubert Humphrey, memo to Herb Waters, July 20, 1957, Box 146, Hubert H. Humphrey Papers, Minnesota Historical Society.

52. Mark Harvey, "Sound Politics: Wilderness, Recreation, and Motors in the Boundary Waters, 1945–1964," *Minnesota History* 58 (Fall 2002): 130–45.

53. Backes, *A Wilderness Within*, 266.

54. Mark W. T. Harvey, *A Symbol of Wilderness: Echo Park and the American Conservation Movement* (Albuquerque: University of New Mexico, 1994).

55. William O. Douglas, *The Autobiography of William O. Douglas*, vol. 2, *The Court Years, 1939–1975* (New York: Random House, 1980), 313.

56. Richard Neuberger and Steve Neal, *They Never Go Back to Pocatello* (Portand: Oregon Historical Society, 1988), 108.

57. Richard A. Baker, "The Conservation Congress of Anderson and Aspinall, 1963–64," *Journal of Forest & Conservation History* 29, no. 3 (1985): 104–19.

58. Quoted in Howard Zahniser, *The Wilderness Writings of Howard Zahniser*, ed. Mark Harvey (Seattle: University of Washington Press, 2014), 155.

59. Linda Luther, "The National Environmental Policy Act: Background and Implementation," Congressional Research Service, February 29, 2008, https:// sgp.fas.org/crs/misc/RL33152.pdf.

60. "The Nature Conservancy History," FundingUniverse, http://www.funding universe.com/company-histories/the-nature-conservancy-history/.

61. Quoted in William D. Blair, Jr., *Katharine Ordway: The Lady Who Saved the Prairies* (Arlington, VA: Nature Conservancy, 1989).

62. Kelly M. Paulson, "'The Lady Who Saved the Prairies'—and Her Brother," honors project, Katharine Ordway Natural History Study Area, Macalester College, Minnesota, 2001.

63. Neal Grove, *Preserving Eden: The Nature Conservancy* (New York: Henry Abrams, 1992), 30–38.

64. Quoted in Anthony Weston, *Back to Earth: Tomorrow's Environmentalism* (Philadelphia: Temple University Press, 1994), 71.

65. Howard Zahniser, "Wilderness: How Much Can We Afford to Lose?," address at Sierra Club Wilderness Conference, 1951, quoted in Scott, *The Enduring Wilderness*, 23.

66. Arnold Krupat, "Chief Seattle's Speech Revisited," *American Indian Quarterly* 35, no. 2 (2011), 197–200.

67. Jack Kerouac, *On the Road* (New York: Viking, 1957).

68. Jack Kerouac, *The Dharma Bums* (New York: Viking, 1958), 65.

69. Frank Zelko, *Make It a Green Peace!* (New York: Oxford University Press, 2013), 184.

70. Steven Watts, *The Magic Kingdom: Walt Disney and the American Way of Life* (Columbia: University of Missouri Press, 2001), 304–5.

71. David Louter, *Windshield Wilderness: Cars, Roads and Nature in Washington's National Parks* (Seattle University Press, 2006), 105.

72. Howard Zahniser to C. Edward Graves, April 25, 1959, in Zahniser, *The Wilderness Writings of Howard Zahniser*, 160–61.

73. Dennis Roth, "The National Forests and the Campaign for Wilderness Legislation," *Journal of Forest History* 28, no. 3 (July 1984), 119–25.

74. Bran Büscher and Robert Fletcher, *The Conservation Revolution: Radical Ideas for Saving Nature Beyond the Anthropocene* (London: Verso, 2020).

75. Matthew J. Lindstrom and Zachary A. Smith, *The National Environmental Policy Act: Judicial Misconstruction, Legislative Indifference, and Executive Neglect* (College Station, TX: A&M University Press, 2001), iv.

76. Davis, *An Everglades Providence*, 409.

77. Quoted in Adam Rome, *The Genius of Earth Day* (New York: Macmillan, 2013), 19–20.

78. Alvin B. Toffler, "Danger in Your Drinking Water," *Good Housekeeping*, January 1960, 41–43, 128–30.

79. Rachel Carson, "The Real World Around Us," Matrix Table Dinner, Theta Sigma Phi, Columbus, Ohio, April 21, 1954, Rachel Carson Papers, Beinecke Rare Book and Manuscript Library, Yale University, New Haven, CT.

Chapter 6: Saving Shorelines

1. National Park Service, Department of the Interior, *A Report on a Seashore Recreation Area Survey of the Atlantic and Gulf Coasts* (Washington, DC: US Government Printing Office, 1955).

2. Hamilton Gray, "Recreation: A Seashore Park for the Nation," *New York Times*, September 5, 1937, A1.

3. Author interview with Jean Kennedy Smith, August 3, 2015.

4. Cornelia Dean, *Against the Tide: The Battle for America's Beaches* (New York: Columbia University Press, 1999), 186.

5. National Park Service, *A Report on the Seashore Recreation Survey of the Atlantic and Gulf Coasts*, 1955, quoted in Francis P. Burling, *The Birth of Cape Cod National Seashore* (Plymouth, MA: Leyden Press, 1978), 6–7.

6. Tom Turner, *David Brower: The Making of the Environmental Movement* (Berkeley: University of California Press, 2015), 111.

7. Quoted in Fredrik Logevall, *JFK: Coming of Age in the American Century, 1917–1956* (New York: Random House, 2020), 593.

8. Quoted in James W. Graham, *Victura: The Kennedys, a Sailboat, and the Sea* (Lebanon, NH: University Press of New England, 2004), 105.

9. Rachel Carson, *The Edge of the Sea* (New York: Houghton Mifflin Harcourt, 1955), 1.

10. Ibid., 140.

11. Henry B. Bigelow to Rachel Carson, October 14, 1955, Rachel Carson Papers, Yale University.

12. "Marine Demimonde," *Time*, November 7, 1955, https://content.time.com /time/subscriber/article/0,33009,807979,00.html.

13. William Souder, *On a Farther Shore: The Life and Legacy of Rachel Carson* (New York: Crown, 2012), 217.

14. Rachel Carson to Dorothy Freeman, November 27, 1955, Rachel Carson Papers, Beinecke Library, Yale University.

15. Paul Brooks, *The House of Life: Rachel Carson at Work* (Boston: Houghton Mifflin, 1989), 211.

16. Rachel Carson, quoted in "Walking in Rachel Carson's Footsteps," March 24, 2021, The Nature Conservancy website.

17. Brooks, *The House of Life*, 197–99.

18. Rachel Carson, *Lost Woods: The Discovered Writing of Rachel Carson*, ed. Linda Lear (Boston: Beacon Press, 1998), 237.

19. Earl Swift, *The Big Roads: The Untold Story of the Engineers, Visionaries, and Trailblazers Who Created the American Superhighways* (New York: Houghton Mifflin, 2011), 296–97.

20. Steven Watts, *JFK and the Masculine Mystique: Sex and Power on the New Frontier* (New York: St. Martin's Press, 2016), 50.

21. Jeff Broadwater, *Adlai Stevenson and American Politics: The Odyssey of a Cold War Liberal* (New York: Twayne, 1994), 172–73.

22. Harry S. Truman, "Press Release of Speech Delivered by Harry S. Truman Before the Democratic National Convention, August 17, 1956," Harry S. Truman Library & Museum, https://www.trumanlibrary.gov/library/research

-files/press-release-speech-delivered-harry-s-truman-democratic-national
-convention?documentid=NA&pagenumber=3.

23. Douglas Brinkley, *The Quiet World: Saving Alaska's Wilderness Kingdom, 1879–1960* (New York: HarperCollins, 2011), 464–66.

24. Richard L. Neuberger, "Plan for Shoreline Parks: U. S. Senate Bills Would Set Aside Recreational Areas on Seacoasts and in the Great Lakes Region," *New York Times*, August 30, 1959, X19, https://timesmachine.nytimes.com/times machine/1959/08/30/89233322.html?pageNumber=440.

25. *Seattle Post-Intelligencer*, August 8, 1958, quoted in Paula Becker, "Conservationists William O. Douglas, Polly Dyer, and Others Begin a 22-Mile Hike Along the Olympic Coastline to Protest Proposed Road Construction on August 19, 1958," HistoryLink.org, December 29, 2010, https://www.history link.org/File/9672.

26. William O. Douglas, *My Wilderness: The Pacific West* (Garden City, NY: Doubleday, 1960), 40.

27. Harvey Manning, *Wilderness Alps: Conservation and Conflict in Washington's North Cascades* (Bellingham, Washington: Northwest Wild Books, 2007), 112.

28. Newton B. Drury, "He Left a Heritage of Beauty," *Sierra Club Bulletin*, April–May 1960, 83.

29. Author interview with Douglas R. Home, April 19, 2020.

30. Ronald Foresta, *America's National Parks and Their Keepers* (Washington, DC: Resources for the Future, 1984), 171; Hal K. Rothman, *The Park That Makes Its Own Weather: A History of Golden Gate National Recreation Area* (San Francisco: Golden Gate National Recreation Area, 2002).

31. "Practical Ecologist: Laurance Spelman Rockefeller," *New York Times*, February 1, 1966, 22, https://www.nytimes.com/1966/02/01/archives /practical-ecologist-laurance-spelman-rockefeller-man-of-many-parts .html.

32. The Outdoor Recreation Resources Review Commission was created by the act of June 28, 1958 (Public Law 85-470, 72 Stat. 238).

33. "Nation and State Take Inventory," *Sierra Club Bulletin*, January 1957, 3.

34. "Laurance S. Rockefeller, 1910–2004," Rockefeller Archive Center, Kykuit Estate, Mt. Pleasant, New York. Also, author interview with Jean Kennedy Smith, August 3, 2015.

35. Michael P. Cohen, *The History of the Sierra Club, 1892–1970* (New York: Random House, 1988), 278; "Point Reyes Park Proposed," *Sierra Club Bulletin*, September 1958.

36. Archie Carr, *The Windward Road: Adventures of a Naturalist on Remote Caribbean Shores* (Gainesville: University of Florida Press, 1956).

37. Carson, *The Edge of the Sea*, 237.

38. Rachel Carson, "Our Ever-Changing Shore," *Holiday* 24 (1958): 71, 117–20.

39. "Remarks of Senator John F. Kennedy Before the California Legislature, Sacramento, California, May 1, 1959," John F. Kennedy Presidential Library and Museum, https://www.jfklibrary.org/archives/other-resources/john-f-kennedy-speeches/sacramento-ca-19590501.

40. Associated Press, "Potomac Is Termed Most Polluted River West of the Nile," June 9, 1959 (North Adams, MA), *Transcript*, June 10, 1959, 1. Also, author interview with Edward Kennedy, July 3, 2007.

41. "Statement of Senator Kennedy before the Select Committee on Water Resources, Boston, Massachusetts, December 8, 1959," John F. Kennedy Presidential Library and Museum, https:www.jfklibrary.org/asset-viewer/archives/JFKCAMP1960/1031/JFKCAMP1960-1031-017.

42. Cape Cod Study Group, Region Five Office, National Park Service, US Department of the Interior, *A Field Investigation Report on a Proposed National Seashore, Cape Cod, Barnstable County, Massachusetts: Report of a Biological Investigation on a Portion of Cape Cod, Massachusetts; Report of the Geologic Features on a Portion of Cape Cod, Massachusetts; Report on the History of Cape Cod, Massachusetts; Report on the Archeology of Cape Cod, Massachusetts* (Washington, DC: US Government Printing Office, 1958); Henry David Thoreau, *Cape Cod* (Boston: Ticknor and Fields, 1866), 227.

43. Quoted Burling, *The Birth of Cape Cod National Seashore*, 18.

44. Don Wilding, *Henry Beston's Cape Cod: How "The Outermost House" Inspired a National Seashore* (Eastham, MA: Henry Beston Society, 2013).

45. Burling, *The Birth of Cape Cod National Seashore*, 30.

46. Caroline Bates, "Walking in Thoreau's Footsteps on Cape Cod," *New York Times*, October 11, 1959, X33, https://www.nytimes.com/1959/10/11/archives/walking-in-thoreaus-footsteps-on-cape-cod.html.

47. William O. Douglas, *The Autobiography of William O. Douglas*, vol. 2, *The Court Years, 1939–1975* (New York: Random House, 1980), 303.

48. Ibid.

Chapter 7: Protesting Plastics, Nuclear Testing, and DDT

1. E. G. Vallianatos with McKay Jenkins, *Poison Spring: The Secret History of Pollution and the EPA* (New York: Bloomsbury, 2014), 1–5. See also Seth M. Siegel, *Troubled Water: What's Wrong with What We Drink* (New York: Thomas Dunne, 2019), 8.

2. Barry Commoner, *Scientific Statesmanship in Air Pollution Control* (Washington, DC: US Public Health Service, 1964), 5.

3. Jeffrey L. Meikle, "Material Doubts: The Consequences of Plastic," *Environmental History* 2, no. 3 (July 1997): 278.

4. Ibid. Also see Laura Parker, "How the Plastic Bottle Went from Miracle Container to Hated Garbage," *National Geographic*, August 23, 2019, https://www.nationalgeographic.com/environment/article/plastic-bottles.

5. John Updike, *Rabbit, Run* (New York: Fawcett Crest, 1969), 87.

6. Barry Commoner, *The Closing Circle: Nature, Man and Technology* (Mineola, NY: Dover Publications, 2020), 47–53.

7. Michael Egan, *Barry Commoner and the Science of Survival* (Cambridge, MA: MIT Press, 2007), 53.

8. Allen Pietrobon, "The Role of Norman Cousins and Track II Diplomacy in the Breakthrough to the 1963 Limited Test Ban Treaty," *Journal of Cold War Studies* 18, no. 1 (Winter 2016): 60–79.

9. Aleksandr Fursenko and Timothy Naftali, *Khrushchev's Cold War: The Inside Story of an American Adversary* (New York: Norton, 2006), 508.

10. "5,000 March Here After Atom Rally," *New York Times*, May 20, 1960, 11.

11. Dr. Martin Luther King, Jr., quoted in Vincent Intondi, "Martin Luther King on Non-violence and Disarmament," *Boston Review*, January 16, 2015, https://bostonreview.net/us/vincent-intondi-martin-luther-king-nuclear-weapons-civil-rights.

12. Edward Teller, *The Legacy of Hiroshima* (Garden City, NY: Doubleday, 1962), 108.

13. Barry Commoner, "The Fallout Problem," *Science* 127 (May 2, 1958): 1023–26.

14. Barry Commoner, *Science and Survival* (New York: Viking, 1966), 120.

15. "Dr. Louise Reiss," *Nation*, June 1959.

16. Dennis Hevesi, "Dr. Louis Reiss, Who Helped Ban Atomic Testing, Dies at 90," *New York Times*, January 10, 2011, https://www.nytimes.com/2011/01/10/science/10reiss.html.

17. Walter Schneir, "Strontium-90 in U.S. Children," *Nation*, April 25, 1959, 355–57.

18. Amy Swerdlow, "Ladies' Day at the Capitol: Women Strike for Peace Versus HUAC," *Feminist Studies* 8, no. 3 (Fall 1982): 496.

19. Norman Cousins, *Dr. Schweitzer of Lambaréné* (Westport, CT: Greenwood Press, 1960), 166.

20. Ibid.

21. Albert Schweitzer, "A Declaration of Conscience," broadcast on Radio Oslo, Oslo, Norway, April 24, 1957. See also Schweitzer, "A Declaration of Conscience," *Saturday Review*, May 18, 1957.

22. Erich Gräßer, "The Significance of Reverence for Life Today," in *Reverence for Life: The Ethics of Albert Schweitzer for the Twenty-First Century*, ed. Marvin Meyer and Kurt Bergel (Syracuse, NY: Syracuse University Press, 2002), 160–64.

23. Schweitzer, "A Declaration of Conscience." See also Schweitzer, "A Declaration of Conscience," *Saturday Review*.

24. Ann Cottrell Free, "Since *Silent Spring*: Our Debt to Albert Schweitzer & Rachel Carson," An International Albert Schweitzer Symposium, August 13, 1992, http://www.anncottrellfree.org/uploads/1/6/7/1/16715062/since_silent_spring_address.pdf.

25. Robert K. Musil, *Rachel Carson and Her Sisters: Extraordinary Women Who Have Shaped America's Environment* (New Brunswick, NJ: Rutgers University Press, 2014), 116.

26. Rachel Carson, letter to the editor, *Washington Post*, April 10, 1959, A12, reprinted in *Lost Woods: The Discovered Writing of Rachel Carson*, ed. Linda Lear (Boston: Beacon Press, 1998), 189–91.

27. William O. Douglas, "Dissent in Favor of Man," *Saturday Review*, May 5, 1960, 59; *US News & World Report*, November 23, 1959, 143. See also *Murphy v. Butler*, 362, US 929, 1960, 929–35. This quote is from Douglas's dissent from denial of certiorari.

28. George DeWan, "The Fight to Ban DDT," *Newsday*, January 18, 2000, 28.

29. Marjorie Spock, quoted in Linda Lear, *Rachel Carson: Witness for Nature* (New York: Henry Holt, 1992), 318.

30. Clarence Dean, "Cranberry Sales Curbed; U.S. Widens Taint Check," *New York Times*, November 11, 1959, 1, 29, https://www.nytimes.com/1959/11/11/archives/cranberry-sales-curbed-45-million-loss-feared-cranberry-crop-facing.html.

31. Quoted in Peter Matthiessen, *Courage for the Earth: Writers, Scientists, and Activists Celebrate the Life and Writing of Rachel Carson* (New York: Houghton Mifflin Harcourt, 2007), 11.

32. John O'Reilly, "The Deadly Spray," *Sports Illustrated*, May 2, 1960, 20–21.

33. Quoted in a bulletin of the International Union for Conservation of Nature and Natural Resources, December 1956. See also Lear, *Rachel Carson*, 321–22.

34. Rachel Carson to Paul Brooks, March 23, 1960, Rachel Carson Papers, Beinecke Library, Yale University.

35. Paul Brooks, *The House of Life: Rachel Carson at Work* (Boston: Houghton Mifflin, 1972), 228.

36. Edward O. Wilson to Rachel Carson, October 7, 1958, RCP/BLYU.

37. Rachel Carson to Edward O. Wilson, October 18, 1958, RCP/BLYU.

38. Musil, *Rachel Carson and Her Sisters*, 90.

39. Quoted in Intondi, "Martin Luther King on Non-Violence and Disarmament." See also Dr. Martin Luther King, Jr., "Address at the Thirty-sixth Annual Dinner of the War Resisters League," New York, New York, February 2, 1959, Martin Luther King, Jr., Research and Education Institute, Stanford University, https://kinginstitute.stanford.edu/king-papers/documents/address-thirty-sixth-annual-dinner-war-resisters-league.

Chapter 8: Forging the New Frontier

1. "John F. Kennedy Statement Announcing Candidacy for President, 2 January 1960," John F. Kennedy Presidential Library and Museum, https://www.jfklibrary.org/asset-viewer/archives/JFKSEN/0905/JFKSEN-0905-021.

2. Doug Scott, *The Enduring Wilderness: Protecting Our Natural Heritage Through the Wilderness Act* (Golden, CO: Fulcrum, 2004), 51.

3. Robert Caro, *The Years of Lyndon Johnson: The Passage of Power* (New York: Vintage, 2012), 59.

4. William O. Douglas, *The Three Hundred Year War: A Chronicle of Ecological Disaster* (New York: Random House, 1972), 51.

5. Author interview with Kenneth Brower, March 16, 2018.

6. William O. Douglas, Oral History Interview, JFK no. 1, November 9, 1967, John F. Kennedy Presidential Library and Museum, 39.

7. John Kenneth Galbraith, *The Affluent Society* (New York: Houghton Mifflin Harcourt, 1958), 253.

8. Author interview with Stewart Udall, September 6, 2009.

9. Author interview with Tom Udall, February 18, 2011.

10. Author interview with Burr Udall, March 3, 2011.

11. Stewart Udall, "Human Values and Hometown Snapshots: Early Days in St. Johns," *American West*, April 1982, library.arizona.edu/exhibits/sludall/articlespages/article/.html.

12. "Stewart L. Udall," American Academy for Park and Recreation Administration, https://aapra.org/Awards/Pugsley-Medal/Recipient-Biography/Id/24.

13. Stewart L. Udall, *To the Inland Empire: Coronado and Our Spanish Legacy* (Garden City, NY: Doubleday, 1987), 3–7.

14. L. Boyd Finch, *Legacies of Camelot: Stewart and Lee Udall, American Culture, and the Arts* (Norman: University of Oklahoma Press, 2008), 10–11.

15. Mo Udall, *Too Funny to Be President* (New York: Henry Holt, 1988).

16. Author interview with Stewart Udall, September 6, 2009.

17. Author interview with Tom Udall, October 4, 2019.

18. "Stewart L. Udall," American Academy for Park and Recreation Administration, https://aapra.org/Awards/Pugsley-Medal/Recipient-Biography/Id/24.

19. "Potomac Timeline," Interstate Commission on the Potomac River Basin, https://www.potomacriver.org/potomac-basin-facts/potomac-timeline/.

20. William V. Shannon, "Out West, Too, First Families Lean Toward Public Service," *New York Post*, December 8, 1960.

21. "Text of Kennedy News Conference Here Appointing Udall," *New York Times*, December 8, 1960, 27, https://timesmachine.nytimes.com/timesmachine/1960/12/08/issue.html.

22. Central Arizona Project, SUP-NA, A2372, box 10, folder 8.

23. Douglas Cornell, "Kennedy Charges G.O.P. Wasted Natural Resources," Greeley (Colorado) *Daily Tribune*, June 18, 1960, 8.

24. "Remarks of Senator John F. Kennedy at Durango, Colorado, June 18, 1960," John F. Kennedy Presidential Library and Museum, https://www.jfklibrary .org/Research/Research-Aids/JFK-Speeches/Durango-CO_19600618.aspx.

25. "Kennedy Plans Western Secretary," (Grand Junction) *Daily Sentinel*, June 18, 1960, 1.

26. "Nixon and Trudeau to Sign an Agreement to Fight Great Lakes Pollution," *New York Times*, April 9, 1972, 49, https://www.nytimes.com/1972/04/09 /archives/nixon-and-trudeau-to-sign-an-agreement-to-fight-great-lakes.html.

27. John H. Hartig, "The Return of the Detroit River's Charismatic Megafauna," Center for Humans and Nature, November 17, 2014, http://www.humans andnature.org/return-detroit-rivers-charismatic-megafauna.

28. Marcy Jane Knopf-Newman, *Beyond Slash, Burn and Poison* (New Brunswick, NJ: Rutgers University, 2004), 38.

29. Nancy Koehn, *Forged in Crisis: The Power of Courageous Leadership in Turbulent Times* (New York: Simon & Schuster, 2017), 401.

30. Steve Chase and Mark Madison, "The Expanding Ark: 100 Years of Wildlife Refuges," *Wild Earth*, Winter 2003–04, 25.

31. Quoted in *Cape Cod National Seashore Park: Hearings Before the Subcommittee on Public Lands* (Washington, DC: US Government Printing Office, 1960), 349.

32. Arthur M. Schlesinger, Jr., *Robert Kennedy and His Times* (New York: Houghton Mifflin Harcourt, 2002), 195.

33. *Henry M. Jackson: Late Senator from Washington* (Washington, DC: US Government Printing Office, 1983), 408.

34. David Shribman, "Senator Henry M. Jackson Is Dead at 71," *New York Times*, September 3, 1983, 10, https://www.nytimes.com/1983/09/03/obituaries /senator-henry-m-jackson-is-dead-at-71.html.

35. Robert G. Kaufman, *Henry M. Jackson: A Life in Politics* (Seattle: University of Washington Press, 2011), 164.

36. Senator Frank Church, Senate debate on the Wilderness bill, *Congressional Record*, September 5, 1961.

37. Author interview with Bethine Clark Church, November 22, 2010.

38. LeRoy Ashby and Rod Gramer, *Fighting the Odds: The Life of Senator Frank Church* (Pullman: Washington State University Press, 1994), 23.

39. Author interview with Bethine Clark Church, November 22, 2010.

40. William O. Douglas, *The Autobiography of William O. Douglas*, vol. 2, *The Court Years, 1939–1975* (New York: Random House, 1980), 314.

41. "Kennedy's Idea Men Map 'New Frontier' Apart from Campaign," *Washington Post*, August 7, 1960, A1.

42. Democratic Party Platforms, 1960 Democratic Party Platform Online by Gerhard Peters and John T. Woolley, The American Presidency Project, https://www.presidency.ucsb.edu/node/273234.

43. Tricia Nixon Cox, email to author, October 27, 2020.

44. "1960 Democratic Party Platform," July 11, 1960, The American Presidency Project, https://www.presidency.ucsb.edu/documents/1960-democratic-party-platform.

45. John F. Kennedy, "We Must Climb to the Hilltop," *Life*, August 22, 1960, 75. See also Richard M. Nixon, "Our Resolve Is Running Strong," *Life*, August 22, 1960, 94. The Kennedy and Nixon articles were included in Oscar Handlin, ed., *American Principles and Issues: The National Purpose* (New York: Holt, Rinehart and Winston, 1960), 3–17.

46. Author interview with Ethel Kennedy, August 14, 2009.

47. Rose Houk, *Heart's Home: Lyndon B. Johnson's Hill Country* (Tucson, AZ: Southwest Parks and Monuments Association, 1986), 28–29.

48. Sherrod Brown, *Desk 88: Eight Progressive Senators Who Changed America* (New York: Farrar, Straus and Giroux, 2019), 247.

49. Jeff Shesol, *Mutual Contempt: Lyndon Johnson, Robert Kennedy, and the Feud that Defined a Decade* (New York: W. W. Norton, 1997), 213.

50. Ibid., 468.

51. Dana Rubin, "The Lake No One Knows," *Texas Monthly*, November 1992, 130–34, https://www.texasmonthly.com/travel/the-lake-no-one-knows/.

52. Lady Bird Johnson, "Remarks at the Dedication in 1976 of the LBJ Grove at the Lady Bird Johnson Park in Washington, D.C.," *Congressional Record*, May 7, 1976.

53. Lewis L. Gould, *Lady Bird Johnson: Our Environmental First Lady* (Lawrence: University Press of Kansas, 1999), 3.

54. Ruth Montgomery, "Selling the Nation on Beauty," *New York Journal American*, May 30, 1965. See also Gould, *Lady Bird Johnson*, 5.

55. Gould, *Lady Bird Johnson*, 5.

56. Jan Jarboe, "Lady Bird Looks Back," *Texas Monthly*, December 1994, 148.

57. Quoted in Gould, *Lady Bird Johnson*, 6.

58. Ibid., 7.

59. Monroe Billington, "Lyndon B. Johnson and Blacks: The Early Years," *Journal of Negro History* 62, no. 1 (January 1977): 1.

60. Tom Wicker, "Lyndon Johnson Is 10 Feet Tall," *New York Times Magazine*, May 23, 1965, 324, 382–86, https://www.nytimes.com/1965/05/23/archives/lyndon-johnson-is-10-feet-tall-johnson-is-10-feet-tall.html.

61. Ronnie Dugger, *The Politician: The Life and Times of Lyndon Johnson* (New York: W. W. Norton, 1982), 140.

62. Quoted in Michael R. Beschloss, *The Crisis Years: Kennedy and Khrushchev, 1960–1963* (New York: HarperCollins, 2008), 666.

63. Michael Beschloss, ed., *Taking Charge: The Johnson White House Tapes, 1963–1964* (New York: Touchstone Books, 1997), 12.

64. Ruth Teiser and Catherine Harroun, oral history, *Conservation with Ansel Adams* (Berkeley: University of California, 1978), 456, InternetArchive.org.

65. Mary Street Alinder, *Ansel Adams: A Biography* (New York: Henry Holt, 1996), 243.

66. William O. Douglas, statement on *This Is the American Earth*, October 1960, Sierra Club Papers, University of California, Berkeley.

67. Charles Poore, "Books of the Times," *New York Times*, July 9, 1960, 17, https://www.nytimes.com/1960/07/09/archives/books-of-the-times.html.

68. Michael P. Cohen, *The History of the Sierra Club, 1892–1970* (New York: Random House, 1988), 259.

69. Linda Lear, *Rachel Carson: Witness for Nature* (New York: Henry Holt, 1992), 367.

70. Douglas Brinkley, "Rachel Carson and JFK, an Environmental Tag Team," *Audubon Magazine*, May–June 2012, https://www.audubon.org/magazine /may-june-2012/rachel-carson-and-jfk-environmental-tag-team.

71. Lear, *Rachel Carson*, 376–77.

72. Rachel Carson to Dorothy Freeman, October 12, 1960, Rachel Carson Papers, Beinecke Library, Yale University.

73. Diana Post and Munro Meyersburg, "Rachel Carson and the House Where She Wrote Silent Spring on the 60th Anniversary of its Construction," Rachel Carson Landmark Alliance, May 28, 2018, https://rachelcarsonlandmarkalliance .org/rachel-carson-and-the-house-where-she-wrote-silent-spring/.

74. Rachel Carson, preface to the revised edition of *The Sea Around Us* (1961), reprinted in ibid. (New York: Oxford University Press, 1991), xxv.

75. Tim Palmer, *The Wild and Scenic Rivers of America* (Washington, DC: Island Press, 1993), 167–68.

76. Stewart Udall with Joseph Stoker, *This Week*, February 12, 1961.

77. William O. Douglas, *My Wilderness: East to Katahdin* (Garden City, NY: Doubleday, 1961), 237.

78. David Nasaw, *The Patriarch: The Remarkable Life and Turbulent Times of Joseph Kennedy* (New York: Penguin, 2012), 749–52.

79. "Conservation Drive Pledged," *New York Times*, November 17, 1960, 17, https://www.nytimes.com/1960/11/17/archives/conservation-drive-pledged .html.

80. Author interview with Tom Udall, February 18, 2011.

81. Arthur M. Schlesinger, *Jacqueline Kennedy: Historic Conversations on Life with John F. Kennedy* (New York: Hyperion, 2011), 121.

82. Quoted in Lawrence Thompson and R. H. Winnick, *Robert Frost: The Late Years, 1938–1963* (New York: Holt, Rinehart and Winston, 1976), 277.

83. Thurston Clarke, *Ask Not: The Inauguration of John F. Kennedy and the Speech That Changed America* (New York: Penguin, 2010), 139–40.

Chapter 9: Wallace Stegner's "Wilderness Letter"

1. "City-Bred Outdoorsman: Laurance Spellman Rockefeller," *New York Times*, February 1, 1962, 16, https://www.nytimes.com/1962/02/01/archives /citybred-outdoorsman-laurance-spelman-rockefeller-33000acre.html.

2. James P. Gilligan, "The Development of Policy and Administration of Forest Service Primitive and Wilderness Areas in the Western United States," PhD diss., University of Michigan, 1953.

3. Wallace Stegner, *Beyond the Hundredth Meridian: John Wesley Powell and the Second Opening of the West* (Boston: Houghton Mifflin, 1953).

4. Author interview with David Pesonen, October 5, 2011.

5. Wallace Stegner, "Wilderness Letter," in *Marking the Sparrow's Fall: Wallace Stegner's American West*, ed. Page Stegner (New York: Henry Holt, 1998), 111–17.

6. Wallace Stegner, "The Geography of Hope," *Living Wilderness*, December 1980, https://web.stanford.edu/~cbross/Ecospeak/wildernessletterintro .html, quoted in Elia T. Ben-Ari, "Defender of the Voiceless: Wallace Stegner's Conservation Legacy," *Bioscience* 50, no. 3 (March 2000): 253.

7. T. H. Watkins, "Letter to Mary," in *The Geography of Hope: A Tribute to Wallace Stegner*, ed. Page Stegner and Mary Stegner (San Francisco: Sierra Club Books, 1996), 67.

8. Wallace Stegner to David E. Pesonen, "The Wilderness Letter," Outdoor Recreation Resources Review Commission, December 3, 1960.

9. Wallace Stegner, *The Sound of Mountain Water: The Changing American West* (New York: Vintage Books, 2017), 114.

10. Nelson G. Hairston, Frederick E. Smith, and Lawrence B. Slobodkin, "Community Structure, Population Control, and Competition," *American Naturalist* 94, no. 879 (November–December 1960): 421–25, https://www .academia.edu/2937962/Community_structure_population_control_and _competition.

11. Stewart L. Udall Journal, December 7, 1960, AZ372, Box 80, Stewart L. Udall Papers, University of Arizona at Tucson.

12. W. H. Lawrence, "Kennedy Chooses Udall of Arizona for Interior Job," *New York Times*, December 8, 1960, 1, 26, https://www.nytimes .com/1960/12/08/archives/kennedy-chooses-udall-of-arizona-for-interior -job-representative-40.html.

13. L. Boyd Finch, *Legacies of Camelot: Stewart and Lee Udall, American Culture, and the Arts* (Norman: University of Oklahoma Press, 2008), 3.

14. "Text of Kennedy News Conference Here Appointing Udall," *New York Times*, December 8, 1960, 27.

15. Edward F. Woods, "Udall Expected to Halt Trend, Restore Interior Department as Champion of Natural Resources," *St. Louis Post-Dispatch*, December 25, 1960, 23.

16. "Kennedy Picks Representative Udall as Secretary of Interior," *Battle Creek Enquirer*, December 7, 1960, 2.

17. *New York Post*, December 9, 1960, University of Arizona Special Collections, Stewart L. Udall Papers, AZ 372, B242 scrapbook.

18. J. Edgar Hoover to Stewart Udall, December 8, 1960; Arthur J. Goldburg to Stewart Udall, January 2, 1961; and Mike Mansfield to Stewart Udall, December 7, 1960, AZ372, Box 81, Udall Papers, University of Arizona at Tucson.

19. Paul Douglas to Stewart Udall, December 8, 1960, University of Arizona Special Collections, Stewart L. Udall Papers, AZ 372, box 81.

20. Barry Goldwater to Stewart Udall, December 6, 1960, University of Arizona Special Collections, Stewart L. Udall Papers, AZ 372, box 81, folder 2.

21. Byron W. Daynes and Glen Sussman, *White House Politics and the Environment: Franklin D. Roosevelt to George W. Bush* (College Station: Texas A&M University Press, 2010), 52.

22. Theodore C. Sorensen, *Kennedy* (New York: Harper & Row, 1965), 276–77.

23. Author interview with Stewart Udall, September 6, 2009.

24. Cory Hatch, "Following Historic Footprints: Group Traces Muries' 1956 Trip up the Sheenjek River in Alaska," Jackson Hole *News & Guide*, August 9, 2006.

25. John F. Kennedy, "Inaugural Address," January 20, 1961, The American Presidency Project, https://www.presidency.ucsb.edu/node/234470.

26. Lawrance Roger Thompson and R. H. Winnick, *Robert Frost: The Later Years, 1938–1963* (New York: Holt, Rinehart and Winston, 1966), 280–82.

27. Scott Raymond Einberger, *With Distance in His Eyes: The Environmental Life and Legacy of Stewart Udall* (Reno: University of Nevada Press, 2018), 196–97.

28. Charles Wilkinson, interview with Stewart Udall, Center of the American West, University of Colorado Boulder, September 24, 2003, http://centerwest.org/wp-content/uploads/2011/01/udall.pdf, 7.

29. "Interior Secretary," *New York Times*, December 8, 1960, 34.

30. Ibid.

31. Einberger, *With Distance in His Eyes*, 68.

32. Stewart Udall, "National Parks for the Future," *Atlantic*, January 1961, 81–84, https://speccoll.library.arizona.edu/online-exhibits/items/show/1487.

33. Ibid.

34. Author interview with Sharon Francis, June 4, 2014.

35. Paula Becker, "Dyer, Pauline (Polly) (1920–2016)," HistoryLink.org, December 22, 2010, https://www.historylink.org/File/9673.

36. Edward A. Whitesell ed., *Defending Wild Washington* (Seattle: Mountaineers Books, 2004), 148–56.

37. Author interview with Sharon Francis, June 4, 2014.

38. University of Washington, "Polly Dyer: A Sweeping Legacy", accessed May 20, 2022, https://www.sos.wa.gov/_assets/legacy/polly-dyer-profile.pdf.

39. Oral history transcript, Sharon Francis interview 1 (1), 512011969, by Dorthy Pierce (McSweeny), LBJ Library Oral History, LBJ Presidential Library, 2, https://www.discoverlbj.org/item/oh-francis-19690520-1-8-68.

40. Author interview with Sharon Francis, June 4, 2014.

41. Peter Hannaford, *Presidential Retreats: Where the Presidents Went and Why They Went There* (New York: Threshold Editions, 2012), 202.

42. Ted Sorensen, *Counselor: A Life at the Edge of History* (New York: Harper-Collins, 2008), 405.

43. John F. Kennedy, "Special Message to the Congress on Natural Resources," February 23, 1961, The American Presidency Project, https://www.presidency.ucsb.edu/documents/special-message-the-congress-natural-resources.

44. Sorensen, *Counselor*, 208.

45. Ibid.

Chapter 10: The Green Face of America

1. "National Park Chiefs Turn to Workshops," (Flagstaff) *Arizona Daily Sun*, April 26, 1961, 2.

2. Author interview with Thomas Udall, August 14, 2018.

3. Author interview with Stewart Udall, September 7, 2009.

4. Author interview with Sharon Francis, June 4, 2014.

5. Abe Chanin, "Setting the Record Straight on the Udall Boys," *Arizona Daily Star*, 53.

6. Richard Severo, "Morris K. Udall, Fiercely Liberal Congressman, Dies at 76," *New York Times*, December 14, 1998, B9, http://www.nytimes.com/1998/12/14/nyregion/morris-k-udall-fiercely-liberal-congressman-dies-at-76.html.

7. Stewart L. Udall, introduction to George B. Hartzog, Jr., *Battling for the National Parks* (Mount Kisco, NY: Moyer Bell, 1988), xii.

8. "Douglas to Lead Reunion Hikers down Towpath; Public Invited," *Washington Post*, May 2, 1961, A17.

9. William O. Douglas, *My Wilderness: The Pacific West* (Garden City, NY: Doubleday, 1960), 168.

10. Patricia Sullivan, "Old Angler's Inn Proprietor Olympia Reges Dies," *Washington Post*, September 4, 2005, https://www.washingtonpost.com/archive/local/2005/09/04/old-anglers-inn-proprietor-olympia-reges-dies/00dc3c43-cd01-4073-98ad-fe7be8780a00/.

11. "Minister Taken for a Tramp," *Guardian*, May 8, 1961, 9.

12. William O. Douglas, *Muir of the Mountains* (Boston: Houghton Mifflin, 1961).

13. Author interview with Stewart Udall, August 6, 2007.

14. William O. Douglas to Conrad L. Wirth, December 4, 1961, William O. Douglas Papers, Library of Congress, Washington, DC.

15. Tim Palmer, *The Wild and Scenic Rivers of America* (Washington, DC: Island Press, 1993), 18.

16. Author interview with Jeanne Halberstam, March 6, 2011.

17. Author interview with Stewart Udall, September 6, 2009.

18. National Park Service, *Economic Report: Proposed Cape Cod National Seashore Park, National Park Service, Region Five, U.S. Dept. of the Interior* (Boston: Economic Development Associates, 1960).

19. *Cape Cod National Seashore Park: Hearings Before the United States Senate*, March 9, 1961 (Washington, DC: US Government Printing Office, 1961), 34.

20. Sigurd F. Olson, *The Meaning of Wilderness: Essential Articles and Speeches*, ed. David Backes (Minneapolis: University of Minnesota Press, 1958), 374.

21. Sigurd F. Olson, *The Singing Wilderness* (New York: Alfred A. Knopf, 1956), 82.

22. Letter from Frank E. Masland, Jr., to Sigurd Olson, March 13, 1961, quoted in David Backes, *A Wilderness Within: The Life of Sigurd F. Olson* (Minneapolis: University of Minnesota Press, 1997), 295.

23. Backes, *A Wilderness Within*, 297.

24. Ansel Adams, "The Artist and the Ideals in Wilderness," in *Wilderness: America's Living Heritage*, ed. David Brower (San Francisco: Sierra Club, 1961), 49–59.

25. Ibid., 44–59.

26. Joseph Wood Krutch, *More Lives than One* (New York: William Sloane Associates, 1962), 3.

27. Joseph Wood Krutch, *Henry David Thoreau* (Westport, CT: Greenwood Press, 1948).

28. Quoted in Char Miller, *On the Edge: Water, Immigration, and Politics in the Southwest* (San Antonio: Trinity University Press, 2013), 168–69.

29. Paul Horgan, "In the Clear, Dry Light of the Desert; THE DESERT YEAR" (book review), *New York Times*, March 16, 1952, BR3, https://www.nytimes.com/1952/03/16/archives/in-the-clear-dry-light-of-the-desert-the-desert-year-by-joseph-wood.html.

30. "Joseph Wood Krutch," *New York Times*, May 23, 1970, 22.

31. Joseph Wood Krutch, *Grand Canyon: Today and All Its Yesterdays* (Tucson: University of Arizona Press, 1989), 275.

32. Joseph Wood Krutch, *The Voice of the Desert: A Naturalist's Interpretation* (New York: Morrow Quill Paperbacks, 1980), 131.

33. Joseph Wood Krutch, "Human Life in the Context of Nature," in *Wilderness: America's Living Heritage*, ed. David Brower (San Francisco: Sierra Club, 1961), 67–79.

34. Krutch, *More Lives than One*.

35. Robert Rowley, "Joseph Wood Krutch: The Forgotten Voice of the Desert," *American Scholar* 64, no. 3 (Summer 1995), 443.

36. Doug Scott, *The Enduring Wilderness* (Golden, CO: Fulcrum, 2004), 153.

37. Quoted in United States Department of the Interior, National Park Service, "Quotes: Conservation, Parks, Natural Beauty," National Park Service, 1966, https://irmaservices.nps.gov/datastore/v4/rest/DownloadFile/446406?access Type=DOWNLOAD.

38. Robert Frost, "Dust of Snow," in *New Hampshire* (New York: Henry Holt, 1923).

39. Wallace Stegner, "Wilderness Letter," in *The Sound of a Mountain Water* (New York: Penguin Random House, 1969), 146.

40. Louis Cassels, "Man in a Hurry Who Loves His Work," *Deseret News*, May 13, 1961.

41. Jack Kerouac, *The Dharma Bums* (New York: Viking Press, 1958); Jack Kerouac, *Desolation Angels* (New York: Coward-McCann, 1965).

42. Stewart Udall, "Notes en Route to the NW," June 7, 1961, Folder 2, Box 90, Stewart L. Udall Papers, Special Collection, University of Arizona, Tucson, AZ.

43. Clinton Anderson, "This We Hold Dear," *American Forests*, July 1963, 24–25.

44. Orville Freeman, diary, vol. 1, July 22, 1961, Freeman Papers, Minnesota Historical Society, St. Paul, Minnesota.

45. Representative John P. Saylor, "Minority Views," in *Providing for the Preservation of Wilderness Areas for the Management of Public Lands and for Other Purposes*, House Report 87-2521, October 3, 1962, 118–32.

46. "Nuclear Test Ban Treaty," John F. Kennedy Presidential Library and Museum, https://www.jfklibrary.org/learn/about-jfk/jfk-in-history/nuclear-test-ban-treaty#:~:text=In%20August%201961%2C%20the%20Soviet, the%20bomb%20dropped%20on%20Hiroshima.

47. "Udall Dedicates Park at Cape Cod," *New York Times*, March 31, 1966, 45.

48. John F. Kennedy, "Remarks upon Signing Bill Authorizing the Cape Cod National Seashore Park," August 7, 1961, The American Presidency Project, https://www.presidency.ucsb.edu/documents/remarks-upon-signing-bill-authorizing-the-cape-cod-national-seashore-park.

49. Paul Schneider, *The Enduring Shore: A History of Cape Cod, Martha's Vineyard, and Nantucket* (New York: Henry Holt, 2000), 304.

50. Horace M. Albright, "To Preserve Seashore; Warning Sounded That Steps Must Be Taken to Save Areas," *New York Times*, September 11, 1961, 26, https://www.nytimes.com/1961/09/11/archives/to-preserve-seashore-warning-sounded-that-steps-must-be-taken-to.html.

51. Michael W. Giese, "A Federal Foundation for Wildlife Conservation of the National Wildlife Refuge System, 1920–1968," PhD diss., American University, Washington, DC, 2008, 347–49.

52. Stewart Udall, tribute to Wallace Stegner at Western History Association, Albuquerque, NM, October 22, 1994. See also Jackson Benson, *Wallace Stegner: His Life and Works* (New York: Penguin, 1997), 278–79.

53. Wallace Stegner to Stewart Udall, June 7, 1963, box 215, folder 2, Stewart L. Udall Papers, University of Tucson, Tucson, AZ.

54. John F. Kennedy, "Proclamation 3443—Establishing the Buck Island Reef National Monument in the Virgin Islands of the United States," December 28, 1961, The American Presidency Project, https://www.presidency.ucsb.edu/documents/proclamation-3443-establishing-the-buck-island-reef-national-monument-the-virgin-islands.

55. John F. Kennedy, "Proclamation 3439—Enlarging the Saguaro National Monument, Arizona," November 15, 1961, The American Presidency Project, https://www.presidency.ucsb.edu/documents/proclamation-3439-enlarging-the-saguaro-national-monument-arizona.

56. Betty Leavengood, *Tucson Hiking Guide*, 4th ed. (Portland, OR: West Winds Press, 2014), 16–17.

57. Stewart L. Udall, *America's Natural Treasures: National Monuments and Seashores* (Waukesha, WI: Country Beautiful Corporation, 1971), 126.

58. Jackson Benson, *Wallace Stegner: His Life and Work* (Lincoln: University of Nebraska Press, 1996), 279.

59. Wallace Stegner to Stewart Udall, December 30, 1961, in Wallace Stegner, *The Selected Collections of Wallace Stegner*, ed. Page Stegner (New York: Shoemaker Hoard, 2001), 367–68.

Chapter 11: Rachel Carson, the Laurance Rockefeller Report, and Kennedy's Science Curve

1. John F. Kennedy, "State of the Union Address," January 11, 1962.
2. William Souder, *On a Farther Shore: The Life and Legacy of Rachel Carson* (New York: Crown, 2012), 318.
3. Souder, *On a Farther Shore*, 321.
4. Quoted in Alex MacGillivray, *Rachel Carson's Silent Spring*, ed. Neil Turnbull (New York: Barron's Educational Series, 2004), 61.
5. William O. Douglas, lecture at National Parks Association, Washington, DC, May 21, 1962, William O. Douglas Papers, Library of Congress, Washington, DC.

6. James M. O'Fallon, *Nature's Justice: Writings of William O. Douglas* (Corvallis: Oregon University Press, 2000), 291.

7. Ibid.

8. Ibid.

9. Ibid.

10. William O. Douglas, lecture, William O. Douglas Papers, Library of Congress, Washington, DC.

11. Rachel Carson to Dorothy Freeman, May 20, 1962, in Rachel Carson and Dorothy Freeman, *Always, Rachel: The Letters of Rachel Carson and Dorothy Freeman, 1952–1964—The Story of a Remarkable Friendship*, ed. Martha Freeman (Boston, MA: Beacon Press, 1994), 404–05.

12. Souder, *On a Farther Shore*, 323–26.

13. Rachel Carson, *Silent Spring* (Boston: Houghton Mifflin, 1994), 3.

14. Ibid., 6.

15. William O. Douglas, *My Wilderness: East to Katahdin* (Garden City, NY: Doubleday, 1961).

16. Ibid., 32.

17. Linda Lear, *Rachel Carson: Witness for Nature* (New York: Henry Holt, 1992), 419.

18. Carson, *Silent Spring*.

19. Quoted in Lear, *Rachel Carson*, 419.

20. Nancy Koehn, "Rachel Caron's Lessons, 50 Years After 'Silent Spring,'" *New York Times*, October 27, 2012, 5.

21. John F. Kennedy, "Annual Message to the Congress on the State of the Union," January 11, 1962, The American Presidency Project, https://www.presidency.ucsb.edu/documents/annual-message-the-congress-the-state-the-union-4.

22. David E. Pesonen, "Outdoor Recreation for America: An Analysis," *Sierra Club Bulletin*, May 1962, 6–12.

23. John F. Kennedy, "Special Message to the Congress on Conservation," March 1, 1962, The American Presidency Project, https://www.presidency.ucsb.edu/documents/special-message-the-congress-conservation.

24. Ibid.

25. Pesonen, "Outdoor Recreation for America: An Analysis," 6–12.

26. Nate Schweber, *This America Is Ours: Bernard and Avis DeVoto and the Forgotten Fight to Save the Wild* (Boston: Mariner Books, 2022).

27. Scott Raymond Einberger, *With Distance in His Eyes: The Environmental Life and Legacy of Stewart Udall* (Reno: University of Nevada Press, 2018).

28. Einberger, *With Distance in His Eyes*, 212–16.

29. Phyllis Paulsell, "Udall Goes Too Far on Federal Parks," *Arizona Republic*, December 12, 1961, 6.

30. Rebecca Conard, "Tough as the Hills: The Making of the Tallgrass Prairie National Preserve," *Kansas History*, Summer 2006, 72.

31. "Rancher Irate; Secretary Ordered off Land," *Wichita Beacon*, December 14, 1961, 1.

32. Charles Wilkinson, interview with Stewart Udall, September 24, 2003. Center for American West, University of Colorado at Boulder, 3, https://drive .google.com/field/1x99fjwxPeLosOVYb4zi4xs5NBL2H/view.

33. Thomas G. Smith, "1962: The Year That Changed the Redskins," *Washingtonian*, October 10, 2011, https://www.washingtonian.com/2011/10/10/1962 -the-year-that-changed-the-redskins/.

34. Author interview with Robert Stanton, August 18, 2020.

35. Ibid.

36. Robert Stanton to author, January 17, 2022. Stanton sent me a fat folder of relevant information that I hope to use in a future book project on national parks.

37. Quoted in introduction to Editors of American Heritage, *The American Heritage Book of Indians* (New York: American Heritage, 1961).

38. Quoted in "The Life of Alvin M. Josephy Jr., Authoritative Interpreter of History," Indian Country Today, September 12, 2018, https://indiancountry today.com/archive/the-life-of-alvin-m-josephy-jr-authoritative-interpreter -of-history.

39. Barbara L. Allen, *Uneasy Alchemy: Citizens and Experts in Louisiana's Chemical Corridor Disputes* (Cambridge, MA: MIT Press, 2003), 28.

40. Kristin Shrader-Frechette, *Environmental Justice: Creating Equality, Reclaiming Democracy* (New York: Oxford University Press, 2002), 9.

41. Marc Reisner, *Cadillac Desert: The American West and Its Disappearing Water* (New York: Viking Penguin, 1986), 233.

42. "Men of the Year: U.S. Scientists," *Time*, January 2, 1961, 40, https://content .time.com/time/subscriber/article/0,33009,895239,00.html.

43. Theodore C. Sorensen, "A View from the White House," in *Jerry Wiesner: Scientist, Statesman, Humanist: Memories and Memoirs*, ed. Walter A. Rosenblith and Judy F. Rosenblith (Cambridge, MA: MIT Press, 2003), 267–68.

44. "President Emeritus Jerome Wiesner Is Dead at 79," MIT News, October 26, 1994, https://news.mit.edu/1994/weisner-obit-1026.

45. Linus Pauling to John F. Kennedy, March 1, 1962, Special Collections and Archives and Research Center, Oregon State University Library, Corvallis, Oregon.

46. R. Revelle and H. Suess, "Carbon Dioxide Exchange Between Atmosphere and Ocean and the Questions of an Increase of Atmospheric CO_2 During the Past Decade," *Tullus* 9, no. 1, (February 1957).

47. Clinton P. Anderson to John F. Kennedy, February 14, 1961, John F. Kennedy, Roosevelt Library, Boston, Massachusetts.

48. Quoted in Scott McVay, *Surprise Encounters with Artists and Scientists, Whales and Other Living Things* (Wild River Books, 2015), 148.

49. John C. Lilly, *Man and Dolphin* (New York: Doubleday, 1961).

50. E. O. Wilson, *Sociobiology: The New Synthesis* (Cambridge, MA: Harvard University Press, 1975), 220.

51. Bob Thomas, "Producer Makes Stars of Dolphone, Rhinoceros," San Bernardino *County Sun*, October 21, 1963, 19.

52. Christopher Riley, "The Dolphin Who Loved Me," *Guardian* (London), June 8, 2014.

Chapter 12: The White House Conservation Conference

1. Sharon Francis, unpublished memoir, private collection, Charlestown, New Hampshire.

2. William O. Douglas, "Thoreau," speech at Dumbarton Oaks, Washington, DC, May 11, 1962, William O. Douglas Papers, Library of Congress, Washington, DC.

3. Joseph A. Loftus, "Udall Urges Aid on Conservation," *New York Times*, May 25, 1962, 30.

4. Robert K. Musil, *Rachel Carson and Her Sisters: Extraordinary Women Who Have Shaped America's Environment* (New Brunswick, NJ: Rutgers University Press, 2014), 117–18.

5. David Backes, *A Wilderness Within: The Life of Sigurd F. Olson* (Minneapolis: University of Minnesota Press, 1997), xiii.

6. Stewart L. Udall, "Address by Secretary of the Interior Stewart L. Udall at White House Conference on Conservation," West Auditorium, Department of State, Washington, DC, May 24, 1962.

7. Ibid., 84.

8. Ibid., 100.

9. Ibid., 99.

10. Ibid., 100.

11. Thomas W. Ottenad, "Grumbling over White House Conference on Conservation," *St. Louis Post-Dispatch*, May 27, 1962, 35.

12. Donald Worster, "The Highest Altruism," *Environmental History*, October 2014, 716.

13. Mark W. T. Harvey, *Wilderness Forever: Howard Zahniser and the Path to the Wilderness Act* (Seattle: University of Washington Press, 2005), 230–31.

14. "Day in the Capitol for One of Nation's Top Congressmen, Representative Wayne Aspinall," (Grand Junction, Colorado) *Daily Sentinel*, October 22, 1962.

15. Denis Collins, "Acting for Wilderness: The Word Is Vigilance," *Washington Post*, December 23, 1984.

16. "The Conservation Conference," *New York Times*, May 26, 1962, 24, https://www.nytimes.com/1962/05/26/archives/the-conservation-confer ence.html.

17. Author interview with Sharon Francis, June 4, 2014.

18. Francis, unpublished memoir.

19. Stephen Tying Mather, *Report on the Proposed Sand Dunes National Park, Indiana* (Washington, DC: US Government Printing Office, 1917), 3.

20. Carl Sandburg, "Chicago," in *Chicago Poems* (New York: Henry Holt, 1916).

21. North Callahan, *Carl Sandburg: Lincoln of Literature* (New York: University Press, 1970), 45, 97–98.

22. Stewart Udall to President John F. Kennedy, October 27, 1961, Stewart L. Udall Papers, University of Arizona, Tucson, AZ.

23. Carl Sandburg, letter to Paul Douglas, 1958, quoted in Ron Cockrell, *A Signature of Time and Eternity: The Administrative History of Indiana Dunes National Lakeshore, Indiana* (Omaha, NE: United States Department of the Interior, National Park Service, Midwest Regional Office, Office of Planning and Resource Preservation, Division of Cultural Resource Management, 1988); National Park Service, "Authorization of the Indiana Dunes National Lakeshore, 1965–1966," https://www.nps.gov/parkhistory/online_books /indu/adhi4a.htm.

24. Paul Douglas, *In the Fullness of Time: The Memoirs of Paul H. Douglas* (New York: Harcourt Brace Jovanovich, 1972), 538.

25. Paul Douglas, "Statement of Senator Paul H. Douglas on a National Park in the Indiana Dunes," 1959, Papers of Paul H. Douglas, Part 2, subseries 17, Chicago History Museum.

26. Jerry Denis, *The Living Great Lakes: Searching for the Heart of the Inland Islands* (New York: Thomas Dunne, 2003), 58.

27. Keith Schneider, "Lee Botts, Environmentalist and Champion of the Great Lakes, Is Dead at 91," *New York Times*, October 17, 2019, https://www.ny times.com/2019/10/17/science/lee-botts-dead.html.

28. Tim Palmer, *The Wild and Scenic Rivers of America* (Washington, DC: Island Press, 1993), 16–22.

29. Quoted in Tim Palmer, *Endangered Rivers and the Conservation Movement* (Berkeley: University of California Press, 1986), 90–93.

30. Quoted in Joe David Rice, "Arkansas Backstories: Buffalo River," *AY Magazine*, May 29, 2019, https://www.aymag.com/arkansas-backstories-buffalo -river/.

31. Author interview with Stewart Udall, September 6, 2009.

32. Quoted in Kenneth L. Smith, *Buffalo River Handbook* (Little Rock, AR: Ozark Foundation, 2018), 107.

33. Chad Montrie, *Making a Living: Work and Environment in the United States* (Chapel Hill: University of North Carolina Press, 2008), 106.

34. Nelson Lichtenstein, *The Most Dangerous Man in Detroit: Walter Reuther and the Fate of American Labor* (New York: Basic Books, 1995), 437.

35. Quoted in Scott Dewey, "Working for the Environment: Organized Labor and the Origins of Environmentalism in the United States, 1948–1970," *Environmental History* (January 1998), 45–63.

36. "The Refuge Recreation Act of 1962, with subsequent amendments, authorizes the Secretary of the Interior to administer refuges, hatcheries and other conservation areas for recreational use, when such uses do not interfere with the primary purpose for which these areas were established." Also, see "Refuge Recreation Act," US Fish & Wildlife Service, https://www.fws.gov /law/refuge-recreation-act#-text=The%20Refuge%20Recreation%20Act%20 of,which%20%these20%areas%20were%established. See also https://www .govinfo.gove/content/pkg/USCODE-2017-title16/pdf/USCODE-2017-title 16-chap1-subchapLXVIII.pdf.

37. "Land Conservation Fund," hearing, 87th Congress, 2nd Session, July 11, 1962, 2–7.

38. United States Department of the Interior, Fish and Wildlife Service, Bureau of Sport Fisheres and Wildlife, "1962 Public Use of National Wildlife Refuges," n.d. (stamped May 20, 1964), https://archive.org/details/1962public useofn449usfi.

39. "Udall Hails Establishment of New Wildlife Refuges as 'Banner Day,'" US Fish and Wildlife Service, June 28, 1962, https://www.fws.gov/news/Historic /NewsReleases/1962/19620628a.pdf.

40. Laura and William Riley, *Guide to The National Wildlife Refuges* (New York: Macmillan, 1992), 241.

41. Riley and Riley, *Guide to the National Wildlife Refuges*, 68–69.

42. John F. Kennedy to Governor Elbert Carvel, June 18, 1963, Douglas Brinkley private papers, Rice University.

43. Eisenhower had dedicated Theodore Roosevelt's Sagamore Hill home in bucolic Oyster Bay, Long Island, as a national shrine on June 14, 1953, but the home, which was open to the public as a museum, was privately run by the Theodore Roosevelt Association (TRA). Roosevelt's birthplace at 28 East 20th Street in New York City was likewise run by the TRA; now, thanks to Kennedy, both became part of the National Park Service's portfolio. See "An Act to Authorize Establishment of the Theodore Roosevelt Birthplace and Sagamore Hill National Historic Sites, New York, and for Other Purposes," Public Law 87-547, 76 July 25, 1962, https://www.govinfo.gov/content/pkg /STATUTE-76/pdf/STATUTE-76-Pg217.pdf.

44. Stewart Udall, speech, July 8, 1963, quoted in Bill Bleyer, *Sagamore Hill: Theodore Roosevelt's Summer White House* (Charleston, SC: History Press, 2016), 113–14.

45. "Legal Footwork in the Wilderness," *New York Times*, August 11, 1962, 28, https://www.nytimes.com/1962/08/11/archives/legal-footwork-in-the-wil derness.html.

46. William O. Douglas, *The Autobiography of William O. Douglas*, vol. 2, *The Court Years, 1939–1975* (New York: Random House, 1980), 308–9.

Chapter 13: Rachel Carson's Alarm

1. Elena Conis, "DDT Disbelievers: Health and the New Economic Poisons in Georgia After World War II," *Southern Spaces*, October 28, 2016, https:// southernspaces.org/2016/ddt-disbelievers-health-and-new-economic-poisons -georgia-after-world-war-ii/.

2. "Washington Wire," *Wall Street Journal*, August 3, 1962, 1.

3. "On Controlling Pests," *Washington Post*, July 13, 1962, A18.

4. https:ww.fws.gov/news/historic/NewsReleases/1961/19611004.pdf.

5. David K. Hecht, "Constructing a Scientist: Expert Authority and Public Im- ages of Rachel Carson," *Historical Studies in Natural Sciences* 41, no. 3 (sum- mer 2011), 277–302.

6. Paul Knight, "A Case Study in Environmental Contamination," quoted in Paul Brooks, *The House of Life: Rachel Carson at Work* (Boston: Houghton Mifflin, 1989), 354.

7. Phil Casey, "Biologist Warns Against Wide Use of Insecticides," *Washington Post*, May 29, 1962, A1. See also Tatiana Schlossberg, *Inconspicuous Con- sumption: The Environmental Impact You Don't Know You Have* (New York: Grand Central, 2019), 117.

8. William Souder, *On a Farther Shore: The Life and Legacy of Rachel Carson* (New York: Crown, 2012), 317.

9. Ibid.

10. Ibid.

11. Danny Heitman, "'The Sea Trilogy' Review: Waves of Wonder," *Wall Street Journal*, March 3, 2022, https://www.wsj.com/articles/the-sea-trilogy-review -waves-of-wonder-11646348073.

12. Arlene R. Quaratiello, *Rachel Carson: A Biography* (Westport, CT: Green- wood Press, 2004), 94.

13. Rachel Carson, "Of Man and the Stream of Time," Scripps College, Clare- mont, California, June 12, 1962, Rachel Carson Papers, Yale Univesity, https://loa-shared.s3.amazonaws.com/static/pdf/Carson_Stream_Time.pdf.

14. Souder, *On a Farther Shore*, 338.

15. Ibid., 319.

16. Ellen Levine, *Up Close: Rachel Carson* (New York: Penguin, 2008), 167.

17. Rachel Carson, *Silent Spring* (Boston: Houghton Mifflin, 1994), 1.

18. Souder, *On a Farther Shore*, 324.

19. Ibid.

20. Ibid., 12.

21. Brooks, *The House of Life*, 267.

22. Quaratiello, *Rachel Carson*, 97.

23. Quoted in Paul Brooks, *Speaking for Nature: How Literary Naturalists from Henry Thoreau to Rachel Carson Have Shaped America* (Boston: Houghton Mifflin, 1960), 285–86.

24. Quaratiello, *Rachel Carson*, 106.

25. Carson, *Silent Spring*, 42.

26. Ibid., 12.

27. "Rachel Carson's Warning," *New York Times*, July 2, 1962, 28, https://times machine.nytimes.com/timesmachine/1962/07/02/82052400.html?page Number=28.

28. Linda Lear, *Rachel Carson: Witness for Nature* (New York: Henry Holt, 1992), 207.

29. Robert A. Caro, "Pesticides," *Newsday*, August 21, 22, 23, and 24, 1962.

30. Author interview with Robert Caro, August 11, 2020.

31. John F. Kennedy, "News Conference 42," August 29, 1962, John F. Kennedy Presidential Library and Museum, https://www.jfklibrary.org/archives/other -resources/john-f-kennedy-press-conferences/news-conference-42. Also quoted in Brooks, *The House of Life*, 305.

32. Souder, *On a Farther Shore*, 4.

33. Lear, *Rachel Carson*, 420.

34. Jack E. Davis, *The Bald Eagle: The Improbable Journey of America's Bird* (New York: W. W. Norton, 2022), 261.

35. Author interview with Edward Kennedy, July 3, 2007.

36. John F. Kennedy, "Commencement Address at Yale University," June 11, 1962, John F. Kennedy Presidential Library and Museum, https://www.jfk library.org/archives/other-resources/john-f-kennedy-speeches/yale-university -19620611.

37. Souder, *On a Farther Shore*, 332.

38. Robert C. Cowen, "The Chemical War on Pests: Is It Getting Out of Hand?," *Christian Science Monitor*, August 10, 1962, 9.

39. Carson, *Silent Spring*, 99.

40. Nancy Koehn, *Forged in Crisis: The Power of Courageous Leadership in Turbulent Times* (New York Scribner, 2017), 427–28.

41. "Silent Spring By Rachel Carson," *Economist*, February 23, 1963, 711.

42. William Darby, "Silence, Miss Carson!," *Chemical and Engineering News* 40 (October 1, 1962): 60–62.

43. "Pesticides: The Price for Progress," *Time*, September 28, 1962, 45, https://content.time.com/time/subscriber/article/0,33009,940091-1,00.html.

44. "The Gentle Storm Center," *Life*, October 12, 1962, 105–10.

45. Souder, *On a Farther Shore*, 345.

46. Robert D. McFadden, "Frances Oldham Kelsey, Who Saved U.S. Babies from Thalidomide, Dies at 101," *New York Times*, August 7, 2015, https://www.nytimes.com/2015/08/08/science/frances-oldham-kelsey-fda-doctor-who-exposed-danger-of-thalidomide-dies-at-101.html.

47. Quoted in Lear, *Rachel Carson*, 412.

48. Loren Eiseley, "Using a Plague to Fight a Plague," *Saturday Review*, September 29, 1962, 18–19.

49. Quaratiello, *Rachel Carson*, 114.

50. Tom Turner, *David Brower: The Making of the Environmental Movement* (Berkeley: University of California Press, 2015), 105–7.

Chapter 14: Point Reyes (California) and Padre Island (Texas) National Seashores

1. W. W. Moss, "Stewart L. Udall Oral History Interview," JFK #6, June 2, 1970, John F. Kennedy Presidential Library and Museum, https://www.jfklibrary.org/sites/default/files/archives/JFKOH/Udall%2C%20Stewart%20L/JFKOH-SLU-06/JFKOH-SLU-06-TR.pdf, 109.

2. Andrea G. Stillman, *Looking at Ansel Adams: The Photographs of the Man* (New York: Little, Brown, 2012), 83.

3. Moss, "Stewart L. Udall Oral History Interview," 112–13.

4. Charles T. Morrissey, "Wayne N. Aspinall Oral History Interview," November 10, 1965, John F. Kennedy, Presidential Library and Museum, https://www.jfklibrary.org/sites/default/files/archives/JFKOH/Aspinall%2C%20Wayne%20N/JFKOH-WNA-01/JFKOH-WNA-01-TR.pdf, 3.

5. Jedediah S. Rogers, *Fryingpan-Arkansas Project* (Washington, DC: Bureau of Reclamation, 2006), 2, 6.

6. John F. Kennedy, "Remarks at the Dedication of the Oahe Dam, Pierre, South Dakota," August 17, 1962, The American Presidency Project, https://www.presidency.ucsb.edu/documents/remarks-the-dedication-the-oahe-dam-pierre-south-dakota.

7. John F. Kennedy, "Remarks in Los Banos, California, at the Ground-Breaking Ceremonies for the San Luis Dam," August 18, 1962, The American Presidency Project, https://www.presidency.ucsb.edu/documents/remarks-los-banos-california-the-ground-breaking-ceremonies-for-the-san-luis-dam.

8. Mark W. T. Harvey, *A Symbol of Wilderness: Echo Park and the American Conservation Movement* (Albuquerque: University of New Mexico, 1994), 56.

9. John F. Kennedy, "Remarks in Pueblo, Colorado Following Approval of the Fryingpan-Arkansas Project," August 17, 1962, The American Presidency

Project, https://www.presidency.ucsb.edu/documents/remarks-pueblo-colo
rado-following-approval-the-fryingpan-arkansas-project.

10. Stewart Udall, Oral History, John F. Kennedy Presidential Library and Museum, 110.

11. Ansel Adams to Stewart Udall, August 14, 1962, in *Ansel Adams: Letters and
Images, 1916–1984*, ed. Mary Street Alinder and Andrea Gray Stillman (Boston: Little, Brown, 1988), 292.

12. Yosemite National Park Facebook post, August 17, 2014.

13. Eli Setencich, "Friendliness, Mutual Interest Mark Kennedy Yosemite Tour,"
Fresno Bee, August 19, 1962, 8.

14. Edward T. Folliard, "15,000 See Kennedy Trigger Start of Work at Dam Site,"
Washington Post, August 19, 1962, A1.

15. Setencich, "Friendliness, Mutual Interest Mark Kennedy Yosemite Tour."

16. Stewart Udall to Ansel Adams, August 14, 1962, Stewart Udall Papers, University of Arizona, Tucson, Arizona.

17. John F. Kennedy, "Remarks at San Luis Dam Ground-Breaking, Los Banos,
California," August 18, 1962, John F. Kennedy Presidential Library and Mu-
seum, https://www.jfklibrary.org/asset-viewer/archives/JFKPOF/039/JFK
POF-039-039.

18. Robert Cahn, "Ansel Adams, Environmentalist," *Sierra Club Bulletin*,
May–June 1979, 31–49.

19. Michael W. Giese, "A Federal Foundation for Wildlife Conservation: The
Evolution of the National Wildlife Refuge System, 1920–1968," PhD diss.,
American University, 2008, 363.

20. Harold Gilliam, *Island in Time: The Point Reyes Peninsula* (San Francisco: Sierra Club, 1962).

21. Stewart Udall, foreword, in Harold Gilliam, *Island in Time: The Point Reyes
Peninsula* (San Francisco: Sierra Club, 1962), 8.

22. Lois Crisler, *Arctic Wild: The Remarkable True Story of One Couple's Adventures Living Among Wolves* (New York: Lyons Press, 1999), 38.

23. Thomas R. Vale, "Conservation Strategies in the Redwoods," *Yearbook of the
Association of Pacific Coast Geographers* 36 (1974): 103.

24. W. Dwayne Jones, *Padre Island National Seashore: An Administrative History*,
http://npshistory.com/publications/pais/adhi/chap4.htm.

25. Ibid.

26. Jack E. Davis, *The Gulf: The Making of an American Sea* (New York: Liveright, 2018), 322–23.

27. Quoted in "The Padre Island National Seashore Recreation Area; Introduction of Bill for Its Creation," *Congressional Record*, January 5, 1961, 159.

28. Jones, *Padre Island National Seashore*.

29. Jones, *Padre Island National Seashore*.

30. "Seashore Bill for Padre Island Virtual 'Giveaway' Says Sadler, UPI Report," (Harlingen, Texas) *Valley Morning Star,* September 21, 1962, 1.

31. "An Act to Provide for the Establishment of the Padre Island National Seashore," Public Law 87-711, September 27, 1962, GovTrack, https://www.gov track.us/congress/bills/87/s4/text.

32. Patrick Cox, *Ralph W. Yarborough: The People's Senator* (Austin: University of Texas Press, 2002).

33. William O. Douglas, *Farewell to Texas: A Vanishing Wilderness* (New York: McGraw-Hill, 1967), 169.

34. Cox, *Ralph W. Yarborough,* 237–38.

35. Edward M. Kennedy, preface to Cox, *Ralph W. Yarborough,* x.

Chapter 15: Campaigns to Save the Hudson River and Bodega Bay

1. "Fast Facts About John F. Kennedy," John F. Kennedy Presidential Library and Museum, https://www.jfklibrary.org/Research/Research-Aids/Ready -Reference/JFK-Fast-Facts/Honey-Fitz.aspx.

2. Amy Swerdlow, *Women Strike for Peace: Traditional Motherhood and Radical Politics in the 1960s* (Chicago: University of Chicago Press, 1993).

3. Associated Press, "Demonstrators Busy," *York Dispatch,* October 24, 1962, 1.

4. Dennis Hevesi, "Dagmar Wilson, Anti-Nuclear Leader, Dies at 94," *New York Times,* January 23, 2011, https://www.nytimes.com/2011/01/24/us/24wilson .html.

5. Alvin Shuster, "Close-up of a 'Peace Striker,'" *New York Times,* May 6, 1962, 251, https://www.nytimes.com/1962/05/06/archives/closeup-of-a-peace -striker-she-is-dagmar-wilson-a-washington.html.

6. "'Human Survival Is Problem,' Mrs. M. L. King," *Pittsburgh Courier,* April 7, 1962, 30.

7. "Newsmakers: Witness," *Spokane Chronicle,* December 12, 1962, 17.

8. "Huge Power Plant Planned on Hudson," *New York Times,* September 27, 1962, 1, https://www.nytimes.com/1962/09/27/archives/huge-power-plant -planned-on-hudson-big-power-plant-to-rise-on.html.

9. David Schuyler, *Embattled River: The Hudson and Modern American Environmentalism* (Ithaca, NY: Cornell University Press, 2018), 11.

10. "Our Legacy," Scenic Hudson, https://www.scenichudson.org/about-us/our -legacy/.

11. Alexandra Zissu, "How Franny Reese fought ConEd, saved Storm King Mountain," Albany *Times-Union,* March 3, 2022.

12. Carl Carmer, "Testimony Before the Federal Power Commission: In the Matter of Consolidated Edison," May 1964, in David Stradling, *The Environmental Movement: 1968–1972* (Seattle, WA: University of Washington Press, 2012), 33.

13. John Cronin and Robert F. Kennedy, Jr., *The Riverkeepers: Two Activists Fight to Reclaim Our Environment as a Basic Human Right* (New York: Scribner, 1997), 33.

14. David Stradling, *The Nature of New York: An Environmental History of the Empire State* (Ithaca, NY: Cornell University Press, 2010), 189.

15. Robert H. Boyle, "An Absence of Wood Nymphs," *Sports Illustrated*, September 14, 1959, E5, https://vault.si.com/vault/1959/09/14/an-absence-of-wood-nymphs.

16. "Is Tule Too Good for Duck?," *Sports Illustrated*, December 14, 1959, 38.

17. Robert H. Boyle, "'I Am a Bit of a Fanatic,'" *Sports Illustrated*, October 21, 1963, 64, https://vault.si.com/vault/1963/10/21/i-am-a-bit-of-a-fanatic.

18. Ibid.

19. Cronin and Kennedy, *The Riverkeepers*, 30.

20. Robert H. Boyle, "A Stink of Dead Stripers," *Sports Illustrated*, April 26, 1965, 81, https://vault.si.com/vault/1965/04/26/a-stink-of-dead-stripers.

21. Robert Boyle, "From a Mountaintop to 1,000 Fathoms Deep," *Sports Illustrated*, August 17, 1964, 76.

22. Robert H. Boyle, testimony, *Hearings Before the Subcommittee on Fisheries and Wildlife Conservation of the Committee on Merchant Marine and Fisheries on Hudson River Spawning Grounds*, May 10–11, 1965 (Washington, DC: US Government Printing Office), 77.

23. Stradling, *The Nature of New York*, 189.

24. Milo Mason and Gaylord Nelson, "Interview: Gaylord Nelson," *Natural Resources & Environment*, Summer 1995, 73–74.

25. Donald Janson, "A National Lakeshore is Proposed for Wisconsin," *New York Times*, September 3, 1967, 83.

26. Michael Brenes, "Here's What Happened to the Last Green New Deal," Politico, December 19, 2018, https://www.politico.com/magazine/story/2018/12/19/green-new-deal-congress-history-mcgovern-223315/.

27. "A.M.A. Gives Plan on Air Pollution," *New York Times*, December 13, 1962, 7.

28. "National Conference on Air Pollution," *Public Health Reports*, vol. 78, no. 5 (May 1963), 424.

29. Luther L. Terry, "Let's Clear the Air," speech delivered at National Conference on Air Pollution, Washington, DC, December 10–12, 1962, 11.

30. Scott Turner to John F. Kennedy, n.d.; Scott Turner to Stewart Udall, December 10, 1962, and undated, Udall Papers, University of Arizona, quoted in Thomas G. Smith, "John Kennedy, Stewart Udall, and New Frontier Conservation," *Pacific Historical Review*, August 1, 1995, 347.

31. Julius Duscha, "Unspoiled Nature Is Udall's Passion," *Washington Post*, December 9, 1962, E3.

32. Gilbert Gude, "Presidents and the Potomac," *White House History*, June 1997, White House Historical Association, https://www.whitehousehistory.org/presidents-and-the-potomac.

33. Robert F. Kennedy, Jr., *American Values: Lessons I Learned from My Family* (New York: HarperCollins, 2018), 106–7.

34. Quoted in Loren Eiseley, "An Appreciation of Robinson Jeffers," *Sierra Club Bulletin* (December 1965), 60.

35. David E. Pesonen, "The Battle of Bodega Bay," *Sierra Club Bulletin*, June 1962, 9.

36. David E. Pesonen, "A Visit to the Atomic Park," the collected *Sebastopol Times* articles from September 27, October 4, 11, and 18, 1962, reprinted from the *Sebastopol Times*, 1962.

37. Philip Flint, "Struggle on the Seacoast," *Sierra Club Bulletin*, April 1961, 9.

38. Don Engdahl, "Earthquake Expert Hits Bodega A-Plant Site," *Santa Rosa Press Democrat*, August 29, 1963, 6.

39. Harrold Gilliam, "Atom Versus Nature at Bodega," *San Francisco Chronicle*, February 11, 1963.

40. Brian Balogh, *Chain Reaction: Expert Debate and Public Participation in American Commercial Nuclear Power, 1945–1975* (Cambridge, UK: Cambridge University Press, 1991), 246; Thomas Raymond Wellock, *Critical Masses: Opposition to Nuclear Power in California, 1958–1978* (Madison: University of Wisconsin Press, 1998), 17–67.

41. James Daly, "Nuclear Fault Line—Bodega Bay," *Sonoma Magazine*, February 3, 2015.

42. Malvina Reynolds, "Little Boxes," Schroder Music Company, 1962.

43. Malvina Reynolds, "Take It Away," Schroder Music Company, 1963.

44. Tom Lehrer, *Too Many Songs by Tom Lehrer with Not Enough Drawings by Ronald Searle* (New York: Pantheon, 1991), 125.

45. Ibid., 113–14.

Chapter 16: The Tag Team of John F. Kennedy, Stewart Udall, and Rachel Carson

1. Richard L. Miller, *Under the Cloud: The Decades of Nuclear Testing* (Woodlands, TX: Two-Sixty Press, 1991), 354.

2. Rachel Carson, speech in acceptance of the Schweitzer Medal, Animal Welfare Institute, January 7, 1963. See also Paul Brooks, *The House of Life: Rachel Carson at Work* (Boston: Houghton Mifflin, 1989), 315–16.

3. Carson, Ibid.

4. Albert Schweitzer, letter to Rachel Carson, Lambaréné, March 16, 1963. The original letter was written in French and is in the Rachel Carson Council archives at the National Conservation Training Center, US Fish & Wildlife Service, Shepherdstown, West Virginia.

5. Ann Cottrell Free, "In Memoriam," *Defenders of Wildlife News Bulletin*, May–July 1964.

6. Rachel Carson, preface to Ruth Harrison, *Animal Machines: The New Factory Farming Industry* (New York: Ballantine, 1964). The preface is reprinted in Rachel Carson, *Lost Woods: The Discovered Writing of Rachel Carson*, ed. Linda Lear (Boston: Beacon Press, 1998), 194–96. See also Zoe Ackerman et al., *Pork and Pollution: An Introduction to Research and Action on Industrial Hog Production*, Rachel Carson Council, 2015, https://rachelcarsoncouncil.org/publications.

7. Carol McKenna, "Ruth Harrison Obituary," *Guardian*, July 5, 2000.

8. Thomas Merton, *The Seven Storey Mountain* (New York: Harcourt, Brace, 1948).

9. Monica Weis, *The Environmental Vision of Thomas Merton* (Lexington: University of Kentucky Press, 2011), 14.

10. Thomas Merton to Rachel Carson, January 12, 1963, Rachel Carson Papers, Yale Collection of American Literature, Beinecke Rare Book and Manuscript Library, Yale University.

11. Author interview with Stewart Udall, June 1, 2008.

12. Niall Ferguson, *The War of the World: Twentieth-Century Conflict and the Descent of the West* (New York: Penguin, 2006), 600–601.

13. John F. Kennedy, "Statement by the President on the Death of Robert Frost," January 29, 1963, The American Presidency Project, https://www.presidency.ucsb.edu/documents/statement-the-president-the-death-robert-frost.

14. Wade Van Dore, "Robert Frost and Wilderness," *Living Wilderness*, Summer 1970, 48.

15. Robert Frost, "In Winter in the Woods Alone," in *In the Clearing* (New York: Holt, Rinehart & Winston, 1962), 101.

16. Doug Scott, *The Enduring Wilderness: Protecting Our Natural Heritage Through the Wilderness Act* (Golden, CO: Fulcrum, 2004), 53.

17. Jack M. Hession, "The Legislative History of the Wilderness Act," master's thesis, San Diego State College, 1967, 207–8.

18. Robert G. Kaufman, *Henry M. Jackson: A Life in Politics* (Seattle: University of Washington Press, 2011), 167–68.

19. David Brower, acknowledgment, in Harvey Manning, *The Wild Cascades, Forgotten Parkland* (San Francisco: Sierra Club, 1965), 901–58.

20. Brigham Daniels, Andrew P. Follett, and Joshua Davis, "The Making of the Clean Air Act," *Hastings Law Journal* 71, no. 4 (2020), 918–19, https://repository.uchastings.edu/hastings_law_journal/vol71/iss4/3.

21. Ibid.

22. Leslie Kemp Poole, *Saving Florida: Women's Fight for the Environment in the Twentieth Century* (Tallahassee: University of Florida Press, 2016), 110.

23. Quoted in Joel K. Goldstein, "Edmund S. Muskie: The Environmental Leader and Champion," *Maine Law Review* 67, no. 2 (2015): 227, https://

digitalcommons.mainelaw.maine.edu/cgi/viewcontent.cgi?article=1027&con text=mlr.

24. Bill McKibben, *Falter: Has the Human Game Begun to Play itself Out?* (New York: Henry Holt, 2019), 4.

25. Memo from Jerome Wiesner to John F. Kennedy, April 3, 1963, John F. Kennedy Presidential Library and Museum.

26. Eric Sevareid and Jay McMullen, "The Silent Spring of Rachel Carson," *CBS Reports*, CBS, April 3, 1963.

27. Eric Sevareid, *Canoeing with the Cree* (New York: Macmillan, 1935).

28. Sevareid and McMullen, "The Silent Spring of Rachel Carson."

29. Ibid.

30. "Robert White-Stevens, Biologist, 66," (White Plains, NY) *Journal News*, September 7, 1978, 17.

31. "Environmentalist Robert White-Stevens Dies," *Central New Jersey Home News*, September 6, 1978, 29.

32. Robert White-Stevens, "Letters: DDT Ban: A Judgment of Emotion and Mystique," *Science* 170, no. 3961 (November 27, 1970): 928.

33. Arlene R. Quaratiello, *Rachel Carson: A Biography* (Westport, CT: Greenwood Press, 2004), 112–13.

34. Sevareid and McMullen, "The Silent Spring of Rachel Carson."

35. Don Kirkley, "Look and Listen," *Baltimore Sun*, April 3, 1963, 12.

36. Raymond Lowery, "Opera Staged Here Tonight; Pesticide Use Protocol by TV," *Raleigh News and Observer*, April 3, 1963, 15.

37. Quoted in "Rachel Carson," *American Experience*, PBS, season 29, episode 3, originally aired January 24, 2017, https://www.pbs.org/wgbh/americanexpe rience/films/rachel-carson/.

38. William M. Blair, "Pesticide Use Will Be Studied in Laboratory," *Arizona Daily Star*, April 26, 1963, 19.

39. "Heavy Price Paid for Pesticide Use, Udall Asserts," *St. Louis Post-Dispatch*, April 25, 1963, 46.

40. Rachel Carson to Stewart Udall, May 3, 1963, Rachel Carson Papers, Bienecke Library, Yale University, New Haven, CT.

41. Rachel Carson, "Remarks" at the Garden Club of America (50th Anniversary) Annual Meeting, Philadelphia, May 6–10, 1963, *GCA Bulletin*, September 1963, 75–76.

42. McKibben, *Falter*, 73.

43. Carson, "Remarks."

44. Naomi Oreskes and Erik M. Conway, *Merchants of Doubt: How a Handful of Scientists Obscured the Truth on Issues from Tobacco Smoke to Climate Change* (New York: Bloomsbury, 2011), 220.

45. *Use of Pesticides: A Report of the President's Science Advisory Committee* (Washington, DC: White House, 1963), 20.

46. President's Science Advisory Committee (PSAC), Pesticides Report, May 15, 1963, Papers of John F. Kennedy, Presidential Papers, President's Office Files, Departments and Agencies.

47. President's Science Advisory Committee, *Use of Pesticides, A Report of the President's Science Advisory Committee* (Washington, DC: US Government Printing Office, 1963), 4.

48. Zuoyue Wang, *In Sputnik's Shadow: The President's Science Advisory Committee and the Cold War* (New Brunswick, NJ: Rutgers University Press, 2008), 200.

49. Rachel Carson, *Silent Spring* (Boston: Houghton Mifflin, 1994), 269–70.

50. Oreskes and Conway, *Merchants of Doubt*, 225–30.

51. Author interview with Ethel Kennedy, March 15, 2018.

52. Robert Kennedy, Jr., *American Values: Lessons I Learned from My Family* (New York: HarperCollins, 2018), 104.

53. Ibid., 105.

54. Ibid.

55. Scott Raymond Einberger, *With Distance in His Eyes: The Environmental Life and Legacy of Stewart Udall* (Reno: University of Nevada Press, 2018), 92–95.

56. Barry Mackintosh, *Assateague Island National Seashore: An Administrative History* (Washington, DC: National Park Service, 1982).

57. Stephen E. Nordlinger, "Goldstein Backs Udall on Assateague Issue," *Baltimore Sun*, June 25, 1963, 24.

58. Richard Homan, "Udall, Landowners Argue over Assateague's Future," *Washington Post*, June 24, 1963.

59. "Assateague," (Salisbury, MD) *Daily Times*, June 25, 1963, 8.

60. US Department of the Interior, *The Department of the Interior During the Administration of President Lyndon B. Johnson, LBJL*, November 1963–January 1969, 76–78.

61. Byron Porterfield, "Udall Aide Asserts Fire Island Road Will Bar Park Plan," *New York Times*, October 23, 1962, 38.

62. Stewart L. Udall Papers, University of Arizona, Tucson, AZ. See also Mackintosh, *Assateague Island National Seashore*, 5–20.

63. Aldo Starker Leopold et al., *Leopold Report: Wildlife Management in the National Parks* (Washington, DC: US Department of the Interior, National Park Service, 1963), 4.

64. Dan Flores, *Coyote America: A Natural and Supernatural History* (New York: Basic Books, 2016), 160.

65. Ibid., 159–63.

66. John F. Kennedy, "Remarks on Signing Outdoor Recreation Bill," May 28, 1963, John F. Kennedy Presidential Library and Museum, https://www.jfkli brary.org/asset-viewer/archives/JFKPOF/044/JFKPOF-044-030.

67. Roger L. Moore and B. L. Driver, *Introduction to Outdoor Recreation: Providing and Managing Natural Resource Based Opportunities* (State College, PA: Venture, 2005), 16.

68. Tom Turner, *David Brower: The Making of the Environmental Movement* (Berkeley: University of California Press, 2015), 118.

Chapter 17: The Limited Nuclear Test Ban Treaty

1. J. F. Rothermel, "Rachel Carson's *Silent Spring* Is Too Emotional," *Birmingham News*, October 21, 1962.

2. Richard G. Hewlett and Jack M. Hall, *Atoms for Peace and War, 1953–1961: Eisenhower and the Atomic Energy Commission* (Berkeley: University of California Press, 1989), 529.

3. A. Constandina Titus, *Bombs in the Backyard: Atomic Testing and American Politics* (Reno: University of Nevada Press, 1986), 65.

4. Quoted in Norman Solomon, "50 Years Later, the Tragedy of Nuclear Tests in Nevada," Fairness & Accuracy in Reporting, January 4, 2001, https://fair .org/media-beat-column/50-years-later-the-tragedy-of-nuclear-tests-in -nevada/.

5. Nikita Khrushchev to SANE, February 25, 1961, Swarthmore College Peace Collection (SCPC), Swarthmore College, Philadelphia, SANE records, Correspondence and Related Papers, 1961, 1962.

6. "Thousands Rally, Parade in New York in Support of 'Sane' Nuclear Policy," Great Falls (Montana) *Tribune*, May 20, 1960, 2.

7. Norman Cousins, *The Improbable Triumvirate: John F. Kennedy, Pope John, Nikita Khrushchev* (New York: W. W. Norton, 1972), 24–25.

8. Author interview with Dr. Candis Cousins Kerns, April 4, 2021.

9. Allen Pietrobon, "The Role of Norman Cousins and Track II Diplomacy in the Breakthrough to the 1963 Limited Test Ban Treaty," *Journal of Cold War Studies* 18, no. 1 (Winter 2016): 60–70.

10. Ibid.

11. Arthur M. Schlesinger, Jr., *A Thousand Days: John F. Kennedy in the White House* (New York: Houghton Mifflin, 1965).

12. *Congressional Record*, 88th Congress, 1st session, 1963, 9415.

13. John F. Kennedy, "Commencement Address at American University, Washington, D.C.," June 10, 1963, John F. Kennedy Presidential Library and Museum, https://www.jfklibrary.org/archives/other-resources/john-f-kennedy -speeches/american-university-19630610.

14. John F. Kennedy, "Televised Address on Nuclear Test Ban Treaty," July 26, 1963, John F. Kennedy Presidential Library and Museum, https://www.jfkli

brary.org/learn/about-jfk/historic-speeches/televised-address-on-nuclear
-test-ban-treaty.

15. Quoted in Vincent J. Intondi, *African Americans Against the Bomb: Nuclear Weapons, Colonialism, and the Black Freedom Movement* (Stanford, CA: Stanford University Press, 2015), 72.

16. Andreas Wagner and Marcel Gerber, "John F. Kennedy and the Limited Test Ban Treaty: A Case Study of Presidential Leadership," *Presidential Studies Quarterly* 29, no. 2 (January 1999), 481.

17. *Woman's Day*, November 1963, 37–39, 141–42.

18. Author interview with Ted Sorensen, July 14, 2008.

19. Theodore Sorensen, *Kennedy* (New York: Harper and Row, 1965), 817.

20. Barry Commoner, *The Closing Circle: Nature, Man and Technology* (New York: Alfred A. Knopf, 1971), 52.

21. Terrence R. Fehner and F. G. Gosling, *Atmospheric Nuclear Weapons Testing, 1951–1963* (Washington, DC: US Department of Energy, 2006), 199.

22. US Atomic Energy Commission, *Annual Report to Congress* (Washington, DC: N.p., 1963), 159.

23. Ethel Taylor, *We Make a Difference: My Personal Journey with Women Strike for Peace* (Philadelphia: Camino Books, 1998).

24. Dennis Hevesi, "Dagmar Wilson, Anti-Nuclear Leader, Dies at 94," *New York Times*, January 23, 2011.

Chapter 18: JFK's Last Conservation Journey

1. E. W. Kenworthy, "Kennedy to Make a 10-State Tour of U.S. Projects," *New York Times*, September 1, 1963, 1, https://www.nytimes.com/1963/09/01/archives/kennedy-to-make-a-10state-tour-of-us-projects-fiveday-inspection.html.

2. Char Miller, *Gifford Pinchot and the Making of Modern Environmentalism* (Washington, DC: Island Press, 2001), 36.

3. William Souder, *On a Farther Shore: The Life and Legacy of Rachel Carson* (New York: Crown, 2012), 376–77; Alvin Shuster, "Close Up of a 'Peace Striker': She Is Dagmar Wilson, a Washington Housewife and Political Neophyte," *New York Times Magazine*, May 6, 1962, 66.

4. Julius Duscha, "JFK Zooms In on 'Quiet Crisis,'" *Washington Post*, September 22, 1963, E3.

5. Michael R. Beschloss, *The Crisis Years: Kennedy and Khrushchev, 1960–1963* (New York: HarperCollins, 2008), 636.

6. "Address at Pinchot Institute for Conservation Studies," Milford, PA, September 24, 1963, Papers of John F. Kennedy, President's Office Files, Speech Files.

7. Tom Wicker, "President Tours 3 States in West," *New York Times*, September 26, 1963, 1, 27.

8. "President Gets Good Look at Islands' Bald Eagles," *Milwaukee Sentinel*, September 25, 1963; Bill Christofferson, *The Man from Clear Lake: Earth Day Founder Senator Gaylord Nelson* (Madison: University of Wisconsin Press, 2004), 306.

9. Christofferson, *The Man from Clear Lake*, 214.

10. Gaylord Nelson to John F. Kennedy, May 16, 1963, Gaylord Nelson Papers, Clear Lake Area Historical Museum, Clear Lake, WI.

11. Gaylord Nelson, recorded interview by Edwin R. Bayley, July 1, 1964, 8–10, John F. Kennedy Library Oral History Program.

12. Gaylord Nelson to John F. Kennedy, May 16, 1963.

13. Christofferson, *The Man from Clear Lake*, 301.

14. John F. Kennedy, "Remarks," Ashland, WI, September 24, 1963, appendix B, in North Central Field Committee, *Proposed Apostle Islands National Lakeshore, Bayfield and Ashland Counties, Wisconsin* (Washington, DC: United States Department of the Interior, 1965).

15. James W. Feldman, *A Storied Wilderness: Rewilding the Apostle Islands* (Seattle: University of Washington Press, 2011), 176.

16. "Kennedy Vows Support for L. Superior Region," *Green Bay Press-Gazette*, September 25, 1963, 2. See also John F. Kennedy, *Public Papers of the Presidents of the United States: John F. Kennedy* (Washington, DC: US Government Printing Office, 1964), 707–09.

17. Quoted in Christofferson, *The Man from Clear Lake*, 307.

18. John F. Kennedy, "Address to the Delegates to the Northern Great Lakes Region Land and People Conference, Duluth, Minnesota, 24 September 1963," John F. Kennedy Presidential Library and Museum, https://www.jfklibrary.org/asset-viewer/archives/JFKWHA/1963/JFKWHA-221-001/JFKWHA-221-001.

19. "Upstate Area Deserves Park Site, Says Kennedy," *Oshkosh Daily Northwestern*, September 25, 1963, 1.

20. "Apostle Islands Are Favored by President," Associated Press, September 25, 1963.

21. Arthur Edson, "JFK Happy with Red Split, Lauds Atomic Test Ban Pact," *Casper Morning Star*, September 26, 1963, 1.

22. Author interview with George McGovern, December 23, 2007.

23. John F. Kennedy, "Remarks at the Hanford, Washington Electric Generating Plant, 26 September 1963," John F. Kennedy Presidential Library and Museum, https://www.jfklibrary.org/asset-viewer/archives/JFKPOF/047/JFKPOF-047-002.

24. Ibid.

25. Cassandra Tate, "President Kennedy Participates in Groundbreaking Ceremonies for Construction of N Reactor at Hanford on September 26, 1963," HistoryLink.org, https://www.historylink.org/file/10640.

26. Author interview with Stewart Udall, September 6, 2009.

27. Roxanne Dunbar-Ortiz, *An Indigenous People's History of the United States* (Boston: Beacon Press, 2014), 181.

28. Tom Wicker, "President Is Hailed by Crowds on Coast," *New York Times*, September 28, 1963, 1, https://www.nytimes.com/1963/09/28/archives/president-is-hailed-by-crowds-on-coast-crowds-on-coast-cheer.html.

29. John F. Kennedy to Mr. and Mrs. Donau, September 28, 1963, John F. Kennedy Presidential Library and Museum.

30. "Kennedy's Visit to Lassen," National Park Service, https://www.nps.gov/lavo/learn/historyculture/jfk.htm.

31. John F. Kennedy, "Remarks at the Convention Center in Las Vegas, Nevada," September 28, 1963, The American Presidency Project, https://www.presidency.ucsb.edu/documents/remarks-the-convention-center-las-vegas-nevada.

32. Stephen Fox, *The American Conservation Movement: John Muir and His Legacy* (Madison: University of Wisconsin Press, 1981), 317.

33. Kennedy, "Remarks at the Convention Center in Las Vegas, Nevada."

34. Ibid.

35. A. Constandina Titus, *Bombs in the Backyard: Atomic Testing and American Politics* (Reno: University of Nevada Press, 1986), 99–100.

36. Ibid., 99.

37. "The 'Conservation' Trip," *New York Times*, September 24, 1963, 38, https://www.nytimes.com/1963/09/24/archives/the-conservation-trip.html.

38. Quoted in John Hart, *Muir Woods National Monument* (San Francisco: Golden Gate National Park Conservancy, 2011), 138.

39. David Brower with Steve Chapple, *Let the Mountains Talk, Let the Rivers Run* (New York: HarperCollins West, 1995), 144.

40. Rachel Carson to Dorothy Freeman, October 26, 1963, in Rachel Carson and Dorothy Freeman, *Always, Rachel: The Letters of Rachel Carson and Dorothy Freeman, 1952–1964—The Story of a Remarkable Friendship*, ed. Martha Freeman (Boston: Beacon Press, 1994), 484.

41. Ibid.

42. Arlene R. Quaratiello, *Rachel Carson: A Biography* (Westport, CT: Greenwood Press, 2004), 113.

43. Rachel Carson, *Lost Woods: The Discovered Writing of Rachel Carson*, ed. Linda Lear (Boston: Beacon Press, 1998), 231–32.

44. Rachel Carson to Dorothy Freeman, October 26, 1963, in ibid., 484–85. Also, author interview with Stewart Udall, September 6, 2009.

45. Quoted in Fox, *The American Conservation Movement*, 298.

46. James T. Yenckel, "Wild, Scenic and Protected," *Washington Post*, July 26, 1962, E1, E9.

47. Doug Scott, *The Enduring Wilderness: Protecting Our Natural Heritage Through the Wilderness Act* (Golden, CO: Fulcrum, 2004), 53–54.

48. UPI wire, November 22, 1963, https://www.tsl.texas.gov/sites/default/files/public/tslac/landing/documents/jfk-upi_feed-1.pdf.

49. Stewart Udall, handwritten letter following news of Kennedy assassination, A2 372, Box 109, Folder 3, Stewart L. Udall Papers, University of Arizona, Tucson, AZ.

50. John Oakes, "John Fitzgerald Kennedy," *New York Times*, November 23, 1963, https://www.nytimes.com/1963/11/23/john-fitzgerald-kennedy.html.

Chapter 19: The Mississippi Fish Kill, the Clean Air Act, and American Beautification

1. Rachel Carson to Dorothy Freeman, November 3, 1963, in Carson and Freeman, *Always, Rachel*, 488–89.

2. Rachel Carson to Dorothy Freeman, November 27, 1963, in ibid., 487–489.

3. Michael J. Hogan, *The Afterlife of John Fitzgerald Kennedy: A Biography* (Cambridge, UK: Cambridge University Press, 2017), 65.

4. Penelope Green, "Carol Johnson, 91, a Trailblazing Landscape Architect," *New York Times*, January 9, 2021, A19.

5. "Pesticide Fatal to Fish Is Traced," *New York Times*, April 5, 1964, 70, https://www.nytimes.com/1964/04/05/archives/pesticide-fatal-to-fish-is-traced-endrin-is-identified-in-tests-at.html.

6. "Pesticides Fatal to Gulf Shrimp," *New York Times*, March 26, 1964, 26.

7. David Zwick and Marcy Benstock, *Water Wasteland* (Chicago: Grossman Publishers, 1971), 48.

8. Linda Lear, *Rachel Carson: Witness for Nature* (New York: Houghton Mifflin Harcourt, 2009), 470.

9. David Zwick and Marcy Benstock, *Water Wasteland: Ralph Nader's Study Group Report on Water Pollution* (New York: Grossman, 1971), 46–47.

10. "Rachel Carson Gets '63 Audubon Medal for 'Silent Spring,'" *New York Times*, December 4, 1963, 49.

11. Rachel Carson to Lois Crisler, March 19, 1963, Rachel Carson Papers, Beinecke Library, Yale University, New Haven, CT.

12. Election of Rachel Carson to the American Academy of Arts and Letters, December 6, 1963, RCP/BLYU.

13. Quoted in Hal K. Rothman, *LBJ's Texas White House: "Our Heart's Home"* (College Station: Texas A&M Press, 2001), 9.

14. Ansel Adams to Lyndon B. Johnson, September 4, 1964, Box 190, Folder 1, Udall Papers, Tucson, AZ.

15. Author interview with Stewart Udall, September 6, 2009.

16. Stewart L. Udall, *The Quiet Crisis and the Next Generation* (Layton, UT: Gibbs Smith, 1991), 182.

17. Charles Stoddard to Stewart Udall, November 12, 1963, Stewart L. Udall Papers, University of Arizona, Tucson, AZ.

18. Alfred Edwards to Stewart Udall, October 25, 1963, Folder 1, Box 213, Stewart L. Udall Papers, University of Arizona, Tucson, AZ.

19. Rachel Carson to Stewart Udall, November 12, 1963, University of Arizona Special Collections Department, Special Exhibition, September 2014.

20. Udall, *The Quiet Crisis*, 174, 179.

21. L. Boyd Finch, *Legacies of Camelot* (Norman: University of Oklahoma Press, 2008), 135; Kathleen Shull, "Stewart Udall Reflects on the Mistakes of This Century," Wildcat Online News, November 15, 1999, https://wc.arizona.edu /papers/93/59/08_1_m.html.

22. Edythe Scott Bagley, *Desert Rose: The Life and Legacy of Coretta Scott King* (Tuscaloosa: University of Alabama Press, 2012), 158–59.

23. Vincent Intondi, "W.E.B. Du Bois to Coretta Scott King: The Untold History of the Movement to Ban the Bomb," Zinn Education Project, July 30, 2015, https://www.zinnedproject.org/if-we-knew-our-history/web-dubois-coretta -scott-king-ban-the-bomb/.

24. Souder, *On a Farther Shore*, 344.

25. Quoted in Ben Bradlee, *A Good Life: Newspapering and Other Adventures* (New York: Simon & Schuster, 1995), 273–74.

26. Quoted in Harold Holzer, *The Presidents vs. the Press: The Endless Battle Between the White House and the Media—from the Founding Fathers to Fake News* (New York: Penguin Random House, 2020).

27. Michael L. Gillette, interview with George Reedy, December 20, 1983, Lyndon Baines Johnson Library Oral History Collection, 11, 49–50.

28. Stewart Udall, Memorandum to the President, "The Administration and Conservation—a Look at Programs and Priorities," December 9, 1963, Folder 8, Box 115, Stewart L. Udall Papers, University of Arizona, Tucson, AZ.

29. Mark Z. Jacobson, *Air Pollution and Global Warming: History, Science, and Solutions* (New York: Cambridge University Press, 2012), 176.

30. Charles Komanoff, "IN MEMORIAM: 'Do All the Good You Can'—The Life of Urban Ecology Pioneer Carolyn Konheim," Streetsblog NYC, December 2, 2019, https://nyc.streetsblog.org/2019/12/02/in-memoriam-do-all -the-good-you-can-the-life-of-urban-ecology-pioneer-carolyn-konheim/. In the late 1960s, Mr. Komanoff was a junior economist working with Konheim in the Lindsay administration environmental offices.

31. Ibid.

32. Arthur C. Stern, "History of Air Pollution Legislation in the United States," *Journal of the Air Pollution Control Association* 32, no. 1 (1982): 44–61, https://www.tandfonline.com/doi/abs/10.1080/00022470.1982.10465369.

33. William L. Rathje, "Rubbish," *Atlantic Monthly*, December 1989, 99.

34. Martin V. Melosi, *Effluent America: Cities, Industry, Energy, and the Environment* (Pittsburgh: University of Pittsburgh Press, 2001), 78–81.

35. Stewart Udall, journal, December 23, 1963, Folder 1, Box 117, Stewart L. Udall Papers, University of Arizona, Tucson, AZ.

36. Udall, quoted in Thomas G. Smith, *Stewart L. Udall: Steward of the Land* (Albuquerque: University of New Mexico Press, 2017), 200.

37. Lyndon B. Johnson, comments at cabinet meeting, February 25, 1965, Special Files, Cabinet Papers, Box 2, LBJ Presidential Library.

38. Lyndon B. Johnson to Stewart L. Udall, October 17, 1965, Stewart Udall file 1/1/1965 to 12/31/1965, LBJL.

39. Wallace Stegner, dust jacket quote for George B. Hartzog, Jr., *Battling for the National Parks* (Mount Kisco, NY: Mayer Bell, 1988).

40. Stewart L. Udall, oral history interview, September 19, 2005.

41. Lady Bird Johnson and Carlton B. Lees, *Wildflowers Across America* (New York: National Wildflower Research Center/Abbeville Press, 1988), 8.

42. Quoted in Rose Houk, *A Biography of Lady Bird Johnson: Legacy of Beauty* (Tucson, AZ: Western National Parks Association, 2006), 31.

43. Lewis L. Gould, "Lady Bird Johnson and Beautification," *The Johnson Years*, vol. 2, *Vietnam, the Environment and Science*, ed. Robert A. Divine (Lawrence: University Press of Kansas, 1987), 152–53.

44. Robert Caro, *The Power Broker: Robert Moses and the Fall of New York* (New York: Alfred A. Knopf, 1974), 870.

45. Ibid., 850.

46. Kenneth T. Jackson, "Robert Moses and the Rise of New York: The Power Broker Perspective," in *Robert Moses and the Modern City: The Transformation of New York*, ed. Hilary Ballon and Kenneth T. Jackson (New York: W. W. Norton, 2008), 69.

47. Kerri Arsenault, *Mill Town: Reckoning with What Remains* (New York: St. Martin's Press, 2020), 282.

48. Caro, *The Power Broker*, 860.

49. "Diminishing Park Areas Opposed," *New York Times*, April 18, 1961, 36, https://www.nytimes.com/1961/04/18/archives/diminishing-park-areas-opposed.html.

50. Thomas J. Campanella, *Brooklyn: The Once and Future City* (Princeton, NJ: Princeton University Press, 2019), 403.

51. Lewis Mumford, *My Works and Days: A Personal Chronicle* (New York: Harcourt Brace Jovanovich, 1979); Lewis Mumford, *The Highway and the City* (New York: Harcourt, Brace & World, 1963), 235.

52. Johnson and Lees, *Wildflowers Across America*, 12.

Chapter 20: The Great Society

1. Jonathan Norton Leonard, "Rachel Carson Dies of Cancer: 'Silent Spring' Author was 56," *New York Times*, April 15, 1964, https://archive.nytimes.com/www.nytimes.com/books/97/10/05/reviews/carson-obit.html?_r=1&scp=1&sq=rachel%2520carson&st=cse.

2. William Souder, *On a Farther Shore: The Life and Legacy of Rachel Carson* (New York: Crown, 2012), 388–89.

3. Arlene R. Quaratiello, *Rachel Carson: A Biography* (Amherst, NY: Prometheus, 2010), 121.

4. Rachel Carson, *The Sense of Wonder* (New York: HarperCollins, 2017), 95.

5. Carol B. Gartner, *Rachel Carson (Literature & Life)* (New York: Frederick Ungar, 1983), 135–36.

6. Norman Boucher, "The Legacy of Silent Spring," *Boston Globe Magazine*, March 15, 1987, 17, 37–47.

7. Rachel Carson, *Silent Spring* (Boston: Houghton Mifflin, 1994), 277.

8. Priscilla Coit Murphy, *What a Book Can Do: The Publication and Reception of "Silent Spring"* (Amherst: University of Massachusetts Press, 2005), 8.

9. "Walking in Rachel Carson's Footsteps: the Icon Who Launched the Contemporary Environmental Movement Also Helped Create the Nature Conservancy in Maine," *Nature Conservancy*, March 24, 2021, accessed May 23, 2022, https://www.nature.org/en-us/about-us/where-we-work/united-states/maine/stories-in-maine/the-nature-conservancy-in-maine-rachel-carson-walking-in-her-footsteps/.

10. Author interview with Allison Rockefeller, May 6, 2021. The Rachel Carson Award is presented annually by the National Audubon Society's Women in Conservation. The Rachel Carson Awards Council was founded in 2004 by Allison Whipple Rockefeller.

11. Margaret Foster, "Carrying on Rachel Carson's Work," *Washington Beacon* 34, no. 3 (March 2022), 16.

12. Quoted in Paul Brooks, *House of Life: Rachel Carson at Work* (Boston: Houghton Mifflin, 1989), 317.

13. Jamie Rappaport Clark to author, June 5, 2001, Defenders of Wildlife, Washington, DC.

14. Gaylord Nelson, "The Effects of Pesticides on Sports and Commercial Fisheries," statement before the Senate Subcommittee on Energy, Natural Resources, and the Environment of the Committee on Commerce, May 19, 1969 (Washington, DC: US Government Printing Office, 1969), 36.

15. Frank Graham, Jr., *Since Silent Spring* (Boston: Houghton Mifflin, 1970), 107.

16. Souder, *On a Farther Shore*, 348.

17. Quoted in Frank Church, "The New Conservation," speech delivered at Idaho Parks and Recreation Convention, Coeur d'Alene, Idaho, April 22, 1969.

18. Charles F. Wurster, *DDT Wars: Rescuing our National Bird, Preventing Cancer, and Creating the Environmental Defense Fund* (New York: Oxford University Press, 2015), 8.

19. "DDT Wars and the Birth of EDF," Special Report, Environmental Defense Fund, Summer 2015, https://www.edf.org/sites/default/files/specialreport _summer2015.pdf, 2.

20. Charles F. Wurster, "Letter on DDT," *Long Island Press*, May 6, 1966.

21. "Conservation: A New Day in Court," *Time*, October 24, 1969.

22. "DDT Wars and the Birth of EDF," 3.

23. Author interview with Victor Yannacone, March 11, 2020.

24. Stewart Udall, "The Legacy of Rachel Carson," *Saturday Review*, May 16, 1964, 23.

25. MarineBio, "Interview with Dr. Sylvia Earle," Marine Bio Conservation Society, April 29, 2010, https://marinebio.org/interview-with-dr-sylvia-earle/.

26. Beth Baker, *Sylvia Earle: Guardian of the Sea* (Minneapolis: Lerner Pub Group, 2001).

27. World Science Festival, "Sylvia Earle: Oceanographer, Explorer, Pioneer, Remarkable Woman," YouTube, October 9, 2013, https://www.youtube.com /watch?v=d06i4isc9fw.

28. Ibid.

29. Sandra Steingraber, ed., *Rachel Carson: The Sea Trilogy* (New York: Library of America, 2021), 1.

30. Rachel Carson to Dorothy Freeman, in Rachel Carson and Dorothy Freeman, *Always, Rachel: The Letters of Rachel Carson and Dorothy Freeman, 1952–1964—The Story of a Remarkable Friendship*, ed. Martha Freeman (Boston: Beacon Press, 1994), 231.

31. Quoted in Eliza Griswold, "How 'Silent Spring' Ignited the Environmental Movement," *New York Times Magazine*, September 21, 2012, 36, https:// www.nytimes.com/2012/09/23/magazine/how-silent-spring-ignited-the-en vironmental-movement.html.

32. Al Gore, introduction to Carson, *Silent Spring*.

33. Zuoyue Wang, "Responding to Silent Spring," *Science Communication* 19, no. 2 (December 1997), 142.

34. Andrew Kenny, "The Green Terror," quoted in Naomi Oreskes and Erik M. Conway, *Merchants of Doubt: How a Handful of Scientists Obscured the Truth on Issues from Tobacco Smoke to Global Warming* (New York: Bloomsbury, 2010), 222.

35. John Tierney, "Fateful Voice of a Generation Still Drowns Out Real Science," *New York Times*, June 5, 2007, https://www.nytimes.com/2007/06/05/sci ence/earth/05tier.html.

36. Tomas Sowell, *Controversial Essays* (Standford: Hoover Institution Press Publication, 2002).

37. Ibid., 227–28.

38. Lyndon B. Johnson, "Remarks at the University of Michigan," May 22, 1964, The American Presidency Project, accessed May 23, 2022, https://www.pres idency.ucsb.edu/documents/remarks-the-university-michigan.

39. Lyndon B. Johnson, "Remarks at the University of Michigan," May 22, 1964, The American Presidency Project, https://www.presidency.ucsb.edu/docu ments/remarks-the-university-michigan.

40. Ibid. See also Randall B. Woods, *Prisoners of Hope: Lyndon B. Johnson, the Great Society, and the Limits of Liberalism* (New York: Basic Books, 2016), 1–4.

41. Author interview with Richard Goodwin, April 9, 2008.

42. Johnson, "Commencement Address at the University of Michigan."

43. Callum Beals, "4 Surprisingly Green Presidents," *Sierra*, February 2014, https://www.sierraclub.org/sierra/green-life/2014/02/4-surprisingly-green -presidents.

44. Tashima Goodwin, judicial opinion, *Wetlands Water Dist. v. U.S. Dep't of Interior*, United States Court of Appeals, Ninth Circuit, 376.F.3rd 853 (9th Cir. 2004).

45. Lyndon B. Johnson, "Remarks on the Transfer to New Jersey of Lands for the Sandy Hook State Park," June 23, 1964, The American Presidency Project, https://www.presidency.ucsb.edu/documents/remarks-the-transfer-new-jer sey-lands-for-the-sandy-hook-state-park.

46. Author interview with Richard Goodwin, April 9, 2008.

47. "Life and Death of a Primeval Empire," *American Heritage*, February 1967, 18–22.

48. Stewart Udall, memorandum to President Lyndon Johnson, May 27, 1964, Folder 8, Box 115, Stewart L. Udall Papers, University of Arizona, Tucson, AZ. See also Thomas G. Smith, *Stewart L. Udall: Steward of the Land* (Albu- querque: University of New Mexico Press, 2017), 195.

49. Stewart L. Udall, foreword, in Philip Hyde and François Leydet, *The Last Redwoods: Photographs and Story of a Vanishing Scenic Resource* (San Fran- cisco: Sierra Club Books, 1963), 11.

50. *The Redwoods: A National Opportunity for Conservation and Alternatives for Action* (Washington, DC: n.p., 1964).

51. Author interview with Stewart Udall, September 6, 2009.

52. "Remarks by Mrs. Lyndon B. Johnson, Park City, Utah, August 15, 1964," Mrs. Johnson—Speeches, Reference File, LBJ Presidential Library.

53. Draft Flaming Gorge speech, White House Social Files, Liz Carpenter, Subject File, Western Trip, Box 9, August 14–17, 1964, LBJ Presidential Library.

54. Lewis L. Gould, *Lady Bird Johnson and the Environment* (Lawrence: University Press of Kansas, 1988), 39. Also "Land and People Tour of

Mrs. Lyndon B. Johnson to Montana, Wyoming and Utah," August 14–17, 1964, Box 6, Stewart Udall Papers, University of Arizona, Tucson, AZ.

55. Stewart Udall to Lyndon Johnson, August 19, 1964; LBJ to Udall, August 24, 1964, WHCF, EX/PP5/LBJ, July 15, 1964–October 1964, LBJ Presidential Library.

56. Paul S. Sutter, *Driven Wild: How the Fight Against Automobiles Launched the Modern Wilderness Movement* (Seattle: University of Washington Press, 2002), 256.

57. Paul Vitello, "Maurice Barbash, Who Saved Fire Island's Terrain, Dies at 88," *New York Times*, March 21, 2013, https://www.nytimes.com/2013/03/22/ny region/maurice-barbash-a-builder-who-fought-for-fire-island-dies-at-88.html.

58. Alan Eysen, "Optimistic Senators Tour Fire Island," *Newsday*, June 29, 1964, 4.

59. LBJ signed Public Law 88-587 to establish Fire Island National Seashore on September 11, 1964.

60. Hal K. Rothman, *LBJ's Texas White House: "Our Heart's Home"* (College Station: Texas A&M Press, 2001), 4.

61. David A. Smith, *Cowboy Presidents: The Frontier Myth and U.S. Politics Since 1900* (Norman: University of Oklahoma Press, 2021), 73.

62. Harold Holzer, *The Presidents vs. the Press: The Endless Battle Between the White House and the Media—from the Founding Fathers to Fake News* (New York: Penguin Random House, 2020).

63. Bob Brister, "LBJ: Outdoor Sportsman," *Argosy*, October 1964, 39–40, 80–88.

64. George Reedy, *Lyndon B. Johnson: A Memoir* (Ann Arbor: University of Michigan Press, 1982), 154.

65. David A. Smith, *Cowboy Presidents: The Frontier Myth and U.S. Politics Since 1900* (Norman: University of Oklahoma Press, 2021), 63.

66. James Morton Turner, *The Promise of Wilderness: American Environmental Politics Since 1964* (Seattle: University of Washington Press, 2012), 2.

67. Lyndon B. Johnson, "Remarks at the Signing of the Highway Beautification Act of 1965," October 22, 1965, https://www.presidency.ucsb.edu/docu ments/remarks-the-signing-the-highway-beautification-act-1965.

68. Irving Bernstein, *Guns or Butter: The Presidency of Lyndon Johnson* (New York: Oxford University Press, 1996), 277.

69. Claude J. Desautels, interview by William Hartigan, February 16, 1977, John F. Kennedy Library Oral History Program, 20.

70. United States Congress, House Committee on Interior and Insular Affairs, *Bills to Establish a National Wilderness Preservation System for the Permanent Good of the Whole People, and for other Purposes* (Washington, DC: US Government Printing Office, 1964), 1205.

71. Kevin R. Marsh, *Drawing Lines in the Forest: Creating Wilderness Areas in the Pacific Northwest* (Seattle: University of Washington Press, 2007), 87.

72. Howard Zahniser, *The Wilderness Writings of Howard Zahniser*, ed. Mark Harvey (Seattle: University of Washington Press, 2014), 198.

73. Steven C. Schulte, *Wayne Aspinall and the Shaping of the American West* (Boulder: University Press of Colorado, 2002), 76.

74. Quoted in Bernstein, *Guns or Butter*, 270–77.

75. Quoted in Zahniser, *The Wilderness Writings of Howard Zahniser*, 198.

76. Howard Zahniser, "Wildlands, a Part of Man's Environment," quoted in Mark W. T. Harvey, *Wilderness Forever: Howard Zahniser and the Path to the Wilderness Act* (Seattle: University of Washington Press, 2007), 253.

Chapter 21: The Wilderness Act of 1964

1. Aldo Leopold, "Wilderness as a Form of Land Use," *Journal of Land and Public Utility Economics* 1, no. 4 (1925): 400.

2. Doug Scott, *The Enduring Wilderness: Protecting Our Natural Heritage Through the Wilderness Act* (Golden, CO: Fulcrum, 2004), 54.

3. United States Congress, Public Law 88-577, An Act to Establish a National Wilderness Preservation System for the Permanent Good of the Whole People, and for Other Purposes, September 3, 1964, quoted in ibid., 127.

4. Hubert Humphrey to Lawrence O'Brien, August 4, 1964, Legislative Background, Wilderness Act File, LBJ Presidential Library.

5. Kevin R. Marsh, *Drawing Lines in the Forest* (Seattle: University of Washington Press, 2007), 5.

6. Scott, *The Enduring Wilderness*, 55.

7. Lyndon B. Johnson, "Remarks upon Signing the Wilderness Bill and the Land and Water Conservation Fund Bill," September 3, 1964, American Presidency Project, https://www.presidency.ucsb.edu/documents/remarks-upon -signing-the-wilderness-bill-and-the-land-and-water-conservation-fund-bill.

8. Johnson, "Remarks upon Signing the Wilderness Bill and the Land and Water Conservation Fund Bill."

9. Gary Snyder, *The Gary Snyder Reader: Prose, Poetry, and Translations* (Berkeley: Counterpoint, 1999), 175.

10. Dylan Zaslowsky and T. H. Watkins, *These American Lands: Parks, Wilderness, and the Public Lands* (Washington, DC: Island Press, 1994), 214.

11. Scott, *The Enduring Wilderness*, 63.

12. Clinton Anderson, "Protection of the Wilderness," *Living Wilderness* 27, no. 78 (Autumn–Winter 1962): 14.

13. Clinton P. Anderson, *Wilderness in a Changing World* (San Francisco: Sierra Club, 1966).

14. Orville Freeman to Bill Worf (cofounder of Wilderness Watch), May 16, 1995.

15. Roderick Nash, *Wilderness and the American Mind* (New Haven, CT: Yale University Press, 1982), xii.

16. Hubert H. Humphrey, "Remarks of Vice President Hubert H. Humphrey," University of Chicago, January 14, 1966, Hubert H. Humphrey Collection, Minnesota Historical Society, http://www2.mnhs.org/library/findaids /00442/pdfa/00442-01779.pdf, 6.

17. Brian Allen Frake, "The Skeptical Environmentalist: Senator Barry Goldwater and the Environmental Management State," *Environmental History* 15, no.4 (October 2010): 594.

18. Tom Miller (photographs) and Harvey Manning (text), *The North Cascades* (Seattle: Mountaineers, 1964).

19. Paul Brooks, "Wilderness in Western Culture," in *Voices for the Wilderness*, ed. William Schwartz (New York: Ballantine, 1969), 44.

20. Lyndon B. Johnson, "Remarks at the Signing of the Highway Beautification Act of 1965," October 22, 1965, The American Presidency Project, https:// www.presidency.ucsb.edu/documents/remarks-the-signing-the-highway -beautification-act-1965.

21. Schwartz, *Voices for the Wilderness*, 26. See also "From the Sierra Club Wilderness Conferences."

22. Alan E. Watson et al., "Wilderness Managers, Wilderness Scientists and Universities: A Partnership to Protect Wilderness Experiences in the Boundary Waters Canoe Area Wilderness," *International Journal of Wilderness* 19, no. 1 (April 2013): 41–42.

23. Sigurd F. Olson, *The Singing Wilderness* (New York: Alfred A. Knopf, 1956), 6.

24. Paul W. Weiblen, "It's Written in the Rocks: The BWCA History," The Conservation Volunteer, January–February 1971, 21–29, https://conservancy.umn .edu/bitstream/handle/11299/93796/bwca.pdf?sequence=1&isAllowed=y.

25. Sigurd Olson, *Reflections from the North Country* (New York: Alfred A. Knopf, 1976).

26. Quoted in David Backes, *A Wilderness Within: The Life of Sigurd F. Olson* (Minneapolis: University of Minnesota Press, 1997), 312–13.

27. Brochure, "Protect Minnesota's Boundary Waters Wilderness" (Washington, DC: Earthworks, 2016).

28. James M. Glover, *A Wilderness Original: The Life of Bob Marshall* (Seattle: Mountaineers, 1986), 224.

29. Marcus B. Simpson, Jr., *Bird Life of North Carolina's Shining Rock Wilderness* (Raleigh: North Carolina Biological Survey and North Carolina State Museum of Natural Sciences, 1994).

30. Charles Jones and Klaus Knab, *American Wilderness: A Goushā Weekend Guide: Where to Go in the Nation's Wilderness, on the Wild and Scenic Rivers and Along the Scenic Trails* (San Jose: Goushā Publications, 1973), 117.

31. "Philip Hyde Artist's Statement," in Philip Hyde, *Range of Light, Slickrock, Drylands and Other Books, Articles, Posters, Interviews and Portfolios*, ed. David Leland Hyde (San Francisco: Sierra Club, 2010), https://vault.sierraclub.org/history/philip-hyde/default.aspx.

32. William O. Douglas, *My Wilderness: The Pacific West* (Garden City, NY: Doubleday, 1960), 165.

33. Michael McCloskey, *In the Thick of It: My Life in the Sierra Club* (Washington, DC: Island Press, 2005), 30.

34. William Cronon, ed., *Uncommon Ground: Rethinking the Human Place in Nature* (New York: W. W. Norton, 1996).

35. Cronon, who joined Yale University's history faculty in 1981, became the leading light in the robust field of wilderness history. A series of books he edited for the University of Washington Press produced three award-winning works related to the wilderness movement: *Wilderness Forever: Howard Zahniser and the Path to the Wilderness Act* by Mark Harvey; *Drawing Lines in the Forest: Creating Wilderness Areas in the Pacific Northwest* by Kevin R. Marsh; and *The Promise of Wilderness: American Environmental Politics Since 1964* by James Morton Turner.

36. David Quammen, foreword, in Stephen Gorman, *The American Wilderness: Journeys into Distant and Historic Landscapes* (Bloomington, IN: Universe, 1999), 9.

37. Land and Water Conservation Fund, September 3, 1964 (Public Law 578).

38. Johnson, "Remarks upon Signing the Wilderness Bill and the Land and Water Conservation Fund Bill."

39. Quoted in Susan R. Schrepfer, *The Fight to Save the Redwoods: A History of the Environmental Reform, 1917–1978* (Madison: University of Wisconsin Press, 1983), 124.

40. Ibid., 120.

41. William O. Douglas, speech before the Governor's Conference for California Beauty, January 11, 1966, Berkeley, University of California, Sierra Club Collection, Mike McCloskey Papers.

42. Gary Ferguson, "Guardians of the Giants," in *The Once and Future Forest: California's Iconic Redwoods* (Berkeley, CA: Heyday, 2018), 88–89.

43. Walter Reuther to President Lyndon Johnson, February 11, 1966, LE/PA3 11/22/63–3/9/66.

44. Author interview with Sharon Francis, May 3, 2021.

45. Ferguson, "Guardians and Giants," 89.

46. Ibid.

47. Michael L. Gillette, *Lady Bird Johnson: An Oral History* (New York: Oxford University Press, 2021), 360.

48. William O. Douglas, foreword, in Harvey Manning, *The Wild Cascades: Forgotten Parkland* (San Francisco: Sierra Club Books, 1965), 18.

49. "Our Wilderness Alps," *Sunset*, June 1965, 14.

50. Edward Abbey, *Desert Solitaire: A Season in the Wilderness* (New York: McGraw-Hill, 1968), 131.

51. William O. Douglas, *A Wilderness Bill of Rights* (Boston: Little, Brown, 1965), 109.

52. Ibid., 166.

53. Ibid., 35.

54. Quoted in William O. Douglas, *The Three Hundred Year War: A Chronicle of Ecological Disaster* (New York: Random House, 1972), 152.

55. Timothy Zick, *Speech Out of Doors: Preserving First Amendment Liberties in Public Places* (New York: Cambridge University Press, 2009), 282.

56. Douglas, *A Wilderness Bill of Rights*, 128–29.

57. Quoted in ibid., 167.

58. M. Margaret McKeown, *Citizen Justice: The Environmental Legacy of William O. Douglas—Public Advocate and Conservation Champion* (Washington, DC: Potomac Books, 2022), 111.

59. William O. Douglas to Marian G. Laurie, March 11, 1965, Box 550, William O. Douglas Papers, Library of Congress, Washington, DC.

60. William O. Douglas to Meyer Lefkowitz, December 15, 1961, Box 548, William O. Douglas Papers, Library of Congress, Washington, DC.

Chapter 22: Ending the Bulldozing of America

1. Stewart L. Udall, "Canyonlands National Park," *Western Gateways*, Autumn 1964.

2. Author interview with Stewart Udall, June 1, 2008.

3. "New National Park in Utah," *New York Times*, September 20, 1964, XX1.

4. Douglas Brinkley, *Rightful Heritage: Franklin D. Roosevelt and the Land of America* (New York: Harper, 2016), 196.

5. Lloyd M. Pierson, "The First Canyonlands New Park Studies: 1959 and 1960," *Canyon Legacy* 1, no. 3 (Fall 1989): 9–14.

6. "Lyndon B. Johnson Signs Public Law 88-590 Establishing Canyonlands National Park," 88th Congress, S. 27, LBJL, September 12, 1964.

7. Author interview with Stewart Udall, July 12, 2007.

8. "The Conservation Congress," *New York Times*, September 20, 1964, E10, https://timesmachine.nytimes.com/timesmachine/1964/09/20/105019488 .html?pageNumber=199.

9. Drew Babb, "LBJ's 1964 Attack Ad 'Daisy' Leaves a Legacy for Modern Campaigns," *Washington Post*, September 5, 2014, https://www.washingtonpost .com/opinions/lbjs-1964-attack-ad-daisy-leaves-a-legacy-for-modern-cam paigns/2014/09/05/d00e66b0-33b4-11e4-9e92-0899b306bbea_story.html.

10. Quoted in Frank Graham, Jr., *Since Silent Spring* (Boston: Houghton Mifflin, 1970), 207.

11. Julia Sweig, *Lady Bird Johnson: Hiding in Plain Sight* (New York: Random House, 2021), 110–22.

12. Lady Bird Johnson and Carlton B. Lees, *Wildflowers Across America* (New York: Abbeville, 1988), 11.

13. Stewart Udall to Lady Bird Johnson, December 10, 1964, Folder 1, Box 145, Stewart L. Udall Papers, University of Arizona, Tucson, AZ.

14. "America the Beautiful?," *New York Times*, February 9, 1965, 36, https://www.nytimes.com/1965/02/09/archives/america-the-beautiful.html.

15. Mary McGrory, "Beauty Blooms in Politics," *Washington Evening Star*, February 10, 1965.

16. Michael Beschloss, *Reaching for Glory: Lyndon Johnson's Secret White House Tapes, 1964–1965* (New York: Simon & Schuster, 2002), 201.

17. Ibid.

18. Bill Moyers to President Lyndon B. Johnson, Febuary 23, 1965, Moyers Papers, LBJL.

19. LeRoy Ashby and Rod Gramer, *Fighting the Odds: The Life of Senator Frank Church* (Seattle: Washington State University Press, 1994), 344.

20. A. Dan Tarlock and Roger Tippy, "The Wild and Scenic Rivers Act of 1968," *Cornell Law Review* 55 (1970): 707, 710.

21. Tim Palmer, *Endangered Rivers and the Conservation Movement* (Berkeley: University of California Press, 1986), 145–46.

22. Lyndon B. Johnson, "The President's Inaugural Address," January 20, 1965, The American Presidency Project, https://www.presidency.ucsb.edu/documents/the-presidents-inaugural-address.

23. Lyndon B. Johnson, "Special Message to the Congress on Conservation and Restoration of Natural Beauty," February 8, 1965, The American Presidency Project, https://www.presidency.ucsb.edu/documents/special-message-the-congress-conservation-and-restoration-natural-beauty.

24. Ibid.

25. Ibid.

26. "The President's Great Message," *Washington Post*, February 9, 1965.

27. Johnson, "Special Message to the Congress on Conservation and Restoration of Natural Beauty."

28. Bureau of Outdoor Recreation, *Trails for America: Report on the Nationwide Trails Study* (Washington, DC: Department of the Interior, December 1966), https://www.nps.gov/parkhistory/online_books/trails/trails.pdf.

29. Charles F. Randall, "White House Conference on Natural Beauty," *Journal of Forestry* (August 1965), 609–61.

30. Lyndon B. Johnson, foreword, in Ansel Adams and Nancy Newhall, *A More Beautiful America* (New York: American Conservation Association, 1965).

31. Martin Arnold, "Kennedy Puts Flag atop Mt. Kennedy," *New York Times*, March 25, 1965, 1.

32. Martin Arnold, "Kennedy Puts Flag atop Mt. Kennedy," *New York Times*, March 25, 1965, 1.

33. "RFK Down from Peak; Family Flag on Crest," *Tacoma News Tribune*, March 25, 1965, 10.

34. Terrianne K. Schulte, "Citizen Experts: The League of Women Voters and Environmental Conservation," *Frontiers: A Journal of Women Studies* 30, no. 3 (2009): 1–29.

35. League of Women Voters Education Fund, *The Big Water Fight: Trials and Tribulations in Citizen Action on Problems of Supply, Pollution, Floods, and Planning Across the U.S.A.* (Brattleboro, VT: Stephen Greene Press, 1966).

36. "Water Pollution Films Offered Local Audience," *Shreveport Times*, November 27, 1966, 16.

37. Joseph B. Frantz, oral history interview with Stewart Udall, 1969, tape 1, LBJL, 32–36.

38. Martin V. Melosi, "Lyndon Johnson and Environmental Policy," in *The Johnson Years*, vol. 2, *Vietnam, the Environment, and Science*, ed. Robert A. Devine (Lawrence: University Press of Kansas, 1987), 132.

39. Author interview with Lynda Bird Johnson Robb, October 23, 2021.

40. Lady Bird Johnson and Carlton B. Lees, *Wildflowers Across America* (New York: Abbeville, 1988), 15.

41. Helen Thomas, UPI, "Lady Bird Works for Beautification," *Charlotte Observer*, February 3, 1965, 11.

42. Quoted in Robert H. Boyle, "America Down the Drain," *Sports Illustrated*, November 16, 1964, 88.

43. "The Lady Bird Johnson Wildflower Center: A Legacy of Beauty," Landscape Notes, November 6, 2013, https://landscapenotes.com/2013/11/06/the-lady-bird-johnson-wildflower-center-a-legacy-of-beauty/.

44. Eric F. Goldman, *The Tragedy of Lyndon Johnson* (New York: Knopf, 1974), 372–73.

45. Author interview with Ethel Kennedy, August 4, 2014.

46. Joe Califano to Lyndon B. Johnson, August 31, 1965, LE/PA31, 11/22/63–3/9/66, LBJL.

47. Office Files of Joseph Califano, "Beautification and Conservation Measures Enacted by the 88th and 89th Congress," Box 28 (1736), LBJL.

48. Lyndon B. Johnson, "Preserving the Hudson Riverway" (Public Law 89-605).

49. Lyndon B. Johnson, "Water Quality Act of 1965," October 2, 1965, LBJL.

50. Lyndon B. Johnson, "Statement by the President in Response to Science Advisory Committee Report on Pollution of Air, Soil, and Waters," November 6, 1965, The American Presidency Project, https://www.presidency.ucsb.edu /documents/statement-the-president-response-science-advisory-committee -report-pollution-air-soil-and.

51. John Cloud, "The 200th Anniversary of the Survey of the Coast," *Prologue Magazine* 39, no. 1 (Spring 2007), https://www.archives.gov/publications /prologue/2007/spring/coast-survey.html.

52. Author interview with Ralph Nader, February 4, 2021.

53. Stephen Fox, *The American Conservation Movement: John Muir and His Legacy* (Madison: University of Wisconsin Press, 1985), 305.

54. Author interview with Ralph Nader, February 4, 2021.

55. Paul Sabin, *Public Citizens: The Attack on Big Government and the Remaking of American Liberalism* (New York: W. W. Norton), xi.

56. Lyndon B. Johnson, special message to Congress, "Restoring the Quality of Our Environment," PSAC, 1965.

57. Lyndon B. Johnson, "Statement by the President in Response to Science Advisory Committee Report on Pollution of Air, Soil, and Waters," November 6, 1965, American Presidency Project, https://www.presidency.ucsb.edu/doc uments/statement-the-president-response-science-advisory-committee-re port-pollution-air-soil-and.

58. Stewart Udall, notes, June 9, 1966, Folder 2, Box 129, Stewart L. Udall Papers, University of Tucson, Tucson, AZ.

59. Stewart Udall to Lyndon Johnson, October 10, 1966, FG 145, Box 204, LBJ Papers.

60. Jan Jarboe Russell, *Lady Bird: A Biography of Mrs. Johnson* (New York: Scribner, 1999), 280.

61. Author interview with Stewart Udall, September 6, 2009.

62. Lyndon B. Johnson, "Special Message to the Congress Transmitting Reorganization Plan 2 of 1966: Water Pollution Control," February 28, 1966, in *Public Papers of the Presidents of the United States*, 1966, vol. 2 (Washington, DC: US Government Printing Office, 1967), 229–30.

63. "Tributes Eulogize Doctor," *Times* (Shreveport, LA), September 6, 1965, 2.

64. James Brabazon, *Albert Schweitzer: A Biography* (Syracuse, NY: Syracuse University Press, 2000), 496.

65. *Sierra Club Bulletin* (December 1965), 81–88.

66. "Lady Bird's Lost Legacy," *New York Times*, July 20, 2007.

67. Mark K. Updegrove, *Indomitable Will: LBJ in the Presidency* (New York: Crown, 2012), 165–66.

68. Lyndon B. Johnson, "Remarks on the Signing of the Highway Beautification Act of 1965," October 22, 1965, accessed May 26, 2022, https://www.pres

idency.ucsb.edu/documents/remarks-the-signing-the-highway-beautification
-act-1965.

69. Lyndon B. Johnson, "Remarks at the Signing of the Highway Beautification
    Act of 1965," October 22, 1965, The American Presidency Project, https://
    www.presidency.ucsb.edu/documents/remarks-the-signing-the-highway
    -beautification-act-1965.

70. Ibid.

71. Robert Dallek, *Flawed Giant: Lyndon Johnson and His Times, 1961–1973*
    (New York: Oxford University Press, 1998), 229.

72. Johnson, "Remarks at the Signing of the Highway Beautification Act of
    1965."

Chapter 23: America's Natural Heritage

1. Lyndon B. Johnson, "Special Message to the Congress Proposing Measures
   to Preserve America's Natural Heritage," February 23, 1966, The American
   Presidency Project, https://www.presidency.ucsb.edu/documents/special
   -message-the-congress-proposing-measures-preserve-americas-natural-heritage.

2. Ibid.

3. Water Pollution Act (Public Law 89-7-53), November 3, 1966.

4. Lyndon B. Johnson, "Special Message to the Congress Proposing Measures to
   Preserve America's Natural Heritage."

5. Author interview with William Ruckelshaus, August 4, 2011.

6. Lyndon B. Johnson, "Special Message to the Congress Proposing Measures to
   Preserve America's Natural Heritage."

7. William M. Blair, "New National Park Urged for Seashore of North Carolina,"
   New York Times, May 6, 1964, 49, https://www.nytimes.com/1964/05/06
   /archives/new-national-park-urged-for-seashore-of-north-carolina.html.

8. "Cape Lookout Bill Signed by Johnson," *New York Times*, March 11, 1966, 15.

9. Lyndon B. Johnson, "Remarks at the Signing of the Cape Lookout National
   Seashore Bill," March 10, 1966, The American Presidency Project, https://
   www.presidency.ucsb.edu/documents/remarks-the-signing-the-cape-lookout
   -national-seashore-bill.

10. Ibid.

11. Ibid.

12. Rachel Carson National Wildlife Refuge, https://www.fws.gov/refuge/rachel
    -carson.

13. "Allagash Waterway in Maine Becomes National 'Wild River,'" *New York
    Times*, July 20, 1970.

14. Lady Bird Johnson, *A White House Diary* (New York: Holt, Rinehart and
    Winston, 1970), 372–79.

15. Quoted in Thomas G. Smith, *Stewart L. Udall: Steward of the Land* (Albuquerque, University of New Mexico Press, 2017), 215.

16. Johnson, *A White House Diary*, April 2, 1966, 376.

17. Ibid.

18. Ibid., 377.

19. Michael Welsh, *Big Bend National Park: Mexico, the United States, and Ecosystems* (Reno: University of Nevada Press, 2021), 109–11.

20. Johnson, *A White House Diary*, April 3, 1966, 381.

21. Ibid., April 2, 1966, 379.

22. Ibid., 379–80.

23. George B. Hartzog, Jr., *Battling for the National Parks* (Mount Kisco, NY: Moyer Bell, 1988), 178–79.

24. Michael Welsh, *Big Bend National Park: Mexico, the United States, and the Borderland Ecosystem* (Reno: University of Nevada Press, 2021), 108–12.

25. Joe Frantz, Oral History Interview, February 25, 1969, LBJ Library, 36–43.

26. Johnson, *A White House Diary*, April 3, 1966, 383.

27. Ibid., 380–81.

28. Joe Frantz, Oral History Interview, 29–30; Johnson, *A White House Diary*, April 3, 1966, 381–83; Smith, *Stewart L. Udall*, 219.

29. Lyndon B. Johnson, "Remarks upon Signing Order Establishing the President's Council and the Citizens' Advisory Committee on Recreation and Natural Beauty," May 4, 1966, The American Presidency Project, https://www.presidency.ucsb.edu/documents/remarks-upon-signing-order-establishing-the-presidents-council-and-the-citizens-advisory.

30. "Watchful Citizen; Norman Cousins," *New York Times*, May 10, 1966, 38, https://www.nytimes.com/1966/05/10/archives/watchful-citizen-norman-cousins.html.

31. United States Congress, House Committee on Public Welfare, Subcommittee on Migratory Labor, 89th Congress, 1st and 2nd sessions, *Amending Migratory Labor Laws* (Washington, DC: US Government Printing Office, 1966), 217–18.

32. Laura Pulida and Devon Peña, "Environmentalism and Positionality: The Early Pesticide Campaign of the United Farm Workers' Organizing Committee, 1965–71," *Race, Gender & Class* 6, no. 1 (1998): 33–50.

33. Author interview with Marc Grossman and Paul Chavez, January 21, 2022.

34. Randy Shaw, *Beyond the Fields: Cesar Chavez, the NFWA, and the Struggle for Justice in the 21st Century* (Berkeley: University of California Press, 2008), 80.

35. Patrick J. Sullivan, *Blue Collar—Roman Collar—White Collar: U.S. Catholic Involvement in Labor Management Controversies, 1960–1980* (Lanham, MD: University Press of America, 1987), 66.

36. "Cultural Realignment & Economic Recession: 1970s: United Farm Workers," Picture This: California Perspectives on American History (Oakland Museum of History), accessed May 26, 2022, http://picturethis.museumca.org /pictures/united-farm-workers-poster-viva-la-huelga; see also Viva Chavez, viva la causa, viva la huelga / this poster was produced by Darien House; painting by Paul Davis; designed by [Richard] Hess and/or Antupit, Library of Congress, https://www.loc.gov/item/90716478/.

37. Author interview with Ethel Kennedy, August 4, 2014.

38. Susan Sontag, *Styles of Radical Will* (New York: Farrar, Straus, Giroux, 1966), 203.

39. Julia Sweig, *Lady Bird Johnson: Hiding in Plain Sight* (New York: Random House, 2021), 235.

40. Quoted in Michael P. Cohen, *The History of the Sierra Club* (San Francisco: Sierra Club, 1988), 358–59.

41. Tom Turner, *David Brower: The Making of the Environmental Movement* (Oakland: University of California Press, 2015), 118–29. Also see https://digitalas sets.lib.berkeley.edu/rohoia/ucb/text/environmentalact00browrich.pdf.

42. "Choice for Grand Canyon," *New York Times*, January 17, 1966, 37.

43. Quoted in Hugh Nash, "Dams in Grand Canyon—A Necessary Evil?" *Sierra Club Bulletin* (December 1965), 49.

44. Hal Wingo, "Knight Errant to Nature's Rescue," *Life*, May 27, 1966, 15.

45. Quoted in Mark Reisner, *Cadillac Desert: The American West and Its Disappearing Water* (New York: Viking, 1986).

46. The series of Grand Canyon Dam ads is found on the Sierra Club website at http://content.sierraclub.org/brower/grand-canyon-ads.

47. Bill McKibben, ed., *American Earth: Environmental Writing Since Thoreau* (New York: Library of America, 2008).

48. Morris Udall, *Too Funny to Be President* (New York: Henry Holt, 1988), 20.

49. Byron E. Pearson, *Still the Wild River Runs: Congress, the Sierra Club, and the Fight to Save Grand Canyon* (Tucson: University of Arizona Press, 2002), xvii.

50. David Brower to Lyndon B. Johnson, September 15, 1966, General Legislation: Le/NR/71, Box 145, LBJL.

51. Donald Worster, *Rivers of Empire: Water, Aridity, and the Growth of the American West* (New York: Oxford University Press, 1985), 276.

Chapter 24. Defenders

1.   Author interview with Kerry Kennedy, January 11, 2022.
2.   Author interview with Robert F. Kennedy, Jr., February 3, 2021.
3.   Author interview with Harry Benson, February 12, 2022.
4.   Author interview with Kerry Kennedy, August 6, 2021.
5.   Delaware Water Gap National Recreation Area (Public Law, 89-66-7).
6.   Lyndon B. Johnson, "Remarks at the Signing Ceremony for Seven Conservation Bills," October 15, 1966, The American Presidency Project, https://www.presidency.ucsb.edu/documents/remarks-the-signing-ceremony-for-seven-conservation-bills.
7.   Charles Layng, "Potential National Park for Texas," *New York Times*, March 28, 1965, XX25.
8.   Public Law 89-668, 89th Congress, H.R. 8678, An Act to Establish in the State of Michigan the Pictured Rocks National Lakeshore, October 15, 1966.
9.   "Pictured Rocks National Lakeshore," Michigan in the World, michiganintheworld.history.lsa.umich.edu/environmentalism/exhibits/show/main_exhibit/origins/wilderness-act/pictured-rocks.
10. Johnson, "Remarks at the Signing Ceremony for Seven Conservation Bills."
11. National Historic Preservation Act of 1966 (Public Law 89-665).
12. Mrs. Lyndon B. Johnson, foreword, in *With Heritage So Rich: A Report of a Special Committee on Historic Preservation under the Auspices of the United States Conference of Mayors with a Grant from the Ford Foundation* (New York: Random House, 1966), vii.
13. Robert E. Stipe, *A Richer Heritage: Historic Preservation in the Twenty-first Century* (Chapel Hill: University of North Carolina Press, 2003), 480.
14. Lyndon B. Johnson, "Remarks at the Signing Ceremony for Seven Conservation Bills."
15. Joe Roman, *Listed Dispatches from America's Endangered Species Act* (Cambridge, MA: Harvard University Press, 2011), 22–23.
16. US Department of the Interior, Office of the Secretary, *Native Fish and Wildlife Endangered Species*, Fed. Res. Doc 67-2758, February 27, 1967. See also US Department of the Interior, "40 Years of the Endangered Species Act: Remarkable Origin, Resilience, and Opportunity," January 14, 2003, https://www.doi.gov/sites/doi.gov/files/migrated/ppa/upload/40-Years-of-the-Endangered-Species-Act.pdf.
17. "Harold Jefferson Coolidge, a Leader in International Conservation and . . . ," February 16, 1985, United Press International, https://www.upi.com/Archives/1985/02/16/Harold-Jefferson-Coolidge-a-leader-in-international-conservation-and/9807477378000/.

18. US Department of the Interior, *Native Fish and Wildlife Endangered Species*.

19. Frank Graham, Jr., *Since Silent Spring* (Boston: Houghton Mifflin, 1970), 206–10.

20. Quoted in *Defenders*, January 1966, 1.

21. Ibid., 208.

22. Mary Hazell Harris testimony, Defenders of Wildlife Archive, Washington, DC.

23. Defenders of Wildlife Archive, Washington, DC, https://www.google
.com/books/edition/Predatory_Mammal/EMZFAQAAMAAJ?gbpv
=1&bsq=%22Defenders%20of%20Wildlife%20objects%20to%20the%20
use%20of%20poisons%20in%20controlling%20predatory%20mam
mals%22.

24. Harris and Pimlott quoted in Defenders of Wildlife File Research Results (June 5, 2021), Washington, DC. Jamie Rappaport Clark graciously assisted in sorting through *Defenders* articles.

25. James M. Cahalan, *Edward Abbey: A Life* (Tucson: University of Arizona Press, 2001), 144.

26. Edward Abbey, *The Journey Home: Some Work in Defense of the American West* (New York: Plume, 1991), 223–38.

27. Archie F. Carr, "In Praise of Snakes," *Audubon* 73, no. 4 (1971): 18–27.

28. Archie Carr, *A Naturalist in Florida: A Celebration of Eden* (New Haven, CT: Yale University Press, 1994), xv.

29. Edward Hoagland, *Red Wolves and Black Bears: Nineteen Essays* (New York: Random House, 1976), 9.

30. Quoted in Jeanne L. Clark, *America's Wildlife Refuges: Lands of Promise* (Portland, OR: Carpe Diem, 2003), 134.

31. "A Conservation Success Story: Aleutian Canada Goose Wings Its Way Back from Brink of Extinction," US Fish and Wildlife Service, Anchorage, Alaska, July 30, 1999, https://www.fws.gov/pacific/news/1999/9948.htm.

32. Ibid.

33. Quoted in Clark, *America's Wildlife Refuges*, 83.

34. Kat Eschner, "The Hopeful Mid-Century Conservation Story of the (Still Endangered) Whooping Crane," *Smithsonian Magazine*, April 24, 2017, https://
www.smithsonianmag.com/smithsonianmag/unbelievable-mid-century-con
servation-story-still-endangered-whooping-crane-180962943/.

35. Quoted in Eschner, "The Hopeful Mid-Century Conservation Story of the (Still Endangered) Whooping Crane."

36. William O. Douglas, *The Three Hundred Year War: A Chronicle of Ecological Disaster* (New York: Random House, 1972), 128–29.

37. Scott Raymond Einberger, *With Distance in His Eyes: The Environmental Life and Legacy of Stewart Udall* (Reno: University of Nevada Press, 2018), 48–49.

38. US Department of the Interior, "Udall Calls for Cooperative Effort to Save Wildlife," July 21, 1968, www.fws.gov/news/Historic/NewsReleases /1968/19680721a.pdf.

39. Patricia Mazzei, "Trying Everything, Even Lettuce, to Save Florida's Beloved Manatees," *New York Times*, April 9, 2022.

40. Dean E. Biggins and Max Schroeder, "Historical and Present Status of the Black-Footed Ferret," *Great Plains Wildlife Damage Control Workshop Proceedings*, April 1987, 50.

41. "Black-Footed Ferret," South Dakota Ecological Field Office, September 9, 2013.

42. Quoted in Tim Palmer, *Endangered Rivers and the Conservation Movement* (New York: Rowman & Littlefield, 2004), 166.

43. Euell Gibbons, *Stalking the Wild Asparagus* (Philadelphia: D. McKay, 1962). See also "Euell Gibbons Dies at 64," *New York Times*, December 30, 1975, 28, https://www.nytimes.com/1975/12/30/archives/euell-gibbons-dies-at -64-wrote-books-about-natural-foods.html.

44. Author interview with Paul Chavez (Cesar Chavez's son), April 14, 2018.

45. Daniel Stone, *The Food Explorer: The True Adventures of the Globe-Trotting Botanist Who Transformed What America Eats* (New York: Penguin, 2018), xiii–xiv.

46. Arthur H. Purcell, "Lake Michigan Dunes: Wonders from the Ice Age," *Los Angeles Times*, June 7, 1987, https://www.latimes.com/archives/la-xpm-1987 -06-07-tr-844-story.html.

47. Nicole Barker, "Saving the Dunes: Then and Now," *Singing Sands* 37, no. 1 (2016): 1, http://npshistory.com/publications/indu/newspaper/2016.pdf.

48. Lady Bird Johnson, *A White House Diary*, April 10, 1964, 107.

49. Ibid.

50. Barry Mackintosh, "Living History," in *Interpretation in the National Park Service: A Historical Perspective* (Washington, DC: National Park Service, 1986), 60–61.

Chapter 25: "Sue the Bastards!" and Environmental Justice

1. Linda Nash, "The Fruits of Ill-Health: Pesticides and Workers' Bodies in Post–World War II California," *Osiris* 19, no. 1 (2004): 203–19.

2. Quoted in Drew Dellinger, "Martin Luther King Jr: Ecological Thinker," https://drewdellinger.org/martin-luther-king-jr-ecological-thinker/.

3. Drew Dellinger, "Dr. King's Interconnected World," *New York Times*, December 22, 2017.

4. Quoted in Dellinger, "Martin Luther King Jr: Ecological Thinker."

5. Lynton Keith Caldwell, *The National Environmental Policy Act: An Agenda for the Future* (Bloomington: Indiana University Press, 1999), 28.

6. Richard J. Lazarus, *The Making of Environmental Law* (Chicago: University of Chicago Press, 2004), 53.

7. Mark Wexler, "75 Years of Conservation," nwf.org/en/naturalwildlife/2011/nwf-75yearstimeline.

8. Bob Deans, "The Environmental Movement's Debt to Martin Luther King, Jr.," Livescience, August 27, 2013, https://www.livescience.com/39210-the-environmental-movement-debt-to-martin-luther-king-jr.html.

9. Harriet A. Washington, *A Terrible Thing to Waste: Environmental Racism and Its Assault on the American Mind* (New York: Little, Brown, 2019), 142–44.

10. *City of Houston v. George, Supreme Court of Texas*, March 22, 1972, https://law.justia.com/cases/texas/supreme-court/1972/b-2726-0.html.phen.

11. Robert D. Bullard, "The Mountains of Houston: Environmental Justice and the Politics of Garbage," *Cite* 93 (Winter 2014): 28–33.

12. Stephen L. Klineberg, *Prophetic City: Houston on the Cusp of a Changing America* (New York: Avid Reader Press, 2020).

13. "Letter Initiating the Down River Anti-Pollution League," David Stradling, ed., *The Environmental Moment, 1968–1972* (Seattle: University of Washington Press, 2012), 52–53.

14. Larry Canter and Ray Clark, eds., *Environmental Policy and NEPA: Past, Present, and Future* (Boca Raton, FL: St. Lucie Press, 1997), 28–31.

15. Phil Dougherty, "Van Ness, William J. 'Bill' Jr. (1938–2017)," History Link, July 20, 2011, https://historylink.org/File/9882. Also see https://www.badc.org/.

16. *Bill Van Ness: Creation of the National Environmental Policy Act* (video), Van Ness Feldman LLP, February 2, 2017, https://www.vnf.com/bill-van-ness-creation-of-the-national-environmental.

17. Ibid.

18. Lynton Keith Caldwell, *The National Environmental Policy Act: An Agenda for the Future* (Bloomington: Indiana University Press, 1999), xvi.

19. Lynton K. Caldwell, "Perspective: An Interview with Lynton Caldwell on the National Environmental Policy Act (NEPA)," *Environmental Practice* 5, no. 4 (December 2003): 281–86.

20. Ibid.

21. *Udall v. FPC*, 387 U.S. 428 (1968), https://caselaw.findlaw.com/us-supreme-court/387/428.html.

22. Paul Sabin, *Public Citizens: The Attack on Big Government and the Remaking of American Liberalism* (New York: W. W. Norton, 2021), 98.

23. "Justice Douglas Leads Wilderness Protest Hike," *Spokesman-Review* (Spokane, WA), August 6, 1967, 3.

24. A. W. Eipper, C. A. Carlson, and L. S. Hamilton, "Impacts of Nuclear Power Plants on the Environment," *Living Wilderness* (Autumn 1970), 6.

25. Rachel Smolkin and Brenna Williams, "How LBJ Scared Visitors at His Ranch," CNN, https://www.cnn.com/interactive/2015/10/politics/lbj-ranch-history/.

26. James M. O'Fallon, *Nature's Justice: Writings of William O. Douglas* (Corvallis: Oregon State University Press, 2000), 265.

27. Author interview with Cathy Douglas Stone, January 9, 2020.

28. Andy Flynn, "Robert Kennedy's Hudson River Kayak at Local Museum," *Sun Community News*, August 19–8, 2011.

29. David A. Schaller, "The 1967 Kennedy River Trip—Looking Back," *Celebrating 100 Years of Grand Canyon National Park: A Gathering of Grand Canyon Historians: Ideas, Arguments, and First Person Accounts*, ed. Richard D. Quartaroli (Grand Canyon, Arizona: Grand Canyon Conservancy, 2020).

30. "Concrete and Canyons: Senator Robert Kennedy's 1967 Family Vacation," National Archives at Denver, Boxes 10 and 302, Photographs of Project Sites, 1966–1983, Public Relations Photographs 1981–1983, National Archives identifier 562813, Engineering and Research Center, Bureau of Reclamation, Department of the Interior, Record Group 115, Records of the Bureau of Reclamation.

31. Author interview with Stewart Udall, March 6, 2006.

32. Author interview with Kerry Kennedy, August 6, 2021.

33. Ibid.

34. Charles F. Wurster, *DDT Wars: Rescuing our National Bird, Preventing Cancer, and Creating the Environmental Defense Fund* (New York: Oxford University Press, 2015), 17.

35. "Yannacone 2-Way," *Living on Earth*, radio broadcast transcript, World Media Foundation, August 20, 1993, https://www.loe.org/shows/segments.html?programID=93-P13-00034&segmentID=3.

36. Victor John Yannacone, Jr., to author, February 25, 2021.

37. Wurster, *DDT Wars*, 18.

38. Ibid., 21–22.

39. Victor Yannacone, Jr., "Origins of the Environmental Defense Fund," https://yannalaw.com/about/about-victor-yannacone-biography-short-history-environmental-movement/environmental-defense-fund-edf/.

40. Victor John Yannacone, "Science, Ethics, and Scientific Ethics in the Modern World," *Environmental Geosciences* 6 (1999): 164–65.

41. Yannacone, "Origins of the Environmental Defense Fund."

42. Quoted in Gilbert Rogin, "All He Wants to Save Is the World," *Sports Illustrated*, February 3, 1969.

43. Wurster, *DDT Wars*, 206.

44. Ibid., 29–30.

45. Frank Graham, Jr., *Since Silent Spring*, 258.

46. Quoted in Stephen Fox, *The American Conservation Movement: John Muir and His Legacy* (New York: Little, Brown, 1981), 304. See also Frank Graham, Jr., "Taking Polluters to Court," *New Republic* 158, January 13, 1968, 9.

47. Quoted in Rogin, "All He Wants to Save Is the World."

48. Quoted in Patrick Anderson, "Ralph Nader, Crusader; or, the Rise of a Self-Appointed Lobbyist," *New York Times Magazine*, October 29, 1967, SM25.

49. Tiffany Dyer, "Pesticides and the United Farm Workers: An Extension of the Struggle for Social Justice," senior thesis, History 400, University of Puget Sound, Fall 2004.

50. Drew Dellinger, "Dr. King's Interconnected World," *New York Times*, December 22, 2017.

51. Steven V. Roberts, "Charge of Peril in Pesticides Adds Fuel to Coast Grape Strike," *New York Times*, March 16, 1969, 46.

52. Harriet A. Washington, *A Terrible Thing to Waste: Environmental Racism and Its Assault on the American Mind* (New York: HarperCollins, 2019), 9.

53. Robert Gottlieb, "The U.S. Air Force Was the Principal Dumper of the Site," *Forcing the Spring*, 163.

54. Lenore Marshall, "The Nuclear Sword of Damocles," *Living Wilderness*, Spring 1971, 17–19.

55. Author interview with Stewart Udall, March 6, 2006.

56. Author interview with Ethel Kennedy, August 4, 2014.

57. Author interview with Cathy Douglas, February 3, 2022.

58. William O. Douglas to Robert Strange McNamara, July 3, 1967, William O. Douglas Papers, Library of Congress, Washington, DC.

59. William O. Douglas to Robert Strange McNamara, July 28, 1967, William O. Douglas Papers, Library of Congress, Washington, DC.

60. William O. Douglas to Lyndon B. Johnson, August 12, 1967, William O. Douglas Papers, Library of Congress, Washington, DC.

61. Lyndon B. Johnson to William O. Douglas, August 28, 1967, William O. Douglas Papers, Library of Congress, Washington, DC.

62. William O. Douglas to Henry M. Jackson, February 1, 1968, William O. Douglas Papers, Library of Congress, Washington, DC.

63. Tom Turner, *David Brower: The Making of the Environmental Movement* (Oakland: University of California Press, 2015), 175–78.

64. Louise Chawla, oral history interview with Oscar H. Geralds, Jr., Louis B. Nunn Center for Oral History, University of Kentucky Libraries, July 27, 1989, http://kentuckyoralhistory.org/ark:/16417/xt7wpz51k30t.

65. "A Resolution Honoring the 50th Anniversary of the Protest Hike," Kentucky Senate Resolution 157, March 13, 2018, Kentucky Legislature, https://apps .legislature.ky.gov/recorddocuments/bill/18RS/sr157/orig_bill.pdf.

66. Ralph W. Derickson, "Justice Douglas Leads Five-Mile Hike Through Gorge of Red River," *Lexington Herald-Leader*, November 19, 1967, 1.

67. Ben A. Franklin, "Conservationists Rallying Against a Dam in Kentucky," *New York Times*, November 20, 1967, 49.

68. Berry, *The Unforeseen Wilderness*, 105.

69. Author interview with Cathy Stoneman Douglas, February 3, 2022.

70. Franklin, "Conservationists Rallying Against a Dam in Kentucky."

71. Britta B. Harris, "The Drama of Oakley Dam: To Build or Not to Build, That Was the Question," *Illinois Issues* 7 (September 1976), https://www.lib.niu .edu/1976/ii760903.html.

72. Trent T. Gilliss, "Martin Luther King's Last Christmas Sermon," The On Being Project, December 25, 2015, https://onbeing.org/blog/martin-luther -kings-last-christmas-sermon/.

Chapter 26: The Unraveling of America, 1968

1. Thomas G. Smith, *Stewart L. Udall: Steward of the Land* (Albuquerque: University of New Mexico Press, 2017), 273–74.

2. L. Boyd Finch, *Legacies of Camelot: Stewart and Lee Udall, American Culture, and the Arts* (Norman: University of Oklahoma Press, 2008), 111–13. See also Stewart Udall journal, February 5, 1968, Folder 2, Box 140, Stewart L. Udall Papers, University of Tucson, Tucson, AZ.

3. Finch, *Legacies of Camelot*, 112.

4. Doris Kearns Goodwin, *Lyndon Johnson and the American Dream* (New York: Harper and Row, 1976), 359.

5. Quoted in Brigham Daniels, Andrew P. Follett, and Joshua Davis, "The Making of the Clean Air Act," *UC Hastings Law Journal* 71, no. 4 (May 10, 2020).

6. Author interview with Robert F. Kennedy, Jr., March 6, 2017.

7. Ibid.

8. Quoted in Steven C. Schutte, *Wayne Aspinall and the Shaping of the American West* (Boulder: University Press of Colorado: 2002), 227.

9. Lyndon B. Johnson, "Special Message to the Congress: Protecting Our Natural Heritage," January 30, 1967, The American Presidency Project, https:// www.presidency.ucsb.edu/documents/special-message-the-congress-protect ing-our-natural-heritage.

10. "Reagan Shows Peculiar Attitude Toward Trees," *Sacramento Bee*, March 15, 1966, 50.

11. "Reagan Aides Will Visit Washington to Push for Redwoods Compromise," *Modesto Bee*, April 10, 1967, 14.

12. Ibid.

13. Jared Farmer, *Trees in Paradise: A California History* (New York: W. W. Norton, 2013), 82.

14. Susan Schrepfer, "Conflict in Preservation: The Sierra Club, Save-the-Redwoods League, Redwood National Park," *Journal of Forest History* 24, no. 2 (1988).

15. Joseph B. Frantz, oral history interview with Wayne Aspinall, June 14, 1974, LBJ Presidential Library, 11.

16. Horace Albright to Stewart Udall, January 27, 1968, Folder 2, Box 190, Stewart L. Udall Papers, University of Tucson, Tucson, AZ.

17. Lyndon B. Johnson, "Special Message to the Congress on the Problems of the American Indian: 'The Forgotten American,'" March 6, 1968, https://www.presidency.ucsb.edu/documents/special-message-the-congress-the-problems-the-american-indian-the-forgotten-american.

18. Lyndon B. Johnson, "Special Message to the Congress on Conservation: 'To Renew a Nation,'" March 8, 1968, The American Presidency Project, https://www.presidency.ucsb.edu/documents/special-message-the-congress-conservation-renew-nation.

19. Dominic Sandbrook, *Eugene McCarthy: The Rise and Fall of Postwar American Liberalism* (New York: Alfred A. Knopf, 2004).

20. Author interview with Stewart Udall, September 6, 2009.

21. Stewart Udall to President and Mrs. Johnson, April 1, 1968, Box 2016, Alphabetical File, Stewart Udall Papers, University of Arizona, Tucson, Arizona.

22. Author interview with Stewart Udall, September 6, 2009. Also see Drew Dellinger, "Martin Luther King Jr: Ecological Thinker," *Common Ground Magazine*, April 2014, 49–50.

23. Author interview with John Lewis, March 7, 2010.

24. Author interview with Andrew Young, January 2, 2022.

25. Tom Turner, *David Brower: The Making of the Environmental Movement* (Oakland: University of California Press, 2015), 192.

26. Jeanne Theoharis, "Coretta Scott King and the Civil-Rights Movement's Hidden Women," *Atlantic*, February 2018, https://www.theatlantic.com/magazine/archive/2018/02/coretta-scott-king/552557. Originally published in the special MLK issue print edition with the headline "Women Have Been the Backbone of the Whole Civil Rights Movement."

27. Lady Bird Johnson and Carlton B. Leer, *Wildflowers Across America* (New York: Abbeville Press, 1988), 263.

28. Ralph Yarborough, foreword, in Lois Williams Parker, *Big Thicket of Texas: A Comprehensive Annotated Bibliography* (Arlington, TX: Sable Publishing, 1977), 4.

29. Michael L. Gillette, *Lady Bird Johnson: An Oral History* (New York: Oxford University Press, 2012), 306.

30. Lyndon B. Johnson, "Statement by the President upon Signing Bill Establishing the National Park Foundation," December 19, 1967, American Presidency Project, https://www.presidency.ucsb.edu/documents/statement-the-president-upon-signing-bill-establishing-the-national-park-foundation.

31. Ibid.

32. Author interview with Mary Elizabeth "Liz" Carpenter, March 1, 2007.

33. "Remarks of Mrs. Lyndon B. Johnson at the Dedication of Padre Island National Seashore," April 8, 1968, https://discoverlbj.org/item/ref-ctj speeches-19680408-1100.

34. Ibid.

35. Lady Bird Johnson, *A White House Diary* (New York: Holt, Rinehart and Winston, 1970), 658–60.

36. Jocelyn Sherman, "NFWA Mourns the Passing of Dr. Marion Moses Who Turned a Weekend in Delano into a Lifetime of Service Helping Cesar Chavez, Farm Workers Combat the Perils of Pesticides," United Farm Workers, August 29, 2020, https://NFWA.org/drmoses.

37. Marion Moses, *Harvest of Sorrow: Farm Workers and Pesticides* (San Francisco: Pesticide Education Center, 1992); Marion Moses, *Designer Poisons: How to Protect Your Health and Home from Toxic Pesticides* (San Francisco: Pesticide Education Center, 1995).

38. "Cesar Chavez Breaks Hunger Strike with Robert F. Kennedy," UPI, https://www.upi.com/News_Photos/view/upi/f67b32c9c1f71f91546f75cf9c7b59ff/Cesar-Chavez-breaks-hunger-strike-with-Robert-F-Kennedy/.

39. Author interview with Paul Chavez, April 3, 2022.

40. Randy Shaw, *Beyond the Fields: Cesar Chavez, the NFWA, and the Struggle for Justice in the 21st Century* (Berkeley: University of California Press, 2010), 134–40.

41. Larry Tye, *Bobby Kennedy: The Making of a Liberal Icon* (New York: Random House, 2017).

42. Pete Hamill, "June 5, 1968: The Last Hours of RFK," *New York*, May 16, 2008, https://nymag.com/news/politics/47041/.

43. Author interview with Stewart Udall, July 12, 2007.

44. John Cronin and Robert F. Kennedy, Jr., *The Riverkeepers: Two Activists Fight to Reclaim Our Environment as a Basic Human Right* (New York: Scribner, 1997), 88–89.

45. Author interview with Robert F. Kennedy, Jr., May 6, 2021.

46. Tricia Nixon Cox, email to author, October 27, 2020.

47. Ralph Nader, "They're Still Breathing," *New Republic*, February 3, 1968, 15.

48. William S. Becker, *The Creeks Will Rise: People Coexisting with Floods* (Chicago: Chicago Review Press, 2021), 152.

49. Robert Gottlieb, *Forcing the Spring: The Transformation of the American Environmental Movement* (Washington, DC: Island Press, 1993), 280–81.

50. Tricia Nixon Cox, email to author, October 27, 2020.

51. "Ehrlichman, Remembered for Watergate, Dies at Age 73—Public Perception of Washington Native Forever Linked to Break-In," *Seattle Times*, February 16, 1999.

52. Author interview with John Ehrlichman, May 29, 2022.

53. "Richard Nixon: Environmental Hero," *Living on Earth*, August 9, 1996.

54. Jack E. Davis, *The Bald Eagle: The Improbable Journey of America's Bird* (New York: W. W. Norton, 2022), 306.

55. John McPhee, *Encounters with the Archdruid* (New York: Farrar, Straus, & Giroux, 1971), July 3, 2006.

56. Author interview with Russell Train, July 3, 2006.

57. Nancy Johnston et al., *Central Park Country: A Tune Within Us* (New York: Sierra Club, 1968).

58. "Marianne Moore Preaches Gently Against War," *New York Times*, June 20, 1968, 47, https://www.nytimes.com/1968/06/20/archives/marianne-moore-preaches-gently-against-war.html.

59. Quoted in Susan R. Schrepfer, *The Fight to Save the Redwoods: A History of the Environmental Reform, 1917–1978* (Madison: University of Wisconsin Press, 1983), 165.

60. Norman Mailer, *St. George and the Godfather* (New York: New American Library, 1972), 3.

61. A. J. Langguth, *Our Vietnam: The War 1954–1975* (New York: Simon & Schuster, 2000), 515–16.

62. "1968 Democratic Party Platform," The American Presidency Project, https://www.presidency.ucsb.edu/documents/1968-democratic-party-platform.

63. Hubert Humphrey to F. Herbert Bormann, August 9, 1968.

64. Edwin Way Teale, "Making the Wild Scene," *New York Times*, January 28, 1968, BR7, https://www.nytimes.com/1968/01/28/archives/making-the-wild-scene-desert-solitaire-a-season-in-the-wilderness.html.

65. Edward Abbey, *Desert Solitaire: A Season in the Wilderness* (Tucson: University of Arizona Press, 1988), 1.

66. Ibid., 9.

67. Ibid., 50.

68. Paul Ehrlich, *The Population Bomb* (New York: Sierra Club–Ballantine, 1968), 34.

69. David Webster, "The Population Bomb," *New York Times*, February 8, 1969.

70. Charles C. Mann, "The Book That Incited a Worldwide Fear of Overpopulation," *Smithsonian Magazine*, January–February 2018.

71. Paul R. Ehrlich and Anne H. Ehrlich, "The Population Bomb Revisited," *The Electronic Journal of Sustainable Development* 1, no. 3 (2009): 63–71.

72. Quoted in LeRoy Ashby and Rod Gramer, *Fighting the Odds: The Life of Frank Church* (Pullman: Washington State University Press, 1994), 346–47.

73. N. Scott Momaday, *House of Dawn* (New York: Harper and Row, 1968).

74. These Native leaders all were introduced to the public during the Long Sixties via articles and books. See especially Chief Seattle, "The Land is Sacred," *Counseling and Values* 18 (Summer 1974): 275–77; Hyemeyohsts Storm, *Seven Arrows* (New York: Random House, 1972).

75. *Ecotactics: The Sierra Club Handbook for Environmental Activists* (New York: Touchstone, 1970).

76. Stewart Brand, *Whole Earth Discipline: An Ecopragmatist Manifesto* (New York: Viking, 2009).

Chapter 27: Lyndon Johnson

1. The bills signed by President Johnson were as follows: Wild and Scenic Rivers Act (Public Law 90-542, 82 Stat. 906); National Trails System Act (Public Law 90-543, 82 Stat. 919); An Act to Establish the North Cascades National Park and Ross Lake and Lake Chelan National Recreation Areas, to Designate the Pasayten Wilderness and to Modify the Glacier Peak Wilderness, in the State of Washington, and for Other Purposes (Public Law 90-544, 82 Stat. 926); and An Act to Establish a Redwood National Park in the State of California (Public Law 90-545, 82 Stat. 931).

2. Lyndon B. Johnson, "Remarks upon Signing Four Bills Relating to Conservation and Outdoor Recreation," October 2, 1968, The American Presidency Project, https://www.presidency.ucsb.edu/documents/remarks-upon-signing-four-bills-relating-conservation-and-outdoor-recreation.

3. Author interview with George McGovern, March 4, 2000.

4. Lyndon B. Johnson, "Special Message to the Congress Transmitting Report on the National Wilderness Preservation System," February 8, 1965, The American Presidency Project, https://www.presidency.ucsb.edu/documents/special-message-the-congress-transmitting-report-the-national-wilderness-preservation.

5. Doris Kearns Goodwin, *Lyndon Johnson and the American Dream* (New York: Harper and Row, 1976)

6. The National Wild and Scenic Rivers Act of 1968, 16 U.S.C., P.L. 90-541, Section 2(b)(1).

7. Tim Palmer, *Rivers of America* (New York: Harry N. Abrams, 2006).

8. Nancy M. Germans, "Negotiating for the Environment," *Federal History* 9 (April 2017): 48–68.

9. US Committee on Interior and Insular Affairs, 90th Congress, 2nd session, HR, Rep. No. 1623, "Providing for a National Scenic Rivers System

and Other Purposes," July 3, 1968, 1–3; United States Senate Committee on National Water Resources, *Water Recreation Needs in the United States, 1960–2000* (Washington, DC: US Government Printing Office, 1960).

10. Johnson, "Remarks upon Signing Four Bills Relating to Conservation and Outdoor Recreation."

11. Lyndon B. Johnson, "Remarks upon Signing Four Bills Relating to Conservation and Outdoor Recreation," online by Gerhard Peters and John T. Wooley, The American Presidency Project, https://www.presidency.ucsb.edu.node /237295.

12. Tim Palmer, *Endangered Rivers and the Conservation Movement* (Lanham, MD: Rowman & Littlefield, 2004), 147–48.

13. Dyan Zaslowsky and T. H. Watkins, *These American Lands: Parks, Wilderness, and the Public Lands* (Washington, DC: Island Press, 1994), 232.

14. Tim Palmer, *Wild and Scenic Rivers: An American Legacy* (Corvallis: Oregon State University, 2017), 39, 230.

15. Germans, "Negotiating for the Environment." Also, author interview with Steward Udall, September 6, 2009.

16. Tim Palmer, *The Wild and Scenic Rivers of America* (Washington, DC: Island Press, 1993), 26.

17. Lyndon B. Johnson, "Remarks upon Signing Four Bills Relating to Conservation and Outdoor Recreation," online by Gerhard Peters and John T. Wooley, The American Presidency Project https://www.presidency.ucsb.edu .node/237295.

18. Author interview with Mary Elizabeth "Liz" Carpenter, March 1, 2007; see also Palmer, 286–69.

19. Quoted in Palmer, *The Wild and Scenic Rivers of America*, 26.

20. Jeff Rennicke, "Of Time and Rivers Flowing: Forty Years Downstream for the National Wild and Scenic Rivers System," *National Parks* 82, no. 3 (Summer 2008): 46–50, http://npshistory.com/npca/magazine/summer-2008.pdf.

21. Ibid.

22. Adam Rome, *The Genius of Earth Day* (New York: Macmillan, 2013), 34.

23. Tim Palmer, *Wild and Scenic Rivers*, 22.

24. "Nelson Gets Enthusiastic Endorsement of Lombardi," *Capital Times* (Madison, WI), October 4, 1968, 1.

25. Palmer, *Endangered Rivers and the Conservation Movement*, 163.

26. Greg Seitz, "50th Anniversary: Read Senator Gaylord Nelson's Fiery 1965 Speech Calling for St. Croix River Conservation," St. Croix 360, https://www.stcroix360.com/2018/03/50th-anniversary-read-senator-gaylord-nelsons-fiery-1965-speech-calling-for-st-croix-river-conservation/.

27. Tim Palmer, *Wild and Scenic Rivers*, 19.

28. "Two Nelson Bills Signed by Johnson," *Capital Times*, October 4, 1968.

29. Bill Christofferson, *The Man from Clear Lake: Earth Day Founder Senator Gaylord Nelson* (Madison: University of Wisconsin Press, 2009), 249.

30. Frank Church, "The New Conservation," speech at Idaho Parks and Recreation Convention, Coeur D'Alene, Idaho, Frank Church Papers, Boise State University, Boise, Idaho, April 22, 1969.

31. Thomas G. Smith, *Green Republican: John Saylor and the Preservation of America's Wilderness* (Pittsburgh, PA: University of Pittsburgh Press, 2006).

32. Edward Abbey, *One Life at a Time, Please* (New York: Henry Holt, 1988), 112.

33. Quoted in Tim Palmer, *Wild and Scenic Rivers*, 206.

34. Gaylord Nelson, "Statement in Support of a Bill to Facilitate the Management, Use, and Public Benefits from the Appalachian Trail," Box 114, Gaylord Nelson Papers, Wisconsin Historical Society, Madison, Wisconsin.

35. *Trails for America: Report on the Nationwide Trail Study*, Department of the Interior, Bureau of Outdoor Recreation, Washington, DC, December 1966, https://www.nps.gov/parkhistory/online_books/trails/trails.pdf.

36. Larry Anderson, *Benton MacKaye: Conservationist Planner and Creator of the Appalachian Trail* (Baltimore: Johns Hopkins University Press, 2002), 358.

37. Samuel P. Hays, *Beauty, Health, and Permanence: Environmental Politics in the United States, 1955–1985* (Cambridge, UK: Cambridge University Press, 1987), 117.

38. Edward P. Cliff, oral history, June 6, 1969, LBJL.

39. Ibid., 115.

40. Ibid., 117.

41. Christofferson, *The Man from Clear Lake*, 360.

42. Jared Farmer, *Trees in Paradise: A California History* (New York: W. W. Norton, 2013), 83.

43. Gary Ferguson, "Guardians of the Giants," in *The Once and Future Forest: California's Iconic Redwoods* (Berkeley, CA: Heyday, 2018), 90–91.

44. Edwin C. Bearss, *History Basic Data: Redwood National Park* (Washington, DC: US Department of the Interior, 1982), https://www.nps.gov/parkhistory/online_books/redw/index.htm.

45. Johnson, "Remarks upon Signing Four Bills Relating to Conservation and Outdoor Recreation."

46. William O. Douglas to David Brower, October 25, 1968, in William O. Douglas, *The Douglas Letters: Selections from the Private Papers of Justice William O. Douglas*, ed. Melvin I. Urofsky (Bethesda, MD: Adler and Adler, 1987), 252.

47. Quoted in Michael McCloskey, "The Battle of the Redwoods," *American West*, September 1969, 55.

48. Susan R. Schrepfer, *The Fight to Save the Redwoods: A History of the Environmental Reform, 1917–1978* (Madison: University of Wisconsin Press, 1983), 186.

49. Ibid., 160.

50. Richard Nixon, "Remarks at the Dedication of Lady Bird Johnson Grove in Redwood National Park in California," August 27, 1969, The American Presidency Project, https://www.presidency.ucsb.edu/documents/remarks-the -dedication-lady-bird-johnson-grove-redwood-national-park-california.

51. Public Law No. 90-544, 82 Stat. 926, 1968.

52. Author interview with Gary Snyder, January 22, 2022.

53. Lyndon B. Johnson, "Remarks upon Signing Four Bills Relating to Conservation and Outdoor Recreation."

54. David Louter, *Windshield Wilderness: Cars, Roads, and Nature in Washington's National Parks* (Seattle: University of Washington Press, 2006).

55. "New Law Creates Two Vast Parks," *Alabama Journal* (Montgomery), November 28, 1968, 4.

56. Lyndon B. Johnson, "Remarks upon Signing Bill to Establish the Biscayne National Monument," October 18, 1968, The American Presidency Project, https://www.presidency.ucsb.edu/documents/remarks-upon-signing-bill-es tablish-the-biscayne-national-monument. That bill is Public Law 90-606, 82 Stat. 1188.

57. Ibid.

58. "For the Future," *Pittsburgh Press*, October 3, 1968, 20.

Chapter 28: Taking Stock of New Conservation Wins

1. Scott Raymond Einberger, *With Distance in His Eyes: The Environmental Legacy of Stewart Udall*, 217–18.

2. Lyndon B. Johnson, "Remarks on Signing Bill to Establish the Biscayne National Monument," October 18, 1968, The American Presidency Project, https://www.presidency.ucsb.edu/documents/remarks-upon-signing-bill-es tablish-the-biscayne-national-monument.

3. Stewart L. Udall, *America's Natural Treasures: National Nature Monuments and Seashores* (Waukesha, WI: Country Beautiful Corporation, 1971), 23.

4. Lyndon B. Johnson, "Remarks upon Signing Bill to Establish the Biscayne National Monument," October 18, 1968, The American Presidency Project, https://www.presidency.ucsb.edu/documents/remarks-upon-signing-bill-es tablish-the-biscayne-national-monument.

5. Stewart Udall, *America's Natural Treasures*, 23.

6. Ibid.

7. Johnson, "Remarks upon Signing Bill to Establish the Biscayne National Monument."

8. Einberger, *With Distance in His Eyes*, 218–19.

9. Thomas G. Smith, *Stewart L. Udall: Steward of the Land* (Albuquerque: University of New Mexico Press, 2017), 286–87.

10. John P. Crevelli, "The Final Act of the Great Conservation President," *Prologue* 12, no. 4 (Winter 1980): 176–77.

11. Stewart Udall, "Notes re the Monument Proclamations," February 5, 1969, Folder 2, Box 182, Stewart L. Udall Papers, University of Arizona, Tucson, AZ.

12. Smith, *Stewart L. Udall*, 288.

13. Ibid, 289.

14. Smith, *Stewart L. Udall*, 289.

15. Stewart Udall, notes, August 1 and 3, 1968, Folder 3, Box 140, Stewart L. Udall Papers, University of Arizona, Tucson, AZ.

16. Lyndon B. Johnson to Stewart L. Udall, December 19, 1968, Udall Personal Files, LBJL.

17. "Transcript of Earthrise in 4K," NASA, https://svs.gsfc.nasa.gov/vis/a000000/a004500/a004593/earthrise_in_4k_transcript.html.

18. Quoted in Robert Poole, *Earthrise: How Man First Saw the Earth* (New Haven, CT: Yale University Press, 2010), 195.

19. Smith, *Stewart L. Udall*, 290.

20. William M. Blair, "Johnson Rebuffs Udall on Plan to Set Aside Vast Park Acreage," *New York Times*, January 21, 1969, 1.

21. Transcript, Stewart L. Udall Oral History Interview IV, October 31, 1969, by Joe B. Frantz, Internet Copy LBJ Library, 12.

22. Jeff Shesol, *Mutual Contempt: Lyndon Johnson, Robert Kennedy, and the Feud That Defined a Decade* (New York: W. W. Norton, 1998), 171–74.

23. Quoted in Smith, *Stewart L. Udall*, 291.

24. Quoted in "President Kennedy's Conservation Message—1962," *Sierra Club Bulletin*, May 1962, 12.

25. Author interview with Stewart Udall, July 12, 2007.

26. Einberger, *With Distance in His Eyes*, 68–96.

27. Author interview with Stewart L. Udall.

28. A. W. Eipper, C. A. Carlson, and L. S. Hamilton, "Impacts of Nuclear Power Plants on the Environment," *Living Wilderness* 34, no. 111 (Autumn 1970): 5–12.

29. Author interview with Stewart Udall, July 12, 2007. See also Barry Commoner, *Science and Survival* (New York: Viking Press: 1966).

30. Stephen Fox, *The American Conservation Movement: John Muir and His Legacy* (Boston: Little, Brown, 1981), 304.

31. Barry Commoner, *The Closing Circle: Nature, Man, and Technology* (New York: Alfred A. Knopf, 1971), 179.

32. Barry Commoner, "Super Technology . . . Will It End the Good Life?," *Field & Stream*, June 1970, 40, 59.

33. Commoner, *The Closing Circle*, 98–99.

34. Joe B. Frantz, "Stewart L. Udall Oral History Interview I," April 18, 1969, LBJ Library, http://www.lbjlibrary.net/assets/documents/archives/oral_histo ries/udall/UDALL01.PDF, 19–20.

35. J. Brooks Flippen, *Nixon and the Environment* (Lawrence: University of Kansas Press, 2000), 52.

36. Russell E. Train, *Politics, Pollution, and Pandas: An Environmental Memoir* (Washington, DC: Island Press, 2003), 53.

37. "Conversation with a Conservative: Russell Train," *Mother Jones*, October 21, 2004.

38. Ibid.

39. Stewart Udall, "The West's Defender of Wild Places," *Los Angeles Times*, July 12, 2005.

40. Stewart L. Udall, *1976: Agenda for Tomorrow* (New York: Harcourt, Brace & World, 1968), 114.

41. Smith, *Stewart L. Udall*, 291.

42. Doris Kearns, *Lyndon Johnson and the American Dream* (New York: St. Martin's Griffin, 1991).

Chapter 29: Santa Barbara, the Cuyahoga River, and the National Environmental Policy Act

1. Leo Janos, "The Last Days of the President," *Atlantic*, July 1, 1973.

2. Lady Bird Johnson and Carlton B. Lees, *Wildflowers Across America* (New York: Abbeville Press, 1988), 12.

3. David and Kay Scott, "Exploring the Parks: Tracing LBJ's Footsteps," National Parks Traveler, November 21, 2013, https://www.nationalparkstraveler .org/2013/11/exploring-parks-tracing-lbjs-footsteps24268.

4. Stuart Long, "L.B.J. Country Revisited," *New York Times*, December 14, 1975, p. XX7.

5. Evan Thomas, *Being Nixon: A Man Divided* (New York: Random House, 2016), 253.

6. "Russell E. Train: Oral History Interview," US Environmental Protection Agency, https://archive.epa.gov/epa/aboutepa/russell-e-train-oral-history-in terview.html.

7. Kate Wheeling and Max Ufberg, "'The Ocean Is Boiling': The Complete Oral History of the 1969 Santa Barbara Oil Spill," *Pacific Standard*, November 7,

2018, https://psmag.com/news/the-ocean-is-boiling-the-complete-oral-history
-of-the-1969-santa-barbara-oil-spill.

8. Russell E. Train, *Politics, Pollution, and Pandas: An Environmental Memoir* (Washington, DC: Island Press, 2003), 55.

9. Christine Mai-Duc, "The 1969 Santa Barbara Oil Spill That Changed Oil and Gas Exploration Forever," *Los Angeles Times*, May 20, 2015.

10. Wheeling and Ufberg, "'The Ocean Is Boiling.'"

11. Author interview with Denis Hayes, January 17, 2022.

12. Adam Rome, *The Genius of Earth Day: How a 1970 Teach-in Made the First Green Generation* (New York: Hill and Wang, 2013), 42.

13. John C. Whitaker, *Striking a Balance: Environment and National Resources Policy in the Nixon-Ford Years* (Washington, DC: American Enterprise Institute for Public Policy Research, 1976), 264.

14. Wheeling and Ufberg, "'The Ocean Is Boiling.'"

15. "Hickel, Walter J. 'Wally,' 1919–2010, Businessman, Governor of Alaska, and U.S. Secretary of the Interior," Anchorage 1910–1940 Legends & Legacies, https://www.alaskahistory.org/biographies/hickel-walter-j-wally/.

16. Dennis Hevesi, "Walter Hickel, Nixon Interior Secretary, Dies at 90," *New York Times*, May 9, 2010.

17. Train, *Politics, Pollution, and Pandas*, 54.

18. "State's Rules Held Better Designed to Prevent Oil Leaks," *Los Angeles Times*, February 6, 1969, 3.

19. Harry Trimborn, "Battle Shaping Up over Offshore Oil," *Los Angeles Times*, February 2, 1969.

20. "Oil-Slick Fighters Hopeful of Curbing Damage to Beaches," *New York Times*, February 3, 1969.

21. Gladwin Hill, "One Year Later, Impact of Great Oil Slick Is Still Felt," *New York Times*, January 25, 1970.

22. "Santa Barbara Disaster," *New York Times*, February 4, 1969.

23. David Dominick, *The Nixon Environmental Agenda: An Insider's View of Republican Decision-Making, 1968–72* (Conneaut Lake, PA: Page Publishing, 2020), 41.

24. Ibid.

25. Whitaker, *Striking a Balance*, 268.

26. Timothy Naftali, "Walter Hickel Interview Transcription," April 25, 2008, Richard Nixon Presidential Library and Museum, https://www.nixonlibrary.gov/sites/default/files/forresearchers/find/histories/hickel-2008-04-25.pdf.

27. Greg Lucas, "Ecological Disaster Creates Impetus for a New Ethos," Cal@170 by the California State Library, January 28, 1969, https://cal170.library.ca.gov/january-28-1969-an-ecological-disaster-and-an-impetus-for-a-new-ethos/.

28. Author interview with Russell Train, July 3, 2006.

29. Richard Nixon, "Remarks Following Inspection of Oil Damage at Santa Barbara Beach," online by Gerhard Peters and John T. Woolley, The American Presidency Project, https://www.presidency.ecsb.edu/node/239715.

30. Ibid.

31. Get Oil Out!, https://getoilout.org/.

32. Dwight Chapin to author, November 12, 2021. Also, author interview with Dwight Chapin, November 14, 2021.

33. Roderick Nash, "Santa Barbara Environmental Declaration," read January 28, 1970, at the January 28 Conference to Mark the First Anniversary of the Santa Barbara Oil Spill, https://es.ucsb.edu/santa-barbara-environmental -declaration#:~:text=We%20have%20contaminated%20the%20air,brought %20others%20close%20to%20annihilation.

34. Roderick Nash, *Wilderness in the American Mind* (New Haven, CT: Yale University Press, 1967).

35. Robert H. Boyle, "The Nukes Are in Hot Water," *Sports Illustrated*, January 20, 1969, 24–28, https://vault.si.com/vault/1969/01/20/the-nukes-are-in -hot-water.

36. Barry Commoner, *The Poverty of Power* (New York: Alfred A. Knopf, 1976), 90.

37. "Nixon's Nuclear Energy Vision," Richard Nixon Foundation, October 20, 2016, https://www.nixonfoundation.org/2016/10/26948/. Also see David Baker, "Nuclear Power's Last Stand in California: Will Diablo Canyon Die?" *San Francisco Chronicle*, November 14, 2015.

38. David Stradling and Richard Stradling, *Where the River Burned: Carl Stokes and the Struggle to Save Cleveland* (Ithaca, NY: Cornell University Press, 2015). Also see Meir Rinde, "Richard Nixon and the Rise of Environmentalism," *Distillations*, June 2, 2017.

39. Ibid., x.

40. "America's Sewage System and the Price of Optimism," *Time*, August 1, 1969, https://content.time.com/time/subscriber/article/0,33009,901182,00.html.

41. Stradling and Stradling, *Where the River Burned*, 150–51.

42. Dr. Seuss, *The Lorax* (New York: Random House, 1971).

43. "There's Nothing Smeary About Lake Erie Anymore," Ohio Sea Grant, March 28, 2019, https://ohioseagrant.osu.edu/news/2019/abfir/lorax-lake -erie.

44. "President Clinton: Celebrating America's Rivers," American Heritage Rivers, July 30, 1998, https://clintonwhitehouse3.archives.gov/CEQ/Rivers/.

45. Laura Johnston, "Cuyahoga Named River of the Year," *Cleveland Plain Dealer*, April 6, 2012.

46. Bruce Allen Murphy, *Wild Bill: The Legend and Life of William O. Douglas* (New York: Random House, 2003), 454.

47. Ibid.

48. Barry Commoner, *The Closing Circle: Nature, Man, and Technology* (New York: Alfred A. Knopf, 1971).

49. Stradling and Stradling, *Where the River Burned*, 150–51.

50. Christina Jarvis, "Tilting the Axis: Kurt Vonnegut and the Environment," The Daily Vonnegut, June 21, 2021, https://thedailyvonnegut.com/tilting-the-axis-kurt-vonnegut-and-the-environment-an-interview-with-christina-jarvis/.

51. Joshua P. Howe, *Behind the Curve: Science and the Politics of Global Warming* (Seattle: University of Washington Press), 47–55.

52. Quoted in Adam Rome, *The Genius of Earth Day* (New York: Macmillan, 2013), 57–54.

53. Robert Gottlieb, *Forcing the Spring: The Transformation of the American Environmental Movement* (Washington, DC: Island Press, 1993), 78.

54. Rome, *The Genius of Earth Day*, 67–72.

55. Janos, "The Last Days of the President."

56. Richard M. Nixon, "A Statement from President Nixon," *Fortune*, February 1970, 92.

57. James R. Hansen, *Dear Neil Armstrong: Letters to the First Man from All Mankind* (West Lafayette, IN: Purdue University Press, 2020), 6.

58. LeRoy Ashby and Rod Gramer, *Fighting the Odds: The Life of Senator Frank Church* (Pullman: Washington State University Press, 1994), 346.

59. René Dubos, *So Human an Animal: How We Are Shaped by Surroundings and Events* (New York: Scribner, 1968), 192.

60. Norman Mailer, *Of a Fire on the Moon* (Boston: Little, Brown, 1978), 186.

61. Grace Hood and Jim Hill, *CPR News*, Colorado Public Radio, September 6, 2019, https://www.cpr.org/2019/09/06/remember-the-first-time-colorado-tried-fracking-with-a-nuclear-bomb/.

62. "'The Ground Went Crazy': Activists Remember Project Rulison 50 Years Later," CBS Denver, September 10, 2019, https://denver.cbslocal.com/2019/09/10/project-rulison-nuclear-bomb-detonation/.

63. "Claims of Minor Damage Follow Underground Blast," *New York Times*, September 12, 1969.

64. Jennifer Thomson, "Surviving the 1970s: The Case of Friends of the Earth," *Environmental History* 22, no. 2 (April 1, 2017): 235–56.

65. David Brower, "David Brower's Farewell," *Wild Cascades*, June–July 1969, http://npshistory.com/newsletters/the-wild-cascades/june-july-1969.pdf.

66. Lawrence E. Davies, "Naturalists Get a Political Arm; Ex–Sierra Club Chief Gives Details on Voters League," *New York Times*, September 17, 1969.

67. "Projects," Friends of the Earth, foei.org (March 2015).

68. John McPhee, *Encounters with the Archdruid* (New York: Farrar, Straus and Giroux, 1980), 5–244.

69. Author interview with Steve Chapple, April 7, 2022.

70. Quoted in M. Margaret McKeown, *Citizen Justice: The Environmental Legacy of William O. Douglas, Public Advocate and Conservation Champion* (Dulles, VA: Potomac Books, 2022).

71. Michael McCloskey, *In the Thick of It: My Life in the Sierra Club* (Washington, DC: Island Press, 2005), 104.

72. Christopher D. Stone, *Should Trees Have Standing?: Law, Morality and the Environment* (New York: Oxford University Press, 2010), xiv.

73. Bill McKibben, *Falter: Has the Human Game Begun to Play Itself Out?* (New York: Macmillan, 2019), 73.

74. Stewart Udall, *1976: Agenda for Tomorrow* (New York: Harcourt, Brace & World, 1968), vii–viii, 30–31.

75. Irving Bernstein, *Guns or Butter: The Presidency of Lyndon Johnson* (New York: Oxford University Press, 1996), 542.

76. "Gaylord Nelson," The Wilderness Society, https://www.wilderness.org/articles/article/gaylord-nelson.

77. Daniel P. Moynihan, memorandum to John Ehrlichman, September 17, 1969, Richard Nixon Presidential Library and Museum, https://www.nixonlibrary.gov/sites/default/files/virtuallibrary/documents/jul10/56.pdf.

78. Hubert Heffner, memorandum to Daniel P. Moynihan, January 26, 1970, Richard Nixon Presidential Library and Museum, https://www.nixonlibrary.gov/sites/default/files/virtuallibrary/documents/jul10/55.pdf.

79. Phil Dougherty, "Van Nexx, William J. 'Bill' Jr., 1938–2017," HistoryLink.org, July 20, 2011, https://historylink.org/File/9882.

80. Ibid.

81. Author interview with Russell Train.

82. Richard Nixon, "Statement About the National Environmental Policy Act of 1969," The American Presidency Project, https://www.presidency.ucsb.edu/documents/statement-about-the-national-environmental-policy-act-1969.

83. Matthew J. Lindstrom and Zachary A. Smith, *The National Environmental Policy Act: Judicial Misconstruction, Legislative Indifference, and Executive Neglect* (College Station, TX: A&M University Press, 2001), 139.

84. National Environment Policy Act of 1969, 16 CFR, 1.82, section 4331, "Congressional Declaration of National Environmental Policy," https://uscode.house.gov/view.xhtml?path=/prelim@title42/chapter55&edition=prelim.

85. Quoted in Jack E. Davis, *The Bald Eagle: The Improbable Journey of America's Bird* (New York: W. W. Norton, 2022), 300.

86. Flippen, *Nixon and the Environment*, 53.

87. Danny C. Reinke, Lucy Swartz, and Lucinda Low Swartz, eds., *The NEPA Reference Guide* (Columbus, OH: Battelle Press, 1999), 88–90.

88. National Environmental Policy Act, as amended, https://www.fsa.usda.gov /Internet/FSA_File/nepa_statute.pdf.

89. Quoted in Wheeling and Ufberg, "The Ocean Is Boiling."

90. Sharon Buccino, "Trump Imperils Key Environmental Law," *New York Times*, January 13, 2020, A27.

91. Evan Thomas, *Being Nixon: A Man Divided* (New York: Random House, 2015), 252–53.

92. John Ehrlichman, White House notes, October 23, 1969, Ehrlichman Papers, Hoover Institution, Stanford University.

93. Michael Schaller and George Rising, *The Republican Ascendancy: American Politics, 1968–2001* (Wheeling, IL: Harlan Davidson, 2002), 29.

94. Thomas, *Being Nixon*, 253.

95. Lindstrom and Smith, *The National Environmental Policy Act*, 3–4.

Chapter 30: Generation Earth Day, 1970–1971

1. Richard Nixon, "Annual Message to the Congress on the State of the Union," January 22, 1970, The American Presidency Project, https://www .presidency.ucsb.edu/documents/annual-message-the-congress-the-state -the-union-2.

2. Ibid.

3. Quoted in Meir Rinde, "Richard Nixon and the Rise of American Environmentalism," *Distillations* (June 2, 2017).

4. John A. Farrell, *Richard Nixon: The Life* (New York: Doubleday, 2017), 380.

5. Douglas Brinkley, *Cronkite* (New York: HarperCollins, 2012), 432–36.

6. Ken Kaye, "Grand Vision of Jetport Unrealized," *South Florida Sun-Sentinel*, July 6, 1997.

7. James C. Clark, "Politics Moved Nixon, but Fla. Reaped Environmental Benefits," *Orlando Sentinel*, September 7, 2014.

8. Harold K. Steen, "An Interview with Russell E. Train," Forest History Society, 1993, https://foresthistory.org/wp-content/uploads/2016/12/Train_Russell _E.ohi_.pdf, 31–32.

9. Robert B. Semple, Jr., "Everglades Jetport Barred by a U.S.-Florida Accord," *New York Times*, January 16, 1970.

10. "The Saving of the Everglades," *New York Times*, January 18, 1970.

11. William Ruckelshaus, "A New Shade of Green," *Wall Street Journal*, April 17, 2010.

12. Lynton Keith Caldwell, *The Nation's Environmental Policy Act: An Agenda for the Future* (Bloomington: Indiana University Press, 1998), 32.

13. Quoted in Steve Johnson, *World in Their Hands: Original Thinkers, Doers, Fighters, and the Future of Conservation* (Lanham, MD: Rowman & Littlefield, 2021), 151.

14. Nixon, "Annual Message to the Congress on the State of the Union," January 22, 1970.

15. Richard Nixon, "Special Message to the Congress on Environmental Quality," February 10, 1970, The American Presidency Project, https://www.presidency.ucsb.edu/documents/special-message-the-congress-environmental-quality.

16. Michael E. Gorn, "Russell E. Train: Oral History Interview," May 5, 1992, US Environmental Protection Agency, https://archive.epa.gov/epa/aboutepa/russell-e-train-oral-history-interview.html.

17. "Gaylord Nelson," The Wilderness Society, https://www.wilderness.org/articles/article/gaylord-nelson.

18. Adam Rome, *The Genius of Earth Day* (New York: Macmillan, 2013), 77.

19. Author interview with Denis Hayes, January 12, 2022.

20. "Angry Coordinator of Earth Day," *New York Times*, April 23, 1970, 30, https://www.nytimes.com/1970/04/23/archives/angry-coordinator-of-earth-day-denis-allen-hayes.html.

21. Author interview with Denis Hayes, January 12, 2022.

22. Ibid.

23. Rome, *The Genius of Earth Day*, 91.

24. Rinde, "Richard Nixon and the Rise of American Environmentalism."

25. Bill Christofferson, *The Man From Clear Lake: Earth Day Founder Senator Gaylord Nelson* (Madison: University of Wisconsin Press, 2009), 310.

26. "History of Earth Day," CNN, http://edition.cnn.com/EVENTS/1996/earth_day/stories/history/index.html.

27. David Bird, "Udall Says Nation Must Curb Growth to Spare Environment," *New York Times*, January 15, 1970, 6, https://www.nytimes.com/1970/01/16/archives/udall-says-nation-must-curb-growth-to-spare-environment.html.

28. Author interview with Stewart Udall, September 6, 2004.

29. Stephen Fox, *The American Conservation Movement: John Muir and His Legacy* (Boston: Little, Brown, 1981), 313.

30. J. Brooks Flippen, *Nixon and the Environment* (Albuquerque: University of New Mexico Press, 2000), 9.

31. Author interview with William Ruckelshaus, August 4, 2011.

32. Quoted in J. Brooks Flippen, *Conservative Conservationist: Russell E. Train and the Emergence of American Environmentalism* (Baton Rouge: Louisiana State University Press, 2006), 97.

33. Joseph Lelyveld, "Mood Is Joyful as City Gives Its Support; Millions Join Earth day Observations," *New York Times*, April 23, 1970, 1.

34. Gladwin Hill, "Activity Ranges from Oratory to Legislation," *New York Times*, April 23, 1970, https://www.nytimes.com/1970/04/23/archives/activity-ranges-from-oratory-to-legislation-oratory-and-legislation.html.

35. William Ruckelshaus, "New Solutions for New Environmental Problems," *Wall Street Journal*, April 17, 2010, https://www.google.com/search?q=William+Ruckelshaus+%22EPA%22+wall+street+journal&oq=William+Ruckelshaus+%22EPA%22+wall+street+journal&aqs=chrome.69i57.11209j0j9&sourceid=chrome&ie=UTF-8.

36. Michael McCloskey, *In the Thick of It: My Life in the Sierra Club* (Washington, DC: Island Press, 2005), 105.

37. John G. Parsons, *Chesapeake & Ohio Canal National Historical Park, District of Columbia/Maryland: General Plan* (Washington, DC: US Department of the Interior, National Park Service, 1976), 4.

38. Christofferson, *The Man from Clear Lake*, 3–6.

39. Quoted in Tim Palmer, *Endangered Rivers and the Conservation Movement* (Berkeley: University of California Press, 1986), 147.

40. LeRoy Ashby and Rod Gramer, *Fighting the Odds: The Life of Senator Frank Church* (Seattle: Washington State University Press, 1994), 346.

41. Author interview with Mary Elizabeth "Liz" Carpenter, March 1, 2007.

42. Martin V. Melosi, "Lyndon Johnson and Environmental Policy," in *The Johnson Years*, vol. 2, *Vietnam, the Environment, and Science*, ed. Robert A. Divine (Lawrence: University of Kansas Press, 1987), 113.

43. White House Press Office, 6311-11a, April 22, 1970, Richard Nixon Presidential Library and Museum.

44. Author interview with Denis Hayes, April 22, 2022.

45. Becky Oskin, "Earth Day, 1970: How President Nixon Spied on Earth Day," *Christian Science Monitor*, April 22, 2013, https://www.csmonitor.com/USA/2013/0422/Earth-Day-1970-How-President-Nixon-spied-on-Earth-Day.

46. Elisabeth Reuther Dickmeyer, email to author, April 29, 2022.

47. John H. Adams and Patricia Adams, *A Force for Nature: The Story of the National Resources Defense Council and Its Fight to Save Our Planet* (San Francisco: Chronicle Books), 12–30.

48. Patricia Sullivan, "Senator Founded Earth Day in 1970," *Washington Post*, July 4, 2005, A8.

49. Marvin Gaye, "Mercy, Mercy Me (The Ecology)," on *What's Going On*, 1971.

50. Three Dog Night, "Out in the Country," on *It Ain't Easy*, 1970.

51. Joni Mitchell, "Big Yellow Taxi," on *Ladies of the Canyon*, 1970.

52. Rome, *The Genius of Earth Day*, 240.

53. Ibid., 259–60.

54. Barry Weisberg, ed., *Ecocide in Indochina: The Ecology of War* (San Francisco: Harper & Row, 1970).

55. Quoted in Meir Rinde, "Richard Nixon: The Rise of American Environmentalism."

56. Ed O'Keefe, "Ex–Interior Secretary Walter Hickel Dies at 90," *Washington Post*, May 8, 2010, http://voices.washingtonpost.com/federal-eye/2010/05/ex-interior_secretary_walter_h.html.

57. James M. Naughton, "Nixon Proposes 2 New Agencies on Environment," *New York Times*, July 10, 1970, https://www.nytimes.com/1970/07/10/archives/nixon-proposes-2-new-agencies-on-environment-major-reshuffling.html.

58. Daniel A. Farber, "The Conservative as Environmentalist: From Goldwater and the Early Reagan to the 21st Century," *Arizona Law Review* 59 (2017), https://papers.ssrn.com/sol3/papers.cfm?abstract_id=2919633.

59. Robert T. Nelson, "Nixon's Environmental Hero," *Seattle Times*, February 16, 1999, https://archive.seattletimes.com/archive/?date=19990216&slug=2944577.

60. E. W. Kenworthy, "Sierra Club and Muskie Accuse the Administration of Disregarding New Environmental Policy Act," *New York Times*, February 22, 1970, https://www.nytimes.com/1970/02/22/archives/sierra-club-and-muskie-accuse-the-administration-of-disregarding.html.

61. Author interview with Russell Train, July 3, 2006.

62. Richard Nixon, "Special Message to the Congress About Reorganization Plans to Establish the Environmental Protection Agency and the National Oceanic and Atmosphere Administration," July 9, 1970, The American Presidency Project, https://www.presidency.ucsb.edu/documents/special-message-the-congress-about-reorganization-plans-establish-the-environmental.

63. James Nestor, *Deep: Freediving, Renegade Science, and What the Ocean Tells Us About Ourselves* (New York: Houghton Mifflin Harcourt, 2014), 3.

64. Richard D. Lyons, "5 Women Named Aquanaut Team," *New York Times*, March 3, 1970, https://www.nytimes.com/1970/03/03/archives/5-women-named-aquanaut-team-hickel-says-foreign-group-will-join.html.

65. "Aquanauts Jolted by Quake but Habitat Is Undamaged," *New York Times*, July 9, 1970, 75, https://www.nytimes.com/1970/07/09/archives/aquanauts-jolted-by-quake-but-habitat-is-undamaged.html.

66. Sylvia A. Earle, "The Sweet Spot in Time: Why the Ocean Matters to Everyone," *Virginia Quarterly Review* 88, no. 4 (Fall 2012): 54–77, https://www.vqronline.org/essay/sweet-spot-time.

67. Richard West Sellars, *Preserving Nature in the National Parks* (New Haven, CT: Yale University Press, 1997), 208.

68. Alfred Runte, *Yosemite: The Embattled Wilderness* (Omaha: University of Nebraska Press, 1990), 202.

69. Michael W. Childers, "The Stoneman Meadow Riots and Law Enforcement in Yosemite National Park," *Forest History Today*, Spring 2017, 28–34, https://foresthistory.org/wp-content/uploads/2017/10/Childers_Stoneman.pdf.

70. Michael W. Childers, "The Stoneman Meadow Riots and the Problem of Law Enforcement," blogwest.org.

71. Robert A. Jones, "National Parks: A Report on the Range War at Generation Gap," *New York Times*, July 25, 1971, XX1.

72. John Fisher to Richard Nixon, July 22, 1970, Yosemite National Park Archive (YNPA), El Portal, California.

73. Jones, "National Parks: A Report on the Range War at Generation Gap."

74. Childers, "The Stoneman Meadow Riots and Law Enforcement in Yosemite National Park," 32.

75. "The Rise of Anti-Ecology," *Time*, August 3, 1970.

76. Rinde, "Richard Nixon and the Rise of American Environmentalism."

77. "William D. Ruckelshaus: Oral History Interview," Environmental Protection Agency, 1993, https://archive.era.gov/epa/aboutepa/William-d-ruck elshaus-oral-history-interview.html/.

78. Robert W. Collin, *The Environmental Protection Agency: Cleaning Up America's Act* (Westport, CT: Greenwood Press, 2006).

79. "William D. Ruckelshaus: Oral History Interview," Environmental Protection Agency, 1993, https://archive.epa.gov/epa/aboutepa/william-d-ruck elshaus-oral-history-interview.html.

80. William D. Ruckelshaus, "Choosing our Common Future: Democracy's True Test," Fifth Annual John H. Chafee Memorial Lecture on Science and the Environment, February 3, 2005, National Council for Science and the Environment, https://www.gcseglobal.org/sites/default/files/inline-files/Chafee MemorialLecture_2005.pdf.

81. "As EPA Turns 40, IU Professor Recalls Its Creation," IU Newsroom, https:// newsinfo.iu.edu/news/page/normal/16660.html.

82. "Celebrating 35 Years of the EPA with William D. Ruckelshaus," Indiana University, https://oneill.indiana.edu/special/images/epa-35.pdf, 11–12.

83. Author interview with Dennis Hayes, April 22, 2022.

84. "Enter Rogers Morton," *New York Times*, November 27, 1970.

85. Author interview with Russell Train, July 3, 2006.

86. Diana Rico, "The Final Battle: How the Taos Pueblo Indians Won Back Their Blue Lake Shrine," http://newmexicohistory.org/2012/06/27/1970-taos-blue -lake-returned-to-pueblo/.

87. "The Return of Blue Lake," https://taospueblo.com/blue-lake/.

88. Matthew Van Buren, "Taos Pueblo Celebrates 40th Anniversary of Blue Lake's Return," *Taos News*, September 18, 2010.

89. Winthrop Griffith, "The Taos Indians Have a Small Generation Gap," *New York Times*, February 21, 1971.

90. Robert Gottlieb, *Forcing the Spring: The Transformation of the American Environmental Movement* (Washington, DC: Island Press, 1993), 128.

91. Allan Hall, "Interview with Barry Commoner," *Scientific American*, June 23, 1997, https://www.scientificamerican.com/article/interview-with-barry -comm/.

92. "Dear Mr. Ruckelshaus" (Sierra Club advertisement), *Los Angeles Times*, March 28, 1973, 18.

93. Charles W. Colson (Special Counsel to the President), memorandum to Dwight Chapin (Special Assistant to the President), January 12, 1971, Richard Nixon Presidential Library and Museum. Also see Brigham Daniels, Andrew P. Follett, and Joshua Davis, "The Making of the Clean Air Act," *Hastings Law Journal* 71, no. 4 (2020), https://repository.uchastings.edu /hastings_law_journal/vol71/iss4/3.

94. William O. Douglas, *The Three Hundred Year War: A Chronicle of Ecological Disaster* (New York: Random House, 1972), 98.

95. William O. Douglas, *Farewell to Texas: A Vanishing Wilderness* (New York: McGraw-Hill, 1967), vii–ix.

96. John McPhee, *Encounters with the Archdruid* (New York: Farrar, Straus and Giroux, 1971), 158–59.

97. Quoted in David Dominick, *The Nixon Environmental Agenda: An Insider's View of Republican Decision Making 1968–1972* (Conneaut Lake, PA: Page, 2020).

98. Dina Gilio-Whitaker, *As Long as Grass Grows: The Indigenous Fight for Environmental Justice, from Colonization to Standing Rock* (Boston: Beacon Press, 2019), 19.

99. Richard Nixon, "Annual Message to the Congress on the State of the Union," January 22, 1971, The American Presidency Project, https://www.presidency .ucsb.edu/documents/annual-message-the-congress-the-state-the-union-1.

100. History, Art & Archives, US House of Representatives, Office of the Historian, *Black Americans in Congress, 1870–2007* (Washington, DC: US Government Printing Office, 2008); "Creation and Evolution of the Congressional Black Caucus," United States House of Representatives, https://history .house.gov/Exhibitions-and-Publications/BAIC/Historical-Essays/Permanent -Interest/Congressional-Black-Caucus/.

101. Representative Ronald V. Dellums to the Norwegian Nobel Committee, January 20, 1978, quoted in Tom Turner, *David Brower: The Making of the Environmental Movement* (Berkeley: University of California Press, 2015), 192.

102. "Open Letter to President Nixon," May 14, 1971, Box 27, Folder 10, David Brower Papers, Bancroft Library, University of California, Berkeley. See also Jennifer Thompson, "Surviving the 1970s: The Case of Friends of the Earth," *Environmental History* (2017), 236.

103. Barry Commoner, *The Closing Circle: Nature, Man, and Technology* (New York: Alfred A. Knopf, 1971), 12–185.

104. Wendell Berry, *Wendell Berry: Essays, 1969–1990*, ed. Jack Shoemaker (New York: Library of America, 2019), 142.

105. Ginger Strand, "The Crying Indian," *Orion*, https://orionmagazine.org/article /the-crying-indian/.

106. "Iron Eyes Cries," United Press International, October 20, 1984, https://www .upi.com/Archives/1984/10/20/IRON-EYES-CRIES/4658467092800/.

107. Strand, "The Crying Indian."

108. "Iron Eyes Cries," United Press International, October 20, 1984, https:// www.upi.com/Archives/1984/10/20/IRON-EYES-CRIES/4658467092800/.

109. Bartow J. Elmore, "The American Beverage Industry and the Development of Curbside Recycling Programs; 1950–2000," *Business History Review* (August 2012), 477–501.

110. Richard Nixon Foundation, "President Nixon's Legacy of Parks," April 24, 2020, https://www.nixonfoundation.org/2020/04/president-nixons-legacy -parks/.

111. Richard Nixon, "Statement About the Legacy of Parks Program, January 7, 1974," The American Presidency Project, https://www.presidency.ucsb.edu /documents/statements-about-the-legacy-parks-program-0.

112. Cynthia Barnett, *Mirage: Florida and the Vanishing Water of the Eastern U.S.* (Ann Arbor: University of Michigan Press, 2007), 77.

113. Flippen, *Nixon and the Environment*, 156.

Chapter 31: Nixon's Environmental Activism of 1972

1. Derra Davis, *When Smoke Ran like Water: Tales of Environmental Deception and the Battle against Pollution* (New York: Basic Books, 2002), 95.

2. H. R. Haldeman, "Unpublished Diaries," Richard M. Nixon Presidential Library and Museum, Yorba Linda, CA.

3. Paul Charles Milazzo, *Unlikely Environmentalists: Congress and Clean Water, 1945–1972* (Lawrence: University Press of Kansas, 2006), 229.

4. Benjamin Kline, *First Along the River: A Brief History of the Environmental Movement* (Lanham, MD: Rowman & Littlefield, 2011), 89.

5. "U.S. Seeks to Curb Leaded Gasoline," *New York Times*, February 23, 1972, 81.

6. Robert B. Semple, Jr., "Great Lakes Pact Signed in Ottawa By Nixon, Trudeau," *New York Times*, April 16, 1972.

7. Gladwin Hill, "Nixon and Trudeau to Sign an Agreement to Fight Great Lakes Pollution," *New York Times*, April 9, 1972.

8. Author interview with William Ruckelshaus, August 4, 2011.

9. Gerald M. Stern, *The Buffalo Creek Disaster: How the Survivors of One of the Worst Disasters in Coal-Mining History Brought Suit Against the Coal Company—and Won* (New York: Vintage Books, 2008).

10. *Disaster on Buffalo Creek: A Citizens' Report on Criminal Negligence in a West Virginia Mining Community* (Charlestown, WV: np, 1972).

11. Stern, *The Buffalo Creek Disaster*, 11.

12. "Buffalo Creek Flood of '72: Why Environmental Disasters Are Nothing New in West Virginia," Nomadic Politics, January 24, 2014, https://nomadicpoli tics.blogspot.com/2014/01/buffalo-creek-flood-of-72-why.html.

13. *Disaster on Buffalo Creek: A Citizens' Report on Criminal Negligence in a West Virginia Mining Community* (Charleston, WV: np, 1972), 10.

14. Neil Compton to J. William Fulbright, March 16, 1972, quoted in Compton, *The Battle for the Buffalo River: A Twentieth-Century Conservation Crisis in the Ozarks* (Fayetteville: University of Arkansas Press, 1992), 465.

15. Marilyn Weeks, "Grande Dame of Conservation," *Fort Lauderdale News*, June 8, 1979, 23.

16. Dyan Zaslowsky, "Along Arkansas's Pristine Waterway," *New York Times*, September 12, 1993.

17. E. W. Kenworthy, "Environmental Act Irks House Panel," *New York Times*, April 30, 1972.

18. Stewart Udall and Jeff Stansbury, "Ecology Agency Head Has Blotted His Copy Book," *Bangor Daily News*, April 1, 1971, 23.

19. United States Congress, Senate Committee on Environment and Public Works, *The Role of Science in Environmental Policy Making*, vol. 4 (Washington, DC: Government Printing Office, 2005), 128, https://archive.org/de tails/gov.gpo.fdsys.CHRG-109shrg38918.

20. Timothy R. Smith, "William D. Ruckelshaus, Who Refused to Join in Nixon's 'Saturday Night Massacre,' Dies at 87," *Washington Post*, November 27, 2019.

21. William Ruckelshaus, oral interview, https://archive.epa.gov/epa/aboutepa /william-d-ruckelshaus-oral-history-interview.html.

22. Nixon's handwritten comments on Daily News Summaries, June 15, 1972, in folder "Annotated News Summaries June 7–22, 1972," 1 of 2, Box 40, President's office files, WHSF, Richard Nixon Presidential Library and Museum.

23. Gladwin Hill, "Planning Completed for U.N. Environment Conference," *New York Times*, March 12, 1972.

24. M. A. Farber, "Environmental Congress Head Asks U.N. to Implement Plans," *New York Times*, October 20, 1972, 10.

25. J. Brooks Flippen, *Nixon and the Environment* (Albuquerque: University of New Mexico Press, 2000).

26. John C. Esposito, *Vanishing Air: The Ralph Nader Study Group Report on Air Pollution* (New York: Grossman, 1970), 259–98.

27. Robert N. Rickles, "Issues 1972," *New York Times*, October 14, 1972, 33, https://www.nytimes.com/1972/10/14/archives/article-1-no-title.html.

28. Jack Rosenthal, "Times Study: Debates Hurt McGovern," *New York Times*, June 8, 1972, 1.

29. Zygmunt J. B. Plater, *The Snail Darter and the Dam: How Pork-Barrel Politics Endangered a Little Fish and Killed a River* (New Haven, CT: Yale University Press, 2013), 64.

30. "To Protect Dolphins," *New York Times*, May 30, 1972, 36.

31. "Senators Are Given Ratings for Environmental Voters," *New York Times*, October 22, 1972.

32. Donald Janson, "The 1972 Campaign," *New York Times*, October 20, 1972.

33. Ibid.

34. Ellen Simon, "The Bipartisan Beginnings of the Clean Water Act," Waterkeeper, January 30, 2019, https://waterkeeper.org/news/bipartisan-beginnings-of-clean-water-act/.

35. Flippen, *Nixon and the Environment*, 182.

36. Quoted in Simon, "The Bipartisan Beginnings of the Clean Water Act."

37. Quoted in ibid.

38. E. W. Kenworthy, "Clean-Water Bill Is Law Despite President's Veto," *New York Times*, October 19, 1972, 26.

39. Adam Rome, *The Genius of Earth Day* (New York: Macmillan, 2013), 73.

40. Mark Simon, "The Co-Founder of Earth Day Recounts 50-Year Movement," *Daily Journal* (San Mateo, CA), April 22, 2020.

41. Kevin M. Benck, Kathy Allen, et al., *Fossil Butte National Monument: Natural Resource Condition Assessment* (Fort Collins, CO: US Department of the Interior, 2017), 5–17.

42. Emil W. Haury, *The Hohoham: Dersert Farmers and Craftsmen* (Tucson: University of Arizona Press, 1976), ix, 150–55. The Hohokam Pima Monument is not open to the public.

43. Joe Roman, *Listed Dispatches from America's Endangered Species Act*, 50–51.

44. "Coastal Zone Management Act," Office for Coastal Management, National Oceanic and Atmospheric Administration, https://coast.noaa.gov/czm/act/.

45. Flippen, *Nixon and the Environment*, 178.

46. Richard Walker, *The Country in the City: The Greening of the San Francisco Bay Area* (Seattle: University of Washington Press, 2008), 94.

47. Ibid., 95.

48. Quoted in John Jacobs, *A Rage for Justice: The Passion and Politics of Phillip Burton* (Berkeley: University of California Press, 1995), 351.

49. Author interview with Russell Train, July 3, 2006.

50. John McPhee, *Encounters with the Archdruid* (New York: Farrar, Straus and Giroux, 1971).

51. Mary Bullard, *Cumberland Island: A History* (Athens: University of Georgia Press, 2003), 12–13; see also Lary M. Dilsaver, *Cumberland Island National*

Seashore: A History in Conservation Conflict (Charlottesville: University of Virginia Press, 2004).

52. J. Brooks Flippen, *Nixon and the Environment* (Albuquerque: University of New Mexico Press, 2000), 184.

Epilogue: Last Leaves on the Tree

1. Quoted in Rose Houk, *Heart's Home: Lyndon B. Johnson's Hill Country* (Tucson: Western National Parks Association, 1986), 31.

2. Lyndon Baines Johnson, *The Vantage Point: Perspectives of the Presidency, 1963–1969* (New York: Holt, Rinehart and Winston, 1971), 336.

3. Lady Bird Johnson and Carlton B. Lees, *Wildflowers Across America* (New York: Abbeville Press, 1988), 12.

4. Lewis C. Gould, "Lady Bird Johnson and Beautification," in *The Johnson Years*, vol. 2, *Vietnam, the Environment, and Science*, ed. Robert A. Divine (Lawrence: University of Kansas Press, 1987), 172.

5. "Environmental First Lady," Lady Bird Johnson Wildflower Center at University of Texas at Austin, August 1, 2007, https://www.wildflower.org/magazine/people/environmental-first-lady.

6. Richard M. Nixon, "Second Inaugural Address of Richard Milhous Nixon," January 20, 1973, Lillian Goldman Law Library, Yale Law School, https://avalon.law.yale.edu/20th_century/nixon2.asp.

7. Tricia Nixon to author, October 27, 2020.

8. *Endangered Species Act of 1973, as Amended Through 2018* (Washington, DC: Department of the Interior, 2018), 1.

9. Author interview with Russell Train, July 3, 2006.

10. Mark V. Barrow, *Nature's Ghosts: Confronting Extinction from the Age of Jefferson to the Age of Ecology* (Chicago: University of Chicago Press, 2009), 338–40.

11. "Lee M. Talbot, Ph.D.," Endangered Species Coalition, https://www.endangered.org/campaigns/wild-success-endangered-species-act-at-40/lee-m-talbot/.

12. Author interview with Russell Train, July 3, 2006.

13. Lupi Saldana, "California's Military Bases 'Last Frontiers for Wildlife,'" *Los Angeles Times*, January 30, 1973, 31.

14. "Learn About Richard Nixon's Legacy of Parks Program!" Richard Nixon Library and Museum, April 27, 2020, https://www.nixonlibrary.gov/news/learn-about-president-nixons-legacy-parks-program#:~:text=Celebrating%20National%20Park%20Week%3A&text=That%20was%20a%20time%20when,the%20Legacy%20of%20Parks%20program.

15. Quoted in Zygmunt J. B. Plater, *The Snail Darter and the Dam: How Pork-Barrel Politics Endangered a Little Fish and Killed a River* (New Haven, CT: Yale University Press, 2013), 65.

16. Richard Nixon, "State of the Union Message to the Congress on Natural Resources and the Environment," February 15, 1973, The American Presidency Project, https://www.presidency.ucsb.edu/documents/state-the-union-message-the-congress-natural-resources-and-the-environment.

17. Will Lissner, "Environmentalists Fear a Shift to Coal," *New York Times*, December 9, 1973, 70, https://www.nytimes.com/1973/12/09/archives/environmentalists-fear-a-shift-to-coal-litter-problem-found-wasted.html.

18. Gladwin Hill, "There Have Been Some Setbacks, But the Damage Isn't Permanent," *New York Times*, December 23, 1973, 106.

19. Tommy Winston, "Science, Practice, and Policy: The Committee on Rare and Endangered Wildlife Species Policy, 1956–1973" (PhD, Arizona State University, May 2011).

20. Richard Nixon, "Statement on Signing the Endangered Species Act of 1973," December 28, 1973, The American Presidency Project, https://www.presidency.ucsb.edu/documents/statement-signing-the-endangered-species-act-1973.

21. "Saving the World's Wildlife," *Washington Post*, February 19, 1973; see also Shannon Peterson, *Acting for Endangered Species: The Statutory Ark* (Lawrence: University of Kansas Press, 2002), 30.

22. Faith McNulty, *The Whooping Crane: The Bird That Defies Extinction* (New York: E. P. Dutton, 1966); McNulty, *Must They Die?: The Strange Case of the Prairie Dog and the Black-Footed Ferret* (Garden City, NY: Doubleday, 1971).

23. "An Endangered Act," *Washington Post*, December 29, 2003, A16.

24. Barrow, *Nature's Ghosts*, 339–42.

25. Robin W. Doughty, *Return of the Whooping Crane* (Austin: University of Texas Press, 1989), 84.

26. Jim Kushlan, director of Patuxent Wildlife Resarch Center, to Douglas Brinkley, May 2, 2022.

27. "U.S. Fish and Wildlife Service Proposes Delisting 23 Species from Endangered Species Act Due to Extinction," US Department of the Interior, September 29, 2021, https://www.doi.gov/pressreleases/us-fish-and-wildlife-service-proposes-delisting-23-species-endangered-species-act-due.

28. Quoted in J. Brooks Flippen, *Conservative Conservationist: Russell E. Train and the Emergence of American Environmentalism* (Baton Rouge: Louisiana State University Press, 2006), 156.

29. Gerald R. Ford, "President Gerald R. Ford's Proclamation 4311, Granting a Pardon to Richard Nixon," September 8, 1974, Gerald R. Ford Presidential Library & Museum, https://www.fordlibrarymuseum.gov/library/speeches/740061.asp.

30. Author interview with Peter Ehrlichman, May 29, 2022

31. Leonard Garment, "John Ehrlichman: Other Legacy," *New York Times*, February 17, 1999.

32. Richard Nixon, *RN: The Memoirs of Richard Nixon* (New York: Grosset and Dunlap, 1978).

33. Monica Crowley, *Nixon in Winter* (New York: Random House, 1998).

34. Flippen, *Conservative Conservationist*, 89.

35. James E. Bishop, "Russell Train," in George A. Cevasco and Richard P. Hammond, *Modern American Environmentalists: A Biographical Encyclopedia* (Baltimore: Johns Hopkins University Press, 2009), 505–9.

36. Author interview with William Ruckelshaus, August 4, 2011.

37. Robert D. McFadden, "William Ruckelshaus, Who Quit in 'Saturday Night Massacre,' Dies at 87," *New York Times*, November 27, 2019.

38. "Biography of William D. Ruckelshaus: First Term," US Environmental Protection Agency, https://archive.epa.gov/epa/aboutepa/biography-william-d-ruckelshaus-first-term.html.

39. Quoted in Otis L. Graham, Jr., *Presidents and the American Environment* (Lawrence: University Press of Kansas, 2015), 243.

40. Linda Charlton, "Officials and Nature Join to Hail Justice Douglas," *New York Times*, May 18, 1977, 27.

41. Richard Cohen, "The Man Who Saved the C&O Canal," *Washington Post*, May 22, 1977, 25.

42. Barry Commoner, *The Closing Circle: Nature, Man, and Technology* (New York: Alfred A. Knopf, 1971), 206.

43. Author interview with Denis Hayes, April 6, 2022.

44. LeRoy Ashby and Rod Gramer, *Fighting The Odds: The Life of Senator Frank Church* (Pullman: Washington State University Press, 1984).

45. Graham, *Presidents and the American Environment*, 242–43.

46. Monica Crowley to author, April 29, 2022.

47. Julia Sweig, *Lady Bird Johnson: Hiding in Plain Sight* (New York: Random House, 2021), 424.

48. Scott Raymond Einberger, *With Distance in His Eyes: The Environmental Life and Legacy of Stewart Udall* (Reno: University of Nevada Press, 2018).

49. Stewart L. Udall, *The Quiet Crisis and the Next Generation* (Layton, UT: Gibbs Smith, 1991), 275–62.

50. Stewart L. Udall, *The Myths of August: A Personal Exploration of Our Tragic Cold War Affair with the Atom* (New York: Pantheon, 1994), 30.

51. Ibid., chapter 11 title.

52. Thomas G. Smith, *Stewart L. Udall: Steward of the Land* (Albuquerque: University of New Mexico Press, 2017), 316–17.

53. Einberger, *With Distance in His Eyes*, 240–51.

54. Smith, *Stewart L. Udall*, 318.

55. "Radiation Exposure Compensation Act," Public Law 101-426, October 15, 1990.

56. Author interview with Stewart Udall, September 7, 2009.

57. Ibid.

58. Stewart and Lee Udall, "A Message to Our Grandchildren," *High Country News*, March 31, 2008, https://www.hcn.org/issues/367/17613.

Bibliography

The following are the secondary sources I found of exceptional value when writing *Silent Spring Revolution*. I profited from many other titles, but this list constitutes the essential ones. Likewise, stories found in the *New York Times, Wall Street Journal, Washington Post, Christian Science Monitor,* and *Los Angeles Times* (published between 1945 to 1973) were indispensable. In addition, relevant issues of *Time, Newsweek,* the *New Yorker,* the *Atlantic, Harper's, Sports Illustrated, National Geographic, Sierra Club Bulletin,* the *Living Wilderness,* and *Defenders of Wildlife News* helped anchor my research.

Abbey, Edward. *Desert Solitaire: A Season in the Wilderness.* New York: Ballantine Books, 1968.

———. *One Life at a Time, Please.* New York: Henry Holt, 1978.

Adams, Ansel. *In the National Parks.* New York: Little, Brown and Company, 2010.

———. *These We Inherit: The Parkland of America.* San Francisco: Sierra Club Books, 1962.

———. *This Is the American Earth.* San Francisco: Sierra Club Books, 1960.

Adams, John H., and Patricia Adams. *A Force for Nature: The Story of NRDC and the Fight to Save Our Planet.* San Francisco: Chronicle Books, 2010.

Alinder, Mary Street. *Ansel Adams: A Biography.* New York: Bloomsbury, 2014.

Alonso, Harriet Hyman. *Peace as a Woman's Issue: A History of the U.S. Movement for World Peace and Women's Rights.* Syracuse, NY: Syracuse University Press, 1993.

Anderson, Clinton P. *Outsider in the Senate: Senator Clinton Anderson's Memoirs.* New York: World Publishing, 1970.

Anderson, Larry. *Benton MacKaye: Conservationist, Planner, and Creator of the Appalachian Trail.* Baltimore: Johns Hopkins University Press, 2002.

Ashby, LeRoy, and Rod Gramer. *Fighting the Odds: The Life of Senator Frank Church.* Pullman: Washington State University Press, 1994.

Backes, David. *A Wilderness Within: The Life of Sigurd F. Olson.* Minneapolis: University of Minnesota Press, 1997.

Baker, Richard A. *Conservation Politics: The Senate Career of Clinton P. Anderson.* Albuquerque: University of New Mexico Press, 1985.

Ball, Howard. *Justice Downwind: America's Atomic Testing Program in the 1950s.* New York: Oxford University Press, 1986.

Bardacke, Frank. *Trampling Out the Vintage: Cesar Chavez and the Two Souls of the NFWA.* New York: Verso, 2011.

Barnett, Cynthia. *Blue Revolution: Unmaking America's Water Crisis.* New York: Beacon Press, 2011.

———. *Mirage: Florida and the Vanishing Water of the Eastern U.S.* Ann Arbor: University of Michigan Press, 2008.

———. *Rain: A Natural and Cultural History.* Studio City: Crown Media, 2015.

Barrow, Mark V., Jr. *Nature's Ghosts: Confronting Extinction from the Age of Jefferson to the Age of Ecology.* Chicago: University of Chicago Press, 2009.

Becker, William S. *The Creeks Will Rise: People Coexisting with Floods.* Chicago: Chicago Review Press, 2021.

Benson, Jackson J. *Wallace Stegner: His Life and Work.* New York: Viking, 1996.

Bernstein, Irving. *Guns or Butter: The Presidency of Lyndon Johnson.* New York: Oxford University Press, 1996.

Berry, Wendell. *New Collected Poems.* Berkeley, CA: Counterpoint, 2012.

———. *Think Little: Essays.* Berkeley, CA: Counterpoint, 2019.

Berry, Wendell, and Gene Meatyard. *The Unforeseen Wilderness: An Essay on Kentucky's Red River Gorge.* Lexington: University Press of Kentucky, 1971.

Beschloss, Michael R. *The Crisis Years: Kennedy and Khrushchev, 1960–1963.* New York: HarperCollins, 1991.

Beston, Henry. *The Outermost House: A Year of Life on the Great Beach of Cape Cod.* New York: Henry Holt, 1928.

Boyd, Karl Brooks. *Before Earth Day: The Origins of Environmental Law, 1945–1970.* Lawrence: University Press of Kansas, 2009.

Brabazon, James. *Albert Schweitzer: A Biography.* 2nd ed. Syracuse, NY: Syracuse University Press, 2000.

Brinkley, Douglas. *The Quiet World: Saving Alaska's Wilderness Kingdom, 1879–1960.* New York: HarperCollins, 2011.

———. *Rightful Heritage: Franklin D. Roosevelt and the Land of America.* New York: HarperCollins, 2016.

———. *The Wilderness Warrior: Theodore Roosevelt and the Crusade for America.* New York: HarperCollins, 2010.

Brooks, Karl B. *Public Power, Private Dams: The Hells Canyon High Dam Controversy.* Seattle: University of Washington Press, 2009.

Brooks, Paul. *The House of Life: Rachel Carson at Work.* Boston: Houghton Mifflin, 1972.

Brower, David R. *For Earth's Sake: The Life and Times of David R. Brower.* Salt Lake City: Peregrine Smith Books, 1990.

———. *Work in Progress.* Salt Lake City: Peregrine Smith Books, 1991.

Brower, David R., ed. *Gentle Wilderness: The Sierra Nevada.* San Francisco: Sierra Club Books, 1964.

———, ed. *The Meaning of Wilderness to Science.* San Francisco: Sierra Club Books, 1960.

———, ed. *Not Man Apart.* San Francisco Books, 1965.

———, ed. *The Sierra Club Wilderness Handbook.* San Francisco: Sierra Club Books, 1961.

———, ed. *Wildlands in Our Civilization.* San Francisco: Sierra Club Books, 1964.

Brower, David, with Steve Chapple. *Let the Mountains Talk, Let the Rivers Run: A Call to Those Who Would Save the Earth*. San Francisco: HarperCollins West, 1995.

Brower, Kenneth. *The Wildness Within: Remembering David Brower*. Berkeley, CA: Heyday, 2012.

Brugge, Doug, Timothy Bannaly, and Esther Yazzie-Lewis, eds. *The Navajo People and Uranium Mining*. Albuquerque: University of New Mexico Press, 2007.

Bullard, Robert D. *Dumping in Dixie: Race, Class, and Environmental Quality*. Boulder, CO: Westview Press, 1990.

Burns, Ken, and Dayton Duncan. *National Parks: America's Best Idea*. New York: Alfred A. Knopf, 2009.

Cahalan, James M. *Edward Abbey: A Life*. Tucson: University of Arizona Press, 2001.

Caldwell, Keith Lynton. *The National Environmental Policy Act: An Agenda for the Future*. Bloomington: University of Indiana Press, 1998.

Califano, Joseph A., Jr. *Governing America: An Insider's Report from the White House and the Cabinet*. New York: Simon & Schuster, 1981.

Caro, Robert. *The Power Broker: Robert Moses and the Fall of New York*. New York: Vintage, 1975.

———. *The Years of Lyndon Johnson: Master of the Senate*. New York: Vintage, 2002.

———. *The Years of Lyndon Johnson: The Passage of Power*. New York: Vintage, 2012.

———. *The Years of Lyndon Johnson: The Path to Power*. New York: Vintage, 1982.

Carpenter, Liz. *Ruffles and Flourishes*. College Station: Texas A&M University Press, 1993.

Carr, Archie. *A Naturalist in Florida: A Celebration of Eden*. New Haven, CT: Yale University Press, 1994.

———. *The Reptiles*. New York, Time, 1963.

Carson, Rachel. *The Edge of the Sea*. Boston: Mariner Books, 1955.

———. *The Sea Around Us*. Oxford, UK: Oxford University Press, 1951.

———. *The Sense of Wonder*. New York: Harper and Row, 1965.

———. *Under the Sea-Wind: A Naturalist's Picture of Ocean Life*. New York: Simon & Schuster, 1941.

Clarke, Thurston. *The Last Campaign: Robert F. Kennedy and 82 Days That Inspired America*. New York: Henry Holt, 2008.

Coatsworth, Elizabeth. *Especially Maine: The Natural World of Henry Beston from Cape Cod to the St. Lawrence*. Brattleboro, VT: Stephen Green Press, 1970.

Cohen, Michael P. *The History of the Sierra Club, 1892–1970*. San Francisco: Sierra Club Books, 1988.

Commoner, Barry. *The Closing Circle: Nature, Man & Technology*. New York: Alfred A. Knopf, 1971.

Conis, Elena. *How to Sell a Poison: The Rise, Fall, and Toxic Return of DDT*. New York: Bold Type Books, 2022.

Cornell, Virginia. *Defender of the Dunes: The Kathleen Goddard Jones Story*. Carpinteria, CA: Manifest, 2001.

Cousins, Norman. *The Improbable Triumvirate: John F. Kennedy, Pope John, and Nikita Khrushchev*. New York: W. W. Norton, 1972.

———. *Modern Man Is Obsolete*. New York: Viking Press, 1945.

Cox, Patrick. *Ralph W. Yarborough: The People's Senator*. Austin: University of Texas Press, 2001.

Cronon, William, ed. *Uncommon Ground: Rethinking the Human Place in Nature.* New York: W. W. Norton, 1996.

Dallek, Robert. *Flawed Giant: Lyndon Johnson and His Times, 1961–1973.* New York and Oxford, UK: Oxford University Press, 1998.

Davis, Frederick Rowe. *The Man Who Saved Sea Turtles: Archie Carr and the Origins of Conservation Biology.* Oxford, UK: Oxford University Press, 2007.

Davis, Jack E. *The Bald Eagle: The Improbable Journey of America's Bird.* New York: W. W. Norton, 2022.

———. *An Everglades Providence: Marjory Stoneman Douglas and the American Environmental Century.* Athens: University of Georgia Press, 2009.

———. *The Gulf: The Making of an American Sea.* New York: Liveright Publishing, 2017.

De Steiguer, J. E. *The Age of Environmentalism.* New York: McGraw-Hill, 1997.

DeBell, Garrett, ed. *The Environmental Handbook.* New York: Ballantine/Friends of the Earth Books, 1970.

Desmann, Raymond F. *The Destruction of California.* New York: Macmillan, 1965.

Douglas, William O. *Beyond the High Himalayas.* New York: Doubleday & Company, 1952.

———. *The Court Years, 1939–1975: The Autobiography of William O. Douglas.* New York: Random House, 1980.

———. *Farewell to Texas: A Vanishing Wilderness.* New York: McGraw-Hill, 1967.

———. *Go East, Young Man: The Early Years: The Autobiography of William O. Douglas.* New York: Random House, 1974.

———. *A Living Bill of Rights.* New York: Doubleday, 1961.

———. *Muir of the Mountains.* New York: Houghton Mifflin, 1961.

———. *My Wilderness: East to Katahdin.* New York: Doubleday, 1961.

———. *My Wilderness: The Pacific West.* New York: Doubleday, 1960.

———. *Of Men and Mountains.* New York: Harper, 1950.

———. *Points of Rebellion.* New York: Random House, 1970.

———. *The Three Hundred Year War: A Chronicle of Ecological Disaster.* New York: Random House, 1972.

———. *A Wilderness Bill of Rights.* Boston: Little Brown, 1965.

Dunbar-Ortiz, Roxanne. *An Indigenous Peoples' History of the United States.* Boston: Beacon Press, 2014.

Dunlap, Thomas R., ed. *DDT, Silent Spring, and the Rise of Environmentalism.* Seattle: Washington University Press, 2008.

Earle, Sylvia A. *Sea Change: A Message of the Oceans.* New York: Fawcett Books, 1995.

———. *The World is Blue: How Our Fate and the Ocean's Fate Are One.* Washington, DC: National Geographic, 2009.

Egan, Dan. *The Death and Life of the Great Lakes.* New York: W. W. Norton & Company, 2017.

Egan, Michael. *Barry Commoner and the Science of Survival: The Remaking of American Environmentalism.* Cambridge: MIT Press, 2007.

Egan, Timothy. *Lasso the Wind: Away to the New West.* New York: Vintage Books, 2009.

———. *The Worst Hard Time.* New York: Houghton Mifflin, 2006.

Ehrlich, Gretel. *The Solace of Open Spaces.* New York: Penguin Books, 1985.

Ehrlich, Paul. *The Population Bomb.* San Francisco: Sierra Club Books, 1969.

Einberger, Scott Raymond. *With Distance in His Eyes: The Environmental Life and Legacy of Stewart Udall.* Reno: University of Nevada Press, 2018.

Elmore, Bartow J. *Citizen Coke: The Making of Coca-Cola Capitalism.* New York: W. W. Norton, 2016.

Farmer, Jared. *Glen Canyon Dammed: Inventing Lake Powell and the Canyon Country.* Tucson: University of Arizona Press, 1999.

Farrell, John A. *Richard Nixon: The Life.* New York: Doubleday, 2017.

Feldman, James W. *A Storied Wilderness: Rewilding the Apostle Islands.* Seattle: University of Washington Press, 2011.

Feldman, Noah. *Scorpions: The Battles and Triumphs of FDR's Great Supreme Court Justices.* New York: Hachette Book Group, 2010.

Fiege, Mark. *The Republic of Nature: An Environmental History of the United States.* Seattle: University of Washington Press, 2012.

Finch, L. Boyd. *Legacies of Camelot: Stewart and Lee Udall, American Culture, and the Arts.* Norman: University of Oklahoma Press, 2008.

Flippen, J. Brooks. *Conservative Conservationist: Russell E. Train and the Emergence of American Environmentalism.* Baton Rouge: Louisiana State University Press, 2006.

———. *Nixon and the Environment.* Albuquerque: University of New Mexico Press, 2000.

Flores, Dan. *Coyote America: A Natural and Supernatural History.* New York: Basic Books, 2016.

Foreman, Dave. *Confessions of an Eco-Warrior.* New York: Harmony Books, 1991.

Fox, Stephen. *The American Conservation Movement: John Muir and His Legacy.* Boston: Little, Brown, 1981.

Fradkin, Philip L. *Fallout: An American Nuclear Tragedy.* Tucson: University of Arizona, 1989.

———. *A River No More: The Colorado River and the West.* New York, Alfred A. Knopf, 1981.

———. *Wallace Stegner and the American West.* New York: Alfred A. Knopf, 2008.

Franklin, Kay, and Norma Schaeffer. *Duel for the Dunes: Land Use Conflict on the Shores of Lake Michigan.* Urbana: University of Illinois Press, 1983.

Free, Ann Cotrell. *Animals, Nature, and Albert Schweitzer.* Washington, DC: Flying Fox Press, 1982.

Freeman, Martha, ed. *Always Rachel: The Letters of Rachel Carson and Dorothy Freeman, 1952–1964.* New York: Beacon Press, 1995.

Frome, Michael. *Strangers in High Places: The Story of the Great Smoky Mountains.* Knoxville: University of Tennessee Press, 1966.

Galbraith, John Kenneth. *The Affluent Society.* Boston: Mariner Books, 1958.

Gessner, David. *All the Wild That Remains: Edward Abbey, Wallace Stegner, and the American West.* New York and London: W. W. Norton, 2015.

Gilio-Whitaker, Dina. *As Long as Grass Grows: The Indigenous Fight for Environmental Justice from Colonization to Standing Rock.* Boston: Beacon Press, 2019.

Gillam, Harold. *Island in Time: The Point Reyes Peninsula.* San Francisco: Sierra Club, 1962.

Gioielli, Robert R. *Environmental Activism and the Urban Crisis.* Philadelphia: Temple University Press, 2014.

Glover, James M. *A Wilderness Original: The Life of Bob Marshall.* Seattle: Mountaineers, 1986.

Goodwin, Doris Kearns. *Lyndon Johnson and the American Dream*. New York: Harper and Row, 1976.

Goodwin, Richard N. *Remembering America: A Voice from the Sixties*. New York: Open Road Integrated Media, 1988.

Gottlieb, Robert. *Forcing the Spring: The Transformation of the American Environmental Movement*. Washington, DC: Island Press, 1993.

Gould, Lewis L. *Lady Bird Johnson: Our Environmental First Lady*. Lawrence: University Press of Kansas, 1988.

Graham, Frank, Jr. *Since Silent Spring*. Boston: Houghton Mifflin, 1970.

Grant, Kim. *Cape Cod, Martha's Vineyard, and Nantucket*. Woodstock, VT: Countryman Press, 1995.

Graves, John. *Goodbye to a River: A Narrative*. New York: Curtis Publishing Company, 1960.

Hacker, Barton C. *Elements of Controversy: The Atomic Energy Commission and Radiation Safety in Nuclear Weapons Testing, 1947–1984*. Berkeley: University of California Press, 1994.

Hamilton, Nigel. *JFK: Reckless Youth*. New York: Random House, 1992.

Hartig, John H. *Burning Rivers: Revival of Four Urban-Industrial Rivers that Caught Fire*. Ontario, Canada: Aquatic Ecosystem Health & Management Society, 2010.

Hartzog, George B., Jr. *Battling for the National Parks*. Mt. Kisco, NY: Moyer Bell, 1988.

Harvey, Mark W. T. *A Symbol of Wilderness: Echo Park and the American Conservation Movement*. Albuquerque: University of New Mexico Press, 1994.

———. *Wilderness Forever: Howard Zahniser and the Path to the Wilderness Act*. Seattle: University of Washington Press, 2005.

Harvey, Mark W. T., ed. *The Wilderness Writings of Howard Zahniser*. Seattle: University of Washington Press, 2014.

Hays, Samuel P. *A History of Environmental Politics Since 1945*. Pittsburgh: University of Pittsburgh Press, 2000.

Hiltzik, Michael. *Colossus*. New York: Free Press, 2010.

Hirt, Paul W. *A Conspiracy of Optimism: Management of the National Forests Since World War Two*. Lincoln: University of Nebraska Press, 1994.

Hoagland, Edward. *Hoagland on Nature*. Guilford, CT: Lyons Press, 2003.

———. *Red Wolves and Black Bears: Nineteen Essays*. New York: Random House, 1976.

Hogan, Michael J. *The Afterlife of John Fitzgerald Kennedy: A Biography*. Cambridge, UK: Cambridge University Press, 2017.

Holzer, Harold. *The Presidents vs. the Press: The Endless Battle Between the White House and the Media—From the Founding Fathers to Fake News*. New York: Penguin Random House, 2020.

Huser, Verne. *Rivers of Texas*. College Station, TX: Texas A&M University Press, 2000.

Hyde, Philip, and François Leydet. *The Last Redwoods*. San Francisco: Sierra Club Books, 1964.

Hynes, H. Patricia. *The Recurring Silent Spring*. Elmsford, NY: Pergamon Press, 1989.

Jett, Stephen C., and Philip Hyde. *Navajo Wildlands: As Long as the Rivers Shall Flow*. San Francisco: Sierra Club Books, 1967.

Johnson, Lady Bird. *A White House Diary*. New York: Holt, Rinehart & Winston, 1970.

Johnson, Lady Bird, and Carlton B. Lees. *Wildflowers Across America*. New York: Abbeville Press, 1988.

Johnson, Lyndon Baines. *The Vantage Point: Perspectives of the Presidency, 1963–1969*. New York: Holt, Rinehart & Winston, 1971.

Johnson, Rich. *The Central Arizona Project, 1918–1968*. Tucson: University of Arizona Press, 1977.

Jones, Charles O. *Clean Air: The Policies and Politics of Pollution Control*. Pittsburgh: University of Pittsburgh Press, 1975.

Josephy, Alvin M., Jr., Joane Nagel, and Troy Johnson, eds. *Red Power: The American Indians' Fight for Freedom*, 2nd ed. Lincoln: University of Nebraska Press, 1999.

Kaufman, Scott. *Project Plowshare: The Peaceful Use of Nuclear Explosives in Cold War America*. Ithaca, NY: Cornell University Press, 2013.

Kennedy, Robert F., Jr. *American Values: Lessons I Learned from My Family*. New York: HarperCollins, 2018.

Kerouac, Jack. *The Dharma Bums*. New York: Viking Press, 1958.

Kline, Benjamin. *First Along the River: A Brief History of the U.S. Environmental Movement*. Lanham, MD: Rowman & Littlefield, 2011.

Koehn, Nancy. *Forged in Crisis: The Making of Five Courageous Leaders*. New York: Scribner, 2017.

Kolbert, Elizabeth. *The Sixth Extinction: An Unnatural History*. New York: Henry Holt, 2014.

Krutch, Joseph Wood. *The Desert Year*. New York: Viking Press, 1951.

———. *Grand Canyon: Today and All Its Yesterdays*. New York: William Morrow and Company, 1957.

———. *The Voice of the Desert: A Naturalist's Interpretation*. New York: William Morrow and Co., 1955.

Krutch, Joseph Wood, and Eliot Porter. *Baja California and the Geography of Hope*. San Francisco: Sierra Club Books, 1967.

Kurlansky, Mark. *Salmon: A Fish, the Earth, and the History of Their Common Fate*. Ventura, CA: Patagonia Press, 2020.

Kushlan, James A. *Seeking the American Tropics: South Florida's Early Naturalists*. Gainesville: University of Florida Press, 2020.

Larson, Derek R. *Keeping Oregon Green: Liveability, Stewardship, and the Challenges of Growth, 1960–1980*. Corvallis: Oregon State University, 2016.

Lazarus, Richard J. *The Making of Environmental Law*. Chicago: University of Chicago Press, 2004.

Lear, Linda. *Rachel Carson: Witness for Nature*. New York: Henry Holt, 1997.

Lear, Linda, ed. *Lost Woods: The Discovered Writing of Rachel Carson*. Boston: Beacon Press, 1998.

Levine, Ellen. *Up Close: Rachel Carson*. New York: Puffin Books, 2007.

Lewis, Michael, ed. *American Wilderness: A New History*. New York: Oxford University Press, 2007.

Leydet, François. *Time and the River Flowing: Grand Canyon*. San Francisco: Sierra Club, 1964.

Lichtenstein, Nelson. *Walter Reuther: The Most Dangerous Man in Detroit*. Urbana: University of Illinois Press, 1995.

Lien, Carsten. *Olympic Battleground: The Power Politics of Timber Preservation*. San Francisco: Sierra Club Books, 1991.

Lifset, Robert D. *Power on the Hudson: Storm King Mountain and the Emergence of Modern American Environmentalism*. Pittsburgh: University of Pittsburgh Press, 2014.

Lindstrom, Matthew J., and Zachary A. Smith. *The National Environmental Policy Act: Judicial Misconstruction, Legislative Indifference, and Executive Neglect*. College Station: Texas A&M Press, 1995.

Logevall, Fredrik. *JFK: Coming of Age in the American Century, 1917–1956*. New York: Random House, 2020.

Louter, David. *Windshield Wilderness: Cars, Roads, and Nature in Washington's National Parks*. Seattle: University of Washington Press, 2006.

Manning, Harvey. *The Wild Cascades: Forgotten Parkland*. San Francisco: Sierra Club Books, 1965.

Martin, Russell. *A Story That Stands Like a Dam: Glen Canyon and the Struggle for the Soul of the West*. New York: Henry Holt, 1989.

Maynard, Barksdale. *Walden Pond: A History*. New York: Oxford University Press, 2004.

McCay, Mary A. *Rachel Carson*. New York: Twayne Publishers, 1993.

McCloskey, Michael. *In the Thick of It: My Life in the Sierra Club*. Washington, DC: Island Press, 2005.

McKibben, Bill. *American Earth: Environmental Writing Since Thoreau*. New York: Library of America, 2006.

———. *The Bill McKibben Reader: Pieces From an Active Life*. New York: Henry Holt, 2008.

———. *The End of Nature*. New York: Random House, 1989.

———. *Falter: Has the Human Game Begun to Play Itself Out?* New York: Henry Holt, 2019.

McPhee, John. *Annals of the Former World*. New York: Farrar, Straus and Giroux, 1981.

———. *Encounters with the Archdruid*. New York: Farrar, Straus and Giroux, 1971.

Meine, Curt. *Aldo Leopold: His Life and Work*. Madison: University of Wisconsin Press, 1988.

Melosi, Martin V. *Fresh Kills: A History of Consuming and Discarding in New York City*. New York: Columbia University Press, 2020.

Melosi, Martin V., and Joseph A. Pratt. *Energy Metropolis: An Environmental History of Houston and the Gulf Coast*. Pittsburgh: University of Pittsburgh Press, 2007.

Meyer, Marvin, and Kurt Bergel, eds. *Reverence for Life*. Syracuse, NY: Syracuse University Press, 2002.

Meyers, Jeffrey. *Robert Frost: A Biography*. Boston and New York: Houghton Mifflin, 1996.

Milazzo, Paul Charles. *Unlikely Environmentalists: Congress and Clean Water, 1945–1972*. Lawrence: University of Kansas Press, 2006.

Miller, Char. *Gifford Pinchot and the Making of Modern Environmentalism*. Washington, DC: Island Press, 2001.

———. *On the Edge: Water, Immigration, and Politics in the Southwest*. San Antonio, TX: Trinity University Press, 2013.

Miller, Char, and Hal Rothman, eds. *Out of the Woods: Essays in Environmental History*. Pittsburgh: University of Pittsburgh Press, 1997.

Miller, Richard L. *Under the Cloud: The Decades of Nuclear Testing*. Woodlands, TX: Two-Sixty Press, 1991.

Mittlefehldt, Sarah. *Tangled Roots: The Appalachian Trail and American Environmental Politics*. Seattle: University of Washington Press, 2013.

Montrie, Chad. *Making a Living: Working and Environment in the United States*. Chapel Hill: University of North Carolina Press, 2009.

Murphy, Priscilla Coit. *What a Book Can Do: The Publication and Reception of* Silent Spring. Amherst: University of Massachusetts Press, 2005.

Musil, Robert K. *Rachel Carson and Her Sisters: Extraordinary Women Who Have Shaped America's Environment*. New Brunswick, NJ: Rutgers University Press, 2014.

Nash, Roderick. *The Rights of Nature: A History of Environmental Ethics*. Madison: University of Wisconsin Press, 1989.

———. *Wilderness and the American Mind*. 5th ed. New Haven, CT: Yale University Press, 2014.

Nash, Roderick, ed. *Grand Canyon of the Living Colorado*. San Francisco: Sierra Club and Ballantine Books, 1970.

Newhall, Nancy. *Ansel Adams*. Vol. 1, *The Eloquent Light*. San Francisco: Sierra Club Books, 1963.

O'Fallon, James M., ed. *Nature's Justice: Writings of William O. Douglas*. Corvallis: Oregon State University Press, 2000.

Oreskes, Naomi, and Erik M. Conway. *Merchants of Doubt: How a Handful of Scientists Obscured the Truth on Issues From Tobacco Smoke to Climate Change*. New York: Bloomsbury Publishing, 2010.

Palmer, Tim. *The Wild and Scenic Rivers of America*. Washington, DC: Island Press, 1993.

Pearson, Byron E. *Still the Wild River Runs: Congress, the Sierra Club, and the Fight to Save Grand Canyon*. Tucson: University of Arizona Press, 2002.

Peters, Greg M. *Our National Forests: Stories from America's Most Important Public Lands*. Portland, OR: Timber Press, 2021.

Peterson, Shannon. *Acting for Endangered Species: The Statutory Ark*. Lawrence: University of Kansas Press, 2002.

Poole, Leslie Kemp. *Saving Florida: Women's Fight for the Environment in the Twentieth Century*. Tallahassee: University of Florida Press, 2010.

Porter, Eliot. *The Place No One Knew: Glen Canyon on the Colorado*. David Brower, ed. San Francisco: Sierra Club Books, 1963.

Quammen, David. *E. O. Wilson: Biophilia, the Diversity of Life, Naturalist*. New York: Literary Classics of the United States, 2021.

Quaratiello, Arlene R. *Rachel Carson: A Biography*. Westport, CT: Greenwood Press, 2004.

Reisner, Marc. *Cadillac Desert: The American West and Its Disappearing Water*. New York: Viking Penguin, 1986.

Remnick, David, and Henry Finder. *The Fragile Earth: Writing From the* New Yorker *on Climate Change*. New York: Ecco, 2020.

Rhodes, Richard. *The Making of the Atom Bomb*. New York: Simon & Schuster, 1968.

Rogers, Heather. *Gone Tomorrow: The Hidden Life of Garbage.* New York: The New Press, 2005.

Roman, Joe. *Listed: Dispatches From America's Endangered Species Act.* Cambridge, MA: Harvard University Press, 2011.

Rome, Adam. *The Bulldozer in the Countryside: Suburban Sprawl and the Rise of American Environmentalism.* Cambridge, UK: Cambridge University Press, 2001.

———. *The Genius of Earth Day: How a 1970 Teach-in Unexpectedly Made the First Green Generation.* New York: Hill and Wang, 2013.

Rosenblith, Walter A. *Jerry Wiesner: Scientist, Statesman, Humanist.* Cambridge, MA: MIT Press, 2003.

Ross, Benjamin, and Steven Amter. *The Polluters: The Making of Our Chemically Altered Environments.* Oxford, UK: Oxford University Press, 2010.

Rothman, Hal K. *LBJ's Texas White House.* College Station: Texas A&M University Press, 2001.

Russell, Jan Jarboe. *Lady Bird: A Biography of Mrs. Johnson.* New York: Scribner, 1999.

Sabin, Paul. *Public Citizens: The Attack on Big Government and the Remaking of American Liberalism.* New York: W. W. Norton, 2021.

Sachs, Jeffrey D. *To Move the World: JFK's Quest for Peace.* New York: Random House, 2013.

Schrepfer, Susan R. *The Fight to Save the Redwoods: A History of Environmental Reform, 1917–1978.* Madison: University of Wisconsin Press, 1983.

Schulte, Steven C. *Wayne Aspinall and the Shaping of the American West.* Boulder: University of Colorado Press, 2002.

Schuyler, David. *Embattled Water: The Hudson and Modern American Environmentalism.* Ithaca, NY: Cornell University Press, 2018.

Schwartz, William, ed. *Voices for the Wilderness.* New York: Ballantine Books, 1969.

Scott, Doug. *The Enduring Wilderness: Protecting Our National Heritage Through the Wilderness Act.* Golden, CO: Fulcrum, 2004.

Seaborg, Glenn T. *Kennedy, Khrushchev, and the Test Ban.* Berkeley: University of California Press, 1981.

Sellars, Richard West. *Preserving Nature in the National Parks: A History.* New Haven, CT: Yale University Press, 1997.

Shaw, Randy. *Beyond the Fields: Cesar Chavez, the NFWA, and the Struggle for Justice in the 21st Century.* Berkeley: University of California Press, 2008.

Simpson, Marcus B., Jr. *Bird Life of North Carolina's Shining Rock Wilderness.* Raleigh: North Carolina Biological Survey and North Carolina State Museum of Natural Sciences, 1994.

Smith, David A. *Cowboy Presidents: The Frontier Myth and U.S. Politics Since 1900.* Norman: University of Oklahoma Press, 2021.

Smith, Thomas G. *Green Republican: John Saylor and the Preservation of America's Wilderness.* Pittsburgh: University of Pittsburgh Press, 2006.

———. "John Kennedy, Stewart Udall, and New Frontier Conservation." *Pacific Historical Review* 64 (August 1995): 329–62.

———. "Robert Frost, Stewart Udall, and the 'Last Go Down.'" *New England Quarterly* 70 (March 1997): 3–32.

———. *Showdown: JFK and the Integration of the Washington Redskins.* Boston: Beacon Press, 2011.

———. *Stewart L. Udall: Steward of the Land.* Albuquerque: University of New Mexico Press, 2017.

Snyder, Gary. *Dimensions of Life.* San Francisco: Sierra Club Books, 1991.

Sollen, Robert. *An Ocean of Oil: A Century of Political Struggle Over Petroleum Off the California Coast.* Juneau, AK: Denali Press, 1998.

Souder, William. *On a Farther Shore: The Life and Legacy of Rachel Carson, Author of "Silent Spring."* New York: Crown, 2012.

Sowards, Adam M. *The Environmental Justice, William O. Douglas and American Conservation.* Corvallis: Oregon State University Press, 2009.

Spence, Mark David. *Dispossessing the Wilderness: Indian Removal and the Making of the National Parks.* New York: Oxford University Press, 1999.

Stegner, Wallace. *Beyond the Hundredth Meridian: John Wesley Powell and the Second Opening of the West.* New York: Penguin, 1953.

———. *The Sound of Mountain Water: The Changing American West.* New York: Penguin, 1946.

Stegner, Wallace, ed. *This Is Dinosaur: Echo Park Country and Its Magic Rivers.* New York: Alfred A. Knopf, 1955.

Steinberg, Ted. *Gotham Unbound: The Ecological History of Greater New York.* New York: Simon & Schuster, 2014.

Sterling, Philip. *Sea and Earth: The Life of Rachel Carson.* New York: Thomas Y. Crowell Company, 1970.

Stern, Gerald, M. *The Buffalo Creek Disaster: How the Survivors of One of the Worst Disasters in Coal-Mining History Brought Suit against the Coal Company—and Won.* New York: Vintage, 2008.

Stone, Christopher D. *Should Trees Have Standing? Law, Morality, and the Environment.* 3rd ed. Oxford, UK: Oxford University Press, 2010.

Stradling, David. *The Nature of New York: An Environmental History of the Empire State.* Ithaca, UK: Cornell University Press, 2010.

Sturgeon, Stephen C. *The Politics of Western Water: The Congressional Career of Wayne Aspinall.* Tucson: University of Arizona Press, 2002.

Suiter, John. *Poets and Peaks: Gary Snyder, Philip Whalen and Jack Kerouac in the North Cascades.* Washington, DC: Counterpoint, 2002.

Sutter, Paul. *Driven Wild: How the Fight against Automobiles Launched the Modern Wilderness Movement.* Seattle: University of Washington Press, 2002.

Svensson, Patrik. *The Book of Eels: Our Enduring Fascination with the Most Mysterious Creature in the Natural World.* New York: HarperCollins, 2019.

Sweig, Julia. *Lady Bird Johnson: Hiding in Plain Sight.* New York: Random House, 2021.

Swerdlow, Amy. *Women Strike for Peace: Traditional Motherhood and Radical Politics in the 1960s.* Chicago: University of Chicago Press, 1993.

Tarr, Joel A. *Devastation and Renewal: An Environmental History of Pittsburgh and its Region.* Pittsburgh: University of Pittsburgh Press, 2003.

Thomas, Evan. *Being Nixon: A Man Divided.* New York: Random House, 2015.

Thoreau, Henry David, with photographs by Eliot Porter. *In Wildness Is the Preservation of the World.* San Francisco: Sierra Club Books, 1962.

Titus, Constandina A. *Bombs in the Backyard: Atomic Testing and American Politics.* Reno: University of Nevada Press, 1986.

Train, Russell, E. *Politics, Pollution, and Pandas: An Environmental Memoir.* Washington, DC: Island Press, 2003.

Treuer, David. *The Heartbeat of Wounded Knee: Native America From 1890 to the Present*. New York: Riverhead Books, 2019.

Turner, James Morton. *The Promise of Wilderness: American Environmental Politics Since 1964*. Seattle: University of Washington Press, 2012.

Turner, Tom. *David Brower: The Making of the Environmental Movement*. Berkeley: University of California Press, 2015.

———. *Sierra Club: 100 Years of Protecting Nature*. New York: Abrams, 1991.

Udall, Stewart L. *The Myths of August: A Personal Exploration of Our Tragic Cold War Affair with the Atom*. New York: Pantheon Books, 1994.

———. *The National Parks of America*. New York: G. P. Putnam's Sons, 1966.

———. *1976: Agenda for Tomorrow*. New York: Harcourt, Brace & World, 1963.

———. *The Quiet Crisis*. New York: Avon Books, 1963.

———. *The Quiet Crisis and the Next Generation*. Layton, UT: Gibbs Smith, 1988.

Vogt, William. *Road to Survival*. New York: William Sloane Associates, Inc., 1948.

Walker, Richard A. *The Country in the City: The Greening of the San Francisco Bay Area*. Seattle: University of Washington Press, 2007.

Washington, Harriet A. *A Terrible Thing to Waste: Environmental Racism and its Assault on the American Mind*. New York: Little, Brown Spark, 2019.

Wells, Christopher. *Car Country: An Environmental History*. Seattle: University of Washington Press, 2012.

Welsh, Michael. *Big Bend National Park: Mexico, the United States, and a Borderland Ecosystem*. Reno: University of Nevada Press, 2021.

Whitaker, John C. *Striking a Balance: Environment and Natural Resources Policy in the Nixon-Ford Years*. Washington, DC: American Enterprise Institute for Public Policy Research, 1976.

Wilkinson, Alec. *The Protest Singer: An Intimate Portrait of Pete Seeger*. New York: Vintage Books, 2010.

Williams, Terry Tempest. *Red: Passion and Patience in the Desert*. New York: Vintage Books, 2001.

———. *Refuge: An Unnatural History of Family and Place*. New York: Pantheon, 1991.

Wilson, Robert. H., Norman J. Glickman, and Laurence E. Lynn, Jr., eds. *LBJ's Neglected Legacy: How Lyndon Johnson Reshaped Domestic Policy and Government*. Austin: University of Texas Press, 2015.

Winkler, Allan M. *Life Under a Cloud: American Anxiety About the Atom*. Chicago: University of Illinois Press, 1993.

Winks, Robin W. *Laurance S. Rockefeller: Catalyst for Conservation*. Washington, DC: Island Press, 1997.

Wirth, Conrad L. *Parks, Politics, and the People*. Norman: University of Oklahoma Press, 1980.

Wittner, Lawrence S. *The Struggle Against the Bomb*. Vol. 1, *One World or None: A History of the World Nuclear Disarmament Movement Through 1953*. Stanford, CA: Stanford University Press, 1993.

———. *The Struggle Against the Bomb*. Vol. 2, *Resisting the Bomb: A History of the World Nuclear Disarmament Movement, 1954–1970*. Stanford, CA: Stanford University Press, 1997.

Woods, Randall B. *Prisoners of Hope: Lyndon B. Johnson, the Great Society, and the Limits of Liberalism*. New York: Basic, 2016.

Worster, Donald. *Rivers of Empire: Water, Aridity, and the Growth of the American West*. New York: Oxford University Press, 1985.

Wurster, Charles. *DDT Wars: Rescuing Our National Bird, Preventing Cancer, and Creating the Environmental Defense Fund*. Oxford, UK: Oxford University Press, 2015.

Wyss, Robert. *The Man Who Built the Sierra Club: A Life of David Brower*. New York: Columbia University Press, 2016.

Zelko, Frank. *Make It a Green Peace!: The Rise of Counterculture Environmentalism*. Oxford, UK: Oxford University Press, 2013.

Image Credits

Interior Image Credits

Interior board: Untitled © Susan Tusa

xii: The Estate of Jacques Lowe / Getty Images

3: Courtesy of the John F. Kennedy Presidential Library and Museum, Boston

9: Courtesy of the Estate of Henry Beston

17: Willard Hatch / Pix / Michael Ochs Archives / Getty Image

36: National Park Service / NP Gallery

38: Everett Collection Historical / Alamy Stock Photo

43: © Erich Hartmann / Magnum Photos

50: Jean-Philippe Charbonnier / Gamma-Rapho / Getty Images

65: © Mark Shaw / mptvimages.com

80: Douglas Chevalier / The Washington Post / Getty Images

92: Arthur Schatz / The LIFE Picture Collection / Shutterstock

104: Alice Zahniser / Courtesy of the Estate of Howard Zahniser and the University of Washington Press

119: Hy Peskin / Getty Images

125: © Erich Hartmann / Magnum Photos

138: Barry Commoner Papers; Manuscript Division, Library of Congress, Box 526, Folder 11

142: Ray Fisher / Getty Images

157: U.S. Congressional Collection; courtesy of Special Collections, the University of Arizona Libraries

173: Norman Dietal / LBJ Museum of San Marcos, the Portal to Texas History

183: Baron Wolman / Getty Images

190: U.S. Congressional Collection; courtesy of Special Collections, the University of Arizona Libraries

196: Frank Church Papers; Special Collection, The Burns Studio, Boise State University

203: © Corbis / Corbis via Getty Images

218: Bettmann / Getty Images

229: Courtesy of Frank Stanton

623: Dev O'Neal / Courtesy National Archives, photo no. 306-PSE-78-241

626: Everett Collection / Alamy Stock Photo

629: National Archives / public domain

652: National Archives / Newsmakers / Getty Images

Insert Image Credits

1 (top): Courtesy of William O. Douglas Films

1 (middle): Universal History Archive / Universal History Group / Getty Images

1 (bottom): *Eliot Porter, Pool in a Brook*, Pond Brook, New Hampshire, October 4, 1953. Dye imbibition print, $10^{11}/_{16}$ x $8^{5}/_{16}$ inches, Amon Carter Museum of American Art, Fort Worth, Texas, Bequest of the artist P.1990.51.4046.1 © 1990 Amon Carter Museum of American Art

2 (top): Keystone-France / Gamma-Keystone / Getty Images

2 (middle): Ansel Adams Archive, Collection Center for Creative Photography, University of Arizona, © The Adams Publishing Rights Trust

2 (bottom): Ethan Daniels / Alamy Stock Photo

3 (top): Alfred Eisenstaedt / The LIFE Picture Collection/Shutterstock

3 (middle, right): *Winter Trees*, © Susan Tusa

3 (middle, left): *Sleeping Bear Dunes*, © Susan Tusa

3 (bottom): © Philip A. Harrington, courtesy of Evan Harrington

4 (top): Rick Friedman / Corbis / Getty Images

4 (middle): Wolfgang Kaehler / Lightrocket / Getty Images

4 (bottom): Courtesy of the US National Park Service

5 (top): Jack E. Boucher / US National Park Service / NPS Historic Photograph Collection

5 (middle): Kevin Fleming / Corbis / Getty Images

5 (bottom): ullstein bild Dtl. / ullstein bild / Getty Images

6 (top): © Philip Hyde; Courtesy of the Estate of Philip Hyde

6 (bottom): Ansel Adams Archive, Collection Center for Creative Photography, University of Arizona, © The Adams Publishing Rights Trust

7 (top): Cecil Stoughton; US National Parks, NP Gallery

7 (middle): © Philip Hyde; Courtesy of the Estate of Philip Hyde

7 (bottom): Courtesy of William O. Douglas Films

8 (top, left): © Philip Hyde; Courtesy of the Estate of Philip Hyde

8 (top, right): Robert Knudson; LBJ Presidential Library

8 (middle): Robert Knudson; LBJ Presidential Library / NP Gallery

8 (bottom): LBJ Presidential Library

9 (top): *Aerial View of the North Cascades in Winter*, © Stephen Matera

9 (middle, left): National Park Service / NP Gallery

9 (middle, right): ZUMA Press Inc. /Alamy Stock Photo

9 (bottom): Bettmann / Getty Images

10 (top): Bettmann / Getty Images

10 (bottom): Maidun Collection / Alamy Stock Photo

11 (top, right): Hulton Archive / Getty Images

11 (top, left): Clair Milanovich / Alamy Stock Photo

11 (bottom, left): Hulton Archive / Getty Images

11 (bottom, right): John Olson / TIME and LIFE Pictures / Shutterstock

12 (top): Bettmann / Getty Images

12 (middle): Jasper Chamber / Alamy Stock Photo

12 (bottom): © Robert Rauschenberg Foundation; Licensed by VAGA at Artists Rights Society (ARS), New York

13 (middle): Bettmann / Getty Images

13 (top, left): Boyd Norton / Alamy Stock Photo

13 (bottom, right): © Roger Ressmeyer / Corbis / VCG / Getty Images

13 (bottom, left): Courtesy of the Stanley Mouse Studio with thanks to the Sierra Club's William E. Colby Memorial Library, for use of its historical archives

14 (top): © Luisa Dörr

14 (bottom): © Mark Vergari/USA Today Network

15 (top): Andy Warhol, *Bald Eagle: Endangered Species, 1983*, screen print on Lenox Museum Board, 38 x 38 inches, Crystal Bridges Museum of American Art, Bentonville, Arkansas, 2015.38.3. Photography by Edward C. Robinson III. © 2022 The Andy Warhol Foundation for the Visual Arts, Inc. / Licensed by Artists Rights Society (ARS), New York; Courtesy Ronald Feldman Fine Arts, New York

15 (middle): Arthur Schatz / The LIFE Picture Collection / Shuttertstock

15 (bottom): © Buddy Mays

16 (top): *Fire in the Sky*, © Michael DeWitt

16 (middle): *God's Window*, © Michael DeWitt

16 (bottom): Steve Northup / Getty Images

Index

Page numbers with illustrations are in *italics*.

A

Hudson River, 159, 201, 285–90, 320, 423–24, 431, 458, 516, 534, 574. *See also* Storm King Mountain
Hudson River's White Water Derby, *349*, 487
Huerta, Dolores, 446–47
Humphrey, Hubert, 107–9, 148, 168, 327, 333, 390–91, 394, 521–22, 600, 643
Hyde, Philip, 100, 380, 398, 454, *455*, 526

I

Ickes, Harold L., 25–26, 30, 33–34, 37, 93, 243, 410–11, 552
Indiana Dunes National Lakeshore, 188, 214, 237, 241–44, 410–11, 429–31, 474–76, 521
Indian Point Nuclear Power Plant, 285, 289
"inhumanism," 346
In Sputnik's Shadow (Wang), 313
Integrity (Neuberger), 109
International Union for Conservation of Nature, 151, 657
In the Thick of It: My Life in the Sierra Club (McCloskey), 399
Iron Eyes Cody, 624–26, *626*
"iron triangle," 40
Island in Time (Gilliam), 274
"islands of poverty," 479
Izaak Walton League, 107, 116, 126, 164, 188, 386, 420, 517, 589, 656

J

Jackson, Henry M. "Scoop," 168, *537*; Chesapeake & Ohio Canal National Historical Park, 601; on dams, 454; Earth Day, 600–601; Ehrlichman and, 519; Environmental Impact Statement, 615; EPA, 608; Federal Water Project Recreation Act of 1965, 420; "impeachment politics,"

662; legacy of, 665; National Environmental Policy Act (NEPA), 482–84, 585–89, 591; "New Frontiersman," 168; Nixon, 564, 589, 596; North Cascades National Park, 194, 305, 485–86, 520; NTSA, 418, 536; ORRRC, 132; overconsumption, combat of, 591; Redwood National Park bill, 505; Senate Committee on Interior and Insular Affairs, 304; trail system, 418; Wilderness bill, 304; wild-river legislation, 414
Jamaica Bay National Wildlife Refuge, 648–49
Japanese beetles, 492
Javits, Jacob, 383, 424
Jeffers, Robinson, 295, 346
John Brown's Body (Benèt), 502
Johnson, Lady Bird, *382, 404*; American beautification campaign, 363–365; Austin, 653; beautification effort, 381; Big Bend National Park, 441; Boyle, 421; California trip, 448–50; children's education, 363; Committee for a More Beautiful Capital, 421–22; "Crossing the Trails of Texas" tour, 511–13; death of, 670; Earth Day, 602; Sharon Francis, 413; Head Start, 363; Highway Beautification Act, *408*, 431; highways, 363–65; Lady Bird Lake, 670; "Lady Bird Special," 412; "Land and People" tour, 381; legacy of, 560; NASA, 412; National Historic Preservation Act of 1966, 461; National Wildlife Research Center, 653–54; Padre Island National Seashore, 512; Point Reyes National Seashore, 448; Rachel Carson Award, 670; Redwood National Park, 541; redwoods, 403–4; Rio Grande River, 444; roadside America, natural beauty, 422–23; Sandburg, 475; Texas Highway

Udall, Stewart, *157*, 161–65; Adams on, 355; *Agenda for Tomorrow*, 549; Allagash, 200; *America's Natural Treasures: National Nature Monuments and Seashores*, 672; Anderson, 210; Antiquities Act and, 545–51, 547; articles, 191; Assateague, 316–18; backers, 188; Baker, 201; Big Bend National Park, 441; Bodega Head, 295; bomb tests, crusader against, 670–71; Brandborg, 355; Brower, 355; *Cadillac Desert*, 230; Canyonlands, 197, 211, 408–10; Cape Cod, 201; Carson, 367, 599; Central Arizona Project, 245, 450; civil rights, 227; Colorado River dams, 321; DDT, editorial on, 639; Department of Health, Education, and Welfare (HEW), 358; Department of the Interior headquarters, 672; *Desert News* interview, 209; Douglas, 200; Earth Day, 599; endangered species, 463; environmentalists and, 528; ESPA, 473; Fire Island, 318; Ford's Theatre, 501; Fort Davis, 217; Fort Smith, 217; Francis, 192; Freeman, 188–89; Frost visit, *190*, 215, 303; Galbraith, 160; Gilliam, 215; Glen Canyon Dam, apology for, 599; global warming, 583; Grand Canyon National Park conference, 196–97; Grand Canyon rafting trip, 488; Great Smoky Mountains National Park, 191; Hartzog, 362; HBCs, 227; high-poverty pockets, 229; Indiana Dunes National Lakeshore, 410–11, 475; *Island in Time: The Point Reyes Peninsula*, 274; Jacqueline Kennedy, 187; JFK, 210, 267, 333–34, 348; Josephy, Jr., 215; Bobby Kennedy, earth feeling of, 267; "Know America First," 361; Krutch, 206;

Lady Bird Johnson, 187, 381, 441; "Land and People" tour, 381; LBJ, 357–58, 361, 509, 557–58; legacy of, 552–55, 670, 672–73; "Leopold Report," 319; Maine North Woods, 440; Matthiessen, 215; media and, 553; Mission 66, 191; Murie, 355; Muskie, 200; *Myths of August: A Personal Exploration of Our Tragic Cold War Affair with the Atom*, 672; national monument wish list, 558–59; National Park Service, Black employees, 227; "National Parks for the Future," 191; national parks signing ceremony, 527; Native American sovereignty, 187; Nelson, Wolf River raft trip, 534; 1965 conservation agenda, 412; *1976: Agenda for Tomorrow*, 583; NPS, 191, 552; nuclear power, 340, 553; Oakes, 201, 355; office of, 191; Old Angler's Inn, 199; Olson, 203–4; Olympia Reges, 199; Ozark National Scenic Riverways, 199, 377; Pacific Southwest Water Plan, 321; Padre Island, 278; Patuxent Wildlife Research Center, 311; personal injury lawyer, 670–71; pesticides, federal land and, 375; pollution, 554; Potomac, 164; Prairie National Park, 226; *The Quiet Crisis*, 215, 217, 250, 334; RECA, 671–72; Redskins, 227; Redwood National Park, 379–80; Reisner criticism of, 230; Revelle, 231; RFK, 513, 515–16; Russell Cave, 216–17; Saguaro National Monument, 217; Sandburg, 475–76; Save the Grand Canyon, 452–53; secretary of interior, 182, 187, 191–92; *Seguoia sempervirens*, 380; *Silent Spring*, 262, 372; Sonoran Desert National Park, 547; "spite cutting," 403; St. Croix River, 245, 533; Stegner, staff, 215; Thoreau anniversary party,

ABOUT THE AUTHOR

DOUGLAS BRINKLEY is the Katherine Tsanoff Brown Chair in Humanities and Professor of History at Rice University, a contributing editor at *Vanity Fair*, and the New-York Historical Society's Presidential Historian. His most recent books are *American Moonshot, The Great Deluge, Cronkite, The Quiet World, The Wilderness Warrior,* and *Rightful Heritage*. Six of his books have been selected as *New York Times* Notable Books of the Year. He lives in Texas with his wife and three children.

READ MORE BY
Douglas Brinkley

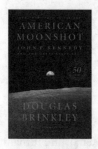

"One of our most brilliant chroniclers of the American past."

—MICHAEL ERIC DYSON,
author of *Entertaining Race: Performing Blackness in America*

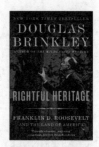

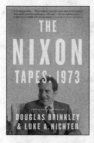

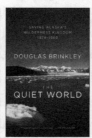

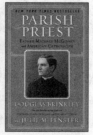

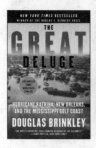

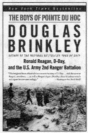

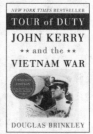